To Keren,

Keep making Northwestern proud.

Greg O

TRANSFORMATIONS

SELECTED WORKS OF

G.B. OLSON

ON MATERIALS MICROSTRUCTURE AND DESIGN

EDITED BY C.E. CAMPBELL, M.V. MANUEL, AND W. XIONG

TMS | ASM INTERNATIONAL

ASM INTERNATIONAL MATERIALS PARK, OH 44073-0002 WWW.ASMINTERNATIONAL.ORG

First printing, October 2017

Prepared under the direction of the ASM International Technical Book Committee (2016–2017), Y. Zayna Connor, Chair.

ASM International staff who worked on this project include Scott Henry, Director, Content and Knowledge-Based Solutions; Karen Marken, Senior Managing Editor; Sue Sellers, Content Development and Business Coordinator; Madrid Tramble, Manager of Production; and Diane Whitelaw, Production Coordinator.

Library of Congress Control Number: 2017950657

ISBN-13: 978-1-62708-137-5
EISBN: 978-1-62708-138-2

SAN: 204-7586

ASM International®
Materials Park, OH 44073-0002
www.asminternational.org

Printed in the United States of America

Contents

Editors' Preface

This collection presents the selected works of Greg B. Olson in honor of his 70th birthday in 2017. It highlights Olson's influential contributions to the understanding of martensite transformations and the development and application of a systems design approach to materials.

Part I: Martensite emphasizes Olson's work to develop a dislocation theory for martensite transformations, improve the understanding of the statistical nature of martensite nucleation, and expand use of quantitative microscopy to characterize phase transformations. Part II: Materials Design focuses on the application of a systems design approach to materials and the development of integrated computational design curriculum for undergraduate education. This section includes several examples of the systems design approach to a variety of applications.

The papers chosen for this volume were selected by the editors after several discussions with Olson. The papers' format (but not content) has been adjusted for reprint in this collection.

In the introductions to Parts I and II, one of his longtime collaborators, Sir Harry Bhadeshia, and a former student and business partner, Dr. Charles Kuehmann, provide perspectives on these papers and the impact they have had on them and will have on future scientists and engineers. The honoree provides a perspective on the events, mentors, and colleagues that have and continue to influence his work in the Author's Preface. We hope this volume not only provides perspective on the significance of Olson's work but also serves as a reference and teaching resource for those wishing to become practitioners of the Materials by Design (Materials by Design is a trademark of QuesTek Innovations) approach.

As we celebrate Greg Olson's achievements on the occasion of his birthday, we hope that the Materials by Design approach continues to permeate the engineering community and industry, expanding the success of this approach to new material classes and applications. We wish that one day designers will be able to specify their performance requirements and computational materials design tools will design the desired material for the specific application.

Expect victory,
Carelyn Campbell
Michele V. Manuel
Wei Xiong

Author's Preface—Reflections on Some Transformations

I have lived in interesting times. I am deeply grateful for a 23-year education during an age of greatness in metallurgy at MIT. The combination of Morris Cohen, Cyril Stanley Smith, and John Cahn, with a sabbatical visit by Jack Christian of Oxford, provided an educational opportunity unparalleled before or since. Combining the perspectives of Cohen's and Christian's critical thinking with Smith's and Cahn's creative thinking gave me the tools to make something of the vast knowledge provided by MIT's infamous "information fire hose." Smith's systems vision encouraged me to approach the science of martensitic transformations with realistic complexity, driving an early interest in interdisciplinary collaboration and providing a conceptual foundation for a systems engineering approach to materials. I am especially grateful to have enjoyed the mentorship of Morris Cohen for as long as I did. Through the Cohen network, I was introduced to the developing CALPHAD computational thermodynamics of Larry Kaufman and its emerging commercial form through Mats Hillert's group at the KTH Royal Institute of Technology in Stockholm. Early collaboration in the late 1970s with Sir Harry Bhadeshia of Cambridge took transformation complexity to a next level with the contemplation of coupled diffusional/displacive transformations, the discussions founding a shared vision of the potential of materials design grounded in the predictive power of CALPHAD thermodynamics.

Trained to pursue mechanistic understanding, I continue to try to interpret the forces driving the dynamic world I have experienced. It seems the elevation of science to the stature of religion by the development of the atom bomb drove a misinterpretation of the 1955 Grinter Report on engineering education, fostering the replacement of engineering with engineering science at research universities. Consequently, despite being in a department in the School of Engineering, my formal MIT education of the 1960s through the 1980s was all science. My later career in design hung from a delicate thread, which is a long interest in the study of music composition. At the time, Cohen and I had no appreciation of the significance of his approval of my doctoral minor in music composition. My continued study culminated with the MIT jazz band director, Herb Pomeroy, allowing me to sit in on his most advanced composition courses at Berklee School of Music (now Berklee College) across the river. Learning an algorithmic approach to the systematic creation of dynamic multilevel structures in the style of Duke Ellington complexity, it hit me one day that this must be what engineering is. As it turns out, Berklee was founded by an MIT electrical engineering alumnus from the days when the institute taught engineering. I had backed in to a complete education by both MITs. The principles of music composition gave me the framework to apply my materials science knowledge in design. I am very grateful to Duke Ellington for a vision of what engineering design can be.

Perhaps a bigger bomb than the atomic one was the Riksbank financing of an Economics Prize "in Memory of Alfred Nobel" that became broadly accepted as a Nobel Prize in Economics. This new stature gave economists the ear of governments, unleashing the notion of the "service economy," and fostering decades of destruction of domestic manufacturing capability. The move away from the selling of actual things to the selling of ideas pervaded all sectors, moving research agencies to pursue novelty at the expense of utility. In this environment, metals were seen as old. MIT pulled the plug. Metallurgy went West.

My early work applying principles of martensitic transformations to biological processes in primitive organisms gave me a unique perspective as I watched research universities follow the money. It struck me as corresponding to the intellectual level of a single-cell organism highly tuned to move in the direction of maximum nutrients. Arriving at Northwestern University, I was introduced as the "Man of Steel" to the university president, who immediately lectured me on the societal importance of materials like steel and concrete, to which the university was strongly committed. Societal importance—an interesting concept. I seemed to have found a university with a soul. Morrie Fine, who had also been strongly influenced by Cyril Smith's generalist vision when he worked for him on the Manhattan Project, showed me essays from the 1950s revealing a vision of a science of all materials through the generalization of the principles of dynamic multilevel microstructure already established in physical metallurgy. At the same time that other departments were rapidly abandoning metals, Fine, aided by Jerry Cohen as dean, was able to maintain a unique commitment to physical metallurgy, understanding its central role in creating a new science. Shortly after hiring David Seidman, Peter Voorhees and I were hired the same year, followed soon after by David Dunand. Building on the considerable metals expertise already present at Northwestern, notably including the inspiring Hans and Julia Weertman, a critical mass balancing theoretical and experimental approaches to the core science of microstructure was maintained, making some great things possible. Equally important was the long-standing tradition of uniquely genuine interdisciplinary collaboration in materials research that Fine and his colleagues established. I doubt we could have achieved such a successful implementation on any other campus of Cyril Smith's systems vision as we so productively integrated materials science, quantum physics, and applied mechanics within the framework of systems engineering. This most notably included pioneering applications of the great Art Freeman's uniquely accurate all-electron density functional theory (DFT) quantum-mechanical full-potential linearized–augmented plane-wave (FLAPW) code, and an evolving series of advanced numerical continuum mechanics methods by Toshio Mura, Brian Moran, and Wing-Kam Liu. It set the foundation for a very productive multi-university collaboration through our "SRG network," notably including Jim Rice and John Hutchinson of Harvard, Dave Parks of MIT, George Smith of Oxford, Mark Eberhart of CSM, Zi-Kui Liu of Penn State, Gary Harlow of Lehigh, Dave McDowell of Georgia Tech, and Martin Blackburn and Dave Furrer of Pratt & Whitney.

I am extremely grateful to have lived long enough to see the pendulum swing back with the destruction of the global service economy, fueling the rapid rediscovery by developed nations of the vital importance of domestic manufacturing to enable genuine wealth creation. In this environment, materials return to the stuff things are actually made of. In the immediate postwar years, U.S. university research was led by the Office of Naval Research (ONR), and the intellectual frontier of materials was metals. After decades of missionless agencies and novelty research, the rediscovery of manufacturing brings us to a national Materials Genome Initiative with the overarching goal of accelerating the full materials development cycle. Today the leading success stories of this endeavor stem from ONR and its affiliated mission-oriented agencies. The intellectual frontier of this acceleration technology is metals—progress indeed. Here we must give a special thanks to the Family Christodoulou, Julie at

ONR, and Leo at DARPA, for their unique leadership role in keeping U.S. metallurgy alive.

Founding a company 20 years ago to offer a service of computational materials design that no one believed was actually possible was a tough slog. Having the president of the United States announce a decade-long national initiative that acknowledged our technology was real and a beautiful thing. Silicon Valley beat a path to our door. I remain extremely grateful to the dedication of our many QuesTekers in making our technology a reality. I am also grateful to the ASM network for bringing Ray Decker and Aziz Asphahani during a critical company reboot. I owe a special thanks to a long synergy with the Thermo-Calc Software company of Stockholm, chaired by John Agren of KTH.

The dominance of science at our research universities has succeeded in creating a great research environment for the teaching of science, but not so for engineering. The example of bringing actual materials design on campus has shown what would be possible if we could "legalize" engineering in academia. Doctoral research at Northwestern that is genuinely focused on the science-based engineering of materials has allowed a hierarchical coaching system where doctoral and postdoctoral researchers can serve as coaches to undergraduate design class teams, enabling experiential learning at an otherwise unattainable technical level. Building on this experience in the materials domain, my experience in the role of music composition and my understanding of design gave me no hesitation to join forces with colleagues at Northwestern in the 1990s in both engineering and the humanities to create a revolutionary integrated "techmanities" approach to the teaching of "design and communication" at the first-year level.We benefited greatly when Clive Dym of Harvey Mudd College helped us build on his expertise in the teaching of first-year-level design. Our highly successful version of first-year design that continues today established a strong foundation for a multiyear design education that has been especially effective in our materials curriculum. It is gratifying to see these pioneering achievements at Northwestern acknowledged by the 2017 National Academy of Engineering (NAE) Gordon Prize for engineering education.

I am very grateful for the resilience of my wife, Jane, and daughter, Elise, in tolerating the burden my career has imposed on our family. Fortunately, they don't seem to harbor serious grudges. I am very grateful for the extended family of the many students at MIT and Northwestern, with whom it has been my great privilege to collaborate and follow their successes. Although it is certainly exhilarating to see the rocketing career of Charlie Kuehmann, I take particular pride in following the success of my women students. As we continue to gradually restore the ancient principle of gender balance, it is especially gratifying to witness the tangible benefits a special female wisdom brings to our enterprise. I recall watching Carrie Campbell receive an award while Morrie Fine turned to me and commented, "It is the achievements of our students that give us the deepest satisfaction." Morrie knew what's what.

G.B. Olson

About the Author

G.B. Olson

Greg Olson grew up in the town of Huntington on Long Island, New York, where his interest in science and engineering was piqued at an early age. While attending grade school he eagerly awaited science time, when new discoveries could be made with beakers and Bunsen burners. In high school, he continued his science experiments in his parent's basement, where he set up a chemistry lab. This lab provided the opportunity to construct model rockets, study fluorescent crystals, and discover extraction metallurgy, converting rock from the backyard into metal. Olson went on to attend MIT to pursue his interest in materials science and engineering and music composition. In 1970, he earned his bachelor's and master's of science in materials science and engineering. He continued his studies and music making, playing lead trumpet in the jazz band at MIT, pursuing his doctorate of science with Morris Cohen, and developing breakthrough insights into martensite transformation behavior.

From 1974 to 1988, Olson continued his research career and collaboration with Cohen at MIT as a research professor. During this time he developed the theory of barrierless classical heterogeneous martensitic nucleation via defect dissociation, concepts explaining the mobility of martensitic interfaces due to coherency dislocations in interphase boundaries, and a theory of transformation plasticity and its influence on flow stability with application to ductility and fracture toughness. This work included detailed experimental analysis of the crystallography and interfacial structure of the martensitic interfaces using then-developing high-resolution microanalysis techniques.

Influenced by Cyril Stanley Smith, Olson also began to explore the interactive structural hierarchy in materials. He integrated Smith's and Cohen's ideas of "reciprocity" within the processing-structure-property paradigm with the systems approach outlined by Jenkins to set the foundation for a systems approach to materials design. This approach became the foundation for the Steel Research Group (SRG), founded in 1985 at MIT. The mission of the SRG was to develop methods, tools, and databases for a systems-based, multiscale, multidisciplinary computational design of materials approach. The initial focus of the SRG was ultrahigh-strength ferrous alloys that required high strength and toughness and hydrogen embrittlement resistance. Designing these new materials incorporated many of the insights into steel transformation behavior that Olson had previously developed. An essential component of this materials design approach was the CALPHAD approach to constructing composition and temperature dependent thermodynamic databases, which could be used to calculate equilibrium and constrained equilibrium quantities. The development of CALPHAD software tools, such as Thermo-Calc, provided an initial computational design platform. Coupled with the CALPHAD approach, Olson began collaborations to use first-principles quantum-mechanical calculations to provide insights into the embrittling potency of impurity segregants. These insights led to significant improvements in grain boundary cohesion and improved fracture toughness.

In 1988, Olson was enticed to Northwestern University to pursue his goals of materials design. The SRG was reorganized at Northwestern and gradually expanded its applications to include aluminum, titanium, nickel, and refractory metals, including shape-memory alloys, high-temperature superalloys, and bulk metallic glasses, with exploratory activities in ceramics, polymers, and biomimetic smart composites. As Olson expanded the concepts of the system design approach to materials design beyond steels, he began to develop a curriculum to teach materials design to undergraduates. In 1989, he taught the first Materials Design course at Northwestern. The first part of the course taught students the concepts of systems design, how to use Ashby plots for materials selection, the basics of a CALPHAD approach, and software tools. With this background, the students tackled various materials design challenges. Within a few years, the course became a part of the Nothwestern Materials Science and Engineering Department core classes required for graduation, and companies were solicited to submit their materials challenges, which ranged from ultrahigh-strength corrosion-resistant bearing steels to bubble gum. Inspired by one of Olson's other hobbies, auto racing, one of the initial design challenges was to improve the wear and fatigue resistance of performance-driven gear steels. As captured in the PBS video *The Infinite Voyage: Miracles by Design* (WQED Pittsburgh in association with the National Academy of Sciences, 1991), Olson was pleased to tell Michael Andretti that he just "happened to have a new alloy up his sleeve" to improve Andretti's race car performance. Eventually, the Ferrium (Ferrium is a trademark of QuesTek Innovations) C61 alloy was designed to improve drive gears for Baja 1000 off-road racing cars.

With the initial success of this systems approach to a variety of materials, there was a clear need to take this technology directly to industry. In 1996, with protégé Charles Kuehmann and two other Northwestern affiliates, Olson founded QuesTek Innovations, the first computational materials design company providing both materials design services and licensing opportunities of designed alloys. Built on the success of the materials design approach, Olson's next step was to consider the materials optimization and qualification at the component level and to design for scale. The U.S. Defense Advanced Research Projects Agency: Accelerated Insertion of Materials (DARPA-AIM) initiatives in the 2000s enabled the development of robust design approaches to reduce intrinsic variations within allowed composition and process tolerances, and to determine minimum design-allowable properties with limited data. The first alloy successfully developed using this approach was Ferrium S53, which reduced the number of required heats by one third and the number of experimental observations by a factor of 10. The first computationally designed steel took flight in December 2010.

Industry took notice of Olson's success, and in fall 2012 a Silicon Valley consumer electronics company acquired the core design team of QuesTek Innovations, along with QuesTek's proprietary software and databases. After this acquisition, Olson rebuilt his company, and today the company is stronger than ever, recently expanding to include an overseas affiliate.

Olson's pioneering work on materials design demonstrates a seamless combination of materials theory and design applications that has transformed the field of materials science and engineering. For more than 30 years, Greg Olson has inspired his students by challenging them to open new regions in design space. Thanks to his efforts, engineers are no longer limited to the materials currently available; they now have the tools to design the materials they need.

Personal History

- Born in Brooklyn, New York, on April 10, 1947.

Academic Degrees

- Bachelor's of science in Metallurgy, MIT, 1970
- Master's of science in Metallurgy, MIT, 1970: "The Design and Construction of an Ultra-high Vacuum Field-ion Microscope"
- Doctor of science in Metallurgy, MIT, 1974: "A General Mechanism of Martensitic Nucleation"
- Honorary doctorate, Kungliga Tekniska Högskolan (KTH) Stockholm, Sweden 2014

Academic Positions

- 1974–1988 Research Professor, MIT
- 1988–1999 Professor, Northwestern University
- 1999–2009 Wilson-Cook Inaugural Endowed Professorship in Engineering Design, Northwestern University
- 2010–present, Walter P. Murphy Professor, Northwestern University

Membership in Learned Societies

- ASM International, Fellow 1990
- The Minerals, Metals and Materials Society (TMS), Fellow 2001
- National Academy of Engineering, 2010
- American Academy of Arts and Sciences, Member, 2012
- Royal Swedish Academy of Engineering Science (IVA), Foreign Associate, 2013
- Alpha Sigma Mu, Distinguished Life Member, 2016

Honors

- AMAX Foundation Fellowship 1972–1974
- Jacob Kurtz Exchange Scientist at Technion–Israel Institute of Technology 1979
- National Science Foundation Creativity Extension Award 1983–1985
- Army Materials Technology Laboratory Special Service Award 1988
- SERC Visiting Professor, University of Cambridge 1992
- Jacob Wallenberg Foundation Award—Sweden 1993
- M.R. Tenenbaum Award—ISS-AIME 1993
- Listed in *Who's Who in America* 1998–present
- Listed in *American Men and Women of Science* 1994–present
- NASA Technology Recognition Award 1994

- Technology of the Year Award, Industry Week 1998
- SERDP Pollution Prevention Project of the Year Award 2002 and 2003
- *Sun-Times* Chicago Innovation Award 2003
- TMS-AIME Structural Materials Distinguished Materials Scientist/Engineer Award 2004
- ASM Gold Medal 2010
- TMS Morris Cohen Award 2013
- IP Lawyers Association of Chicago Innovator of the Year Award 2013
- Evanston/Northwestern Catalyst Award for Entrepreneurship 2014
- SBA Tibbetts Award for SBIR Achievement (QuesTek) 2016

Distinguished Lectures

- New Mexico Distinguished Lecturer, Materials 1983
- ASM Boston Chapter Sauveur Memorial Lecturer 1986
- ASM Alpha Sigma Mu Lecturer 1996
- Northwestern University Inland Steel Lecturer 1997
- ASM Philadelphia Chapter Sauveur Lecturer 1998
- Speich Memorial Lecturer, ASM Chicago Chapter 2000
- ASM Edward DeMille Campbell Memorial Lectureship 2006
- China Distinguished Materials Scientist Lecture 2007
- Kelly Lecture, University of Cambridge 2008
- Golick Lecturer, Missouri University of Science and Technology 2009
- Winegard Lecturer, University of Toronto 2015

PART I

MARTENSITE

Transformations—Selected Works of G.B. Olson on Materials, Microstructure, and Design
C.E. Campbell, M.V. Manuel, and W. Xiong, editors

Introduction to Part I—Martensite

Harry Bhadeshia

Martensitic transformations in steels are thermodynamically of first order in the Ehrenfest classification. This means that the product and parent can coexist in a two-phase mixture, a fact that is apparent from the myriad of direct observations. In almost all known circumstances, the nucleation of martensite occurs heterogeneously at defects present in the austenite. The first paper by Greg was coauthored with Morris Cohen in 1972 (Ref 1) in a festschrift honoring W. G. Burgers, a name surely familiar to every materials scientist in the context of the displacement vector associated with a dislocation. This work identified a clear distinction between stress-assisted and strain-induced nucleation, the former being a simple manifestation of the change in thermodynamic force activating already existing sites, whereas the latter involving the creation of additional nucleation sites by plastic deformation, thus permitting the cascades of nucleation events described generically by the term "autocatalytic nucleation."

Morris Cohen, Greg's thesis advisor at MIT, had previously published a number of papers on martensitic nucleation (Ref 2–4), based on classical theory involving heterophase fluctuations, some of which may be quenched in from the austenitic state. However, such fluctuations lead to an inverse functional dependence between the activation energy for nucleation and a power of the driving force, which is not observed experimentally (e.g.. Ref 5). By 1976, Olson and Cohen had published a series of seminal papers (Ref 6–8) in which they generalized the dislocation mechanism for martensitic nucleation that previously had been used in the context of the conceptually simpler austenite to epsilon martensitic transformation (Ref 9). In that particular transformation, when the volume change is neglected, the partial dislocations (Shockley partials) that accomplish the lattice deformation involve simple shear by glide on a close-packed plane, so the nucleus has the final crystallography of the visible martensite plate that evolves. However, the austenite lattice cannot be changed into a body-centered one by a shear—the deformation is more complex and leads to a crystallography with irrational parameters, one that generates a substructure in the final product. Chapter 1 presents a detailed review of Olson and Cohen's dislocation-based martensitic transformation theory. The basis for this dislocation-based theory is outlined in Chapter 2, which presents the hierarchy of interfacial dislocation structures.

Olson and Cohen abandoned some of the features of the macroscopically observed crystallography when proposing the nucleus, arguing that the macroscopic form develops as the nucleus evolves, an assumption that permitted a description of the detailed structure of the dislocations and the nucleus/matrix interface. These ideas, reviewed in Chapters 3 through 6, were implemented in a thermodynamic framework that naturally led to a linear dependence of the (small) activation energy for nucleation on the magnitude of the chemical driving force (pointed out by Ref 5 in the derivation of the classical Koistinen and Marburger equation in Ref 10). Classical heterophase fluctuation theory, in contrast, gives an inverse power relationship. I was enthralled by the

1976 publications, which I studied repeatedly. In later years, when I observed a particular dependence of the bainite-start temperature of each alloy on temperature, I was able to explain this quantitatively in terms of this linear dependence of activation energy on driving force (Ref 11) (Chapter 7). The 1976 papers nurtured our subsequent collaborations on theory (Ref 12, 13), with some of the work done at the beach in Boston, with peanut butter and jelly sandwiches provided by Greg's wife, Jane. There is some fascinating unpublished work with Carrie Campbell and Greg that develops the theme further to encompass a wider range of transformations, with the genesis done in Cambridge when Greg spent an extended period. There is much more to be said about the nucleation of martensite, a subject reviewed at length in Greg and Morris Cohen's (Ref 14) contribution to Nabarro's epic compilation on dislocations (Chapters 1 and 3). Chapters 1 and 5 include a charming discourse on biological martensitic transformations, work I quote to the delight of students in my annual lectures for the Natural Sciences Tripos in Cambridge.

I recall a period of gentle confusion in martensite, partly because there was a concerted effort to make the subject interdisciplinary. Soft modes and solitons became fashionable, and the term *martensite* began to be used in circumstances that were not entirely appropriate. I vividly remember two publications that were seminal in resolving the issues; both were in the proceedings of the International Conference on Martensitic Transformations (ICOMAT) '79, a conference organized at MIT by Greg and his co-workers. This was my first international conference. What a wonderful experience it turned out to be, with every paper that was presented pregnant with meaning and presented in the presence of giants in the subject. The plenary lecture (Ref 15) by Cohen, Olson, and Clapp demonstrated that the term *martensite* should be reserved for those transformations dominated by homogeneous deformations that lead to shape deformations with large shear components. This distinguished martensite proper from shuffle-dominated transformations and from those where the strain energy is not the determining factor. The second influential paper was by Christian (Ref 16); the idea that the genesis of martensite can be described by a lattice instability in which some elastic constant vanishes (soft mode, strain spinodal, etc.) does not make sense, he argued, because we can observe martensite and austenite to coexist. The transformation is thermodynamically of first order. To obviate this difficulty, some were arguing for a soft mode to occur at a dislocation strain field, leading to the nucleation of martensite. However, such localization, he said, is not formally different from classical heterophase fluctuation theory because it will involve an interface (a soft one) between the nucleus and matrix (Ref 16).

Greg and I had not connected when ICOMAT '79 was organized, but the conference was formative for my career and in stark contrast to modern meetings that lack depth. The 1979 paper on interphase boundary dislocations (Ref 17) inspired me to contact Greg and Morris Cohen. This paper clarified for me how single-atom steps can be described as "coherency dislocations," whose motion accomplishes the lattice change (including dilatation) and yet are glissile! The long-range strain fields due to the coherency dislocations are then accommodated by a set of "anticoherency" dislocations, which have Burgers vectors that are lattice vectors. Many years later at a meeting in Bariloche, Argentina, Aaronson suggested that the fcc to bcc transformation in brass must involve diffusion because of the volume change. I immediately responded that coherency dislocations can do the job without diffusion. He asked me how the Burgers vector of the so-called "coherency dislocation" could be described, and I was able to explain it in terms of the generalized Burgers circuit defined by Bilby and Christian. Greg was in the audience, and I could see him smile out of the corner of my eye.

Greg has been impressive in applying his knowledge to the design of strong steels (Ref 18–20), and Chapter 8 lays the foundation for this design approach, outlining the application of transformation plasticity models to predict the mechanical behavior of advanced steels. Charlie Kuehmann discusses the commercial success in creating a design-based industry in the introduction to Part II of this book, the latest achievement being the blast-resistant steels for the navy (Ref 21, 22) (Chapters 18–19). I have a feeling that his design philosophy was kindled by the work on lanthanum additions to rapidly solidified steel (Ref 23), where the purpose was to getter the impurities that exacerbate stress corrosion cracking. Steel design never involves the optimization of a single property; Greg worked hard to educate undergraduates on what design actually means (Ref 24) (Chapter 29), illustrating his ideas with the systems approach that he so loved, apparently inspired by C.S. Smith. At one point, I had dreams about his flow charts. Greg went on to develop a friendship with first-principles calculations, beginning, I think, in the 1990s. Iron has the most odd magnetic properties in all its manifestations, and the first paper (Ref 25) revealed the stabilities of the ferromagnetic and antiferromagnetic states of austenite. With Krasko (Ref 26), he published a rather important paper based on ab initio methods on the role of species such as boron, carbon, and phosphorus on intergranular embrittlement, commenting in particular on the effect of these "impurities" on the anisotropy of bonding across the boundary. More detailed studies on this and associated subjects followed later in a close collaboration with Freeman and his team at Northwestern University (Ref 27–29) and are reviewed in Chapters 13–15. I often refer to these publications in my own work on impurity effects (Ref 30) and cite them in an undergraduate text (Ref 31).

There is so much to say about Greg's contributions to science that cannot fit into this introduction, but a few items must get an honorable mention. Ling and Olson accounted for anisotropic elasticity in dealing with the strain field around martensite (Ref 32). The work on the early stages of tempering (Ref 33) is important in showing that the vast majority of martensite in high-M_S steels is unlikely to contain all of its carbon in the solid solution as defined in thermodynamics. Why is the lattice invariant shear by twinning in some ferrous martensites and by slip in others? Reference 34 suggested that the mobility of twinned interfaces is much greater than when the lattice invariant deformation involves the glide of dislocations. Therefore, for rapid transformations the twinning mode is favored. Indeed, it is observed experimentally (Ref 35) that when the heat of transformation cannot be dissipated fast enough, the resulting rise in temperature causes the interface to slow down, so a plate that is internally twinned in the center will change to being internally slipped at the peripheries.

I'll finish with some personal comments. Greg has been a great friend and inspiration, a true scholar and a gentleman. I was able to discuss science with him. I sometimes felt inadequate afterward, though never disheartened. Greg and Morris Cohen were both there when, unknown to me, I needed support during an era of controversy about bainite. I congratulate him wholeheartedly on his 70th birthday, and I am certain he will continue to surprise and delight.

REFERENCES

1. G.B. Olson and M. Cohen, Mechanism for strain-induced nucleation of martensitic transformations. *J. Less-Common Met.* 28, 1972, 107–18
2. L. Kaufman and M. Cohen, Homogeneous versus Heterogeneous Nucleation in the Martensite Transformation, *The Mechanism of Phase Transformations in Metals*, Institute of Metals, 1995, p 187–92
3. L. Kaufman and M. Cohen, Thermodynamics and kinetics of martensitic transformations. *Prog. Metal Phys.* 7, 1958, 165–246
4. V. Raghavan and M. Cohen, Nucleation Model for Martensitic Transformations in Iron-Base Alloys. *Acta Metall.* 20, 1972, 333–38
5. C. Magee, The Nucleation of Martensite, *Phase Transformations*, eds. H.I. Aaronson and V.F. Zackay, ASM International, 1970, p 115–56, .
6. G.B. Olson and M. Cohen, General Mechanism of Martensitic Nucleation. 1. General Concepts and FCC->HCP Transformation. *Metal. Trans. A.* 7, 1976, 1897–904
7. G.B. Olson and M. Cohen, General Mechanism of Martensitic nucleation. 2. FCC->BCC and Other Martensitic Transformations, *Metal. Trans. A.* 7, 1976, 1905–14
8. G.B. Olson and M. Cohen, General Mechanism of Martensitic Nucleation. 3. Kinetics of Martensitic Nucleation, *Metal. Trans. A.* 7, 1976, 1915–23
9. J.W. Christian, A theory of the Transformation in Pure Cobalt, *Proc. R. Soc. London*, Ser. 1 206, 1951, 51–64
10. D.P. Koistinen and R.E. Marburger, A General Equation Prescribing the Extent of the Austenite-Martensite Transformation in Pure Iron-Carbon Alloys and Plain Carbon Steels. *Acta Metall.* 7, 1959, 59–60
11. H. Bhadeshia, A Rationalization of Shear Transformations in Steels. *Acta Metall.* 29: 1117–30, 1981

12. G.B. Olson, H. Bhadeshia, and M. Cohen, Coupled Diffusional Displacive Transformations. *Acta Metall.* 37, 1989, 381–89
13. G.B. Olson, H. Bhadeshia, and M. Cohen, Coupled Diffusional Displacive Transformations. 2. Solute Trapping. *Metal. Trans. A.* 21, 1990, 805–9
14. G.B. Olson and M. Cohen, Dislocation Theory of Martensitic Transformations, Dislocations in Solids, ed. F.R.N. Nabarro, North-Holland, 1986, p 297–407
15. M. Cohen, G.B. Olson, and P.C. Clapp, On the Classification of Displacive Transformations (What is Martensite?), *International Conference on Martensitic Transformations ICOMAT '79*, eds. Olson G.B. and M. Cohen, Alpine Press, 1979, p 1–11
16. J.W. Christian, Thermodynamics and Kinetics of Martensite, *International Conference on Martensitic Transformations ICOMAT '79*, eds. G.B. Olson G.B. and M. Cohen, Alpine Press, 1979, p 220–34
17. G.B. Olson and M. Cohen, Interphase-Boundary Dislocations and the Concept of Coherency, *Acta Metall.* 27, 1979, 1907–18
18. G.B. Olson, Advances in Theory: Martensite by Design. *Maer. Sci. Eng.*, A 438–40, 2006, 48–54
19. C.J. Kuehmann and G.B. Olson, Computational Materials Design and Engineering. *Mater. Sci. Technol.* 25, 2009, 472–78
20. G.B. Olson and C.J. Kuehmann, Materials Genomics: from CALPHAD to Flight. *Scr. Mater.* 70, 2014, 25–30
21. A. Saha and G.B. Olson, Computer-Aided Design of Transformation Toughened Blast Resistant Naval Hull Steels: Part I. *J. Comput. Aided. Mater. Des.* 14, 2007, 177–200
22. A. Saha, J. Jung, and G.B. Olson, Prototype Evaluation of Transformation Toughened Blast Resistant Naval Hull Steels: Part II. *J. Comput. Aided. Mater. Des.* 14, 2007, 201–33
23. J.F. Watton, G.B. Olson, and M. Cohen, A Novel Hydrogen-Resistant UHS Steel, *Innovations in Ultrahigh-Strength Steel Technology*, eds. E.S. Wright, G.B. Olson, and M. Azrin, US Army Materials Technology Laboratory, 1987, p 705–37
24. G.B. Olson, New Steels by Design. *J. Mat. Educ.* 11, 1989, 515–28
25. G.L. Krasko and G.B. Olson, Ferromagnetism and Crystal-Lattice Stability of BCC and FCC Iron, *J. Appl. Phys.* 67, 1990, 4570–72
26. G.L. Krasko and G.B. Olson, Effect of Boron, Carbon, Phosphorus and Sulfur on Intergranular Cohesion in Iron. *Solid State Commun.* 76, 1990, 247–51
27. R.Q. Wu, A.J. Freeman, and G.B. Olson, First Principles Determination of the Effects of Phosphorus and Boron on Iron Grain-Boundary Cohesion. Science 265, 1994, 376–80
28. R.Q. Wu, A.J. Freeman, and G.B. Olson, Effects of Carbon on Iron-Grain-Boundary Cohesion: First-Principles Determination. *Phys. Rev. B.* 53, 1996, 7504–9
29. I.G. Kim, G. Rahman, J.H. Jang, Y.Y. Song, S.W. Seo, H. Bhadeshia, A.J. Freeman, and G.B. Olson, A Systematic Study on Iron Carbides from First-Principles, *Pricm 7*, pts 1–3, eds. J.F. Nie and A. Morton, Trans Tech Publications, 2010, p 47–50
30. H. Bhadeshia and D.W. Suh, Is Low Phosphorus Content in Steel a Product Requirement? *Ironmaking Steelmaking* 42, 2015, 259–67
31. H. Bhadeshia and R. Honeycombe, Steels—Microstructure and Properties (4th ed.), Elsevier, 2017
32. H.C. Ling and G.B. Olson, Strain Energy of a Martensitic Nucleus, *Proceedings of an International Conference on Solid to Solid Phase Transformations*, Metallurgical Society of AIME, 1982, p 1171–75
33. G.B. Olson and M. Cohen, Early Stages of Aging and Tempering of Ferrous Martensites. *Metal. Trans. A.* 14, 1983, 1057–65
34. M. Grujicic, G.B. Olson. and W.S. Owen, Mobility of Martensitic Interfaces. *Metal. Trans. A.* 16, 1985, 1713–22
35. R.L. Patterson and C.M. Wayman, Crystallography and Growth of Partially-Twinned Martensite Plates in Iron-Nickel Alloys. *Acta Metall.* 14, 1966, 347–69

Transformations—Selected Works of G.B. Olson on Materials, Microstructure, and Design
C.E. Campbell, M.V. Manuel, and W. Xiong, editors

CHAPTER 1

Dislocation Theory of Martensitic Transformations*

G.B. Olson and Morris Cohen
Massachusetts Institute of Technology

Introduction

A martensitic phase transformation can be considered as the spontaneous plastic deformation of a crystalline solid in response to internal chemical forces. It is no surprise, then, that the theory of dislocations, after incorporating the generalized forces of chemical thermodynamics and the special constraints of crystalline interfaces, has provided the most useful basis for understanding these transformations. Martensitic transformations have recently been defined as a subset of diffusion-less/displacive solid-state transformations in which the strain energy arising from a shear-dominant lattice distortion controls the kinetics and product morphology during transformation (Ref 1). Such transformations are first-order and heterogeneous in nature, proceeding by the propagation of relatively sharp interfaces. These characteristics make the approximations of traditional linear-elastic dislocation theory particularly appropriate, and it is for this reason that the application of the theory to martensitic transformations has been so successful.

The current state of the general theory of martensitic transformation crystallography and nucleation mechanisms has been reviewed fairly recently (Ref 2, 3). We here address martensitic transformation theory from a dislocation perspective with special emphasis on the most recent advances; this includes the dislocation theory of martensitic interfacial structure and mobility, and its application to the mechanism and kinetics of nucleation and growth.

Crystallography and Interfacial Structure

One approach to the structure of interphase boundaries is the generalization of the theory of grain-boundary structure which emphasizes the *topological* characteristics of interfacial dislocations and their role in surface-pattern matching (Ref 4). In contrast, theory of martensitic interfacial structure has emphasized the *kinematic* characteristics of interfacial dislocations, that is, the deformations accompanying their motion, based on a separately evolved deformation theory of interphase crystallography. Originally phenomenological in nature (Ref 5, 6), the theory of martensitic transformation crystallography was put on a fundamental basis via the principles of elastic strain-energy minimization by Christian, (Ref 7), who also applied these basic concepts to nonmartensitic transformations. In comparison with grain-boundary theory, which evolved from the consideration of planar interfaces in

*G.B. Olson and M. Cohen, Dislocation Theory of Martensitic Transformations, *Dislocations in Solids*, Vol. 7, F.R.N. Nabarro, Ed., North-Holland, 1986, p 295–407

long-range stress-free infinite bicrystals, this approach is particularly useful for the problem of enclosed second-phase particles of importance to phase transformations. After examining the deformation approach to transformation crystallography in its most general form, we will discuss the prediction of the dislocation structure of interphase boundaries of varying states of coherency, and then explore its application to the structure and properties of martensitic interfaces.

Kinematic Approach to Transformation Crystallography

Crystal Deformations and Particle Coherency

In any theory of crystalline interfacial structure, it is necessary to identify a *lattice correspondence* which relates the identity of lattice planes and directions in the two crystal structures which meet at the interface. This correspondence then defines a *lattice deformation* which is an affine transformation describing the homogeneous distortion of a unit cell of one structure to that of the other. In matrix notation, a lattice deformation $\boldsymbol{S}$ can be factored into a pure strain $\boldsymbol{B}$ and a rigid-body rotation $\boldsymbol{R}$:

$$\boldsymbol{S} = \boldsymbol{RB} \qquad \text{(Eq 1)}$$

When one or both of the crystal structures involves a basis, additional inhomogeneous atom displacements, termed *shuffles*, may occur within the unit cell which, as such, do not contribute to the distortion of lattice vectors. The distinction between the homogeneous lattice deformation and inhomogeneous shuffle displacements is illustrated in Fig. 1.

The lattice deformation accompanying a phase transformation can be regarded as a "plastic" deformation in that it relates structural states which are stable or metastable, as distinct from the intrinsically unstable states associated with an "elastic" deformation. For the case of an enclosed particle in a matrix, if a particle has undergone no plastic deformation other than the lattice deformation (with or without shuffles), it is considered *fully coherent*, and its elastic strain energy per particle volume can be calculated according to the procedure prescribed by Ref 8 (for ellipsoidal shapes) using the "stress-free transformation strain", ε_{ij}^{T}, defined by the lattice deformation, $\boldsymbol{S}$. While shuffle displacements will contribute

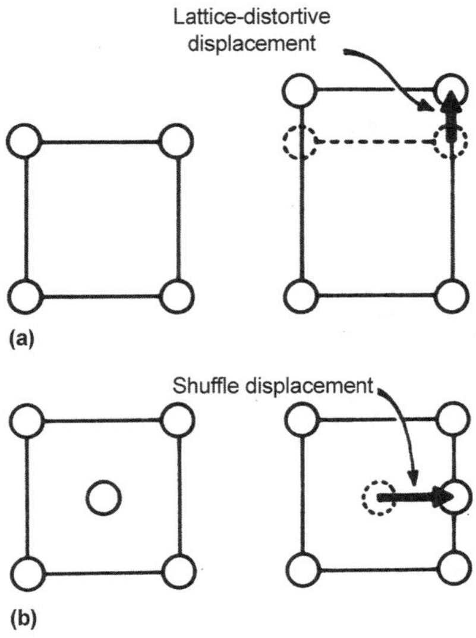

Fig. 1 Displacements of atomic sites in a crystal transformation. (a) Lattice-distortive displacements associated with a homogeneous lattice deformation, $\boldsymbol{S}$; (b) shuffle displacements within the unit cell. Source: Ref 1

to the coherent-particle interfacial energy, only the homogeneous deformation $\boldsymbol{S}$ contributes to the long-range elastic strains which determine the particle volume strain energy.

As will be discussed further in the section "Thermodynamic and Kinetic Constraints on Morphology and Orientation Relations," the character of the deformation $\boldsymbol{S}$ will determine the particle morphology which minimizes the strain energy of a fully coherent particle. Further strain-energy reduction can be achieved by decreasing the net transformation strain through the superposition of other internal plastic deformations within the particle. (The influence of plastic deformation of the surrounding matrix will be discussed in the section "Nonthermoelastic Growth" in this chapter.) Once the lattice deformation has produced the correct structure of the transformation product, any further deformation must consist of a *lattice-invariant* deformation which produces a change of shape without altering the lattice. This can occur by any of the usual mechanisms of conventional plastic deformation in crystals, such as the glide and climb of crystal lattice dislocations or mechanical twinning, and will be subject to the usual constraints of crystal plasticity. Macroscopic (long-range) cancellation of a component of the homogeneous lattice deformation is most effectively accomplished by a uniform periodic array of lattice dislocations or twins.

To understand the character of the lattice-invariant deformations associated with interphase boundaries, it is useful at this point to consider the specific nature of the deformations generated by periodic arrays of interfacial dislocations. An *invariant-plane strain* (IPS) is a deformation which leaves a particular plane macroscopically undistorted and unrotated, and therefore invariant. Such a deformation of an initially rectangular body is illustrated in the upper portion of Fig. 2. The combination of a simple shear γ_{12} and a uniaxial strain ε_{22} leaves the x_2 plane invariant. As indicated by the vector in Fig. 2, the resulting displacement of a point at the upper right corner of the rectangle includes components along both x_1 and x_2. Dislocations in a periodic array at the interface of an enclosed particle are in the form of closed loops defined by the intersection of the particle interface with the composition plane (i.e., plane of motion) of each dislocation. As depicted for such an array in the lower portion of Fig. 2, the two components of the dislocation Burgers vector in the loop plane (b_1) and normal to the plane (b_2) produce the two macroscopic strain components γ_{12} and ε_{22} of an invariant-plane strain with the dislocation loop plane as the invariant plane. For an array of dislocation loops of spacing D_0, $\gamma_{12} = b_1/D_0$ and $\varepsilon_{22} = b_2/D_0$. As these strains arise from discrete displacements at intervals of D_0, they are not truly homogeneous, but are actually inhomogeneous deformations on the scale of D_0. Expansion or

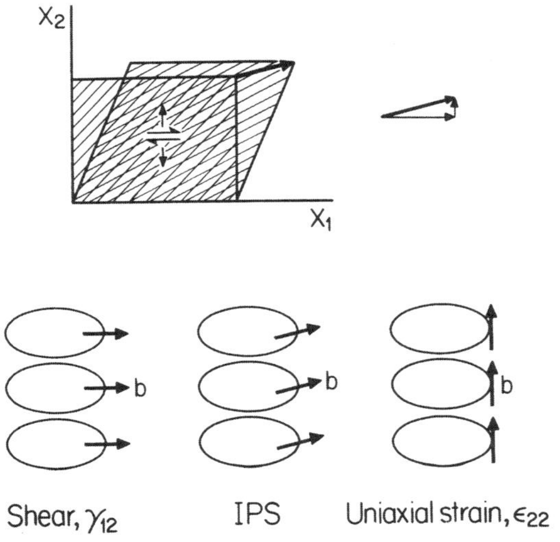

Fig. 2 Invariant-plane strain (IPS) deformation of an initially rectangular body, and arrays of dislocation loops producing cases of simple shear, general IPS, and uniaxial strain deformations.

contraction of the dislocation loops during motion of the particle interface generally requires glide of the b_1 component and climb of the b_2 component of the dislocation Burgers vector. Hence, a lattice-invariant deformation arising from dislocation Burgers vectors with only a b_1 glide component is conservative, while a deformation involving a b_2 climb component will be nonconservative. In an interface involving the superposition of several such periodic dislocation arrays, each array can be considered individually to contribute an invariant-plane strain defined by its Burgers vector, loop plane, and dislocation spacing.

The superposition of several invariant-plane strains can produce (or accommodate) an arbitrary uniform deformation. Denoting an invariant-plane lattice-invariant deformation as $\boldsymbol{P}_i$, the net transformation strain or macroscopic shape strain $\boldsymbol{E}$ of a transformation product arising from the superposition of a lattice deformation (Eq 1) and n invariant-plane lattice-invariant deformations can be described in matrix notation by:

$$\boldsymbol{E} = \boldsymbol{RBP}_n \cdots \boldsymbol{P}_2\boldsymbol{P}_1 \qquad \text{(Eq 2)}$$

While the notation suggests a sequence of operations beginning with $\boldsymbol{P}_1$ (which is often most convenient mathematically for calculations), the processes can be regarded as physically simultaneous during a phase transformation. The occurrence of a lattice-invariant deformation within the transformed volume reduces the degree of overall coherency of the transformation product. For a *semicoherent* enclosed particle, the volume elastic strain energy is equivalent to that of a coherent particle with the net transformation strain defined by $\boldsymbol{E}$. As will be developed in the section "Martensitic Interfaces," the inhomogeneous nature of the lattice-invariant deformation produces a short-range interfacial stress field which contributes to the interfacial (surface) energy of the particle.

Thermodynamic and Kinetic Constraints on Morphology and Orientation Relations

As Eq 2 indicates, the most general phase transformation can involve a complex set of deformations. Which deformations actually occur and the product morphology that results will depend on the thermodynamic and kinetic constraints which apply during the transformation. Kinetic factors include restrictions on the lattice-invariant deformations (e.g. conservative versus nonconservative, twinning versus slip) and the morphological influence of relative interfacial mobilities (e.g. ledge structures versus smooth interfaces). Once kinetic constraints can be specified for a given phase transformation, prediction of the extent and specific nature of lattice-invariant deformation and its influence on morphology can be sought from the thermodynamic constraints embodied in the minimization of strain energy and/or surface energy. While competition between strain-energy and surface-energy constraints (or between thermodynamic and kinetic constraints) can be complex, the behavior of simple systems suggests some principles of energy minimization which may be fairly general. Before discussing these, we examine a relation between product morphology and crystallographic orientation relations which should have broad applicability.

Orientation relations. In addressing the constraints that apply to the crystallography of interphase boundaries, it is helpful to distinguish two classes of problem which can be termed the *interphase boundary* (IPB) and the *interphase grain boundary* (IPGB). The IPGB problem concerns the prediction of the structure of an interface between two crystals related by both an arbitrary rotation and an arbitrary lattice deformation. This is a generalized grain boundary corresponding to the interface produced by the growth impingement of two randomly oriented crystals of differing structure, or the formation of a second phase at a grain boundary. In contrast, an IPB accompanying the formation of a phase of given structure (lattice deformation) in a crystal matrix involves a rotation, or orientation relation, which is not arbitrary. A function of the deformation-crystallographic theory, as it has been applied to martensitic transformations, is to predict the orientation relations for a phase transformation involving a specified lattice deformation.

While the function of grain-boundary dislocations is to produce a rotation, the primary function of the interphase boundary dislocations accounting for the lattice-invariant deformation of Eq 2 is to produce a strain (canceling a component of $\boldsymbol{B}$). As has been established for the influence of grain shape on crystallographic texture development (Ref 9, 10), the rigid-body rotation that accompanies a finite strain will depend on the boundary conditions imposed by the morphology of the deformed (transformed) region. This can be illustrated by comparison of the rotation produced by the same set of twinning dislocations when they are arranged to form a plate-shaped twin with a habit on the K_1 or on the K_2 conjugate twinning planes. In both instances, the pure deformation (principal strains) is identical, but the habit plane of the twin is always unrotated, giving different rigid-body rotations in the two cases. In the same manner that a plate shape constrains its habit plane against rotation, a rod or lath shape will constrain its habit direction against rotation. These considerations suggest the following principle influencing the orientation relations in crystalline phase transformations:

Principle 1. *The rigid-body rotation accompanying the transformation of an enclosed particle will tend to minimize rotation of directions of maximum dimension.*

While additional conditions will generally be required to uniquely specify the orientation relations of a given transformation, the operation of this principle sets an important restriction on the range of possible orientation relations, which has not been widely recognized. The principle is expected to hold regardless of the nature of the factors (thermodynamic or kinetic) controlling product morphology. When thermodynamic factors such as strain-energy or surface-energy minimization prevail, some restrictions on particle shape and habit can then be predicted from the nature of the pure strain produced by the combined lattice deformation and lattice-invariant deformation.

Strain-energy minimization. Although the strain energy of a particle of arbitrary shape and transformation strain will depend on the elastic properties of the product and matrix phases, certain energy extrema can be identified from the character of the net deformation $\boldsymbol{E}$ of Eq 2 and its relation to particle morphology, independent of elastic properties. Clearly there is no strain energy when there is no net transformation strain, $\boldsymbol{E} = \boldsymbol{I}$, whatever the morphology. The slow motion of an interphase boundary at high temperatures, where unrestricted multiple nonconservative lattice-invariant deformations occur, could achieve this condition, resulting in a particle which is macroscopically fully incoherent. Restricting the lattice-invariant deformations to conservative deformations, but allowing unlimited shear systems to operate (including interface sliding) corresponds to a "dilating incoherent" particle, the strain energy of which has been treated by Nabarro (Ref 11). While these notions of incoherent particles have provided convenient approximations for the treatment of the energetics of diffusional phase transformations, it is not clear that such extremes ever occur in solid-state reactions, and there is increasing evidence of partial coherency in a wide range of crystalline phase transformations (Ref 12).

It was noted in the section, "Crystal Deformations and Particle Coherency," that a linear-elastic estimate of the strain energy of a coherent or semicoherent particle can be obtained from the method of Eshelby (Ref 8) using the pure transformation strain $\varepsilon_{ij}^{\mathrm{T}}$ defined by the net deformation $\boldsymbol{E}$. As discussed by Christian (Ref 7), after due consideration of the limitations inherent in applying linear-elastic infinitesimal strain calculations to finite strain problems (Ref 13), one can identify from the general Eshelby solutions some important energy extrema in the relation between $\varepsilon_{ij}^{\mathrm{T}}$ and particle morphology consistent with deformation compatability. A particularly important result is the case of a deformation which leaves a plane undistorted. For an $\varepsilon_{ij}^{\mathrm{T}}$ consisting of components γ_{12} and ε_{22} which leave the x_2 plane undistorted (as in the deformation of Fig. 2), the strain energy per particle volume of a thin ellipsoidal plate of radius r and semithickness c lying in the x_2 plane is expressed by:

$$E_{\mathrm{v}} = Kc/r, \quad K = k_1\gamma_{12}^2 + k_2\varepsilon_{22}^2 \qquad \text{(Eq 3)}$$

where k_1 and k_2 are functions of the anisotropic elastic constants; for isotropic elasticity with the same shear modulus μ and Poisson's ratio v for both phases, $k_1 = \pi\mu(2 - v)/8(1 - v)$ and $k_2 = \pi\mu/4(1 - v)$. In the limit $c/r \rightarrow 0$, the strain energy vanishes regardless of the magnitude of the elastic constants. In contrast, a plate with the same $\varepsilon_{ij}^{\mathrm{T}}$ but a habit in the x_3 plane has a large strain energy which is insensitive to c/r. This arises from the poor elastic accommodation (compatability) of strains which distort vectors in the plate habit plane. From the general form of the Eshelby solutions, it is evident that elastic accommodation is generally poorest for strains which distort directions of largest dimension of an enclosed particle. For a given $\varepsilon_{ij}^{\mathrm{T}}$ then, this suggests a principle for the prediction of a particle morphology (shape and habit) which minimizes the strain energy per particle volume:

Principle 2. *A transformed particle which minimizes strain energy per volume will maximize its dimensions in directions of minimum distortion.*

For an $\varepsilon_{ij}^{\mathrm{T}}$ which leaves a plane undistorted, strain energy will be minimized by a plate shape with the undistorted plane being adopted as its habit, as for the case of Eq 3. Principle 1 then requires that this plane be unrotated, and so $\boldsymbol{E}$ must be an invariant-plane strain. The latter condition, together with the restriction of conservative lattice-invariant deformations in diffusionless transformations, provides the fundamental assumption underlying the crystallographic theory commonly applied to martensitic transformations. When the lattice deformation $\boldsymbol{B}$ leaves no plane undistorted, but leaves a line (or cone of lines) undistorted, the superposition of a single lattice-invariant shear $\boldsymbol{P}$ is sufficient to produce a macroscopically undistorted plane. Equation 2 then becomes (Ref 5, 6):

$$\boldsymbol{E} = \boldsymbol{RBP} \qquad \text{(Eq 4)}$$

When a shear system of $\boldsymbol{P}$ is selected (or inferred from experimental observation), the magnitude of $\boldsymbol{P}$ can be determined such that $\boldsymbol{BP}$ leaves a plane undistorted. The $\boldsymbol{R}$ which leaves this plane unrotated can then be determined, making $\boldsymbol{E}$ an IPS (Ref 5). The complete orientation relation is usually specified by the additional condition that the undistorted line of $\boldsymbol{B}$ which lies in the invariant plane of $\boldsymbol{E}$ is an invariant line. This method has successfully accounted for the observed habit planes, sub-structures (twinning systems), and macroscopic shape strains for semicoherent plate martensites in a wide variety of thermoelastic alloys and low-temperature f.c.c. → b.c.c. transformations in ferrous alloys (Ref 14).

There is also evidence that diffusional f.c.c. → b.c.c. transformations involving plate-shaped products, such as the Widmänstatten ferrite transformation, represent semicoherent IPS transformations (Ref 15), and high-resolution microscopy of interfacial dislocation structures indicates a nonconservative lattice-invariant deformation (Ref 16). Adaptation of the IPS calculational methods to obtain a quantitative solution of the crystallography of this transformation, using observed interfacial dislocation line directions and Burgers vectors, is under investigation (Ref 17).

In the case of a fully coherent particle for which $\boldsymbol{B}$ gives an undistorted line but not an undistorted plane, principle 2 predicts a rod or lath shape lying along this line, and principle 1 then requires that $\boldsymbol{E} = \boldsymbol{RB}$ is an invariant-line strain. There is some evidence for coherent precipitation of b.c.c. Cr from f.c.c. Cu involving such invariant-line rods (Ref 18). The f.c.c. → b.c.c. lattice deformation actually defines a cone of undistorted lines, and another condition in addition to principle 2 is required to predict which line is selected as the rod habit. A detailed crystallographic analysis of rod-shaped b.c.c-f.c.c. precipitation at free surfaces in Cu-Zn alloys has identified that the rod axis lies along the line of intersection of the undistorted cone with the free surface, and that the orientation relations are consistent with an invariant-line strain as required by principle 1 (Ref 19).

The importance of an invariant line established by the lattice deformation in determining the orientation relations in solid-state phase transformations is receiving increasing recognition (Ref 20–22). As in the case of the simple "single shear" martensitic transformations mentioned earlier, only a single IPS lattice-invariant deformation $\boldsymbol{P_1}$ is required to make $\boldsymbol{E}$ a low strain-energy IPS when $\boldsymbol{RB}$ is itself an invariant-line strain. This can be equivalently demonstrated by application of the Bollmann O-lattice theory (Ref 13, 23): an invariant-line lattice deformation defines an O-line lattice requiring only a single set of primary interfacial dislocations to match the two structures. This gives a relatively simple interfacial structure and is the expected result when strain-energy minimization is the only consideration.

Surface-energy minimization. The simplest interfacial structure that minimizes strain energy does not necessarily minimize the particle surface energy. As the interfacial energy of a semicoherent particle arises in part from the elastic energy of its short-range interfacial stress field, we distinguish here between particle strain energy and particle surface (interfacial) energy by separating those energy contributions which scale with the particle volume from those which scale with the interfacial area. For guidance in minimizing surface energy, we look to generalized grain-boundary theory (Ref 4). Although it is more difficult to quantitatively predict interfacial energies compared to particle strain energies, theoretical modeling and accumulated experimental evidence indicate that energy is minimized for interfaces involving a high density of coincidence between lattice sites of the two structures, giving rise to relatively short wavelength patterns of good matching. This then provides a third principle which influences interphase-boundary crystallography.

Principle 3. *A particle which minimizes surface energy will adopt a morphology which maximizes the coincidence of lattice sites in its interface.*

For a given lattice deformation, the probability of site coincidence along a particular plane or direction scales with the packing density. Thus, a corollary of this principle is that surface-energy minimization will favor habits near the closer-packed low-index planes and directions.

Under kinetic conditions where lattice-invariant deformations are very restricted, surface-energy minimization may not be possible. This is likely the case for low-temperature single lattice-invariant-shear plate martensites where the observed irrational habits can be accounted for entirely on the basis of strain-energy minimization, as for the (3, 10, 15)-type habit in ferrous alloys. When the transformation temperature is increased (or the transformation rate decreased), more complex lattice-invariant deformations such as multiple shears become possible, and may then allow the development of higher-coincidence interfacial patterns according to principle 3. Hence, a tendency toward closer-packed low-index habits might underlie the transition in ferrous f.c.c. → b.c.c. transformations to habits near (225), containing the [110] close-packed direction, and ultimately toward the close-packed (111) habit. Correlation of habit transitions with plastic-flow properties confirm that these habits form under conditions where lattice-invariant deformations are less restricted (Ref 24).

In the application of generalized grain-boundary theory to planar interfaces, it is usually assumed that the interface possesses no long-range stress field such that volume strain energy is zero, consistent with principle 2. Similarly, one might expect that the operation of principle 3 in an enclosed particle will be subjected to the constraint of principle 2. There are many phase transformations for which some degree of strain-energy reduction (semi-coherency) is a primary requirement in order that the total volume free-energy change (chemical + strain energy) can become negative for the transformation to be thermodynamically possible. Absolute strain-energy minimization will not always be necessary, however, and due to the relative contributions of strain energy and surface energy for a finite particle, a compromise may be reached between principles 2 and 3.

A clear example of such a compromise is the coherent f.c.c.-h.c.p. martensitic transformation occurring in many Co- and Fe-based alloys. The transformation takes place fully coherently with a close-packed habit plane, and the matching of the close-packed planes is so close that the lattice deformation is very nearly an invariant-plane strain. The matching is not exact, however, so that a small isotropic elastic distortion exists in the habit plane. Thus, the habit is neither a true invariant plane, nor does it contain an undistorted line of the lattice deformation, in violation of the usual assumptions of martensite crystallographic theory. On the other hand, undistorted lines do exist out of the habit plane, and so strain-energy minimization could be achieved by the operation of a single lattice-invariant shear resulting in an irrational true invariant-plane habit. That this does not actually occur is undoubtedly a consequence of the very low surface energy of the coherent close-packed interface. The possibility of such elastic distortions in martensitic interfaces was considered by Bowles and Mackenzie (Ref 6), thus relaxing the invariant-plane concept to one of "near invariance." Similarly, it should be noted that, referred to the two stress-free lattices, the coincidence of lattice sites in the low-energy close-packed interface is not exact either. A modification of exact coincidence-lattice concepts to admit "near coincidence" has also been found necessary to treat such cases (Ref 4).

While it is difficult to quantify a compromise between surface and strain energies with the same precision as the minimization of either factor alone, it can be expected that the relative importance of these factors will be influenced by the size-dependent surface-to-volume ratio of an enclosed particle. Therefore, the morphology and degree of coherency of a transformation product will usually be most influenced by surface-energy minimization at small sizes, particularly at nucleation.

Coherency of Particles and Interfaces

The state of coherency of enclosed particles, as discussed thus far, is usually defined in terms of the *continuum* concept of continuity of displacements and tractions across the particle interface (Ref 8). From a deformation viewpoint, this is controlled by the extent to which lattice-invariant deformations occur, relaxing the displacements and tractions which arise from an otherwise elastically accommodated (coherent) lattice deformation. In contrast, the coherency of interfaces is usually defined in terms of the *discrete lattice* concept of continuity of corresponding crystal planes and lines across the interface (Ref 7). If we employ the corresponding crystal planes and lines to define a continuous "network" across the interface, the preservation of this network guarantees continuity of displacements and tractions; accordingly, these two notions of coherency are physically equivalent and any interface of a coherent particle is a coherent interface (Ref 25).

Just as the state of coherency of a given particle from a continuum viewpoint will depend on the choice of the lattice deformation relating the two phases, its coherency from the discrete-lattice viewpoint likewise depends on the choice of lattice correspondence relating crystal planes and lines. In assessing the coherency of a particle or interface, therefore, it is essential to specify a correspondence (Ref 26). This is particularly important in examining the properties of interfacial dislocations.

Interfacial Dislocations

Coherency versus Anticoherency Dislocations

The concepts of lattice deformations and lattice-invariant deformations identify two roles of interphase-boundary dislocations which, in turn, define two types of dislocations with markedly different properties. The first type is illustrated in Fig. 3. Figure 3(a) shows an ellipsoidal coherent continuum particle which has undergone a lattice deformation consisting of a simple shear γ_{xz} on its habit plane. The same particle in a discrete lattice is depicted in Fig. 3(b) wherein the continuity of two sets of crystal planes a and b is indicated. Due to the discrete nature of the crystal planes, the homogeneous shear of the material inside the particle actually corresponds to finite relative displacements of each a plane in the x direction. These discrete displacements relative to the material outside define the Burgers vectors of an array of partial dislocations shown in Fig. 3(c). Since all corresponding crystal planes and lines are continuous across the interface even though the dislocations are present, these dislocations are not lattice

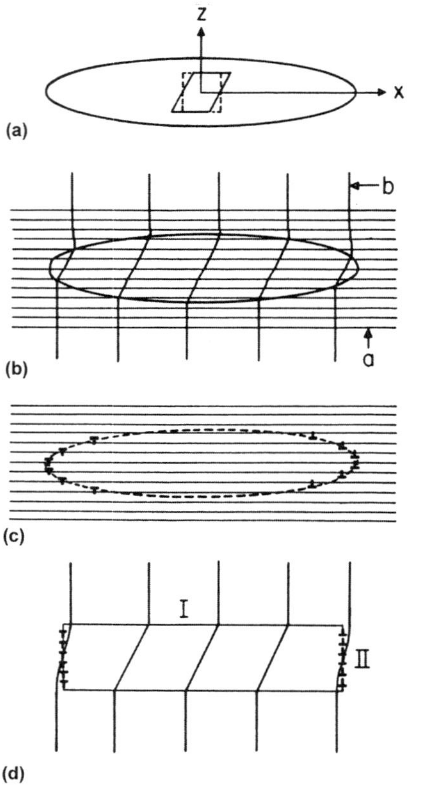

Fig. 3 Coherent shear particle. (a) Ellipsoidal continuum-elastic particle; (b) same particle in discrete lattice; (c) position of coherency dislocations in interface; (d) squared-off particle distinguishing stress-free coherent interface I from forced-elastic coherent interface II. Source: Ref 25

discontinuities. Such dislocations, which accomplish lattice deformations, are often called transformation dislocations. In view of their special property of maintaining lattice continuity, the name *coherency* dislocation has been suggested (Ref 25). Moving the dislocations in their glide planes to define a rectangular particle in Fig. 3(d) helps to illustrate two types of coherent interface. Dislocation-free interface I represents "stress-free" coherency, whereas interface II represents "forced-elastic" coherency.

Lattice continuity through a single coherency dislocation from the upper left of Fig. 3(c) is demonstrated in Fig. 4 for the case where γ_{xz} is a twinning shear. Motion of the dislocation of Fig. 4(b) to the left advances the stress-free coherent twin interface of Fig. 4(a) by one plane (Fig. 4(c)). As long as the corresponding planes are followed, a one-to-one correspondence of planes exists across the interface, even through the dislocation. This is also shown by a similar construction for planes initially orthogonal to the interface in Fig. 5.

A second basic type of coherency dislocation is illustrated in Fig. 6 displaying a coherent particle of the same shape as that of Fig. 3 but with a lattice deformation consisting of a uniaxial strain in the *z* direction, ε_{zz}. Again the discrete relative displacements of the crystal planes define the finite Burgers vectors of the coherency dislocations indicated. Moving these dislocations to form a rectangular particle shape as in Fig. 6(b) again produces a stress-free coherent interface I and a forced-elastic coherent interface II.

The coherency dislocations of Fig. 3 and 6 are partial dislocations of both the parent and product crystal lattices and consequently their motion is constrained by fault energy. This forces the dislocations to remain in the interface and determines the nature of their motion as individual defects. For three-dimensional enclosed particles, the individual dislocations are in the form of closed loops defined by the line of intersection of the crystal planes (*a* planes of Fig. 3) with the interface. To produce a homogeneous (fault-free) lattice deformation requires one such dislocation per lattice plane transformed. This means that the motion of each coherency dislocation is restricted to the plane of its loop. For the simple shear transformation of Fig. 3 where the Burgers vector lies in the loop plane, this motion corresponds to ordinary glide. For the uniaxial strain of Fig. 6, the Burgers vector lies normal to the loop plane, and the required motion corresponds to dislocation climb. The climb motion in this case is a conservative process, however, inasmuch as it does not create or destroy lattice points, but simply alters an interplanar spacing. This behavior represents another special property of coherency dislocations.

The two special types of coherency dislocations of Fig. 3 and 6 can be called "glide coherency dislocations" and "climb coherency dislocations," respectively. As for the lattice dislocation arrays of Fig. 2, the general case can involve Burgers vectors with components of both types, each such array generating an invariant-plane strain. The superposition of several such arrays, each lying on rational crystal planes, can account for any lattice deformation.

In contrast to the coherency dislocations, the crystal-lattice dislocations which accomplish lattice-invariant deformations possess the usual properties of lattice dislocations. Climb motion

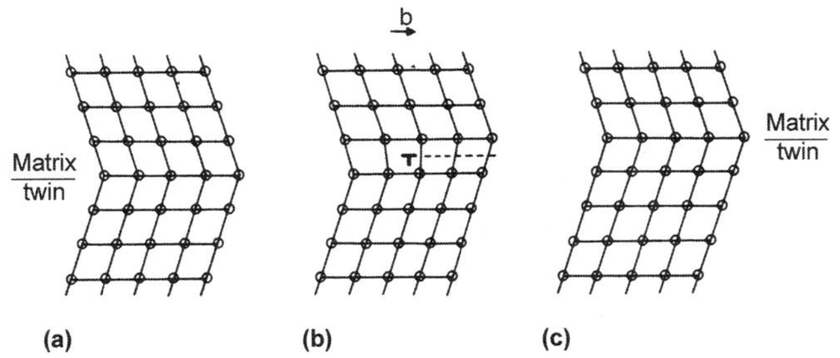

Fig. 4 Atom positions across coherent twin boundary. (a) Stress-free coherent boundary; (b) boundary with twinning dislocation; (c) stress-free boundary after twinning advance by one plane. (Source: Ref 25)

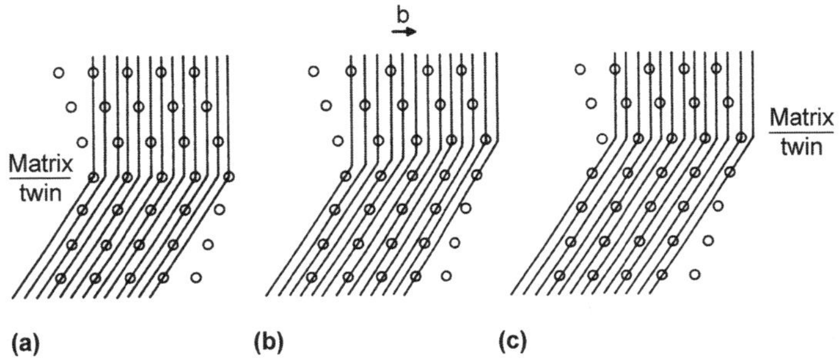

Fig. 5 Same atom positions as in Fig. 4 showing continuity of planes initially orthogonal to twin boundary. Source: Ref 25

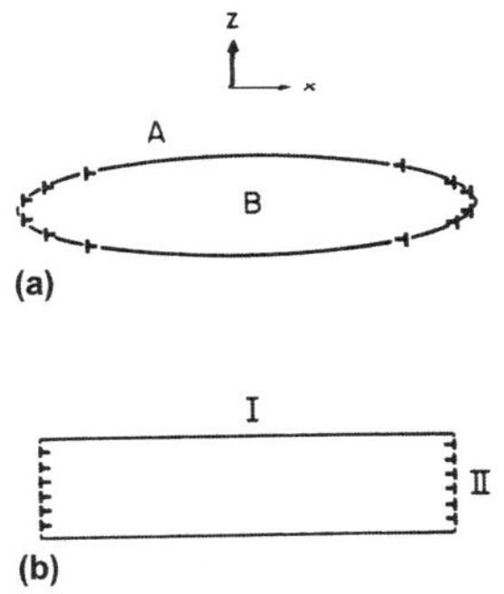

Fig. 6 Enclosed coherent particle of crystal B in matrix A with lattice deformation consisting of a uniaxial distortion. (a) Particle shape corresponding to Fig. 3(c); (b) particle shape corresponding to Fig. 3(d), again distinguishing stress-free coherent interface I from forced-elastic coherent interface II. (Source: Ref 25)

of these dislocations is nonconservative and would thus require diffusion. Instead of producing a homogeneous lattice deformation and maintaining lattice continuity, they disrupt the uniformity of the lattice correspondence and act as true lattice discontinuities. Sometimes called "misfit" dislocations, these dislocations which reduce the coherency of an interface can also be termed *anticoherency* dislocations. The function and properties of coherency and anticoherency dislocations are contrasted in Table 1.

Extending the forced-elastic coherent interface II of the uniaxial-strain particle of Fig. 6(b) indefinitely in the *z* direction produces the familiar case of the "epitaxial" misfitting interface illustrated in Fig. 7. This provides a convenient example for examining the interrelation of coherency and anticoherency dislocations. Figure 7(a) depicts two stress-free crystals which differ in interplanar spacing normal to the *z* direction. Joining the crystals coherently along a plane normal to *x* requires the relative uniaxial distortions shown in Fig. 7(b). The discrete relative displacements of the crystal planes then define the Burgers vectors of the coherency dislocation array of Fig. 7(c), corresponding to the climb coherency dislocations of Fig. 6. The long-range stress field of this array can be canceled by the superposition of an inhomogeneous array of anticoherency lattice dislocations resulting in an equal and opposite long-range stress field as in the semicoherent interface of Fig. 7(d). The anticoherency dislocations act as discontinuities, and a short-range elastic stress field remains from the superposition of the two

Table 1 Characteristics of interphase-boundary dislocations

	Coherency dislocations	Anticoherency dislocations
Alternate names	Transformation	Misfit: accommodation
Function	Lattice deformation (accomplish transformation strain)	Lattice-invariant deformation (reduce strain energy)
Effect on coherency	Maintain coherency and lattice correspondence	Reduce coherency; disrupt uniformity of lattice correspondence
Burgers vector	Partial *b*	Lattice *b*(a)
Motion	Conservative climb and/or glide, but restricted to plane of dislocation loop	Conservative glide; climb by diffusion

(a) Except special case of twinning dislocations accomplishing a lattice-invariant deformation.

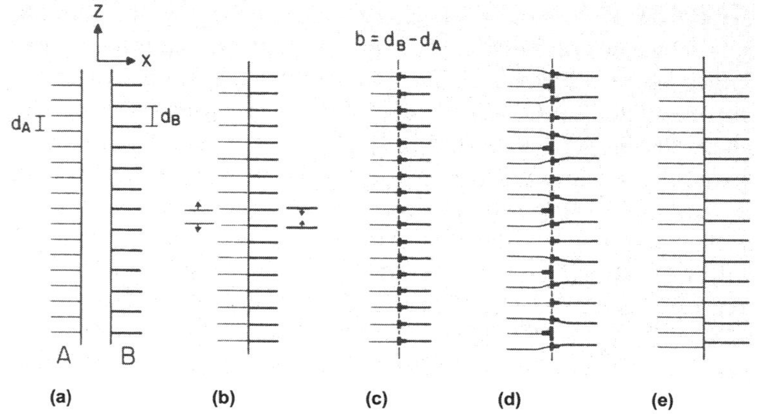

Fig. 7 Planar interphase boundary with uniaxial distortion. (a) Two crystals, A and B, with different interplanar spacings, d_A and d_B, in the *z* direction; (b) position of planes across coherent interface: (c) partial dislocations required in discrete lattice to maintain configuration of (b): (d) semicoherent boundary with anticoherency dislocations superimposed on coherency dislocations of (c); (e) incoherent boundary. Source: Ref 25

arrays. This short-range field can be eliminated if fault energy permits the dissociation of the anticoherency dislocations in the interface to completely cancel the coherency dislocations. This gives the fully incoherent interface of Fig. 7(e) in which interfacial energy arises entirely from chemical fault energy.

Interfacial Topography

Just as a crystal can be equivalently described as a stacking of any *hkl* crystal planes, a given lattice deformation can be decomposed into any number of invariant-plane strains generated by coherency dislocation arrays on rational crystal planes. A means of assessing the relative significance of potential coherency dislocation arrays can be sought from the fine-scale topography of a crystalline interface (Ref 27). As illustrated by the example of Fig. 4, an interfacial step can be associated with each coherency dislocation. The relative step character of these dislocations then provides the basis for a comparison of their potential physical (or mechanistic) significance in a given interface.

The crystal step structure of a (locally) atomically smooth interface of normal *n* is exhibited by Fig. 8. The interfacial steps associated with a particular crystal plane are conveniently described using a step vector *d* which is normal to the crystal plane and equal to the interplanar spacing *d* in magnitude. The separation *s* of the interfacial steps associated with a given set of planes is:

$$s = d/\sin\alpha \qquad \text{(Eq 5)}$$

where α is the angle between *n* and *d*. The coherency-dislocation Burgers vector associated with a given interfacial step can be determined from the transformation lattice deformation as illustrated by the construction of Fig. 9. If the median reference lattice at the right is transformed to lattice 1 by the lattice deformation $\boldsymbol{S_1}$ and to lattice 2 by the lattice deformation $\boldsymbol{S_2}$, then the representations of the step vector $\boldsymbol{d}$ in lattices 1 and 2 are transformed to their associated vectors in the reference lattice by the inverse deformations S_1^{-1} and S_2^{-1}. A Burgers-circuit closure failure in the reference lattice then results from the difference of these two vectors, identifying the Burgers vector $\boldsymbol{b}$ depicted in Fig. 9 and expressed by:

$$\boldsymbol{b} = \left(S_1^{-1} - S_2^{-1}\right)\boldsymbol{d} \quad \text{(Eq 6a)}$$

Similarly, if lattice 1 is adopted as the reference lattice and is transformed to lattice 2 by the lattice deformation $\boldsymbol{S}$, then b is given by:

$$\boldsymbol{b} = \left(\boldsymbol{1} - \boldsymbol{S}^{-1}\right)\boldsymbol{d} \quad \text{(Eq 6b)}$$

where $\boldsymbol{I}$ is the identity matrix.

If a surface atom is defined as one which lies within a distance δ_0 from a mathematical reference surface, then each crystal plane will contribute surface atoms which lie within a ribbon of width δ as designated in Fig. 8. The density of atoms ρ_L along a surface step (dislocation) line is then:

$$\rho_L = d\delta/\Omega \quad \text{(Eq 7)}$$

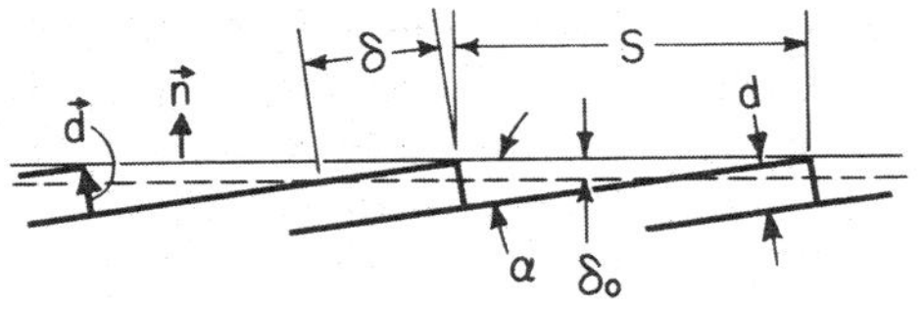

Fig. 8 Step structure of a crystal surface. Source: Ref 27

where Ω is the atomic volume in the reference lattice. Analysis of crystalline free-surface atomic structures, both computed and directly observed by field-ion microscopy, reveals that ρ_L is a useful index of the absolute intensity of a step line, and this is expected to hold for internal surfaces as well. A convenient parameter for assessing the extent to which a set of steps is resolvable as discrete step lines is the ratio of the step separation s to the average atomic spacing along the step lines ρ_L^{-1}:

$$\beta = \rho_L s \quad \text{(Eq 8)}$$

A set of steps is considered resolvable when $\beta \geq 1$.

For coherency dislocations in an internal interface, the width parameter δ can be taken as the dislocation core size $r_c = Ab$ with A between 1 and 3 (Ref 27). The parameters ρ_L and β then become:

$$\rho_L = Abd/\Omega \quad \text{(Eq 9)}$$

$$\beta = \frac{Abd^2}{\Omega \sin \alpha} \quad \text{(Eq 10)}$$

For a given lattice deformation, the magnitude of $\boldsymbol{b}$ given by Eq 6 will be proportional to d. Hence, ρ_L will scale with d^2 and β with d^3. This leads to the general result that the crystal step and coherency-dislocation structure of an interphase boundary will be dominated by crystal planes of relatively high packing density.

For constructing the coherency-dislocation/step structure of an interface, it is essential that lattice deformations and vectors be referred to

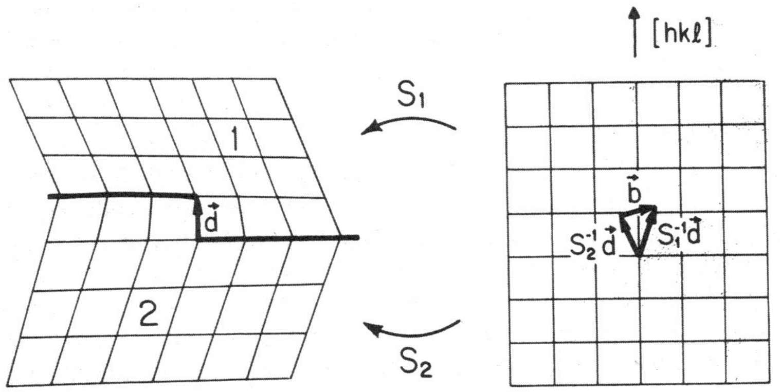

Fig. 9 Interphase boundary with an interfacial step. Associated Burgers vector is defined in median reference lattice. Source: Ref 27

the same reference lattice. When two crystals involve different numbers of atoms per lattice point (i.e., different bases), the reference lattice must be constructed from only those points which correspond to lattice points in *both* structures. In the structure with the smaller basis, this then defines a superlattice representing the finest scale to which the lattice deformation can be regarded as homogeneous. Relative to the crystal lattice, this superlattice will define larger step vectors and coherency-dislocation Burgers vectors.

DSC Dislocations

The state of coherency of a given interface and its dislocation description are functions of the lattice correspondence (and the lattice deformation defined by this correspondence) chosen to define the relation between the two crystals. In discussing the interrelation of various descriptions, it is helpful to identify certain "standard" correspondences. The crystallography of martensitic transformations is traditionally described adopting a correspondence and lattice deformation which relate primitive lattice cells with minimum principal strains. This provides a useful basis for describing the overall state of strain relaxation, and amounts to the interphase analogue of the "identity correspondence" ($\sum 1$) adopted to describe low-angle grain boundaries. As in generalized grain-boundary theory, an O-lattice can be constructed from this lattice deformation to identify possible arrays of "primary" interfacial dislocations, which are equivalent in the interphase case to the anticoherency lattice dislocations described in the section "Coherency versus Anticoherency Dislocations." Such a correspondence can be termed a *primary correspondence*, and its associated lattice deformation a *primary lattice deformation.*

Due to the requirement for invariance of the two crystal structures, all other correspondences and lattice deformations can be derived from the primary ones via the superposition of lattice-invariant deformations. To relate finite lattice cells, these lattice-invariant deformations will be rational in magnitude, involving lattice-vector displacements spaced at integer multiples of lattice planes. Such a combination of a primary lattice deformation with a rational lattice-invariant deformation can establish a larger near-coincidence lattice cell of importance to surface-energy minimization, and defining a new lattice deformation with smaller principal strains. The new correspondence and lattice deformation form the basis of DSC-lattice and O2-lattice constructions to prescribe possible arrays of "secondary" interfacial dislocations. A coincidence lattice then defines a *secondary correspondence* and a *secondary lattice deformation.*

Adopting a coincidence-lattice reference frame, a DSC lattice is constructed from the (DSC) shuffle displacements relating lattice points of the two crystals within the coincidence cell. An O2-lattice construction can then define the spacing of potential interfacial DSC-lattice dislocations which accommodate deviations from exact coincidence in an interface. As the DSC dislocations generally reduce the long-range stress field of an interface, their function is similar to anticoherency dislocations. However, since their Burgers vectors are not lattice vectors of either crystal, they do not, strictly speaking, generate a lattice-invariant deformation. Referred to the primary correspondence, DSC dislocations can be recognized as local perturbations in the spacing of the primary (anticoherency) lattice dislocations producing an otherwise rational lattice-invariant deformation that defines the coincidence cell (Ref 28). DSC dislocations can be described by the superposition of an anticoherency lattice dislocation with one or more coherency dislocations defined by the primary lattice deformation.

While the coherency/anticoherency-dislocation description of interphase-boundary structure based on the primary correspondence provides a convenient account of the overall state of (long-range) strain relaxation, the coincidence-lattice/secondary-dislocation representation is very helpful for understanding surface-energy minimization. Both can furnish useful descriptions of interfacial dislocation structure, provided the nature of the lattice correspondence being adopted is clearly stated.

Martensitic Interfaces

As discussed in the section "Thermodynamic and Kinetic Constraints on Morphology and Orientation Relations," the diffusionless nature of martensitic transformations imposes the constraint that the lattice-invariant deformation must be conservative, and therefore composed of simple shear deformations. We here illustrate the essential characteristics of martensitic interfaces using the simple case where the lattice-invariant deformation consists of a single shear as represented by Eq 4.

Interfacial Structure

Theoretical predictions. The relation of Eq 4 can be equivalently expressed by factoring the homogeneous lattice deformation $\boldsymbol{RB}$ into two components (Ref 6).

$$\boldsymbol{RB} = \boldsymbol{EP}^{-1} \quad \text{(Eq 11)}$$

Here, as in Eq 4, $\boldsymbol{E}$ represents the portion of $\boldsymbol{RB}$ that is observable macroscopically, while $\boldsymbol{P}^{-1}$ is a homogeneous "complementary" or "invisible" shear constituting the portion of $\boldsymbol{RB}$ which is canceled macroscopically by the inhomogeneous lattice-invariant shear $\boldsymbol{P}$ of Eq 4. Each of these deformations can be represented by arrays of interfacial dislocations as illustrated schematically in Fig. 10 (Ref 29). The homogeneous lattice deformation is represented by the uniform arrays of coherency dislocations producing the deformations $\boldsymbol{E}$ and $\boldsymbol{P}^{-1}$ shown in Figs. 10(a) and (b), respectively. Consistent with the IPS character of $\boldsymbol{E}$, the Burgers vector of the first array is denoted as inclined to the dislocation-loop plane,

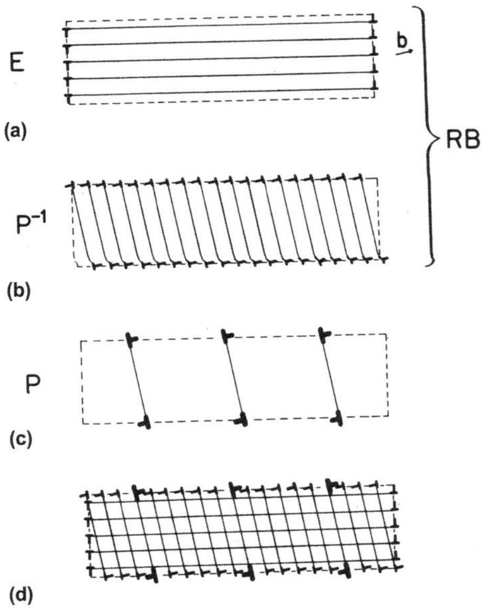

Fig. 10 Discrete-dislocation model of martensitic interface. (a) Coherency dislocations producing macroscopic shape strain $\boldsymbol{E}$; (b) coherency dislocations producing complementary shear $\boldsymbol{P}^{-1}$; (c) anticoherency dislocations producing lattice-invariant deformation $\boldsymbol{P}$; (d) total interfacial structure. Source: Ref 29

reflecting both glide and (conservative) climb components. The Burgers vector of the array generating the $\boldsymbol{P}^{-1}$ shear has only a glide component. Figure 10(c) illustrates a periodic array of anticoherency lattice dislocations producing the inhomogeneous lattice-invariant shear $\boldsymbol{P}$ by slip. Superposition of the three arrays gives a complete dislocation description of the interfacial structure of an enclosed semicoherent martensitic particle as shown in Fig. 10(d).

Decomposition of the deformations according to Eq 11 and the dislocation arrays of Fig. 10 provides a convenient basis for calculation of interfacial stress fields, as will be demonstrated in the section "Interfacial Stress Field and Stability." The $\boldsymbol{E}$ array of Fig. 10(a) produces the long-range stress field of the particle, accounting for the volume strain energy. Superposition of the $\boldsymbol{P}^{-1}$ and $\boldsymbol{P}$ arrays of Fig. 10(b) and (c) produces a short-range elastic-stress field, the energy of which scales with the interfacial area and thus accounts for the elastic contribution to the (surface) interfacial energy. When the lattice-invariant shear $\boldsymbol{P}$ occurs by twinning, the interfacial structure can be modeled by substituting a periodic array of groups of twinning dislocations in place of the slip dislocations of Fig. 10(c). As illustrated in Fig. 11, the superposition of such a twinning dislocation array on the uniform $\boldsymbol{P}^{-1}$ coherency-dislocation array of Fig. 10(b) yields a periodic array of groups of twin-related coherency dislocations. Motion of this interface generates twin-related martensitic variants, and can be equivalently described by the operation of twin-related lattice deformations in the two regions.

While representation of the homogeneous lattice deformation in terms of the $\boldsymbol{E}$ and $\boldsymbol{P}^{-1}$ coherency dislocation arrays is most convenient for analyzing the interfacial stress fields, the invariant plane of $\boldsymbol{E}$ will not generally be a rational low-index crystal plane and therefore this description will not provide the coherency dislocation arrays of greatest potential mechanistic significance according to the considerations of the section "Interfacial Topography." The interfacial-structure parameters defined by Eq 5–10 have been calculated for some specific martensitic transformations. Table 2 summarizes the computed parameters (Ref 27) for the *hkl* crystal planes nearest the habit plane for the f.c.c.(γ) $\rightarrow$ b.c.c.(α) transformation in Fe-31Ni, using the crystallographic data of Wayman (Ref 30). The f.c.c. γ-phase is

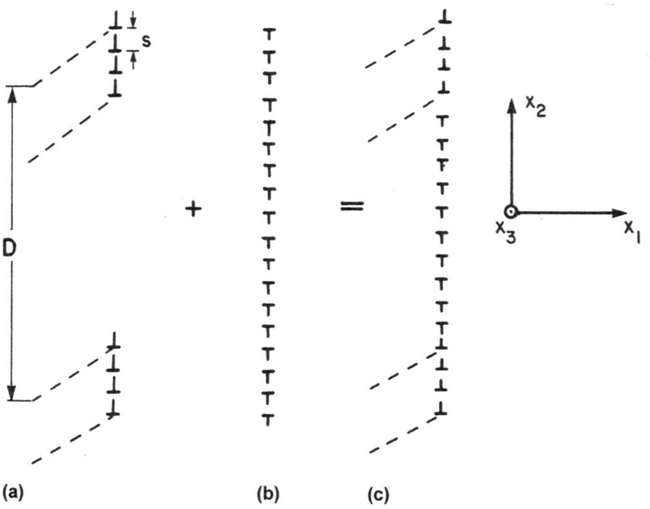

Fig. 11 Discrete-dislocation model of martensitic interface with lattice-invariant deformation by twinning. (a) Inhomogeneous array of twinning dislocations producing lattice-invariant shear, ***P***; (b) coherency dislocations producing complementary shear, $\boldsymbol{P}^{-1}$; (c) array of twin-related coherency dislocations arising from superposition of (a) and (b). Source: Ref 34

Table 2 Coherency dislocations in $(31015)_\gamma$ austenite-martensite interface

hkl	α	*d/a*	*d* (Å)	*b/a*	*b* (Å)	ρ_L (Å^{-1})	*s* (Å)	β
111	25.61°	$\begin{bmatrix} 1/3 \\ 1/3 \\ 1/3 \end{bmatrix}$	2.07	$\begin{bmatrix} 0.074615 \\ -0.175744 \\ 0.081243 \end{bmatrix}$	0.745	0.266	4.79	1.274
002	38.52°	$\begin{bmatrix} 0 \\ 0 \\ 1/2 \end{bmatrix}$	1.80	$\begin{bmatrix} -0.0082252 \\ -0.067728 \\ 0.061039 \end{bmatrix}$	0.329	0.102	2.89	0.295
022	13.15°	$\begin{bmatrix} 0 \\ 0 \\ 1/4 \end{bmatrix}$	1.27	$\begin{bmatrix} 0.024837 \\ -0.091461 \\ 0.053724 \end{bmatrix}$	0.391	0.0858	5.58	0.479
113	19.49°	$\begin{bmatrix} 1/11 \\ 1/11 \\ 3/11 \end{bmatrix}$	1.08	$\begin{bmatrix} 0.017349 \\ -0.072559 \\ 0.044353 \end{bmatrix}$	0.312	0.0582	3.24	0.188
133	8.02°	$\begin{bmatrix} 1/19 \\ 3/19 \\ 3/19 \end{bmatrix}$	0.822	$\begin{bmatrix} 0.022239 \\ -0.066364 \\ 0.035449 \end{bmatrix}$	0.282	0.0401	5.89	0.236
024	15.04°	$\begin{bmatrix} 0 \\ 1/10 \\ 1/5 \end{bmatrix}$	0.804	$\begin{bmatrix} 0.008284 \\ -0.050230 \\ 0.033698 \end{bmatrix}$	0.219	0.0304	3.10	0.094
135	6.25°	$\begin{bmatrix} 1/35 \\ 3/35 \\ 1/7 \end{bmatrix}$	0.607	$\begin{bmatrix} 0.011129 \\ -0.043767 \\ 0.026219 \end{bmatrix}$	0.187	0.0196	5.58	0.109

adopted as the reference lattice and the dislocation core size parameter is taken as $A = 2$ (Eq 9). For the habit plane near $(3\ 10\ 15)_\gamma$ the coherency dislocations of greatest topographical significance, in terms of both ρ_L and β, are those associated with the close-packed (111) planes giving a dislocation separation of $s \approx 5$ Å. Other coherency dislocations of possible significance are associated with the (022) and (002) planes. The arrangement of these potential coherency-dislocation arrays in the $(3\ 10\ 15)_\gamma$ interface is illustrated in Fig. 12(a) with the thickness of the lines reflecting the relative β parameters. The corresponding atomic step structure in the nearby $(135)_\gamma$ plane is depicted by a ball model in Fig. 13 (Ref 31), illustrating

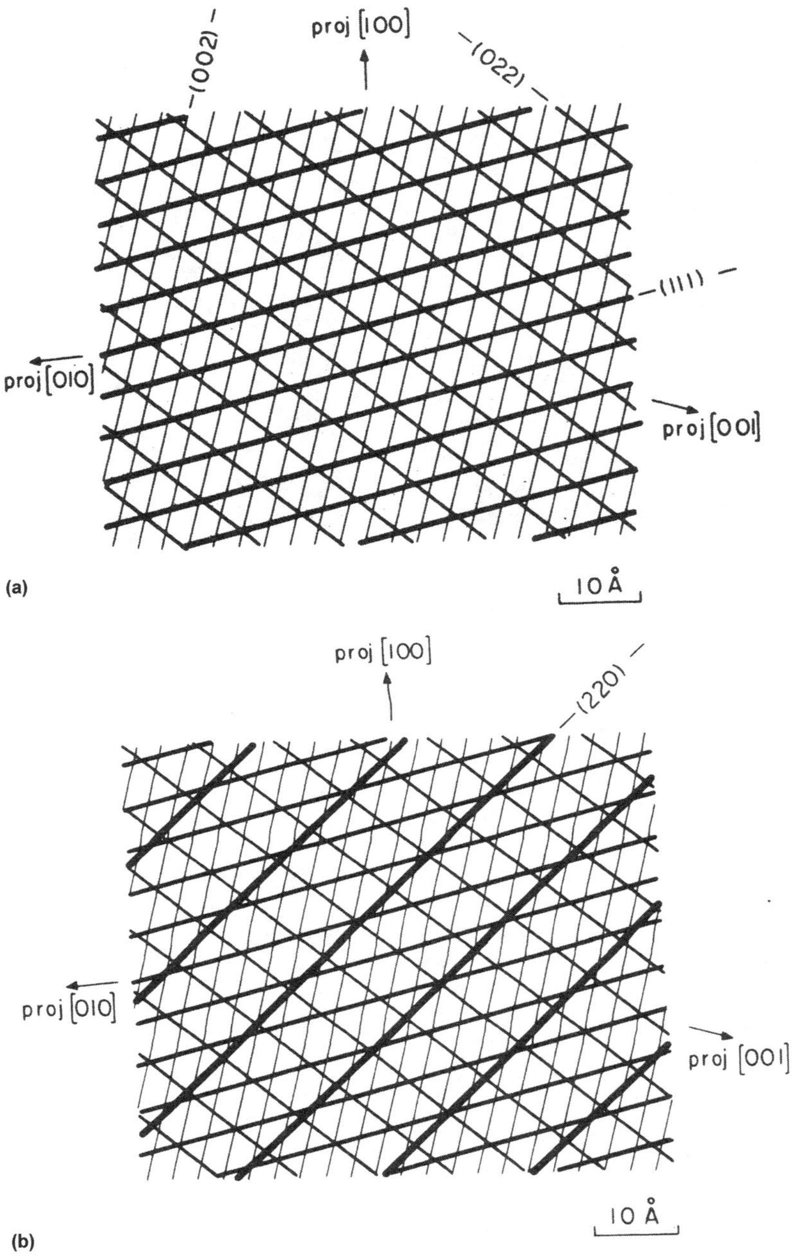

Fig. 12 Computed dislocation structure of f.c.c.-b.c.c. $(3\ 10\ 15)_\gamma$ martensitic interface in Fe-31Ni. (a) Most significant coherency dislocations of Table 2; (b) complete interfacial structure from superposition of anticoherency lattice dislocations on (a). Source: Ref 27

the higher topographical significance of the (111) steps. While these steps lie along rational atomic rows in the rational $(135)_\gamma$ plane illustrated, they will possess occasional kinks in order to lie along irrational directions in the actual irrational habit plane described by Fig. 12. To make this habit an invariant plane, the lattice-invariant shear ***P*** must occur by slip or twinning on $(112)_\alpha$ corresponding to $(220)_\gamma$. For the slip case, interfacial anticoherency dislocations with unit Burgers vectors $\frac{1}{2}a[110]_\gamma$ and a separation of 11.7 Å must be superimposed on the coherency-dislocation arrays to give the interfacial dislocation structure represented in

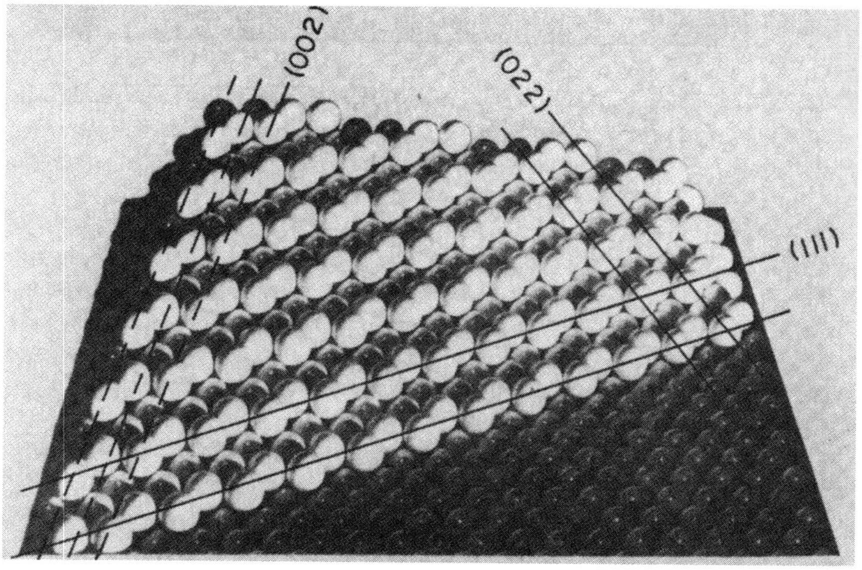

Fig. 13 Ball model of step structure of $(135)_\gamma$ plane corresponding to coherency dislocation array of Fig. 12(a). Source: Ref 31

Fig. 12(b). Such an interfacial-structure analysis has also been applied recently to ferrous lath martensites with a habit near $(111)_\gamma$ by Sandvik and Wayman (Ref 32).

Calculation of the interfacial coherency-dislocation structure of internally twinned martensite can be illustrated by the example of the b.c.c. (β_1) $\rightarrow$ $2H(\gamma'_1)$ transformation in thermoelastic Cu-Al-Ni alloys, a case of particular interest due to available interfacial mobility measurements to be discussed in the section "Observations in β-Cu-Al-Ni Alloys." The martensitic product exists in two orientations, a primary variant I and a second twin-related variant II. The $\beta_1 - \gamma'_1$ interfacial structure of each can be calculated separately. Although the β_1 and γ'_1 phases exhibit atomic ordering, the lattice deformation is essentially that for transformation between disordered b.c.c.(β) and 2H(γ') phases (the orthorhombic 2H phase "inheriting" the atomic order of the parent) so that atomic order can be neglected from the viewpoint of deformation mechanisms. We therefore adopt a reference lattice based on the fundamental b.c.c. lattice of the β-phase. For flat interfaces on the observed habit plane near $(33\bar{1})_\beta$, the calculated parameters (based on available crystallographic data [Ref 33]) for $\beta - \gamma'$ interfacial coherency dislocations of γ' variants I and II are given in Tables 3A and 3B (Ref 34), again taking $A = 2$. The b.c.c. $\rightarrow$ 2H transformation involves planar atomic shuffles, and this must be taken into account in constructing the reference lattice as discussed in the section "Interfacial Topography." Referred to the b.c.c. coordinate system, this shuffling converts reference lattice planes from (101) to $(\frac{1}{2}0\frac{1}{2})$ in variant I and from (110) to $(\frac{1}{2}\frac{1}{2}0)$ in variant II. The effect is a doubling of b and s for the coherency dislocations associated with these particular planes. The most significant coherency dislocations in terms of the β parameter are those associated with (110) in variant I and $(\frac{1}{2}\frac{1}{2}0)$ in variant II, with separations of 8 Å and 16 Å, respectively. Also potentially significant, particularly in terms of ρ_L, are the coherency dislocations associated with the (011) and (101)-type planes in both variants.

The arrangement of the computed coherency dislocation arrays of Tables 3A and 3B is shown in Fig. 14 with variants I and II in the lower and upper portions; the relative amounts of I and II reflect the twin fraction for this transformation. The two interfacial structures meet at an internal martensitic twin boundary, the trace of which is horizontal in the figure. Again the thicknesses of the lines are drawn to reflect relative values of the β parameter. The separation of the most significant coherency dislocations associated with (110)-type planes is seen to double in variant II relative to variant I due to the atomic shuffling on these planes in variant II. Similarly shuffling on (101)-type planes in variant I doubles the

Table 3A Coherency dislocations in $\beta_1 - \gamma_1'$ interface: Variant I

hkl	α	d/a_0	d (Å)	b/a_0	b (Å)	ρ_L (Å^{-1})	s (Å)	β
110	15.15°	$\begin{bmatrix}1/2\\1/2\\0\end{bmatrix}_\beta$	2.063	$\begin{bmatrix}0.044620\\-0.071246\\-0.051746\end{bmatrix}_\beta$	0.262	0.086	7.894	0.688
200	44.33°	$\begin{bmatrix}1/2\\0\\0\end{bmatrix}_\beta$	1.459	$\begin{bmatrix}0.020852\\0.023742\\-0.012295\end{bmatrix}_\beta$	0.103	0.024	2.088	0.050
$21\bar{1}$	17.38°	$\begin{bmatrix}1/3\\1/6\\-1/6\end{bmatrix}_\beta$	1.191	$\begin{bmatrix}0.025929\\-0.043454\\-0.019988\end{bmatrix}_\beta$	0.156	0.030	3.988	0.120
310	27.86°	$\begin{bmatrix}3/10\\1/10\\0\end{bmatrix}_\beta$	0.923	$\begin{bmatrix}0.017273\\-0.025281\\-0.009105\end{bmatrix}_\beta$	0.093	0.014	1.975	0.028
$32\bar{1}$	8.27°	$\begin{bmatrix}3/14\\1/7\\-1/14\end{bmatrix}_\beta$	0.780	$\begin{bmatrix}0.017487\\-0.028372\\-0.013067\end{bmatrix}_\beta$	0.104	0.012	5.418	0.070
$33\bar{2}$	10.65°	$\begin{bmatrix}3/22\\3/22\\-1/11\end{bmatrix}_\beta$	0.622	$\begin{bmatrix}-0.014045\\-0.024619\\-0.006172\end{bmatrix}_\beta$	0.091	0.008	3.365	0.030
$66\bar{2}$	3.13°	$\begin{bmatrix}6/76\\6/76\\1/38\end{bmatrix}_\beta$	0.335	$\begin{bmatrix}0.007692\\-0.012751\\-0.006172\end{bmatrix}_\beta$	0.047	0.002	6.127	0.016
$0\bar{1}1$	129.9°	$\begin{bmatrix}0\\-1/2\\1/2\end{bmatrix}_\beta$	2.063	$\begin{bmatrix}-0.0.7101\\0.071517\\0.041414\end{bmatrix}_\beta$	0.271	0.086	2.689	0.242
$\frac{1}{2}0\frac{1}{2}$	87.8°	$\begin{bmatrix}1\\0\\1\end{bmatrix}_\beta$	4.126	$\begin{bmatrix}0.008272\\0.001356\\0.014200\end{bmatrix}_\beta$	0.048	0.128	4.128	0.266

Table 3B Coherency dislocations in $\beta_1 - \gamma_1'$ interface: Variant II

hkl	α	d/a_0	d (Å)	b/a_0	b (Å)	ρ_L (Å^{-1})	s (Å)	β
$\frac{1}{2}0\frac{1}{2}$	15.15°	$\begin{bmatrix}1\\1\\0\end{bmatrix}_\beta$	4.216	$\begin{bmatrix}0.020188\\-0.024921\\-0.004210\end{bmatrix}_\beta$	0.094	0.704	15.788	2.828
200	44.33°	$\begin{bmatrix}1/2\\0\\0\end{bmatrix}_\beta$	1.459	$\begin{bmatrix}0.026797\\-0.001976\\0.037549\end{bmatrix}_\beta$	0.134	0.022	2.088	0.046
$21\bar{1}$	17.38°	$\begin{bmatrix}1/3\\1/6\\-1/6\end{bmatrix}_\beta$	1.191	$\begin{bmatrix}0.014173\\0.003241\\0.025801\end{bmatrix}_\beta$	0.086	0.016	3.988	0.066
310	27.80°	$\begin{bmatrix}3/10\\1/10\\0\end{bmatrix}_\beta$	0.923	$\begin{bmatrix}0.013435\\-0.010364\\0.005825\end{bmatrix}_\beta$	0.052	0.008	1.975	0.016
$32\bar{1}$	8.27°	$\begin{bmatrix}3/14\\1/7\\-1/14\end{bmatrix}_\beta$	0.780	$\begin{bmatrix}0.009802\\0.001107\\0.016471\end{bmatrix}_\beta$	0.056	0.006	5.418	0.038
$33\bar{2}$	8.27°	$\begin{bmatrix}3/22\\3/22\\-1/11\end{bmatrix}_\beta$	0.622	$\begin{bmatrix}0.007157\\0.005036\\0.016471\end{bmatrix}_\beta$	0.059	0.006	3.366	0.020
$66\bar{2}$	3.13°	$\begin{bmatrix}6/76\\6/76\\-1/38\end{bmatrix}_\beta$	0.355	$\begin{bmatrix}-0.004351\\-0.000891\\-0.007092\end{bmatrix}_\beta$	0.027	0.002	6.127	0.010
$01\bar{1}$	129.90°	$\begin{bmatrix}0\\-1/2\\1/2\end{bmatrix}_\beta$	2.063	$\begin{bmatrix}-0.005160\\-0.056061\\-0.008602\end{bmatrix}_\beta$	0.305	0.102	2.689	0.272
101	87.80°	$\begin{bmatrix}1/2\\0\\1/2\end{bmatrix}_\beta$	2.063	$\begin{bmatrix}0.015748\\-0.084717\\-0.055244\end{bmatrix}_\beta$	0.234	0.078	2.064	0.164

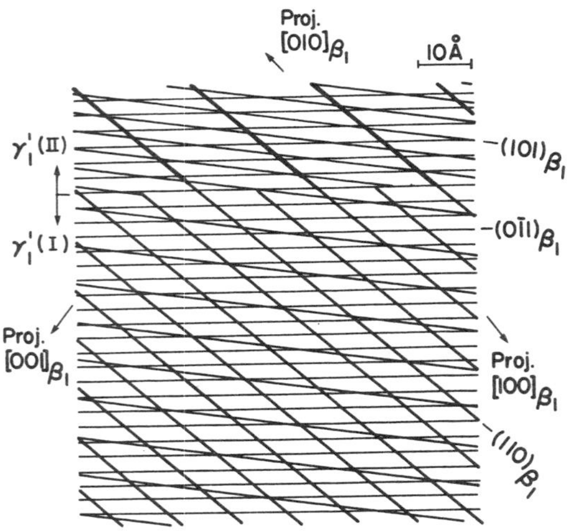

Fig. 14 Computed dislocation structure of $\beta - \gamma'$ $(33\bar{1})_\beta$ interface of internally twinned martensite in Cu-Al-Ni. Most significant coherency dislocations of Table 3 are shown for variant I in lower portion and twin-related variant II in upper portion. Source: Ref 34

separation of these coherency dislocations relative to those of variant II. Also indicated are the coherency dislocations associated with (011) corresponding to the plane of twinning; these dislocations thus represent the arrays depicted in Fig. 11. Coherency dislocation structures very similar to Fig. 14 are also obtained when the γ' phase is adopted as the reference lattice (Ref 34).

The $\beta - \gamma'$ interface in Cu-Al-Ni alloys is observed to be smooth and generally parallel to the macroscopic habit plane as assumed in the foregoing calculations. The martensitic transformation between very similar structures in Ti-Mn alloys is sometimes observed to exhibit a "corrugated" or "faceted" interface in which the local interface plane differs for the martensitic twin variants I and II (Ref 35). The effect has been explained for this system via minimization of the net coherency-dislocation Burgers-vector content of each interfacial segment (Ref 36). Calculation of the detailed coherency-dislocation/step structures of such faceted interfaces can be done using the local interface plane of each segment.

Experimental observations. Direct experimental observations of discrete coherency dislocations in martensitic interfaces are rare due to the difficulty of resolving closely spaced dislocations with small Burgers vectors. An exceptional case is the coherent f.c.c. → h.c.p. transformation with a close-packed habit plane such that a gradually tapering particle has well-spaced coherency dislocations corresponding to the ***E*** array of Fig. 10(a). This simple transformation constitutes the first case for which coherency (transformation) dislocations were proposed (Ref 37, 38). Observation of these dislocations in the interface of a tapering particle, using conventional transmission electron microscopy diffraction-contrast techniques, is shown in Fig. 15 (Ref 39). Similar coherency dislocations have been observed in the closely related problem of mechanical twinning in both f.c.c. (Ref 40) and b.c.c. alloys (Ref 41). While the twinning dislocations represent pure glide coherency dislocations, the Burgers vectors of the f.c.c. → h.c.p. coherency dislocations of Fig. 15 include a climb component due to a small difference in the spacing of the close-packed planes of the order of 1%. This additional component has not been directly detected by conventional dislocation Burgers vector contrast experiments, but its presence can be inferred from precise measurements of the displacement contrast of faults associated with the same dislocations (Ref 42). As mentioned in the section "Thermodynamic and Kinetic Constraints on Morphology and Orientation Relations," this transformation also involves a small dilatation in the habit plane of about 0.1%. Hence, the IPS produced by the dislocations imaged in Fig. 15 cannot account for the entire lattice deformation and the remaining portion must be ascribed to a small additional component of unresolvable Burgers-vector content. As follows from the analysis of the section "Interfacial Topography," the total coherency dislocation Burgers-vector content of an interface will generally involve both resolvable and unresolvable components.

Observation of the discrete-crystal interfacial steps associated with the coherency dislocations can be sought from high-resolution lattice-fringe imaging of martensitic interfaces. The best such observation to date is the edge-on view of a b.c.c.(β)-h.c.p.(α) martensitic interface in Ti-5Mn shown in Fig. 16 left (Ref 43), viewing the structure normal to the h.c.p. basal plane, $[001]_\alpha \| [110]_\beta$. Three sets of planes are simultaneously imaged in the h.c.p. α-phase, and computer-aided image analysis allows the determination that the dark-imaging areas correspond to the atomic planes in this structure. It could not be similarly determined whether the light or dark fringes represent the

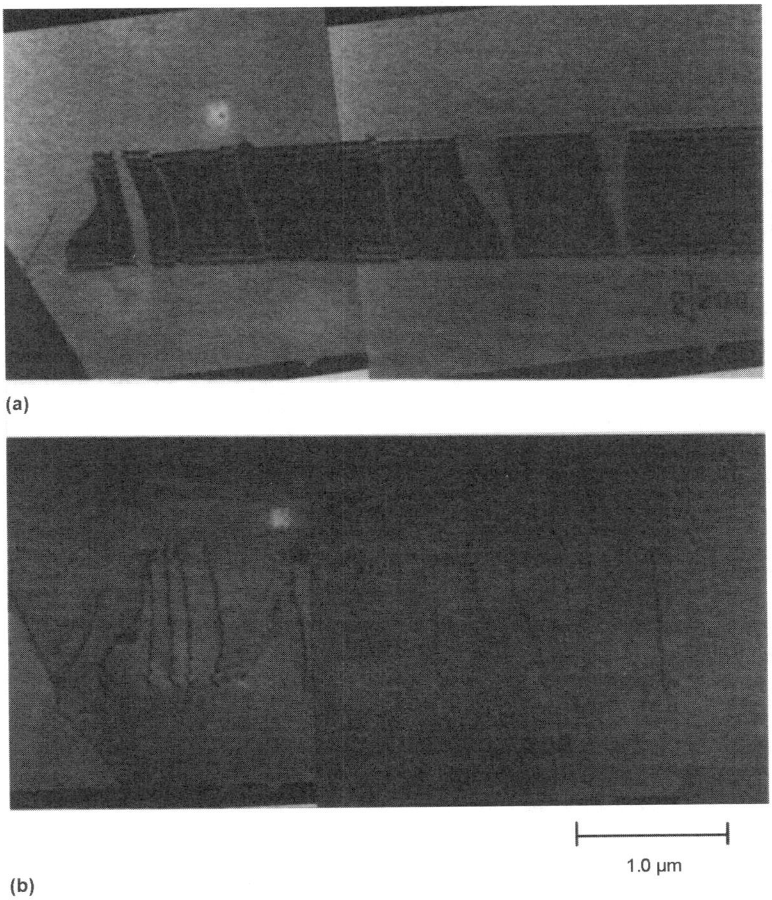

Fig. 15 Tapered h.c.p. martensitic particle in f.c.c. Co-Fe alloy. (a) Displacement contrast image of particle; (b) strain contrast image of interfacial coherency dislocations. Source: Ref 39

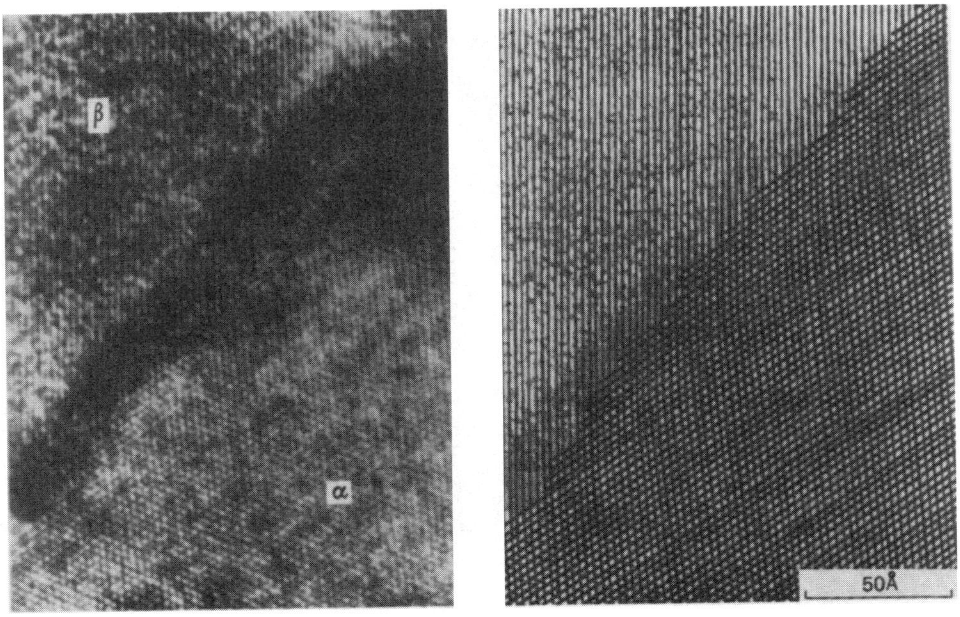

Fig. 16 Lattice-fringe image of b.c.c.(β)-h.c.p.(α) martensitic interface in Ti-5Mn. Electron micrograph at left (Ref 43), with superimposed construction of corresponding planes at right.

single set of (110) atomic planes imaged in the b.c.c. β-phase, but the information obtained is sufficient to verify a one-to-one correspondence of planes across the interface, attesting to the local coherency of the area of interface examined. This is demonstrated by the set of nearly vertical lines drawn through the image of the $(110)_\beta$ planes and corresponding α-prism planes at the right in Fig. 16.

The topographically most significant coherency dislocations for this interface; according to the analysis of the section "Interfacial Topography," correspond to steps associated with $(112)_\beta$ planes which are not imaged in this micrograph. However, the corresponding prism planes of the α are imaged and these have also been drawn in for clarity in the figure at the right. The intersection of these planes with the interface defines the array of steps corresponding to the coherency-dislocation array.

The planes imaged in Fig. 16 lie at a relatively high angle to the interface so that the coherency dislocations are rather closely spaced. The relative smoothness and sharpness of the interface is a common feature of high-resolution images of martensitic interfaces obtained thus far (Ref 44, 45), supporting the underlying assumptions of the dislocation/step analysis. Similar observations on a martensitic system with the interface plane closer to a close-packed crystal plane should allow the detection of well-separated crystal steps. The coherency-dislocation Burgers vectors of the steps may then be discernible with the weak-beam dark-field diffraction-contrast technique (Ref 46). The coherent b.c.c. $\rightarrow$ 9R martensitic transformation in β-Cu-Zn alloys, having habits near $(110)_\beta$, appears to be well suited for such combined lattice-fringe and diffraction-contrast experiments.

The weak-beam diffraction-contrast technique has proved most useful in the high-resolution electron microscopy investigation of interphase-boundary dislocation structures. A recent observation of the anticoherency lattice dislocations in the $\gamma - \alpha'$ interface of lath martensites in a Fe-Ni-Mn alloy with a habit near $(111)_\gamma$ is shown in Fig. 17 (Ref 47). The dislocations have a separation of ≈ 45 Å. Although $(111)_\gamma$ steps in the interface of similar lath martensites have been found by lattice-fringe imaging (Ref 45), the associated coherency dislocations have not yet been imaged by diffraction contrast in this case.

Interfacial Stress Field and Stability

The elastic stress field of an interphase boundary can be calculated by summing the stress fields of the component dislocation arrays represented in Fig. 10. For isotropic elasticity, the stress field of individual periodic arrays in infinite planar interfaces of the type corresponding to Fig. 10(b) and (c) have been computed analytically by Hirth and Lothe (Ref 47). The stress field of a planar semicoherent martensitic interface with a lattice-invariant shear by slip can be calculated by combining the stress fields of the two periodic arrays of Fig. 10(b) and (c), with periods of s and D, respectively. For pure screw dislocations lying along the z direction in an interface lying in the x plane, the stress fields of the individual and combined arrays are plotted in Fig. 18 (Ref 29). Parameters appropriate to the Frank model (Ref 48) for the ferrous $\gamma - \alpha'$ interface with a $(225)_\gamma$ habit have been chosen. Referred to the α'-martensite lattice, the anticoherency lattice dislocations of Fig. 10(c) then have a Burgers vector of $\boldsymbol{b} = \frac{1}{2}a[11\bar{1}]$ and a periodic spacing D corresponding to six (112) planes. The coherency array of Fig. 10 (b) has a Burgers vector $\boldsymbol{b}_\mathrm{c} = -\boldsymbol{b}/6 = (a/12)[\bar{1}\bar{1}1]_\alpha$ and a separation s corresponding to one dislocation for each (112) plane. This interfacial model neglects a slight misfit in the $[11\bar{1}]_\alpha$ close-packed direction which requires a second lattice-invariant shear to operate in the actual $(225)_\gamma$ transformation, but the pure screw dislocation array provides a convenient example to illustrate the essential characteristics of martensitic interfaces.

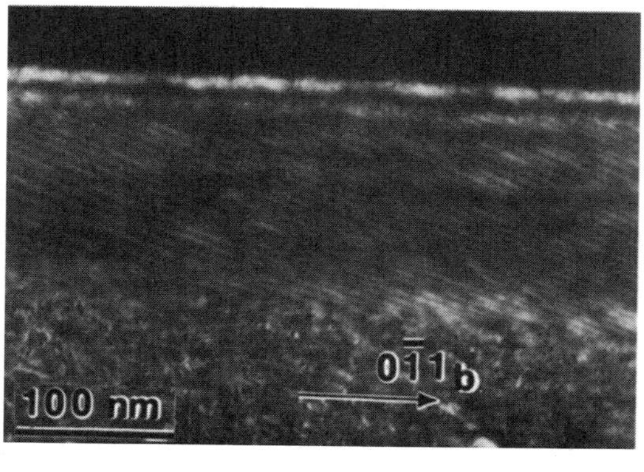

Fig. 17 Weak-beam diffraction contrast image of anticoherency lattice dislocations in f.c.c.-b.c.c. martensitic interface in Fe-Ni-Mn alloy. Source: Ref 47

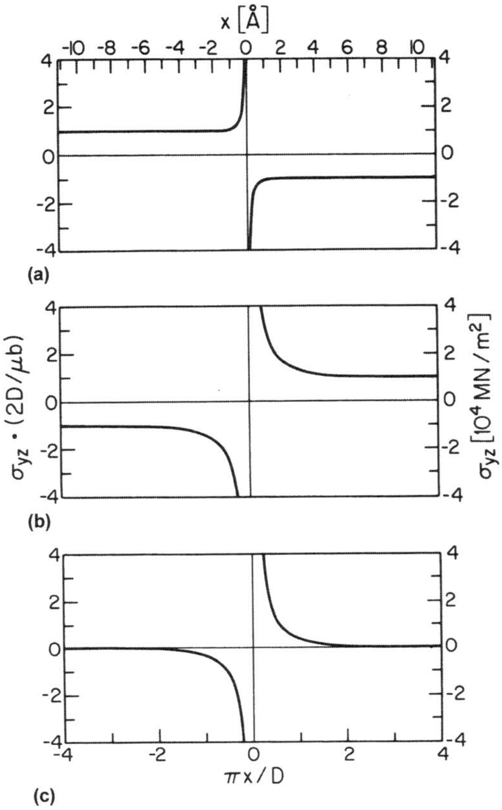

Fig. 18 Stress fields of interfacial dislocation arrays for f.c.c.-b.c.c. (225) interface (y = 0). (a) Coherency dislocation array of Fig. 10(b); (b) anticoherency dislocation array of Fig. 10(c); (c) combined arrays. Source: Ref 29

The stress fields of Fig. 18 (evaluated at $y = 0$ centered on a lattice dislocation core) disclose the equal and opposite long-range stress fields of the coherency and anticoherency arrays, which cancel upon superposition leaving a short-range stress field of range $\approx D$ similar to that of a low-angle grain boundary. For $x \geq s/\pi \approx 0.5$ Å, the stress field of the coherency-dislocation array is constant, equivalent to a continuous dislocation distribution or "surface dislocation" (Ref 49). Adopting the approximation of continuously distributed coherency dislocations canceling the long-range stresses of the anticoherency array, the stress components of a more general interface with anticoherency Burgers-vector components b_x, b_y, and b_z can be expressed as (Ref 29, 34) (see equation 12a to 13c at bottom of page):

$$\sigma_{xz}(b_z) = \frac{\mu b_z}{2D}\left[\frac{\sin 2\pi y/D}{\cosh 2\pi x/D - \cos 2\pi y/D}\right] \quad \text{(Eq 14a)}$$

$$\sigma_{yz}(b_z) = \frac{\mu b_z}{2D}\left[\frac{\sinh 2\pi x/D}{(\cosh 2\pi x/D - \cos 2\pi y/D)} - 1\right] \quad \text{(Eq 14b)}$$

The edge component b_x corresponds to a simple symmetric tilt boundary as treated by Read and Shockley (Ref 50) while the edge component b_y represents the "misfit" dislocation array treated by Frank (Ref 48) and van der Merwe

$$\sigma_{xx}(b_x) = -\frac{\mu b_x}{2D(1-v)}\times\frac{(\sin 2\pi y/D)[\cosh 2\pi x/D - \cos 2\pi y/D + (2\pi x/D)\sinh 2\pi x/D]}{(\cosh 2\pi x/D - \cos 2\pi y/D)^2} \quad \text{(Eq 12a)}$$

$$\sigma_{yy}(b_x) = -\frac{\mu b_x}{2D(1-v)}\times\frac{(\sin 2\pi y/D)[\cosh 2\pi x/D - \cos 2\pi y/D - (2\pi x/D)\sinh 2\pi x/D]}{(\cosh 2\pi x/D - \cos 2\pi y/D)^2} \quad \text{(Eq 12b)}$$

$$\sigma_{xy}(b_x) = \frac{\mu b_x}{2D(1-v)}\frac{(2\pi x/D)[(\cosh 2\pi x/D)(\cos 2\pi y/D) - 1]}{(\cosh 2\pi x/D - \cos 2\pi y/D)^2} \quad \text{(Eq 12c)}$$

$$\sigma_{xx}(b_y) = \frac{\mu b_y}{2D(1-v)}\frac{(2\pi x/D)[(\cosh 2\pi x/D)(\cos 2\pi y/D) - 1]}{(\cosh 2\pi x/D - \cos 2\pi y/D)^2} \quad \text{(Eq 13a)}$$

$$\sigma_{yy}(b_y) = \frac{\mu b_y}{2D(1-v)}\times\left\{\frac{(2\sinh 2\pi x/D)(\cosh 2\pi x/D - \cos 2\pi y/D) - (2\pi x/D)[(\cosh 2\pi x/D)(\cos 2\pi y/D)] - 1}{(\cosh 2\pi x/D - \cos 2\pi y/D)^2} - 1\right\} \quad \text{(Eq 13b)}$$

$$\sigma_{xy}(b_y) = -\frac{\mu b_y}{2D(1-v)}\times\frac{(\sin 2\pi y/D)[\cosh 2\pi x/D - \cos 2\pi y/D - (2\pi x/D)(\sinh 2\pi x/D)]}{(\cosh 2\pi x/D - \cos 2\pi y/D)^2} \quad \text{(Eq 13c)}$$

(Ref 51) and shown in Fig. 7(c). The screw component b_z is the Frank interfacial array represented by Fig. 18; Fig. 18(c) corresponds to σ_{yz} of Eq 14b evaluated for $y = 0$.

It is apparent from the stress fields plotted in Fig. 18 for the pure screw-dislocation case that a strong attractive interaction exists between the coherency and anti-coherency dislocation arrays. Because each array contributes a long-range σ_{yz} stress of magnitude $\mu b/2D$, separation of the two arrays would require a force per area of interface of:

$$F_A = \tfrac{1}{2}\mu\left(\frac{b}{D}\right)^2 \qquad \text{(Eq 15)}$$

The force required to remove a single anticoherency dislocation from the interface can be calculated from the interaction of the dislocation with the stress field of the remaining interface; the latter corresponding to the superposition of the stress field of Eq 14b with that of an opposite-sign "image" dislocation at $(x, y) = (0, 0)$. For removal of the dislocation orthogonally from the interface (along x), the interaction force per length of dislocation is then:

$$F_\ell = b\sigma_{yz} = \frac{\mu b^2}{2D}\left[\coth(\pi x/D) - (\pi x/D)^{-1} - 1\right] \qquad \text{(Eq 16)}$$

giving a maximum attractive force at $x = 0$ of $\mu b^2/2\ D$. Force-distance curves similarly computed for removal of the dislocation at various angles from the interface (as should be required for glide of a dislocation with an edge component) are plotted in Fig. 19. The angle of 90° corresponds to Eq 16. At lower angles, the force curve oscillates due to the periodic stress field of the interface. At the rather shallow angle of 15°, this gives rise to a metastable configuration indicated by the solid point. Whether there exists a martensitic transformation with a ***P*** shear system defining such a shallow angle is not known. Apart from the remote possibility of such metastable configurations, the large interaction forces demand that the interfacial arrays are very stable, and it is quite unlikely that an anticoherency dislocation could be removed from the interface during forward or reverse motion; the possibility of other accommodation dislocations being generated in the interface stress field will be discussed in the section "Thermoelastic-Nonthermoelastic Transition." This interfacial stability accounts for the intrinsic mechanical reversibility of martensitic interfacial motion, in contrast to the irreversible nature of conventional plastic deformation by the disorganized motion of slip dislocations. Moreover, via the strong attractive interaction, chemical forces acting on the coherency dislocations are transmitted to the anticoherency dislocations and thus act on the interface as a whole.

The interfacial stress field of internally twinned martensite can be obtained by summing the stress fields of the periodic coherency-

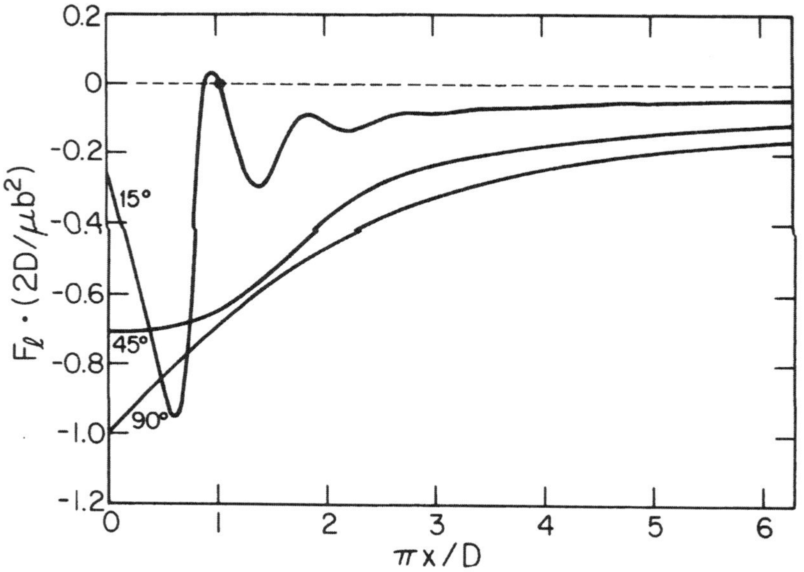

Fig. 19 Force-distance curves for removal of an anticoherency dislocation from interface of Fig. 18 along glide planes inclined to interface by angles indicated. Solid point indicates metastable configuration.

dislocation arrays shown in Fig. 11. A twinned interface with the same period D as the slip-dislocation interface represented in Fig. 18 can be obtained by spreading the cores of the $\frac{1}{2}a[11\bar{1}]$ slip dislocations to produce $\frac{1}{6}a[11\bar{1}]$ twinning dislocations bounding three-layer twins. Superposition on the coherency dislocation array results in alternating three-layer segments of $\frac{1}{12}a[11\bar{1}]$ and $\frac{1}{12}a[\bar{1}\bar{1}1]$ coherency dislocations. Summing the stress fields of the six coherency dislocation arrays (of period D) so defined gives the interfacial stress-field contours which are compared against the slip case in Fig. 20. Except for the regions very close to the slip-dislocation cores, the stress fields are quite similar. The stress fields of twinned interfaces of larger period can be similarly computed by increasing the number of coherency dislocations of each sign. The range of the stress field increases in proportion to D.

Interfacial Energy

Following the method of Hirth and Lothe (Ref 47), the elastic contribution to the interfacial energy can be calculated from the interaction force between two interfaces of opposite sign integrated from a cut-off size r_0 to ∞, with $r_0 = b/\alpha$; α is a core-energy parameter taken between 1 and 2 for metals. For $r_0 \geq s/\pi$ the interfacial stress fields of Eq 12–14 are appropriate for the slip dislocation case, giving the following contributions, γ, to the energy per area from each of the Burgers vector components (Ref 29, 34):

$$\gamma(b_x) = \frac{\mu b_x^2}{4\pi D(1-\nu)}\left[\frac{\pi r_0}{D}\coth\frac{\pi r_0}{D} - \ln\left(2\sinh\frac{\pi r_0}{D}\right)\right] \quad \text{(Eq 17)}$$

$$\gamma(b_y) = \frac{\mu b_y^2}{4\pi D(1-\nu)}\left[\frac{2\pi r_0}{D} - \frac{\pi r_0}{D}\coth\frac{\pi r_0}{D} - \ln\left(2\sinh\frac{\pi r_0}{D}\right)\right] \quad \text{(Eq 18)}$$

$$\gamma(b_z) = \frac{\mu b_z^2}{4\pi D}\left[\frac{\pi r_0}{D} - \ln\left(2\sinh\frac{\pi r_0}{D}\right)\right] \quad \text{(Eq 19)}$$

Within the isotropic elastic approximation, the three stress field components are non-interacting, and hence their energy contributions can be simply summed, taking $r_0 = b/\alpha$ with b the total Burgers vector of the anticoherency dislocations. The interfacial energy for the pure screw case defined by Eq 19 is plotted versus b/D for α-values of 1 and 2 in Fig. 21 (Ref 29). While γ is sensitive to b/D at low values, it is rather insensitive for the large lattice-invariant shears involved in ferrous transformations (the (31015) case actually involves a slight edge component). The high sensitivity to α for the

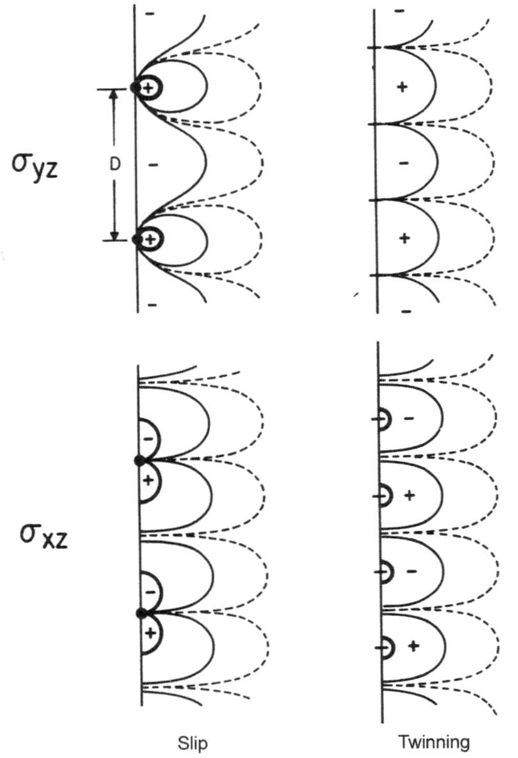

Fig. 20 Constant-stress contours for martensitic interfaces with lattice-invariant deformations by slip and by twinning. Source: Ref 29

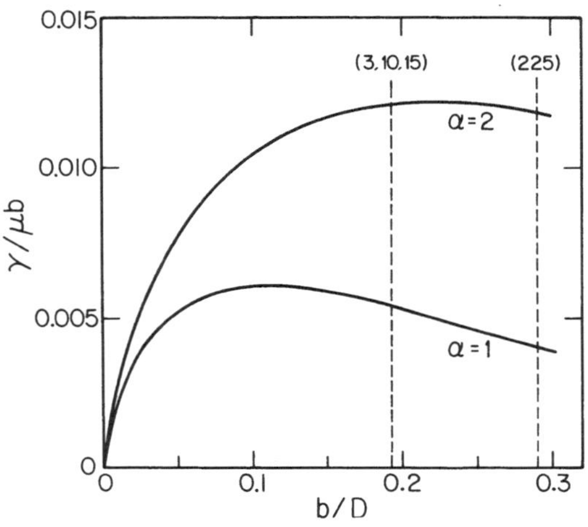

Fig. 21 Interfacial energy versus b/D for lattice-invariant deformation by slip in screw orientation. Values of b/D for ferrous martensites with (3, 10, 15) and (225) habits are indicated. Source: Ref 29

latter cases reflects a substantial contribution from the anticoherency dislocation core energy; despite the core-energy uncertainty, the predicted elastic-energy contribution of 0.1–0.2 J/m^2 sets a lower limit to the semicoherent interfacial energy that is substantially greater than typical coherent interfacial energies in metallic systems. Similar interfacial energies have been obtained using anisotropic elasticity with the same calculational techniques (Ref 52).

The same interaction-energy method can be applied to the twinned interface summing the interactions of all the component coherency-dislocation arrays. For the screw-dislocation case represented in Fig. 20 with N-atomic-layer twins, the summation is of the form:

$$\gamma = \frac{b_c}{2D}\int_{r_0}^{\infty}\left[\sum_{p=1}^{N}\sum_{q=1}^{N}\sigma_0 - \sum_{p=1}^{N}\sum_{q=N+1}^{2N}\sigma_0 - \sum_{p=N+1}^{2N}\sum_{q=1}^{N}\sigma_0 + \sum_{p=N+1}^{2N}\sum_{q=N+1}^{2N}\sigma_0\right]\mathrm{d}x \qquad \text{(Eq 20)}$$

with

$$\sigma_0 = \frac{\mu b_c}{2D}\left[\frac{\sinh 2\pi x/D}{\cosh 2\pi x/D - \cos 2\pi y/D}\right]$$

where $y = (p - q)s$ and $D = 2Ns$. On integrating, long-range terms cancel, giving:

$$\gamma = \frac{\mu b_c^2}{4\pi D}\left[\sum_{p=1}\sum_{q=1}\phi - \sum_{p=1}\sum_{q=N+1}\phi\right] \qquad \text{(Eq 21)}$$

with

$$\phi = -\ln[\cosh 2\pi r_0/D - \cos 2\pi(p-q)s/D]$$

Computer summing of Eq 21 and taking b_c = 0.42 Å, $\mu = 7 \times 10^{10}$ N/m^2, and s = 1.46 Å appropriate to the $(225)_\gamma$ transformation in steels gives the γ curves plotted versus N in Fig. 22. Comparison with the slip-interface case is made both at the same α values (1 and 2) and at the same r_0 (α = 1/6 and 1/3); the latter is likely the more realistic comparison. The interfacial energies are similar at N = 3 corresponding to identical D. The energy for the twinned case increases approximately linearly with N reflecting the greater range of the interfacial stress field. Values are very high when extrapolated to experimentally observed twin thicknesses of $N \approx 10^2$. In such instances, it is expected that the stress fields will induce accommodation slip, creating a local structure

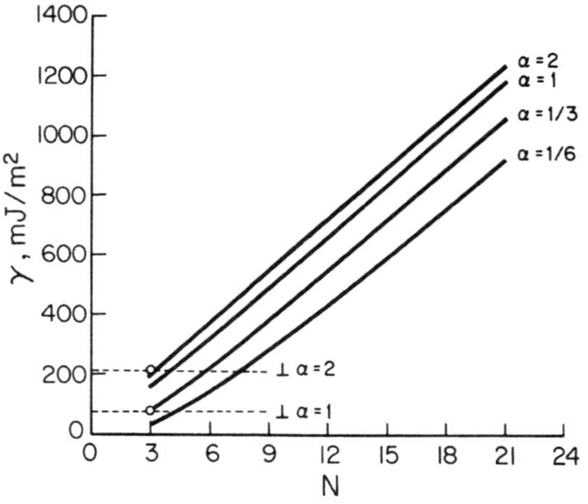

Fig. 22 Twinned-martensite interfacial energy versus twin thickness. Open points indicate values for slip interface (denoted by ⊥ symbol). Source: Ref 29

resembling the slip interface. Combining the high interfacial energy with the additional energy of internal twin boundaries, it appears that the observed internal twinning in ferrous martensites will never be *thermodynamically* favored relative to slip, arising instead from the inhibition of slip through *kinetic* factors as embodied in relative interfacial mobilities to be discussed in the section "Defect Interactions."

When $r_0 \geq s/\pi$ the stress field and elastic energy of the discrete coherency-dislocation arrays are equivalent to those of continuous surface dislocations and the sums of Eq 20 and 21 can be replaced by integrals of continuously distributed Burgers-vector density. Methods for such continuum-elastic calculations have been developed and applied to internally twinned martensites (Ref 53).

Interfacial Mobility

Conservative versus Nonconservative Motion

The general solid-state (crystalline) transformation can involve a change of composition, atomic order, and crystal structure. When the structural change involves a *nonconservative* lattice-invariant deformation, diffusion plays a role in all three of these processes. It is customary, however, to describe a transformation as "diffusion controlled" when the kinetics are governed by the atom transport accomplishing a composition (or order) change, and "interface controlled" when governed by the structural change occurring at the interface (Ref 54). Hillert (Ref 54) has proposed that the interface-controlled velocity υ of an interphase boundary can be related to the volume free-energy change driving it. Δg, by a mobility relation of the form:

$$\upsilon = M(-\Delta g)^m \qquad \text{(Eq 22)}$$

where M is a mobility coefficient and m a fixed exponent. Although it should be possible to model nonconservative interfacial motion on the basis of climb of the anticoherency dislocations of Fig. 2, interfacial mobility data for nonconservative transformations are scarce, and mobility parameters are typically estimated from grain-boundary kinetic data for which m is taken as unity.

When the structural change, as embodied in the processes represented by Eq 2, involves lattice-invariant deformations which are *conservative*, interfacial mobility does not involve diffusion and the intrinsic mobility behavior for such instances can be sought from the special case of martensitic transformations which, due to their diffusionless nature, represent the prototypical interface-controlled conservative structural transformation. The intrinsic mobility of martensitic interfaces can be assessed by application of the theory of dislocation dynamics to the interfacial dislocation arrays developed in the section "Martensitic Interfaces". After examining theoretical prediction and experimental measurements of martensitic interfacial mobility, the application of the developed mobility relations to the case of coupled diffusional/displacive transformations will be explored in the section, "Coupled Diffusional/Displacive Transformation," in this chapter.

Interfacial Driving Force

The interfacial force per unit area acting in the x direction arising from a transformational volume free-energy change Δg can be expressed as:

$$F_A = -\frac{1}{A}\frac{dV}{dx}\Delta g \qquad \text{(Eq 23)}$$

with V being the transformed volume and A the interfacial area. For a planar interface, $(1/A)dV/dx$ is unity, and $-\Delta g$ can then be taken as the "driving force" (Ref 55), as assumed in Eq 22. When applied mechanical forces act on the interface, it is convenient to define a generalized Gibbs free energy, as employed in the theory of dislocation motion (Ref 56), which takes into account the work done by both shear and hydrostatic components of the applied stress. In addition to a chemical free-energy change Δg^{ch}, the interface will then respond to a mechanical free-energy contribution defined by:

$$-\Delta g^\sigma = \sum \sigma_{ij}^A \varepsilon_{ij}^T \qquad \text{(Eq 24)}$$

where σ_{ij}^A is the applied stress and ε_{ij}^T is the transformation strain discussed in the section "Crystal Deformations and Particle Coherence."

Patel and Cohen (Ref 57) have demonstrated that the thermodynamic effect of applied stress on semicoherent martensitic transformations under various stress states can be predicted satisfactorily using the ε_{ij}^T defined by the macroscopic deformation $\boldsymbol{E}$. When $\boldsymbol{E}$ is an IPS with shear and normal stresses components γ_T and ε_n, Eq 24 becomes:

$$-\Delta g^\sigma = \tau\gamma_T + \sigma_n\varepsilon_n \qquad \text{(Eq 25)}$$

where τ and σ_n are the resolved shear and normal stresses in the planes and directions of γ_T and ε_n. This defines the net macroscopic force acting on the interface. For a highly flexible interface, the local force acting on a coherent interfacial segment would be defined by the ε_{ij}^T of the local lattice deformation. However, the surface tension associated with the high interfacial energies estimated in the section "Interfacial Energy" will make martensitic interfaces behave in a relatively rigid manner for which the macroscopic force defines the force acting everywhere in the interface. This would account for the success of the Patel-Cohen assumption, and this assumption will be employed throughout the treatment of interfacial mobility developed here.

As the driving force acting on individual line dislocations is normally expressed as a force per unit length F_ℓ, it is useful to establish a common driving force basis for comparing the behaviors of interfacial arrays and isolated slip dislocations. The F_ℓ acting on each dislocation in an array of separation s will be $F_\ell = sF_A$. When an interface acts as a continuously distributed array or surface dislocation, it is more convenient for comparison purposes to define an effective F_A acting on the isolated slip dislocation. An effective Δg^σ for this case can be derived from a modification of the Peach-Kohler formula, using the dyadic notation (Ref 47):

$$-\Delta g^\sigma = (\boldsymbol{\sigma} \cdot b) \cdot d/d^2 \quad \text{(Eq 26)}$$

where $\boldsymbol{d}$ is a vector normal to a slip plane as defined in the section "Interfacial Topography." This is geometrically equivalent to treating the line dislocation as a ribbon defect of height d which deforms a volume element of this height through a strain of b/d.

Continuous Motion

When a slip dislocation moves in response to a force F_ℓ arising from an applied resolved stress τ^A, $F_\ell = \tau^A b$, the dependence of the crystal energy on dislocation, position can give rise to an internal friction stress τ_f providing an opposite force $\tau_f b$ which resists the motion. The τ_f will vary markedly with location in the crystal, with occasional peak values defining a mechanical threshold stress $\hat{\tau}$. When the applied force exceeds $\hat{\tau}b$, continuous motion of the dislocation ensues with a velocity governed by the net driving force $F_\ell^{net} = (\tau^A - \tau_f)b$. For a curved dislocation the net force is further modified by line tension. When the net force is positive, the dislocation will tend to accelerate at a rate governed by its inertial mass, m_ℓ (Ref 58), but will be damped by dissipative drag mechanisms. For motion in the x direction at nonrelativistic velocities, a Newtonian equation of motion can be employed of the form:

$$F_\ell^{net} = (\tau^A - \tau_f)b + \gamma\frac{d^2x}{dz^2} = m_\ell\frac{d^2x}{d\tau^2} + B_\ell\frac{dx}{dt} \quad \text{(Eq 27)}$$

with γ the line tension, d^2x/dz^2 the dislocation curvature, and B_ℓ a drag coefficient describing the interaction of the moving dislocation with phonons and (at low temperatures) with electrons (Ref 56). Proposed phonon drag mechanisms typically lead to a drag coefficient of the form:

$$B_\ell = kT/\Omega v \quad \text{(Eq 28)}$$

with k the Boltzmann constant, T the absolute temperature, Ω the atomic volume, and v the atomic vibration frequency. Under most circumstances of interest, inertial effects can be neglected and continuous dislocation motion will occur at a drag-controlled steady-state velocity of:

$$\upsilon = F_\ell/B_\ell \quad \text{(Eq 29)}$$

These same equations of motion should be applicable to the continuous motion of a well-separated array of interfacial dislocations, incorporating the chemical force $-\Delta g^{ch}s$ in F_ℓ; Eq 29 is then equivalent to Eq 22 with $m = 1$. However, the phonon drag force is based on the concept that the dislocation encounters more phonons from the forward direction than from the backward direction. The drag coefficient B_ℓ may be significantly modified for interphase-boundary dislocations as it has been suggested that the dislocation may either gain or lose energy during motion due to the differing phonon modes in the two structures (Ref 59, 60).

For the continuous motion of an interface acting as a continuously distributed array or surface dislocation, Eq 27 can be replaced by a similar relation using analogous parameters defined for a two-dimensional defect:

$$F_A^{net} = -(\Delta g - \Delta g^f) + \gamma(\kappa_1 + \kappa_2) = m_a\frac{d^2x}{dt^2} + B_a\frac{dx}{dt} \quad \text{(Eq 30)}$$

Here Δg^1 represents a frictional force, γ the (isotropic) surface tension, κ_1 and κ_2 principal curvatures, m_a the interfacial inertial mass, and B_a the interfacial drag coefficient. From Eq 26, B_a can be estimated as B_ℓ/d. Again continuous phonon-drag controlled motion will ensue when the driving force exceeds a threshold level $-\Delta\hat{g} \approx \hat{\tau}\gamma_T$ corresponding to the mechanical threshold stress $\hat{\tau}$ of the crystal.

Thermally Activated Motion

General Concepts

When the acting driving force does not exceed the mechanical threshold level, a dislocation reaches a state of static equilibrium where the local driving force F^{net} falls to zero. Further motion requires thermal activation to overcome the local barrier of the internal friction force. Once the dislocation overcomes the barrier, it accelerates to a drag-controlled velocity which it maintains until another barrier is encountered. The overall rate of motion is governed by the waiting time at the obstacles, and this forms the basis of the kinetic theory of thermally activated dislocation motion as recently reviewed by Kochs, Argon, and Ashby (Ref 56).

The essential concepts of thermally activated dislocation motion can be illustrated by the schematic dislocation/obstacle interaction force-distance profile depicted in Fig. 23. Representing the position of a dislocation segment by the area a swept during its motion, the variation of internal friction stress with position can define a τ versus a profile of the form of Fig. 23(a). For an applied stress τ^A, the dislocation is at equilibrium at point a_1. To reach the unstable point a_2 requires an activation energy defined by the shaded area under the curve. During the thermally activated event, the dislocation line sweeps through an "activation area" a^*. From the shape of the obstacle profile in Fig. 23(a), a^* will be a function of τ^A, which can be represented by the "sheared" profile (Ref 56) in Fig. 23(b). The activation energy can be determined by integration of either profile:

$$Q(\tau^A) = b \int_{a_1}^{a_2} (\tau - \tau^A)\, da = b \int_{\tau^A}^{\hat{\tau}} a^* d\tau \quad \text{(Eq 31)}$$

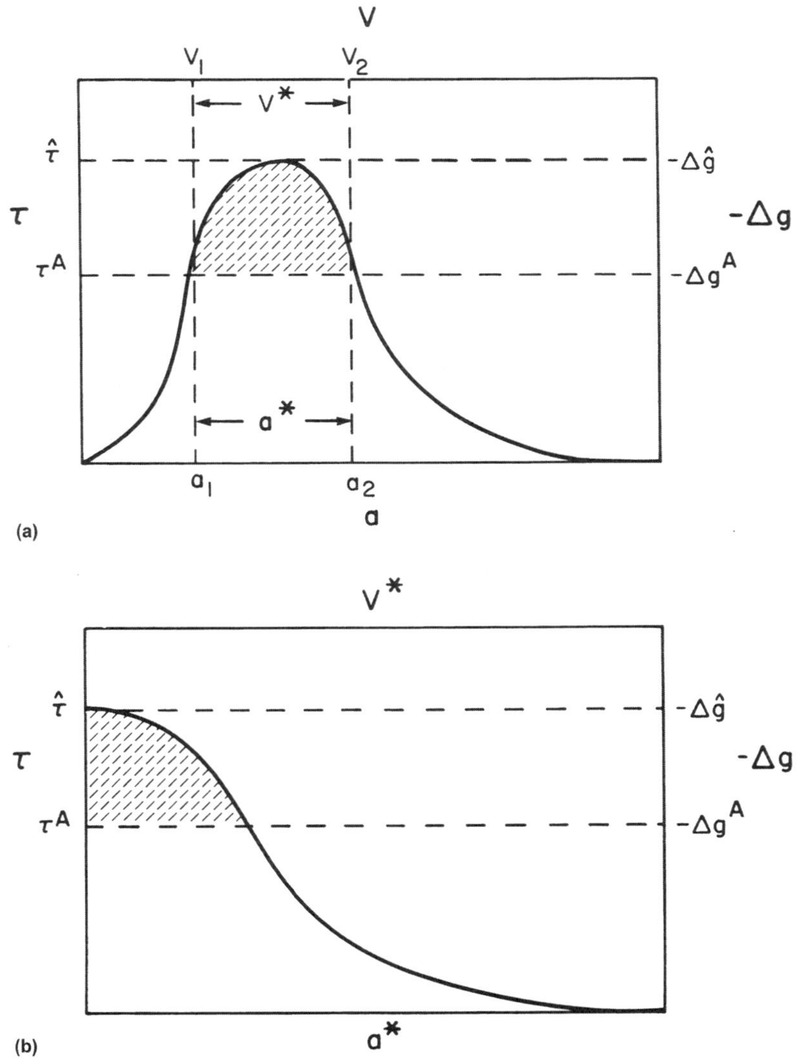

Fig. 23 Dislocation/obstacle interaction profile. (a) Internal stress versus dislocation position; (b) "sheared" profile showing stress versus activation area. Scales at top and right represent corresponding quantities for interface/obstacle interaction.

The overall velocity of dislocation motion is then:

$$\upsilon = \upsilon_0 \exp(-Q/KT) \quad \text{(Eq 32)}$$

where the pre-exponential factor υ_0 is proportional to an attempt frequency. Operational parameters corresponding to these model activation parameters can be derived from experimental mobility data via the relations:

$$Q = \left.\frac{\partial \ln \upsilon}{\partial(-1/kT)}\right|_{\tau^A} \quad \text{(Eq 33)}$$

$$a^* = -\frac{1}{b}\frac{\partial Q}{\partial \tau^A} = \left.\frac{kT}{b}\frac{\partial \ln \upsilon}{\partial \tau^A}\right|_T \quad \text{(Eq 34)}$$

As discussed in the section "Interfacial Driving Force," it is most convenient for the problem of interfacial motion to express the driving force in terms of Δg as defined in Eq 23–26. One can then define an "activation volume" V^* analogous to a^*:

$$V^* = \left.\frac{\partial Q}{\partial \Delta g}\right|_T \quad \text{(Eq 35)}$$

The values of Δg and V^* corresponding to τ and a^* are also shown in Fig. 23 and the activation analysis can be equivalently formulated in terms of these variables. As will be demonstrated in the next two sections, this is particularly appropriate when an interface behaves as a continuous surface and the activated configuration is truly three-dimensional in character. For the case of individual line dislocations, V^* is related to a^* through Eq 26 and 34 as:

$$V^* = da^* \quad \text{(Eq 36)}$$

For slip dislocations, this quantity is similar to the commonly employed "stress activation volume" defined by:

$$V_\tau = -\left.\frac{\partial Q}{\partial \tau^A}\right|_T = ba^* \quad \text{(Eq 37)}$$

However, the latter quantity does not correspond to a physical volume.

When the obstacle profile of Fig. 23 involves a very long-range interaction such that V^* is very large, thermal activation is of negligible importance; then $\tau^A \approx \hat{\tau}$ for dislocation motion and the obstacle can be regarded as contributing an athermal back-stress τ_μ retarding dislocation motion. In the presence of τ_μ, the "thermal" component of τ^A which can contribute to activation over short-range obstacles is $\tau^A - \tau_\mu$. When different obstacles of comparable range impede motion, the net τ^A required for a given velocity can be approximately represented by the root sum of the squares of their individual contributions (Ref 56). Again, these contributions can be equivalently expressed in terms of the Δg driving force. We next consider the interaction of interfacial dislocation arrays with some specific types of obstacles.

Motion in a Perfect Crystal

When a dislocation moves in a perfect crystal, its core energy varies in a periodic manner due to the translational symmetry of the lattice. The resulting periodic resisting force provides a "Peierls barrier" to dislocation motion which can provide a rate-controlling thermal obstacle at very low temperatures. If the separation s of an array of significant coherency dislocations is substantially greater than the lattice period λ_0 (as for the $(110)_\beta$ dislocations of Fig. 14), the dislocations may individually interact with the periodic lattice potential in much the same manner as isolated slip dislocations. The dislocation energy as a function of position x for an edge dislocation is often approximated by an expression of the form (Ref 47, 56):

$$\varepsilon(x) = \frac{\mu b^2}{2\pi(1-\nu)}[1 - \alpha_1 \cos 2\pi x/\lambda_0] \quad \text{(Eq 38)}$$

with

$$\alpha_1 = \frac{\exp(-4\pi\delta_c/\lambda_0)}{\ln(R/\delta_c)}$$

where δ_c is the effective dislocation core width, and R is the range of the dislocation stress field, usually taken as the average dislocation spacing in the crystal. The parameter α_1 which determines the amplitude of the energy variation with x/λ_0 is very sensitive to δ_c. Using Eq 38 to determine the activation energy for the formation of critical dislocation-kink pairs gives an activation energy proportional to b^2 and an activation volume proportional to $\alpha_1^{1/2}$. For comparable δ_c, coherency dislocation motion would involve similar activation volumes but a substantially smaller Peierls barrier compared to slip dislocations because of the smaller Burgers vectors.

When s is very small so that the interface acts as a continuous surface dislocation, the critical configuration may be a three-dimensional bulge. Approximating the interface as a membrane with isotropic surface tension γ_0, a similar analysis of the activation parameters arising from a periodic variation of the interfacial energy with position x/λ_0 gives (Ref 34):

$$Q = \tfrac{1}{4}\pi^3\gamma_0\lambda_0^2(1 - \Delta g/\Delta\hat{g})^2 \qquad \text{(Eq 39)}$$

$$V^* = \frac{\pi^3}{16}\frac{\lambda_0 3}{(\Delta\gamma/\gamma_0)}(1 - \Delta g/\Delta\hat{g}) \qquad \text{(Eq 40)}$$

where $\Delta\gamma/\gamma_0$ represents the amplitude of the energy variation, analogous to α_1 of Eq 38. Again, this amplitude will be sensitive to the interfacial core width δ_c (Ref 61). Using a nonlinear elastic model and incorporating a strain gradient energy, the equilibrium interfacial core width is:

$$\delta_c = (\kappa/\Phi_0)^{1/2} \qquad \text{(Eq 41)}$$

where κ is a gradient-energy coefficient and Φ_0 is the strain-energy density barrier for a homogeneous lattice deformation. For the special case of a lattice on the verge of mechanical instability, the effective Φ_0 could be quite low, giving a high-δ_c diffuse interface with a greatly reduced Peierls barrier. Observation of interfacial cores (Fig. 16) suggests, however, that in normal circumstances δ_c is comparable to that for lattice dislocations. Taking $\Delta\gamma/\gamma_0$ to be equivalent to the α_1 estimated for slip dislocations, Eq 40 predicts relatively large activation volumes of $\approx 10^4\ \Omega$. However, γ_0 values appropriate to martensitic interfaces give maximum Q values (Eq 39) of only $\approx 10^{-5}$eV. It is thus unlikely that the Peierls barrier associated with continuous-interface behavior will be rate-controlling in metals under experimentally accessible conditions.

In contrast to the motion of individual line dislocations, the motion of an interface as in the growth of the enclosed particle of Fig. 10 requires not only dislocation motion but the regular generation of discrete dislocation loops. When significant coherency dislocations lie in crystal planes at a shallow angle to a sharp interface, such as for the f.c.c.-h.c.p. transformation of Fig. 15, the activation energy for nucleation of discrete dislocation loops producing long-range stress fields can provide a substantial anisotropic interfacial Peierls barrier which opposes particle thickening. A simple calculation (Ref 62) of this activation energy shows that it is indeed significant for the f.c.c.-h.c.p. transformation and of potential importance to the f.c.c.-b.c.c. transformation with a habit near $(111)_\gamma$, although the results are typically very sensitive to ill-defined non-linear-elastic core energies. Similar calculations have been performed for the generation of anticoherency dislocation loops, treating their interaction with the coherency-dislocation arrays in an approximate way through a modification of their Burgers vector (Ref 55, 63); the predicted barrier is potentially rate-controlling for the edge-wise growth of very small particles but is easily surmountable at moderate sizes.

Defect Interactions

While the intrinsic interfacial mobility in a perfect crystal is of potential import in the growth of microscopic particles, the thermally activated motion of an interface of more macroscopic dimensions will be dominated by interaction with crystal defects. For the case of slip, the cutting of forest dislocations can represent an appreciable thermal obstacle for the deformation of metals at moderate temperatures. Similar "line-line" interactions might occur for well-spaced anticoherency dislocations when the lattice-invariant deformation takes place by slip. Forest-dislocation cutting by closely spaced coherency dislocations will, however, constitute more of a "line-surface" interaction; although this process is of potential importance in the generation of coherency dislocations for the thickening of small martensitic particles via a pole mechanism (Ref 62), its contribution to the interfacial friction stress will be predominately long-range and athermal in nature. Similar considerations apply to the "surface-surface" interactions of the interface with other interfaces such as antiphase boundaries.

Defects which are most likely to play a rate-controlling role in thermally activated interfacial motion are those which contribute to "point-surface" interactions of the interface with a three-dimensional array of discrete obstacles. Such "point" obstacle interactions involve long-range contributions from modulus and strain interactions of the particles with the interfacial stress field, as well as short-range interactions with the interfacial core. These interactions have recently been modeled analytically by (Grujicic Ref 64), and we next examine each in turn.

Strain interaction. An array of partially coherent particles can interact with the interfacial stress field via their particle transformation strain, ε_{ij}^{T}. Adopting the approximation that the interfacial stress field is uniform over the particle volume, the strain interaction energy for a spherical particle of radius r_p is:

$$W_s = -\tfrac{4}{3}\pi r_p^3 \sigma_{ij}\varepsilon_{ij}^{T} \quad \text{(Eq 42)}$$

where σ_{ij} is the interfacial stress field evaluated at the center of the particle. The derivative of W_s with respect to distance x of the particle from the interface defines a strain interaction force:

$$F = -\frac{dW_s}{dx} = \tfrac{4}{3}\pi r_p^3 \varepsilon_{ij}^{T} \frac{d\sigma_{ij}}{dx} \quad \text{(Eq 43)}$$

A particle transformation strain ε'_{ij} defined in the particle coordinate system can be expressed in the interface coordinate system using appropriate direction cosines, a_{ik}:

$$\varepsilon_{ij}^{T} = a_{ik}a_{il}\varepsilon'_{kl} \quad \text{(Eq 44)}$$

Taking the case of a simple "tetragonal" distortion $\varepsilon'_{ij} = \varepsilon'_{33}$, the interaction force for an interfacial dislocation array in pure screw orientation can be obtained from Eq 43 using the σ_{ij} of Eq 14:

$$F = \tfrac{4}{3}\frac{\pi^2 a_{33}\varepsilon'_{33}\mu b r_p^3}{D^2} \times \{a_{13}\sinh(2\pi x/D)\sin(2\pi y/D) + (1 - a_{13}^2 - a_{33}^2)^{1/2}[1 - \cosh(2\pi x/D)\cos(2\pi y/D)]\} \times [\cosh(2\pi x/D) - \cos(2\pi y/D)]^{-2} \quad \text{(Eq 45)}$$

The force is a maximum when $a_{33} = 2^{-1/2}$ for which the particle distortion direction is at 45° to the line direction, z, of the screw dislocations. Close to an anticoherency dislocation ($x, y \approx 0$), the stress field is very similar to that of an isolated screw dislocation for which the interaction force is:

$$F = \tfrac{4}{3}a_{33}\varepsilon'_{33}\mu b r_p^3 \frac{2a_{13}xy + (1 - a_{13}^2 - a_{33}^2)^{1/2}(y^2 - x^2)}{(x^2 + y^2)^2} \quad \text{(Eq 46)}$$

The strongest interaction occurs for particles that intersect the interface along the $y = 0$ plane in which the anticoherency dislocations reside. The maximum force, $\hat{F}$, for these particles can be estimated from Eq 46, evaluated at a cut-off distance $x = r_0$ taken as either the dislocation core size r_c or the particle size r_p:

$$\hat{F} = \tfrac{4}{3}a_{33}(1 - a_{13}^2 - a_{33}^2)^{1/2}\varepsilon'_{33}\mu b \frac{r_p^3}{r_0^2}. \quad \text{(Eq 47)}$$

The force/distance behavior thus defined can be used to determine driving force/activation-volume profiles as in Fig. 23(b), taking into account the three-dimensional nature of the interface/particle interaction. The geometry of such an interaction with an array of particles of mean spacing λ is illustrated schematically in Fig. 24. The interfacial driving force necessary to exert a force F on the center particle is:

$$-\Delta g = F/\lambda^2 \quad \text{(Eq 48)}$$

and the activation volume associated with an increment of interfacial motion Δx is:

$$V^* = \tfrac{2}{3}\lambda^2 \Delta x \quad \text{(Eq 49)}$$

In contrast, the corresponding quantities for the interaction of a single line dislocation with the same obstacles are:

$$-\Delta g = F/\lambda d \quad \text{(Eq 50)}$$

$$V^* = \lambda d \Delta x \quad \text{(Eq 51)}$$

Compared to slip, then, the same obstacle interaction involves an interfacial driving force which is smaller by a factor of d/λ, and an activation volume which is larger by a factor of $2\lambda/3d$.

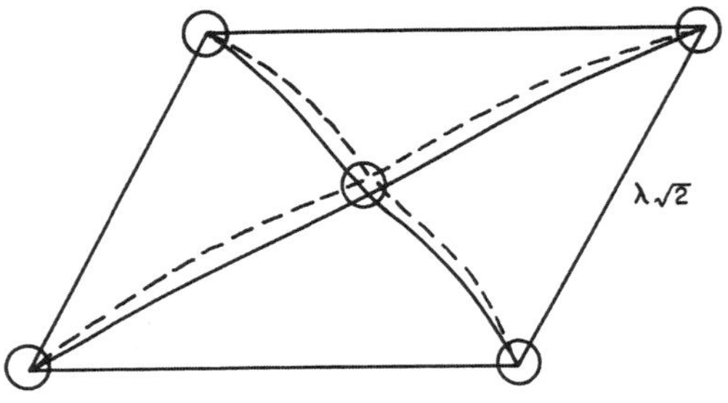

Fig. 24 Schematic representation of the interaction between an interface and regularly distributed obstacle particles of average planar spacing λ.

Taking F to be proportional to x in the region $x < r_0$, the force/distance behavior then defines a driving-force/activation-volume relation for $y = 0$ of the form:

$$V^*/V_0^* = \frac{D}{\pi r_0}\sinh^{-1}\left[\sinh(\pi r_0/D)\cdot\left(\frac{\Delta g}{\Delta \hat{g}}\right)^{-1/2}\right] - \frac{\Delta g}{\Delta \hat{g}} \qquad \text{(Eq 52)}$$

with $V_0^* = \frac{2}{3}\lambda^2 r_0$ and $\Delta\hat{g} = -\hat{F}/\lambda^2$. The long-range exponential tail of the interfacial stress field gives a logarithmic dependence of V^* on Δg in the region of very low Δg where Eq 52 can be approximated by:

$$V^*/V_0^* = -\frac{1}{2\pi}\left(\frac{D}{r_0}\right)\ln\left[\frac{1}{4}\left(\frac{D}{\pi r_0}\right)^2\frac{\Delta g}{\Delta \hat{g}}\right] \qquad \text{(Eq 53)}$$

For large Δg, the behavior is controlled by the short-range stress field which is equivalent to an isolated slip dislocation:

$$V^*/V_0^* = (\Delta g/\Delta \hat{g})^{-1/2} - \Delta g/\Delta \hat{g} \qquad \text{(Eq 54)}$$

The latter is compared with the interfacial behavior, Eq 52, for D/r_0 values of 10^1 and 10^3 in Fig. 25(a). The behaviors are quite similar overall but differ at very low Δg due to the

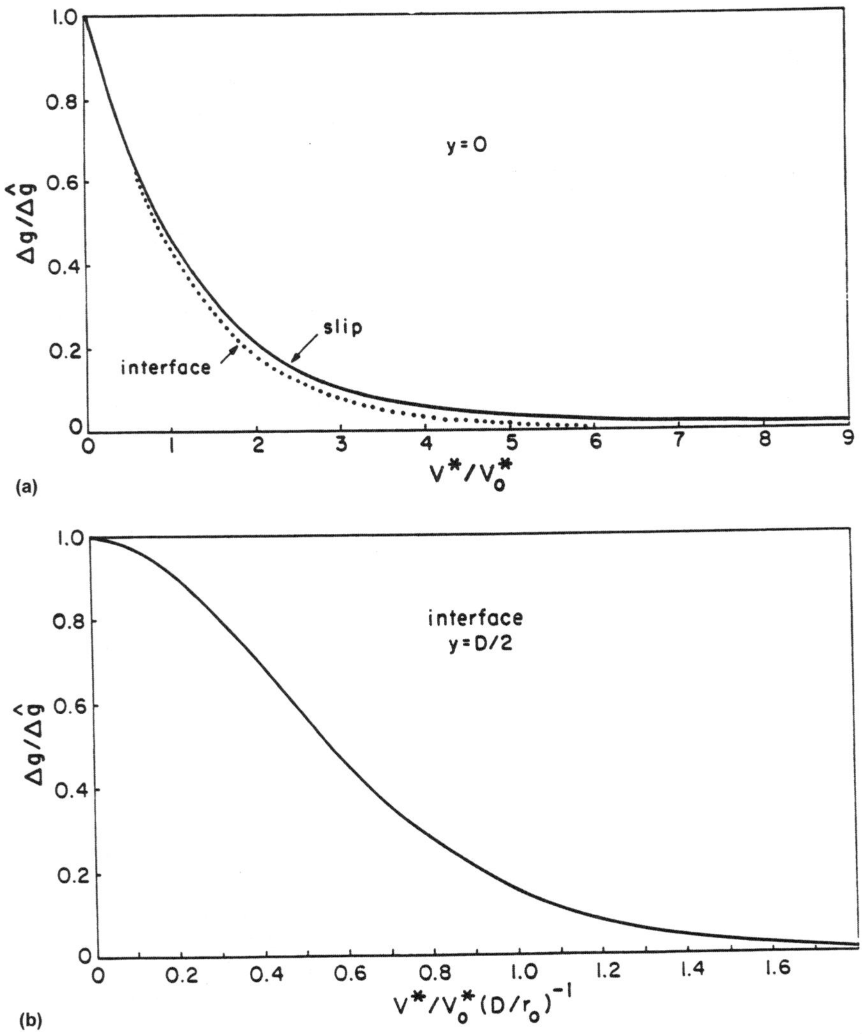

Fig. 25 Interface/particle strain interaction. (a) Force-distance profiles at $y = 0$. Dotted curve represents interface with $D/r_0 = 10$; solid curve depicts both $D/r_0 = 10^3$ interface and individual slip dislocation; (b) behavior at $y = D/2$ corresponding to twinned interface. Source: Ref 64

shorter range of the interfacial stress field which scales with D/r_0.

Integration of Eq 52 as in Eq 31 provides the dependence of the activation energy Q on Δg. This is compared with the slip-dislocation case in Fig. 26. Complete integration gives a total activation energy for $\Delta g = 0$ of $Q_0 = \hat{F}r_0$. The high Δg (short-range) behavior is again identical to that of the isolated dislocation which can be represented by:

$$Q = Q_0\left[1 - \tfrac{4}{3}(\Delta g/\Delta\hat{g})^{1/2} + \tfrac{1}{3}(\Delta g/\Delta\hat{g})^2\right] \quad \text{(Eq 55)}$$

At low Δg the behavior again differs depending on the magnitude of D/r_0, and can be approximated by:

$$Q = Q_0 + \frac{V_0^* D}{2\pi\, r_0}\left(\frac{\Delta g}{\Delta\hat{g}}\right)\left\{\ln\left[\frac{1}{4}\left(\frac{D^2}{\pi r_0}\right)\left(\frac{\Delta g}{\Delta\hat{g}}\right)\right] - 1\right\} \quad \text{(Eq 56)}$$

It is convenient to approximate the overall $Q - \Delta g$ behavior by a relation of the form (Ref 56):

$$Q = Q_0[1 - (\Delta g/\Delta\hat{g})^p]^q \quad \text{(Eq 57)}$$

where the exponents p and q describe the long-range (small Δg) and short-range (large Δg) behaviors, respectively. Both the slip-dislocation behavior defined by Eq 54 and 55 and the interfacial behavior defined by Eq 52, 55, and 56, despite some long-range differences, can be described approximately by $p = 1/2$ and $q = 3/2$.

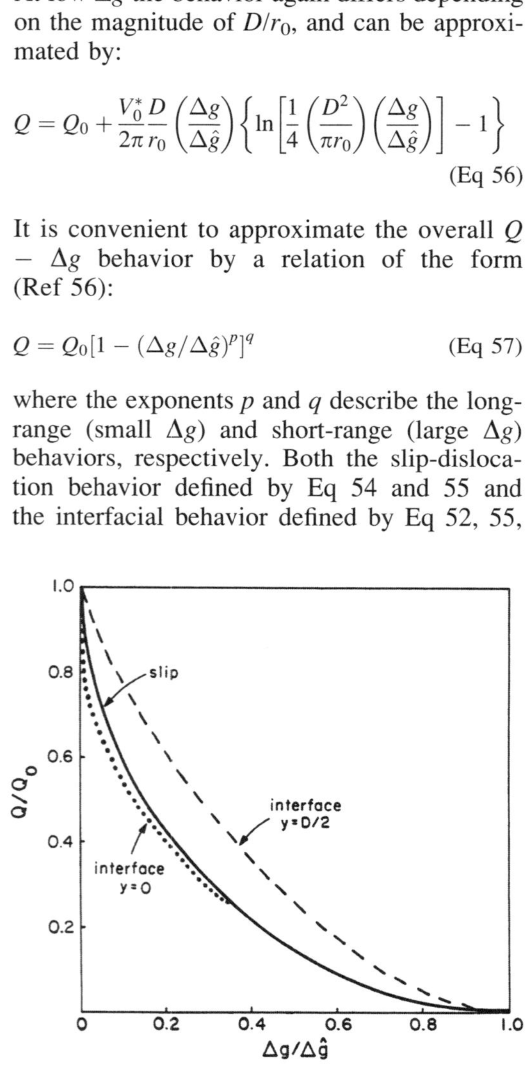

Fig. 26 Activation-energy/driving-force plots for strain interaction. Dotted curve depicts interface with $D/r_0 = 10$; solid curve represents both $D/r_0 = 10^3$ interface and individual slip dislocation; dashed curve ($y = D/2$) corresponds to twinned interface. Source: Ref 64

While the maximum interaction occurs for $y = 0$, another local maximum exists at $y = D/2$. Although the weaker interaction at this position is unlikely to be of rate-controlling significance for the slip-dislocation interface, the stress field here is very similar to that of the internally twinned case, as demonstrated in Fig. 20, and therefore serves as a convenient basis for comparison of the slip and twinning cases. The interaction force is maximum at $x = 0$ and equal to:

$$\hat{F} = \frac{2\pi}{3D} a_{33}\left(1 - a_{13}^2 - a_{33}^2\right)^{1/2} \varepsilon_{33}' \mu r_p^3 (b/D) \quad \text{(Eq 58)}$$

For the twinned interface, b/D would represent the magnitude of the lattice-invariant shear and D the projected twin period. The $F-x$ behavior is symmetric about $x = 0$ and defines a $\Delta g - V^*$ relation of the form:

$$V^*/V_0^* = \frac{2\,D}{\pi\, r_0}\cosh^{-1}\left[(\Delta g/\Delta\hat{g})^{-1/2}\right] \quad \text{(Eq 59)}$$

The overall interaction is substantially weaker than in the case $y = 0$, and integration of Eq 59 gives a significantly lower activation energy of the form:

$$Q = Q_0(\Delta g/\Delta\hat{g})^{1/2}\left\{\sinh\left[\cosh^{-1}(\Delta g/\Delta\hat{g})^{-1/2}\right] - (\Delta g/\Delta\hat{g})^{1/2}\cosh^{-1}(\Delta g/\Delta\hat{g})^{-1/2}\right\} \quad \text{(Eq 60)}$$

with

$$Q_0 = (2/3\pi)\hat{F}D$$

This is denoted by the dashed curve in Fig. 26. In terms of Eq 57 the behavior can be represented by $p = 1$ and $q = 2$, identical to the form obtained for the Peierls barrier, Eq 39.

It is useful to compare the activation parameters for interfacial motion with those for the motion of slip dislocations, inasmuch as knowledge of slip kinetic parameters might then be used to predict interfacial mobility in the same microstructure. Because the shapes of the $Q - \Delta\hat{g}$ relations are very similar, differences reside primarily in the Q_0, $\Delta\hat{g}$, and V_0^* parameters. For the case $y = 0$, $\hat{F}$ is identical for the interface and the isolated slip dislocation and from Eqs 48 and 50, $\Delta\hat{g}$ scales inversely with λ/d.

Due to the shorter range of the interfacial field, integration of the force-distance curves by the same methods gives a Q_0 for the interface which is 2/3 that for the slip dislocation. The same activation energy would then require a larger $\Delta g/\Delta\hat{g}$ and hence a smaller V^*/V_0^* (shorter-range) interaction with a given obstacle. Moreover, a lower vibrational frequency can be expected for the more three-dimensional configurations involved in interfacial motion, and so the same velocity would require a lower Q, tending to reduce V^*/V_0^* even further. Comparing V^*, this may be sufficient to compensate for the factor of $\frac{2}{3}\lambda/d$ relating V_0^* (Eq 49 and 51), tending to make the V^* values comparable in the two cases.

For the twinned interface as represented by the calculations for $y = D/2$, the ratio of $\hat{F}$ for the interface (Eq 58) to $\hat{F}_\perp$ for the slip dislocation (Eq 47) is:

$$\hat{F}/\hat{F}_\perp = \tfrac{1}{2}\pi(D/r_0)^{-2} \qquad \text{(Eq 61)}$$

which diminishes quickly as D/r_0 increases. The threshold driving force is correspondingly reduced:

$$\Delta\hat{g}/\Delta\hat{g}_\perp = \tfrac{1}{2}\pi(\lambda/d)^{-1}(D/r_0)^{-2} \qquad \text{(Eq 62)}$$

Hence, while the slip interface and isolated slip dislocation have similar obstacle strain interactions, the twinned interface interaction is appreciably weaker. Integrating the force/distance behavior to obtain Q_0, the reduced $\hat{F}$ is partly compensated by the greater range of interaction as D/r_0 increases, giving:

$$Q_0/Q_0^\perp = \frac{4}{9\pi}\frac{D}{r_0}\frac{\hat{F}}{\hat{F}_\perp} = \frac{2}{9}\left(\frac{D}{r_0}\right)^{-1} \qquad \text{(Eq 63)}$$

The weaker obstacle interaction means that lattice-invariant shear by twinning rather than by slip provides a much more mobile interface in the thermally activated regime. This may account for the occurrence of internal twinning in martensitic transformations despite the thermodynamic disadvantages discussed in the section "Interfacial Energy."

Modulus interaction. When a particle possesses different elastic constants from those of its matrix, the resulting modification of the strain energy of the interfacial stress field can also provide an interaction force. Using the example of a screw-dislocation interface, the strain-energy density at a point in the interfacial stress field is:

$$W = \frac{1}{2\mu}\left(\sigma_{xz}^2 + \sigma_{yz}^2\right) \qquad \text{(Eq 64)}$$

with σ_{xz} and σ_{yz} given by Eq 14. Adopting similar approximations as in the strain interaction case, the force arising from a spherical particle of radius r_p with a shear-modulus difference $\Delta\mu$ is then:

$$F = -\tfrac{4}{3}\pi r_p^3 \Delta\mu \frac{dW}{dx} \qquad \text{(Eq 65)}$$

From eqs. (14) and (64) we obtain:

$$F = \frac{4}{3}\frac{\pi^2\Delta\mu r_p^3 b^2}{D^3}\frac{\sinh(2\pi x/D)\cos(2\pi y/D)}{[\cosh(2\pi x/D) - \cos(2\pi y/D)]^2} \qquad \text{(Eq 66)}$$

Again, close to an anticoherency dislocation at x, $y = 0$, the interaction is the same as for an isolated dislocation:

$$F = \frac{2\Delta\mu r_p^3 b^2}{3\pi}\frac{x}{(x^2 + y^2)^2} \qquad \text{(Eq 67)}$$

With a cutoff distance $x = r_0$, the maximum force in the $y = 0$ plane is then:

$$\hat{F} = \frac{2\Delta\mu b^2}{3\pi}\left(\frac{r_p^3}{r_0^3}\right) \qquad \text{(Eq 68)}$$

Relating Δg to F and V^* to Δx via Eq 48 and 49, and once more assuming a linear behavior of F for $x < r_0$, the $V^* - \Delta g$ obstacle-interaction profile can be obtained by expressing V^* as the sum of long- and short-range contributions V_1^* and V_2^* which are defined by:

$$\frac{\cosh\left(3\pi V_1^*/2\lambda^2 D\right)}{\sinh^3\left(3\pi V_1^*/2\lambda^2 D\right)} = \frac{\cosh(\pi r_0/D)}{\sinh^3(\pi r_0/D)} \cdot \left(\frac{\Delta g}{\Delta\hat{g}}\right) \qquad \text{(Eq 69a)}$$

$$V_2^*/V_0^* = \Delta g/\Delta\hat{g} \qquad \text{(Eq 69b)}$$

with $V_0^* = \frac{2}{3}\lambda^2 r_0$ and $\Delta g = - F/\lambda^2$ as before. For small Δg, the behavior arising from the exponential tail of the stress field can be described by:

$$V^*/V_0^* = -\frac{1}{2\pi}\left(\frac{D}{r_0}\right)\ln\left[\frac{1}{4}\left(\frac{D}{\pi r_0}\right)^3(\Delta g/\Delta\hat{g})\right] \qquad \text{(Eq 70)}$$

For large Δg, the short-range interaction approaches that of an isolated slip dislocation expressed by:

$$V^*/V_0^* = (\Delta g/\Delta\hat{g})^{-1/3} - (\Delta g/\Delta\hat{g}) \quad \text{(Eq 71)}$$

Figure 27(a) compares the $\Delta g{-}V^*$ profile for the slip dislocation with that of the interface for D/r_0 values of 10^1 and 10^3. The behavior is again quite similar overall but differs at large V^* due to the shorter range of the interfacial stress field.

Integrating Eq 69 gives the dependence of Q on Δg, and this is compared with the slip dislocation case in Fig. 28. Complete integration gives $Q_0 = \frac{2}{3}\hat{F}r_0$. The behavior is very similar to that of the slip dislocation which is described by:

$$Q = Q_0\left[1 - \tfrac{3}{2}(\Delta g/\Delta\hat{g})^{2/3} + \tfrac{1}{2}(\Delta g/\Delta\hat{g})^2\right] \quad \text{(Eq 72)}$$

At small Δg, Q for the interfacial case is reduced depending on the magnitude of D/r_0, and can be approximated by:

$$Q = Q_0 + \frac{V_0^* D}{2\pi\, r_0}\left(\frac{\Delta g}{\Delta\hat{g}}\right)\left\{\ln\left[\frac{1}{4}\left(\frac{D^3}{\pi r_0}\right)\frac{\Delta g}{\Delta\hat{g}}\right] - 1\right\} \quad \text{(Eq 73)}$$

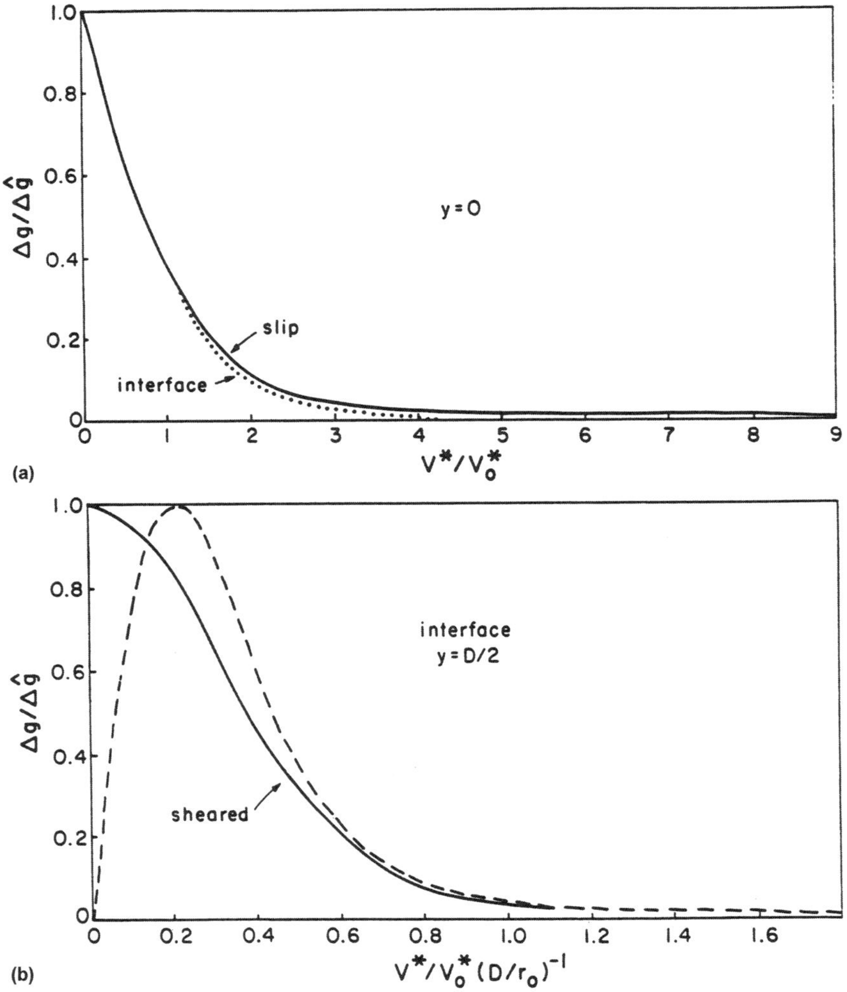

Fig. 27 Interface/particle modulus interaction (a) Force-distance profiles at $y = 0$. Dotted curve represents interface with $D/r_0 = 10$; solid curve depicts both $D/r_0 = 10^3$ interface and individual slip dislocation; (b) behavior at $y = D/2$ corresponding to twinned interface; solid curve shows sheared profile. Source: Ref 64

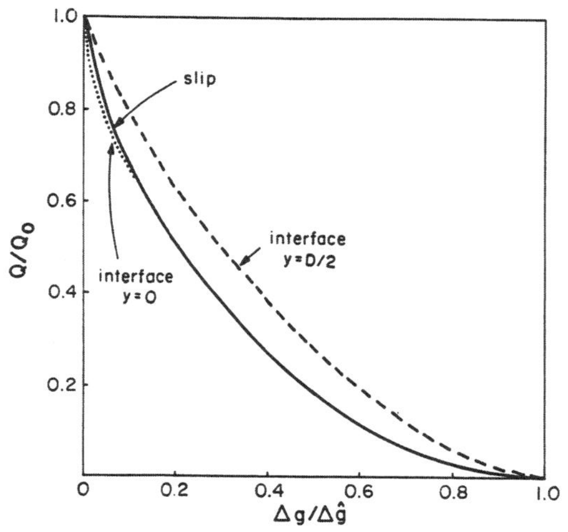

Fig. 28 Activation-energy/driving-force plots for modulus interaction. Dotted curve depicts interface with D/r_0 = 10; solid curve represents both $D/r_0 = 10^3$ interface and individual slip dislocation; dashed curve ($y = D/2$) corresponds to twinned interface. Source: Ref 64

Using the approximation of Eq 57, the overall behavior can be described by the parameters $p = 0.8$ and $q = 2$.

With the stress field of Eq 14 evaluated at $y = D/2$ as a model for the twinned martensitic interface, a maximum modulus interaction occurs at $x_m = 0.2096D$ where the peak force is:

$$\hat{F} = 0.2567\frac{\pi^2\Delta\mu r_p^3}{D}\left(\frac{b}{D}\right)^2 \quad \text{(Eq 74)}$$

An "unsheared" obstacle profile (as in Fig. 23 (a)) is defined by the relation:

$$\Delta g/\Delta\hat{g} = 5.198\frac{\sinh\left(2\pi\frac{V}{V_0^*}\frac{r_0}{D}\right)}{\left[\cosh\left(2\pi\frac{V}{V_0^*}\frac{r_0}{D}\right)+1\right]^2} \quad \text{(Eq 75)}$$

This function and the corresponding sheared profile are depicted in Fig. 27(b) again showing a smoother short-range interaction compared to the $y = 0$ case. Integration gives $Q_0 = 0.0102\,\hat{F}D$ and the $Q - \Delta g$ behavior represented by the dashed curve in Fig. 28. In terms of Eq 57, this behavior is described by the exponents $p = 1$ and $q = 1.8$.

Comparing the interfacial behaviors with that of the isolated slip dislocation, the same relations apply for the $y = 0$ case as for the strain interaction discussed previously, predicting very similar mobility for dislocated interfaces and slip dislocations. For the twinned interface, however, the modulus interaction is even more sensitive to D/r_0 than is the strain interaction, giving a much weaker relative maximum force defined by:

$$\hat{F}/\hat{F}_\perp = 1.1550\pi^3(D/r_0)^{-3} \quad \text{(Eq 76)}$$

with a correspondingly reduced threshold driving force:

$$\Delta\hat{g}/\Delta\hat{g}_\perp = 1.1550\pi^3\left(\frac{\lambda}{d}\right)^{-1}\left(\frac{D}{r_0}\right)^{-3} \quad \text{(Eq 77)}$$

The maximum activation energy is similarly reduced:

$$Q_0/Q_0^\perp = 0.0102\frac{D}{r_0}\hat{F}/\hat{F}_\perp = 0.0118\pi^3\left(\frac{D}{r_0}\right)^{-2} \quad \text{(Eq 78)}$$

Comparison with Eq 61–63 shows a stronger D/r_0 dependence than for the strain interaction. For an obstacle which interacts with a slip dislocation via both strain and modulus effects, interaction with a twinned interface will thus tend to involve a greater relative contribution from the strain interaction.

Short-range interactions. While the interfacial stress fields of Eq 12–14, based on the approximation of continuously distributed coherency dislocations, are adequate to treat the longer-range elastic interactions with defects, the short-range interactions at play when a defect or particle makes direct contact with the interface will be sensitive to the actual discrete structure of the interfacial core. The interaction energetics will be influenced by nonlinear elastic effects that are more difficult to estimate, but the essential characteristics of contact interactions involving modifications of particle shape, structure, and interfacial energy can be anticipated.

When a particle does not undergo the same transformation as its matrix, elastic back stresses will accompany the matrix transformation, just as for the normal plastic deformation of a matrix containing a dispersion of hard particles (Ref 65). Stored energy (including a reversible change of particle interfacial energy) which is eliminated upon reverse interfacial motion does not contribute to a friction stress, per se, but rather alters the thermodynamic equilibrium stress τ_0 at which the martensite and parent phase have equal volume free energies. However, irreversible plastic deformation of the particle in response to the local stresses as the interface crosses the particle will contribute to the interfacial friction stress. The temporary change of net interfacial energy when a particle resides on the martensitic interface gives a contact interaction force equivalent to that treated by Zener (Ref 66) for the case of grain-boundary pinning by second-phase particles. The short-range interaction force, F_{SR}, in the range $x \leq r_p$, with r_p the particle radius, can be approximated by:

$$F_{SR} = 2\pi\Delta\gamma(r_p - x) \quad \text{(Eq 79)}$$

where $\Delta\gamma$ is the change in interfacial energy when the particle resides on the interface, equal to the (local) martensitic interfacial energy when the particle does not transform. For a particle which undergoes a transformation as some component of the martensitic interfacial dislocation array passes through it, an additional force, also of range r_p, can arise from a difference in the lattice friction stress within the particle. Compared to the long-range elastic interactions, these short-range forces will usually provide a relatively weak contribution to the interfacial resistance.

Multiparticle interactions. When the effective spacing λ of obstacles is comparable to their range of interaction, the foregoing analysis in terms of interaction with individual discrete obstacles is no longer appropriate. Instead, the critical events controlling the kinetics of thermally activated motion will involve the interaction with many obstacles simultaneously. Such multiparticle interaction is known to be important in the case of solid-solution strengthening, and statistical methods are being developed to deal with this theoretical problem. The theory of Labusch (Ref 67) has been shown to predict important observed features of solid-solution strengthening, as discussed by Nabarro (Ref 68). The latter discussion implies that the role of thermal activation can in this case still be represented by Eq 57 with $p = 1$, $q = \frac{3}{2}$. Compared to single particle interaction, however, Q_0 is substantially increased, with an attendent increase in V^*.

An analysis of observed solid-solution strengthening behavior by Kocks (Ref 69) supports a mechanism of repeated unlocking from smoothly distributed solute atmospheres. Such a mechanism provides a $Q - \Delta g$ relation of the form:

$$Q = A\frac{[1 - (\Delta g/\Delta\hat{g})^p]^q}{\Delta g/\Delta\hat{g}} \quad \text{(Eq 80)}$$

where A is a constant; available data suggest $p = 1$, $q = \frac{1}{2}$ for this case.

While the models and experimental data have so far been developed only for the motion of slip dislocations, similar forms of behavior can be expected to apply to interfacial motion. However, geometric differences between the motion of line and surface defects might alter the solute-concentration dependence of Q_0 and $\Delta\hat{g}$.

The further complications which can arise when solutes become mobile are discussed in the section "Stabilization Effects."

Observations in β-Cu-Al-Ni Alloys

Stress-assisted single-interface transformations in single crystals of thermoelastic Cu-Al-Ni alloys have allowed direct measurements of the intrinsic interfacial mobility of internally twinned martensites in the absence of significant extrinsic accommodation slip (Ref 34, 70, 71); this has been examined most recently by Grujicic (Ref 34). In a limited temperature range, uniaxial compression testing at constant crosshead speed can impose motion of a $\beta_1 - \gamma_1'$ martensitic interface at a fixed velocity, and this can be directly measured by observation of the interface. An example of the applied stress/crosshead displacement curve obtained in this way is shown in Fig. 29. After nucleation at a specimen corner at stress τ_N, a single interface propagates at a fairly constant stress level; reversing the crosshead at the same speed, the interface moves in the reverse direction at a lower stress level. One-half the resulting stress hysteresis represents the average friction stress for forward and reverse motion at the imposed velocity. If these friction stresses are taken to be equal, the midpoint of the hysteresis loop defines the equilibrium stress τ_0 for which the net interfacial driving force is zero. Based on the crystallographic parameters for the particular martensitic variant which forms, the applied stresses can be converted to interfacial driving forces via Eq 25.

Results obtained in this way for two crystal compositions of differing thermodynamic stabilities to cover a wider temperature range are shown in Figs. 30 and 31 (Ref 34, 70). Alloy A is a Cu-13.9Al-3.9Ni crystal quenched from 1073 K while the more stable alloy B is a Cu-14.2Al-3.5Ni crystal quenched from 1249 K. For isothermal interfacial motion at velocities

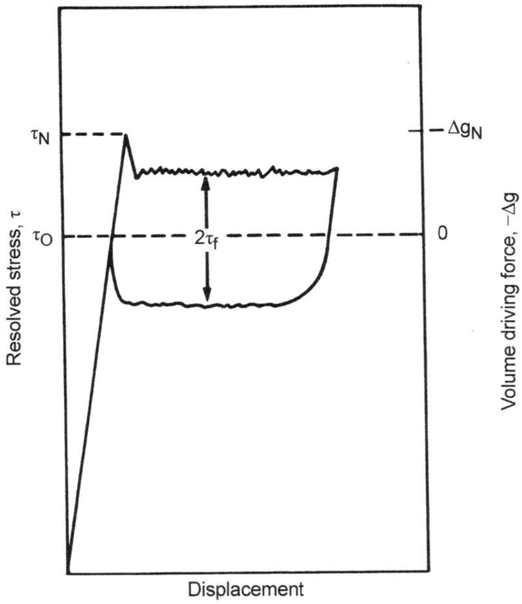

Fig. 29 Typical applied stress/crosshead displacement curve for $\beta_1 - \gamma_1'$ single interface transformation in Cu-Al-Ni crystals. Source: Ref 34

of 10^{-6} to 10^{-2} ms^{-1} over a temperature range of 210 to 410 K, the interfacial friction force shows a normal temperature and rate dependence for thermally activated motion. A fairly well-defined athermal component (Δg_μ) for alloy A is indicated by the dashed line in Fig. 30. A somewhat smaller athermal component (above 210 K) is indicated for alloy B. An anomalous temperature dependence below 210 K is observed for alloy B. The relatively soft C′-elastic constant of the β-phase also shows an anomalous temperature dependence with a steeper slope in this region, suggesting that the reduced interfacial friction force might arise from interface widening associated with incipient lattice instability as discussed in the section "Motion in a Perfect Crystal."

For the behavior above 210 K, a linear plot of ln υ versus $1/kT$ at constant driving force verifies that the thermally activated rate law of Eq 32 holds with $\upsilon_0 \approx 30$ ms^{-1} and that the operational activation energy Q of Eq 33 is a single-valued function of the thermal component of driving force ($\Delta g - \Delta g_\mu$) for both alloys. Deriving activation volumes according to Eq 35, the $\Delta g - V^*$ obstacle profile and activation-energy $Q - \Delta g$ relation obtained are shown in Fig. 32 and 33. The behavior is well represented by Eq 57 with $Q_0 = 0.305$ eV, $\Delta\hat{g} = 0.9\,MJ/m^3$, $p = 0.39$, and $q = 1.35$. This

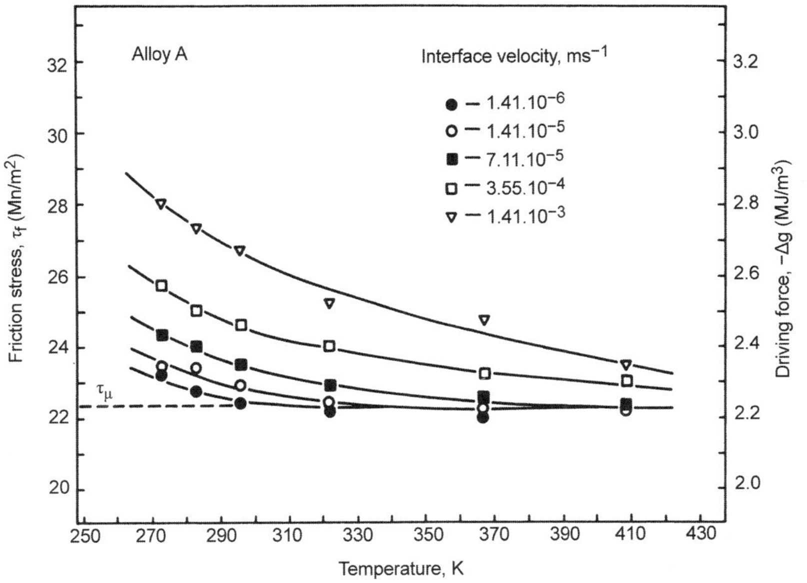

Fig. 30 Temperature and velocity dependence of friction stress (driving force) for Cu-13.9Al-3.9Ni alloy A quenched from 1073 K. Source: Ref 34

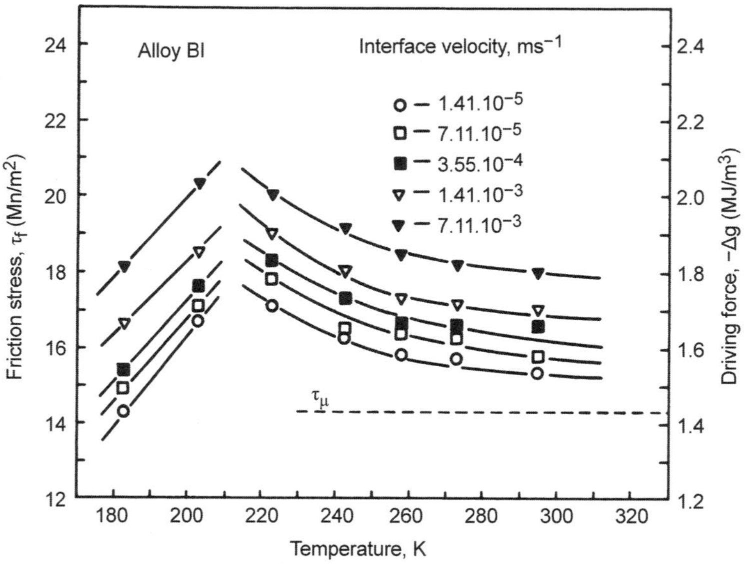

Fig. 31 Temperature and velocity dependence of friction stress (driving force) for Cu-14.2Al-3.5Ni alloy B1 quenched from 1250 K. Source: Ref 34

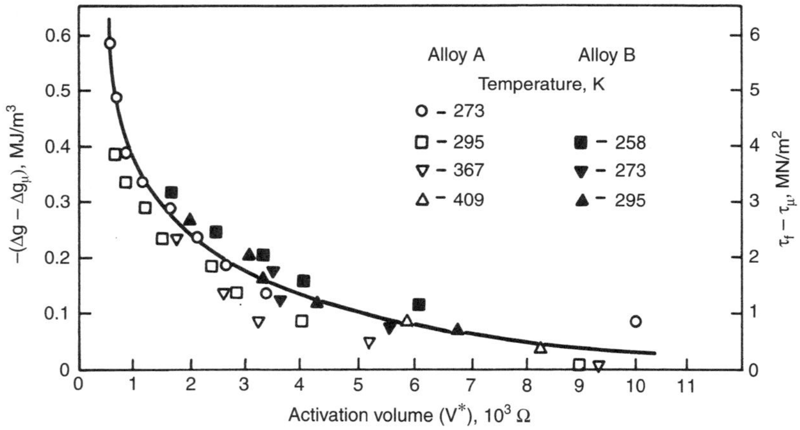

Fig. 32 Thermal component of driving force (friction stress) versus activation volume for alloys of figs. 30 and 31. $T > 250$ K. Source: Ref 34

is very similar to the form of Fig. 25(b) and 26, suggesting that the rate control is by strain interaction with discrete particles.

Transmission electron microscopy of the β-phase crystals revealed a low forest-dislocation density which did not change appreciably with a small number of transformation cycles, but also identified numerous second-phase particles. The features of greatest potential significance to interfacial mobility are shown in Fig. 34. The particles of Fig. 34(a) have a 2H structure. Assuming a similar ε_{ij}^{T} and μ as for the martensitic phase, detailed strain and modulus interaction calculations were performed for the observed 2H-phase particle-size distribution, using the stress field of the $(011)_\beta$ coherency dislocation array of Fig. 14, and integrating the interactions over the finite volumes of the particles. These particles were determined to be on too coarse a size scale to account for the thermal component of the interfacial friction force, but could provide a long-range force of sufficient magnitude to account for the observed athermal component.

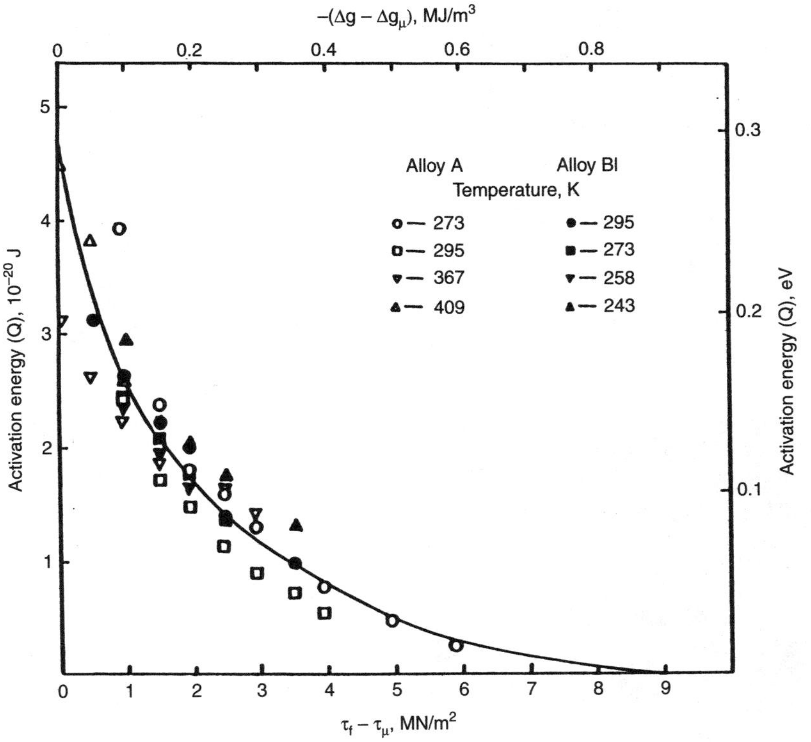

Fig. 33 Driving-force dependence of activation energy for alloys of Fig. 30 and 31. Source: Ref 34

The only observed microstructural feature on a sufficiently fine scale to account for the thermal component was the "tweed" structure shown in Fig. 34(b). Such modulated structures are commonly observed in β-phase alloys with a soft C' shear constant and have been attributed to periodic atomic displacements on this shear system. The origin of the displacements is still uncertain, but has been variously ascribed to small coherent particles or other point centers of predominantly tetragonal distortion such as clusters of interstitials or vacancies. Treating the fine structure as an array of point obstacles with average spacing equal to the modulation period of 3–6 nm, interaction calculations were performed for various assumed cases, including both modulus and strain interaction with vacancy clusters, coherent 2H-phase particles, and tetragonal strain centers. Using a local interfacial twin period from recent electron microscopy observations (Ref 72), the calculated $Q - \Delta g$ behavior gave reasonable agreement with experiment (Ref 34).

While a better understanding of the tweed microstructure is necessary for a quantitative test of the mobility models derived in the section "Defect Interactions," these experiments in Cu-Al-Ni alloys clearly demonstrate that the form of the mobility behavior is consistent with thermally activated interfacial motion, and microstructural features can be identified which provide a plausible basis for both the athermal and thermal components of the resistance to interfacial motion. Further experiments allowing a direct comparison of slip and interfacial mobilities in the same microstructure would also provide an important test of the theoretical models.

Coupled Diffusional/Displacive Transformation

As discussed in the section "Conservative versus Nonconservative Motion," the mobility of martensitic interfaces may share common features with the interfacial mobility in other solid-state phase transformations that involve conservative lattice-invariant deformations. Consider an interface with the dislocation structure and mobility of the martensitic

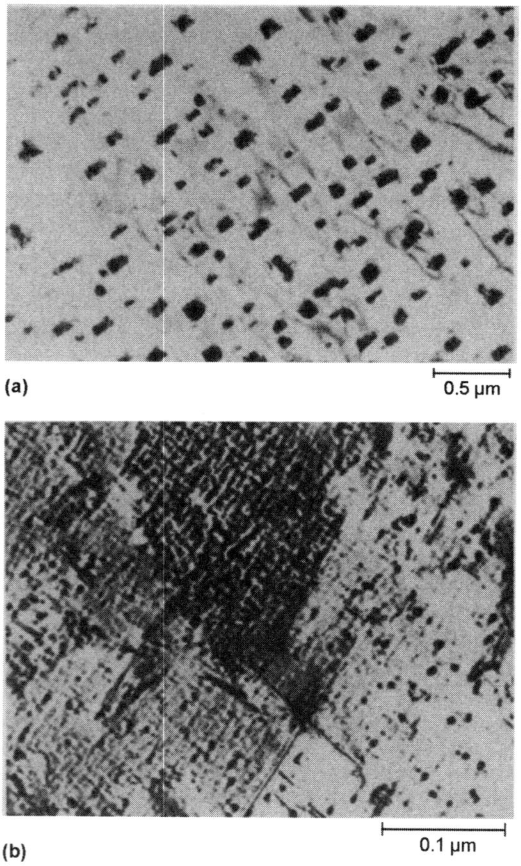

Fig. 34 Transmission electron micrographs of microstructural features in βCu-Al-Ni crystals. (a) 2H-phase particles; (b) tweed structure; modulations on (110) planes viewed along [001] direction. Source: Ref 34

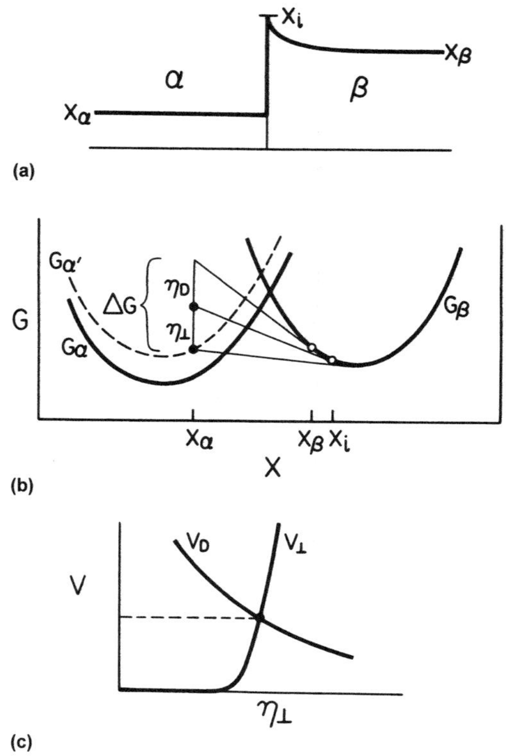

Fig. 35 Coupled diffusional/displacive transformation. (a) Composition profile for conversion of β of composition X_β to α of composition X_α; (b) free-energy/composition diagram defining β composition at interface (X_i) constrained by interfacial dissipation $\eta_\perp$; (c) diffusion-field velocity (for edge-wise growth of plate-shaped particle) and dislocation-interface velocity versus interfacial dissipation, identifying stable steady-state velocity for coupled process.

interfaces just treated, but under thermodynamic conditions where the driving force at constant composition is insufficient to allow interfacial motion (e.g. $\Delta g < \Delta g_\mu$). If the temperature is sufficient to allow some degree of atomic diffusion, solute partitioning between the two phases could increase the operative driving force and allow interfacial motion. For a parent phase β of composition X_β and a product phase α of composition X_α, the diffusion field (composition versus position) that must move with the interface is of the form depicted in Fig. 35(a), where X_i is the composition of the β-phase at the interface. The composition X_i and the total (molar) driving force ΔG acting on the interface can be determined from solution thermodynamics via the construction given by the schematic free-energy/composition diagram of Fig. 35(b) (Ref 73). For the interface of an enclosed semicoherent particle, the thermodynamic contribution of elastic strain energy and interfacial energy raises the effective molar free energy of the product α-phase from the solid G_α curve to the dashed $G_{\alpha'}$ curve. The molar free-energy change accompanying the transformation of β of composition X_β to α of composition X_α is shown by the difference between the upper point on the tangent line from the G_β curve at X_β and the point on the dashed $G_{\alpha'}$ curve at X_α, as indicated in the figure. If no free energy dissipation were to occur at the interface, the above-designated driving force would be entirely dissipated by the motion of the diffusion field, and the local equilibrium β-composition at the interface would be that for which ΔG for the transformation to α of composition X_α is zero. This is defined by the tangent point on the G_β curve of a line from the point $G_{\alpha'}(X_\alpha)$.

As suggested by Hillert (Ref 74), the occurrence of a free-energy dissipation at the interface will cause a deviation from local equilibrium, altering the interfacial composition X_i. If motion of the dislocation interface involves a dissipation $\eta_\perp$ corresponding to the frictional force for thermally activated motion, then the total driving force ΔG is dissipated by both $\eta_\perp$ and the diffusional dissipation η_D; the interface composition is now the β-composition for which ΔG is equal to the dissipation $\eta_\perp$. As represented in Fig. 35 (b), this corresponds to the tangent point on G_β of a line from the point $G_{\alpha'}(X_\alpha) + \eta_\perp$. Accordingly, X_i can be expressed as a function of $\eta_\perp$.

For the motion of a planar interface, a steady-state velocity for the diffusion field of Fig. 35(a) does not exist; instead, the motion follows a parabolic law. On the other hand, for the edgewise growth of a plate-shaped particle, a steady-state diffusion velocity does exist and can be expressed in the form:

$$\upsilon_D = f\left(\frac{X_i - X_\beta}{X_\beta - X_\alpha}\right) \quad \text{(Eq 81)}$$

where the function f may be approximated by the Ivantsov (Ref 75) solution for a parabolic cylinder. As shown in Fig. 35(c), υ_D is a function of $\eta_\perp$ through the dependence of X_i on $\eta_\perp$. For the thermally activated behavior discussed in the section "Thermally Activated Motion," the velocity of the dislocation interface $\upsilon_\perp$ is also a function of $\eta_\perp$. The point of intersection in Fig. 35(c), where $\upsilon_\perp = \upsilon_D$, corresponds to a stable steady-state interfacial velocity for which just sufficient diffusion can occur to supply the necessary driving force for motion of the dislocation interface. Such a process is neither strictly diffusion-controlled nor displacement-controlled but, rather, constitutes a coupled diffusional/displacive mechanism of interfacial motion.

This mechanism may very likely account for the behavior of bainitic transformations which exhibit clear evidence of both displacive characteristics and a necessary kinetic role of diffusion. Preliminary calculations by Bhadeshia (Ref 76) for an Fe-0.4C alloy indicate that this mechanism is quite plausible in the temperature range of observed bainitic transformations. Using the same calculational procedures for X_α product compositions deviating from the equilibrium value, X_α^{eq}, the intriguing result is obtained that when $\eta_\perp$ is small relative to the equilibrium ΔG, a maximum velocity is obtained for intermediate compositions between X_α^{eq} and X_β, predicting incomplete partitioning of solute. When applied to the interface-controlled heterogeneous nucleation mechanism described in the sections "Heterogeneous Nucleation by Defect Association" and "Nucleation Kinetics," the model successfully predicts the C-curve kinetics of the bainite reaction. Another interesting prediction for the lower temperatures is that if the surface- and strain-energy contributions should substantially diminish with increasing particle size, the total driving force could increase sufficiently that final growth of a bainitic subunit may occur by a diffusionless mechanism equivalent to lath martensite, as has often been suggested (Ref 77, 78). Mechanisms of final growth arrest for such a case are discussed in the section "Nonthermoelastic Growth."

Martensitic Nucleation

Classical Nucleation Theory

The application of classical nucleation theory to martensitic transformations, using both continuum-elastic and discrete-dislocation descriptions of a nucleus, has been extensively reviewed (Ref 3, 79); the present discussion will emphasize the contribution of dislocation theory. The classical approach treats the energetics of formation of a nucleus along a "path" of fixed structure and increasing size. The nucleus crystal structure is taken to be that of the fully formed transformation product, with a relatively sharp interface; the specific interfacial energy is customarily assumed to be size independent.

As developed in the section "Interfacial Energy," the specific interfacial energy γ of a fully coherent particle will consist of a "chemical" core energy, while the interfacial energy in the semicoherent case will be dominated by the elastic-energy contribution described by Eq 17–21. If the nucleus transformation strain is an invariant-plane strain, the strain energy per particle volume for a thin oblate spheroid of radius r and semithickness c is of the form Kc/r as described by Eq 3. When the transformation strain deviates from an exact IPS, an additional shape-insensitive strain energy contribution g_p^{str} arises from distortions in the particle habit plane. Following Kaufman and Cohen (Ref 55), the total free energy of an isolated martensitic particle can thus be described by:

$$\Delta G(r,c) = \tfrac{4}{3}\pi r^2 c\left(\Delta g^{ch} + g_p^{str}\right) + \tfrac{4}{3}\pi r c^2 K + 2\pi r^2 \gamma \quad \text{(Eq 82)}$$

and the free-energy barrier for homogeneous martensitic nucleation can be determined from a saddle point in the $\Delta G(r, c)$ surface defined by:

$$r^* = \frac{4\gamma K}{\left(\Delta g^{ch} + g_p^{str}\right)^2} \quad \text{(Eq 83a)}$$

$$c^* = -\frac{2\gamma}{\left(\Delta g^{ch} + g_p^{str}\right)} \quad \text{(Eq 83b)}$$

$$\Delta G^* = \frac{32\pi\gamma^3 K^2}{3\left(\Delta g^{ch} + g_p^{str}\right)^4} \quad \text{(Eq 83c)}$$

Under the thermodynamic conditions for which martensitic transformations are known to occur, this ΔG^* is far too large to be thermally surmountable, and it can be safely concluded that homogeneous classical nucleation is then not possible.

In an imperfect crystal, defect interactions can substantially alter the energy of a classical nucleus such that the free-energy barrier for *heterogeneous* nucleation is greatly diminished relative to the homogeneous case. Consistent with the strain-energy dominant character of martensitic transformations, it is generally assumed that the most important interaction in heterogeneous martensitic nucleation is with the defect stress field, and it is now well established by quantitative calculations that classical nucleation via such interaction is indeed quite plausible at experimentally observed heterogeneous nucleation sites. While this interaction can be described by traditional continuum-elastic methods, the structurally equivalent discrete-dislocation description in terms of a defect-dissociation process leads most directly to the essential behavior, and is actually the method by which the first viable theory of heterogeneous martensitic nucleation was developed (Ref 80, 81).

Heterogeneous Nucleation by Defect Dissociation

Nucleus Model

Adopting the discrete-dislocation description of Fig. 10 for the interfacial structure of a classical martensitic embryo, the long-range stress field and volume strain energy of the embryo arise primarily from the array of significant coherency dislocations lying on crystal planes at a shallow angle to the embryo habit as represented schematically in Fig. 10(a). Clearly the free-energy barrier for nucleation is greatly diminished if the latter dislocation array is derived from the dissociation of an existing defect. A convenient defect model for this purpose is the extrinsic boundary defect represented by the array of "excess" interfacial dislocations in Fig. 36(a), residing in an existing grain boundary or interphase boundary. The defect has a finite length and height, and a net Burgers vector content giving rise to a stress field of range comparable to its length, ℓ. Embryo formation by dissociation of this defect is depicted in Fig. 36(b) and (c). As discussed in the section "Motion in a Perfect Crystal," the interfacial motion producing such a particle involves both the motion of

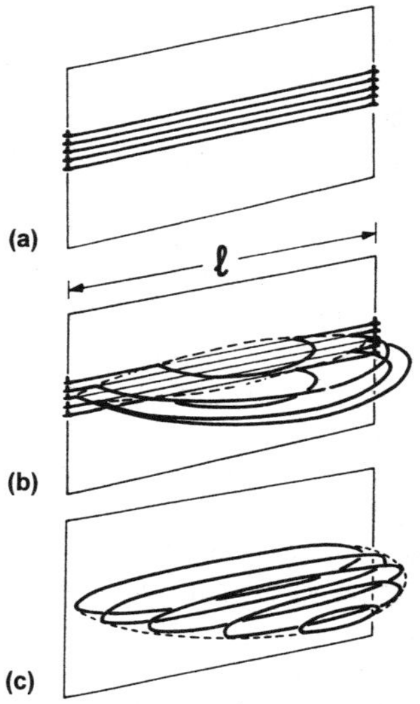

Fig. 36 Heterogeneous martensitic nucleation by defect dissociation. (a) Nucleating defect; (b) dissociation to produce interfacial coherency dislocations of Fig. 10(a) in horizontal planes; (c) simultaneous generation of interfacial dislocation arrays of Fig. 10(b) and (c) in nearly vertical planes. Source: Ref 3

the array of Fig. 10(a) as shown in Fig. 36(b), and the sequential formation of the discrete loops of Fig. 10(b) and (c) under the acting thermodynamic driving force, represented by Fig. 36(c).

As shown in Fig. 36(b), a finite number of coherency dislocations can be directly derived from dissociation of the defect. Further growth of the embryo in a direction normal to the plane of these dislocations requires the generation of additional discrete dislocation loops. In contrast to the arrays of Fig. 10(b) and (c) which together produce no long-range stresses, these coherency dislocations contribute to the long-range stress field of the embryo. As noted already, thermally activated nucleation of the latter dislocations can be difficult depending on the relative significance (ρ_L) of the particular coherency-dislocation arrays. The resulting anisotropic interfacial mobility would then restrict the embryo to adopt a habit parallel to the crystal plane of the coherency dislocations as in the case of Fig. 10(a). For the special case of the coherent f.c.c. → h.c.p. transformation, such a restriction applies to the growth of macroscopic particles even though the resulting rational habit plane is not a true invariant plane, as noted in the section "Thermodynamic and Kinetic Constraints on Morphology and Orientation Relations." For transformations involving less significant coherency dislocations, the restriction may apply only for very small particles where the net thermodynamic driving force for loop formation is relatively weak. Such a size-dependent constraint would lead to the nucleation sequence illustrated in Fig. 37. Once unrestricted loop nucleation can occur, an embryo which is initially constrained to a rational habit plane in Fig. 37(a) can adopt the irrational invariant-plane habit of Fig. 37(b), with reduced elastic strain energy.

Following the prediction of the section "Interfacial Topography" that the most significant coherency dislocations will be associated with planes of closest packing, it is appropriate for the construction of the dislocation arrays of Fig. 10(a) and (b) to decompose the transformation lattice deformation into invariant-plane strains on close-packed planes. For f.c.c. → b.c.c. transformations as in ferrous alloys, such a decomposition has been proposed by Bogers and Burgers (Ref 82) based on a hard-sphere deformation model. This defines the f.c.c.(γ) → b.c.c.(α) semicoherent nucleus model illustrated in Fig. 38, involving coherency dislocation arrays with $(111)[1\bar{2}1]_\gamma$ and $(11\bar{1})[1\bar{2}\bar{1}]_\gamma$ shear components. Assuming an initial nucleus habit restricted to $(111)_\gamma$, $\frac{1}{2}a[011]_\gamma$ and $\frac{1}{2}a[\bar{1}10]_\gamma$ anticoherency lattice dislocations lying on $(11\bar{1})_\gamma$ produce the lattice-invariant shear predicted by crystallographic analyses for this transformation (Ref 83, 84). Having such a specific model for a martensitic nucleus provides the basis for a detailed quantitative treatment of the energetics of heterogeneous classical nucleation.

Nucleation Energetics

Dissociation at fixed thickness. When thickening of the nucleus is restricted by the difficulty of coherency dislocation loop generation, the constraint of embryo formation at constant thickness simplifies the energetics of nucleation, and the essential characteristics of heterogeneous martensitic nucleation become readily apparent by analogy to the familiar problem of stacking-fault formation via simple dislocation dissociation. By this analogy, it is convenient to separate the total energy of defect and nucleus into (a) a "dislocation energy" representing the elastic interaction (including self-energy) of the right-hand coherency-dislocation

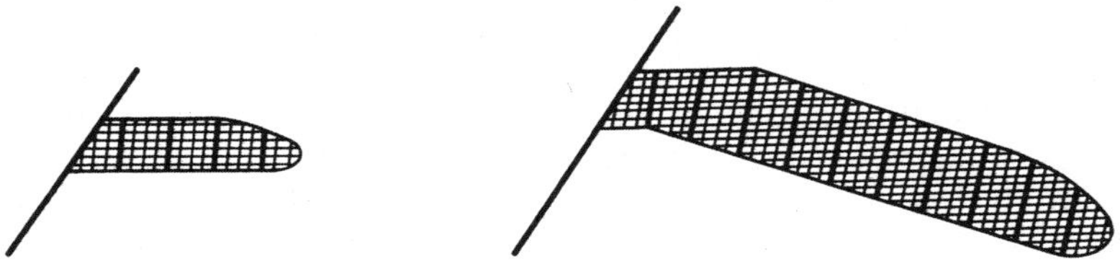

(a) (b)

Fig. 37 Possible morphological sequence for nucleation and early growth. (a) Initial nucleus formation on rational habit constrained by difficult coherency dislocation generation at small size; (b) subsequent growth on irrational habit when interfacial dislocation formation is unrestricted.

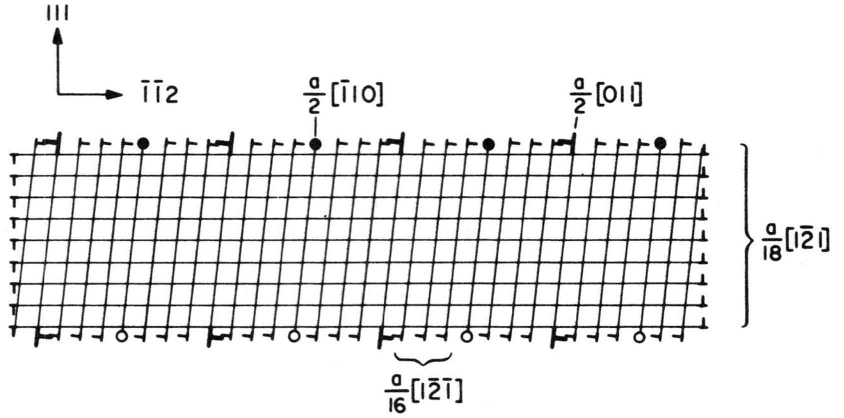

Fig. 38 Nucleus model for f.c.c. → b.c.c. transformation. Coherency dislocations $(a/18)[1\bar{2}1]$ and $(a/16)[1\bar{2}\bar{1}]$ produce transformation lattice deformation, while anticoherency lattice dislocations $\frac{1}{2}a[011]$ and $\frac{1}{2}a[\bar{1}\bar{1}0]$ accomplish lattice-invariant deformation of semicoherent particle. Source: Ref 79

array of Fig. 10(a) and 38 with the array arising from the superposition of the left-hand array with the defect array of Fig. 36, and (b) a "fault energy" which includes all other contributions. As an energy per area (in the plane of dissociation), the fault energy can be expressed by:

$$\gamma_f = nd(\Delta g^{ch} + g_f^{str}) + 2\gamma \qquad \text{(Eq 84)}$$

where n is the number of planes in the thickness of the embryo, γ is the specific interfacial energy of the top and bottom interfaces in Fig. 10(d), and g_f^{str} is the volume elastic energy arising from the component of the transformation strain which is not derived from the dissociation process; the latter energy stems primarily from distortions in the habit plane, and so is approximately equal to g_p^{str} of Eq 82. For transition-metal alloy systems which undergo f.c.c. → h.c.p. transformations, Eq 84 has been extensively tested for the simple case of intrinsic stacking faults where n = 2 (based on the number of planes which undergo the lattice deformation associated with the intrinsic fault displacement). Estimation of Δg^{ch} from solution thermodynamic data gives excellent agreement with the measured temperature and composition dependence of intrinsic fault energy (Ref 80, 85, 86) and consistently defines an f.c.c. → h.c.p. coherent surface energy of γ = 5 to 15 mJ/m^2 (Ref 80, 86).

At the early stages of dissociation when a ribbon-shaped nucleus forms along the linear defect, the bounding coherency-dislocation arrays of Fig. 10(a) and 38 can be treated as straight and parallel. When the ribbon width is much greater than the thickness, the elastic interaction of the arrays can be approximated by that of superdislocations defined by their net Burgers vectors. If the net Burgers vector of the defect is $\boldsymbol{b}$ and that of the right-hand coherency dislocations (Fig. 10(a)) is $\boldsymbol{b_1}$, the interaction will be equivalent to that of two superdislocations of Burgers vectors $\boldsymbol{b_1}$ and $\boldsymbol{b_2} = \boldsymbol{b} - \boldsymbol{b_1}$. Denoting screw and edge components by the subscripts s and e, the dislocation interaction energy per unit length for a nucleus of width a can be expressed by:

$$G_\ell^\perp(a) = -\frac{\mu}{2\pi}\left[b_{1s}b_{2s} + \frac{b_{1e}b_{2e}}{1-\nu}\right]\ln\frac{a}{a_0} \qquad \text{(Eq 85)}$$

where a_0 is a cutoff distance. The total defect/nucleus free-energy change per unit length upon widening of the nucleus to a is then the sum of the dislocation and fault-energy contributions:

$$\Delta G_\ell = G_\ell^\perp(a) + \gamma_f a \qquad \text{(Eq 86)}$$

with γ_f expressed by Eq 84. $G_\ell^\perp$(a) is a monotonic function of a, and so when the net dislocation interaction is repulsive in nature, the stability of the nucleus is controlled entirely by the sign of γ_f as illustrated in Fig. 39. For $\gamma_f > 0$ a long-range restoring force stabilizes the nucleus at a finite a. When $\gamma_f \leq 0$ this force changes sign and the nucleus is unstable with respect to unrestricted widening, leading to barrierless nucleation. From Eq 84, the critical Δg^{ch} for barrierless nucleation at a given defect is expressed by:

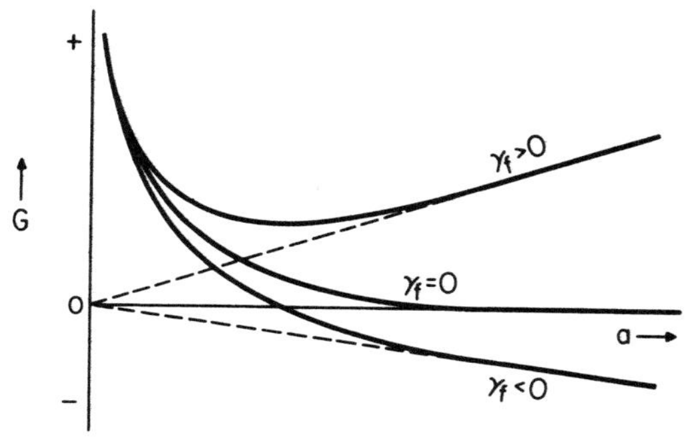

Fig. 39 Schematic free-energy curves for heterogeneous nucleation by defect dissociation. Barrierless nucleation occurs when $\gamma_f \leq 0$. Source: Ref 80

$$\Delta g_c^{ch} = -\left(g_f^{str} + 2\gamma/n_0 d\right) \qquad \text{(Eq 87)}$$

where n_0 is the number of coherency dislocations that can be derived from the defect consistent with a net repulsive dislocation interaction. (For nucleation at a defect of finite length, ℓ, as in Fig. 36, an additional line-tension restraining force inversely proportional to ℓ must be added to the right side of Eq 87. The effect on the condition for barrierless nucleation is negligible for defect lengths comparable to typical grain sizes of polycrystalline metals.) As will be discussed in the section "Nucleation Kinetics," the actual rate of nucleation by such a barrierless process will be controlled by the kinetics of interfacial propagation driven by the negative fault energy.

Calculating γ and g_f^{str} for the nucleus model of Fig. 38, a defect potency corresponding to $n_0 \approx 15$ is sufficient to account for f.c.c. $\rightarrow$ b.c.c. martensitic nucleation at the M_s temperature in ferrous alloys. Taking the γ determined from intrinsic stacking-fault energy measurements, a similar defect potency also accounts for the nucleation of the coherent f.c.c. $\rightarrow$ h.c.p. martensitic transformation in transition-metal alloys (Ref 80, 81). Such a defect is equivalent to a group of four or five lattice dislocations. The nature of these defects and their probability distribution will be discussed further in the section "Nucleation Sites."

Once the conditions are met for the unrestricted widening of a nucleus, subsequent thickening in a material for which coherency-dislocation generation is difficult even at larger particle sizes can be rationalized via interaction with forest dislocations as illustrated in Fig. 40 (Ref 62). When a single partial dislocation intersects a forest dislocation with a screw component equal to the lattice interplanar spacing (Fig. 40(a)), a partial dislocation dipole can be generated as shown in Fig. 40(b), (c). When an array of transformation coherency dislocations intersects the same forest dislocation (Fig. 40(d), (e)), a well-spaced dipole is created as in Fig. 40(f), (g). The maximum attractive glide-force interaction between the partial dislocations in the dipole of Burgers vector b_T is (Ref 62):

$$F_\ell^{max} = -\frac{\mu b_T^2}{8\pi(1-\nu)nd} \qquad \text{(Eq 88)}$$

While this is a formidable force for the small dipole of Fig. 40(c), it is small compared to the net driving force at the M_s temperature for f.c.c. $\rightarrow$ h.c.p. and f.c.c. $\rightarrow$ b.c.c. transformations when $n \approx 15$ as required for nucleus widening. The coherency dislocations can then spiral around the forest dislocation, thickening the embryo via a "pole mechanism" as denoted in Fig. 40(h). Operation of such a thickening mechanism for the f.c.c. $\rightarrow$ h.c.p. transformation is supported by observation of transformation shear in a single direction for thicknesses of hundreds of planes (Ref 87, 88); independent generation of coherency dislocations would otherwise involve three different Burgers vectors (shear directions) for this transformation.

Evidence of difficult coherency dislocation generation, as manifested in constraint of an embryo to a close-packed habit plane, is less clear for the case of f.c.c. $\rightarrow$ b.c.c. transformations. Observations of b.c.c. embryos in ferrous alloys include embryo habits near the close-packed $(111)_\gamma$ plane (Ref 89–91) and near $(112)_\gamma$ (Ref 92, 93). While the latter findings involve embryos of several microns dimension, observations of submicron embryos support the $(111)_\gamma$ habit (Ref 94, 95). Although free-surface relaxation in transmission electron microscopy experiments may remove constraints at smaller embryo sizes than for transformations in the bulk, the presence of embryos with habits near $(112)_\gamma$ suggests that the constant thickness constraint may not necessarily apply at the rate-controlling step which leads to a macroscopic growth event. It is of interest, therefore, to examine next the nucleation behavior when this constraint is removed.

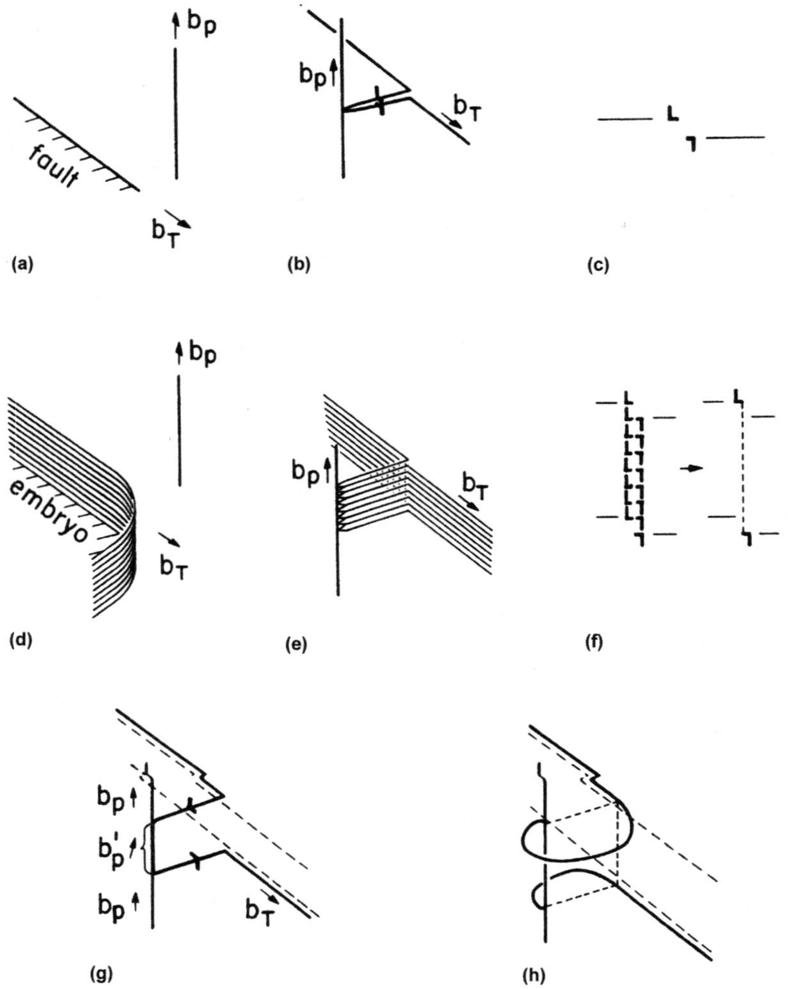

Fig. 40 Thickening of martensitic embryo by pole mechanism. Transformation coherency dislocation b_T and pole dislocation b_p in (a) intersect in (b) to form dipole as in (c). Group of coherency dislocations bounding embryo in (d) intersect pole dislocation in (e) to form dipoles which interact as in (f) to produce single dipole shown in (g). Dipole dislocations spiral around pole dislocation in (h) to thicken embryo. Source: Ref 62

Thickness fluctuations. The energetics of heterogeneous nucleation in which the embryo thickness is allowed to fluctuate can be treated by allowing n in Eq 84 to vary. Retaining the superdislocation approximation of Eq 85, the long-range dislocation interaction can be treated by making the net Burgers vector $\boldsymbol{b}_1$ of the coherency dislocation array proportional to n. Again taking the defect net Burgers vector as $\boldsymbol{b}$ with screw and edge components $\boldsymbol{b}_s$ and $\boldsymbol{b}_e$, the components of $\boldsymbol{b}_1$ of the coherency array are defined by resolving the transformation shear strain γ_T into components γ_s and γ_e in the directions of $\boldsymbol{b}_s$ and $\boldsymbol{b}_e$. The dislocation interaction energy of Eq 85 then becomes:

$$G_\ell^\perp(a,n) = \frac{\mu d^2}{2\pi}\left(\gamma_s^2 + \frac{\gamma_e^2}{1-v}\right)\left(n^2 - nn_0\right)\ln\frac{a}{a_0} \qquad \text{(Eq 89)}$$

where n_0 is again the maximum embryo thickness consistent with a net repulsive interaction, expressed by:

$$n_0 = \frac{b_s\gamma_s + b_e\gamma_e/(1-v)}{d\left[\gamma_s^2 + \gamma_e^2/(1-v)\right]} \qquad \text{(Eq 90)}$$

The total energy ΔG_ℓ can then be obtained by combining Eqs 84 and 89 as in Eq 86.

It is convenient to normalize the net volume driving force, $\Delta g^{ch} + g_f^{str}$, to its critical value for barrierless nucleation at constant thickness, n_0; from Eq 87 this normalized driving force is given by:

$$\tilde{f} = n_0 d(\Delta g^{ch} + g_f^{str})/2\gamma \quad \text{(Eq 91)}$$

Normalizing the energy per volume to $\frac{1}{2}\mu\gamma_T^2$ and distances to d, and taking $a_0 \approx d$, the total free energy per unit length can now be expressed as:

$$\Delta F_\ell = \Delta G_\ell / \left(\tfrac{1}{2}\mu\gamma_T^2 d^2\right) = \gamma_\mu\left(1 - \frac{n}{n_0}\tilde{f}\right)w + k(n^2 - nn_0)\ln w \quad \text{(Eq 92)}$$

with

$$w = a/d$$

$$\gamma_\mu = 4\gamma/\mu\gamma_T^2 d$$

$$k = \left[\gamma_s^2 + \gamma_e^2/(1-\nu)\right]/\pi\gamma_T^2$$

Setting $\partial\Delta F_\ell/\partial n = 0$, the thickness for minimum energy is defined as:

$$n = \frac{1}{2}\left(n_0 + \frac{\gamma_\mu \tilde{f}}{kn_0}\frac{w}{\ln w}\right) \quad \text{(Eq 93)}$$

Combining Eq 92 and 93 then gives $\Delta F_\ell(w)$ for an embryo unconstrained in thickness.

For semicoherent f.c.c. → b.c.c. nucleation in ferrous alloys with an interfacial structure like that of Fig. 38, γ_μ can be taken as unity. For the simple case of a defect in pure edge orientation with $\boldsymbol{b}$ parallel to the direction of γ_T, and n_0 = 15, computed curves of ΔF_ℓ versus w are plotted in Fig. 41 for various values of normalized driving force, $\tilde{f}$. The behavior shares the qualitative feature of the curves of Fig. 39 that barrierless nucleation can occur when the driving force is sufficiently high, as indicated by the $\tilde{f} = 1.25$ case. However, the embryo-thickness variations introduce an inflection in the curves such that, at lower driving forces, a local minimum exists at a position w_0 along with a maximum at a position w^*. This defines a finite nucleation barrier (per unit length) expressed by:

$$\Delta G_\ell^* = \Delta G_\ell(w^*) - \Delta G_\ell(w_0) \quad \text{(Eq 94)}$$

Such a finite barrier suggests that the barrierless nucleation condition could be preceded at lower driving force by thermally activated nucleation via heterophase fluctuations. Analogous to models of the nucleation of precipitation on dislocations (Ref 96), the critical nucleus would involve a local perturbation in the size of a subcritical embryo at size w_0. When w_0 is much smaller than w^*, the energetics of formation of such a critical perturbation is effectively equivalent to that of the formation of a small plate-shaped particle in the stress field of the defect. The latter problem has been treated for both oblate-spheroidal (Ref 97) and square-plate (Ref 98) shapes, giving a ΔG^* which diminishes as the reciprocal of the driving force to a large power. The possible relevance of such a ΔG^* to the nucleation kinetics will be discussed in the section "Nucleation Kinetics."

The curves of Fig. 41 also suggest that nucleation with variable embryo thickness requires a higher driving force for barrierless nucleation, compared to the constant thickness case which is barrierless at $\tilde{f} = 1$ for $n = n_0$. This is partly an artifact of the superdislocation

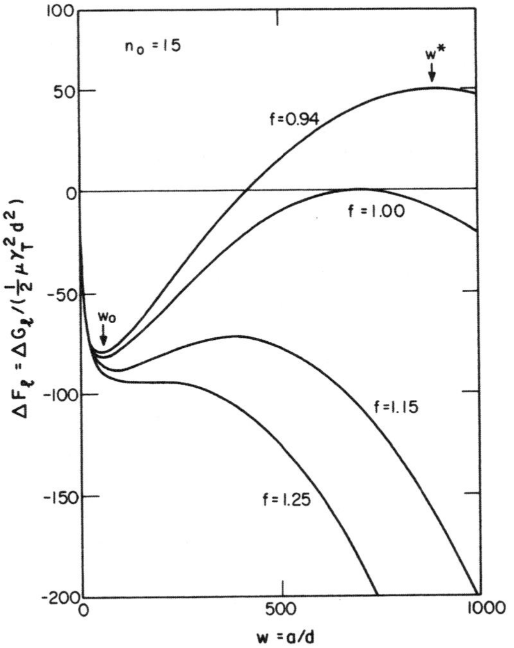

Fig. 41 Heterogeneous nucleation with variable thickness nucleus; computed free energy versus nucleus width for various values of normalized driving force, f. Barrierless nucleation for $f \geq 1.25$ preceded by formation of subcritical embryo of size w_0.

approximation which greatly overestimates the short-range dislocation interaction and, in turn, overestimates the depth of the minimum at w_0. Interaction of a finite array of coherency dislocations with a spatially extended defect as in Fig. 36 will not only provide an overall weaker short-range interaction, but will tend to stabilize subcritical embryos at a thickness comparable to the defect height. For the case $\tilde{f} = 1$ in Fig. 41, the embryo at w^* has a thickness of n_0, while the thickness at w_0 is 3/5 of this value in the superdislocation approximation; if the actual short-range defect interaction stabilizes subcritical embryos at thickness n_0, the local minimum is eliminated and barrierless nucleation can occur at this driving force. An approximate expression for the ΔG_ℓ^* barrier that precedes the barrierless condition, appropriate to this circumstance, has been derived by Roitburd (Ref 99).

Calculations by Ling and Olson (Ref 98) demonstrate that the essential features of the dislocation-dissociation description of heterogeneous martensitic nucleation can be equivalently obtained using a traditional continuum description of the nucleus/defect interaction in which the total elastic energy is described in terms of a nucleus self-energy and a nucleus/defect interaction energy. This method allows a more precise treatment of the elastic energy, particularly for the short-range interaction. Numerical calculations for the f.c.c. → b.c.c. transformation in steels, using the nucleus model of Fig. 38 and a defect (Fig. 36(a)) consisting of five lattice dislocations spaced three planes apart, are shown in Fig. 42 for three levels of the molar chemical free energy change ΔG^{ch}. Curves are plotted for various nucleus thicknesses, the dotted curves corresponding to a thickness of 15 planes, equal to the defect height, h. While the dotted curves substantiate the schematic curves of Fig. 39, the results further demonstrate that the nature of the short-range interaction is such that the thickness giving lowest energy as small sizes is equal to h, as indicated by the energy minimum at the arrow in Fig. 42(a). Therefore, the early stages of nucleation at a defect of finite height will tend to occur at constant thickness regardless of whether the nucleus is constrained against thickness fluctuations. These detailed calculations also verify that treatment of the non-IPS strain energy contributions (e.g., g_p^{str} of Eq 82 and g_f^{str} of Eq 84) as shape-independent constants is a reasonable approximation (Ref 98).

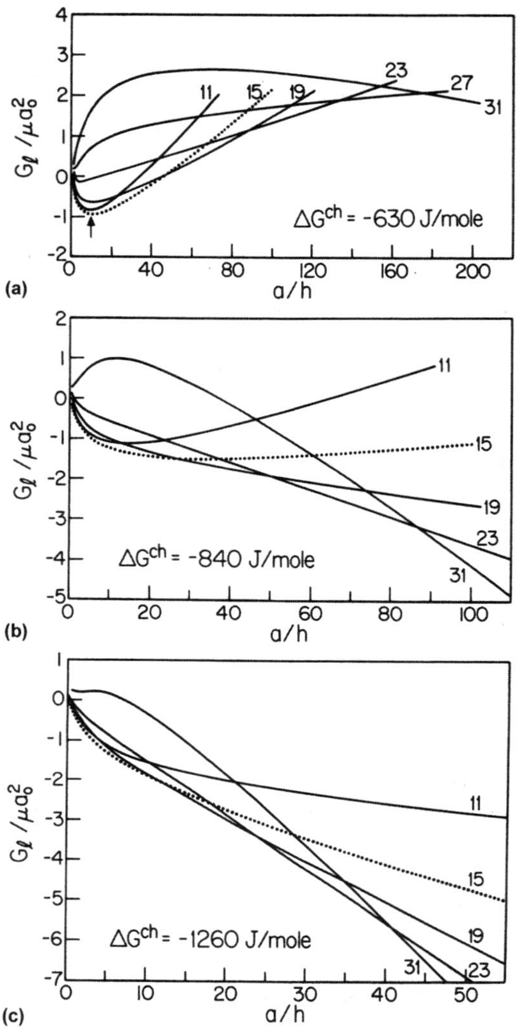

Fig. 42 Free energy per unit nucleus length for heterogeneous f.c.c. → b.c.c. nucleation in steels at three levels of driving force. Numbers identify nucleus thickness in units of crystal close-packed planes. Shear modulus, μ: f.c.c. lattice parameter, a_0. Dislocation defect in mixed orientation. Source: Ref 98

The detailed calculations have also allowed an assessment of the possibility of short-range energy barriers to embryo formation (Ref 98). This problem has been examined analytically by Chen (Ref 100) using the dislocation-dissociation description, and treating the dislocation arrays as continuously distributed. The superdislocations of Burgers vectors $\boldsymbol{b}_1$ and $\boldsymbol{b}_2$ of Eq 85 are then replaced by dislocation walls of height h with Burgers vector densities $\tilde{\boldsymbol{b}}_1 = \boldsymbol{b}_1/h$ and $\tilde{\boldsymbol{b}}_2 = \boldsymbol{b}_2/h$. The glide force per length F_ℓ acting between the arrays is the sum of the forces arising from the edge and screw components:

$$F_{\ell e} = \left[\mu \tilde{b}_{1e}\tilde{b}_{2e}/2\pi(1-\nu)\right]h\left[\xi \ln\left(1+1/\xi^2\right)\right] \quad \text{(Eq 95a)}$$

$$F_{\ell s} = \left[\mu \tilde{b}_{1s}\tilde{b}_{2s}/2\pi\right]h\left[2\tan^{-1}(1/\xi) - \xi \ln\left(1+1/\xi^2\right)\right] \quad \text{(Eq 95b)}$$

with $\xi = a/h$. Integrating these forces with respect to a provides a more precise description of the dislocation energy contribution $G_\ell^{\perp}(a)$ in Eq 86. Since the force arising from the fault energy γ_f is independent of a, and the dislocation forces of Eq 95 are of fixed sign, the existence of a short-range energy maximum will require an inflection in $G_\ell^{\perp}(a)$ where $F'_\ell = 0$. Differentiating the forces of Eq 95 with respect to ξ gives:

$$F'_\ell = \frac{\mu}{2\pi}\left\{\frac{\tilde{b}_{1e}\tilde{b}_{2e}}{1-\nu}\left[\ln(1+\xi)^{-2} - \frac{2}{1+\xi^2}\right] - \tilde{b}_{1s}\tilde{b}_{2s}\ln(1+\xi)^{-2}\right\}, \quad \text{(Eq 96)}$$

which predicts an inflection in the region $0 < \xi < 0.4508$ when $\tilde{b}_{1e}\tilde{b}_{2e}/(1-\nu) > \tilde{b}_{1s}\tilde{b}_{2s}$. Thus a short-range barrier is possible when the dislocation interaction involves a sufficiently strong edge component, but only when the fault energy is sufficiently positive, due to the monotonic nature of $G_\ell^{\perp}(a)$. This barrier will vanish at or before the point where the long-range barrier vanishes ($\gamma_f \leq 0$) and therefore does not alter the condition for nucleation.

A common feature of all heterogeneous martensitic nucleation calculations is the prediction that nucleation is preceded at lower driving forces by the formation of subcritical embryos, in line with the "preexisting embryo" hypothesis originally advanced by Kaufman and Cohen (Ref 55). These embryos are not necessarily "frozen" in size, however, and may respond relatively reversibly to changes in driving force. How far (in driving force) embryo formation can precede nucleation will be sensitive to details of the short-range nucleus/defect interactions, such as the possible short-range energy barriers just considered.

Nucleation Sites

High-energy nonequilibrium defects of the type illustrated in Fig. 36, fulfilling the geometric and energetic requirements for heterogeneous martensitic nucleation, will tend to exist primarily as *extrinsic* interfacial defects at grain boundaries and interphase boundaries where they can be partially stabilized by interfacial interactions. Although individual defects will tend to relax to states of lower energy (and lower nucleation potency) during high-temperature annealing, a rate of potent defect production can be expected from "accidents" of boundary motion during annealing and from differential thermal expansion at interphase boundaries on heating and cooling. Electron microscopy observations of martensitic embryos formed at internal interfaces are now quite numerous (Ref 89–91, 93, 101, 102). Figure 43 shows an in situ observation of the nucleation of a b.c.c. → orthorhombic martensitic transformation in a fine-grained rapidly solidified Au-Cd alloy. A nucleus initially restricted to the closest-packed $(110)_{b.c.c.}$ habit, forming at sub-boundary A in Fig. 43(a), adopts the macroscopic habit in Fig. 43(b), similar to the sequence suggested in Fig. 37. Other defects are seen to be progressively dissociating in boundary A, while more limited faulting at boundary B gives rise to

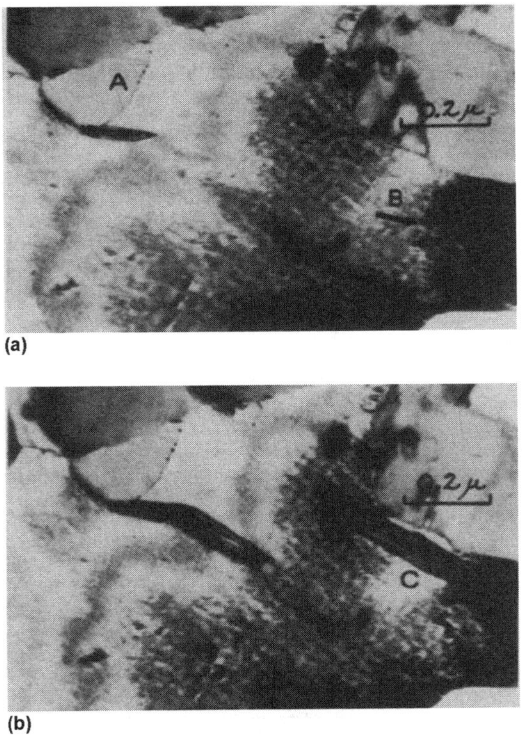

(a)
(b)

Fig. 43 Electron microscopy observation of martensitic nucleation at grain boundary defects in rapidly solidified Au-Cd alloy. Nucleus forming at boundary A in (a) is initially restricted to close-packed habit plane, but adopts macroscopic habit in (b). Source: Ref 101 another martensitic plate at C. Heterogeneous martensitic nucleation has been observed in ferrous alloys at grain boundaries (Ref 93), twin boundaries (Ref 89), inclusion-particle interfaces (Ref 90), and grain-boundary carbides (Ref 102).

A wide distribution of nucleation-site potencies can be expected, and this has been verified by small-particle experiments in which the fraction of particles undergoing transformation to martensite is measured as a function of particle size and cooling temperature. A cumulative potency distribution derived (Ref 103) from the experimental results of Cech and Turnbull (Ref 104) for f.c.c. → b.c.c. transformation in Fe-30Ni is presented in Fig. 44, depicting the number of sites per unit volume, U_V, as a function of the critical driving force for their operation. The solid curve in Fig. 44 represents the prediction of a simple geometric model based on defects of the type shown in Fig. 36. The distribution with respect to defect potency, n_0, is of the form:

$$U_V(n_0) = U_V^0 \exp(-\alpha n_0), \qquad \text{(Eq 97)}$$

with $U_V^0 = 2 \times 10^{11} \text{cm}^{-3}$ and $\alpha = 0.86$ for the data of Fig. 44. The potency parameter n_0 is converted to a critical driving force via Eq 87. From this distribution, the number of highly potent defects which trigger nucleation at the M_s temperature is about 10^6cm^{-3} or one per 100 μm grain. Such defects are too sparse to be directly observed by electron microscopy, and observations such as that of Fig. 43 must represent more plentiful less potent defects from further down in the potency distribution, which will nonetheless share common characteristics with the more potent defects.

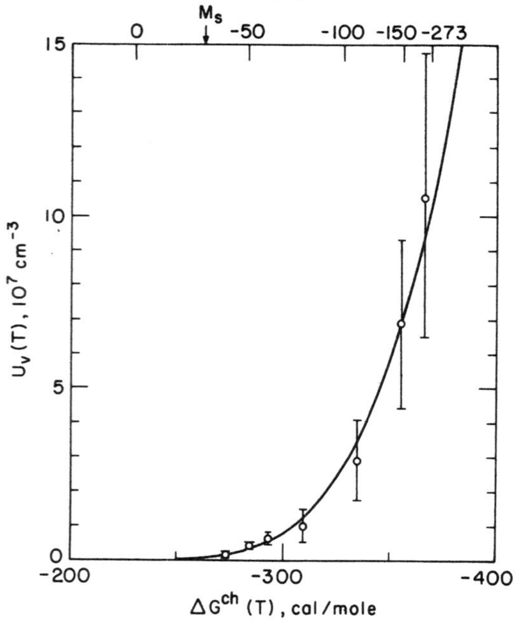

Fig. 44 Cumulative nucleating defect potency distribution obtained from small-particle experiments in Fe-30Ni. Curve represents Eq 97. Source: Ref 103

Applied stress can assist the operation of the same nucleation sites responsible for the transformation on cooling; such *stress-assisted* nucleation can be described by the thermodynamic effect of stress as in Eq 25. However, if the applied stress exceeds the yield stress of the parent phase, *strain-induced* nucleation can occur on new potent nucleation sites created by the plastic strain. Under these circumstances, nucleation sites are plentiful, and this has allowed numerous detailed observations of strain-induced sites, consisting primarily of various types of shear-band intersections, and including intersections with grain boundaries (Ref 105–110). The shear bands may be mechanical twins, hcp ε-martensite, bundles of stacking faults, or slip bands. In each case, the intersection can be interpreted as producing a dislocation stacking of the type illustrated in Fig. 36(a), and the relation of the active displacement to the orientation variant of the martensite being formed is consistent with the defect dissociation model (Ref 108–110). More potent defects can be generated in this way, compared to those existing in the annealed parent phase, and hence strain-induced nucleation can occur at lower driving forces. Based on shear-band intersections as the dominant nucleation site, a quantitative model for the kinetics of strain-induced transformation in austenitic steels has been developed (Ref 111), predicting the volume fraction of martensite versus plastic strain as a function of temperature; temperature influences both the thermodynamic driving force and the rate of shear-band formation through its influence on austenite intrinsic stacking-fault energy.

It is well established that the formation of one martensitic unit stimulates the nucleation of others. The mechanism of such *autocatalysis* can involve both stress-assisted and strain-induced nucleation via the elastic stresses and plastic strains accommodating the transformation shape strain. A third mechanism of *interfacial autocatalysis* has been proposed, in which groups of martensitic interfacial dislocations might directly nucleate another variant by dissociation (Ref 81).

Using thermal- and chemical-gradient specimens, Knorovsky (Ref 112) determined that autocatalytic nucleation of ferrous plate martensites can occur at a chemical driving force 200 J/mole less than that at the M_s temperature. Applying this increment to the site potency distribution of Fig. 44, it was concluded that the amount of observed autocatalytic nucleation cannot be explained via stress-assisted nucleation on preexisting sites. The crystallographic relation between groups of autocatalytically generated plate variants, also found by Bokros and Parker (Ref 113) was then rationalized by strain-induced nucleation via the dissociation model of Fig. 36, using the active slip systems accommodating the transformation shape strain (Ref 79, 112). This relation, involving four variants grouped around a matrix (110) pole and related by 2 mm symmetry, is common to other martensitic transformations in cubic crystals, including thermoelastic transformations in b.c.c. alloys (Ref 114, 115). The relation also applies to the "butterfly" morphology of groups of paired variants in ferrous alloys; a detailed analysis of this case by Clark (Ref 116) has established a sequence of interactions between the four variants which can be accounted for in terms of the strain fields of the units. A primary coupling between individual units meeting at an obtuse angle involves an additive relation between shape strains, favored by the most intense strain field at the edges of each unit. A secondary coupling between pairs of units meeting at an acute angle involves a canceling strain relation favored by the weaker strain field of the broad interfaces of the units. The stress field and plastic zones of martensitic units will be discussed further in the section "Thermoelastic-Nonthermoelastic Transition." Further understanding of autocatalytic nucleation mechanisms would be greatly aided by more detailed knowledge concerning the dynamic stress fields and plastic strain distributions during growth events.

Nonclassical Nucleation

While a classical nucleus forms along a path of fixed structure and increasing size, one of the earliest proposed martensitic nucleation mechanisms involved the continuous distortion of a region of fixed size, taking the nucleus from the parent structure to the martensitic structure via a local mechanical instability of the parent crystal lattice (Ref 117, 118). Interest in the possibility of "nonclassical" nucleation along paths of varying structure has revived in recent years (Ref 119–121), due to numerous observations of elastic-constant anomalies indicative of incipient mechanical instability of crystals which undergo martensitic transformations (Ref 122). Adapting a formalism originally applied to analogous nonclassical nucleation in diffusional decomposition reactions by Cahn and Hilliard (Ref 123), a unified theory of martensitic nucleation has recently been developed which allows for nucleation via both classical and nonclassical paths, and identifies the circumstances favoring each (Ref 124). We will briefly review here the manner in which the nucleation behavior just discussed can be modified by nonclassical effects.

In order to treat the energetics of nonclassical martensitic nucleation, it is first necessary to describe the energetics of large-strain, nonlinear-elastic homogeneous lattice deformations. An electronic pseudopotential calculation for the lowest-energy-path homogeneous f.c.c.-b.c.c. lattice deformation in sodium at 0 K is summarized in Fig. 45 (Ref 125). Calculations were performed at various pressures in order to vary the f.c.c.-b.c.c. free-energy (enthalpy) difference. The state of deviatoric strain of the structures intermediate between the f.c.c. and b.c.c. states is described by the variable η, with $\eta = 0$ representing b.c.c. and $\eta = 1$ f.c.c. The f.c.c.-b.c.c. molar free-energy difference, or the thermodynamic driving force, is the value of ΔG at $\eta = 1$. At the equilibrium pressure $P_0 = 87.6$ kbar where the driving force is zero, there is an energy-density barrier of 150 J/mole for the homogeneous lattice deformation. As the driving force is varied, the curves distort, altering the magnitude of this barrier. Although only a limited range of pressures could be explored in these calculations, extrapolation of the observed behavior to higher driving forces predicts the limiting cases indicated by the dashed curves for which local minima vanish and one phase becomes mechanically unstable. These curves thus define a critical driving force for homogeneous lattice instability.

Expressed as free-energy per volume, Δg, the computed free-energy curves of Fig. 45 can be approximated by a fourth-order expansion:

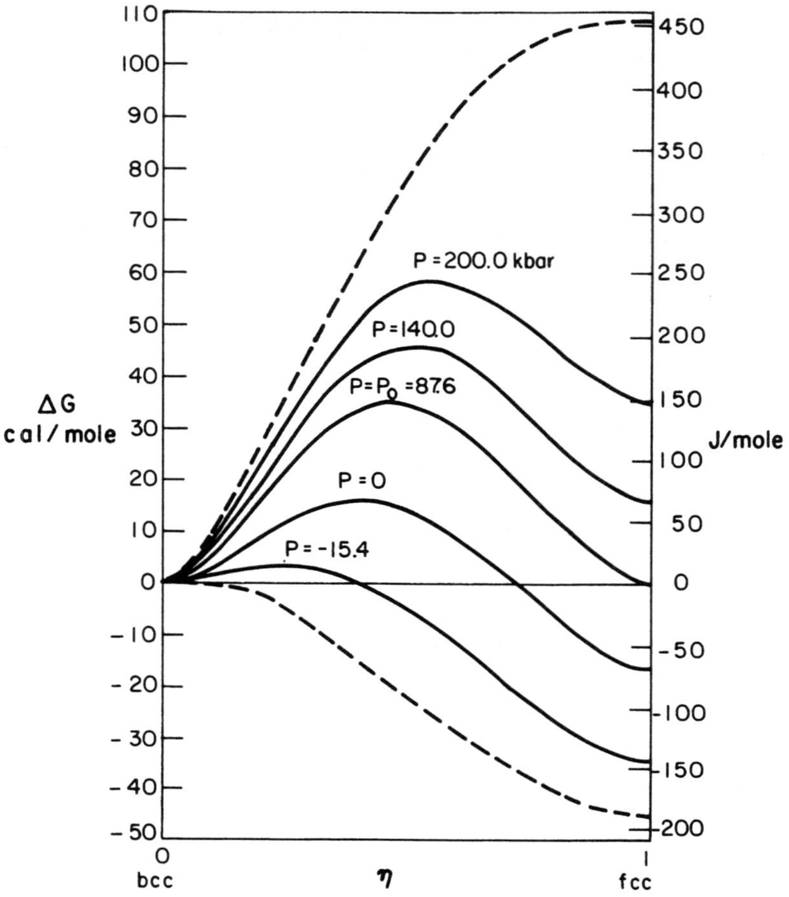

Fig. 45 Computed free energies for homogeneous b.c.c. ⇄ f.c.c. lattice deformation in Na at 0 K for various pressures indicated. The dashed curves are extrapolated. Source: Ref 125

$$\Delta g(\eta) = A\eta^2 - B\eta^3 + C\eta^4 \quad \text{(Eq 98)}$$

where the coefficients A, B, and C correspond to second-, third-, and fourth-order elastic constants. Constraining these coefficients to give minima at $\eta = 0$ and $\eta = 1$, the driving-force dependence of the curve shapes can be approximated by taking A to be linearly related to a normalized driving force:

$$\alpha = \Delta g^{ch}/\Delta g_i^{ch} \quad \text{(Eq 99)}$$

where Δg_i^{ch} corresponds to the driving force at instability. The coefficients are then expressed by:

$$A/\phi_0 = 16(1 - \alpha) \quad \text{(Eq 100a)}$$

$$B/\phi_0 = 32(1 - \alpha) - 4\alpha\beta \quad \text{(Eq 100b)}$$

$$C/\phi_0 = 16(1 - \alpha) - 3\alpha\beta \quad \text{(Eq 100c)}$$

where ϕ_0 is the equilibrium ($\Delta g^{ch} = 0$) barrier height, and $\beta = \Delta g_i^{ch}/\phi_0$. The form of Eq 98 leads to $\beta = -3.08$ for instability under stress at P_0. A similar β value describes the instability of the f.c.c. crystal under pressure in Fig. 45, although a smaller β value describes b.c.c. instability under hydrostatic tension ($P < 0$).

Assuming the homogeneous lattice-deformation energetics can be described by relations of the form of Eq 98–100, description of the energetics of *inhomogeneous* deformations in a nucleation process must incorporate the additional contributions of either interfacial energy or a strain-gradient energy. From the analogy of nonclassical nucleation of diffusional decomposition reactions [123], diffuse interface behavior can be anticipated in the vicinity of an instability. In dislocation terms, the interfacial dislocation cores become wide, and the interface resembles the nonlinear "solitons" of renewed interest to solid-state physicists (Ref 126–128). For such interfaces, a gradient-energy formulation is appropriate. By this formalism, the traditional volume chemical free energy and interfacial-energy terms are replaced by a functional of the form:

$$\int \left[\Delta g + \kappa(\nabla\eta)^2\right] dV \qquad \text{(Eq 101)}$$

where Δg is the homogeneous deformation-energy density just described and κ is a strain-gradient energy density coefficient. Finding the minimum energy variation $\eta(z)$ for a planar interface at equilibrium ($\alpha = 0$), and describing Δg by Eq 98, the gradient-energy coefficient can be defined by the equilibrium coherent surface energy, γ_0:

$$\kappa = 9\gamma_0^2/16\phi_0 \qquad \text{(Eq 102)}$$

and the equilibrium interfacial core width δ_0 is represented by Eq 41:

$$\delta_0 = (\kappa/\phi_0)^{1/2} \qquad \text{(Eq 103)}$$

The latter quantity is directly obtainable in principle from high-resolution microscopy observations of the type shown by Fig. 16. With the form of the energetics given by these relations, a reasonably general description of nonclassical martensitic nucleation can be obtained by normalizing energy densities to ϕ_0 and distances to δ_0.

In the simple case of a fully coherent shear transformation, the minimum energy one-dimensional variation $\eta(z)$ and associated energy per area F_A for an infinite thin plate producing a total displacement $u\gamma_T$, with an η value of η_0 at its center, can be solved analytically (Ref 124). The function $F_A(\eta_0, u)$ thus defined can be applied to the description of the energetics of a thin-plate nucleus if we assume, as in the case of classical nucleation, that the nucleus "interfacial" energy arises primarily from the broad interfaces as opposed to the plate edges, and the elastic energy accommodating the transformation strain resides primarily in the region of the matrix beyond the edges. The energy of a nucleus of radius r can then be approximated as:

$$\Delta G(r, \eta_0, u) = 2\pi r^2 F_A + 2\pi r \xi A u^2 \qquad \text{(Eq 104)}$$

where the second term is identical to the elastic energy of a classical particle producing the same u, with $\xi = \pi(2 - \nu)/4(1 - \nu)$ for a simple shear. Using the displacement parameter u to define an effective nucleus semithickness $c = u/\eta_0$, numerical calculations of the homogeneous nucleation saddle-point energy $\Delta G^*(r^*, c^*)$ as a function of nucleus structure parameter η_0 are presented in Fig. 46 for various values of α (Ref 129).

At low α, the minimum energy η_0^* of the critical nucleus is very close to $\eta_0 = 1$ representing classical behavior. As α increases, ΔG^* diminishes and η_0^* decreases representing progressively more nonclassical behavior as instability is approached. The ΔG^* barrier for homogeneous nucleation vanishes at $\alpha = 1$ as $\eta_0^* \to 0$ where nucleation and instability merge. Hence, homogeneous coherent nucleation can be thermally activated at driving forces sufficiently close to Δg_i^{ch}, and the behavior will be strongly nonclassical in this regime. Note, however, that to achieve the condition $\eta_0^* < 0.5$ for which the critical nucleus structure is closer to the parent phase than the martensite requires $\alpha \approx 0.95$.

Homogeneous nucleation will only be possible in crystals of high perfection, as the heterogeneous nucleation described in the section "Heterogeneous Nucleation by Defect Dissociation" can occur for α of the order of 0.1. Significantly nonclassical behavior is not expected at these low driving forces. An approximate treatment of the extent of possible nonclassical effects in heterogeneous nucleation can be obtained using the superdislocation approximation of Eq 89–93, taking the n^2 term of Eq 89 (corresponding to the nucleus elastic self-energy) to be proportional to η_0^2, and the nn_0 term (corresponding to the defect-interaction energy) proportional to η_0. The detailed calculations of homogeneous nonclassical nucleation summarized by Fig. 46 indicate that the interfacial width of a thin-plate nonclassical nucleus is insensitive to α (at small α) and the effective interfacial energy can be approximated by $\gamma = \gamma_0\eta^m$ with $m \approx 1$. Taking g_f^{str} as approximately constant, the volume chemical free-energy term Δg^{ch} of Eq 84 and 91 can be replaced by $\Delta g(\eta_0)$ which, in turn, may be approximated by a simple function of $(1 - \eta_0)^2$ for η_0 near 1 (Ref 3). Contour

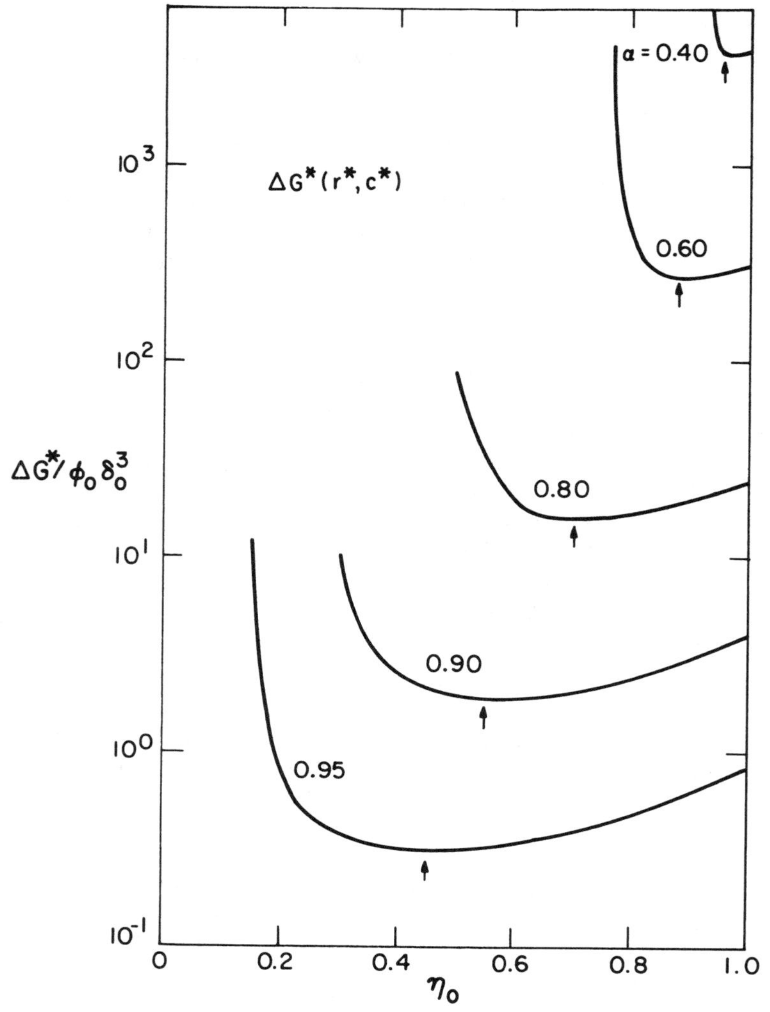

Fig. 46 Nonclassical homogeneous martensitic nucleation. Saddle-point free energy versus nucleus structure parameter η_0, for various values of normalized driving force, α. Source: Ref 129

plots of ΔF_ℓ (as in Eq 92 thus defined are presented in Fig. 47 as a function of η_0 and nucleus width w for two values of driving force corresponding to the cases $f = 0.94$ and $f = 1.15$ presented in Fig. 41, assuming the same defect. For $f = 0.94$ a local minimum exists at point A with a saddle-point barrier at point B. For $f = 1.15$ the barrier vanishes and barrierless nucleation can proceed from point AB. If the actual short-range interaction stabilizes a nucleus of thickness η_0 at small w, as discussed in the section "Nucleation Energetics" the barrierless condition will be met for the case $f = 0.94$. Regardless of the short-range details, η_0^* for such heterogeneous nucleation is found to be within 10% of the classical case, and its influence on the critical driving force for barrierless nucleation will be of a similar magnitude. These are small relative to the uncertainty in any of the estimated quantities in a nucleation calculation, and so the classical approximation and its representation by traditional linear-elastic dislocation theory are quite appropriate for heterogeneous nucleation.

Surface Nucleation

Thus far we have considered only transformation within the bulk. It is well established that martensitic nucleation can be stimulated by free

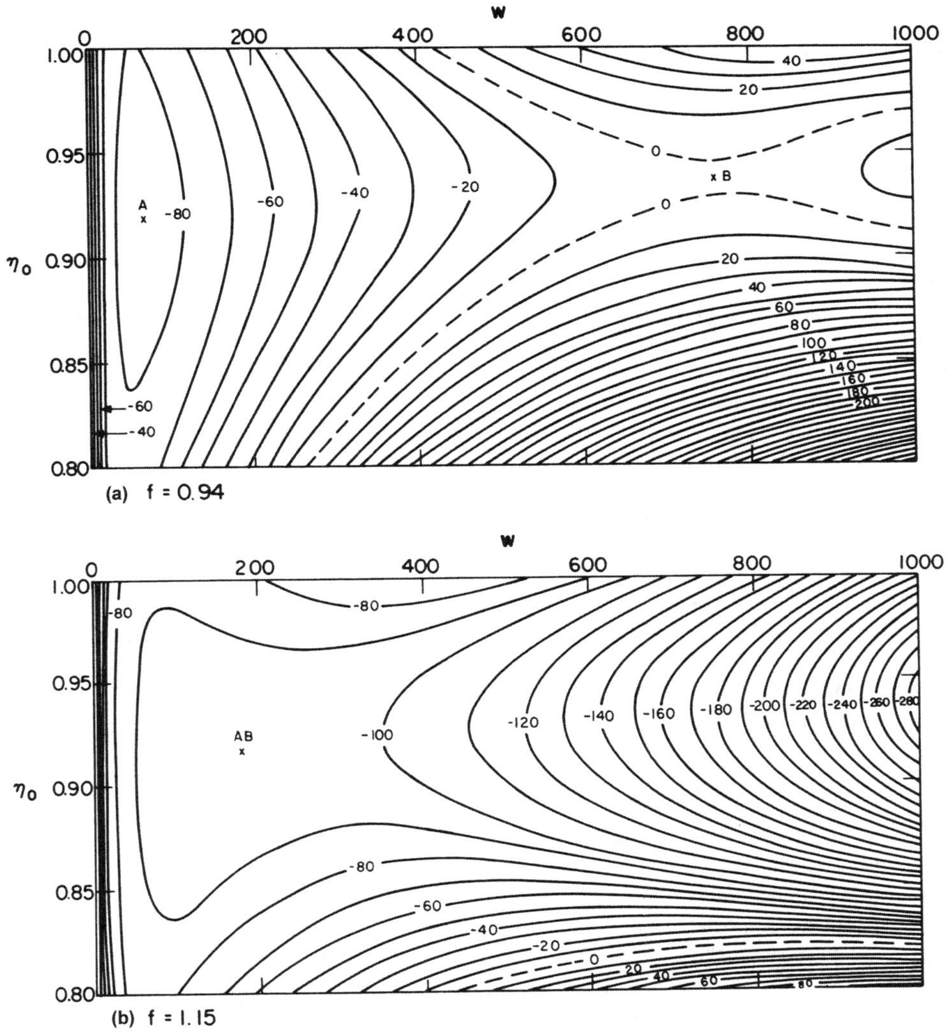

Fig. 47 Nonclassical heterogeneous martensitic nucleation; contour plots of free energy ΔF_ℓ versus nucleus width w and structure parameter η_0. (a) Normalized driving force $f = 0.94$, showing local minimum at A and saddle point at B; (b) $f = 1.15$, depicting barrierless nucleation from inflection point AB.

surfaces. This has been demonstrated in ferrous alloys by comparison of nucleation kinetics with and without platings or surface doping (Ref 130), and similar effects have been found for the $\beta \rightarrow \gamma_1$ thermoelastic transformation in Cu-Al-Ni alloys (Ref 131). Such effects can be particularly pronounced in thin foils where two free surfaces are in close proximity.

A major influence of a free surface on the nucleation energetics can arise from a reduction of the nucleus elastic energy through surface relaxations which can be described approximately by surface "image force" interactions. For an IPS transformation the problem is somewhat similar to the nucleation of a dislocation loop at a surface. A plate of finite thickness lying nearly parallel to a free surface will experience a maximum strain-energy reduction equivalent to that of two oppositely shearing parallel plates in contact; calculations for the latter problem give a reduction by a factor of one-third (Ref 132, 133). The behavior for the more complicated strains that may be involved in fully coherent nucleation can be anticipated from the general elastic calculations of Horner (Ref 134) which demonstrate that surface interaction can locally stabilize a coherent phase when transformation in the bulk requires a semi-coherent product. Specific calculations for f.c.c. $\rightarrow$ b.c.c. transformations in steels, however, indicate a strain energy reduction of only $\approx$ 10% for fully coherent transformation, making surface nucleation unlikely to be coherent in this case (Ref 135).

In addition to image-force effects, an important interaction of a nucleus with a free surface can arise from modification of the nucleus interfacial energy. For a nucleus lying nearly parallel to the free surface, the 2γ interfacial energy term in Eq 82 for homogeneous nucleation and in Eq 84 and 92 for heterogeneous nucleation can be replaced by $\gamma + \Delta\gamma_s$, where $\Delta\gamma_s$ is the free surface-energy difference between the parent and martensitic phases. The latter quantity is sensitive to surface orientation and can be positive or negative. If $\Delta\gamma_s$ is sufficiently negative (such that $\gamma + \Delta\gamma_s < 0$), free-surface wetting can occur to produce a thin layer of the martensitic phase even when $\Delta g^{ch} > 0$. Thin surface films of metastable martensitic phases have been observed in β-phase copper-base alloys for surface orientations near $(110)_\beta$; these films do not appear to interact strongly with subsequent transformation in the bulk (Ref 136). A similar $\Delta\gamma$ term can contribute to heterogeneous nucleation at internal surfaces, in addition to the important role of interfacial stress fields. This is likely to be an important factor in the observed nucleation at carbide particle interfaces in sensitized stainless steels (Ref 102).

While the elastic interaction of a nucleus with a free surface (potentially including the nonlinear effects of the section "Nonclassical Nucleation") can be predicted on a reasonably quantitative basis, quantitative prediction of the potentially overwhelming influence of the surface-energy effects is quite difficult. Despite this inherent theoretical difficulty, surface nucleation may offer a unique experimental opportunity for the direct observation of nucleation atomic displacements by high-resolution surface-imaging techniques, as suggested by preliminary field-ion microscopy experiments on meta-stable Fe_3Pt (Ref 137).

Nucleation Kinetics

Operational Nucleation

In order to express a rate of nucleation, it is necessary to define the end product of the nucleation process, or the "operational" nucleus. The nucleation rate can then be expressed as:

$$J = \sum_i N_i \hat{\chi}_i \hat{\beta}_i \qquad \text{(Eq 105)}$$

where N_i is the number of nucleation sites of the ith type, $\hat{\chi}_i$ is the fraction of these sites occupied (at a given instant) by operational nuclei, and β_i is a frequency factor controlling the rate at which nuclei become operational. In classical nucleation theory, where the state of a nucleus is characterized by its size, the fraction of sites $\hat{\chi}$ occupied by nuclei of operational size $\hat{r}$ is given as:

$$\hat{\chi} = Z\chi^* = Z\exp(-\Delta G^*/kT) \qquad \text{(Eq 106)}$$

Here ΔG^* is again the activation free energy for forming a nucleus of *critical* size r^* having an equal probability of growth or reversion, and Z is a nonequilibrium factor equal to the fraction of critical nuclei which reach size $\hat{r}$. The frequency factor $\hat{\beta}$ is then the rate at which nuclei arrive at $\hat{r}$, and is expressed by:

$$\hat{\beta} = \nu\exp(-Q_m/kT) \qquad \text{(Eq 107)}$$

where Q_m is the activation energy for motion of the nucleus interface, and ν is an associated attempt frequency.

In diffusional transformations, the operational nucleus is defined in terms of a size having negligible probability of dissolution. Inasmuch as the unit process of interfacial motion is regarded as atom-by-atom transfer, Q_m is taken as the activation energy for atomic diffusion. For martensitic transformations, the operational nucleus size has been defined (Ref 138) as that required for rapid growth (leading to a detectable event), and Q_m becomes the activation energy for motion of a glissile martensitic interface. From the analysis of the section "Thermally Activated Motion," Q_m will be sensitive to the net forces acting to drive the interface; operational size may thus correspond to a critical slope of the nucleus energy curves of Fig. 41 for nuclei unconstrained in thickness. In cases where thickening is difficult, the operational size may correspond to the size at which the thickness constraint is effectively removed, as by intersection with a pole dislocation in Fig. 40.

For the barrierless nucleation discussed in the section "Nucleation Energetics," $\Delta G^* = 0$, $\hat{\chi} = 1$, and the nucleation kinetics defined by Eq 105 are controlled entirely by interfacial motion at suboperational (but supercritical) size. Possible rate-controlling mechanisms can then be sought from the various processes which accompany interfacial motion.

Rate-Controlling Mechanisms

As discussed in the section "Motion in a Perfect Crystal," the growth of an enclosed particle of the type illustrated in Fig. 10 and 38 involves the formation and motion of both coherency and anticoherency dislocations. Each of these processes is potentially rate-controlling at the nucleation stage. The process of forming loops of anticoherency dislocations at the edge of a thin nucleus was analyzed in Kaufman and Cohen (Ref 55) and later in Raghavan and Cohen (Ref 63). Treating the interaction of the dislocations with the lattice deformation (coherency dislocations) in an approximate way by a modification of the Burgers vector, the activation energy for the formation of a loop of critical radius ρ^* can be expressed as:

$$Q = \tfrac{5}{16}\mu b^2 \rho^* [\ln(\rho^*/b) - 1.6 + z] \qquad \text{(Eq 108)}$$

where z is a core-energy parameter. The critical loop radius ρ^* is determined by the nucleus radial-growth force $-(\partial\Delta G/\partial r)$ through the relation:

$$-C_n(\partial\Delta G/\partial r) = \tfrac{5}{16}(\mu b^2/\rho^*)[\ln(\rho^*/b) + 0.4 + z] \qquad \text{(Eq 109)}$$

where C_n is a geometric constant dependent on the specific nucleus model (Ref 62, 63). Selecting an embryo size to give an appropriate growth force, Eq 108 and 109 are found to give good agreement with an observed linear dependence of the nucleation activation energy on chemical driving force, as determined from isothermal f.c.c. → b.c.c. nucleation kinetic experiments in ferrous alloys (Ref 130).

The Kaufman-Cohen-Raghavan model for the nucleation kinetics is dependent on the occurrence of a lattice-invariant deformation. Critical nucleation kinetic experiments on the coherent f.c.c. → h.c.p. martensitic transformation, where the lattice-invariant deformation is absent, give essentially the same linear Q versus Δg relation, demonstrating that the lattice-invariant deformation is not necessary to the observed kinetic behavior (Ref 139). As developed in the section "Thermally Activated Motion," the motion of interfacial dislocations, even in the absence of a lattice-invariant deformation, can give rise to thermally activated behavior. The notion of nucleation-rate control via dislocation motion was originally suggested by Magee (Ref 91), and applied to coherency dislocation motion during nucleation via defect dissociation by Olson and Cohen (Ref 62). For a process with a "stress-free" activation energy, Q_0, and a fixed activation volume, V^* (corresponding to $p = 1$, $q = 1$ in Eq 57, coherency-dislocation motion under the influence of a negative fault energy prescribed by Eq 84 gives a linear Q versus Δg relation (Ref 62):

$$Q = Q_0 + V^*\left[(2\gamma/nd) + g_f^{str}\right] + V^*(\Delta g - \Delta g_\mu) \qquad \text{(Eq 110)}$$

where Δg_μ represents the athermal component of frictional resistance discussed in the section "General Concepts." Expressing experimental activation-energy relations in the form:

$$Q = A + B\Delta g \qquad \text{(Eq 111)}$$

kinetic parameters derived from experimental data are summarized in Table 4. The coefficient B corresponding to V^* in Eq 110 is found to be in the range 100 to 300Ω, with Ω being the atomic volume. Estimating the other terms in Eq 110, the experimental A parameters suggest a Q_0 of $\approx$ 1 eV. These V^* and Q_0 values are comparable to those for low-temperature slip in similar alloys. The most likely controlling mechanism is solid solution strengthening. Because it is possible to explore only a limited range of activation energies in a given alloy, the observed variations in B may in part be due to nonlinearity of the Q versus Δg function as depicted in Fig. 26 and 28. Detailed comparisons among alloy systems requires normalization to elastic moduli as implied, for example, by Eq 45.

In both of the kinetic models of Eq 108 and 110, the activation energy is sensitive to embryo size and a distribution of activation energies is thus expected, in line with the distribution of defect potencies discussed in the section "Nucleation Sites." This is verified by nucleation kinetic experiments on small particles where the complication of

Table 4 Kinetic parameters from isothermal nucleation experiments. $Q = A + B\Delta g$

		B	
	A (J)	(m^3)	(Ω)
f.c.c. → b.c.c.			
Bulk specimens (Ref 122)			
Fe-24Ni-3Mn	3.66×10^{-19}	1.05×10^{-27}	90
Fe-29Ni	8.27×10^{-19}	3.69×10^{-27}	315
Small particles (Ref 132)			
Fe-24Ni-4Mn	3.39×10^{-19}	1.20×10^{-27}	103
Fe-22Ni-0.5C	4.48×10^{-19}	1.77×10^{-27}	151
f.c.c. → h.c.p.			
Bulk specimens (Ref 131)			
Fe-17Mn-7Ni	2.04×10^{-19}	1.70×10^{-27}	146

autocatalytic nucleation is circumvented (Ref 140). Whereas the kinetic data from bulk specimens in Table 4 reflect some degree of averaging, the quantities reported for small-particle experiments correspond to the minimum activation energy associated with the most potent embryos available in quantities for detectable transformation. The small-particle results for the Fe-22Ni-0.5C alloy, which exhibits apparent "athermal" behavior in bulk form, demonstrate the intrinsic thermally activated behavior of the initial nucleation events which can be obscured in bulk specimens by extreme autocatalysis.

Alternative proposals for rate-controlling mechanisms in martensitic nucleation include rate control by accommodation slip (Ref 141). However, such slip is unlikely to be necessary to the process of embryo growth, nor is slip necessary to account for the observed kinetic behavior in view of the well-established thermally activated interfacial motion in thermoelastic alloys (section "Observations in β-Cu-Al-Ni Alloys") where accommodation slip is virtually absent. As discussed in the section "Nucleation Energetics," depending on the constraints which apply to a microscopic nucleus, the barrierless condition might be preceded by a thermally surmountable barrier, and it has been proposed that ΔG^* might thus account for observed isothermal nucleation behavior (Ref 97). Although ΔG^* itself is far too sensitive to Δg to account for the observed kinetic behavior of Table 4, Eq 105–107 predict that the nucleation activation energy associated with a finite ΔG^* will be the sum of ΔG^* and Q_m. Hence, ΔG^* might provide an additional contribution to the experimental activation energy at driving forces lower than that for the barrierless condition, increasing the curvature of the experimental Q versus Δg relation at low Δg. This high activation-energy regime has not been experimentally accessible thus far.

It should be noted at this point that although the embryo as a whole is expected to follow a classical nucleation path for the low driving forces at which heterogeneous nucleation occurs, nonclassical effects may be significant for the more microscopic processes of interfacial dislocation production and motion controlling the kinetics of operational nucleation. Therefore, microscopic fluctuations of "homophase" rather than "heterophase" character may be important in surmounting local interfacial barriers determined by short-range interactions which are dominated by nonlinear-elastic core energies. The anomalous low-temperature interfacial mobility depicted in Fig. 31, correlating with anomalous elastic constants, is suggestive of such behavior. It is in this context that further information on nonlinear elasticity and gradient energies may contribute significantly to a quantitative understanding of the nucleation kinetics of martensitic reactions under conditions of practical interest.

Nucleation Attempt Frequency

In early analyses of experimental operational nucleation kinetics, the attempt frequency ν in Eq 107 was taken to be of the order of the atomic vibration frequency, $\approx 10^{13}$ s^{-1} (Ref 130). However, it was emphasized by Magee (Ref 91) that the coordinated motion of large groups of atoms as in thermally activated dislocation motion would involve much lower attempt frequencies of the order of 10^{10}–10^{11} s^{-1}. The type of operational nucleation predicted by the barrierless nucleation curves of the shape depicted in Fig. 41 further implies that many individual thermally activated events are required for an embryo to reach the operational nucleus size at which rapid growth ensues. (This behavior is similar to the phenomenon of transient nucleation (Ref 73) in conventional nucleation with a barrier, but is related to the establishment of *operational* nuclei rather than a steady-state distribution of *critical* nuclei.) This predicts a greatly diminished effective nucleation attempt frequency of the form:

$$\nu = \upsilon_0/\Delta a \quad \text{(Eq 112)}$$

where υ_0 is the pre-exponential velocity in the interfacial mobility law of Eq 32, and Δa is the size difference between the preexisting embryo (w_0 in Fig. 41) and the operational nucleus. A recent analysis (Ref 142) of f.c.c. → b.c.c. nucleation kinetic data from experiments in which the driving force was varied independently of temperature by the application of high magnetic fields (Ref 143) provides an experimental nucleation attempt frequency of the order of 10^7 s^{-1}. Taking $\upsilon_0 = 30$ m/s from the direct interfacial mobility measurements discussed in the section "Observations in β-Cu-Al-Ni Alloys," this low attempt frequency is consistent with Eq 112 if Δa is a few μm. The rather large operational nucleus size that this implies is consistent with the numerous observations of suboperational embryos of μm dimensions in ferrous alloys (Ref 89–93).

Stabilization Effects

The evident control of nucleation kinetics by the motion of dislocation interfaces suggests that any feature of the kinetics of conventional crystal plasticity should have a counterpart in martensitic transformations. It is observed that any matrix-strengthening mechanism which operates on a size scale relevant to operational nucleation, for example, precipitation hardening (Ref 144), can inhibit martensitic nucleation. These mechanisms can interact with the transformation through either thermal or athermal components of the resistance to interfacial motion. Although small pre-strains which form planar dislocation pile-ups can stimulate nucleation (Ref 145) through production of potent nucleation sites, it is often found that the matrix work hardening associated with large plastic deformations inhibits subsequent transformation, termed "mechanical stabilization" (Ref 146, 147). Similarly, an apparent inhibiting influence of grain refinement on initial nucleation has been attributed to an associated increase in dislocation density within the grains (Ref 148). A subtle rate-dependent interaction of embryo interfaces with concurrent matrix slip during stress-assisted transformation has been invoked to account for an observed negative rate sensitivity phenomenon similar to "dynamic strain aging" (Ref 149).

A more direct analogy with strain aging effects in carbon steels can be found in the phenomenon of "thermal stabilization." An inhibiting influence of isothermal holding near room temperature on the kinetics of subsequent continued transformation on cooling has been correlated with the kinetics of carbon diffusion in martensite (Ref 150). It thus appears that carbon segregation to the interface of suboperational martensitic embryos can cause a rate-dependent drag force; the kinetics of such solute drag effects at martensitic dislocation interfaces has been modeled quantitatively in Owen, Schoen, and Srinivasan (Ref 151). Magee (Ref 91) has proposed that steels which transform above room temperature may be "dynamically stabilized" by carbon drag, exhibiting apparent athermal transformation kinetics analogous to the negative rate sensitivity of flow stress observed in materials exhibiting dynamic strain aging. When mechanical and thermal stabilization effects are simultaneously present, complex interactions may occur, as has been suggested to account for unusual kinetic observations at extreme cooling rates (Ref 152).

Martensitic Growth

Thermoelastic Behavior

The interfacial mobility described in the section "Interfacial Mobility", when combined with consideration of the net forces acting on the interface, leads naturally to the prediction of the growth behavior of a martensitic particle. The simplest case to treat is the elastically accommodated oblate spheroidal particle, for which the total free energy as a function of size and shape is described in Eq 82. For an IPS transformation $\left(g_p^{str} = 0\right)$, Raghavan and Cohen (Ref 153) modeled the growth of such a particle using an empirical mobility relation based on dislocation dynamics, and calculating the growth forces acting in the r and c directions

($-\ \partial \Delta G/\partial r$, $-\partial \Delta G/\partial c$) by differentiation of Eq 82. Computer simulations of growth events give results which are insensitive to the geometry of the initial embryo, leading to a steady-state aspect ratio during growth which is controlled by the relative interfacial mobilities for radial growth (r direction) and thickening (c direction). Once radial growth is stopped by impingement with an obstacle (e.g., a grain boundary), a stage of continued particle thickening ensues.

In the absence of a significant athermal component of interfacial friction, the final thickening of an elastically accommodated martensitic particle continues until a "thermoelastic equilibrium" is reached where the thickening force associated with the chemical free-energy change Δg^{ch} is balanced by that arising from the IPS elastic strain energy. Differentiating ΔG of Eq 82 with respect to c, the final shape is then described by the condition (Ref 154, 155):

$$\Delta g^{ch} + 2Kc/r = 0 \qquad \text{(Eq 113)}$$

When this condition is maintained, subsequent changes of Δg^{ch} (as by changing the temperature) produce reversible changes in particle thickness. The temperature dependence of the particle volume and shape for this behavior is shown schematically in Fig. 48 (Ref 155). Nucleating at the M_s temperature at point a, a thin plate grows rapidly, impinges on a grain boundary, and thickens to the thermoelastic equilibrium condition of Eq 113 at point b. On further cooling the plate thickens to point c; on heating it thins according to Eq 113 following line cbd. Differentiation of ΔG of Eq 113 with respect to r reveals a critical Δg^{ch} for which a radial instability occurs expressed by (Ref 154, 155):

$$\Delta g^{ch} = -2(K\gamma/r)^{1/2} \qquad \text{(Eq 114)}$$

This condition is met at point d where the plate vanishes, identifying the final reversion temperature A_f. Since the critical Δg^{ch} of Eq 114 is negative, reversion occurs below the temperature T_0 where $\Delta g^{ch} = 0$.

Modification of thermoelastic behavior by an athermal interfacial friction stress, τ_μ, is illustrated in Fig. 49 (Ref 155). While Fig. 49(a) depicts the "equilibrium" behavior of Fig. 48, modification of the force balances of Eq 113 and 114 by the driving force increment $\Delta g_\mu = \pm$

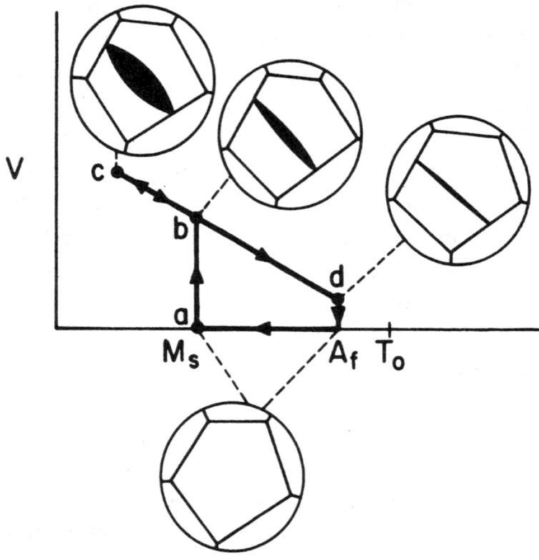

Fig. 48 Thermoelastic growth behavior of an isolated martensitic particle without interfacial friction. Source: Ref 155

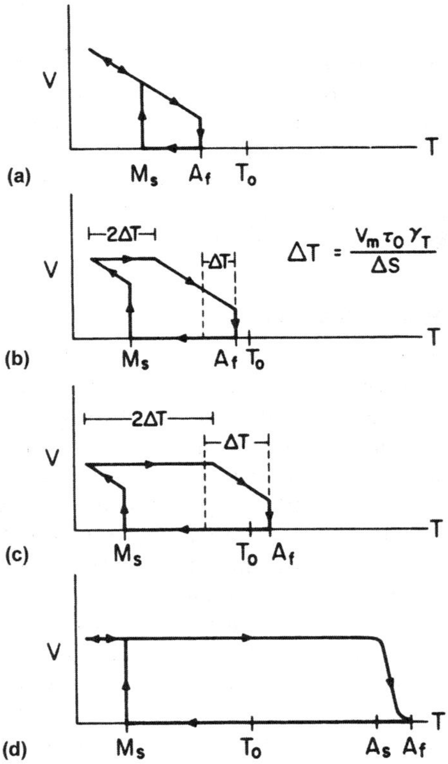

Fig. 49 Influence of interfacial friction stress on the growth behavior of an isolated martensitic particle. (a) Behavior without friction; (b) and (c) behavior with two levels of τ_μ, but with elastic accommodation; (d) nonthermoelastic behavior resulting from plastic accommodation and attendant large τ_μ. Source: Ref 155

$\tau_{\mu}\gamma_T$ for forward and reverse interfacial motion introduces the additional ΔT hysteresis increments shown in Fig. 49(b) and (c) for two levels of τ_{μ}. Depending on the magnitude of τ_{μ}, final reversion can occur above or below T_0.

Figure 49(d) indicates the nonthermoelastic case where, rather than elastic accommodation, substantial plastic accommodation of the transformation shape strain occurs. Plastic accommodation not only reduces the elastic restoring force for reverse interfacial motion but causes thickening to stop in a highly irreversible manner through a greatly increased τ_{μ} via matrix strain hardening. No significant interfacial motion then occurs on further cooling or heating and reversion requires renucleation of the parent phase over the temperature range $A_s - A_f$. If a similar Δg^{ch} is required for nucleation of the forward and reverse transformation, the T_0 temperature is symmetrically bracketed by the M_s and A_s temperatures (Ref 55). These considerations suggest that the single requirement for thermoelastic growth behavior is the relative absence of plastic accommodation of the transformation shape strain (Ref 154).

Thermoelastic–Nonthermoelastic Transition

The relative propensity of a given alloy system to exhibit thermoelastic growth can be assessed by an estimate of the plastic zone size of a martensitic particle in hypothetical thermoelastic equilibrium at the alloy M_s temperature. A simple two-dimensional calculation of the particle long-range stress field can be obtained by summing the stress fields of arrays of straight coherency dislocations of the type depicted in Fig. 10(a) producing the IPS shape strain. Such a calculation for an ellipsoidal martensitic particle in an Fe_3Pt alloy with a c/r ratio defined by Eq 113 evaluated at M_s is shown in Fig. 50 (Ref 156). Contours of the component of shear stress τ_{xy} corresponding to the transformation shear γ_T are plotted in Fig. 50(a) and (b) for interfacial dislocations in edge and screw orientations, respectively.

Solid contours in Fig. 50(a) and (b) correspond to the critical resolved shear stress for slip in disordered Fe_3Pt austenite. Upon ordering, the austenite elastic shear modulus μ diminishes while the flow stress increases. Taking $\Delta g^{ch}(M_s)$ to be proportional to μ (as implied by Eq 87) the same c/r aspect ratio for a particle in hypothetical thermoelastic equilibrium at M_s can be adopted for the disordered and ordered conditions. The dotted lines in Fig. 50(a) and (b) then represent the yield stress contours in the ordered case. While Fig. 50(a) and (b) represent only the stress component τ_{xy}, Fig. 50(c) shows the yield contours for the total stress based on a von Mises yield criterion for the screw orientation case of Fig. 50(b). It is clear that the increased yield strain arising from the combined effect of elastic softening and plastic strengthening via ordering provides a dramatic decrease in the extent of plastic accommodation of the transformation macroscopic shape strain. This then accounts for the observed transition from nonthermoelastic to thermoelastic growth behavior with ordering in Fe-Pt alloys.

The results of the two-dimensional stress field calculations of Fig. 50 have been substantiated by three-dimensional continuum elastic calculations for martensitic plates in Fe_3Pt by Ling and Owen (Ref 157), which also demonstrated that incorporation of anisotropic elasticity does not significantly alter the behavior. In addition, they extended the analysis to pairs of plates with oppositely directed shears in order to assess the role of "self-accommodating" plate group formation in minimizing plastic accommodation. The elastic interaction of such pairs is found to reduce K in Eq 113 by a factor of $\approx 1/3$. Though long-range stresses are greatly reduced, maintenance of the thermoelastic equilibrium condition of Eq 113 then requires a tripling of the aspect ratio, with an intensification of local stresses at the plate edges. Figure 51 shows von Mises yield contours for a pair of oppositely shearing plates in ordered Fe_3Pt with twice the aspect ratio of Fig. 50. Comparison with the ordered Fe_3Pt yield contours in Fig. 50(c) suggests that pair formation is not particularly effective in further reducing local plastic accommodation. Local stress field reduction might be more effective in experimentally observed tapered-plate multivariant geometries (Ref 158), but long-range displacement cancellation is incomplete in such cases, and precise stress field calculations have yet to be performed. It should be borne in mind that the intrinsic mobility of martensitic interfaces in the absence of extensive plastic accommodation gives thermoelastic martensitic plates the ability to "translate" after their formation, so that extensive rearrangement of plates favoring total strain energy reduction can occur during the course of thermoelastic

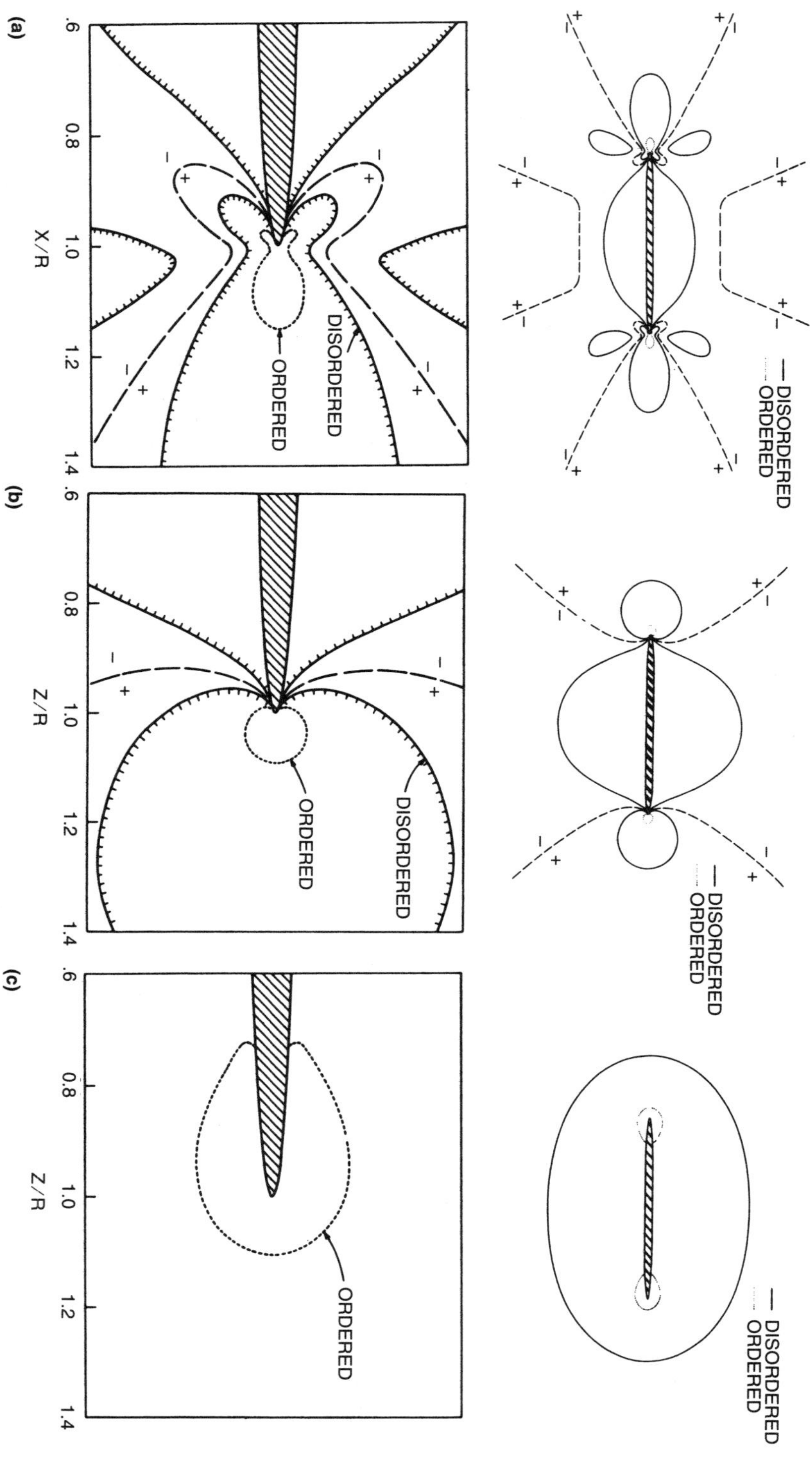

Fig. 50 Stress field contours of a martensitic particle in Fe_3Pt lying in the *xz* plane with transformation shear in the *x* direction; (a) represents a section in the *xy* plane with interface in edge orientation; (b) and (c) are sections in the *yz* plane with interface in screw orientation; (a) and (b) illustrate contours where τ_{xy} reaches the critical stress for slip on the nearest slip system; (c) illustrates contours where maximum shear stress τ_{max} reaches the critical stress. Source: Ref 156

transformation. The "self-accommodating" plate groups characteristic of thermoelastic systems may thus be a consequence of thermoelastic behavior rather than a cause.

While the nature of macroscopic shape-strain accommodation is certainly an important factor in thermoelastic behavior, experimental observations of accommodation dislocations left behind after transformation cycling in partially ordered Fe_3Pt indicate that accommodation of the local interfacial stress field of the internally twinned martensite (Fig. 20) is also important (Ref 159). Kajiwara and Owen (Ref 160) attributed an observed reduction in the internal twin period with ordering in Fe_3Pt to a decrease in twin-boundary energy associated with the tetragonality of the ordered martensitic product. They proposed an attendant reduction in local accommodation slip based on a simple geometric criterion for slip nucleation. A calculation (Ref 161) of local yield contours (as in Fig. 50 and 51) using the interfacial stress field calculational procedures of the section "Interfacial Stress Field and Stability" is shown in Fig. 52. The direction of the lattice-invariant shear ($[1\bar{1}0]_A = [11\bar{1}]_M$) is nearly parallel to the ≈ (3, 10, 15) interface for this transformation so that the pure screw dislocation array can again be used to model the interfacial stress field. Using the same critical resolved shear stress values as in Fig. 50 and 51 and assuming them equal in both phases, computed yield contours are plotted for slip on $(111)[1\bar{1}0]$ and $(11\bar{1})[1\bar{1}0]$ in austenite and $(112)\ [11\bar{1}]$ in the martensite. A substantial reduction of local plastic accommodation arises from both the reduced twin period and the higher yield strain in the ordered case. A further reduction of the extent of accommodation may arise from the requirements for slip nucleation, but a quantitative assessment of this requires a calculation of the energetics of dislocation loop formation in the local interfacial stress field. Such a calculation for slip nucleation at the periphery of a martensite plate is discussed in the section, Thermoelastic-Nonthermoelastic Transition, in this chapter.

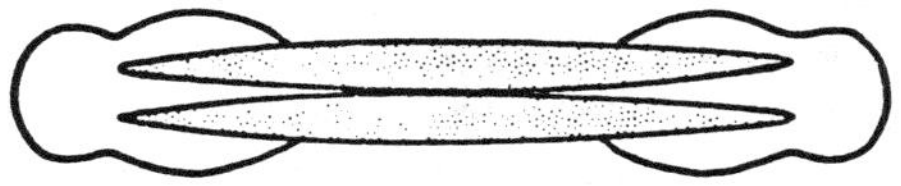

Fig. 51 Yield stress contours for a pair of martensitic particles with oppositely directed shears. Source: Ref 157

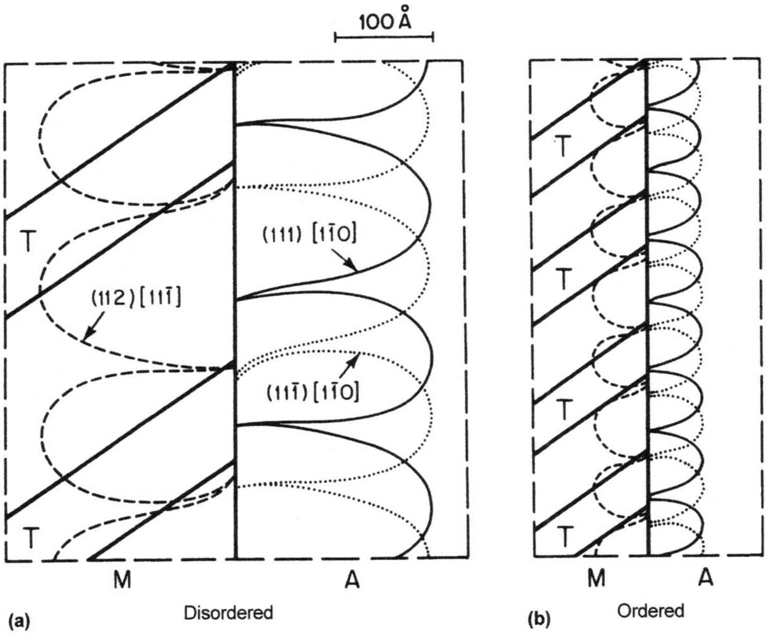

Fig. 52 Plastic accommodation of local interfacial stress field in Fe_3Pt; yield contours plotted for various slip systems in (a) disordered and (b) ordered conditions. M and A denote martensite and austenite; T denotes internal twins in martensite.

It is evident that numerous factors can influence the degree of elastic versus plastic accommodation which underlies thermoelastic behavior. From the considerations embodied in Eq 113, a primary factor is the magnitude of Δg^{ch} during transformation. The nucleation models of the section "Martensitic Nucleation" indicate that the latter is decreased by a small transformation strain and soft elastic shear constants. A second important factor is the matrix plastic flow properties. Strengthening mechanisms which preferentially inhibit slip without interfering with interfacial mobility will be especially effective in reducing plastic accommodation. Based on the interfacial mobility considerations of the section "Interfacial Mobility," long-range order strengthening may be unique in this regard (Ref 154). This may account for the tendency for thermoelastic transformations to occur in ordered alloys.

Nonthermoelastic Growth

When plastic accommodation is substantial, the dynamic interaction of the growing particle and its plastic zone becomes important, and the simple description of growth discussed in the section "Thermoelastic Behavior" is no longer appropriate. The complex interactions which occur during nonthermoelastic growth have been modeled in Haezebrouck (Ref 129) using computer simulations of dynamic growth events in FeNi alloys. The interactions considered are summarized in Fig. 53. As in the thermoelastic case, the driving force for interfacial motion arises from the chemical free-energy change (a), with restraining forces arising from interfacial energy (b) and elastic strain energy (c). Differentiating the total energy with respect to particle radius r and semithickness c, these terms all influence the net driving force per unit area acting on the interface (d) as expressed by Eq 23, for radial growth and thickening. Through the interfacial mobility relations (e) discussed in the section "Interfacial Mobility," these net forces determine the interfacial velocities in the r and c directions. For a first-order phase transformation with a latent heat, the interfacial velocity influences the heat transfer across the interface and so affects the temperature at the interface (f). This in turn affects the interfacial mobility (e), modifies the chemical free-energy change (a) at the interface, and weakly affects the interfacial energy (b). The overall interaction thus far discussed would also apply to the initial radial growth events in thermoelastic transformations.

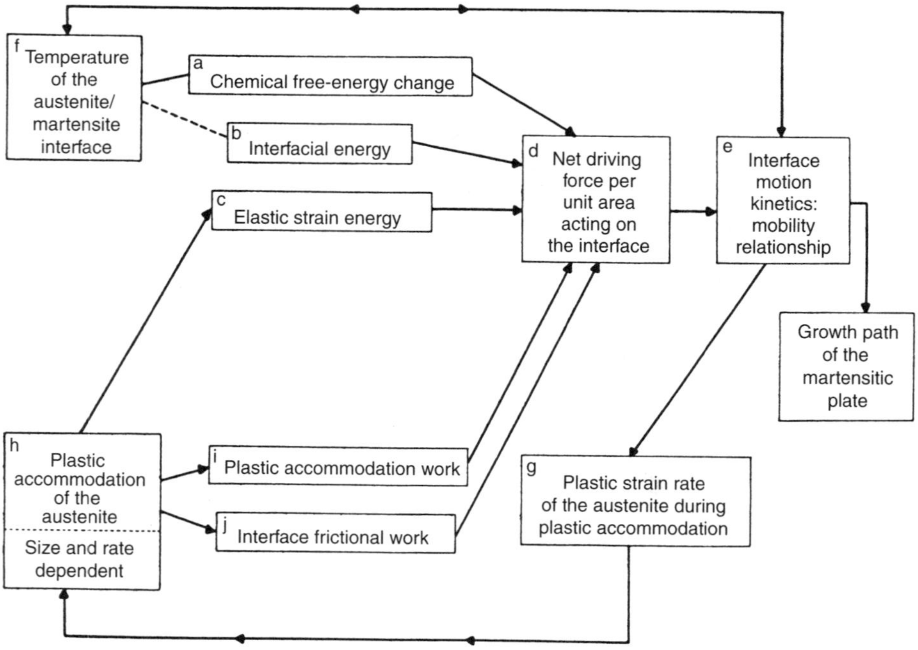

Fig. 53 Dynamic interactions during nonthermoelastic growth of a martensitic particle. Source: Ref 162

In a nonthermoelastic transformation, the interfacial velocity also determines the plastic strain rate (g) imposed on the parent phase. This influences the extent of plastic accommodation (h) which is also dependent on particle size through the conditions for slip nucleation in the particle stress field. Plastic accommodation modifies the elastic strain energy (c) and introduces two important dissipations: the work of creating the plastic zone (i), and the additional interface frictional work (j) for moving the interface through strain-hardened material. These dissipations modify the net interfacial driving forces (d).

As discussed in the section "Rate-Controlling Mechanisms," interfacial mobility parameters for thermally activated behavior in ferrous alloys can be estimated from nucleation kinetic data; phonon drag parameters are estimated as in the section "Continuous Motion." Steady-state heat transfer during radial growth of a martensitic particle has been analyzed in Lyubov and Roitburd (Ref 162), and the transient behavior during thickening has been treated by Nishiyama, et al. (Ref 163). From the form of the behavior obtained for the latter case, the interface temperature at time t can be taken to be a function of $\overline{\upsilon^2}t$ where $\overline{\upsilon^2}$ is the time-averaged mean-square velocity. Using available alloy thermodynamic parameters and elastic constant data, all the necessary parameters can then be defined to describe the uppermost loop of Fig. 53.

The mechanics of plastic accommodation can be treated using available slip kinetic data for the alloys of interest, and adapting concepts from fracture mechanics developed for crack-tip plasticity. Using the plastic zone solution for a mode III shear crack (Ref 165), the radial plastic zone size R_p is estimated as:

$$R_p = r[1 + 2.5(\bar{\tau}_f/\tau_i)]^{-2} \qquad \text{(Eq 115)}$$

where $\bar{\tau}_f$ is the matrix effective shear strength and τ_i is the internal shear stress in the particle as defined by the Eshelby elastic particle calculation (Ref 8). Using the same model for the form of the plastic distribution, the average plastic strain rate imposed in the plastic zone by a radial interfacial velocity υ_r is expressed by:

$$\dot{\gamma} = 2\upsilon_r\bar{\tau}_f/\mu R_p \qquad \text{(Eq 116)}$$

From the form of calculated elastic stress fields in the c direction (Ref 157), the plastic-zone size normal to the particle habit is estimated as:

$$C_p = \tfrac{1}{2}r(1 - \bar{\tau}_f/\tau_i) \qquad \text{(Eq 117)}$$

The dependence of the local plastic strain rate on υ_c and the total plastic displacement across the zone can both be estimated by assuming a constant plastic strain gradient in the zone and fitting to experimental displacement fields based on detailed analysis of surface relief effects from isolated martensite plates in ferrous alloys. Modification of the IPS elastic strain energy of Eq 3 (and the associated particle internal stress parameter τ_i) is taken into account by a radial plastic-zone size correction to the effective particle radius (Ref 165), and by an effective transformation shear strain which incorporates accumulated plastic displacements averaged over the particle volume. Integration of the plastic strain fields defines the work of plastic zone formation. The primary influence of strain hardening is assumed to lie in the athermal resistance to interfacial motion, and this is taken to vary linearly with the average plastic strain in each zone (r and c directions treated separately). The magnitude of the strain hardening effect, within the range of theoretical possibilities, is estimated from the thermodynamic conditions for which radial growth is observed to cease in thermal-gradient and composition-gradient experiments (Ref 112).

Nucleation of slip dislocations in the particle stress field is regarded as a necessary condition for plastic-zone formation. A linear elastic calculation of the conditions for barrierless nucleation of slip dislocations in the stress field at the plate periphery gives a critical plate radius as a function of plate aspect ratio plotted in Fig. 54. For an aspect ratio of a few percent, a plate radius of a few μm is required to nucleate slip. Once this nucleation condition is met, a radial plastic zone size can be calculated by combining Eq 115 and 116 and expressing the matrix strength $\bar{\tau}_f$ as a function of $\dot{\gamma}$. The latter function is denoted by the schematic solid curve in Fig. 55. Following the discussion of dislocation mobility in the sections "Continuous Motion" and "Thermally Activated Motion," region I $(\tau_\mu < \bar{\tau}_f < \hat{\tau})$ corresponds to thermally activated behavior, linear region II represents phonon drag control, and region III

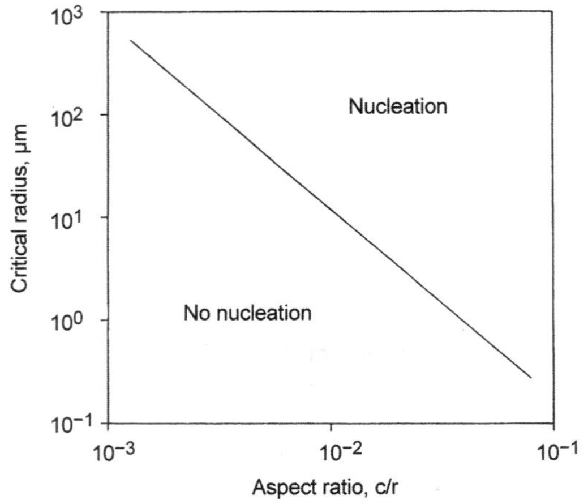

Fig. 54 Critical condition for barrierless nucleation of slip dislocations in stress field of plate periphery. Source: Ref 162

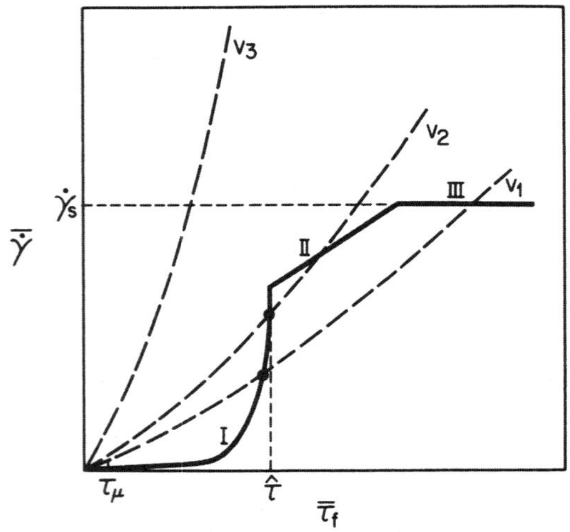

Fig. 55 Conditions for radial plastic zone formation. At a given radial velocity, dashed curve representing imposed plastic strain rate defined by Eq 115 and 116 must intersect solid flow stress curve. Source: Ref 162

is a relativistic limit where $\dot{\gamma}$ is not allowed to exceed $\dot{\gamma}_s$ corresponding to a dislocation velocity equal to the elastic shear wave velocity. The dashed curves indicate the imposed strain rate in the plastic zone as a function of $\bar{\tau}_f$, defined by Eq 115 and 116, depicting three different radial interface velocities. For the velocities v_1 and v_2, physically meaningful plastic zone solutions correspond to the curve intersections shown by the heavy points. At the higher velocity v_3, no solution exists, and a radial plastic zone cannot form. Similar behavior obtains for determination of the plastic zone size in the *c* direction.

Results of computer simulations of growth events at the M_s temperature for two FeNi alloys are summarized by plots of radial velocity versus plate radius in Fig. 56. For the Fe-31Ni case, radial growth rapidly accelerates to a velocity near one-third the shear wave velocity. The condition for slip nucleation is met at the point indicated by the first arrow, but the interface velocity is sufficiently high that a plastic zone cannot form. As the plate grows, its increasing tip radius makes heat transfer less effective and the interface temperature increases, causing a gradual deceleration. At the point marked by the second arrow, a critical velocity is reached where plastic zone formation is possible. Abrupt formation of a plastic zone causes a strong deceleration and erratic growth behavior leading to complete cessation of radial growth at a plate radius of ≈ 80 μm. This is then followed by an additional increment of plate thickening.

For the case of the Fe-24Ni alloy which forms lath martensite, a lower peak velocity is achieved during growth, due primarily to increased phonon drag at is higher M_s temperature (Eq 28). Soon after the slip nucleation condition is achieved, as indicated by the arrow, extensive plastic accommodation brings growth to a halt. In contrast to the Fe31Ni growth involving a macroscopic elastic event, the major part of the growth occurs in a plastic state and stops at a size which is smaller by an order of magnitude.

The behavior described here identifies three basic modes of martensitic growth:

- *Elastic*. When significant plastic accommodation is absent, the behavior described in the section "Thermoelastic Behavior" occurs, corresponding to thermoelastic plate martensite (Fig. 49).
- *Elastic/plastic*. When plastic accommodation is delayed by a high initial growth rate, a macroscopic elastically accommodated "midrib" forms, followed by thickening and growth cessation in an irreversible plastic mode. This corresponds to nonthermoelastic plate martensite (as in Fe-31Ni, Fig. 56).
- *Fully plastic*. When plastic accommodation is not significantly delayed beyond the slip nucleation condition, the entire growth event can be considered to take place in a plastic condition. This corresponds to the lath martensite of Fe-24Ni.

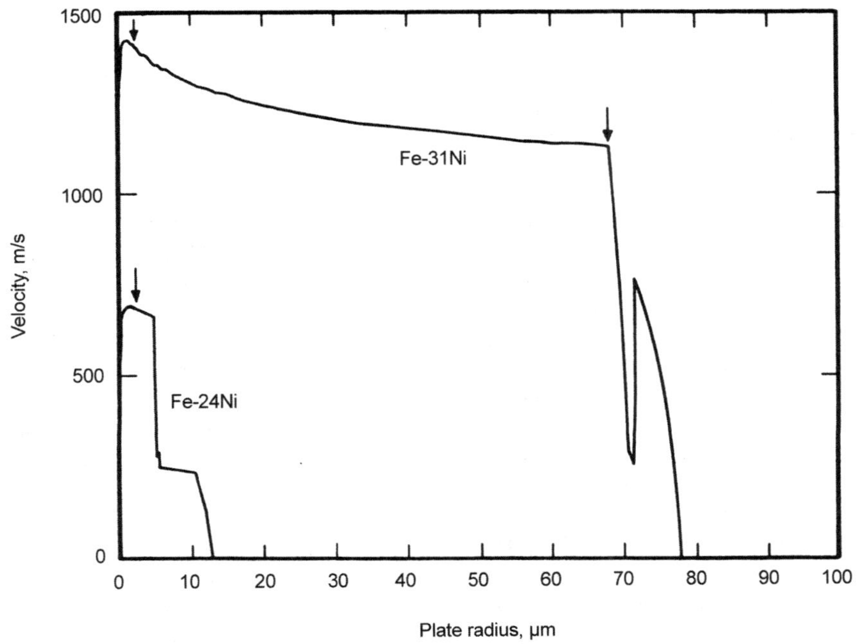

Fig. 56 Summary of computer simulation results for growth events at M_s temperatures of Fe-31Ni and Fe-24Ni alloys. Source: Ref 162

Although the approximations employed in the simulations of Fig. 56 are based on small-scale yielding and cannot accurately describe the highly plastic behavior of the Fe-24Ni alloy, the model does predict a substantial transition in growth mode which correlates with the lath/plate morphological transition, and provides a likely physical basis for it. While radial symmetry was imposed in these calculations, anisotropic plastic zone interactions (e.g., "edge" versus "screw" orientations of Fig. 50(a) and (b) could provide some tendency toward a lath shape during fully plastic growth. The macroscopic elastic growth events common to the thermoelastic and nonthermoelastic plate martensites would be consistent with the success of habit plane analyses for these cases based on elastic strain energy minimization alone. Following the considerations of the section "Surface-energy Minimization," the more complex deformations possible during fully plastic growth may favor interfacial energy minimization, promoting habits near close-packed planes as observed for lath martensites.

Overall Kinetics

While the basic characteristics of individual nucleation and growth events determine many essential features of martensitic transformations, the overall kinetics of the reaction also depend on the nature of the interactions between martensitic particles. As discussed in the section "Thermoelastic-Nonthermoelastic Transition," the strain-energy reducing rearrangements allowed by particle translation will play an important role in the course of thermoelastic transformations. For nonthermoelastic behavior where particles are frozen into the configuration in which they originally form, the nature of autocatalytic nucleation will play a more important role in microstructural development. The intense stress fields at the plate periphery in Fig. 50(a) and (b), where the stresses are of the same sign as the particle shape strain (denoted as + in Fig. 50) provide a rationale for the strong primary autocatalytic coupling between habit variants with "additive" shape strains discussed in the section "Nucleation Sites"; a secondary coupling between variants with canceling strains arises from the weaker negative stresses at the broader interfaces. A special case of primary (additive) coupling is possible for lath-shaped particles when the shape strain direction is nearly normal to the lath axis, corresponding to the two-dimensional stress field calculation of Fig. 50(a). As indicated by the dashed node lines in the figure, the stress field above and below the particle changes to positive again at distances comparable to the particle half-width. Provided plastic growth does not alter this feature of the stress field, it could account for the formation of periodic arrays of martensite laths shearing in the same direction, followed by the formation of oppositely shearing (secondary coupled) units in the regions between them. Such a mechanism for the formation of lath martensite packets has been proposed by Roitburd and Kurdjumov (Ref 166), and the associated sequence has been verified experimentally (Ref 151).

In addition to particle shape-related differences in the character of the stress fields of martensitic laths and plates, the macroscopic elastic growth event predicted for the plate martensite of Fig. 56 suggests a very different type of autocatalysis. With plastic zone formation and autocatalytic nucleation as competitive modes of stress field accommodation, the intense elastic event can provide an autocatalytic nucleation site of extreme potency, in line with the "bursting" form of autocatalysis encountered in plate martensites. Similar to Hall-Petch strengthening, austenitic grain refinement can reduce the intensity of these elastic stress fields and inhibit the grain-to-grain spread of autocatalytic nucleation, further extending the analogy between the kinetics of conventional crystal plasticity and martensitic transformations. The effect is manifested in a grain size dependence of the M_b bursting temperature (Ref 167), and an influence of grain size on the relative roles of the grain-to-grain "spreading" and intragranular "filling-in" modes of autocatalysis in controlling microstructure development (Ref 168, 169). The latter effects are important to the overall transformation kinetics through the dependence of average plate volume $\bar{V}$ on the extent of transformation (Ref 170).

When the dependence of $\bar{V}$ on the transformed volume fraction f is established, the overall kinetics of a nucleation-controlled martensitic transformation can be modeled by a rate equation of the form (Ref 171):

$$\dot{f} = (n_i + pf - N_v)(1 - f)\bar{V}\nu \exp(-Q/RT) \quad \text{(Eq 118)}$$

Here n_i is the density of most-potent initial nucleation sites, p defines the density of autocatalytically generated sites, N_v is the number of plates per unit volume accounting for sites which have operated, the $(1 - f)$ factor accounts for potential sites "swept up" by the transformation, $\bar{V}$ is the instantaneous mean plate volume, and ν and Q are the nucleation attempt frequency and activation energy discussed in the sections "Rate-Controlling Mechanisms" and "Nucleation Attempt Frequency." This model has been successfully fitted to experimental transformation curves of f versus time t. Since N_v is equal to $f/\bar{V}$, the shape of the $f(t)$ function is sensitive to the product $p\bar{V}$, curving upward when $p\bar{V} > 1$. For $p\bar{V} < 1$, $f(t)$ curves downward leading to a finite saturation level of transformation at a given reaction temperature. Although the model is normally applied with the approximation of a singly activated process, it can in principle be extended to incorporate the potency distribution of initial sites $n_i(Q)$ discussed in the sections "Nucleation Sites" and "Rate-Controlling Mechanisms"; however, a distribution $p(Q)$ must also be specified for the autocatalytic sites. A relation for the temperature dependence of the transformation saturation level has been derived based on a simple assumption for the potency distribution of total sites (initial plus autocatalytic) with respect to Δg^{ch} (Ref 91). Further refinement of the overall kinetic models will require more detailed information on the autocatalytic site-potency distribution.

Some Special Applications

Transformation Plasticity

An important objective of crystal plasticity theory is the prediction of constitutive relations for plastic flow. Having developed a transformation kinetic theory from the principles of crystal plasticity, it is natural to consider the prediction of flow behavior when the transformation acts as a deformation mechanism. The quantitative theory of transformation plasticity and its application to the control of mechanical properties has been reviewed (Ref 172). We here briefly examine the essential characteristics of the mechanical behavior associated with martensitic transformation under stress, placing special emphasis on some recent advances.

If martensitic transformation occurs under an applied elastic stress, macroscopic yielding in a constant strain-rate test will be associated with a fixed rate of transformation, $\dot{f}$. From Eq 111 and 118, the critical driving force required to impose this rate will be linearly dependent on temperature:

$$\Delta g_c(\dot{f}) = \frac{1}{B}\left(A + RT\ln\frac{\dot{f}}{n_i V \nu}\right) \qquad \text{(Eq 119)}$$

Computing the total acting driving force from the chemical free-energy change and the work of the applied stress (Eq 25), this linear relation is found to hold for the measured yield stress of high-strength TRIP steels at low temperatures (Ref 174). The predicted transformation stress σ_t is compared against the measured 0.2% yield stress and the stress for 1% transformation in Fig. 57. Below a temperature designated M_s^σ, yielding is seen to be controlled by stress-assisted martensitic transformation. Above M_s^σ, yielding by slip precedes transformation, and transformation is then detected at lower stresses than σ_t; this is consistent with the additional contribution of *strain-induced* transformation on new nucleation sites created by plastic strain as discussed in the section "Nucleation Sites." The unusual temperature dependence of σ_t below M_s, giving a minimum yield stress at 158 K, is associated with the curvature of $\Delta g^{ch}(T)$. Another unusual characteristic of the flow stress observed in this low-temperature regime is an extreme stress-state dependence arising from the effect of the transformation volume change as expressed in Eq 25 (Ref 174). While the temperature and stress-state dependences of the flow stress are unusual, the strain-rate sensitivity predicted by Eq 119 is of the normal magnitude observed for deformation by slip, owing to the close similarity between the B parameter and V^* for slip as discussed in the section "Rate-Controlling Mechanisms."

The kinetic model of Eq 118 can also predict the form of the $\sigma - \varepsilon$ curve at temperatures below M_s^σ, based on the dependence of σ_t on f:

$$\sigma_1(f,\dot{f},T) = -\left(B\frac{\partial\Delta g}{\partial\sigma}\right)^{-1}\left[A + B\Delta g^{ch} + RT\ln\frac{\dot{f}}{(n_i + pf - N_v)(1-f)V\nu}\right] \qquad \text{(Eq 120)}$$

Again, $\partial\Delta g/\partial\sigma$ is taken as a constant defined by Eq 25 for the stress state of interest usually assuming the most optimally oriented transformation variant (Ref 57). With the plastic strain ε proportional to f, Eq 120 then provides a

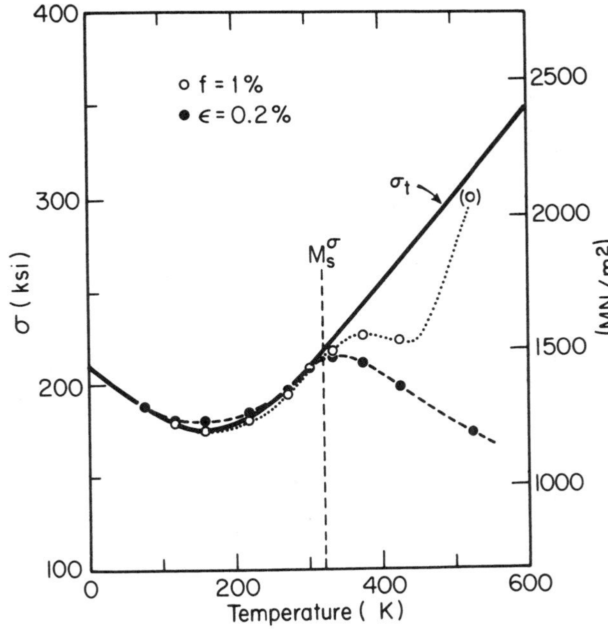

Fig. 57 Observed temperature dependence of 0.2% yield stress and stress at which 1% martensite is detected in high-strength TRIP steel. Solid curve σ_t represents theoretical tensile stress prediction (Eq 119) for stress-assisted transformation. Source: Ref 173

complete constitutive relation $\sigma(\varepsilon, \dot{\varepsilon}, T)$ for plastic flow controlled by stress-assisted martensitic transformation. Using the A and B parameters defined by the $\sigma(T)$ fit of Fig. 57, and typical values of the other kinetic parameters including the initial plate volume $\bar{V}_0$, the predicted $\sigma - \varepsilon$ curve is compared against the true $\sigma - \varepsilon$ curve of the TRIP steel measured at 158 K in Fig. 58. The curve shape is controlled by the behavior of the denominator in the last term of Eq 120. At yielding, where $p\bar{V} > 1$, the initial flow stress is controlled by n_i and a yield drop arises from the autocatalytic nucleation expressed by the pf term. This is followed by flow at a fairly constant stress level in a regime where $p\bar{V} \approx 1$. At higher f where $p\bar{V}<1$, nucleation sites become depleted and extreme strain hardening accompanies the approach to a saturation level of transformation. The model describes the low-strain behavior quite well, but the strain-hardening behavior in the saturation region is too abrupt; this stems primarily from the approximation of a singly activated process.

As discussed in the section "Overall Kinetics," better modeling of the transformation behavior of homogeneous materials at high f requires an activation-energy distribution and this requires further information on the nature of the autocatalytic site potency distribution. However, the manner in which a potency distribution affects strain-hardening behavior can be illustrated by the special case of a well-spaced dispersion of metastable particles in a stable matrix, a problem of relevance to certain ceramic systems such as partially stabilized zirconia. Here, transformation is controlled by the preexisting nucleation sites for which the potency distribution is described by Eq 97. For an average particle volume V_p, the fraction f of particles to transform via sites of a potency with a cumulative number density U_V will be:

$$f = 1 - \exp(-V_P U_V) \quad \text{(Eq 121)}$$

From Eq 87, the stress σ required to operate a nucleation site of potency n_0 is linearly related to a parameter of the form:

$$\begin{aligned}\phi &= \left[\sigma \cdot (-\partial \Delta g/\partial \sigma) - \left(\Delta g^{ch} + g_f^{str} + w_f\right)\right](d/2\gamma) \\ &= n_0^{-1} \end{aligned} \quad \text{(Eq 122)}$$

where w_f represents the frictional work of interfacial motion. For the potency distribution of Eq 97, f and ϕ are related by:

$$\phi = \alpha / \ln\left[-V_P U_V^0 / \ln(1-f)\right] \quad \text{(Eq 123)}$$

Taking the plastic strain ε to be proportional to f with complete transformation giving a strain

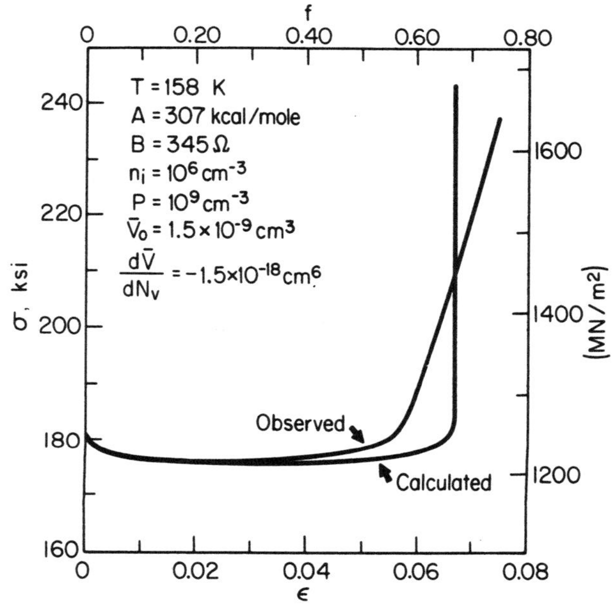

Fig. 58 Comparison of calculated and observed true $\sigma - \varepsilon$ curves for TRIP steel of Fig. 57. Source: Ref 173

ε_0, and using $\alpha = 0.86$ from the distribution of Fig. 44, the form of the $\sigma - \varepsilon$ curves predicted for various values of V_p normalized to $\left(U_v^0\right)^{-1}$ are shown in Fig. 59. For large particles where nucleation sites are plentiful, the flow stress is fairly constant. For small particles, a high strain-hardening rate arises from early exhaustion of the most potent sites, and further strain requires operation of less potent sites at higher stresses. The smoother form of the "exhaustion" hardening predicted by the potency distribution function is more consistent with the observed high-strain behavior in Fig. 58 than is the simple singly activated model.

While the curves of Fig. 59 demonstrate the character of the distribution effects on transformation plasticity, the assumption that $\partial\Delta g/\partial\sigma$ can be treated as a constant is not entirely realistic in view of the distribution of possible nucleation-site orientations with respect to the applied stress system. An analysis (Ref 174) for the case of a random-orientation distribution has demonstrated the manner in which applied stress broadens the thermodynamic potency distribution $U_v(\Delta g)$ associated with a given structural potency distribution $U_v(n_0)$. The model can account for previous discrepancies in the stress dependence of the M_s temperature observed in ferrous alloys, and predicts a smooth yield locus for transformation plasticity under multiaxial stress. Application to the dispersed-phase transformation plasticity problem of Fig. 59 predicts curves which are qualitatively similar, but differ quantitatively due to the effective distribution broadening; the curves are reasonably approximated by applying a constant correction factor to $\partial\Delta g/\partial\sigma$ of Eq 122.

Despite the high strain-hardening rates associated with site exhaustion, transformation plasticity at temperatures below M_s^σ represents a *softening* phenomenon relative to the flow behavior of the parent phase in the absence of transformation. Above M_s^σ, where *strain-induced* nucleation dominates the transformation behavior, transformation and slip act as parallel deformation mechanisms. While a softening effect is still associated with operation of the transformation as a deformation mechanism, a hardening effect also arises from the transformation product as an obstacle to slip. These two effects have been separated by prestrain experiments, providing empirical constitutive relations for this class of transformation plasticity (Ref 175). The hardening and softening effects are monotonic functions of f and $\dot{f}$, respectively. Combined with a sigmoidal $f(\varepsilon)$ curve for strain-induced transformation (Ref 111), they produce an unusual upward curvature to the $\sigma - \varepsilon$ curve which has been shown to be particularly effective in enhancing uniform ductility in single-phase and multiphase materials (Ref 172).

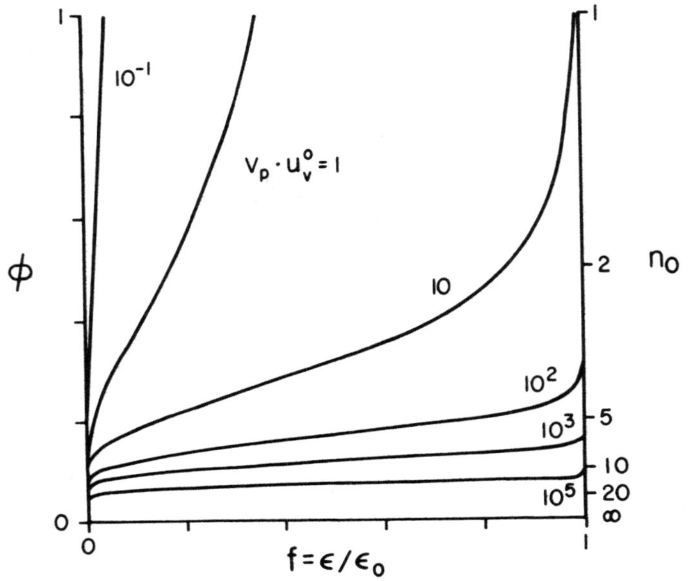

Fig. 59 Normalized $\sigma - \varepsilon$ curve shapes for stress-assisted transformation plasticity from noninteracting dispersed particles of volume V_p.

Recent experiments have demonstrated that the flow-stabilizing influence of transformation plasticity can substantially enhance fracture toughness when the fracture is controlled by plastic shear instability (Ref 176). A series of γ' precipitation-hardened metastable austenitic steels was designed to have (a) a constant 1300 MPa yield strength, (b) varying stability with respect to transformation, and (c) a varying transformation volume change. Figure 60 summarizes the measured increment of J-integral toughness enhancement ΔJ_{IC} relative to the toughness of stable austenite, J_{IC}^{A}, as a function of relative transformation stability. The latter stability is expressed by a normalized temperature interval:

$$\theta = (T - M_d)/(M_d - M_s^{\sigma}) \quad \text{(Eq 124)}$$

where M_d is the maximum temperature at which transformation is induced by deformation. Both the M_s^{σ} and M_d temperatures were experimentally determined for the stress state ahead of the crack tip. Solid points represent shear-instability controlled ductile fracture, while open points depict a flat tensile mode of fracture. The shape of the ΔJ_{IC} versus θ behavior for shear-instability controlled fracture is consistent with the effect of transformation plasticity on flow stability, indicating an optimum amount of crack-tip transformation near the M_s^{σ} temperature.

Curves labeled HV in Fig. 60 represent alloys with a relatively high transformation-volume change of 4%, while LV represents a lower volume change of 2.5%. Toughness enhancement is clearly increased by a larger volume change. An empirical power-law correlation between relative toughening and amount of crack-tip transformation allows a quantitative assessment of the role of dilatation in toughening. Although the rate of toughening increases with ΔV, extrapolation to $\Delta V = 0$ suggests a significant toughening contribution even in the absence of dilatation. This behavior is qualitatively consistent with models of void-induced shear instability (Ref 177). Even in the absence of dilatation, the $\sigma - \varepsilon$ curve-shaping effect of transformation plasticity will delay shear instability. Dilatation will further delay the instability by reducing stress triaxiality, which in turn lowers the critical strain-hardening rate for void-induced instability.

The largest toughening increment shown in Fig. 60 is 60%, corresponding to a total fracture toughness of K_{IC} = 250 MPa m$^{1/2}$. This is unusually high for a 1300 MPa yield-strength material, illustrating the potential technological significance of transformation toughening.

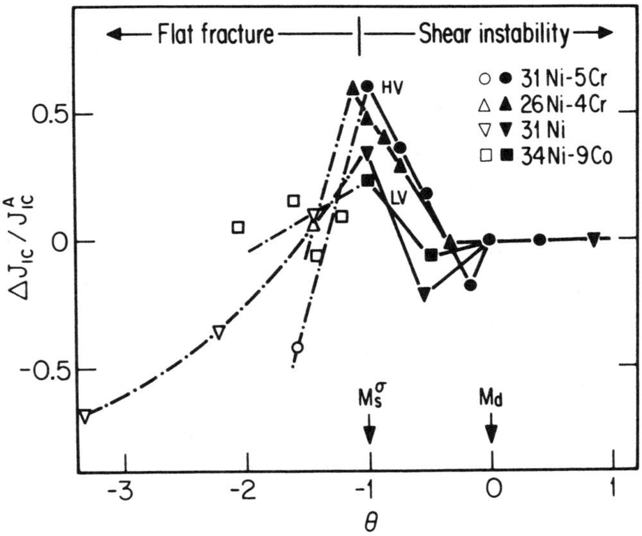

Fig. 60 Fracture toughness enhancement relative to stable austenite plotted versus austenite stability for alloys with high (HV) and low (LV) transformation volume change. Solid points represent shear-instability controlled ductile fracture while open points correspond to flat tensile modes. Source: Ref 176

In the traditional use of phase transformations to obtain microstructures with desirable properties, the role of transformation kinetic theory in structure/property relationships is quite indirect. On the other hand, when the transformation is used as a deformation mechanism, this role becomes a very direct one, and subtleties of transformation kinetics have a profound influence on the shape of the $\sigma - \varepsilon$ curve as well as on its sensitivity to stress state. The examples cited here showing the impact of these effects on macroscopic ductility and fracture toughness indicate that the direct application of transformation kinetic theory to the control of mechanical properties can have far-reaching practical consequences.

Biological Martensitic Transformations

The principles of martensitic transformations examined here are now being tested in an increasingly wide range of materials. This is particularly true, for example, in the case of ceramic materials. In contrast, martensitic transformations have received little attention in polymeric systems, presumably due to the noncrystalline structures typical of many polymeric materials of engineering importance. However, the polymers from which nature constructs its living machines exhibit a surprising preference for periodic crystalline structures, and research indicates that martensitic transformations are well represented in biology (Ref 178).

Figure 61 shows the example of tail-sheath contraction in the T4 bacteriophage, a virus which infects *E. coli* bacteria. The virus consists of a DNA-filled icosahedral head or capsid attached to a tail assembly composed of a rigid tail-core, a cylindrical tail-sheath, and a baseplate assembly with six long and short tail fibers. The tail-sheath is a metastable two-dimensional protein crystal closed to form a cylinder. When the tail fibers attach to the appropriate bacterial membrane, they distort the baseplate which, in turn, triggers a strain-induced single-interface martensitic transformation in the crystalline tail-sheath. The transformation shape strain produces a substantial contraction which drives the rigid tail-core through the bacterial membrane, injecting the virus DNA into the bacterium.

Figure 62 depicts a coherency dislocation model of the martensitic interface transforming the protein crystal between the structures corresponding to the extended and contracted states. The helical nature of the cylindrical crystal provides a built-in pole mechanism

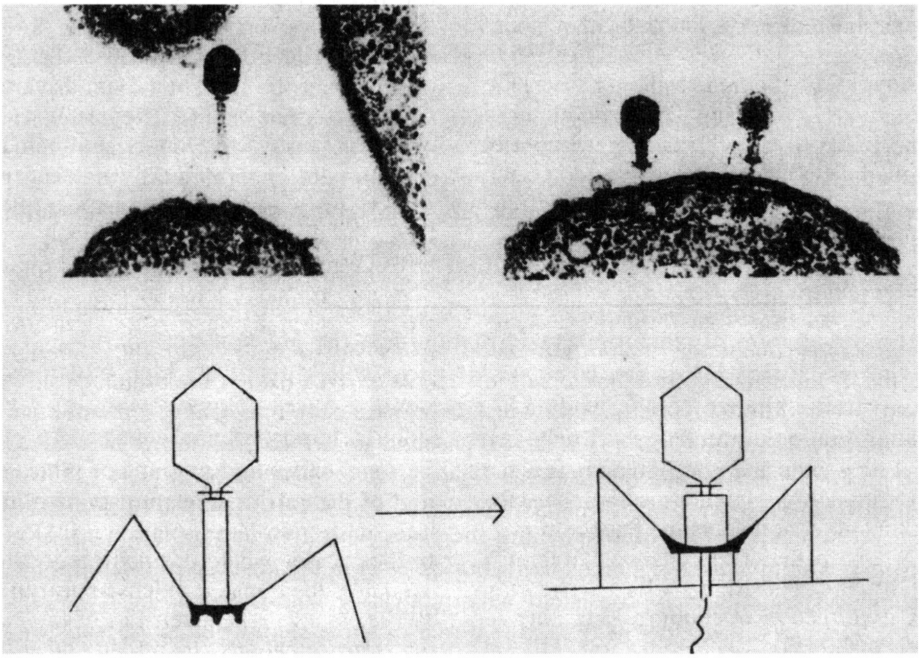

Fig. 61 Bacterial virus tail-sheath contraction. Source: Ref 178

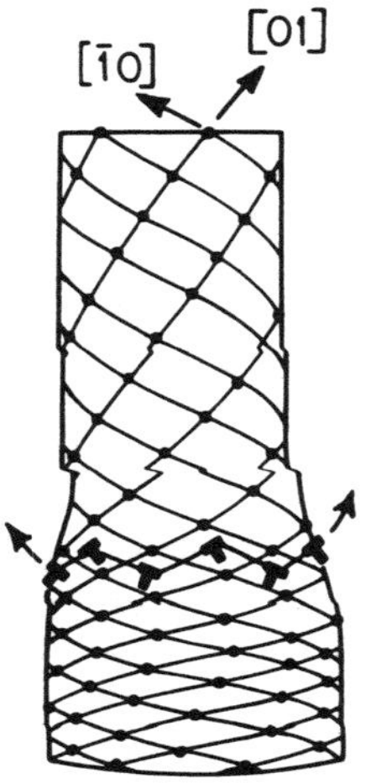

Fig. 62 Cylindrical crystal structure of virus tail-sheath during contraction by martensitic transformation. Interface is described by coherency dislocations which spiral up helical close-packed crystal rows. Source: Ref 178

whereby only a fixed number of dislocations is necessary to spiral up the close-packed rows and completely transform the crystal. The transformation shape strain involves a shear greater than 100%, which not only shortens the sheath but also provides a large twist about the cylinder axis, thus helping the tail-core to "bore" through the bacterial cell wall. Calorimetric measurements on this highly irreversible transformation indicate a latent heat of about −190 kJ/mole.

Smaller-strain thermoelastic martensitic transformations producing a shape-memory effect are found to be important in the control of motion in certain "swimming" bacteria, such as *Salmonella* and *E. coli*. These bacteria propel themselves by rigid rotation of "flagella," which are helical filaments consisting of distorted cylindrical crystals of the protein flagellin. Martensitic transformation shears producing a twist about the crystal axis can convert the helical forms from left-hand to right-hand with differing pitch. When the direction of rotation of the flagella is reversed, viscous forces induce transformation, and the shape strain across the martensitic interface produces a kink that renders the flagellum ineffective as a propeller. This causes a "tumbling" motion of the bacterium which randomly changes its direction of motion. Returning to the original direction of flagellar rotation, the crystals revert to their initial form and "swimming" motion ensues. The bacterium thus controls its motion by alternating between these two states.

Studies of flagellar crystals in isolation have revealed a wide variety of "polymorphic" structures related by martensitic transformations, and phase diagrams have been constructed mapping the stability of the various forms as a function of pH, ionic concentration, and temperature. The transformation shears associated with these thermoelastic transformations are only a few percent, and measured torque-twist hysteresis loops analogous to Fig. 29 have allowed determination of critical driving forces for transformation. The driving force required for nucleation is much higher than that for interfacial motion, and heterogeneous nucleation from distorted regions has been directly observed. Low-pitch helical close-packed rows in these crystals allow the transformation shears to be accomplished by a single coherency dislocation (Ref 178).

The most ancient forms of bacteria move along solid surfaces by a motion termed "gliding," for which a mechanism has yet to be established. Longitudinal arrays of thin filaments or "fibrils" have been observed between flexible and rigid outer membranes of the bacterial cell in some of these species. Assuming that these fibrils possess transformation properties similar to those of the bacterial flagella, and further that such pH-sensitive transformations involve proton absorption, a possible mechanism of gliding motility is being developed on the basis of a proton-driven shape-memory turbine engine (Ref 180). In higher organisms, arrays of cylindrical protein crystals such as the microtubules in cilia produce motion on a much more macroscopic level. Here, the displacements are actively amplified by ATP hydrolysis, but it appears quite likely that the primary microscopic displacements are controlled by martensitic transformations in the microtubules (Ref 180). Similar displacive characteristics of microtubules in the brain may have an important role in information processing (Ref 181).

These examples suggest a significant function of martensitic transformations in the control of motion and shape in living things. In those cases where detailed observations are available, the transformations appear to display all the essential characteristics of the martensitic transformations found in other classes of materials. That this commonality exists despite the detailed bonding differences among metals, ceramics, and polymers supports the underlying role of strain energy associated with the large displacements which define this class of transformation and for which a dislocation-based transformation theory is most appropriate. The behavior of crystalline proteins thus attests to the broad applicability of the theoretical approach to martensitic transformations reviewed here.

Conclusions

Application of the principles of crystal plasticity to the lattice deformations and lattice-invariant deformations accompanying martensitic transformations has provided a unified theoretical approach to this important class of solid-state reactions. Geometric constraints on the interphase crystallography have been identified, and a clear connection is established between the necessary deformations and the dislocation structure of interphase boundaries in discrete crystals, as observed by modern high-resolution microscopy techniques. The dislocation description allows the prediction of numerous interfacial properties, including the interfacial stress field, energy, stability, and perhaps most important, the interfacial mobility. Direct measurement of the predicted mobility behavior in single-interface experiments has not only provided a major advance in our understanding of martensitic kinetics, but constitutes the most precise test of thermally activated deformation theory to date. Application of the dislocation-based approach to the mechanism and kinetics of martensitic nucleation and growth has led to a reasonably quantitative explanation of the essential features of martensitic transformations. Interface-controlled barrierless nucleation at heterogeneous sites with a distribution of potencies accounts for the observed nucleation behavior, consistent with measured kinetics and metallographic observations of suboperational embryos. Analysis of the dynamic interaction of martensitic growth with matrix plasticity has provided a basis for kinetic (e.g., thermoelastic versus nonthermoelastic) and morphological (e.g., lath versus plate) transitions in growth behavior. Numerous parallels between martensitic transformations and conventional plastic deformation are identified, from which more subtle "second-order" effects such as thermal stabilization can be anticipated.

Further important progress in the understanding of martensitic transfomations can be expected from the incorporation of nonlinear-elastic and gradient-energy effects. Development of a nonlinear elastic theory would allow a better treatment of core energies and nonclassical kinetic phenomena in martensitic transformations, and would also expand crystal plasticity theory more generally. The greater complexity of a nonlinear description may warrant the use of atomistic computer simulations, once suitable interatomic potentials can be developed for materials of interest. Another area where further progress would be particularly beneficial is the precise nature of autocatalytic nucleation. The recent progress in modeling of growth dynamics may provide a theoretical basis for prediction of the specific characteristics of this important contribution to the overall transformation kinetics. Taking martensitic reactions as the prototype of a purely structural transformation, an especially fruitful area of further endeavor is the application of the current concepts of martensitic transformations to the intrinsic structural aspects of other first-order solid-state reactions.

ACKNOWLEDGMENTS

The authors are grateful for helpful discussion with Dr. M. Grujicic, Mr. D. M. Haezebrouck, Prof. I-W. Chen, Prof. W. S. Owen, and Prof. A. S. Argon of MIT.

Research at MIT on martensitic transformations has been sponsored by the National Science Foundation under Grant DMR79-15196.

REFERENCES

1. M. Cohen, G.B. Olson, and P.C. Clapp, On the classification of displacive phase transformations, *Proc. Int. Conf. on Martensitic Transformations*, ICOMAT-79 (1979) p 1
2. J.W. Christian and A.G. Crocker, in: *Dislocations in Solids*, Vol. 3 (F.R.N. Nabarro, ed.) (North-Holland, 1980) p 165
3. G.B. Olson and M. Cohen, Theory of martensitic nucleation: a current assessment, *Proc. Int. Conf. Solid-Solid Phase Transf.*, 1981 (H.I. Aaronson, ed.) (AIME, 1982) p 1145
4. R.W. Balluffi, A. Brokman, and A.H. King, Acta Metall. 30 (1982) 1453

5. M.S. Wechsler, D.S. Lieberman, and T.A. Read, Trans. Metall. Soc. AIME 127 (1953) 1503
6. J.S. Bowles and J.K. Mackenzie, Acta Metall. 2 (1954) 129
7. J.W. Christian, *Theory of Transformations in Metals and Alloys* (Pergamon, 1965)
8. J.D. Eshelby, Prog. Solid Mech. 2 (1961) 89
9. U.F. Kocks and G.R. Canova, How many slip systems, and which?, *Deformation of Polycrystals: Mechanisms and Microstructures*, 2nd Riso Int. Symp. Metall. and Mater. Sci., (1981) p 35
10. U.F. Kocks and H. Chandra, Acta Metall. 30 (1982) 695
11. F.R.N. Nabarro, Proc. R. Soc. London Ser. A 175 (1940) 519
12. P.R. Howell and R.W.K. Honeycombe, Crystallographic aspects of diffusional phase transformations in metals and alloys, *Int. Conf. Solid-Solid Phase Transf.* (H.I. Aaronson, ed.), (TMS-AIME, (1982), p 399
13. J.W. Christian, O-lattice, surface dislocation and elastic theories of martensite crystallography and martensitic nuclei, *New Aspects of Martensitic Transformations*, Proc. First JIM Int. Symp., Suppl. Trans JIM 17 (1976)
14. B.C. Muddle, The crystallography and morphology of ferrous martensites, *Int. Conf. Solid-Solid Phase Transf.* (H.I. Aaronson, ed.), (TMS-AIME, 1982), p 1347
15. J.D. Watson and P.G. McDougall, Acta Metall. 21 (1973) 961
16. J.M. Rigsbee and H.I. Aaronson, Acta Metall. 27 (1979) 351
17. P.G. McDougall, G.B. Olson, and M. Cohen, unpublished research
18. C.G. Weatherly, P. Humble, and D. Borland, Acta Metall. 27 (1979) 1815
19. A. Crosky, P.G. McDougall, and J.S. Bowles, Acta Metall. 28 (1980) 1495
20. U. Dahmen and K.H. Westmacott, The role of the invariant line in the nucleation and growth of precipitates, *Int. Conf. Solid-Solid Phase Transf.* (H.I. Aaronson, ed.), (TMS-AIME, 1982), p 433
21. U. Dahmen, Acta Metall. 30 (1982) 63
22. U. Dahmen, P. Ferguson, and K.H. Westmacott, Acta Metall. 32 (1984) 803
23. W. Bollmann, *Crystal Defects and Crystalline Interfaces* (Springer, 1970)
24. R.G. Davies and C.L. Magee, Metall. Trans. 2 (1971) 1939
25. G.B. Olson and M. Cohen, Acta Metall. 27 (1979) 1907
26. G.B. Olson, Surf. Sci. 144 (1984) 302
27. G.B. Olson, Acta Metall. 29 (1981) 1475
28. R.W. Balluffi and G.B. Olson, Metall. Trans. 16A (1985) 529
29. G.B. Olson and M. Cohen, Dislocation structure of martensitic interfaces, *Proc. Int. Conf. Solid-Solid Phase Transf.*, 1981 (H.I. Aaronson, ed.) (AIME 1982), p 1209
30. C.M. Wayman, *Introduction to the Crystallography of Martensitic Transformations* (Macmillan, 1964)
31. J.F. Nicholas, *An Atlas of Models of Crystal Surfaces* (Gordon and Breach, 1965) p 57
32. P. Sandvik and C.M. Wayman, Metall. Trans. 14A (1983) 823, 835
33. K. Otsuka and K. Shimizu, Trans. JIM 15 (1974) 103
34. M. Grujicic, Sc.D. thesis, MIT, 1983
35. C. Hammond and P.M. Kelly, Acta Metall. 17 (1969) 869
36. K.M. Knowles and D.A. Smith, Acta Metall. 29 (1981) 1445
37. J.W. Christian, Proc. R. Soc. London Ser. A 206 (1951) 51
38. W. Bollmann, Acta Metall. 9 (1961) 972
39. S. Mahajan, M.L. Green, and D. Brasen, Metall. Trans. 8A (1977) 283
40. S. Mahajan and G.Y. Chin, Acta Metall. 21 (1973) 1353
41. S. Mahajan, Philos. Mag. 26 (1972) 161
42. J.W. Brooks, M.H. Loretto, and R.E. Smallman, Acta Metall. 27 (1979) 1839
43. K. Knowles, Proc. R. Soc. London Ser. A 380 (1982) 187
44. R. Sinclair, in: *Proc. 35th Annual Meeting EMSA* (Claitor's, 1977) p 110
45. B.V.N. Rao and G. Thomas, in: *Proc. Int. Conf. Mart. Transf.*, ICOMAT-79 (1979) p 12
46. D.J.H. Cockayne, J. Microsc. (Oxford) 98 (1973) 116
47. J.P. Hirth and J. Lothe, *Theory of Dislocations* (McGraw-Hill, 1968)
48. F.C. Frank, Acta Metall. 1 (1953) 15
49. R. Bullough and B.A. Bilby, Proc. Phys. Soc. 69B (1956) 1276

50. W.T. Read and W. Shockley, Phys. Rev. 78 (1950) 275
51. J.H. van der Merwe, Proc. Phys. Soc. A63 (1950) 616
52. H.C. Ling, Sc.D. thesis, MIT (1978)
53. T. Mura, T. Mori, and M. Kato, J. Mech. and Phys. Solids 24 (1976) 305
54. M. Hillert, Metall. Trans. 6A (1975) 5
55. L. Kaufman and M. Cohen, Prog. Metal Physics 1 (1958) 165
56. U.F. Kocks, A.S. Argon, and M.F. Ashby, Prog. Mater. Sci. 19 (1975) 1
57. J.R. Patel and M. Cohen, Acta Metall. 1 (1953) 531
58. F.C. Frank, Proc. Phys. Soc. A62 (1949) 131
59. J.D. Eshelby, Proc. Phys. Soc. B69 (1956) 1013
60. J. Weertman, J. Mech. and Phys. Solids 11 (1963) 197
61. A.L. Roitburd, Kristallografiya 7 (1962) 291
62. G.B. Olson and M. Cohen, Metall. Trans. 7A (1976) 1915
63. V. Raghavan and M. Cohen, Acta Metall. 20 (1972) 333
64. M. Grujicic, unpublished research, MIT
65. M.F. Ashby, Philos. Mag. 11 (1966) 1157
66. C. Zener, quoted by C. S. Smith, Trans. AIME 175 (1948) 47
67. R. Labusch, Acta Metall. 20 (1972) 917
68. F.R.N. Nabarro, Solution hardening, *Dislocations and Properties of Real Materials* (Institute of Metals, 1985) p 152
69. U.F. Kocks, Metall. Trans. A16 (1985) 2109
70. M. Grujicic, G.B. Olson, and W.S. Owen, Kinetics of martensitic interface motion, *Proc. Int. Conf. Mart. Transf.*, ICOMAT-82, 1982, J. de Physique 43 (1982) C4-173
71. K. Otsuka, C.M. Wayman, K. Nakai, H. Sakamoto, and K. Shimizu, Acta Metall. 24 (1976) 207
72. T. Tadaki, T. Kakeshita, and K. Shimizu, *Proc. Int. Conf. Mart. Transf.*, ICOMAT-82, J. de Physique 43 (1982) C4-191
73. K.C. Russell, Adv. Colloid and Interface Sci. 13 (1980) 205
74. M. Hillert, in: *Mechanisms of Phase Transformations in Crystalline Solids*, Inst. of Metals Monograph 33 (Inst. of Metals, 1969) p 231
75. P.V. Ivantsov, Dokl. Akad. Nauk SSSR 58 (1947) 567
76. H.K.D.H. Bhadeshia, G.B. Olson, and M. Cohen, unpublished research
77. H.K.D.H. Bhadeshia, Acta Metall. 29 (1981) 1117
78. R.F. Hehemann, The bainite transformation, *Phase Transformations* (ASM, 1970) p 397
79. G.B. Olson and M. Cohen, Annu. Rev. Mater. Sci. 11 (1981) 1
80. G.B. Olson and M. Cohen, Metall. Trans. 7A (1976) 1897
81. G.B. Olson and M. Cohen, Metall. Trans. 7A (1976) 1905
82. A.J. Bogers and W.G. Burgers, Acta Metall. 12 (1964) 1905
83. M.L. Wechsler, T.A. Read, and D.S. Lieberman, Trans. TMS-AIME 218 (1960) 202
84. F.J. Schoen, J.L. Nilles, and W.S. Owen, Metall. Trans. 2 (1971) 2489
85. L. Remy and A. Pineau, Mater. Sci. and Eng. 36 (1978) 47
86. A.P. Miodownik, CALPHAD 2 (1978) 207
87. J.W. Christian, *Physical Metallurgy* (R.W. Cahn, ed.) (North-Holland, 1965) p 521
88. M. Gedwill, C. Altstetter, and C.M. Wayman, Trans. TMS-AIME 230 (1964) 453
89. S. Dash and N. Brown, Acta Metall. 14 (1966) 595
90. J. Gaggero and D. Hull, Acta Metall. 10 (1962) 995
91. C.L. Magee, The nucleation of martensite, *Phase Transformations* (ASM, 1970) p 115
92. K. Shimizu, M. Oka, and C.M. Wayman, Acta Metall. 18 (1970) 1005
93. S. Kajiwara, *Proc. Int. Conf. Martensitic Transformations*, ICOMAT-79 (1979) p 362
94. C.H. Li, D.A. Porter and K.E. Easterling, On the nucleation and growth of martensite in small iron particles in copper, *Proc. Int. Conf. Solid-Solid Phase Transf.* (H.I. Aaronson, ed.) (TMS-AIME, 1982) p 1177
95. K.E. Easterling and P.R. Swann, Acta Metall. 19 (1971) 117
96. J.W. Cahn, Acta Metall. 5 (1957) 169
97. M. Suezawa and H.E. Cook, Acta Metall. 28 (1980) 423
98. H.C. Ling and G.B. Olson, Strain energy of a martensitic nucleus, *Proc. Int. Conf. Solid-Solid Transf.* (H.I. Aaronson, ed.) (TMS-AIME, 1982) 1171

99. A.L. Roitburd, Dokl. Akad. Nauk SSSR 256 (1981) 80; Sov. Phys. – Dokl. 26 (1981) 92
100. I-W. Chen, unpublished research, MIT, 1983
101. P.L. Ferraglio and K. Mukherjee, Acta Metall. 22 (1974) 835
102. E.P. Butler and M.G. Burke, *Proc. Int. Conf. Mart. Transf.*, ICOMAT-82, J. de Physique 43 (1982) C4-121
103. M. Cohen and G.B. Olson, Martensitic nucleation and the role of the nucleating defect, *New Aspects of Martensitic Transformations*, Proc. First JIM Int. Symp., Suppl. Trans. JIM 17 (1976) 93
104. R. Cech and D. Turnbull, Trans. AIME 206 (1956) 124
105. J.A. Venables, Philos. Mag. 7 (1962) 35
106. R. Lagneborg, Acta Metall. 12 (1964) 823
107. D.L. Manganon and G. Thomas, Metall. Trans. 1 (1970) 1577
108. F. Lecroisey and A. Pineau, Metall. Trans. 3 (1972) 387
109. T. Suzuki, H. Kojima, K. Suzuki, K. Hashimoto, and M. Ichihara, Acta Metall. 25 (1977) 1151
110. J.W. Brooks, M.H. Loretto, and R.E. Smallman, Acta Metall. 27 (1979) 1829
111. G.B. Olson and M. Cohen, Metall. Trans. 6A (1975) 791
112. G.A. Knorovsky, Autocatalysis of Martensitic Transformations, Sc.D. thesis, MIT, 1977
113. J.C. Bokros and E.R. Parker, Acta Metall. 11 (1963) 1291
114. L. Delaey, in: *New aspects of martensitic transformations*, Proc. First JIM Int. Symp., Suppl. Trans. JIM 17 (1976) 135
115. T. Saburi and C.M. Wayman, Acta Metall. 27 (1979) 979
116. H.R. Clark, Paired-variant martensite in Fe-30% Ni, *Proc. Int. Conf. Martensitic Transformations*, ICOMAT-82, J. de Physique 43 (1982) C4-517
117. M. Cohen, E.S. Machlin, and V.G. Paranjpe, in: *Thermodynamics in Physical Metallurgy* (ASM, 1949) p 242
118. E.S. Machlin and M. Cohen, Trans. AIME 194 (1952) 489
119. P.C. Clapp, Phys. Stat. Sol. (b) 57 (1973) 561
120. T. Suzuki, Metall. Trans. 12A (1981) 709
121. G. Guenin and P.F. Gobin, Metall. Trans. 13A (1982) 1127
122. N. Nakanishi, Prog. Mater. Sci. 24 (1980) 143
123. J.W. Cahn and J.E. Hilliard, J. Chem. Phys. 31 (1959) 688
124. G.B. Olson and M. Cohen, Classical and nonclassical mechanisms of martensitic transformations, in: *Proc. Int. Conf. Mart. Transf.*, ICOMAT-82, J. de Physique 43 (1982) C4-75
125. G. Krasko, G.B. Olson, and M. Cohen, unpublished research, MIT, 1983
126. A.R. Bishop and T. Schneider, eds., *Solitons and Condensed Matter Physics* (Springer, 1978)
127. F. Falk, in: *Proc. Int. Conf. Mart. Transf.*, ICOMAT-82, J. de Physique 43 (1982) C4-203
128. G.R. Barsch and J.A. Krumhansl, Phys. Rev. Lett. 53 (1984) 1069
129. D.M. Haezebrouck, MIT thesis research in progress
130. V. Raghavan and M. Cohen, Metall. Trans. 2 (1971) 2409
131. R.J. Salzbrenner and M. Cohen, Acta Metall. 27 (1979) 739
132. H.C. Ling, The Influence of the Invar Effect on the Elastic Properties and the Martensitic Transformation of Fe_3 Pt, Sc.D. thesis, MIT (1978)
133. J.K. Lee and W.C. Johnson, Ser. Metall. 11 (1977) 477
134. H. Horner, The role of strain in phase transformations, *Proc. Int. Conf. Solid-Solid Phase Transf.* (H.I. Aaronson, ed.) (TMS-AIME, 1982) p 87
135. M. Suezawa and S. Ishioka, Mater. Sci. and Eng. 37 (1979) 283
136. F. Lovey, M. Chandrasekaran, R. Rapacioli, and M. Ahlers, Z. Metallkd. 71 (1980) 37
137. C.V. Thompson, Field-Ion Microscopy of Fe_3Pt, SB thesis, MIT (1976)
138. M. Cohen, Metall. Trans. 3 (1972) 1095
139. G.B. Olson and M. Cohen, Kinetics of isothermal FCC-HCP martensitic nucleation, *Proc. Int. Conf. Mart. Transf.*, ICOMAT-79 (1979) p 310
140. C.L. Magee, Metall. Trans. 2 (1971) 2419
141. S. Kajiwara, Philos. Mag. 43 (1981) 1483
142. G.B. Olson, unpublished research (1984)
143. M.K. Korenko, Martensitic Transformations in High Magnetic Fields, Sc.D. thesis, MIT (1973)
144. M. Laverroux and A. Pineau, Ser. Metall. 8 (1974) 351
145. J.F. Breedis, Acta Metall. 13 (1965) 239

146. H.C. Fiedler, B.L. Averbach, and M. Cohen, Trans. ASM 47 (1955) 267
147. J.R.C. Guimaraes and J.C. Shyne, Ser. Metall. 4 (1970) 1019
148. P.J. Brofman and G.S. Ansell, Metall. Trans. 14A (1983) 1929
149. R.H. Richman and G.F. Bolling, Metall. Trans. 2 (1971) 2451
150. K.R. Kinsman and J.C. Shyne, Acta Metall. 14 (1966) 1063
151. W.S. Owen, F.J. Schoen, and G.R. Srinivasan, in: *Phase Transformations* (ASM, 1970) p 157
152. G.S. Ansell, S.J. Donachie, and R.W. Messler Jr, Metall. Trans. 2 (1971) 2443
153. V. Raghavan and M. Cohen, Acta Metall. 20 (1972) 779
154. G.B. Olson and M. Cohen, Ser. Metall. 9 (1975) 1247
155. G.B. Olson and M. Cohen, Ser. Metall. 11 (1977) 345
156. G.B. Olson and W.S. Owen, The stress field of a martensitic particle and the conditions for thermoelastic behavior, *New Aspects of Martensitic Transformations*, Proc. First JIM Int. Symp., Suppl. Trans. JIM 17 (1976) 105
157. H.C. Ling and W.S. Owen, Acta Metall. 29 (1981) 1721
158. M. Umemoto and C.M. Wayman, Acta Metall. 26 (1978) 1529
159. S. Kajiwara and W.S. Owen, Metall. Trans. 4 (1973) 1988; 5 (1974) 2061
160. S. Kajiwara and W.S. Owen, Ser. Metall. 11 (1977) 137
161. G.B. Olson and M. Cohen, unpublished research
162. B.Y. Lyubov and A.L. Roitburd, Dokl. Akad. Nauk SSSR 131 (1960) 809
163. Z. Nishiyama, A. Tsubaki, H. Suzuki, and Y. Yamada, J. Phys. Soc. Jpn. 13 (1958) 1084
164. J.A.H. Hult and F.A. McClintock, in: *Proc. 9th Int. Congress Applied Mechanics*, Vol. 8 (1956) p 51
165. F.A. McClintock and G.R. Irwin, in: *Symp. Fracture Toughness Testing and Applications*, ASTM Spec. Tech. Pub. 381 (1965) p 84
166. A.L. Roitburd and G.V. Kurdjumov, Mater. Sci. and Eng. 39 (1979) 141
167. M. Umemoto and W.S. Owen, Metall. Trans. 5 (1974) 2041
168. J.R.C. Guimaraes, Acta Metall. 26 (1978) 1591
169. J.R.C. Guimaraes and J.C. Gomes, Metall. Trans. 10A (1979) 109
170. P-H. Chang, P.G. Winchell, and G.L. Liedl, Metall. Trans. 14A (1983) 163
171. S.R. Pati and M. Cohen, Acta Metall. 17 (1969) 189, 1327
172. G.B. Olson, in: *Deformation, Processing, and Structure* (G. Krauss, ed.) (ASM, 1984) p 391
173. G.B. Olson and M. Cohen, Metall. Trans. A, 13A (1982) 1907
174. K. Tsuzaki and G.B. Olson, MIT, unpublished research
175. T. Narutani, G.B. Olson, and M. Cohen, in: *Proc. Int. Conf. Mart. Transf.*, ICOMAT 82, J. de Physique 43 (1982) C4-429
176. R.H. Léal, Transformation Toughening of Metastable Austenitic Steels, Sc.D. thesis, MIT (1984)
177. A. Needleman and J.R. Rice, in: *Mechanics of Metal Sheet Forming* (D.P. Koistinen and N.M. Wang, eds.) (Plenum, 1978) p 237
178. G.B. Olson and H. Hartman, in: *Proc. Int. Conf. Mart. Transf.*, ICOMAT 82, J. de Physique 43 (1982) C4-855
179. G.B. Olson and H. Hartman, MIT, unpublished research
180. J.J. Blum and M. Hines, Quart. Rev. Biophysics 12 (1979) 120
181. S.R. Hameroff and R.C. Watt, J. Theor. Biol. 98 (1982) 549

Transformations—Selected Works of G.B. Olson on Materials, Microstructure, and Design
C.E. Campbell, M.V. Manuel, and W. Xiong, editors

CHAPTER 2

On the Hierarchy of Interfacial Dislocation Structure*

R.W. Balluffi and G.B. Olson
Massachusetts Institute of Technology

Abstract

Many different types of dislocations have been defined in dislocation models for grain boundaries and interphase boundaries. It is emphasized that there is no unique dislocation model for a boundary, and that the formal dislocation content depends upon the choice of the lattice correspondence relating the adjoining lattices. However, it is concluded that no problems of real physical significance arise from this lack of uniqueness. "Best," or most useful, descriptions often exist, and these are discussed. A hierarchy consisting of four different types of interfacial dislocations may be distinguished, which is useful in describing the dislocation content of interfaces. These entities are termed: (1) primary interfacial dislocations; (2) secondary interfacial dislocations; (3) coherency interfacial dislocations; and (4) translational interfacial dislocations. While there may be a lack of agreement on terminology in the literature, it is believed that these dislocation types are distinguishable and play unique roles in useful dislocation models for interfaces. Detailed descriptions of these dislocation types are given, and actual examples in real interfaces are presented. It is concluded that dislocation descriptions of interface structures become of purely formal significance in the limit of fully incoherent interfaces since the cores are then delocalized. The utility of various dislocation descriptions therefore depends on the degree to which various types of local coherency exist.

Introduction

We may regard all interfaces in crystalline solids, that is, grain boundaries and interphase boundaries, as being composed of arrays of dislocations (Ref 1–3). However, it has been clearly recognized (Ref 3–5) that the dislocation content of a given interface cannot be specified uniquely, since, as we shall explicitly demonstrate below, the transformation producing the change in crystal structure or orientation associated with the interface cannot be uniquely described. In addition, as we shall also demonstrate, various reference states may be employed to describe the dislocation contents of interfaces. Consequently, many different (or apparently different) types of dislocations and dislocation terminologies have been proposed for dealing with the dislocation structure of grain boundaries and interphase boundaries. These have included, for example:

- Interfacial (grain boundary or interphase boundary) dislocations (Ref 3)
- Lattice dislocations (Ref 1)

*R.W. Balluffi and G.B. Olson, On the Hierarchy of Interfacial Dislocation Structure, *Metall. Mater. Trans. A*, Vol 16 A, April 1985, p 529–541

- Primary interfacial dislocations (Ref 3, 6)
- Surface dislocations (Ref 7)
- Secondary interfacial dislocations (Ref 3, 6)
- DSC-Lattice dislocations (Ref 8)
- Virtual interfacial dislocations (Ref 8, 9)
- Coherency interfacial dislocations (Ref 10)
- Anti-coherency interfacial dislocations (Ref 10)
- Misfit interfacial dislocations (Ref 11)
- Transformation dislocations (Ref 12)
- Partial interfacial dislocations (Ref 13)

The purpose of the present paper is to review a number of topics of current interest concerning dislocation descriptions of crystalline interfaces and to demonstrate a relatively simple hierarchy consisting of four types of interfacial dislocations, which we shall term:

- Primary interfacial dislocations
- Secondary interfacial dislocations
- Coherency interfacial dislocations
- Translational interfacial dislocations

These four dislocation types are sufficient to represent all interfacial dislocation structures. In carrying out this exercise it is recognized that there will probably never be general agreement as to the best nomenclature which should be used to describe these entities. Nevertheless, these entities are distinguishable types of dislocations which play unique roles in dislocation models for interfaces. The nomenclature which we have selected merely represents a compromise between descriptive power and brevity. No attempt is made to carry out a systematic review of the history of this subject. Instead, we simply proceed with the descriptions of the four types of interfacial dislocations listed above. Efforts are made throughout to discuss briefly the roles which these dislocations play in determining the structure of the interfaces of interest.

Grain Boundaries

Primary Interfacial Dislocations

Any grain boundary may be produced by an operation in which an appropriate set of lattice dislocations is collected into a surface array. The resulting array is then taken as the "primary dislocation" content of the boundary. This is illustrated in Fig. 1 for a tilt

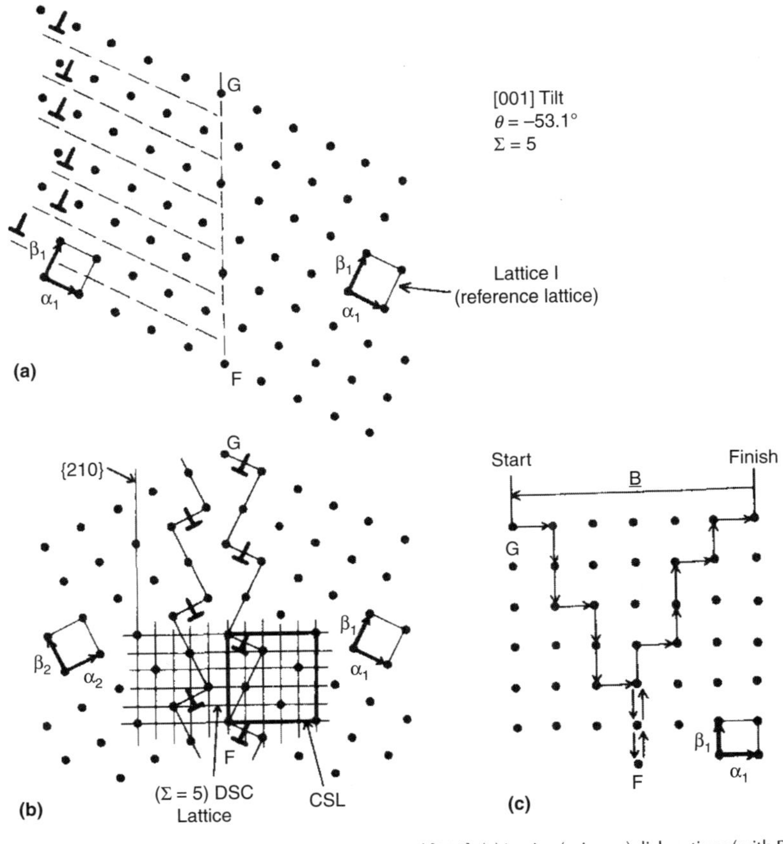

Fig. 1 Construction of $\Sigma = 5$ boundary by −53.1 deg rotation around [001]. (a) Lattice (primary) dislocations (with Burgers vector, $-\alpha_1$) introduced from left. (b) Final boundary configuration. (c) Burgers circuit, G → F → G in (b), mapped in the reference lattice.

boundary in the simple cubic structure. If, as shown in Fig. 1(a), edge dislocations with Burgers vector $-\alpha_1$ are introduced from the left on every successive plane of the reference lattice and collected in a planar array along GF, the result is the large angle boundary shown in Fig. 1(b). Here, the introduction of the lattice dislocations has caused the left side of the reference crystal to rotate by −53.1 deg around the [001] axis (which runs into the plane of the paper). The resulting configuration may be described as a symmetric [001] tilt boundary where the transformation between adjoining crystals corresponds to the $\theta = -53.1$ deg rotation which produced Lattice 2 from Lattice 1 (i.e., the reference crystal). The primary dislocation content of the boundary may be revealed explicitly by a Burgers circuit which closes in the bicrystal and runs from G to F in Lattice 1 and then back to G in Lattice 2. When this circuit is mapped in the reference crystal as shown in Fig. 1(c), a closure failure is obtained which is precisely the sum of the Burgers vectors of the six lattice dislocations arrayed in the boundary between G and F. (We note that our Burgers circuit and vector conventions follow those described in Ref 2.)

We next demonstrate that the identical atomic configuration (i.e., boundary) produced in Fig. 1 can be produced by means of a different array of lattice dislocations. First, as shown in Fig. 2(a), edge lattice dislocations with Burgers vectors α_1 are introduced from the left on every other plane to produce the array along GF illustrated in Fig. 2(b). Then, a second set with Burgers vectors β_1 is introduced from the left on every other plane to produce the final atomic configuration in Fig. 2(c) which is seen to be identical to that in Fig. 1(b). In this case the introduction of the dislocations has caused the left side of the reference crystal to rotate by 36.9 deg around [001]. The resulting boundary may now be described as a symmetric [001] tilt boundary where the transformation which produced Lattice 2 is a $\theta = 36.9$ deg rotation around [001]. When the Burgers circuit G → F → G is mapped in the reference crystal (Fig. 2d), the closure failure is $3\alpha_1 + 3\beta_1$ as expected.

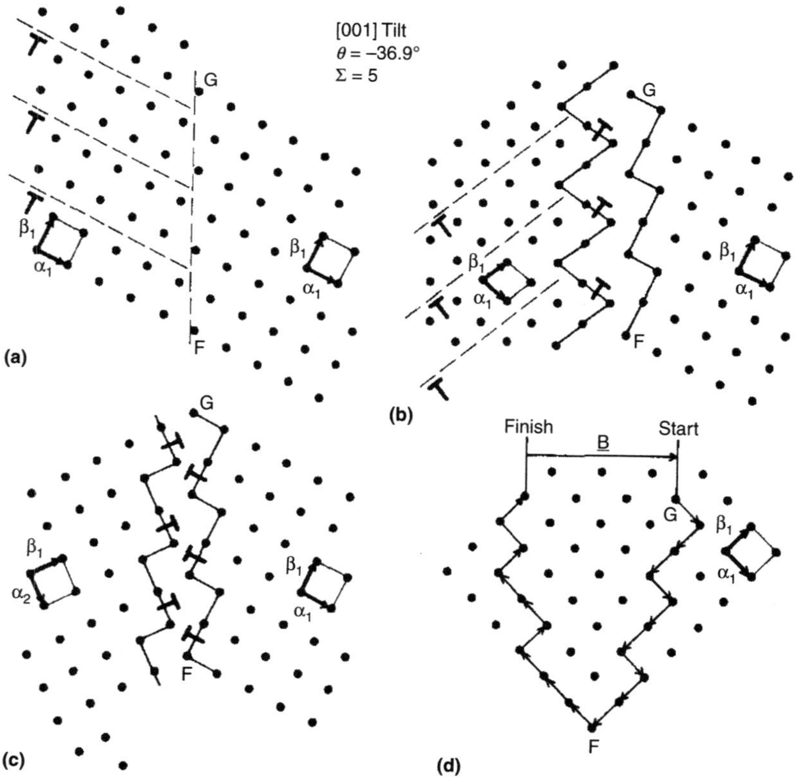

Fig. 2 Construction of $\Sigma = 5$ boundary by 36.9 deg rotation around [001]. (a) Lattice dislocations (with Burgers vector, α_1) introduced from left. (b) Lattice dislocations (with Burgers vector, β_1) introduced from left. (c) Final boundary configuration. (d) Burgers circuit, G → F → G in (c), mapped in the reference lattice.

It is readily shown that the boundary atomic configuration shown in Fig. 1(b) and 2(c) can be produced by still further operations. For example, if four lattice dislocations with Burgers vector β_1 are introduced on every successive plane as in Fig. 3(a), a boundary with the same atomic structure is again produced as shown in Fig. 3(b). In this case the primary dislocation content of the boundary is considerably higher than in the previous two examples, and the crystal rotation on the left side is correspondingly greater, that is, 126.9 deg. The closure failure of the standard Burgers circuit now corresponds to 12 β_1. Finally, we point out that this particular boundary can also be produced by a simple shearing of the reference lattice in a direction parallel to the boundary plane as illustrated in Fig. 4. This can be accomplished by passing a partial dislocation with Burgers vector $1/5 \cdot [\alpha_1 - 2\beta_1]$ on every successive {210} plane as shown in Fig. 4(a). There is therefore no accumulation of lattice dislocations in the resulting boundary (Fig. 4b) which may now be described as a twin boundary where the transformation which produces Lattice 2 is a twinning shear along ⟨210⟩. The zero dislocation content in this description is confirmed by the Burgers circuit construction in the reference lattice (Fig. 4c) which exhibits zero closure failure.

The previous examples demonstrate that the primary (lattice) dislocation content of a given boundary cannot, in general, be specified uniquely. In the examples presented it was shown that precisely the same boundary could be produced by the introduction of different arrays of dislocations which in turn produced different transformations between Lattices 2 and 1 adjoining the boundary. The result that these different transformations produced the same boundary structure is due to the existence of symmetry elements in the simple cubic lattice such as the 4-fold axis around [001] which causes rotations of −53.1 deg, 36.9 deg, and 126.9 deg to produce the same configuration. In addition, the mirror symmetry across {210} produced by the $1/5 \cdot [\alpha_1 - 2\beta_1]$ shear allows the sheared configuration to be described as a {210} twin which is devoid of lattice dislocation content.

As is well known (Ref 3), the result described above can be expressed mathematically by the relation:

$$B = \sum^{i} b_1 = [I - A^{-1}]p \qquad \text{(Eq 1)}$$

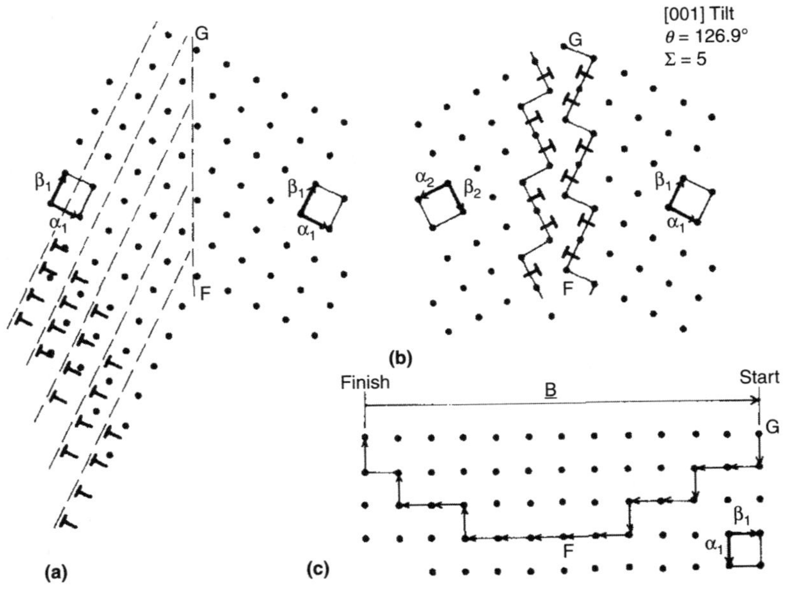

Fig. 3 Construction of Σ = 5 boundary by 126.9 deg rotation around [001]. (a) Lattice dislocations (with Burgers vector, β_1) introduced from left. (b) Final boundary configuration. (c) Burgers circuit, G → F → G in (b), mapped in the reference lattice.

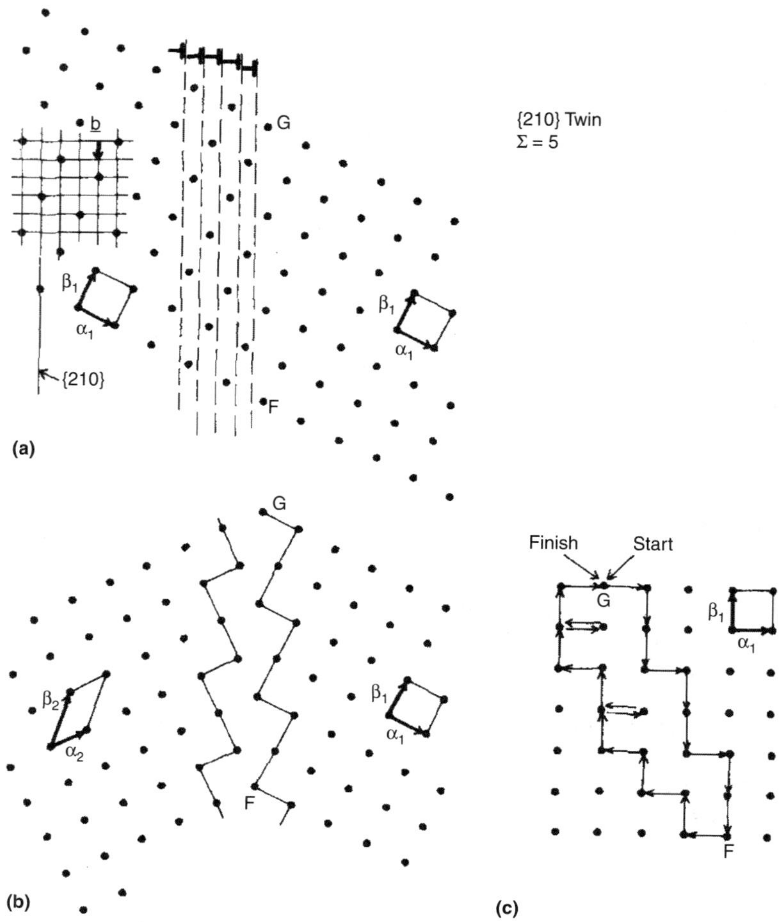

Fig. 4 Construction of $\Sigma = 5$ boundary (twin) by shearing on successive {210} planes by passage of partial dislocations with Burgers vector $b = 1/5[\alpha_1 - 2\beta_1]$ introduced at top/left. (b) Final boundary configuration. (c) Burgers circuit, G → F → G in (b), mapped in the reference lattice.

where B is the sum of the Burgers vectors of the dislocations in the boundary which are cut by the vector p which lies in the boundary. The matrix [A] represents the affine transformation which produces Lattice 2 from Lattice 1 (taken as the reference lattice). Equation 1 establishes the relationship between the Burgers vector content of the boundary and the transformation, **A**. However, since there are usually many different transformations (or paths) which can be used to produce Lattice 2 from Lattice 1 in the same final configuration, there are correspondingly many Burgers vector contents, that is, primary dislocation contents, which can be used to produce the boundary (Ref 3–5).

At first glance, it may seem that this state of affairs should cause serious problems in describing the primary dislocation content of grain boundaries. However, this is not the case. For static boundaries *any* of the various possible descriptions can be used depending upon the nature of the problem at hand and the ease with which the problem can be analyzed through use of a particular description. The physical results obtained through use of any chosen description should be identical to those obtained through use of any other description as long as each analysis is carried out consistently. As an example, consider the determination of the strain (or stress) field in the vicinity of the $\Sigma = 145$ [001] tilt boundary in Fig. 5 which, according to our previous discussion, can be considered to possess either a $\theta = -9.5$ deg or $\theta = 170.5$ deg tilt angle. The corresponding primary dislocation contents would then be those shown in Fig. 5(a) and 5(b), respectively. Presumably the strain (or stress) field could be calculated (via a dislocation model) using either dislocation representation and the result would be the same. However, the calculation of the strain/stress field due to the simple dislocation configuration in Fig. 5 (a) is obviously far simpler than that due to the complex configuration in Fig. 5(b). In addition, the image of the boundary in Fig. 5, as obtained by electron microscope diffraction strain contrast, would consist of a family of straight lines corresponding to the dislocation line structure in Fig. 5(a) rather than Fig. 5(b) even though the strain fields associated with the representation are, of course, identical. (In this respect see Fig. 8, below.) It is therefore more convenient to consider the primary dislocation content to correspond to that of the relatively simple array in Fig. 5(a) and this may therefore be regarded as the preferred description. As might be expected, the preferred primary dislocation description in the general case will usually correspond to the simplest dislocation configuration of largest spacing. Such a description is often obtained by taking the transformation, **A**, to correspond to the "disorientation," that is, the rotation angle/rotation axis pair chosen to minimize the rotation angle (Ref 14). In general, we may reach the important conclusion that no problems of real physical significance appear to arise due to the formal lack of uniqueness in the description of the static equilibrated dislocation structure.

In the case of a migrating boundary where the static configuration is perturbed by an applied force it may be possible to make an unequivocal choice of the transformation **A** (and, therefore, of the dislocation representation) on the basis of the nonequilibrium physical behavior of the boundary. Consider, for example, the Σ = 5 [001] tilt boundary illustrated in Fig. 6. Conceivably, the boundary could migrate by the glissile movement of its primary dislocations by either of the two

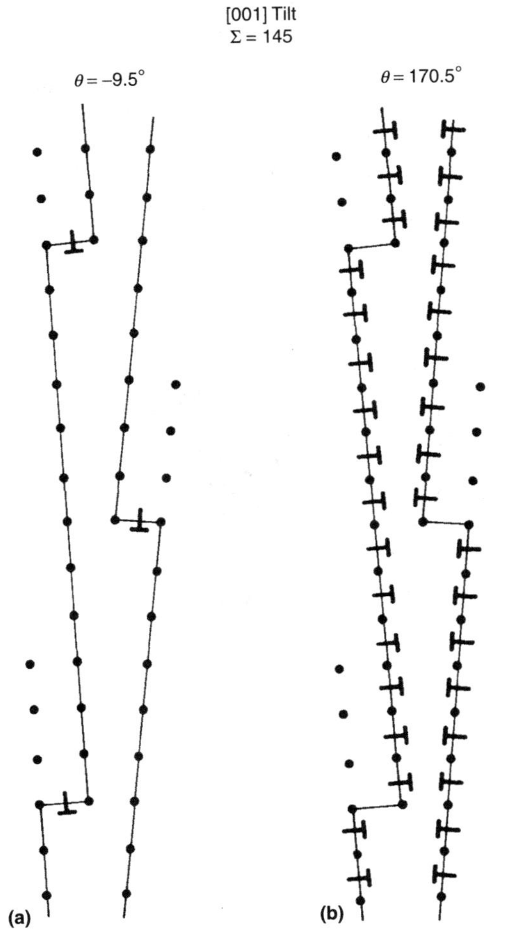

Fig. 5 Σ = 145 boundary. (a) Boundary, with primary dislocation content indicated, considered to possess a θ = −9.5 deg tilt angle. (b) Boundary with primary dislocation content again indicated, considered to possess a θ = 170.5 deg tilt angle.

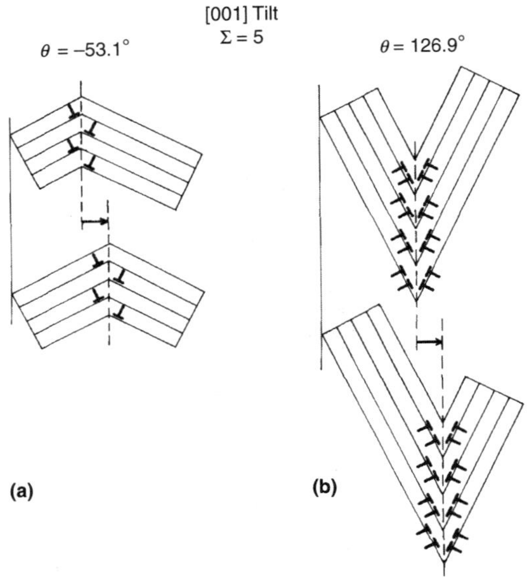

Fig. 6 Change of bicrystal shape due to glissile motion of primary dislocations considered to make up a Σ = 5 tilt boundary. (a) Boundary considered as a −53.1 deg tilt boundary. (b) Boundary considered as a 126.9 deg tilt boundary.

modes illustrated in Fig. 6(a) and 6(b) which correspond to the motion of different primary dislocation representations. In each representation the boundary motion would occur by the coordinated "military" motions of atoms in the dislocation cores in a manner consistent with the glissile motion of the dislocations involved. Different sets of atomic motions would be required for the glissile motion of each representation and, as seen in Fig. 6, quite different changes in the external shape of the bicrystal would then occur. Observation of the change in bicrystal shape could therefore be used to identify the operative **A** in cases where the boundary migration actually occurred by the military motions of atoms corresponding to the glissile motion of a particular dislocation array. On simple physical grounds we would expect such glissile motion to correspond to the simplest dislocation description of lowest dislocation density since that motion would take place with the simplest displacements. For example, the configurations in Fig. 6(a) and 5(a) would be more likely candidates than their counterparts in Fig. 6(b) and 5(b), respectively. This conclusion is confirmed by the observations of Washburn and Parker (Ref 15, 16) who could account for the glissile motion of a simple tilt boundary in zinc in terms of a simple dislocation representation of the type illustrated in Fig. 5(a) rather than 5(b).

While the examples considered thus far involve planar interfaces, the same principles can be applied to curved interfaces, including interfaces of enclosed particles. It is easily demonstrated that the primary dislocation content of an occluded grain surrounded by a grain boundary can be obtained by an elaboration of the dislocation production method used to made the previous planar boundaries. As illustrated in Fig. 7, such a configuration containing a $\Sigma = 5$ boundary can be produced by first generating lattice dislocation dipoles in the interior of the intended occluded grain region (Fig. 7a) and then expanding them on {210} planes. If this is done on appropriate successive planes (by combined glide and climb), the configuration in Fig. 7(b) is achieved, in which a large amount of shear distortion is present. If a second set of dislocation dipoles is now expanded in the transverse direction (Fig. 7b), the final configuration illustrated in Fig. 7(c) is

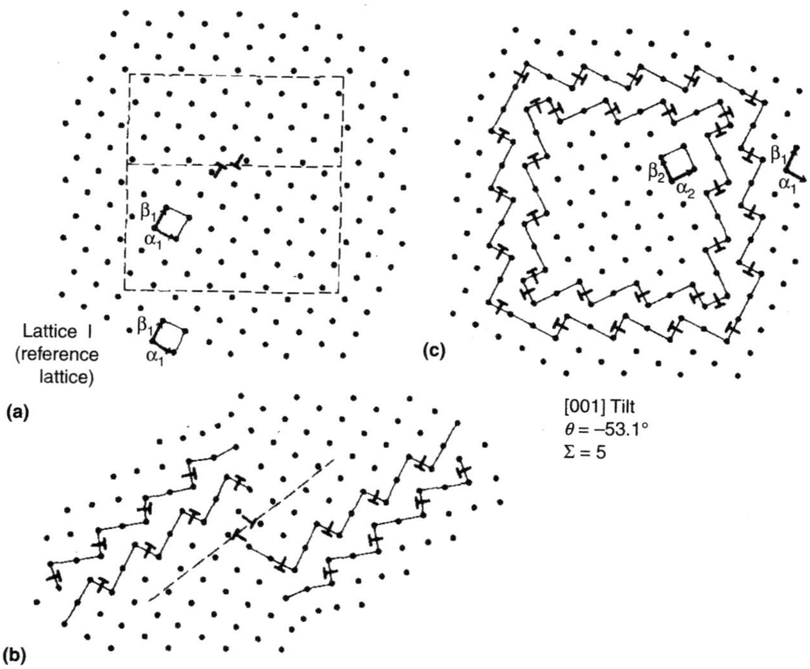

Fig. 7 Construction of an occluded grain surrounded by a $\Sigma = 5$ grain boundary. (a) Lattice dislocation dipoles (with Burgers vector, $-\alpha_1$) generated on {210} planes. (b) Lattice dislocation dipoles (with Burgers vector, $-\beta_1$) generated on second set of {210} planes. (c) Final configuration.

achieved. In this configuration the shear distortion has been eliminated, and the interior region has been rotated to produce an occluded grain surrounded by a $\Sigma = 5$ symmetric tilt boundary. It should be emphasized that, whether or not the interface exhibits faceting, the entire interface of an enclosed particle should be considered one interface, so that a self-consistent description requires a single **A** for the entire particle.

Finally, in Fig. 8 we present (from Ref 16) an electron micrograph of a real symmetric [001] tilt boundary which shows the strain field of its primary dislocation content as imaged by means of diffraction contrast. The boundary can be described as a $\theta = -5.3$ deg tilt boundary or also as, for example, a $\theta = 174.7$ deg tilt boundary. In the former case the primary dislocation representation consists of a simple stack of well-spaced edge dislocations similar to that illustrated in Fig. 5(a). The image shows a family of contrast lines of this geometry with a spacing corresponding to the spacing of the primary dislocations expected in the $\theta = -5.3$ deg tilt representation. For this reason, as well as those pointed out previously, we take this as the preferred representation for this boundary.

Secondary Interfacial Dislocations

When the misorientation angles in all descriptions of a given boundary are relatively large (i.e., $\gtrsim 15$ deg), the spacings between the primary dislocations in the various descriptions become sufficiently small that their cores always overlap. For many purposes such primary dislocation descriptions lose much of their usefulness. It is then often profitable to adopt a "secondary dislocation" model which is based on perturbations in the spacings of the primary dislocations (Ref 1–3, 5, 6, 8, 18). Since the perturbations are more widely spaced than the primaries themselves, such secondaries are more discrete and distinct. However, they are weaker, possessing smaller Burgers vectors. This may be illustrated simply by considering the structures of the $\Sigma = 5$, 17, and 233 boundaries in the series of symmetric [001] tilt boundaries possessing the progressively decreasing tilt angles displayed in Fig. 9. The $\Sigma = 5$ boundary has already been presented in Fig. 1(b) and is seen to possess a uniform array of primary dislocations. The $\Sigma = 17$ boundary (constructed by the same dislocation production method as the $\Sigma = 5$ boundary) is shown in Fig. 10(a) and is seen to possess an array of primary dislocations which exhibit periodic perturbations in spacing at locations indicated by the arrows (at the right). In Fig. 10(b) it is seen that each perturbed region in Fig. 10 (a) is the locus where two extra {210} planes terminate in the boundary. Furthermore, the regions between these perturbations closely resemble patches of the perfect $\Sigma = 5$ boundary in Fig. 1(b). These perturbations may therefore be regarded as the cores of secondary edge dislocations which are embedded in the $\Sigma = 5$ interface. The Burgers vectors correspond to two {210} interplanar spacings. The array of secondary dislocations therefore acts as a low angle tilt boundary superimposed on the $\Sigma = 5$ reference boundary which changes the tilt angle from -53.1 deg to -61.9 deg while preserving patches of the reference boundary. It may be noted that this array acts in the same manner as the array of primary edge dislocations in the low angle tilt boundary in Fig. 5(a) which changed the tilt angle from 0 deg to -9.5 deg while preserving patches

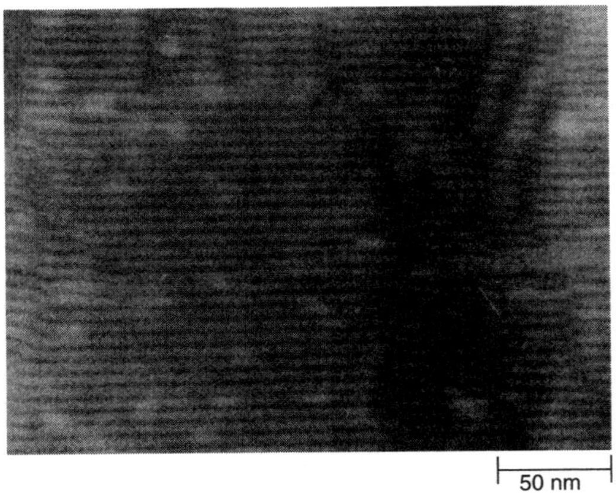

Fig. 8 Array of primary edge dislocations (family of parallel horizontal lines) in symmetric [001] tilt boundary in aluminum. Boundary viewed at normal incidence. Tilt angle is $\theta = -5.3$ deg measured between [010] directions. From Ref 17.

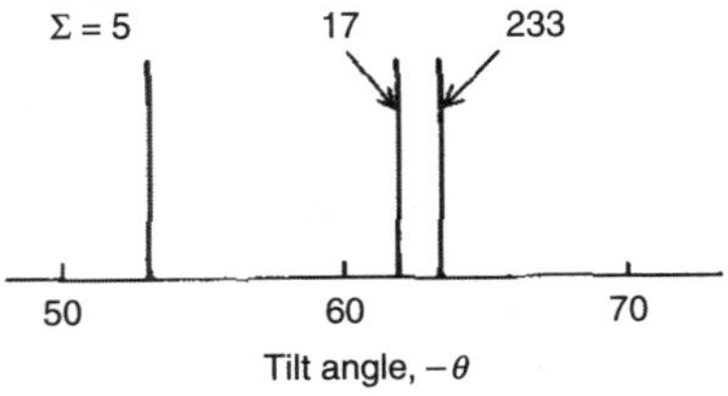

Fig. 9 Diagram displaying tilt angles of $\Sigma = 5$, $\Sigma = 17$, and $\Sigma = 233$ [001] tilt boundaries.

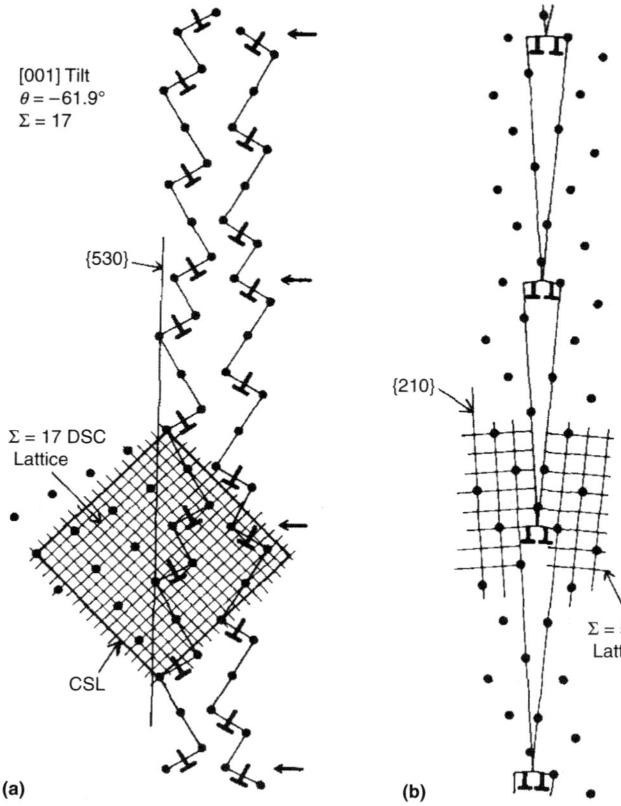

Fig. 10 Primary and secondary dislocation models for Σ = 17 [001] tilt boundary. (a) Primary (lattice) edge dislocation model. (b) Secondary edge dislocation model. Burgers vector of secondary dislocations is vector of Σ = 5 DSC Lattice shown.

of the reference boundary which in that case was the perfect lattice.

As is well known (Ref 3, 5, 6, 18), all of this structural detail can be neatly expressed within the frameworks of the Coincidence Site Lattice and DSC-Lattice. Only a very brief description of this representation will be given here. If the two crystal lattices adjoining the Σ = 5 boundary are extended throughout all space, they form the so-called dichromatic pattern in three dimensions (Ref 19). The three-dimensional array of points in space which have identical atomistic surroundings then forms the Coincidence Site Lattice (CSL). The CSL therefore delineates the periodicities of the dichromatic pattern and hence the periodicity of any planar grain boundary which may be constructed from the two crystal lattices making up the dichromatic pattern. The parameter Σ is defined as the number of Lattice 1 (or Lattice 2) sites per CSL unit cell, and, hence, is a measure of the relative size of the CSL unit cell. The DSC-Lattice is the lattice made up of all translation vectors between Lattices 1 and 2 which preserve the dichromatic pattern (Ref 6) (For the present simple cubic structure the DSC Lattice is the reciprocal lattice of the CSL.) Hence, any vector of the DSC-Lattice can be used as the Burgers vector of a perfect boundary dislocation, that is, a dislocation which lies in the boundary under conditions where the boundary structure on either side of it is the same.

Returning now to Fig. 1(b) and 10, we note in Fig. 1(b) the CSL and DSC-Lattice formed by Lattices 1 and 2. The Σ = 5 boundary is seen to be relatively short period boundary running parallel to a cell face of the CSL. Similar constructions are presented for the longer period Σ = 17 boundary in Fig. 10(a). In Fig. 10(b) the Σ = 5 DSC Lattice is superimposed on the two crystal lattices adjoining the Σ = 17 boundary, and it is evident that the Burgers vector of the secondary edge dislocations is a vector of the Σ = 5 DSC Lattice.

Furthermore, in the DSC Lattice construction of Fig. 10(b) the secondary edge dislocations clearly appear as simple edge dislocations. It is easily shown (Ref 20) that the secondary dislocation content of the boundary in Fig. 10(b) can be revealed formally by carrying out a Burgers circuit construction using the $\Sigma = 5$ DSC Lattice as the reference lattice. This reinforces the concept that the $\Sigma = 17$ boundary may be regarded as a $\Sigma = 5$ boundary containing a superimposed array of $\Sigma = 5$ DSC Lattice secondary dislocations.

Consider next the relatively long period $\Sigma = 233$ tilt boundary ($\theta = 63.2$ deg) shown in Fig. 11. Perturbations in the spacings of the primary dislocations (indicated by the arrows at the left) are again present. However, the spacings of these perturbations are no longer uniform, as in the $\Sigma = 17$ boundary, but are themselves perturbed. If we regard each perturbation in the primary array as a $\Sigma = 5$ DSC Lattice secondary dislocation (as was the case for the $\Sigma = 17$ boundary), we obtain the construction shown in Fig. 11(b) where each perturbation appears as a secondary dislocation in the $\Sigma = 5$ DSC-Lattice. The boundary is therefore considered as a $\Sigma = 5$ boundary containing a superimposed array of $\Sigma = 5$ DSC Lattice secondary edge dislocations which supplies the tilt component necessary to change the tilt angle from the $\Sigma = 5$ value of -53.1 deg to -63.2 deg in order to produce the $\Sigma = 233$ boundary. However, as we have already noted, the spacings of the secondary dislocations (primary dislocation perturbations) in Fig. 11(b) themselves exhibit periodic perturbations, and we may therefore regard these latter perturbations as a set of even weaker more widely spaced secondary dislocations as illustrated in Fig. 11(c). Here, we regard the regions where the perturbations in the spacings of the array of perturbations in the spacings of the primary dislocation array contribute excess Burgers vector strength [see arrows on left in Fig. 11(b)] as the cores of these weaker secondary dislocations. It is seen that these are the loci where extra {530} planes terminate in the boundary. The $\Sigma = 17$ DSC Lattice construction shown in Fig. 11(c) shows that the Burgers vectors of these relatively weak secondary dislocations are vectors of the $\Sigma = 17$ DSC Lattice. This result could be verified formally by means of a Burgers circuit construction using the $\Sigma = 17$ DSC-Lattice as the reference lattice. In the representation shown in Fig. 11(c) the $\Sigma = 233$ boundary is therefore considered as a $\Sigma = 17$ boundary containing a superimposed array of $\Sigma = 17$ DSC Lattice secondary edge dislocations which supply the tilt component necessary to change the tilt angle from the $\Sigma = 17$ value of -61.9 deg to -63.2 deg in order to produce the $\Sigma = 233$ boundary. This example of the $\Sigma = 233$ boundary offers a clear demonstration of the hierarchial nature of secondary dislocation descriptions.

The different dislocation descriptions of, for example, the $\Sigma = 233$ boundary may all be useful in certain contexts. Consider the observation of this boundary by the diffraction contrast method in the electron microscope. The dislocation line structure observed in a

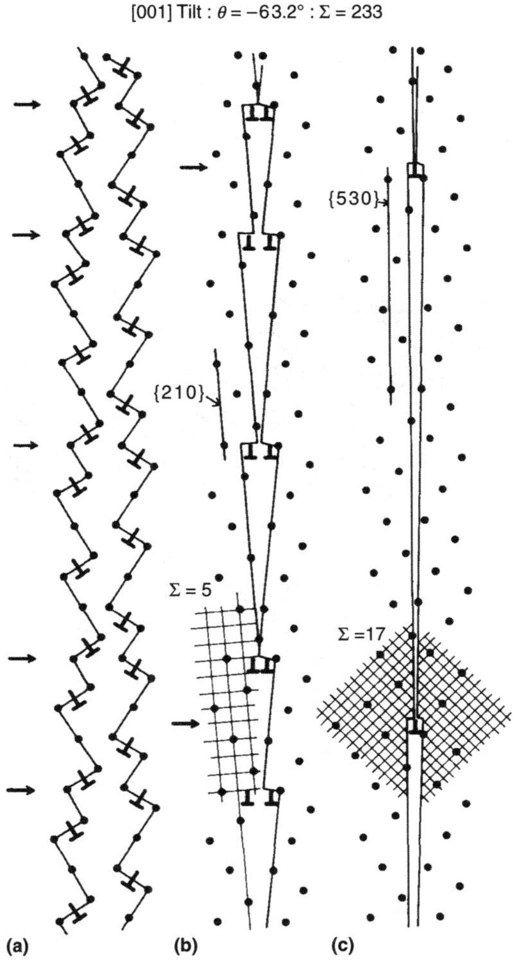

Fig. 11 Primary and secondary dislocation models for $\Sigma = 233$ [001] tilt boundary. (a) Primary (lattice) edge dislocation model. (b) Secondary edge dislocation model. Burgers vector is vector of $\Sigma = 5$ DSC Lattice shown. (c) Further secondary edge dislocation model. Burgers vector is vector of $\Sigma = 17$ DSC Lattice shown.

high-resolution, high-contrast instrument might correspond to the primary dislocation array. On the other hand, a lower resolution instrument might be able to resolve only the far strain field of the boundary corresponding to the $\Sigma = 17$ DSC Lattice secondary dislocation model. The intermediate case where the secondary dislocations in the $\Sigma = 5$ DSC Lattice secondary dislocation array would be visible but the primary array would remain invisible is also possible. Figure 12(a) (from Ref 21) shows an unusual electron micrograph where the strain fields of both primary and secondary interfacial dislocations are apparently imaged simultaneously. Figure 12(b) (from Ref 17) shows a situation where a secondary array is imaged but the primary array remains invisible.

Finally, we mention that the multiplicity of primary dislocation descriptions discussed in the section "Primary Interfacial Dislocations" in this chapter does not play a role in the selection of secondary dislocation models. The symmetry operations described earlier correspond to the superposition of arrays of uniformly spaced primary dislocations and, hence, do not produce any new perturbations in the primary dislocation spacings.

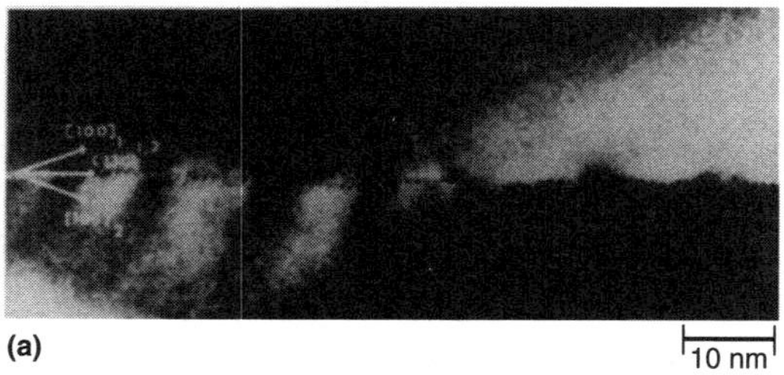

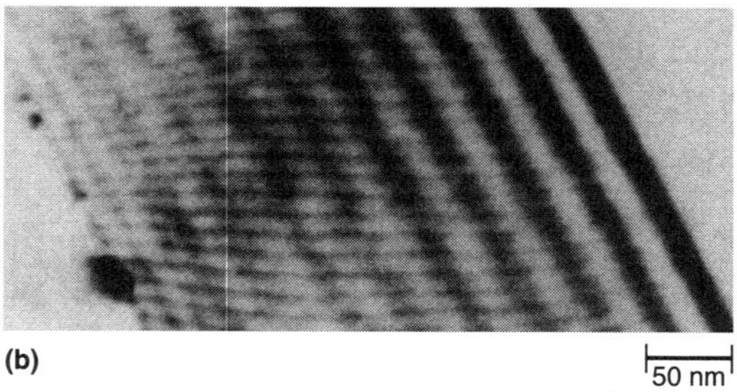

Fig. 12 (a) Electron microscope diffraction contrast image of [001] tilt boundary in gold with misorientation near the $\Sigma = 5$ misorientation. Grain boundary dislocations viewed end-on. Spacing of primary edge dislocations which produce the near displacement field is $\simeq$0.6 nm; spacing of secondary edge dislocations, which account for the deviation from the $\Sigma = 5$ misorientation and produce the farther field, $\simeq$7 nm. From Ref 21. (b) Diffraction contrast image of secondary edge dislocations (family of horizontal lines) in symmetric [001] tilt boundary in aluminum with misorientation near the $\Sigma = 5$ misorientation. Boundary viewed at oblique incidence. Burgers vectors are vectors of the $\Sigma = 5$ DSC Lattice. Deviation from exact $\Sigma = 5$ coincidence is 0.80 deg. From Ref 17.

Translational Interfacial Dislocations

Crystals 1 and 2, adjoining a grain boundary, generally tend to adopt a specific rigid body translational position with respect to each other, in order to allow the core region to reach a minimum energy configuration (Ref 22). There is therefore a variety of situations in which two regions of an interface meet along a line under conditions where the boundary structures in the two regions differ and the translational positions of the adjoining lattices therefore also differ. Such a line possesses dislocation character with a Burgers vector equal to the vector difference between the rigid body translations characterizing the two regions. We term such a partial interfacial dislocation a "translational interfacial dislocation" since it separates two regions of an interface which are in different translational states.

Such dislocations have been observed (Ref 13, 22–24) at the intersections of facets in faceted boundaries, since the rigid body translations for the individual facets generally differ. Translational interfacial dislocations may also occur in flat boundaries between domains which differ in structure (Ref 13, 22–24). The differences in structure may be due to the presence of entirely different atomic configurations or else configurations which are locally the same but lie in different symmetry related arrangements. In the latter case the domains are, of course, energetically degenerate.

An example of a conceivable translational interfacial screw dislocation in a flat boundary is illustrated in Fig. 13 for a [001] $\Sigma = 5$ twist boundary in the fcc structure. Calculations by Bristowe and Crocker (Ref 24) indicate that such a boundary in copper can exist in three different translational states corresponding to three different atomistic boundary structures of almost the same energy. The screw dislocation in Fig. 13 separates two domains which are in two of these different translational states and therefore possess entirely different structures. Its Burgers vector is given by half of one of the primitive vectors of the $\Sigma = 5$ DSC Lattice (see figure caption for details). Finally, an example (from

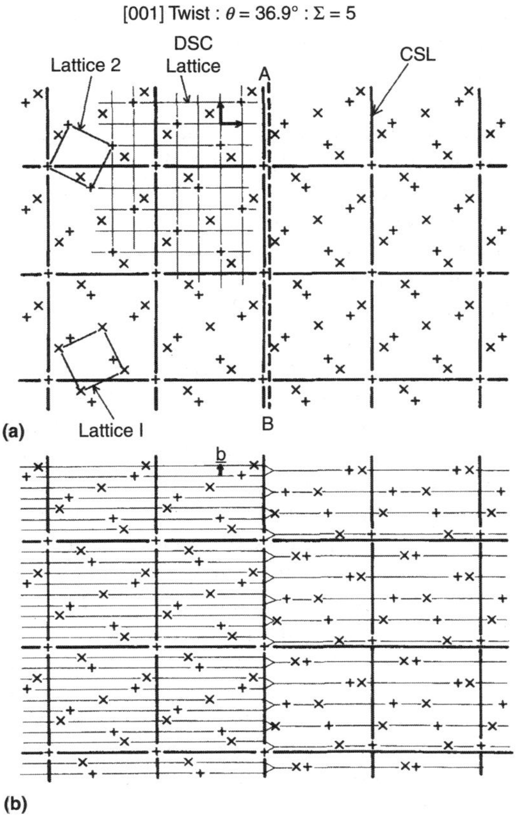

Fig. 13 Introduction of translational interfacial screw dislocation into $\Sigma = 5$ [001] twist boundary in fcc structure. Structures viewed along [001]: first planes of Lattices 1 and 2 adjoining boundary midplane indicated by + and × symbols, respectively. (a) Boundary before introduction of dislocation. (b) Boundary containing translational screw dislocation. Dislocation introduced by making cut at the boundary midplane in (a) from right side up to line AB and then translating Lattice 1 with respect to 2 by b, which is just one-half of a primitive vector of the $\Sigma = 5$ DSC Lattice.

Ref 13) of the strain field image of an actual translational interfacial dislocation separating two energetically degenerate domains in a flat $\Sigma = 3$ $(1\bar{2}1)$ symmetrical tilt boundary in aluminum is shown in Fig. 14.

Interphase Boundaries

Coherency Interfacial Dislocations

The coherency of crystalline interfaces is defined by the continuity of corresponding lattice planes and directions across the interface (Ref 3). The correspondence of lattice vectors which underlies the notion of coherency is in turn defined by the lattice deformation (i.e., the

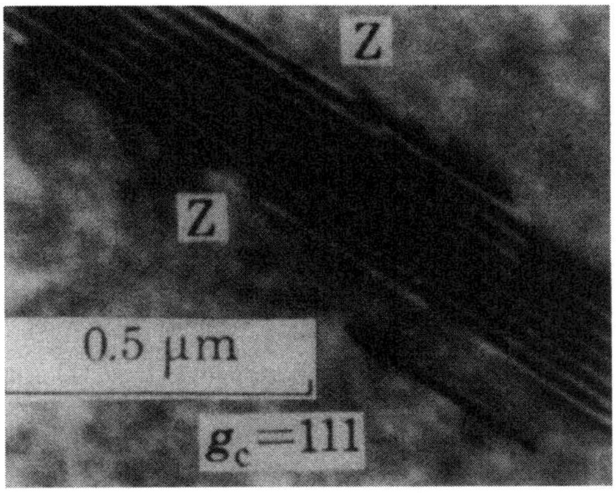

Fig. 14 Electron microscope strain contrast image of translational interfacial dislocation (at ZZ) separating two energetically degenerate domains in a flat $\Sigma = 3$ $(1\bar{2}1)$ symmetric tilt boundary in aluminum. Boundary viewed at oblique incidence. g_c = diffraction vector. From Ref 13.

affine transformation **A** of Eq 1) which relates the two crystals. As the two crystal lattices adjoining an interphase boundary are generally incommensurate, lattice continuity across the boundary generates long-range strain fields which can be represented in a discrete crystal in terms of arrays of interfacial dislocations.

Figure 15 illustrates a fully coherent particle with a lattice deformation **A** consisting of a simple shear on the z plane in the x direction. Figure 15(a) represents the ellipsoidal continuum particle treated by Eshelby (Ref 26) for which all displacements and tractions are continuous across its interface. In a discrete crystal this mechanical continuity arises from the structural continuity of the corresponding crystal planes and directions, and the deformation of the crystal can be represented by superimposing a grid on the continuum of Fig. 15(a), as depicted in Fig. 15(b). The simple shear lattice deformation leaves the crystal planes "a" undistorted but distorts the planes "b" in the manner indicated. Although all corresponding planes and directions are continuous across the interface of this fully coherent particle, the homogeneous lattice deformation in the discrete crystal amounts to a finite shear displacement per crystal plane of type "a" relative to the untransformed material outside; this shear defines the Burgers vectors of the interfacial partial dislocations depicted in Fig. 15(c). These discrete dislocations can be regarded as generating the elastic strain field of the particle (interface), and their motion produces the transformation lattice deformation. In view of their very special property of preserving lattice continuity, the term "coherency dislocation" has been proposed for this distinct class of interfacial dislocations (Ref 10). In the case of diffusionless phase transformations, they have also been termed "transformation dislocations" (Ref 12). Moving the dislocations to produce a faceted particle as in Fig. 15(d) helps distinguish two types of coherent interface: a "stress-free" coherent interface (I) and a "forced-elastic" coherent interface (II).

In order to produce a homogeneous lattice deformation, coherency dislocations must be homogeneously distributed in the sense that one coherency dislocation is required for each lattice plane transformed, and its motion is always restricted to that particular plane; for example, the dislocations of Fig. 15(c) are restricted to the successive "a" planes of Fig. (b). The actual spacing between the dislocations depends on the inclination of the interface with respect to the crystal planes with which they are associated. Coherency dislocations can always be formally associated with an interfacial step, and methods for predicting the more topographically significant coherency dislocation arrays in a given interphase boundary have been developed based on this step character (Ref 27). It is convenient to describe the steps in terms of a step vector d, normal to the lattice plane defining the step, and equal to the interplanar spacing in magnitude. The Burgers vector b of the coherency dislocation associated with such a step is then defined by the manner in which d is transformed by the lattice deformation (Ref 27). Fig. 16 illustrates such a step in a coherent interface. Adopting the median reference lattice shown at the right, which is transformed to Lattice 1 by the deformation $\mathbf{S}_1$ and to Lattice 2 by the deformation $\mathbf{S}_2$, an appropriate Burgers circuit construction (Ref 10) gives the Burgers vector

$$b = \left(\mathbf{S}_1^{-1} - \mathbf{S}_2^{-1}\right)d \qquad \text{(Eq 2)}$$

(When Lattice 1 is adopted as the reference lattice, transforming to Lattice 2 with the deformation **A**, Eq 2 becomes equivalent to Eq 1 with $d = p$.) As depicted in the reference lattice of Fig. 16, coherency dislocation Burgers vectors will generally possess glide and climb components with respect to the motion of the dislocation in its associated crystal plane normal to d. Another unusual property of these special dislocations is that their climb motion

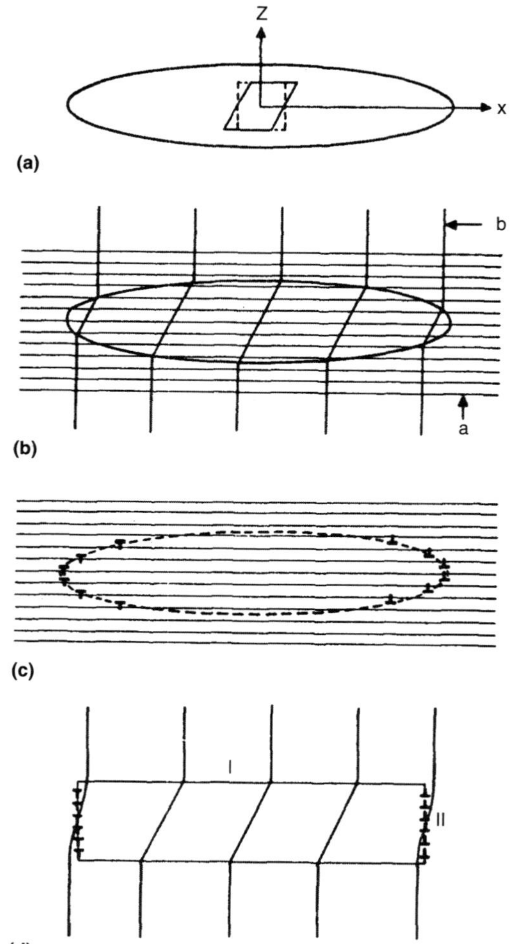

Fig. 15 Coherent shear particle. (a) Ellipsoidal continuum elastic particle. (b) Same particle in discrete lattice. (c) Positions of coherency dislocations in interface. (d) Squared-off particle to distinguish stress-free coherent interface I from forced-elastic coherent interface II. From Ref. 10.

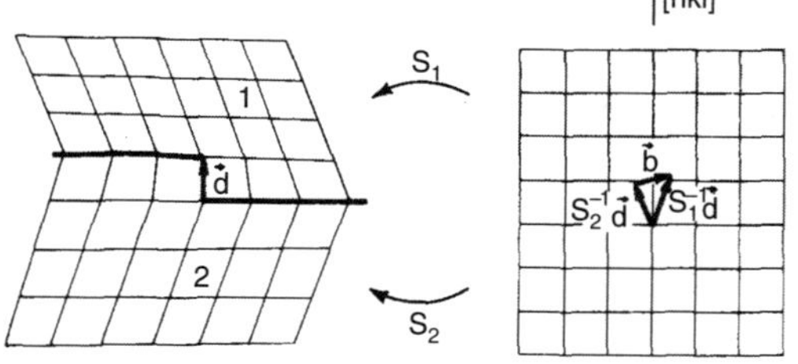

Fig. 16 Coherent interphase boundary with an interfacial step. From Ref 27. Associated Burgers vector of coherency dislocation at step is defined in median reference lattice at right.

is conservative with respect to matter (but not with respect to the volume of the continuum); the motion neither creates nor destroys lattice points, but simply changes an interplanar spacing. With these two Burgers vector components, a single array of coherency dislocations produces an invariant-plane strain (with the crystal plane of d invariant), and the superposition of several such arrays can describe any lattice deformation. In the general case, the arrays can consist of both resolvable and unresolvable components with respect to observation of discrete line defects in the electron microscope.

An observation (Ref 28) of a resolvable array of coherency dislocations in the interface of a tapered hcp particle produced by an fcc → hcp martensitic transformation in a Co-Fe alloy is shown in Fig. 17. The operative lattice deformation is in this case very close to an invariant-plane strain on a close-packed crystal plane, so that a single set of coherency dislocations can account for the major portion of the deformation. The particle is imaged in Fig. 17(a) using contrast from the total particle displacement, while the individual coherency dislocations associated with interfacial steps are imaged in Fig. 17(b) using dislocation strain-field contrast. Although the latter contrast arises primarily

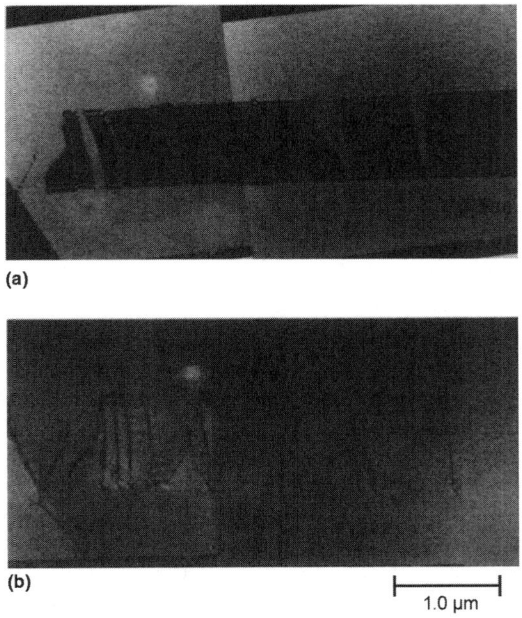

Fig. 17 Tapered hcp martensitic particle in fcc Co-Fe alloy viewed at oblique incidence. (a) Displacement contrast image of particle. (b) Strain contrast image of interfacial coherency dislocations (family of vertical lines). From Ref 28.

from the Burgers vector glide component, the existence of the climb component associated with an ~1% difference in interplanar spacing has been demonstrated for this transformation by a precise analysis of the fault displacement contrast associated with a single dislocation (Ref 29). An additional small unresolvable component of coherency dislocation content (requiring additional arrays) arises from small distortions in the close-packed plane.

The fcc → hcp transformation is a good example of the potential lack of uniqueness in specifying the relation between two crystal lattices. Due to the fcc crystal symmetry, an hcp crystal lying on a particular $(111)_{fcc}$ plane can be generated by numerous combinations of planar displacements in $\langle 11\bar{2}\rangle_{fcc}$ directions, which cannot be distinguished by observation of planar stress-free $(111)_{fcc}$ coherent interface. However, treating the particle as a whole, the strain field associated with interfacial coherency manifested in the coherency dislocation strain fields breaks the crystal symmetry. In a manner similar to the identification of a macroscopic displacement as depicted in Fig. 6, the identification of the local strain field (Burgers vectors) of the coherency dislocations in Fig. 17 defines a distinct and physically meaningful lattice correspondence for this interface.

Primary Interfacial Dislocations

The strain energy associated with the strain fields of a coherent interface can be reduced if the long-range displacements are canceled by another array of interfacial dislocations. Once a lattice deformation has produced the correct crystal structure in a transformed region, any further deformation which reduces strain energy must be lattice invariant, corresponding to the types of operations described the section "Primary Interfacial Dislocations" in this chapter which typically lead to the introduction of primary (lattice) interfacial dislocations.

The roles of coherency dislocations and primary interfacial dislocations in the formation of an enclosed semicoherent particle are depicted schematically in Fig. 18. The lattice deformation converting Lattice 1 to Lattice 2 according to the correspondence defined in Fig. 18(a) can be generated in the rectangular region of Lattice 1 outlined in Fig. 18(b) by the formation and motion of coherency dislocation dipoles (with glide and climb components) along crystal planes of the type

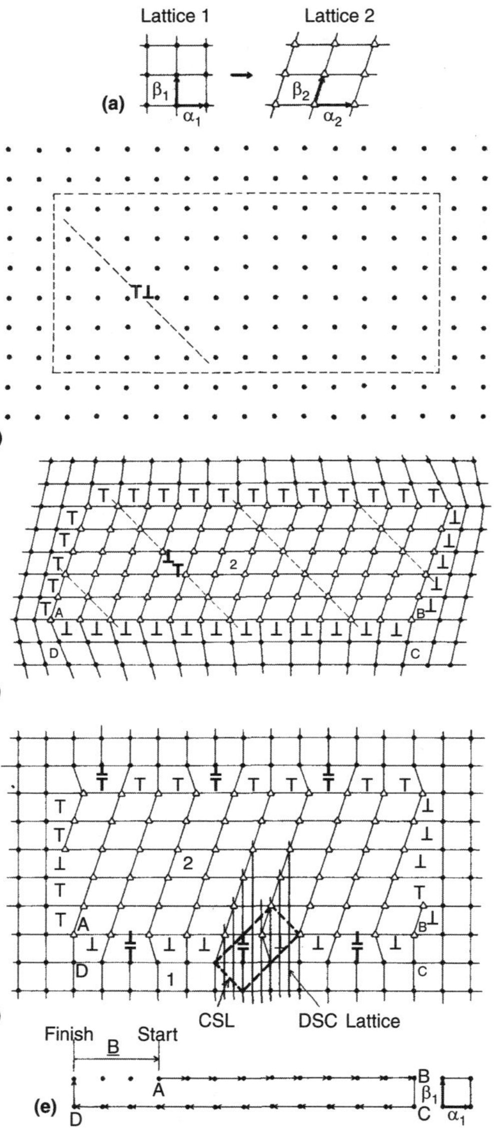

Fig. 18 Formation of occluded particle of Phase 2/Lattice 2 in a large matrix of Phase 1/Lattice 1 by a phase transformation. (a) Lattice transformation between Phases 1 and 2. (b) First step in transformation. Coherency dislocation dipoles are generated in interior of region to be transformed (dashed region) and are expanded (by combined glide and climb) on successive {110} planes (dashed) of Lattice 1. (c) Second step in transformation. Lattice (primary) dislocation dipoles are generated in Phase 2/Lattice 2 region and are expanded (by combined glide and climb) on dashed {110} planes. (d) Final configuration. Interphase boundary contains both coherency and primary dislocations. Also shown are CSL and DSC Lattice formed by Lattices 1 and 2. (e) Burgers circuit, A → B → C → D → A in (d), mapped in the reference lattice (i.e., Lattice 1) to reveal primary interfacial dislocation content.

indicated by the dashed line. This produces the fully coherent particle with an array of coherency interfacial dislocations as shown in Fig. 18(c). The elastic distortions associated with the interfacial coherency can be relaxed in this case by the introduction of a nonconservative lattice-invariant deformation produced by the glide and climb of a periodic array of lattice dislocations along the planes indicated by the dashed lines of Fig. 18(c). This then produces the semicoherent particle of Fig. 18(d) for which the long-range strain field of the coherency dislocation array is exactly canceled by an array of primary interfacial dislocations.

The Burgers vector strength of the primary dislocations in the boundary can be revealed formally by making a Burgers circuit A → B → C → D → A in Fig. 18(d). When this is mapped into Lattice 1 as a reference lattice in Fig. 18(e), the closure failure *B* is then just the sum of the Burgers vectors of the three primary dislocations shown. The net Burgers vector contents of the separate coherency and primary interfacial dislocation arrays are equal and opposite, although different types of Burgers circuits are required to reveal each set, as discussed in Ref 10. As depicted by the corresponding crystal planes drawn in Fig. 18(d), the primary dislocations act as lattice discontinuities and can be termed "anti-coherency dislocations" which reduce the degree of interfacial coherency (Ref 10). The interface is still semicoherent in the sense that local patches of coherency exist between the primary dislocations, giving an interfacial elastic strain field with a range comparable to the spacing of the primary dislocations. If the primary dislocations would dissociate to cancel completely the coherency dislocations, the interface would become incoherent, and the two dislocation arrays would then be of only formal significance. The degree of such dissociation, "spreading," or "delocalization" is controlled by the interplay between the interface core energy and the elastic strain field energy.

An observation of primary interfacial dislocations in a semicoherent fcc-bcc martensitic interface in an FeNiMn alloy is shown in Fig. 19 (Ref 30). Coherency dislocations associated with interfacial steps on $(111)_{fcc}$ are not resolved in this case. In this example the lattice-invariant deformation is conservative so that the primary dislocations can glide as the interface advances. Primary interfacial dislocation arrays indicative of a nonconservative lattice-invariant deformation requiring dislocation climb, as in Fig. 18(d), have been observed

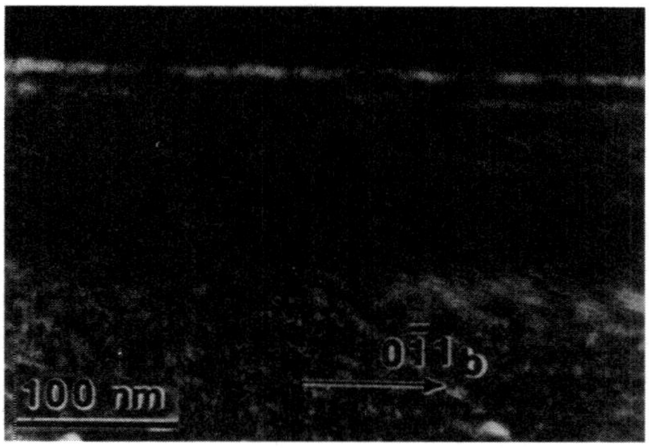

Fig. 19 Primary interfacial dislocations (family of finely spaced inclined lines) imaged by strain contrast in an fcc-bcc martensitic interface in an Fe-Ni-Mn alloy. Interface viewed at oblique incidence. From Ref 30.

for the fcc → bcc Widmanstatten ferrite transformation in steels (Ref 31).

Secondary Interfacial Dislocations

The example constructed in Fig. 18 represents a very special case in which the two lattices are sufficiently commensurate that the lattice deformation is exactly canceled by a "rational" lattice-invariant deformation for which the primary dislocations are uniformly spaced at an integer number of crystal planes. This special circumstance creates a CSL which contains 3 sites of Lattice 1 and 4 sites of Lattice 2 (i.e., $\Sigma = 3:4$). The CSL unit cell is indicated by the dashed rectangle in Fig. 18(d). If, as a more general example, we increase the lattice parameter α_2 of Lattice 2 [Fig. 18(a)] by about 5%, the Burgers vectors of the coherency dislocations of Fig. 18 are slightly increased, but the character of the coherency array is not altered. The primary dislocation array necessary to cancel long-range displacements is, however, significantly changed, and the horizontal interface AB of Fig. 19(d) could now appear as in Fig. 20(a); the primary dislocations are now closer together, on average, and are also spaced nonuniformly.

We now demonstrate that we can construct a secondary interfacial dislocation representation of this generalized "near-coincidence" interface by using the special commensurate $\Sigma = 3:4$ interface as a reference structure. It is easily shown (Ref 8, 32) that a DSC Lattice is generated by the two different Lattices 1 and 2 in the reference structure as illustrated in Fig. 18(d). This DSC Lattice embedded in Lattices 1 and 2 of the generalized configuration is shown in

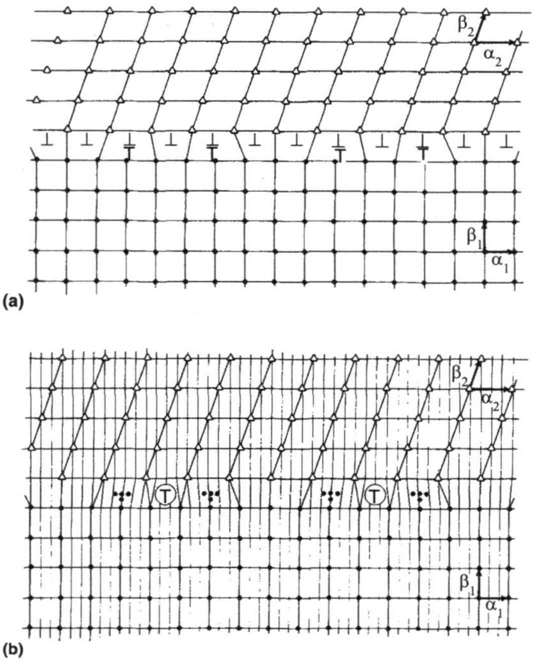

Fig. 20 (a) Final semicoherent interphase boundary produced when lattice parameter of Lattice 2 in Fig. 18 is increased slightly (i.e., by ≃5%). (b) Same as (a) except that the $\Sigma = 3:4$ DSC Lattice is shown embedded in Lattices 1 and 2. Secondary edge dislocations are shown encircled. Primary dislocations are shown dotted.

Fig. 20(b) and is seen to contain an array of secondary edge dislocations at the interface. The Burgers vector of these dislocations is, of course, a vector of the DSC Lattice. Also, these dislocations appear at perturbations in the spacings of the primary dislocations in regions where there is extra Burgers vector strength. Referred to the crystal lattice reference frame of Fig. 20(a), the DSC Lattice vectors can in fact be expressed as the superposition of a lattice vector and some multiple of coherency dislocation vectors. As in the grain boundary examples earlier, the array of secondary dislocations acts as an array of "correction" dislocations superimposed on the reference $\Sigma = 3:4$ boundary in order to produce the more general boundary. We note that the secondary dislocation structure of the boundary in Fig. 20(b) has been derived using somewhat different approaches in Ref 8. However, the results are equivalent.

An example of the diffraction contrast from an actual array of secondary interfacial dislocations in a $\Sigma = 8:10$ interphase boundary between copper and silver (from Ref 17) is shown in Fig. 21 under conditions where no primary dislocation structure is detectable.

The examples developed here illustrate the general interrelation between the reference states or lattice correspondences that underlie primary and secondary dislocation descriptions of interfacial structure in crystals. Each can usefully describe an aspect of an interfacial structure provided the nature of the reference state, or correspondence, being applied is clearly specified. The correspondence between primitive crystal lattice cells defining primary (and coherency) dislocation contents can be termed a "primary correspondence." As in the grain boundary case, a convenient "preferred" primary correspondence can often be specified by minimization of transformation principal strains and rotations, and this is generally found to represent the most physically realistic description when macroscopic displacements or local coherency strain fields can be determined experimentally. The primary, that is, $\Sigma = 1$, correspondence provides a useful framework for gauging the state of overall strain relaxation of a transformed particle which governs the particle-volume-dependent elastic strain energy. The CSL correspondence (with $\Sigma > 1$) defining the secondary dislocation content of an interface relates larger unit cells involving smaller transformation strains, and can be termed a "secondary" correspondence.

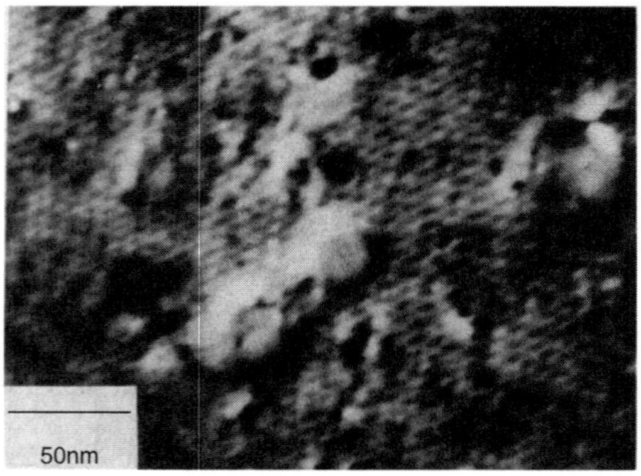

Fig. 21 Network of secondary interfacial dislocations (square array of nearly horizontal and vertical fine lines) in Ag/Cu interphase boundary between α Cu-rich and β Ag-rich phases. Interface viewed at normal incidence. Misorientation corresponds to small [001] twist deviation from special $\Sigma = 8_\beta/10_\alpha$ near-coincidence misorientation produced by 26.6 deg rotation around common [001]. Secondary dislocations possess Burgers vectors which are vectors of the $\Sigma = 8_\beta/10_\alpha$ DSC Lattice, and they serve to accommodate both the twist deviation and the small mismatch from exact coincidence. From Ref 17.

This provides a useful framework for treating the local pattern matching that governs the area-dependent interfacial energy.

Just as the dislocation description of an interface is a function of the chosen lattice correspondence, so is the state of interfacial coherency. Note that the interface of Fig. 18(d) is semicoherent with respect to the primary correspondence but fully coherent with respect to the secondary correspondence. It must be emphasized that any meaningful statement regarding dislocation structure or coherency of a crystalline interface requires that the correspondence or reference state applied be specified.

Final Remarks

All of the boundary structures presented in the present work were merely schematic, and no attempt was made to describe fully relaxed atomic structures of minimum energy which might be characteristic of boundaries in real materials. In actual practice there is an infinite number of detailed ways in which the atomic structure of the boundary core can relax. In general, these depend upon the detailed atomic forces which are present which, of course, vary from system to system, and upon thermodynamic variables such as, for example, temperature. The different atomistic ways in which the core can relax merely represent different ways in which the two crystals adjoining the boundary can be joined together. It is important to point out that these complications at the atomistic level do not affect the present general result that we can describe the dislocation content of boundaries in terms of interface dislocations belonging to the relatively simple hierarchy which we have described.

As discussed in the subsection "Primary Interfacial Dislocations" in the previous section the dislocation description of interfacial structure becomes of purely formal significance in the limit of the fully incoherent interface in which all dislocation cores are delocalized. In such a case the long-range dislocation displacements still account for the transformation relating the two adjoining crystals. However, since the dislocation content is defined by the transformation, this relation is tautological. The utility of the dislocation structural description, therefore, depends on the extent to which cores are localized and local interfacial coherency generates elastic stress fields extending into the crystals adjoining the interface. The aspect of dislocations which is most amenable to quantitative treatment is the linear elastic stress field, and interface dislocation models therefore provide a basis for the prediction of the linear elastic portion of interfacial strain fields, which is the actual feature observed in the electron micrographs presented here. Though the energetic aspects of the nonlinear elastic core region of dislocations in general are currently poorly defined, atomistic models of dislocation cores in terms of structural units (Ref 33–35) can provide a useful description of interfacial dislocation core structures, even in cases where the dislocations are closely spaced. In this context, both long- and short-range aspects of the dislocation paradigm can contribute usefully to our understanding of the structure of crystalline interfaces.

We further note in closing that interfacial dislocations with localized cores have also proved useful in modeling interfacial phenomena such as, for example, source and sink action for point defects (Ref 36) and boundary sliding (Ref 37). Dislocation models of martensitic interfaces have also allowed prediction of observed interfacial mobility in thermoelastic martensitic transformations (Ref 38, 39). Further discussion of these topics is beyond the scope of the present paper.

ACKNOWLEDGMENTS

Helpful discussions were held with Dr. P. D. Bristowe, Dr. J. W. Cahn, Professor M. Cohen, Professor G. Kalonji, and Dr. P. G. McDougall. R. W. B. was supported by the United States Department of Energy under Contract DE-AC02-78ER05002. G. B. O. was supported by the National Science Foundation under Grant DMR-79-15196. This manuscript was presented at the Cyril S. Smith 80th Birthday Symposium on Structure held at the TMS Fall Meeting 1983 on Tuesday, Oct 4, Philadelphia, PA.

REFERENCES

1. W.T. Read and W. Shockley: *Phys. Rev.*, 1950, vol. 78, p 275–289
2. J.P. Hirth and J. Lothe: *Theory of Dislocations*, Wiley, New York, NY, 1982
3. J.W. Christian: *Theory of Transformations in Metals and Alloys: Part I*, 2nd edition, Pergamon Press, Oxford, 1975
4. J.W. Christian and A.G. Crocker: *Dislocations in Solids*, F.R.N. Nabarro, ed., North-Holland Publishing Co., Amsterdam, 1980, vol. 3, p 165–249
5. A.P. Sutton and V. Vitek: *Phil. Trans. R. Soc. Lond.*, 1983, vol. A309, p 1–36
6. W. Bollmann: *Crystal Defects and Crystalline Interfaces*, Springer Verlag, New York, NY, 1970
7. B.A. Bilby: *Report of Conference on Defects in Crystalline Solids*, The Physical Society, London, 1955, p 124–33
8. R.W. Balluffi, A. Brokman, and A. H. King: *Acta Metall.*, 1982, vol. 30, p 1453–70
9. M.J. Marcinkowski and W.F. Tseng: *Metall. Trans.*, 1970, vol. 1, p 3397–401
10. G.B. Olson and Morris Cohen: *Acta Metall.*, 1979, vol. 27, p 1907–18
11. J.W. Matthews: *Phil. Mag.*, 1961, vol. 6, p 1347–49
12. B.A. Bilby: *Phil. Mag.*, 1953, vol. 44, p 782–85
13. R.C. Pond: *Proc. R. Soc. Lond.*, 1977, vol. A357, p 471–83
14. H. Mykura: *Grain Boundary Structure and Kinetics*, ASM, Metals Park, OH, 1980, p 445–56
15. J. Washburn and E.R. Parker: *Metall. Trans.*, 1952, vol. 194, p 1076–78
16. W.T. Read: *Dislocations in Crystals*, McGraw-Hill, New York, NY, 1953
17. J.S. Liu: Ph.D. Thesis, Massachusetts Institute of Technology, Cambridge, MA, 1984
18. R.W. Balluffi, Y. Komem, and T. Schober: *Surf. Sci.*, 1972, vol. 31, p 68–95
19. R.C. Pond and D.S. Vlachavas: *Proc. R. Soc. Lond.*, 1983, vol. A386, p 95–143
20. J.P. Hirth and R.W. Balluffi: *Acta Metall.*, 1973, vol. 21, p 929–42
21. F. Cosandey and C.L. Bauer: *Phil. Mag.*, 1981, vol. A44, p 391–403
22. R.C. Pond and V. Vitek: *Proc. R. Soc. Lond.*, 1977, vol. A357, p 453–70
23. P.H. Pumphrey, T.F. Malis, and H. Gleiter: *Phil. Mag.*, 1976, vol. 34, p 227–33
24. R.C. Pond: *Phil. Mag.*, 1983, vol. A47, p L49–L55
25. P.D. Bristowe and A.G. Crocker: *Phil. Mag.*, 1978, vol. A38, p 487–502
26. J.D. Eshelby: *Prog. Solid Mech.*, 1961, vol. 2, p 89–140
27. G.B. Olson: *Acta Metall.*, 1981, vol. 29, p 1475–84
28. S. Mahajan, M.L. Green, and D. Brasen: *Metall. Trans. A*, 1977, vol. 8A, p 283–93

29. J.W. Brooks, M.H. Loretto, and R.E. Smallman: *Acta Metall.*, 1979, vol. 27, p 1839–47
30. B.P.J. Sandvik and C.M. Wayman: *Metall. Trans. A*, 1983, vol. 14A, p 823–34
31. J.M. Rigsbee and H.I. Aaronson: *Acta Metall.*, 1979, vol. 27, p 351–63
32. R. Bonnet and F. Durand: *Phil. Mag.*, 1975, vol. 32, p 997–1006
33. M.F. Ashby, F. Spaepen, and S. Williams: *Acta Metall.*, 1978, vol. 26, p 1647–63
34. A.P. Sutton and V. Vitek: *Phil. Trans. R. Soc. Lond.*, 1983, vol. A309, p 37–54; *ibid.*, p 55–68
35. R.W. Balluffi and P.D. Bristowe: *Surface Sci.*, 1984, vol. 144, p 28–43
36. R.W. Balluffi: *Grain Boundary Structure and Kinetics*, ASM, Metals Park, OH, 1980, p 297–329
37. G.R. Kegg, C.A.P. Horton, and J.M. Silcock: *Phil. Mag.*, 1973, vol. 27, p 1041–55
38. M. Grujicic, G.B. Olson, and W.S. Owen: Proc. Intl. Conf. Martensitic Transformations—ICOMAT 82, *J. de Physique*, Coll. C4, 1982, vol. 43, p C4-173–78
39. M. Grujicic: Sc.D. Thesis, Massachusetts Institute of Technology, Cambridge, MA, 1983

Transformations—Selected Works of G.B. Olson on Materials, Microstructure, and Design
C.E. Campbell, M.V. Manuel, and W. Xiong, editors

CHAPTER 3

Martensitic Nucleation*

G.B. Olson, Northwestern University
A.L. Roitburd, University of Maryland

Historical Perspective

To set the stage for consideration of the varied factors influencing martensitic nucleation, and to fully appreciate the pervasive innovations of Professor Morris Cohen in the development of this field, it is instructive to review the developments during the first decade in which theoretical concepts of martensitic nucleation were intensively investigated.

In the late 1940s, when Ref 1 and 2 were first introducing the classical theory of homogeneous nucleation into the field of martensitic transformations, Ref 3 applied the then-new concept of dislocations to the *heterogeneous* nucleation of martensite by local mechanical instability of the crystal lattice, corresponding to what is now termed "non-classical" nucleation. Figure 1 from their paper depicts a volume of material of width w and length L interacting with a group of parallel lattice screw dislocations of height h (Fig. 1a). Figure 1(b) represents the shear strain θ as a function of distance from the dislocation group. Figure 2 depicts the energy density of a homogeneously deformed lattice as a function of shear strain θ, with $\theta = 0$ representing the undeformed parent structure and $\theta = \theta_M$ the fully formed martensite structure. Curves for two temperatures are shown, illustrating that as temperature decreases and the relative energy of the martensite ($\theta = \theta_M$) becomes lower, the lattice deformation energy density barrier (ΔF_C) diminishes. Termed the "reaction path" mechanism, Ref 3 proposed that the highly strained region near the dislocation defect of Fig. 1 could reach a condition of local mechanical instability (at a critical level of the parent-

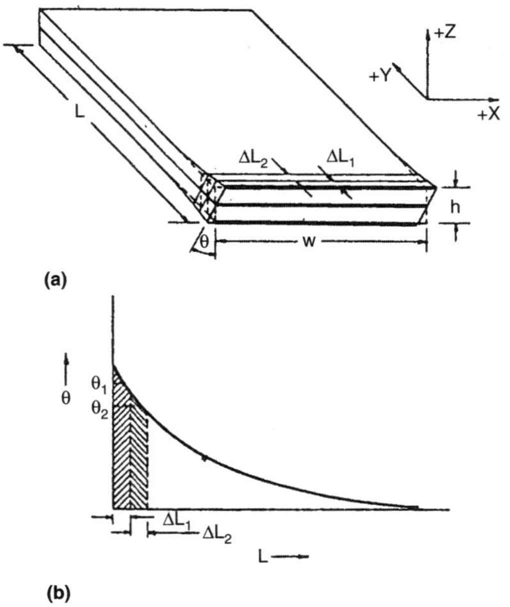

Fig. 1 (a) Rectangular volume of material interacting with a set of lattice screw dislocations denoted by heavy lines. (b) Distribution of strain (θ) versus distance normal to dislocations. Source: Ref 3

*G.B. Olson and A.L. Roitburd, Martensitic Nucleation, *Martensite*, G.B. Olson and W.S. Owen, Eds., ASM International 1992, p 149–174 martensite free energy difference) leading to the formation of a critical nucleus along a path of increasing strain.

In a further refinement of the model, Ref 4 considered that the process could be assisted by thermal fluctuations, and that the progress of overall transformation with cooling would be governed by a potency distribution of the "strain embryos" formed at various defects. Figure 3 from their paper depicts the embryo potency distribution (right side) and the associated temperature dependence of the activation energy of the most potent embryos (left side) governing the start of transformation at the M_s temperature.

In 1953, Ref 5 extended the quantitative treatment of the reaction-path model by incorporating a strain-dependent nucleus interfacial energy. Employing a linear elastic description of the energy of strained martensite, they concluded that the strain θ^* characterizing the structure of a critical nucleus would not depart significantly from the $\theta = \theta_M$ limit (fully formed martensite) under the thermo-dynamic conditions at which martensite is observed to nucleate. Reference 6 concluded that "classical" nucleation ($\theta^* = \theta_M$) could occur *homogeneously* at the observed M_s temperature if the nucleus interfacial energy is sufficiently low (~10 mJ/m^2). It became clear, however, from the subsequent small-particle experiments of Ref 7 that nucleation at M_s occurs *heterogeneously* on sparsely distributed defects.

A major advance in the reconciliation of theory and experiment came with the understanding that the measured macroscopic strains accompanying the formation of martensite plates are associated with a partially relaxed or "semicoherent" state of the martensitic particle; further, the effect of elastic stress on nucleation (Ref 8) is consistent with a critical nucleus in this state. In 1958, Ref 9 adopted such a structure for the oblate-spheroidal nucleus model depicted in Fig. 4, containing an array of lattice dislocations in its interface. Calculating the work of formation ΔW of such a particle as a function of radius r and semi-thickness c, they determined the saddle-point energy barrier ΔW^* for homogeneous classical ($\theta = \theta_M$) nucleation, as depicted in Fig. 5. Estimating the interfacial energy of such a semicoherent nucleus as 200 mJ/m^2 due to the elastic fields of the interfacial dislocations, they demonstrated that the barrier ΔW^* is orders of magnitude too high to allow homogeneous nucleation at the observed M_s temperature.

Reference 9 went on to apply the semicoherent classical nucleus model of Fig. 4 to the problem of *heterogeneous* nucleation, leading to the concepts represented schematically in Fig. 6. They reasoned that the interaction of such a nucleus with a defect like that of Fig. 1 would provide an elastic interaction energy behaving dimensionally like a negative surface energy. At temperatures above T_0, where the martensite is not thermodynamically stable as a bulk phase, this interaction would then lead to the local stabilization of martensitic embryos, as depicted by the minima in the

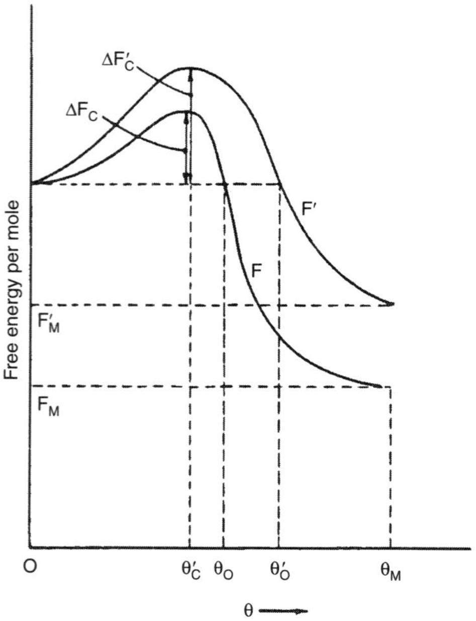

Fig. 2 Schematic plot of free energy density versus strain (θ) for homogeneous lattice deformation from parent phase (θ = 0) to martensitic (θ = θ$_M$), depicted at two levels of transformation driving force. Source: Ref 3

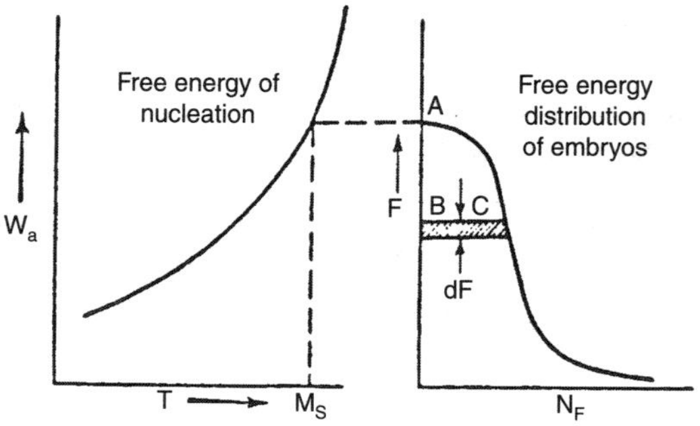

Fig. 3 Schematic potency distribution of martensitic embryos and associated temperature dependence of required potency for nucleation. Source: Ref 4

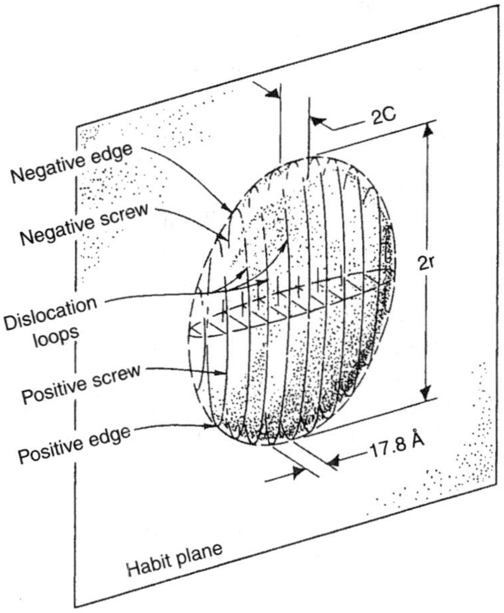

Fig. 4 Model of semicoherent martensitic nucleus with array of interfacial dislocations. (Source: Ref 9)

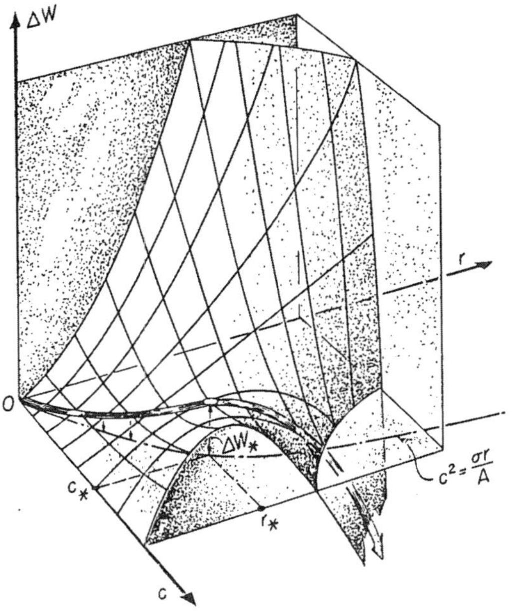

Fig. 5 Work of martensitic embryo formation (ΔW) versus radius (r) and semithickness (c), defining saddlepoint homogeneous nucleation barrier (ΔW*) (Source: Ref 9)

total energy versus size curves of Fig. 6(a). In line with the considerations of Ref 5, they considered such "preexisting" embryos to be of the fully formed martensite structure ($\theta^* = \theta_M$), rather than the earlier proposed "strain embryos" of intermediate structure. On cooling

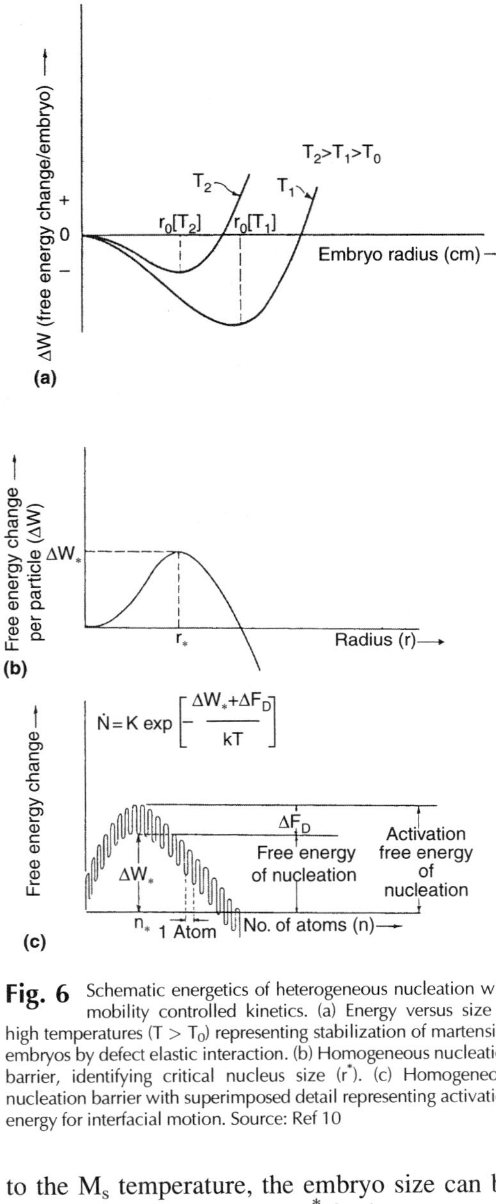

Fig. 6 Schematic energetics of heterogeneous nucleation with mobility controlled kinetics. (a) Energy versus size at high temperatures ($T > T_0$) representing stabilization of martensitic embryos by defect elastic interaction. (b) Homogeneous nucleation barrier, identifying critical nucleus size (r^*). (c) Homogeneous nucleation barrier with superimposed detail representing activation energy for interfacial motion. Source: Ref 10

to the M_s temperature, the embryo size can be larger than the critical size r^* for homogeneous nucleation depicted in Fig. 6(b), leading to a supercritical nucleus capable of unrestricted growth. The defect interaction thus circumvents the large ΔW^* barrier of homogeneous nucleation, equivalent to "barrierless" heterogeneous nucleation. In analogy to the nucleation kinetics of diffusional transformations depicted schematically in Fig. 6(c), the apparent activation energy for nucleation is the sum of ΔW^* and the activation energy for interfacial motion, the latter corresponding to the fine detail superimposed on the $\Delta W(r)$ curve in Fig. 6(c). As also emphasized by Ref 11–13, the kinetics of martensitic nucleation in the barrierless limit would be controlled by interfacial mobility, which Ref 9 modeled in terms of the thermally activated production of the dislocation loops of Fig. 4.

In the ensuing decades, dramatic progress has been made in the quantitative treatment of the competing theoretical possibilities in martensitic nucleation, aided in large measure by the continuing research efforts of Morris Cohen. From our current level of understanding, it is remarkable to look back and recognize that all of the key concepts were in place in the period from 1948 to 1958, stemming largely from Morris Cohen's innovations.

Current Theory of Martensitic Nucleation

Parameters of Nucleation Theory

The key issues of martensitic nucleation theory identified in its first decade of investigation can be organized to define the "nucleation theory space" depicted in Fig. 7. The first axis, x_1, denotes the structure of the nucleus as represented by a strain order parameter η, which can be related to the strain θ of Fig. 1 and 2 by $\eta = \theta/\theta_M$, such that $\eta = 0$ denotes the parent phase and $\eta = 1$ denotes the fully formed martensite. "Classical" nucleation assumes a critical nucleus structure of $\eta^* = 1$, while "nonclassical" nucleation admits nuclei of intermediate structure $\eta^* < 1$, including η^* approaching $\eta^* = 0$ near a lattice mechanical instability.

The second axis, x_2, denotes the state of overall strain relaxation within a martensitic particle, governing the nature of its interface and internal substructure. For a fully *coherent* particle, the net transformation strain is the full homogeneous lattice deformation that converts one Bravais lattice to another. The term *semicoherent* is generally applied to a particle that undergoes sufficient plastic relaxation (by slip or twinning) to make the net transformation strain an invariant plane strain

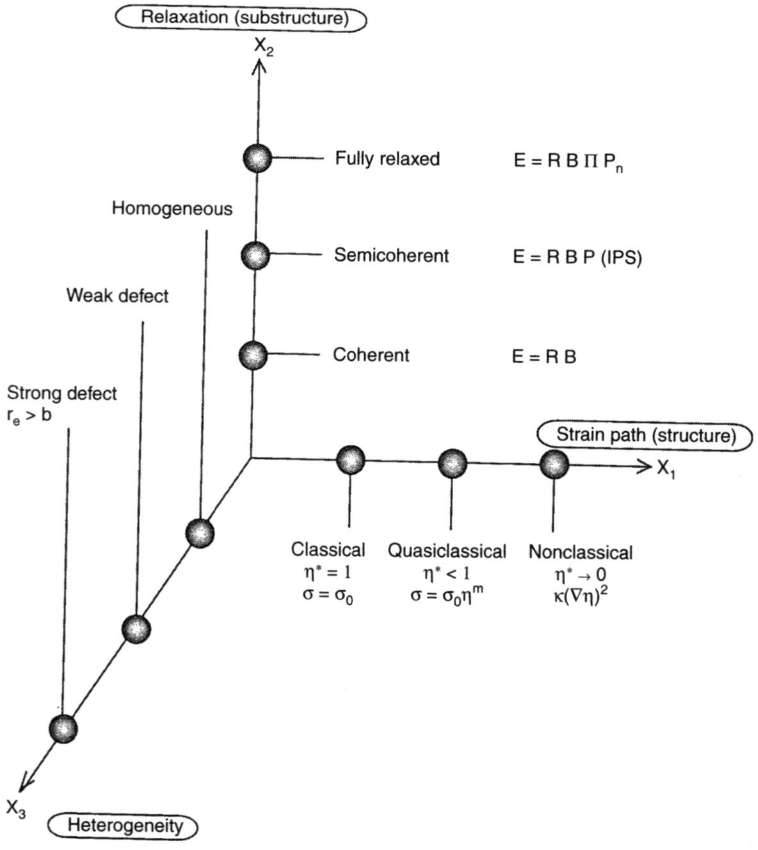

Fig. 7 Principal parameters defining a "theory space" for martensitic nucleation

(IPS), thereby greatly reducing the volume elastic self-energy of the particle. The conceptual limit of a *fully relaxed* particle allows additional plastic relaxation to cancel the shear component of the IPS.

The third axis, x_3, denotes the degree of heterogeneity of nucleation, homogeneous corresponding to nucleation without significant defect interaction. A concept of defect "strength" separating two classes of heterogeneous nucleation can be made on the basis of whether the defect interaction stabilizes a static embryo prior to nucleation.

All combinations of the parameters denoted by the three axes of Fig. 7 define a space in which all of the competing theoretical mechanisms of nucleation can be described. We next review the quantitative description of the role of each of these parameters to allow assessment of the competition among these potential mechanisms.

Strain Path—Classical Versus Nonclassical Nucleation

The issue of critical nucleus structure as represented by the strain parameter η was not only the first issue debated in martensitic nucleation, but has been the longest issue to resolve, requiring recent advances in understanding the energetics of large strain lattice deformations. The issue can be addressed most clearly for the case of homogeneous coherent nucleation, where current understanding defines three regimes of behavior, depicted schematically in Fig. 8.

As in the treatment of homogeneous nucleation by Ref 9 represented in Fig. 5, we consider an oblate spheroidal embryo of radius r and semi-thickness c. For nucleation at sufficiently small driving force Δg (free energy change per unit volume), critical nucleus size will be large, such that c^* is much greater than the interfacial core width δ, and the classical approximation of a constant interfacial energy per area σ_0 can be employed. As will be demonstrated later, we can also treat the nucleus as having the fully formed martensite structure $\eta^* = 1$ in this regime. If the transformation strain is of the form of an IPS, the elastic energy per volume of such a particle can be expressed as (Ref 14):

$$g_{IPS}^{str} = K\frac{c}{r} \qquad \text{(Eq 1)}$$

where $K = [\pi(2-\nu)/8(1-\nu)]\,\mu\gamma_T^2 + [\pi/4(1-\nu)]\,\mu\,\varepsilon_n^2$ for an isotropic shear modulus μ, Poisson's ratio ν, and transformation shear and normal strains γ_T and ε_n. When the transformation strain is not an exact IPS, distortions in the habit plane contribute an additional shape-insensitive strain energy per volume g_p^{str} such that the total strain energy g^{str} is:

$$g^{str} = g_{IPS}^{str} + g_p^{str} \qquad \text{(Eq 2)}$$

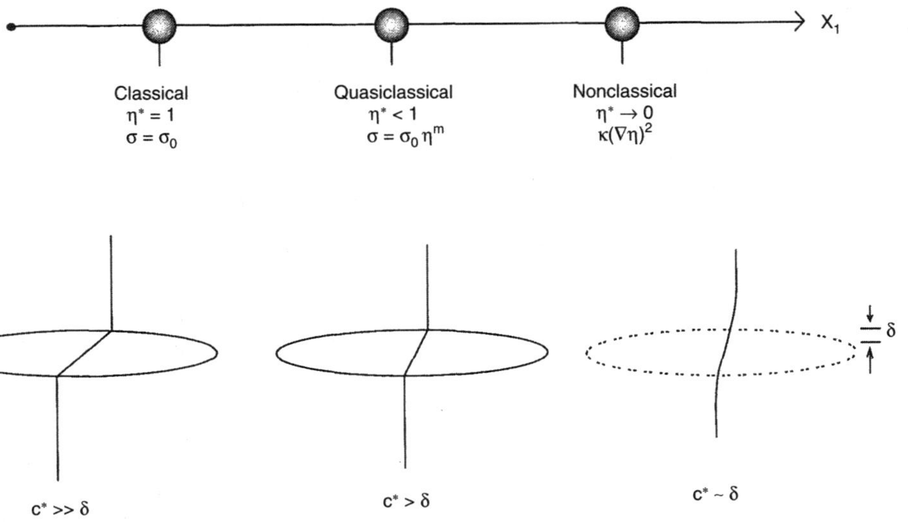

Fig. 8 Nucleation strain path. Schematic illustrations of nucleus structure in three regimes of behavior, with associated theoretical approximations

With these approximations, the work of forming a thin oblate spheroidal coherent particle can be expressed as:

$$\Delta W(r,c) = \frac{4\pi}{3} r^2 c\left(\Delta g + g_p^{str}\right) + \frac{4\pi}{3} r c^2 K + 2\pi r^2 \sigma_0 \quad \text{(Eq 3)}$$

This defines the energy surface depicted in Fig. 5 with a saddlepoint defining a critical nucleus as:

$$r^* = \frac{4\sigma_0 K}{\left(\Delta g + g_p^{str}\right)^2} \quad \text{(Eq 4)}$$

$$c^* = -\frac{2\sigma_0}{\left(\Delta g + g_p^{str}\right)} \quad \text{(Eq 5)}$$

$$\Delta W^* = \frac{32\pi \sigma_0^3 K^2}{3\left(\Delta g + g_p^{str}\right)^4} \quad \text{(Eq 6)}$$

This type of classical critical nucleus is depicted at the left of Fig. 8. For small values of the net driving force $\Delta g + g_p^{str}$ where these approximations are valid, Eq 4 to 6 describe a relatively large critical nucleus, with ΔW^* typically orders of magnitude greater than kT for strongly first-orders transformations.

At larger net driving forces, some departure from the classical $\eta^* = 1$ limit can be expected. If the critical nucleus is still sufficiently large relative to the interfacial core width, that is, $c^* > \delta$, the concept of interfacial energy remains useful, and in this intermediate regime a "quasi-classical" model like that of Ref 5 can be employed, represented at the center of Fig. 8. As mentioned earlier, the strain dependence of the interfacial energy can be treated as a function $\sigma_0\eta^m$. Both the strain-energy parameters K and g_p^{str} can be multiplied by η^2. For small departures from $\eta^* = 1$, the martensite can be treated as linear elastic, giving the Fisher-Turnbull approximation for $\Delta g(\eta)$:

$$\Delta g(\eta) = \Delta g(\eta = 1) + \frac{1}{2}\mu \gamma_T^2 (1-\eta)^2 \quad \text{(Eq 7)}$$

Equation 3 then takes the form:

$$\Delta W(r,c,\eta) = \frac{4\pi}{3} r^2 c\left[\Delta g_1 + \frac{1}{2}\mu\gamma_T^2(1-\eta)^2 + g_p^{str}\eta^2\right] + \frac{4\pi}{3} r c^2 K \eta^2 + 2\pi r^2 \sigma_0 \eta^m \quad \text{(Eq 8)}$$

Choosing a value for m such as m = 2, a saddlepoint with respect to r, c, and η similar to that of Fig. 5 defines a homogeneous nucleation barrier ΔW^* which, as shown by Ref 5, can be about one-half of the classical $\eta^* = 1$ model. At higher driving force where η^* departs further from $\eta^* = 1$, the linear elastic approximation of Eq 7 must be replaced by a nonlinear $\Delta g(\eta)$ function of the type depicted in Fig. 2. This was first treated by Ref 15, showing that, at sufficiently high driving force where $\eta^* \sim 0.5$, ΔW^* can be reduced by an order of magnitude relative to the classical model of Eq 3. It should be noted, however, that such effects require driving forces much higher than those at which martensitic transformations typically occur in bulk materials.

Recent interest in the possibility of strong departures from classical nucleation behavior has been fostered by observations in numerous systems of lattice softening behavior indicative of incipient mechanical instability associated with the vanishing of an elastic shear modulus. It can be anticipated from the analogous problem of nucleation of a diffusional transformation near a spinodal point as treated by Ref 16 that nucleation near an instability will involve a nucleus with a highly diffuse interface (large δ), a small η^*, and a greatly reduced ΔW^*. This is represented by the particle with the smooth displacement field shown at the right in Fig. 8. In this regime, it is necessary to treat not only highly non-linear lattice deformation energetics, but the simple notion of interfacial energy is best replaced by a "nonlocal" strain gradient energy of the form $\kappa(\nabla\eta)^2$ and the structure of the diffuse interface addressed as a variational problem.

This approach was first applied to a coherent martensitic interface by Ref 17, employing a tensor multiple-strain order parameter. Extension of the approach by Ref 18 and 19 has included estimates of nonlinear energetics from higher-order elastic constants, estimates of gradient energy coefficients from phonon dispersion curves, and applications of the method to predict inter-facial structure and energy in specific systems. Within the approximation of the scalar order parameter η discussed here, Ref 20, 21 have applied a rigorous one-dimensional solution for the diffuse martensitic interface to the description of a three-dimensional thin-plate nucleus as in Fig. 8. The approach not only describes the unusual nucleation behavior near a lattice instability, but also provides a unified framework to assess the range of validity of the approximations represented in Fig. 8.

The description of the nonlinear energetics of homogeneous lattice deformations in strongly first-order martensitic transformations has been made possible by recent advances in first-principles electronic total energy calculations. Enthalpy density at 0 K has been calculated as a function of strain along the fcc-bcc Bain deformation path for sodium and iron employing hydrostatic pressure to vary relative phase stability (Ref 22–24). These calculations give curves of the type represented in Fig. 2, defining the detailed nature of the energy density barrier between the two energy minima, and identifying the critical driving force for lattice mechanical instability. While the latter is found to be only 200 to 450 J/mol for a simple metal like sodium, in a transition metal like iron it is found to be 14 kJ/mol. Under hydrostatic tension, sodium shows smooth "precursor" behavior, with the shear modulus monotonically decreasing to zero at the instability. The behavior can be well described in this case by a model similar to a Landau expansion for which the free energy density relative to that of the parent phase ($\eta = 0$) is expressed by:

$$g(\eta) = A\eta^2 - B\eta^3 + C\eta^4 \quad \text{(Eq 9)}$$

where the coefficients A, B, and C correspond to second-, third-, and fourth-order elastic constants. Denoting the transformation driving force, $g(\eta = 1)$, as Δg, the driving force dependence of the curve shapes can be approximated by taking A to be linearly related to a normalized driving force:

$$\alpha = \Delta g/\Delta g_i \quad \text{(Eq 10)}$$

where Δg_i is the critical driving force for instability. Denoting the lattice deformation barrier as $\hat{g}$ at equilibrium ($\alpha = 0$), and constraining g (η) to give minima at $\eta = 0$ and $\eta = 1$, yields the coefficients for Eq 9:

$$A/\hat{g} = 16(1 - \alpha) \quad \text{(Eq 11a)}$$

$$B/\hat{g} = 32(1 - \alpha) - 4\alpha\beta \quad \text{(Eq 11b)}$$

$$C/\hat{g} = 16(1 - \alpha) - 3\alpha\beta \quad \text{(Eq 11c)}$$

where $\beta = \Delta g_i/\hat{g}$. For the cases so far treated, the electronic calculations give values of β between -1 and -3.

As a further simplification to more easily treat the problem of nucleation, we apply the model of Eq 9 to 11 to the special case of a coherent transformation for which the lattice deformation is a simple shear, such that g_p^{str} (Eq 2). Examples are the fcc-hcp and bcc-9R transformations in metals and the orthorhombic-monoclinic transformation in ZrO_2. We then describe the energetics of an inhomogeneously strained system adopting the framework of the Cahn-Hilliard theory of nonclassical nucleation. Assuming a strain-gradient energy of the form $\kappa(\nabla\eta)^2$, the traditional volume chemical free energy and interfacial energy terms of a classical description of a two-phase system are then replaced by a functional of the form:

$$\int \left[g + \kappa(\nabla\eta)^2\right] dV \quad \text{(Eq 12)}$$

with g defined by Eq 9 to 11. The energy of a coherent parent/martensite interface in the plane of the transformation shear at $\alpha = 0$ is then given by:

$$\sigma_0 = \int_{-\infty}^{\infty} \left[g + \kappa(\eta')^2\right] dz \quad \text{(Eq 13)}$$

Solving the Euler equation for the variation $\eta(z)$ which minimizes σ_0, a solution for σ_0 is obtained which defines κ as $\left(9\sigma_0^2/16\right)\hat{g}$ and gives an interfacial width (at $\alpha = 0$) of $\delta_0 = \sqrt{\kappa/\hat{g}}$.

A problem of relevance to martensitic nucleation is that of a thin slab of product phase centered on the $z = 0$ plane, where $\eta = \eta_0$. The total free energy functional of interest is then:

$$F_A = \int_0^{\infty} \left[g + \kappa(\eta')^2\right] dz \quad \text{(Eq 14)}$$

and a fixed total displacement across the particle corresponds to the condition:

$$\int_0^{\infty} \eta \, dz = u \quad \text{(Eq 15)}$$

where u is the total displacement normalized by the transformation shear γ_T. This defines an effective semithickness of the slab of $c = u/\eta_0$. The constraint of Eq 15 introduces a multiplier λ in the Euler equation for the variation $\eta(z)$ which minimizes the F_A. Solution with appropriate boundary conditions gives (Ref 21):

$$F_A(\eta_0, \lambda) = \sqrt{\kappa} \int_0^{\eta_0} \frac{g + \lambda\eta}{\sqrt{g + \lambda\eta}} d\eta \quad \text{(Eq 16)}$$

$$u(\eta_0, \lambda) = \sqrt{\kappa} \int_0^{\eta_0} \frac{\eta}{\sqrt{g + \lambda\eta}} d\eta \quad \text{(Eq 17)}$$

The effective semithickness $c = u/\eta_0$ is varied by the λ parameter and can be used to define an interfacial energy:

$$\sigma(\eta_0, \lambda) = F_A(\eta_0, \lambda) - \frac{g(\eta_0)}{\eta_0} u(\eta_0, \lambda) \quad \text{(Eq 18)}$$

A fairly general description of the energetics of coherent nucleation can be obtained from these relations by normalizing energy density to $\hat{g}$ and distances to δ_0.

The major contribution to the interfacial energy of a thin-plate coherent-shear particle arises from its broad interfaces, while the elastic energy accommodating the transformation strain resides primarily in the matrix beyond the periphery. The total energy ΔW of a thin-plate nucleus of radius r lying in the $z = 0$ plane can then be reasonably described by the energy of a one-dimensional variation $\eta(z)$ within a cylinder of radius r combined with the elastic energy associated with the long-range field of a particle of semithickness c producing the net displacement defined by u. A first-order correction for the effective interfacial energy associated with the plate periphery can be obtained using σ of Eq 18, analogous to the isotropic interfacial energy approximation. The nucleus energy is then expressed as (Ref 21, 23):

$$\Delta W(r, c, \eta_0) = 2\pi r^2 F_A + 2\pi r \xi A u^2 + 4\pi r c \sigma \quad \text{(Eq 19)}$$

where ξ is an accommodation factor equal to $\pi(2 - \upsilon)/4(1 - \upsilon)$ for the simple shear, and υ is Poisson's ratio.

Evaluating ΔW of Eq 19 by numerical integration of Eq 16 to 18 to obtain F_A, u, and σ, Ref 25 has computed the homogeneous nucleation saddlepoint barrier ΔW^* (r^*, c^*) as a function of nucleus structure, η_0, for various values of normalized driving force, α, as depicted in Fig. 9. In parallel with the Cahn-Hilliard results for diffusional nucleation, classical behavior $(\eta_0^* \approx 1)$ is found to prevail at low driving force (e.g., $\alpha = 0.35$) where the homogeneous nucleation barrier is very high. At high driving forces, the nucleation barrier is greatly reduced and the behavior becomes strongly nonclassical $(\eta_0^* \rightarrow 0)$ as the instability is approached. The nucleation barrier vanishes at the instability.

The dependence of critical nucleus structure η_0^* on driving force α is plotted in Fig. 10. The dashed line marks a convenient boundary

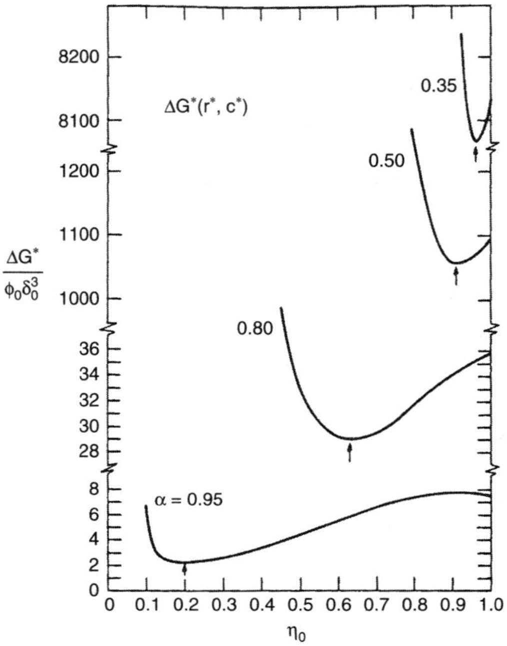

Fig. 9 Normalized homogeneous martensitic nucleation saddlepoint energy barrier, $\Delta W^*/\hat{g}\delta_0^3$ versus nucleus structure (η_0) for various values of driving force (α); $\beta = -3.08$. Source: Ref 25

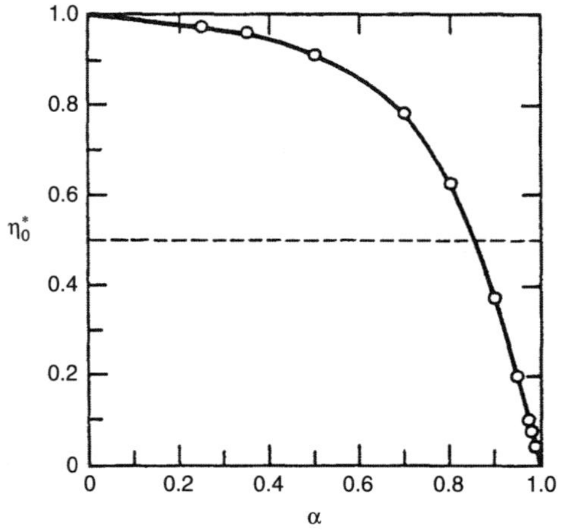

Fig. 10 Dependence of critical nucleus structure (η_0) on driving force (α). Source: Ref 25

where $\eta_0^* = 0.5$. At lower α the nucleus structure is closer to the product phase than the parent and the behavior can be regarded as a modification of classical nucleation. For $\alpha > 0.85$ the nucleus structure is closer to the parent phase than the product and the behavior is strongly nonclassical. In this regime, nucleation can be regarded as a localized lattice instability.

Critical nucleus size as expressed by c^* is plotted versus α in Fig. 11. As in the Cahn-Hilliard case, the critical nucleus size diminishes with increasing driving force in the classical regime, but diverges in the strongly nonclassical regime where nucleation and homogeneous instability smoothly merge. This is accompanied by a divergence of the interfacial width δ, beyond $\alpha = 0.9$. Nucleus c/r aspect ratios denoted in Fig. 11 show that the thin-plate approximations are reasonable up to the boundary of the strongly non-classical regime, but become less valid closer to the instability.

Referring back to the alternative approximations represented in Fig. 8, the model depicted in Fig. 9 to 11 gives insight into their ranges of validity. The fully classical ($\eta^* = 1$, $\sigma = \sigma_0$) model describes the behavior quite well in the range $0 < \alpha < 0.5$, even though the model system of Eq 9 to 11 shows strong instability precursor softening promoting nonclassical effects. The quasi-classical ($\sigma = \sigma_0\eta^m$) model will give an improved description up to $\alpha \sim 0.8$. At higher α, only the nonclassical diffuse interface model can capture the essential features associated with the divergence of δ at the instability.

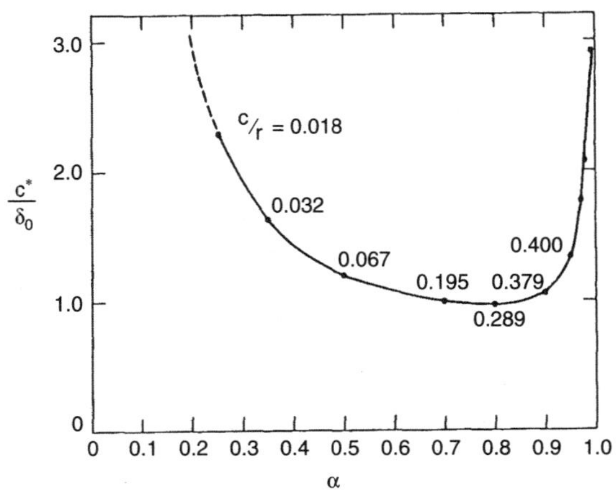

Fig. 11 Critical nucleus size (c^*) versus driving force (α) with c/r ratios indicated. Source: Ref 25

Except for very weak first-order transformations for which $\hat{g}$ is small, the homogeneous nucleation barrier tends to be quite high in the regimes of validity of the classical and quasi-classical models. In the nonclassical regime, however, the barrier can be made arbitrarily small, allowing thermally activated homogeneous nucleation to become possible. We can anticipate, then, that if homogeneous nucleation can be made to occur in a strongly first-order system, it will tend to involve nonclassical behavior.

Strain Relaxation—States of Coherency

Possible states of overall strain relaxation of a critical nucleus are summarized in Fig. 12. As described in the crystallographic or "kinematic" theories of martensitic transformations (Ref 26, 27), the finite strain deformations accompanying martensitic transformations can be represented by deformation matrices. We here use E to represent the net macroscopic shape strain of a martensitic particle, while the homogeneous lattice deformation relating Bravais lattices can be represented by the pure strain B and rigid body rotation R. As depicted at the left of Fig. 12, for a fully coherent martensitic particle, the homogeneous lattice deformation is the only deformation occurring, so that E = RB. When RB is not an IPS, the strain energy associated with distortions in the particle interface (g_p^{str} of Eq 2) can be quite large. For example, for coherent fcc → bcc transformation in iron-base alloys, g_p^{str} has been estimated (Ref 28, 29) as 8×10^8 J/m^3. Coherent transformation is not thermo-dynamically possible unless Δg exceeds this magnitude. In such a case, the volume elastic self-energy of the particle can be greatly reduced by adopting a microscopically in-homogeneous "lattice-invariant" internal deformation P such that E = RBP becomes an IPS and $g_p^{str} \rightarrow 0$. Transformation is then thermodynamically possible at much smaller Δg. The deformation P can occur by either slip or twinning, the latter case resulting in the internally twinned substructure depicted for the "semicoherent" IPS particle of Fig. 12.

While the elastic strain energy per volume is greatly reduced to the shape-sensitive g_{IPS}^{str} of Eq 1, the microscopic inhomogeneity of a slip dislocation array or a periodic twin array gives rise to a short-range interfacial elastic field that significantly increases the particle interfacial energy per area σ_0 compared to the fully coherent case. Further reduction of g_{IPS}^{str} can occur by

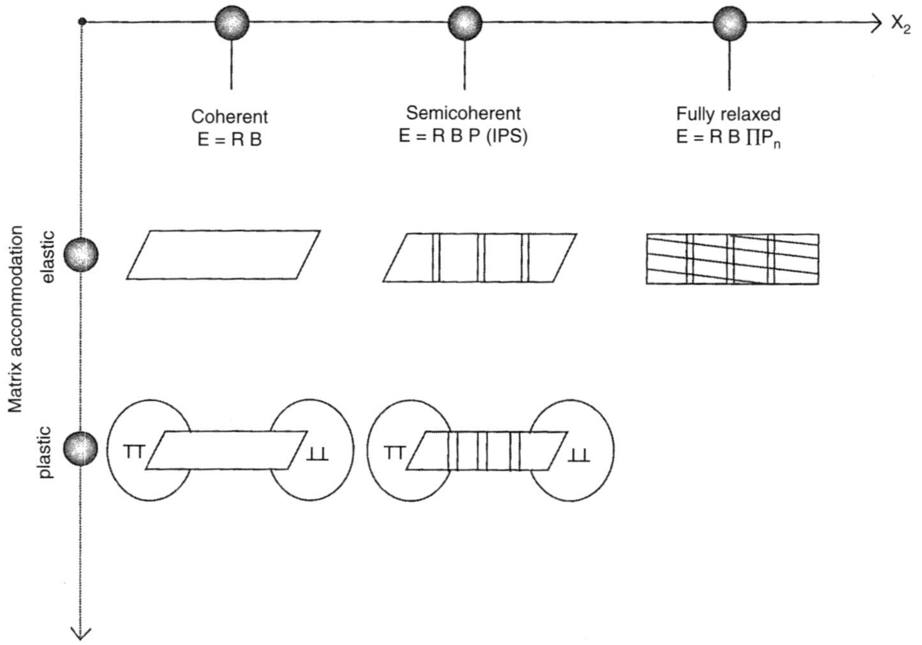

Fig. 12 States of nucleus strain relaxation. Schematic illustration of particle substructures, with corresponding mathematical description of net deformations

the operation of additional P_n deformations, represented by the product ΠP_n at the right of Fig. 12, denoting the conceptual limit of the "fully relaxed" particle where E (and thereby K of Eq 1) is arbitrarily reduced in magnitude. For a thin-plate particle, this constitutes a less significant reduction in strain energy compared to the single P establishing the IPS condition and is attended by a further increase in interfacial energy. The trade-off between volume strain energy reduction and increased interfacial energy in the varying states of strain relaxation is of special importance in nucleation.

The states of relaxation represented by the x_2 axis of Fig. 12 can be described in terms of uniformly periodic deformations within the transformed particle. These deformations can be regarded as elastically accommodated by the matrix. Alternatively, as denoted by a second axis in Fig. 12, the highly inhomogeneous stress field associated with the edge of a plate particle can also give rise to a far less orderly mode of strain relaxation, corresponding to a dislocation "plastic zone." Observation of such dislocation zones at small martensitic particles has led to the suggestion (Ref 30) that matrix plastic accommodation might control nucleation kinetics. Although no quantitative nucleation kinetic theory has been developed along these lines, modeling of dynamic plastic zone interactions during growth of a supercritical particle (Ref 25) suggests that this mode of relaxation constitutes an inhibiting influence. We will here restrict our attention to the more thoroughly studied modes of relaxation represented by the x_2 axis of Fig. 12.

The interfacial energies of semicoherent martensitic interfaces with lattice-invariant deformations by both slip and twinning have been estimated by Ref 31 employing detailed discrete dislocation models. A direct comparison of computed interfacial stress fields for the two cases is shown in Fig. 13. Simpler models have been applied by Ref 28, 32 and 33 in a thorough assessment of the energetic competition of the modes of relaxation depicted in Fig. 12, primarily in the context of classical homogeneous nucleation. For the slip interface of Fig. 4 with lattice dislocations of Burgers vector **b**, the dislocation model of Fig. 13 gives $\sigma_0/\mu b$ in the range of 0.006 to 0.012 for IPS semicoherent fcc-bcc transformation. The relatively wide range is associated with uncertainty in dislocation core energy cutoff parameters. For iron-based alloys, this corresponds to σ_0 = 0.1 to 0.2 J/m^2. Employing such an estimate for σ_0 in Eq 3, with $g^{str}_{p^*} = 0$ for an IPS transformation, gives ΔW^* for homogeneous classical semicoherent nucleation equal to the Kaufman-Cohen estimate discussed earlier.

With equivalent assumptions regarding core cutoff parameters, the discrete dislocation model for the interface of internally twinned martensite of otherwise identical crystallography (as depicted in Fig. 13) gives $\sigma_0 = \beta_T \mu D$, where D is the twin period and the constant β_T is determined as 3×10^{-3}. The semicoherent interfacial energy thus scales with D (which also sets the range of the short-range interfacial stress field) and is comparable to the energy of the slip interface when D is of the order of six lattice planes. For an internal twin boundary energy of γ, fine-scale twinning introduces a significant additional volume energy of $2\gamma/D$, which must be added to the volume Δg term of Eq 3. Substituting the $\sigma_0 = \beta_T \mu D$ expression for the interphase boundary energy in Eq 3 and minimizing total energy with respect to D gives a minimum energy twin spacing (Ref 34) of $D = \sqrt{2\gamma c / \beta_T \mu}$, which thus increases with plate thickness as $c^{1/2}$. This leads to a size-dependent interfacial energy $\sigma_0 = 2\sqrt{\beta_T \mu c} = \Gamma c^{1/2}$. Equation 3 then takes the form:

$$\Delta W(r, c) = \frac{4\pi}{3} r^2 c \Delta g + \frac{4\pi}{3} r c^2 K + 2\pi r^2 c^{1/2} \Gamma \quad \text{(Eq 20)}$$

This gives a nucleation barrier

$$\Delta W^* = 67\pi \Gamma^6 K^2 / \Delta g^7 \quad \text{(Eq 21)}$$

which is more sensitive to Δg than the result of Eq 6 (Ref 28, 29).

The energetic competition between coherent and semicoherent nucleation considered thus far can be represented by the schematic plot of ΔW^* versus $-\Delta g$ in Fig. 14. Curve 3 depicts the nucleation barrier for the fully coherent case, which is only thermo-dynamically possible for $-\Delta g > g_p^{str}$, but has the lowest ΔW^* at extreme Δg due to its low interfacial energy. Curve 2 represents the internally twinned IPS semicoherent case, which is thermodynamically stable for $-\Delta g < g_p^{str}$, but tends to exhibit a large ΔW^* for homogeneous nucleation.

For $-\Delta g$ near g_p^{str}, an alternative "intermediate" non-IPS semicoherent state has been shown to give a lower ΔW^* (Ref 35). If the twin fraction φ is allowed to vary from its

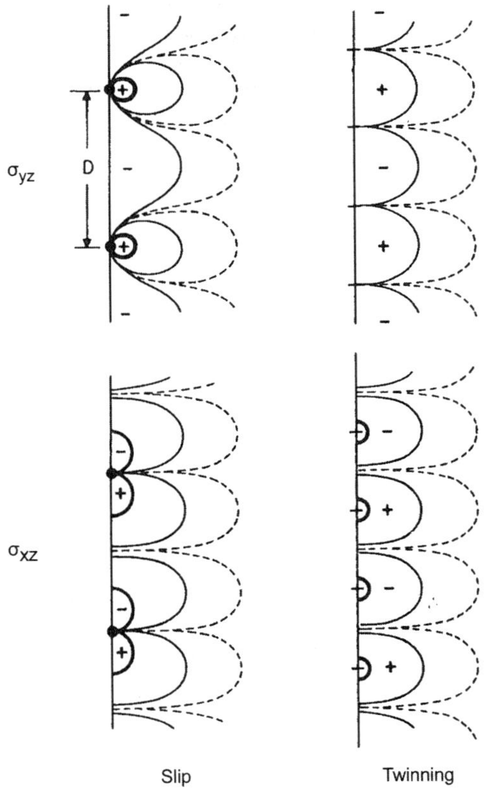

Fig. 13 Computed stress contours for semicoherent martensitic interfaces with lattice-invariant deformations by slip and by twinning. Source: Ref 31

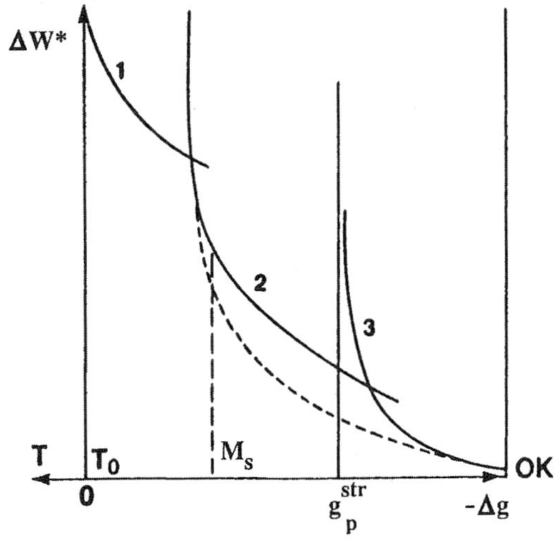

Fig. 14 Homogeneous nucleation barrier (ΔW^*) as a function of driving force ($-\Delta g$): curve 1, fully relaxed martensite; curve 2, internally twinned (IPS) semicoherent martensite; curve 3, fully coherent martensite. Dashed curve denotes non-IPS twinned martensite. Superimposed temperature scale indicates T_0 (Δg = 0) temperature and typical position of M_s. Source: Ref 35

unique value φ_0 required for an exact IPS condition, the interphase boundary energy $\sigma_0 = \beta_T \mu D$ will then scale with $\varphi^2(1 - \varphi)^2$ such that the optimal twin spacing D will now scale with $\sqrt{\gamma c}/\varphi(1 - \varphi)$. The deviation from an exact IPS will contribute a g_p^{str} energy equal to the fully coherent value scaled by $(1 - \varphi/\varphi_0)^2$. The strain-energy coefficient K of Eq 1 will similarly be modified by a factor of the form $(1 - 2\varphi)^2/\varphi_0^2$. Minimizing ΔW^* with respect to φ gives the behavior depicted by the dashed curve in Fig. 14, smoothly connecting to the fully coherent curve 3 in the limit $\varphi^* \rightarrow 0$, and connecting to the IPS semi-coherent curve 2 in the limit $\varphi^* \rightarrow \varphi_0$. The ΔW^* barrier for the non-IPS semicoherent state drops steeply with Δg, and constitutes the lowest barrier path for $-\Delta g$ in the vicinity of g_p^{str}.

An estimate of the relative energetics of the fully relaxed state depicted in Fig. 12 has also been made (Ref 35) corresponding to curve 1 in Fig. 14. Analogous to the dashed non-IPS curve, states of partial relaxation can also be considered to construct an additional curve smoothly connecting cases 1 and 2.

While the estimates discussed thus far have all retained the classical $\eta^* = 1$ approximation, a preliminary calculation employing the quasi-classical approximation (Ref 35) has addressed the possibility of internally twinned semicoherent nucleation following an $\eta^* < 1$ path. Such a path resembles a localized displacive spinodal decomposition in which matrix constraint imposes an effective strain conservation. The calculation suggests further reduction of ΔW^* relative to the dashed curve of Fig. 14.

From these considerations we can expect that achieving a thermally surmountable homogeneous nucleation barrier in a strongly first-order system would require sufficiently high driving force to favor a fully coherent or non-IPS semicoherent nucleus, likely involving some departure from $\eta^* = 1$ classical behavior. Transformation at lower driving force (which will not occur by homogeneous nucleation) will tend to involve states of higher relaxation, such as IPS semicoherent and partially relaxed semicoherent.

Heterogeneity—Homogeneous Versus Heterogeneous Nucleation

Detailed calculations of the elastic interactions of a martensitic nucleus with defects like that of Fig. 1 have shown the range of behaviors depicted schematically in Fig. 15 for four levels of Δg. For the class of "strong" defect depicted at the right, the defect interaction has been shown at low driving force (Δg_1) to stabilize an embryo of size r_e smaller than the critical size r^*. This behavior is directly in line with the Kaufman-Cohen concept depicted in Fig. 6 (a). At a critical driving force Δg_2, the interaction leads to barrierless nucleation where $r_e = r^*$. The value of Δg_2 is inversely related to the total Burgers vector content of the defect. For the "weak" defect case depicted at the center

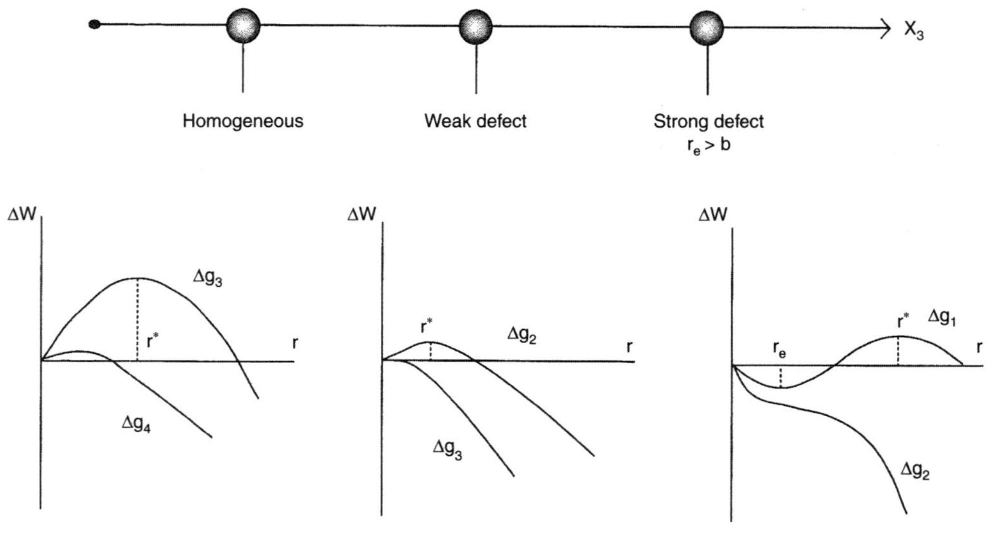

Fig. 15 Role of heterogeneity of nucleation. Schematic curves of energy (ΔW) versus size (r) depict behavior for homogeneous, weak-defect heterogeneous, and strong-defect heterogeneous nucleation at four levels of driving force (Δg).

of Fig. 15, barrierless nucleation is achieved at Δg_3, but is not preceded by the formation of a stable embryo at lower driving force (Δg_2). The barrierless condition allowed by heterogeneous nucleation contrasts sharply with the homogeneous nucleation behavior represented at the left of Fig. 15, for which the barrier can be made thermally surmountable at sufficient driving force (Δg_4), but can only be reduced to zero in the extreme limit of a lattice instability.

The first quantitative demonstration of the strong defect behavior depicted in Fig. 15 was made by Ref 36, which treated the widening of a ribbon-shaped embryo lying along the defect of Fig. 1, analogous to a dislocation dissociation. The model showed both embryo stabilization and barrierless nucleation. Barrierless nucleation was also demonstrated by Ref 37, which considered a circular plate interacting with a linear defect. (As discussed by Ref 38, an earlier treatment of a plate-shaped embryo interacting with a circular dislocation loop that expands with the plate [Ref 39] seriously overestimates the effectiveness of heterogeneous nucleation through omission of the dislocation self-energy.) Numerical calculations by Ref 40 showed that the circular plate is an unrealistic assumption, as the ribbon-shaped embryo lying along the defect has significantly lower energy.

Further insight into the character of heterogeneous nucleation at a linear defect is offered by the analytical model of Ref 41, which considered the same rectangular particle geometry depicted in Fig. 1. In addition to three energy terms analogous to those of Eq 3, elastic interaction with a defect stress field $\tau(x)$ introduces an interaction energy contribution of the form:

$$E^{int} = -hw\gamma_T \int_{r_0}^{L} \tau(x)dx \qquad \text{(Eq 22)}$$

employing the notation of Fig. 1, with τ the resolved stress on the system of γ_T, and r_0 a dislocation core cutoff parameter. The simplest case is to take $\tau(x)$ as the 1/x form of the long-range field of the defect of Fig. 1, corresponding to a "superdislocation" approximation. Normalizing energies to the self-energy parameter K of Eq 1, the four terms describing the total energy of the particle then take the form:

$$f(h, w, L) = -\alpha hwL + h^2(w + L) + \beta wL - khL \qquad \text{(Eq 23)}$$

with α representing the driving force, β the interfacial energy, and k an interaction parameter scaling with the defect total Burgers vector content and including a weak logarithmic dependence on L. Finding the saddlepoint of f (h, w, L) gives the critical nucleus parameters:

$$h^* = \frac{1}{3}\left\{\left(k + \frac{2\beta}{\alpha}\right) + \left[\left(k + \frac{2\beta}{\alpha}\right)^2 - 9k\frac{\beta}{\alpha}\right]^{1/2}\right\} \qquad \text{(Eq 24)}$$

$$f^* = \frac{1}{\alpha}\frac{h^{*3}(h^* - k)}{h^* - \beta/\alpha} \qquad \text{(Eq 25)}$$

$$L^*/w^* = 1 - k/h^* \qquad \text{(Eq 26)}$$

The predicted behavior is shown in Fig. 16. Curve a represents the barrier for homogeneous nucleation, while curve b shows the reduced barrier for heterogeneous nucleation at a defect of finite length. Equation 26 predicts that the nucleus tends to be progressively more ribbon-shaped (lying along the defect) as h^* diminishes with increasing α. From Eq 24 and 25, a special behavior occurs at $\alpha = \beta/k$, where the nucleus "wets" the defect so that W^* is limited only by the defect length. For a defect of infinite length, depicted by curve c of Fig. 16, this condition corresponds to barrierless nucleation. For finite defects, barrierless nucleation requires that α must exceed β/k by an amount that scales inversely with defect length.

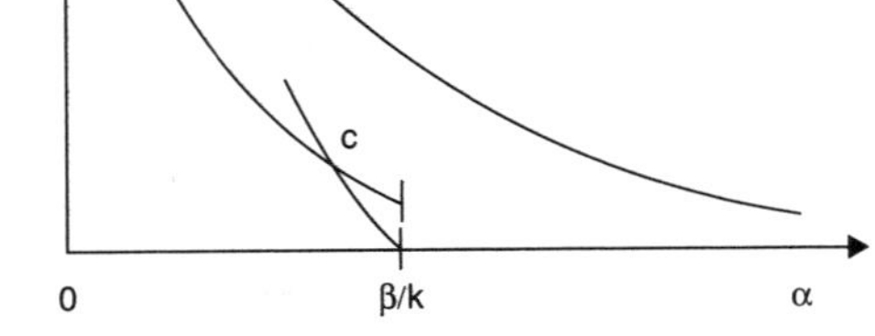

Fig. 16 Work of formation of a critical nucleus as a function of driving force α, for superdislocation model of defect interaction: (a) in absence of dislocation, (b) on a finite dislocation segment, (c) on infinite length dislocation. Source: Ref 41

The Roitburd analysis indicates further that the relatively sharp 1/x stress field of the superdislocation defect approximation promotes the weak defect class of behavior (Fig. 15) for which subcritical embryo stabilization is absent. Treating the longer range $1/x^{1/2}$ field of a dislocation pileup, embryo stabilization is found. More detailed treatment of the finite dislocation group of Fig. 1 (Ref 31), however, shows that, although the defect long-range field has the 1/x form, the milder short-range field of the spatially extended defect can promote embryo stabilization and the "strong" defect behavior of Fig. 15. The issue of prior embryo stabilization will thus depend on local details of the actual defect structure.

The most significant result of heterogeneous nucleation theory is that it can account for nucleation at the relatively modest driving forces at which martensitic transformations are observed to occur. Following the considerations of the previous section, particles will tend to be semicoherent in this regime, and interfacial energies appropriate to this state are generally applied for the β parameter of Eq 23.

The original 1949 proposal of Ref 3 was that strong-defect heterogeneous nucleation would be nonclassical in nature; this proposition has been restated in more modern terms as a "localized soft-mode mechanism" discussed by Ref 42. The concept has been examined quantitatively by Ref 25, applying the nonclassical model of the simple shear transformation of Eq 19 to the geometry of Fig. 1 and incorporating the E^{int} of Eq 22 to obtain:

$$\Delta W(h, w, L, \eta_0) = wLF_A + E^{self} + 2hw\sigma + E^{int} \quad \text{(Eq 27)}$$

where F_A and σ are again defined by Eq 16 and 18, E^{self} is of the form in Eq 23 but scaled to η_0^2, and E^{int} is scaled to η_0. Using the normalized driving force α of Eq 10, Fig. 17 depicts the computed energy ΔW per length w normalized to $\hat{g}\delta_0^2$, plotted versus L for various levels of α, for the case of a defect of sufficient potency to give barrierless nucleation at $\alpha \sim 0.2$. For all conditions represented in this figure, energy minimization with respect to η_0 gives $\eta_0 = 1$. Rather than nucleation by a

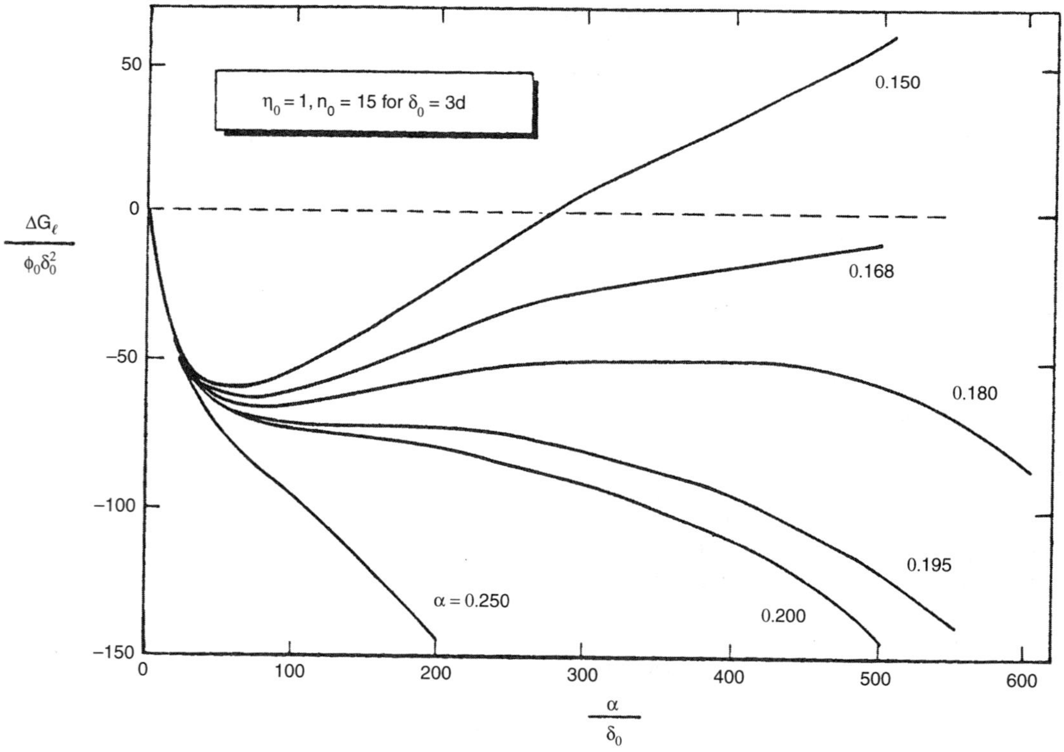

Fig. 17 Energy change per length $(\Delta W_1/\hat{g}\delta_0^2)$ versus nucleus size (L/δ_0) at various values of driving force (α) for case of heterogeneous nucleation at strong defect. Source: Ref 25

localized lattice instability preceded by "homophase" fluctuations of "soft mode" character, the model predicts that heterogeneous nucleation at such low α will occur by the classical $\eta_0^* = 1$ path preceded by the local formation of classical embryos.

The predicted behavior for a weak defect giving barrierless nucleation near $\alpha = 0.7$ is depicted in Fig. 18 and 19. The energy curves of Fig. 18 show that the barrierless condition is not preceded by the formation of an embryo of significant size ($r_e < b$). For $\alpha = 0.7$, the energy contour plot versus L and η_0 (Fig. 19) gives a saddlepoint near $\eta_0 = 0.7$. The character of fluctuations reaching such a configuration is more "heterophase" than "homophase," offering no support for a "localized soft-mode" description of heterogeneous nucleation even in the weak-defect limit.

A recent more rigorous description of a nonclassical nucleus, applied to the less realistic case of one-dimensional nucleation at a planar defect (Ref 43) further supports the strong defect behavior of Fig. 17. The condition for unrestricted nucleus growth is preceded well in advance ($\alpha < 0$) by wetting of the defect by a classical ($\eta \sim 1$) embryo.

It should be emphasized that the model of Eq 9 to 19 upon which the results of Fig. 17 to 19 are based represents a system with strong instability precursor effects in the form of an anomalous temperature dependence of an elastic constant as governed by Eq 11a. An actual system with features closely resembling this model is the bcc $\rightarrow$ 9R transformation in Cu-Zn-Al alloys which exhibit a large softening of the C′ elastic shear constant (Ref 44). The magnitude of the normalized driving force α at the observed M_s temperature can be determined from the relative magnitudes of C′ at M_s and T_0, giving $\alpha \sim 0.2$ corresponding to the conditions modeled in Fig. 17. It is clear then that the heterogeneous nucleation controlling the transformation behavior in Cu-Zn-Al alloys should be classical in nature, despite the large lattice softening effects exhibited in this system.

The theoretical viability of barrierless heterogeneous nucleation lends support to the Kaufman-Cohen proposal of interfacial-mobility-controlled

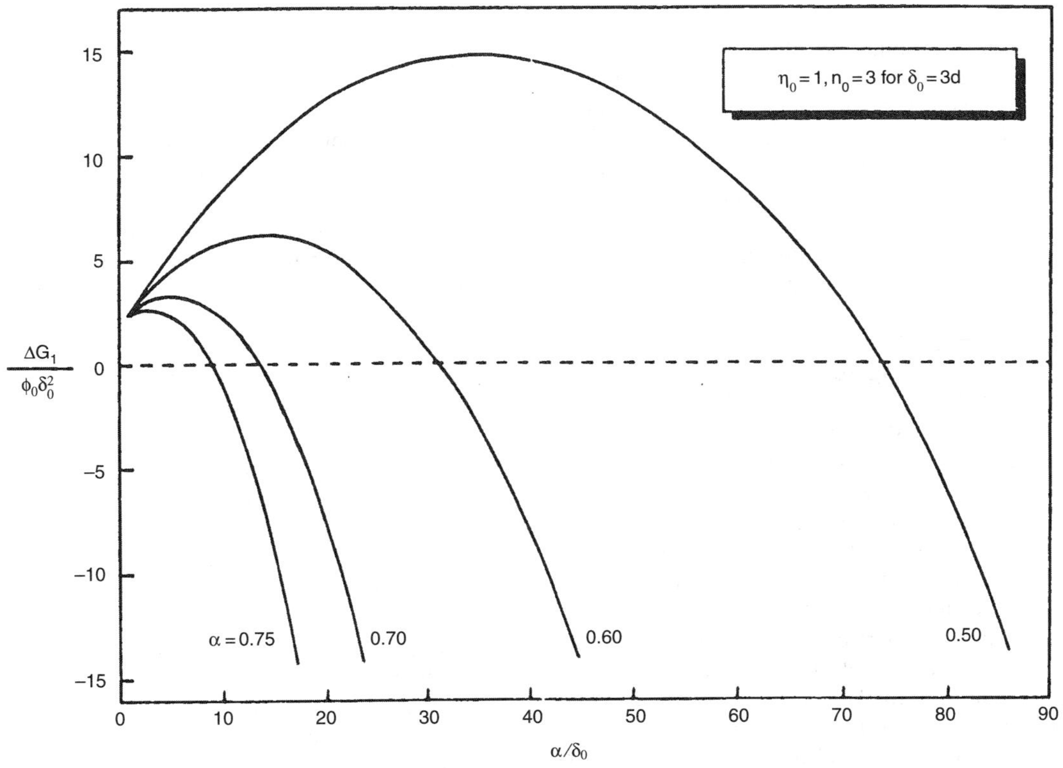

Fig. 18 Energy change per length ($\Delta W_1/\hat{g}\delta_0^2$) versus nucleus size ($L/\delta_0$) at various values of driving force (α) for heterogeneous nucleation at weak defect. Source: Ref 25

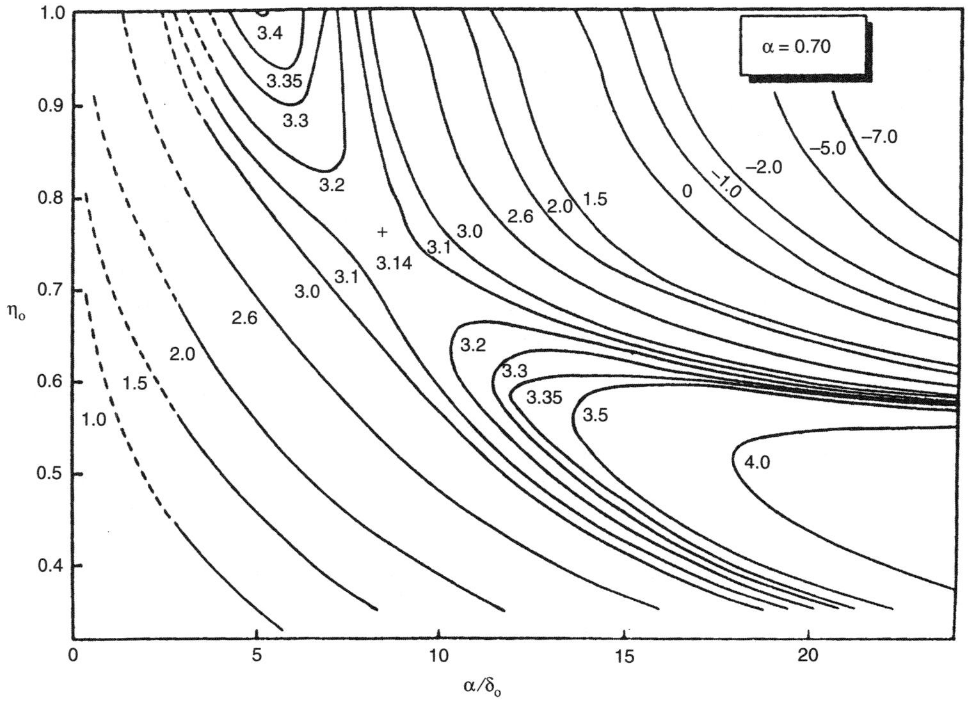

Fig. 19 Contour plot of $\Delta W_1/\hat{g}\delta_0^2$ versus nucleus size (L/δ_0) and strain amplitude (η_0) for weak-defect heterogeneous nucleation case of Fig. 18 at α = 0.7. Source: Ref 25

nucleation kinetics. While the originally proposed rate-controlling step was the formation of the dislocation loops of Fig. 4 associated with the lattice-invariant shear of a semicoherent particle, isothermal nucleation kinetic measurements on the coherent fcc-hcp transformation (where the lattice-invariant shear is absent) show the same kinetic behavior as the fcc-bcc transformation in the same Fe-Ni-Mn system (Ref 45). This, together with a much larger composition dependence of the critical Δg for nucleation than can be accounted for by simple elastic modulus variations, has led to the interpretation of the interfacial mobility as controlled primarily by solution hardening (Ref 31).

A convenient single parameter quantifying the potency of a heterogeneous nucleation site is the thickness h^* of the critical nucleus stabilized by the defect interaction, evaluated at the point of barrierless nucleation. In the Olson-Cohen model (Ref 36), this thickness is normalized to the close-packed interplanar spacing d, giving the defect potency parameter $n = h^*/d$. In this model the critical Δg for barrierless nucleation at a defect of potency n is expressed by:

$$\Delta g_0(n) = -\left(\frac{2\sigma_0}{nd} + g_p^{str} + w_f\right) \quad \text{(Eq 28)}$$

Recognizing that the invariant plane of the IPS is a macroscopic concept, the g_p^{str} term accounts for the strain energy of small distortions in the habit plane associated with a microscopic particle adopting a rational habit. The w_f term accounts for the frictional work of interfacial motion, and consists of athermal and thermal (rate and temperature dependent) components, the latter accounting for isothermal nucleation kinetics. The influence of solution hardening on w_f accounts for the relatively strong composition dependence of Δg_0 in alloys. Neglect of $g_p^{str} + w_f$ can lead to the conclusion that fairly small defects (e.g., single lattice dislocations) can account for heterogeneous nucleation at observed M_s temperatures, but inclusion of these terms gives total Burgers vector contents of four to five lattice vectors for the defects controlling M_s.

Inherent in the nature of heterogeneous nucleation is a defect potency distribution, as outlined in Fig. 3. Relating Δg to defect size n according to Eq 28, an analysis (Ref 46) of the Cech-Turnbull small-particle data mentioned earlier reveals an exponential form for the cumulative number density of defects of size n or greater:

$$N_v(n) = N_v^0 \exp(-\alpha n) \qquad \text{(Eq 29)}$$

with α a constant. The same function describes the defect distribution deduced from similar small-particle experiments conducted on the martensitic transformation in ZrO_2 (Ref 47). Analysis of the defect distribution has been extended to bulk Fe-32Ni poly-crystals by Ref 48 employing highly sensitive acoustic emission experiments to detect the M_s^{AE} corresponding to the first nucleation event, which is plotted versus grain size in Fig. 20. Also shown is the conventional M_b temperature corresponding to the first detection of transformation by electrical resistivity measurements. The behavior indicates that the amplitude of the preexisting defect distribution increases with grain size according to $N_v^0 = k_1 + k_2\bar{D}^{-m}$, with m = 2.6. The reduction of M_b with grain refinement is analogous to Hall-Petch strengthening and can be attributed to a reduction in the characteristic potency of autocatalytic nucleation sites generated by growth events. Further quantitative analysis of the transformation behavior in the same specimens by Ref 49 has separated the potency distribution of autocatalytic defects, revealing a Gaussian form with an amplitude scaling with the fraction transformed. When initial defects are more potent than the mean autocatalytic potency, sparse nucleation events can occur 40 °C above M_b, as shown in the fine-grain region in Fig. 20. For the coarse grain sizes at the right of Fig. 20, the autocatalytic defects are more potent than the most potent preexisting defect, and a single nucleation event triggers macroscopic "bursting." In this regime, a predicted specimen-size dependence of the M_b temperature was experimentally confirmed.

The most quantitative test of heterogeneous nucleation theory thus far is the experiment of Ref 50, in which small defect-free particles of tetragonal ZrO_2 were electron irradiated during direct observation in the electron microscope at 1000 °C below the bulk M_s temperature. The observed critical size of dislocation loop which nucleated martensitic transformation under these conditions was in good agreement with classical theory. No precursor embryo formation was detected, representing an example of the weak-defect case of Fig. 18. A direct electron microscopy observation of the strong-defect case of Fig. 17 has been made by Ref 49. For the cubic →

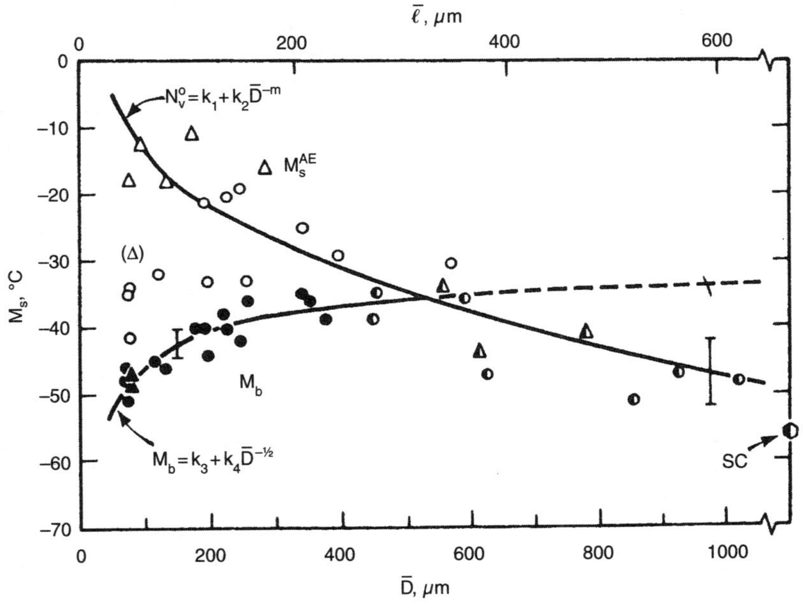

Fig. 20 Grain-size dependence of acoustic-emission M_s^{AE} (open symbols) and electrical-resistivity M_b (closed symbols) measured in Fe-32.3Ni. Circles represent acoustic emission experiments conducted at 70 dB gain; triangles denote more sensitive measurements at 90 dB gain. SC, single-crystal M_b temperature. Source: Ref 48

orthorhombic transformation in a thermoelastic Ti-Ni-Cu alloy, they recorded the smooth evolution of a classical embryo emanating from a superdislocation (similar to that envisioned in Fig. 1) on the interface of an oxide inclusion particle.

A novel experiment by Ref 49 has established the conditions for homogeneous nucleation of the fcc-bcc martensitic transformation in iron-based alloys. Defect-free fcc particles 10 to 20 nm in diameter with compositions near Fe-33Co were formed by coherent precipitation from a copper matrix. The cobalt contents were chosen to achieve transformation driving forces of the order of 10 kJ/mole. The Δg at which transformation on cooling occurred in such particles is plotted versus temperature in Fig. 21. Also plotted in the figure are dashed lines summarizing the observed critical Δg for heterogeneous nucleation in bulk Fe-Ni and Fe-Ni-Mn alloys (Ref 52). The experiment confirms that the Δg necessary for nucleation without aid of significant defects is a factor of ~7 greater than that for heterogeneous nucleation.

From the estimated critical driving force for fcc lattice instability (from the electronic calculations for pure iron mentioned earlier) of $\Delta g_i = -14$ kJ/mol, the Δg levels of Fig. 21 correspond to a normalized $\Delta g/\Delta g_i$ of $\alpha = 0.7$, for which strongly nonclassical effects are not expected (Fig. 10). The Δg does exceed the g_p^{str} strain-energy estimates for coherent fcc-bcc transformation, so that coherent nucleation is viable. The solid curve in Fig. 21 represents a fit of classical homogeneous nucleation theory (Eq 6) for coherent nucleation with $\sigma_0 = 10$ mJ/m^2, taking $\Delta W^* = 32$ kT for detectable nucleation. In contrast to barrierless heterogeneous nucleation, which shows no special behavior at low temperatures, the essential role of thermal activation in homogeneous nucleation causes the critical Δg to diverge near 0 K. In support of this feature, the solid points in Fig. 21 represent particles which did *not* transform on cooling to 4 K. As a further test, a 15 Tesla magnetic field was applied to increase Δg by an additional 320 J/mole. While application of the field did not cause transformation at 4 K, the same particles did transform when the field was applied at 77 K. The experiment thus offers support for the qualitative difference in the role of thermal activation in homogeneous versus heterogeneous nucleation.

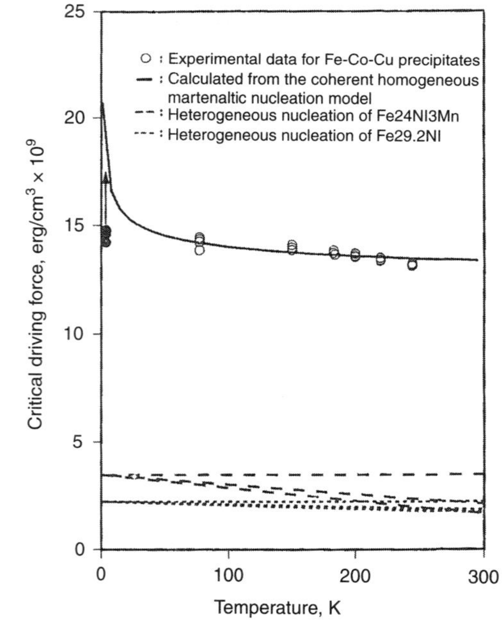

Fig. 21 Measured critical Δg for martensitic nucleation in defect-free coherent Fe-Co particles in a copper matrix. Closed symbols represent Δg at which nucleation did not occur at T = 4 K. Dashed lines summarize observed Δg for heterogeneous nucleation in bulk Fe-Ni and Fe-Ni-Mn alloys. Solid curve is fit of homogeneous classical nucleation theory for fully coherent case. Source: Ref 49

Future Directions

From the developments reviewed here, a clear picture emerges of the interrelations among the parameters of Fig. 7. Strongly nonclassical behavior can only occur at extreme driving forces very near Δg_i for lattice instability. This regime could only be accessed by suppression of heterogeneous nucleation, as in the example of finely dispersed particles. Heterogeneous nucleation governs the behavior of bulk materials and allows transformation at low driving forces where the fully coherent state is typically not stable, thus favoring semicoherent nucleation. Defect interaction leads to mobility-controlled barrierless nucleation, where potency distributions of both initial and autocatalytically generated nucleation sites can dominate macroscopic transformation behavior.

While opportunities abound for further refinement of martensitic nucleation theory, the case can be made that martensite currently represents the best understood heterogeneous phase transformation, and an important direction of future research is continued application of this understanding to other solid-state reactions. Development of universal defect-potency distribution functions, for example, would have broad impact throughout materials science.

From a practical standpoint, the most important issue remaining in the theory of heterogeneous martensitic nucleation concerns the interfacial mobility controlling the kinetics of barrierless nucleation, especially the role of solution hardening, as this appears essential to precise prediction of M_s temperatures in alloy design. Mobility issues are also central to the predictive control of hysteresis in thermoelastic shape-memory alloys.

The Saburi-Nenno direct observation of strong-defect heterogeneous nucleation in Ti-Ni-Cu provides the opportunity for a new level of detail in the three-dimensional modeling of classical martensitic nucleation to compare with the observed evolution of embryo shape. Further detailed modeling was not previously warranted in view of the lack of information on the detailed nature of real nucleating defects.

A major shortcoming of classical nucleation theory has been the lack of a predictive theory of interfacial energy. Developments in electronic total energy calculation methods open the way to prediction of coherent interfacial energies in specific systems, and also allow prediction of the nonlinear elastic energetics of the coherent state. In terms of the detailed experimental information available concerning their nucleation behavior, prime candidates for application of these methods are the tetragonal → monoclinic transformation in ZrO_2 and the coherent fcc → bcc transformation (Fig. 21) in iron. The development of methods to treat the nonlinear energetics of coherent transformation also addresses a broad issue of relevance to precipitation phenomena throughout materials science. Another system amenable to precise electronic total energy calculations is the bcc → 9R transformation in β-Cu-Zn, for which detailed information on interfacial core structure has been obtained by high-resolution electron microscopy (Ref 53). This system can serve as a model for commercial Cu-Zn-Al shape-memory alloys, and a Landau model (Ref 54) has been developed for the bcc → 9R lattice deformation energetics which explicitly treats the role of atomic "shuffle" displacements within the unit cell, supported by electronic calculations performed for the case of sodium (Ref 55).

Further insight into the three-dimensional nature of nonclassical nucleation in displacive transformations can be sought from the special case of a pure dilational transformation, as occurs, for example, in cerium. Modeling of such a transformation would more closely parallel the original Cahn-Hilliard theory.

A fruitful area for further work would be the combined theoretical and experimental investigation of nucleation behavior in weakly first-order systems, where nonlinear nonlocal continuum models could be reasonably applied to predict the parameters of classical nucleation theory. A suitable candidate system may be the weakly first-order fcc → fct transformation in Fe-Pd alloys, where continuum energetic models are already being developed (Ref 56) and applied to prediction of interfacial structure and energy.

In light of recent homogeneous martensitic nucleation experiments in Fe-Co, the question of the dynamics of homogeneous nucleation in displacive transformations becomes a real issue. While interfacial mobility models could treat the dynamics of the classical nucleation path in a manner analogous to the theory of diffusional transformations, the alternative nonclassical (variable-strain) path poses a different class of dynamics which has yet to be developed quantitatively. As originally discussed by Ref 4, large differences in the dynamics of these paths might alter the competition between classical and nonclassical nucleation relative to predictions based on barrier height alone.

Finally, a full understanding of martensitic transformations requires a criterion to assess the point where a sufficiently weak first-order transformation would no longer be classed as martensitic. Toward this end, a theory is under development (Ref 24, 57) for an alternative "quasi-martensitic" mechanism of microstructure evolution by a continuous strain modulation path representing the displacive analog of spinodal decomposition. Analysis suggests that energy conservation can lead to the selection of a fastest growing wavelength governing the dynamics of such a process. A quantitative description of the competition between nucleation and modulation would complete a unified view of first-order phase transformation mechanisms.

ACKNOWLEDGMENTS

Research on martensitic transformations at Northwestern University, including the collaboration of the authors, is sponsored by the National Science Foundation under grant DMR-8820116. G.B. Olson is grateful for many years of inspiring collaboration with Professor Morris Cohen. A.L. Roitburd acknowledges the inspiration of academician G.V. Kurdjumov.

REFERENCES

1. G.V. Kurdjumov, *Dokl. Akad. Nauk SSSR*, Vol 60, 1948, p 1543
2. J.C. Fisher, J.H. Hollomon, and D. Turnbull, *Trans. AIME*, Vol 185, 1949, p 691
3. M. Cohen, E.S. Machlin, and V.G. Paranjpe, *Thermodynamics in Physical Metallurgy*, American Society for Metals, 1949, p 242
4. E.S. Machlin and M. Cohen, *Trans. AIME*, Vol 194, 1952, p 489
5. J.C. Fisher and D. Turnbull, *Trans. AIME*, Vol 197, 1953, p 921
6. J.H. Hollomon and D. Turnbull, *Prog. Met. Phys.*, Vol 4, 1953, p 333
7. R.E. Cech and D. Turnbull, *Trans. AIME*, Vol 206, 1956, p 124
8. J.R. Patel and M. Cohen, *Acta Metall.*, Vol 1, 1953, p 531
9. L. Kaufman and M. Cohen, *Prog. Met. Phys.*, Vol 1, 1958, p 165
10. M. Cohen, *Trans. AIME*, Vol 212, 1958, p 171
11. G.V. Kurdjumov and O.P. Maksimova, *Dokl. Akad. Nauk SSSR*, Vol 61, 1948, p 83
12. B.Y. Lyubov, *Probl. Metalloved. Fiz. Met.*, Vol 3, 1952, p 83
13. G.V. Kurdjumov, *J. Iron Steel Inst.*, Vol 195, 1960, p 26
14. J.D. Eshelby, *Prog. Solid Mech.*, Vol 2, 1961, p 89
15. A.L. Roitburd, *Phys. Met. Metall.*, Vol 10, 1960, p 1
16. J.W. Cahn and J.E. Hilliard, *J. Chem. Phys.*, Vol 32, 1959, p 688
17. A.L. Roitburd, *Kristallografiya*, Vol 1, 1962, p 291
18. F. Falk, *Z. Physik B—Condensed Matter*, Vol 51, 1983, p 177
19. G.R. Barsch and J.A. Krumhansl, *Phys. Rev. Lett.*, Vol 53, 1984, p 1069
20. G.B. Olson and M. Cohen, *Proc. Int. Conf. Solid-Solid Phase Transformations*, H.I. Aaronson, Ed., TMS-AIME, 1982, p 1145
21. G.B. Olson and M. Cohen, *J. Phys.*, Vol 43, 1982, p C4-75
22. G.L. Krasko and G.B. Olson, *Phys. Rev. B*, Vol 40, 1989, p 536
23. G.B. Olson, *Proc. ICOMAT 86*, Japan Institute of Metals, 1985, p 25
24. G.B. Olson, *Proc. ICOMAT 89*, Trans Tech, Zurich, 1989, p 89
25. D.M. Haezebrouck, "Nucleation and Growth of a Single Martensitic Particle," Sc.D thesis, Massachusetts Institute of Technology, 1987
26. M.S. Wechsler, D.S. Lieberman, and T.A. Read, *Trans. AIME*, Vol 127, 1953, p 1503
27. J.S. Bowles and J.K. Mackenzie, *Acta Metall.*, Vol 2, 1954, p 129
28. A.L. Roitburd, *Solid State Physics*, Vol 33, H. Ehrenreich, F. Seitz, and D. Turnbull, Ed., Academic Press, 1978, p 317
29. A.L. Roitburd and G.V. Kurdjumov, *Mat. Sci. Eng.*, Vol 39, 1979, p 141
30. S. Kajiwara, *Proc. ICOMAT 79*, Massachusetts Institute of Technology, 1979, p 362
31. G.B. Olson and M. Cohen, *Dislocations in Solids*, Vol 7, F.R. Nabarro, Ed., North-Holland, 1986, p 295
32. A.L. Roitburd, *New Aspects of Martensitic Transformation*, Japan Institute for Metals, 1976, p 100
33. A.L. Roitburd, *Phys. Stat. Sol. A*, Vol 10, 1977, p 333
34. A.L. Roitburd, *Sov. Phys.—Solid State*, Vol 10, 1969, p 2870
35. A.L. Roitburd, *Mat. Sci. Eng.*, Vol A127, 1990, p 229
36. G.B. Olson and M. Cohen, *Metall. Trans.*, Vol 7A, 1976, p 1897
37. M. Suezawa and H.E. Cook, *Acta Metall.*, Vol 28, 1980, p 423
38. G.B. Olson and M. Cohen, *Ann. Rev. Mat. Sci.*, Vol 11, 1981, p 1
39. K.E. Easterling and A.R. Tholen, *Acta Metall.*, Vol 24, 1976, p 333
40. H.C. Ling and G.B. Olson, *Proc. Int. Conf. Solid-Solid Phase Transformations*, H.I. Aaronson, Ed., TMS-AIME, 1982, p 1171
41. A.L. Roitburd, *Sov. Phys.-Dokl.*, Vol 26, 1981, p 92
42. P.C. Clapp, *Phys. Stat. Sol. (b)*, Vol 57, 1973, p 561
43. W. Cao, J.A. Krumhansl, and R.J. Gooding, *Phys. Rev. B*, Vol 41, 1990, p 11319
44. G. Guenin and P.F. Gobin, *Metall. Trans.*, Vol 13A, 1982, p 1127

45. G.B. Olson and M. Cohen, *Proc. ICOMAT 79*, Massachusetts Institute of Technology, 1979, p 310
46. M. Cohen and G.B. Olson, in *New Aspects of Martensitic Transformation*, Japan Institute of Metals, 1976, p 93
47. I-W. Chen, Y-H. Chiao, and K. Tsuzaki, *Acta Metall.*, Vol 33, 1985, p 1847
48. G.B. Olson, K. Tsuzaki, and M. Cohen, in *The Turnbull Symposium: Phase Transitions in Condensed Systems*, G.S. Cargill, F. Spaepen, and K.N. Tu, Ed., MRS, 1987, p 129
49. M. Lin, "Distributed Activation Kinetics of Heterogeneous and Homogeneous Martensitic Nucleation," Sc.D. thesis, Massachusetts Institute of Technology, 1990
50. I-W. Chen and Y-H. Chiao, *Acta Metall.*, Vol 33, 1985, p 1827
51. T. Saburi and S. Nenno, *Proc. ICOMAT 86*, Japan Institute of Metals, 1986, p 671
52. V. Raghavan and M. Cohen, *Metall. Trans.*, Vol 2, 1971, p 2409
53. F-R. Chen and G.B. Olson, *Proc. XIIth Int. Cong. Electron Microscopy*, San Francisco Press, 1990
54. R.J. Gooding and J.A. Krumhansl, *Phys. Rev.*, Vol 38, 1990, p 1695
55. Y-Y. Ye, C.T. Chan, K-M. Ho, and B.N. Harmon, *Int. J. Supercomputer Appl.*, Vol 4, 1990, p 111
56. G.R. Barsch, private communication, 1991
57. A. Umantsev and G.B. Olson, research in progress, Northwestern University, 1991

CHAPTER 4

Statistical Aspects of Martensitic Nucleation*

G.B. Olson, K. Tsuzaki, and Morris Cohen
Massachusetts Institute of Technology

Abstract

Analysis of the Cech-Turnbull small-particle martensitic transformation experiments in terms of heterogeneous nucleation theory defines an exponential nucleation-site potency distribution. Sensitive acoustic-emission detection of martensitic nucleation events shows that the same form of distribution describes the behavior of bulk polycrystals, and the influence of heat treatment on the distribution amplitude can be identified in this way. On the assumption of a random distribution of preexisting nucleation-site orientations, the effect of applied stress on the effective potency distribution has been calculated, thus accounting for an observed nonlinear stress dependence of the transformation kinetics. The model also predicts a transformation "yield locus" for multiaxial stress.

Introduction

While materials science has much in common with condensed matter science, an important difference is that it must include a science of "dirt" if we are to have any hope of understanding the phenomena governing the behavior of the technologically important subset of condensed matter that we call materials. An excellent example of the pioneering role of Professor David Turnbull in establishing this branch of science is the classic Cech-Turnbull small-particle experiment of 1956, which quantified the heterogeneous nature of martensitic nucleation and established principles of broad relevance to heterogeneous phenomena controlled by sparse defects. We here assess the pivotal role of this experiment in light of recent developments, with particular attention to the emerging technology of transformation plasticity.

The Nucleation-Site Potency Distribution

Small Particle Behavior

The results of the Cech-Turnbull experiment on an Fe-30Ni alloy are summarized in Fig. 1, showing the fraction χ of particles which contain detectable martensite on cooling to various temperatures as a function of particle size. If a number density N_v of nucleation sites is randomly distributed throughout the volume, the probability p that a particle of volume V_p contains at least one such site is (Ref 1):

$$p = 1 - \exp(-N_v V_p) \quad \text{(Eq 1)}$$

*G.B. Olson, K. Tsuzaki, and M. Cohen, Statistical Aspects of Martensitic Nucleation, Turnbull Symposium: Phase Transitions in Condensed Systems, G.S. Cargill, F. Spaepen, and K.N. Tu, Eds., MRS (1987) 129; 2011–1985 MRS Fall Meeting

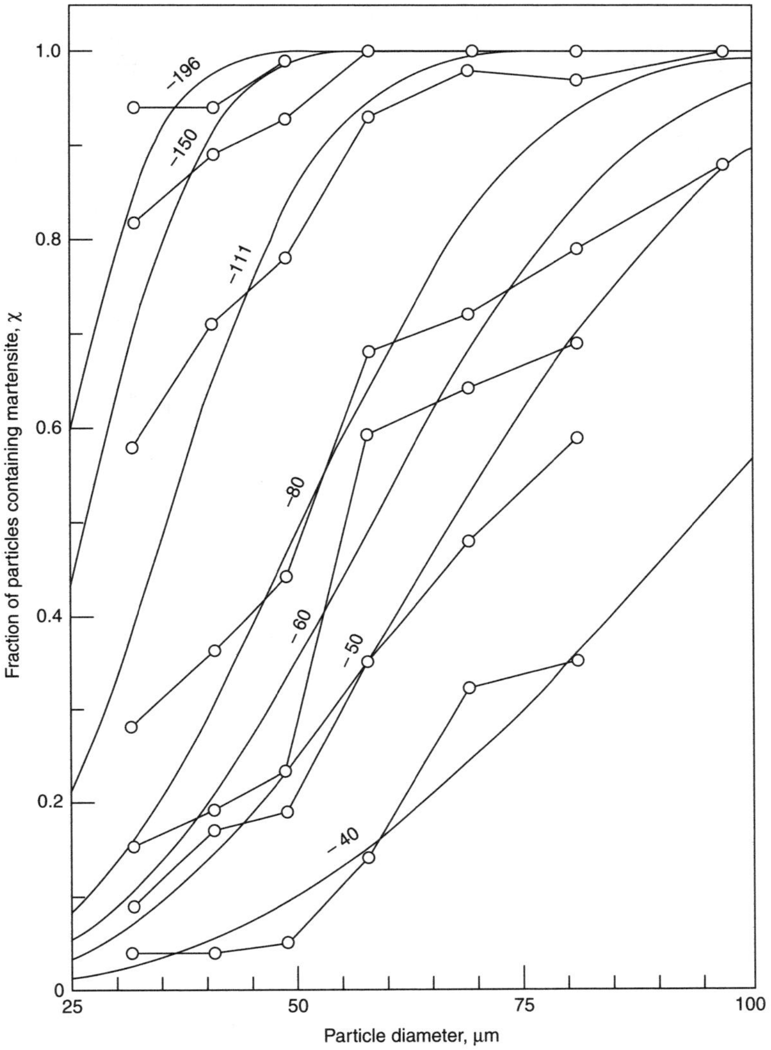

Fig. 1 Fraction χ of Fe-30Ni particles showing detectable martensitic transformation on cooling to various temperatures (°C) vs. particle diameter. The curves represent best fits of Eq 1. Source: Ref 2

If sites occur only on particle surfaces, a similar function of surface area obtains. Since the transformation of bulk materials demands a volume distribution of sites, the two possible alternatives are that nucleation occurs according to a volume distribution or a combination of volume and surface distributions. Of these possibilities, the simple volume distribution best describes the behavior observed in Fig. 1. Identifying χ with p, the best fit of Eq 1 is represented by the solid curves in Fig. 1 and these then define the value of N_v for sites of sufficient potency to nucleate the transformation on cooling to a given temperature T. From knowledge of the fcc-bcc transformation thermodynamics of this alloy, the potency of nucleation sites can be equivalently expressed in terms of the critical volume chemical free energy change Δg^{ch} for their operation, and the cumulative potency distribution thus defined by the Cech-Turnbull data of Fig. 1 is plotted in Fig. 2 (Ref 1).

Based on a simple model of barrierless heterogeneous classical martensitic nucleation via elastic interaction with superdislocation-like linear defects similar to those commonly observed in strain-induced nucleation experiments, the potency of a nucleation site

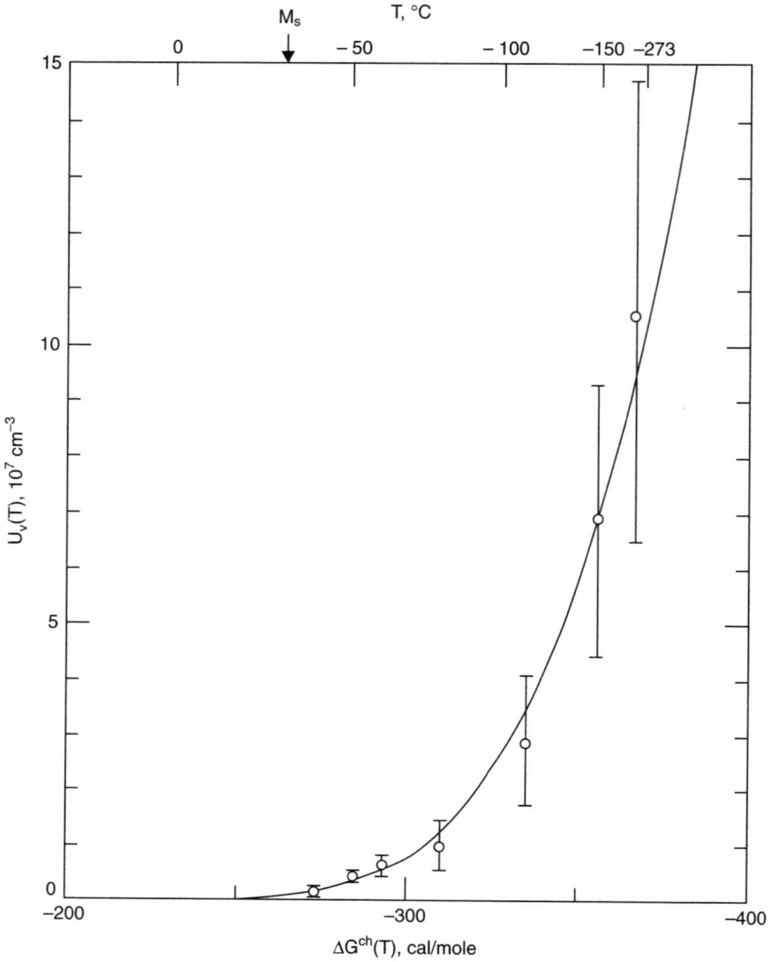

Fig. 2 Cumulative number density $U_v = N_v$ of operational nucleation sites as a function of molar chemical free energy change ΔG^{ch} and corresponding cooling temperature, T. The curve is the best fit of Eqs 2 and 3. Source: Ref 1

operating at a given volume free energy change Δg can be expressed in terms of a defect size parameter of the form (Ref 3, 4):

$$n = \frac{-2\gamma/d}{\Delta g + g^{e\ell} + w_f} \quad \text{(Eq 2)}$$

where γ is the nucleus specific interfacial energy, d is the close-packed interplanar spacing, $g^{e\ell}$ is an elastic coherency strain energy, and w_f is the frictional work of interfacial motion. With estimated values of γ=0.15 J/m^2 and $g^{e\ell} + w_f$ = 6.1×10^7 J/m^3, the thermodynamic potency distribution of Fig. 2 is consistent with a cumulative structural defect potency distribution of the form:

$$N_v(n) = N_v^o \exp(-\alpha n) \quad \text{(Eq 3)}$$

with α a constant shape factor and N_v^o the total number density of defects of all potencies. The solid curve in Fig. 2 is the best fit of Eq 2 and 3 giving α = 0.84 and N_v^o = 2×10^{17} m^{-3}. The form of Eq 3 was theoretically derived (Ref 1) for a simple defect model appropriate to the nucleation mechanism represented by Eq 2, and is typical of experimentally observed distribution functions for sparsely distributed nonequilibrium structural defects as encountered in fracture and fatigue. The Cech-Turnbull experiment thus defines a structural "dirt" function which may be of comparable fundamental significance to heterogeneous phenomena in materials as the Boltzmann function is to thermally activated phenomena. The distribution of Fig. 2 implies that the number of defects initiating transformation at the M_s

temperature of underformed bulk materials can be quite low; the search for the scarce highly potent defects described by Eq 3 seems to have more in common with high-energy particle physics than condensed matter physics. Observations of nucleation at more plentiful low-potency defects in metallic alloys generally provide qualitative support for the classical heterogeneous nucleation mechanism of Eq 2. The most quantitative confirmation of theory has come from recent observations of nucleation at well-characterized low potency defects in small particles of ZrO_2 ceramics (Ref 5). The statistics of transformation behavior in the ZrO_2 particles is found to conform to the same form of distribution function as the Cech-Turnbull experiment, extending the available data to smaller particle sizes and confirming the generality of Eq 3 (Ref 6).

Bulk Polycrystal Behavior

Measurements have now been extended to the extreme high-potency tail of the defect distribution using highly sensitive acoustic emission measurements to detect initial nucleation events in macroscopic polycrystalline specimens of an Fe-32.3Ni-0.004C alloy. The acoustic-emission transformation start temperature M_s^{AE} was measured using a Dunegan S9204 175 kHz resonance transducer with a Panametrics 5660B preamplifier, a 100–300 kHz band pass filter, and a Dunegan 301 totalizer operated at a total system amplification of 50–90 dB. The conventional transformation burst temperature M_b was also monitored by simultaneous electrical resistivity measurements. The grain size of specimens of 5×10^{-8} m^3 volume was varied between 75 and 1000 μm by austenitizing at 850–1350 °C followed by a one-hour stress-relief anneal at 500 °C. Figure 3 shows M_s^{AE} (open symbols) and M_b (closed symbols) measured at a cooling rate of 0.5 °C/min, plotted as a function of grain diameter $\bar{D}$ and corresponding linear intercept $\bar{\ell}$ Circles depict experiments run at an acoustic emission system gain of 70 dB, while triangles represent a 90 dB gain. The M_s^{AE} is insensitive to gain at larger grain sizes, but the highest gain is required to detect the earliest events in the finer grained material. Optical metallography revealed that the first event in the latter case corresponded to a small cluster of plates presumably autocatalytically connected to a single initial nucleation event. Serial sectioning at 50 μm intervals showed that in two instances a single martensitic plate had been detected.

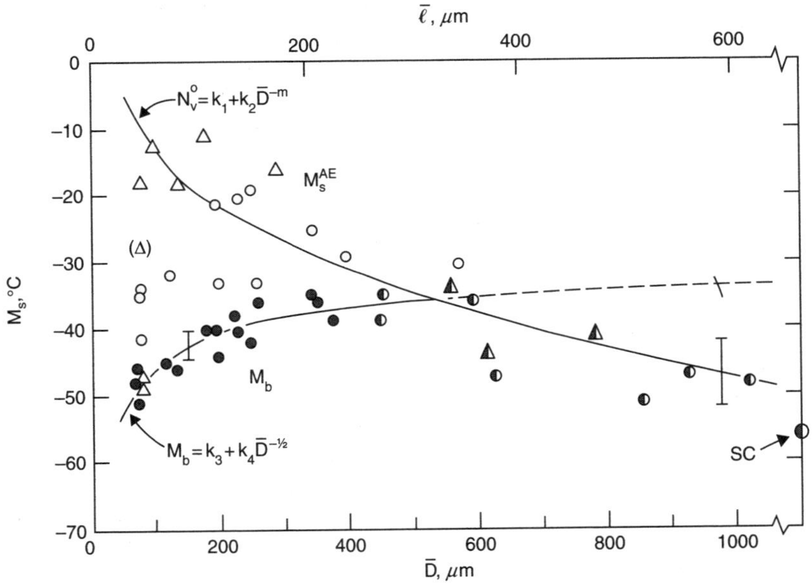

Fig. 3 Grain-size dependence of acoustic-emission M_s^{AE} (open symbols) and electrical-resistivity M_b (closed symbols). Circles represent experiments conducted at 70 dB gain while triangles denote 90 dB gain. SC indicates single crystal value from similar alloy.

Assuming the same defect-potency distribution form ($\alpha = 0.84$) as derived from the Cech-Turnbull data, the solid M_s^{AE} curve in Fig. 3 corresponds to the prediction of Eq 1–3 with a grain-size dependent N_v^o of the form:

$$N_v^o = k_1 + k_2\overline{D}^{-m} \quad \text{(Eq 4)}$$

with $m = 2.6$, $k_1 = 1.73 \times 10^{12}\ m^{-3}$, and $k_2 = 1.4 \times 10^6$ ($\overline{D}$ in meters). The value of k_1 is estimated from the single-crystal M_s of a similar alloy (Ref 7) denoted by SC in Fig. 3. Over the 75–1000 μm grain-size range examined, N_v^o then varies by three orders of magnitude from 9.0×10^{13} to $7.4 \times 10^{16}\ m^{-3}$, extrapolating to the value derived from the Cech-Turnbull small-particle data at $\overline{D} \approx 50 \mu m$. Although observed initial nucleation events were invariably related to grain boundaries, the $\overline{D}$ dependence of N_v^o is stronger than the $\overline{D}^{-1}$ dependence of the boundary area per volume, suggesting that more defective boundaries are associated with the finer grain sizes produced by lower temperature austenitizing.

The M_b temperature shows the form of $\overline{D}$ dependence commonly observed in previous studies and reflects a grain-refinement inhibition of the spread of autocatalytic nucleation across boundaries, analogous to Hall-Petch strengthening. The solid M_b curve in Fig. 3 represents a relation of the form:

$$M_b = k_3 + k_4\overline{D}^{-1/2} \quad \text{(Eq 5)}$$

with $k_3 = -27.9$ °C and $k_4 = -5.67$ °C $\sqrt{mm}$. This reflects the $\overline{D}$ dependence of the effective potency of autocatalytic nucleation. Above a critical $\overline{D}$ of ~500 μm, the macroscopic transformation bursting behavior is triggered by a single nucleation event controlled by M_s^{AE}. Bursting at lower temperatures than the extrapolated $M_b(\overline{D})$ relation leads to a catastrophic overdriven autocatalysis in which 40–50% of the sample transforms in a single burst.

Bars in Fig. 3 depict the standard deviation of the experimental scatter of M_s^{AE} and M_b relative to Eq 4 and 5. The larger scatter of the M_s^{AE} measurements associated with the first nucleation event is in good quantitative agreement with the corresponding confidence limits predicted by Eq 1 to 3. In addition, the statistical nature of the initial nucleation behavior governed by the defect-potency distribution predicts a specimen-size dependence of M_s^{AE}. This was confirmed in specimens with 950 μm grain size where the initial event triggers bursting. Over a range of specimen volumes of $1.7 - 13.3 \times 10^{-8}\ m^3$, the M_s temperature exhibited a systematic 20 °C variation, consistent with prediction. The M_b temperature of the finer 220 μm grain size material was size independent over the same specimen volume range, indicating a deterministic nature of the autocatalytic spreading, in contrast to the statistical nature of initial nucleation.

Effect of Applied Stress

The Mechanical Driving-Force Distribution

When martensitic transformation occurs under an applied stress as originally treated by Ref 8 the total volume free-energy change or "driving force" for transformation, Δg, is the sum of a chemical Δg^{ch} and a mechanical Δg^{σ} contribution. The Δg^{σ} term is orientation dependent and, for a uniaxial stress, σ, can be expressed by:

$$\Delta g^{\sigma} = \frac{\sigma}{2}\{\gamma_0 \sin 2\theta \cos \alpha + \varepsilon_0(1 + \cos 2\theta)\} \quad \text{(Eq 6)}$$

where γ_0 and ε_0 are the transformation shear and normal strains, θ is the angle between the stress axis and the habit normal, and α is the angle between the transformation shear direction and the maximum shear stress direction resolved on the habit. The value of Δg^{σ} lies in the range from

$$\Delta g^{\sigma}_{min} = -\frac{1}{2}\left(-|\sigma|\sqrt{\gamma_0^2 + \varepsilon_0^2} + \sigma\varepsilon_0\right)$$

to

$$\Delta g_{max} = -\frac{1}{2}\left(|\sigma|\sqrt{\gamma^2 + \varepsilon_0^2} + \sigma\varepsilon_0\right)$$

For a distribution of nucleation-site orientations, the effective site-potency distribution expressed in terms of the applied stress for site operation will be broadened relative to the structural potency distribution of Eq 3. In the case of a random orientation distribution, this broadening can be predicted from a mechanical driving-force distribution computed from the fraction of θ and α values in Eq 6 which yield a particular value of Δg^σ. Computed distributions for uniaxial tension and compression are shown in Fig. 4. The normalized total number of sites is plotted as a function of a normalized mechanical driving force $\xi = \Delta g^\sigma/\Delta g^\sigma_{max}$. Due to the ε_0 dilational term, the distribution has two peaks at ξ positions determined by the strain ratio $R = \varepsilon_0/\gamma_0$. The distributions plotted in Fig. 4 are for R = 0.2, appropriate to the transformation in steels. The positions of the distribution centers indicate that nucleation sites with positive ξ will be more plentiful than those with negative ξ in tension and vice versa for compression.

Combining the driving-force distribution of Fig. 4 with the thermodynamic defect-potency distribution defined by Eq 2 and 3, the cumulative operational nucleation-site potency distribution for cooling under constant stress can be expressed as:

$$N_v\left(\Delta g^{ch}, \sigma\right) = \int_{\Delta g^\sigma = \Delta g^\sigma_{min}}^{\Delta g^\sigma_{max}} \frac{dN^o_v(\Delta g^\sigma)}{d\Delta g^\sigma} \times \exp\left(\frac{2\alpha\gamma/d}{\Delta g^{ch} + \Delta g^\sigma + g^{e\ell} + w_f}\right) d\Delta g^\sigma \quad \text{(Eq 7)}$$

For $\gamma_0 = 0.2$, Fig. 5 shows the predicted applied tensile stress for various critical values of N_v for transformation, N_v^c, as a function of the increment in Δg^{ch} relative to its value at the stress-free M_s temperature, M_s^o. The Δg^{ch} increment can be directly related to cooling temperature as in Fig. 2. Also represented are a normalized chemical driving force

$$\phi = -\frac{\Delta g^{ch} + g^{e\ell} + w_f}{2\gamma/d}$$

and a normalized stress

$$\Sigma = -\frac{\Delta g^\sigma_{max}}{2\gamma/d}$$

The predictions are compared with experimental results of Ref 9 for Fe20Ni 0.34C and Ref 8 for Fe20Ni 0.52C obtained for the stress dependence of the M_s temperature. The Fisher data are in reasonable agreement with the nonlinear predictions of the statistical model with $N_v^c/N_v^o \approx 10^{-7}$. The latter value is typical of the N_v^c corresponding to M_b in the finer-grained material in Fig. 3. At high stress, however, the curves of Fig. 5 approach a limiting slope corresponding to $\Delta g^\sigma = \Delta g^\sigma_{max}$ represented by the dashed line. The latter corresponds to the prediction for a fully biased orientation distribution in which all nucleation sites are of the optimum orientation.

The data of Ref 8 appear to be in better agreement with the Δg^σ_{max} prediction. This can be rationalized if the stronger bursting tendency of their higher-carbon alloy provides the supercritical behavior found in the coarse-

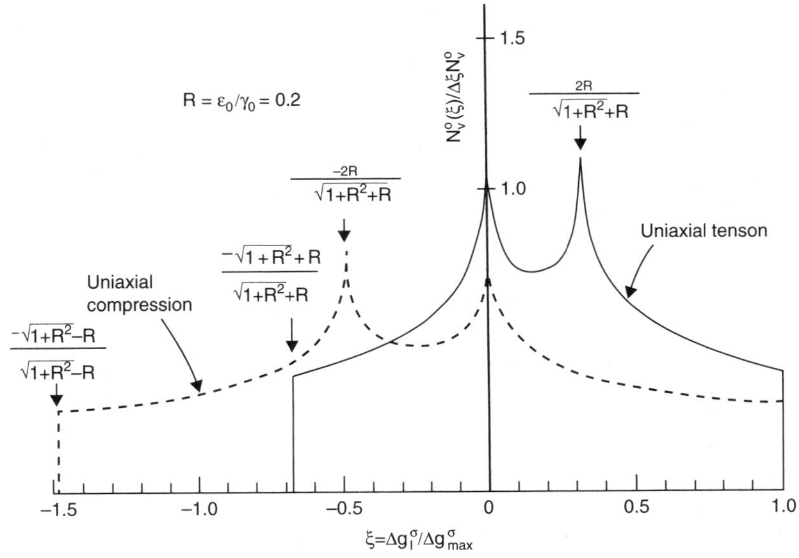

Fig. 4 Mechanical driving force distribution for random orientation distribution of nucleation sites under uniaxial applied stress.

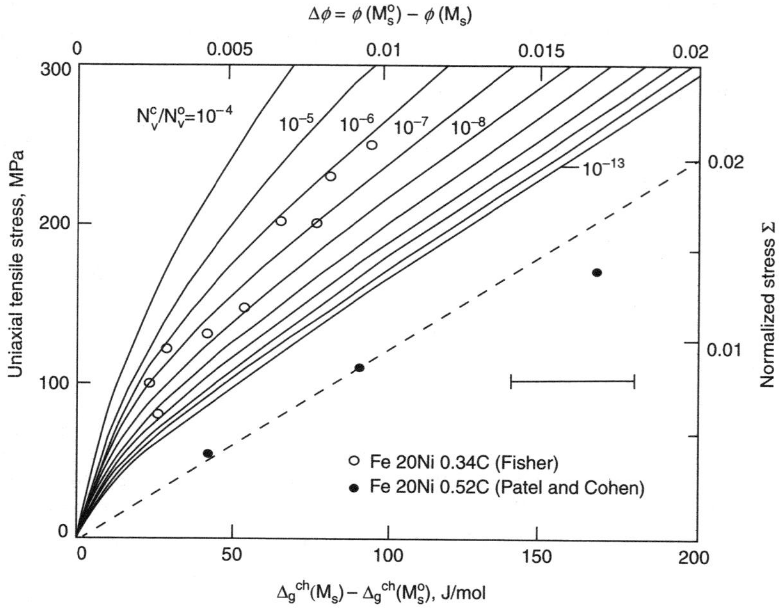

Fig. 5 Stress-driving-force plot corresponding to experimental measurements of the stress-dependence of the M_s temperature. Solid curves depict statistical model for N_v^c/N_v^o from 10^{-4} to 10^{-13}. Dashed curve represents $\Delta g^\sigma = \Delta g^\sigma_{max}$ model.

grained material of Fig. 3 for which a single initial event triggers a burst. The prediction for this case corresponds to $N_v^c/N_v^o = 10^{-13}$ in reasonable agreement with the observed behavior, apart from an offset in Δg^{ch} equivalent to a 4 °C temperature interval. This is smaller than the 95% confidence limit of 8 °C defined by the observed M_s scatter in Fig. 3. This limit is denoted by the bracket in the lower right of Fig. 5. Within such experimental scatter, then, the difference between these two alloys can be rationalized in terms of the dependence of N_v^c/N_v^0 on the severity of autocatalysis.

While the behavior described thus far applies to cooling under constant stress, further calculations based on the statistical model predict enhanced transformation under conditions of isothermal loading to the same level of Δg^{ch} and σ, in good agreement with such experimental comparisons reported by Ref 10. Cooling to temperature before monotonic loading allows the operation of some sites which would otherwise have an unfavorable interaction with applied stress. Extending the predictions of Fig. 5 into compressive stress, the statistical model also has the property of predicting a smooth curve in contrast to the sharp corner associated with a $\Delta g^\sigma = \Delta g^\sigma_{max}$ model.

Applications to Transformation Plasticity

Martensitic transformation under stress can provide an effective deformation mechanism for which constitutive relations can be derived from transformation kinetic theory. The mechanical behavior associated with retained austenite in martensitic steels and the toughening influence of metastable dispersed phases in certain ceramic systems such as partially stabilized zirconia have called attention to the special transformation plasticity behavior of dispersed particles in a solid matrix. When the dispersed metastable particles are sufficiently isolated that autocatalytic nucleation interactions can be neglected, the constitutive relations for stress-assisted transformation plasticity can be predicted directly from the nucleation site potency distribution.

Following the considerations applied to the data of Fig. 1, the fraction of a large number of particles to transform to martensite can be described by the probability p of Eq 1. When deformation is controlled by stress-assisted transformation, the transformed fraction f and the resulting plastic strain ε can be linearly related:

$$f = \varepsilon/\varepsilon_1 \qquad \text{(Eq 8)}$$

where ε_1 is the total strain at f = 1. Combining Eqs 1–3 with the considerations previously described for isothermal monotonic loading, the isothermal σ-ε curve for transformation plasticity from dispersed particles of size V_p can be described by:

$$f = \varepsilon/\varepsilon_1 = 1 - \exp\left\{-V_p\left[\int_{\Delta g^{\sigma}_{min}}^{o} \frac{dN_v^o(\Delta g^{\sigma})}{d\Delta g^{\sigma}} \times \exp\left(\frac{2\alpha\gamma/d}{\Delta g^{ch} + g^{el} + w_f}\right) d\Delta g^{\sigma} + \int_{o}^{\Delta g^{\sigma}_{max}} \frac{dN_v^o(\Delta g^{\sigma})}{d\Delta g^{\sigma}} \times \exp\left(\frac{2\alpha\gamma/d}{\Delta g^{ch} + \Delta g^{\sigma} + g^{el} + w_f}\right) d\Delta g^{\sigma}\right]\right\} \quad \text{(Eq 9)}$$

In the case of the simple $\Delta g^{\sigma} = \Delta g^{\sigma}_{max}$ model the relation becomes

$$f = \varepsilon/\varepsilon_1 = 1 - \exp\left[-V_p N_v^o \times \exp\left(\frac{2\alpha\gamma/d}{\Delta g^{ch} + \Delta g^{\sigma}_{max} + g^{el} + w_f}\right)\right] \quad \text{(Eq 10)}$$

Calculated normalized stress-strain curves (Σ-f) at ϕ = 0.07 for $N_v{}^oV_p$ = 10^2, 10^3, 10^4 are depicted in Fig. 6. Unnormalized stress is shown along the right vertical axis. The stress-strain curve based on the $\Delta g^{\sigma}{}_{max}$ prediction for $N_v{}^oV_p$ = 10^4 is shown for comparison. For large particles where nucleation sites are plentiful, some particles transform to martensite during cooling before deformation (**$N_v{}^oV_p$ = 10^4**) and the transformation flow stress is low. For small particles, a high strain hardening arises from early exhaustion of the most potent sites, and further strain requires operation of less potent sites at higher stresses. The compression case shows a slightly higher strain

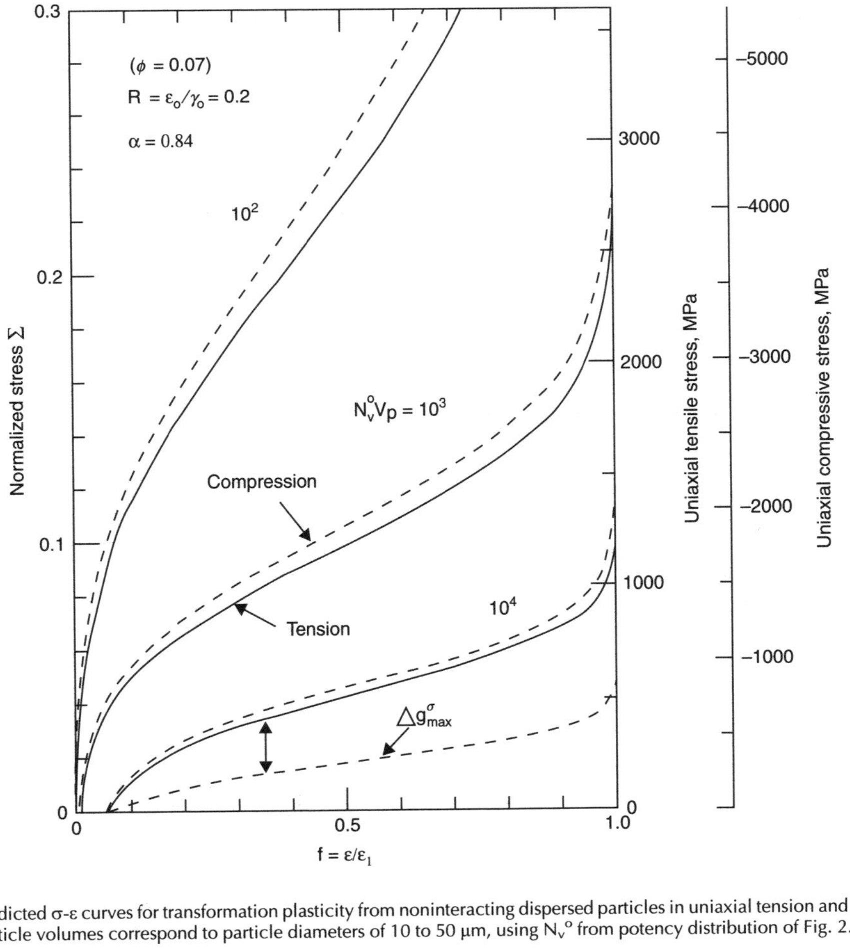

Fig. 6 Predicted σ-ε curves for transformation plasticity from noninteracting dispersed particles in uniaxial tension and compression. Particle volumes correspond to particle diameters of 10 to 50 μm, using $N_v{}^o$ from potency distribution of Fig. 2.

hardening rate than in tension arising from the wider mechanical driving force distribution in compression. The stress level is higher than that of the Δg^{σ}_{max} prediction by about a factor of two.

As the stress-strain curve of the Δg^{σ}_{max} prediction can be obtained analytically, it is useful to express the stress-strain curve of the random orientation prediction by that of the Δg^{σ}_{max} prediction as:

$$\Sigma_{\alpha}(f) = \beta\Sigma_{m}(f) \qquad \text{(Eq 11)}$$

where $\Sigma_{\alpha}(f)$ and $\Sigma_{m}(f)$ are normalized stresses of the random orientation prediction and the Δg^{σ}_{max} prediction, respectively. The coefficient β tends to be slightly smaller with increase in f and with decrease in $N_v^oV_p$, but is almost constant. The value of β in uniaxial tension (β_t) is around 2.2 and the ratio β_c/β_t is around 1.07, where β_c corresponds to compression.

Recent compression experiments performed on dispersed ZrO_2 ceramics under superimposed hydrostatic pressure provide σ-ε curves very similar in shape to those predicted in Fig. 6, but show significant quantitative departures which are rationalized in terms of autocatalytic interactions among particles in the high volume-fraction dispersions investigated (Ref 11).

The statistical model can also be used to calculate a transformation "yield" locus for multiaxial stress. Generalizing Eq 6 and the Δg^{σ} distribution of Fig. 4 to other stress states and applying the type of analysis represented by Eq 9, a computed yield locus for biaxial stress is represented in Fig. 7 for three values of normalized chemical driving force,

$$\phi_{ch} = \frac{\Delta g^{ch} + g^{el} + w_f}{2\gamma/d}$$

$N_v^c/N_v^o = 5 \times 10^{-6}$ was deduced from the yielding behavior of austenitic TRIP steels. Stresses are normalized to the transformation stress in uniaxial tension, σ_o. The smooth "egg shaped" yield locus is compared with the sharp-cornered shape given by the $\Delta g^{\sigma} = \Delta g^{\sigma}_{max}$ model. At low ϕ_{ch}, corresponding to

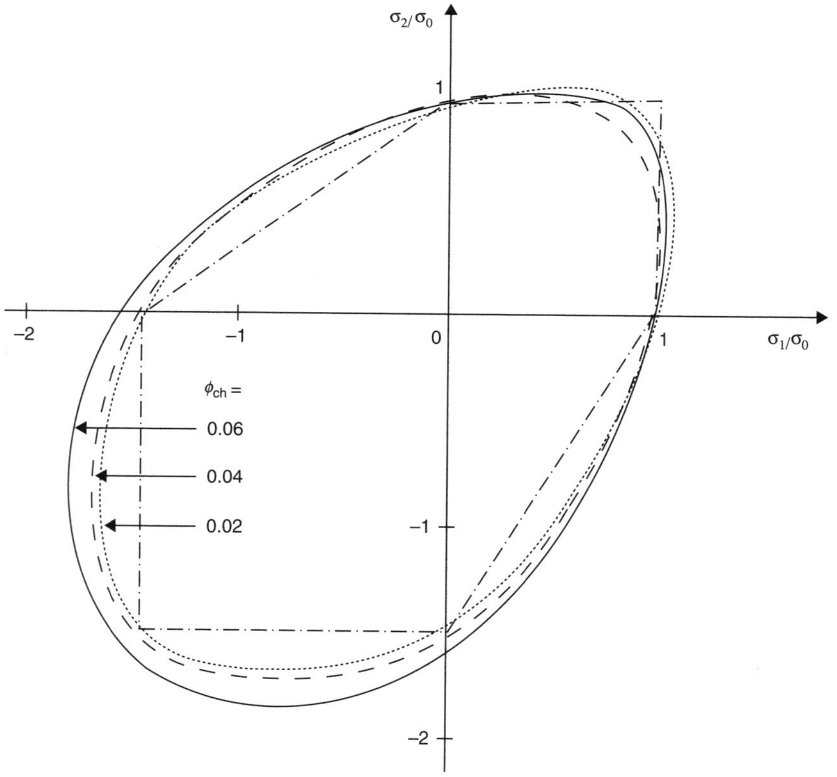

Fig. 7 Biaxial-stress transformation yield-locus shapes at three levels of normalized chemical driving force, ϕ_{ch}. Sharp-cornered locus represents Δg^{σ}_{max} model. $N_v^c/N_v^o = 5 \times 10^{-6}$.

higher transformation stresses, the locus moves toward the $\Delta g^{\sigma}{}_{max}$ limit, consistent with the high-stress limiting slopes depicted in Fig. 5. Further calculations for triaxial stress states indicate that the yield condition in the low-stress limit (high ϕ_{ch}) can be reasonably approximated by a linear combination of tensor invariants, σ_h and $\bar{\sigma}$. For tensile stress states ranging from plane stress to the plane-strain crack-tip case, this can be represented by:

$$\bar{\sigma} = \sigma_s - k\sigma_h \qquad \text{(Eq 12)}$$

with σ_s the yield stress in pure shear and k = 0.794 for the transformation in steels. Predictions of transformation behavior at a crack tip can thus be obtained from observations in uniaxial tension. Such predictions are in good agreement with recent experimental observations of crack-tip transformation behavior and associated toughening in high-strength TRIP steels (Ref 12).

Closure

The martensitic nucleation-site potency distribution provides a rational basis for many statistical aspects of the kinetics of heterogeneous martensitic transformations, including specimen-size effects and the magnitude of experimental scatter. When combined with the mechanical driving-force distribution associated with a random distribution of nucleation site orientations relative to an applied stress, the statistical model also accounts for observed nonlinear stress effects, describes potency-distribution-based strain hardening in dispersed-phase transformation plasticity, and allows prediction of the transformation behavior under multiaxial stress.

A research program is now under way in which the predicted austenite-stabilizing effect of particle-size refinement is exploited to control the stability of dispersed austenite for optimum transformation plasticity behavior during plane-strain tensile deformation in order to provide enhanced stretch form-ability in high-strength microalloyed sheet steels, and enhanced fracture toughness in ultrahigh-strength martensitic alloy steels. These developments are a direct extension of the defect-potency distribution defined by the classic Cech-Turnbull small-particle experiments of 1956.

ACKNOWLEDGMENTS

Research at MIT on the mechanism and kinetics of martensitic transformations is sponsored by NSF under Grant DMR-7915196. Research on transformation plasticity is sponsored by DOE under Grant DE-F602-84ER45154.

REFERENCES

1. M. Cohen, and G.B., Olson, in: *New Aspects of Martensitic Transformation*, Japan Inst. of Metals (1976), p 93
2. R.E., Cech, and D. Turnbull, (1956) *Trans. AIME* 206, 124
3. G.B. Olson, and M. Cohen, (1976) *Metall. Trans.* 7A, 1897, 1905, 1915
4. G.B. Olson, and M. Cohen, in: *Proc. Int'l Conf. Solid-Solid Phase Transf.* (AIME, 1982), p 1145
5. I.-W., Chen, and Y.-H., Chiao, (1985) *Acta Metall.* 33, 1827
6. I.-W., Chen, Y.-H., Chiao, and K., Tsuzaki, (1985) *Acta Metall.* 33, 1847
7. E.S. Machlin, and M. Cohen, (1951) *Trans. AIME* 191, 746
8. J.R., Patel, and M. Cohen, (1953) *Acta Metall.* 1, 531
9. P. Fisher, (1974). Ph.D. thesis, Univ. New South Wales, Australia; also P. Fisher, and D.J.H. Corderoy, (1974) *J. Australian Inst. Met.* 19, 51
10. G.F., Bolling, and R.H. Richman, (1970) *Scripta Metall.* 4, 539
11. I.-W., Chen, and P.E. Reyes Morel, (1986) *Acta Metall.* 34
12. R.H. Léal, Transformation Toughening of Metastable Austenitic Steels, Ph.D. thesis, MIT (1984)

Transformations—Selected Works of G.B. Olson on Materials, Microstructure, and Design
C.E. Campbell, M.V. Manuel, and W. Xiong, editors

CHAPTER 5

New Directions in Martensite Theory*

G.B. Olson, Northwestern University

Abstract

Advances in materials science, applied mechanics, physics, and mathematics offer new opportunities for the science and engineering of martensitic materials. A systems approach to multilevel dynamic microstructure emphasizes the importance of distributed defect hierarchy in predictive control of desired behaviors. Total energy electronic calculations and Landau free energy functions describe nonlinear energetics of lattice deformations with atomic shuffles, and incorporation of nonlocal strain gradient energies in Landau-Ginzburg models has allowed numerical simulation of homogeneous and heterogeneous nucleation as well as semicoherent growth and twin variant conversion. Analysis of multiscale defect interactions addresses growth in a plastic matrix and the nature of nucleation site potency distributions in both thermoelastic and non thermoelastic systems. Predictive design theory incorporates macroscopic constraints in polycrystals and thin films. Biological molecular martensites underlying infection mechanisms in HIV and influenza viruses offer a fertile area for adaptation of martensite theory.

Systems—Dynamic Hierarchy

Modern numerical methods paced by rapidly advancing computational power have opened the way to a far more realistic theoretical treatment of the complexity of materials behavior. The resulting enhanced appreciation of natural complexity has fostered renewed interest in the "systems" view of materials proposed by C.S. Smith (Ref 1), which recognized the inevitable hierarchical nature of materials structure, with strong interactions amongst length scales, and an interplay between perfection and imperfection involving a "duality" whereby structure can be equivalently regarded as a hierarchy of three dimensional space filling domains of relative perfection or a hierarchy of lower-dimensionaldefects, imperfections which bound them. Reference 1 warned of a cultural bias toward perfection, and emphasized the importance of defect hierarchy in the willful control of materials behavior. That view also acknowledged an intrinsic dynamic nature, whereby a spectrum of relaxation times associated with structural hierarchy assures an intrinsic path-dependent nonequilibrium structure of real materials. This inherent complexity sets limits on the degree of predictability of materials behavior from "first principles" approaches, and Morris Cohen (Ref 2) eloquently argued the importance of a "reciprocity" between the deductive cause/effect logic of reductionist analysis and the inductive goal/means logic of systems synthesis whereby useful insights such as structure/properly relations can be distilled from natural complexity by probing structure from the viewpoint of specific desired properties.

This philosophy has formed the basis of a systems approach to computational materials design which has been successfully applied to

*G.B. Olson, New Directions in Martensite Theory, Proc. ICOMAT 98, *Mat. Sci. & Eng.*, A273–275 (1999) p 11–20

martensitic steels over the past decade, and is now being extended to all classes of materials (Ref 3). The pioneering application in martensites is due largely to an early scientific appreciation of complexity in martensitic systems. Current design research aimed at acceleration of the full materials development cycle emphasizes (a) integration of microstructural evolution in process models supporting a design-for-processability approach anticipating process scaleup; and (b) a probabilistic science quantifying the role of distributed defect hierarchy, especially in structure/property relations, for predictive control of variation to accelerate materials qualification. This review assesses the evolution of martensite theory in the context of this changing materials engineering environment.

Consonant with this view, Fig. 1 depicts the spatiotemporal hierarchy of first-order displacive (e.g., marlensitic and bainitic) transformations. Blocks at the left represent interactive levels of structure both preceding and generated by the transformation, while blocks at the right denote a sequence of dynamic phenomena interacting with these structural levels. As denoted by the interconnections, initial nucleation is sensitive to interfacial defects generated by the processing history, determining the initial number density N_i of heterogeneous nucleation sites and the critical driving force at which they operate. The latter determines the dynamics of unit growth events and influences the degree of dislocation substructure generation determining subunit morphology (plate/lath) and the initial average particle volume $\bar{V}_0$. The character of growth in turn establishes the elastic or plastic fields governing the number density N_A and potency of autocatalytic nucleation sites, which evolve with phase fraction, f. The character of autocatalytic nucleation in turn establishes the initial geometry of variant pairs influencing the cooperative multivariant growth of higher level groups (packet/sheaf) determining the evolution of average particle volume $\bar{V}(f)$. These are also influenced by the transgranular spread of transformation in polycrystals, as well as other macroscopic constraints of component shape and external boundary conditions. On the finest scale denoted, solutes and other fine scale defects exert an influence on several levels, primarily through interfacial friction. Progress in martensitc theory has addressed each of these aspects, with due attention to both hierarchies of space-filling domains and space-dividing defects.

An overall assessment of the current state of computational materials science is that simulations have demonstrated remarkable success in reproducing aspects of observed natural complexity. In the spirit of Cohen's reciprocity,

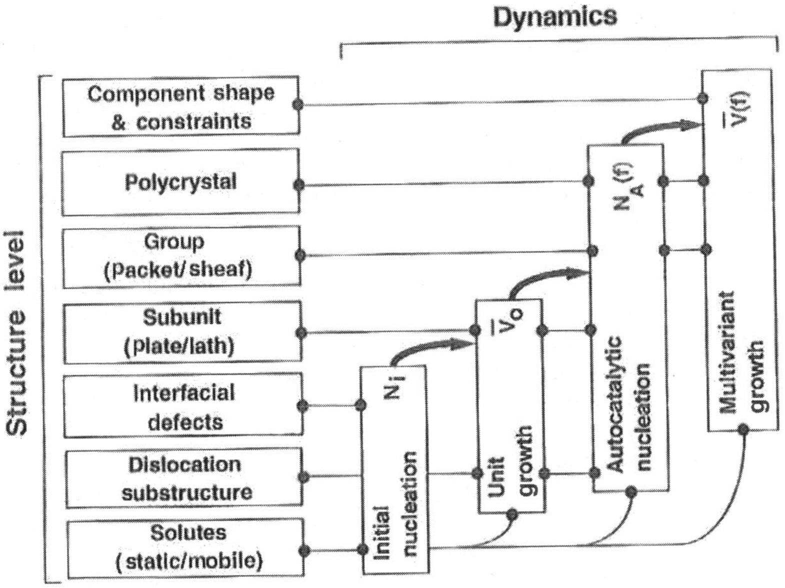

Fig. 1 Flow-block diagram of martensitic transformation as a multilevel dynamic system.

the challenge for the near future is to purposefully exploit the capabilities of these techniques (allowing parametric studies not accessible by experiment) to yield new insights offering predictive control of specific desired behaviors.

Physics of Lattice, Deformations, and Interfaces

A major advance in theoretical capability has been the improved accuracy of ab initio electronic quantum mechanical calculations of the energy differences between parent and product lattices, and the interpretation of their origins in terms of band structure and bonding. The approach offers the potential for prediction of new directions in alloy chemical composition for control of desired phase stability, which has yet to be significantly exploited. A further application of these methods, which has already had significant impact on martensite theory, is the prediction of the nonlinear energetics of homogeneous lattice deformations between structures. For relatively high symmetry strain paths through energy saddle points, calculations have included alkali metals (Ref 4, 5), Fe and it alloys (Ref 6–8), Cu-base alloys (Ref 7), and TiNi alloys (Ref 9, 10). The lattice deformation calculations define energy density barrier heights important to coherent interfacial energy, and identify conditions where energy surfaces are cusped (Ref 6) rather than smooth. Most important to issues of competing transformation mechanisms, calculations as a function of pressure identify critical driving forces for lattice instabilities (Ref 6).

In transformations involving atomic shuffle displacements within the unit cell as well as lattice strains, energy calculations for all possible combinations of shuffle displacements and strains along a lattice deformation path can become prohibitively complex. An especially useful tool in this case is the development of Landau free energy functions which incorporate crystal symmetry constraints and define external configurations. Coefficients of functions thus defined can be calibrated by a reasonable number of electronic total energy calculations. Such Landau functions have now been derived for specific structural transformations involving multiple shuffle-based order parameters (Ref 11–13). An important advance bringing this approach to technologically important martensitic systems is the recent analysis by Barsch (Ref 13) of the Landau order parameters of the transformations from B2 to the R, B19 and B19′ structures in TiNi and AuCd alloys.

Figure 2 summarizes the symmetry defined shuffle and train order parameters for each group → subgroup transformation in these systems. Detailed analysis of available phonon dispersion curves in both the TiNi and AuCd systems has allowed development of analytical Landau free energy functions versus temperature for the B2 → R transformations, and functions are being developed for the larger strain B2→B19 and B2→B19′ martensitic transformations (Ref 14).

A versatile method for the bridging of length scales within the structural hierarchy of Fig. 1 is the application of nonlocal continuum models. Detailed calculations at a fine scale can be "homogenized" by gradient energy coefficients to define size-dependent constitutive laws to treat coarser-scale problems. The approach is now seeing broad application in computational materials science via the "phase field" approach to simulation of phase transformations as pioneered by Cahn and Hilliard (Ref 15). Within this realm, martensitic transformations represent perhaps the most rigorous application of the approach, building on the seminal research of Barseh and coworkers developing Landau-Ginzburg models with symmetry-constrained gradient energy coefficients deduced from measured phonon dispersion curve anomalies in specific martensitic systems (Ref 16–18), recently extended to incorporate gradient contributions of shuttle parameters (Ref 12). In strongly first-order systems without such anomalies, gradient energy coefficients can be deduced from direct observation of interfacial core widths by HREM lattice imaging. This requires special systems with near-rational invariant lines to allow edge-on viewing of a semicoherent interface in a low index direction. This has allowed detailed analysis of interfacial core displacement fields for the B2→18R transformation in CuZn (Ref 19) and more recently the FCC → BCT transformation in an Fe alloy (Ref 20).

Interfacial energies, particularly for the internal twin boundaries of martensitic substructures, have been calculated both from variational solutions of the Landau-Ginzburg models (Ref 16–18), and by direct calculations by quantum electronic methods (Ref 21, 22). Greater accuracy of the latter methods is now

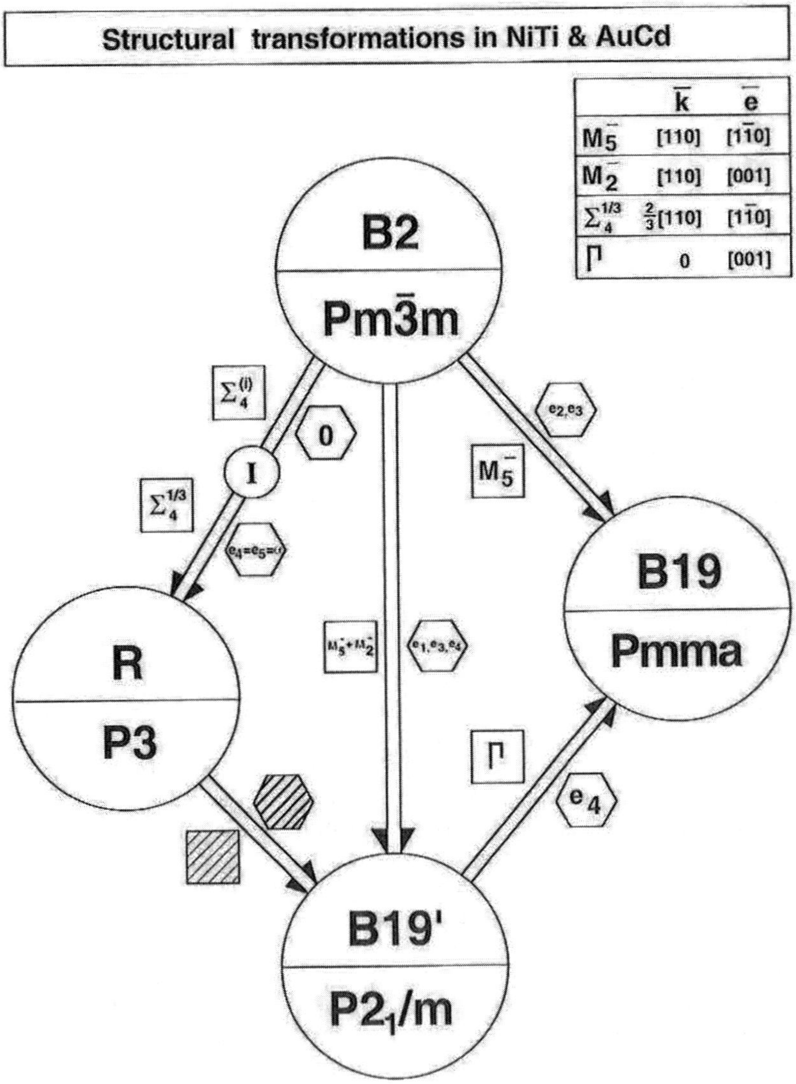

Fig. 2 Summary of shuffle and strain order parameters for group/subgroup transformation in TiNi and AuCd alloys. Source: Ref 13.

being achieved by addressing nonlocality at the electronic level as well. A modification of the full potential augmented plane wave (FLAPW) method replacing the local spin density approximation (LSDA) with a generalized gradient approximation (GGA) has for the first time predicted the correct relative energies and lattice parameters of the FCC and BCC structures of Fe from an ab initio calculation (Ref 23), and is now being applied to BCC incoherent twin boundaries (Ref 24).

The new predictive capabilities in the physics of phase stability and lattice deformations offer a fertile ground for inventing new martensites from a very fundamental perspective. New devices such as microactuators based on traditional thermally induced martensitic transformations suffer limitations of heating and cooling processes. An "electromartensite" switched on and off directly by an electric field would offer greater efficiency, but large strain ferroelectric martensites are constrained by the intrinsic brittleness of insulators. "Magnelomartensite" driven by magnetic fields can offer the better mechanical properties of metallic systems; prototype systems show promise

(Ref 25), and detailed materials design considerations have been explored by R. James (Ref 26).

Nucleation

A topic which has particularly benefited from the nonlocal continuum Landau-Ginzburg models is the treatment of martensitic nucleation, the first process in Fig. 1. Compared to finite element methods, new meshless numerical methods such as the element-free Galerkin (EFG) method (Ref 27) have proved to be more efficient in the computation of strain gradients to find variational solutions within the nonlocal models. A first problem to test this approach has been the modeling of homogeneous nucleation at high driving forces where an issue of principal interest is the degree of nonclassical behavior, characterized by critical nuclei of reduced strain amplitude. Chu and Moran (Ref 28) rigorously treated the case of a simple 2D coherent shear transformation and demonstrated reduced strain amplitude and divergence of critical nucleus size and interfacial width on approaching a lattice instability, similar to the Cahn-Hilliard treatment of nonclassical nucleation in decomposing fluids (Ref 15). Based on these solutions, an approximate treatment was developed to describe the energetics of nonclassical homogeneous nucleation in 3D systems including the FCC-BCT transformation in steels (Ref 29). For the latter case, the model was fit to measured critical driving force data for homogeneous martensitic nucleation in small particle experiments conducted in the Fe-Co system (Ref 30). The solid curve in Fig. 3 represents the model prediction for fully coherent homogeneous nucleation with a coherent interfacial energy corresponding to 20 mJ/m^2 in the classical limit. Based on the electronic calculations of lattice stability limits (Ref 6), the measured nucleation critical driving forces are at approximately 70% of the critical driving force for instability, and the new model here predicts that departures from fully classical behavior reduce the nucleation critical driving force by 20% from the classical limit (upper dashed curve). The previous rigorous calculations indicate that the critical nucleus strain amplitude is still within 90% of classical at this level of driving force.

The homogeneous nucleation critical driving forces of Fig. 3 are seven times higher than for nucleation under normal conditions, which is well known to be controlled by heterogeneous nucleation. Observed nucleation sites are generally linear defects with dislocation character, typically partially stabilized by interaction with interfaces (e.g., Ref 31). Groups of lattice dislocations were first proposed to model such defects in 1949 (Ref 32), and have been employed in detailed dislocation models of heterogeneous nucleation Kaufman and Cohen (Ref 33). The latter have demonstrated that barrierless nucleation at such defects is preceded at lower driving force by the formation of local martensite embryos, consistent with the "preexisting embryo" hypothesis proposed by Kaufman and Cohen in 1958 (Ref 34). Linear elastic dislocation models are inadequate to treat the earliest stage of the process, and this has now been addressed by Reid (Ref 35) using the nonlocal Landau-Ginzburg models and EFG method. A group of three lattice dislocations in the form of a finite tilt boundary segment (equivalent to a wedge disclination dipole) is embedded in the continuum, representing the dislocations as displacement gradient source functions continuously distributing the Burgers vector content over the scale of a lattice vector. The computed strain field at the upper right of Fig. 4 is obtained for a linear elastic material with the parabolic energy-strain curve shown at the upper left. The lower figures show the computed behavior for the nonlinear Landau-Ginzburg model of a simple coherent

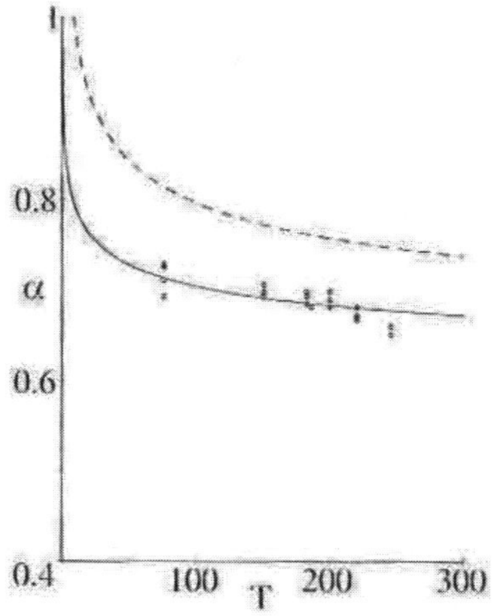

Fig. 3 Critical driving force tor homogeneous nucleation (normalized to instability driving force) for FCC-BCT martensitic transformation in defect-free small particles. Solid curve is fit of nonclassical coherent nucleation; dashed curve is classical limit. Source: Ref 29, 30

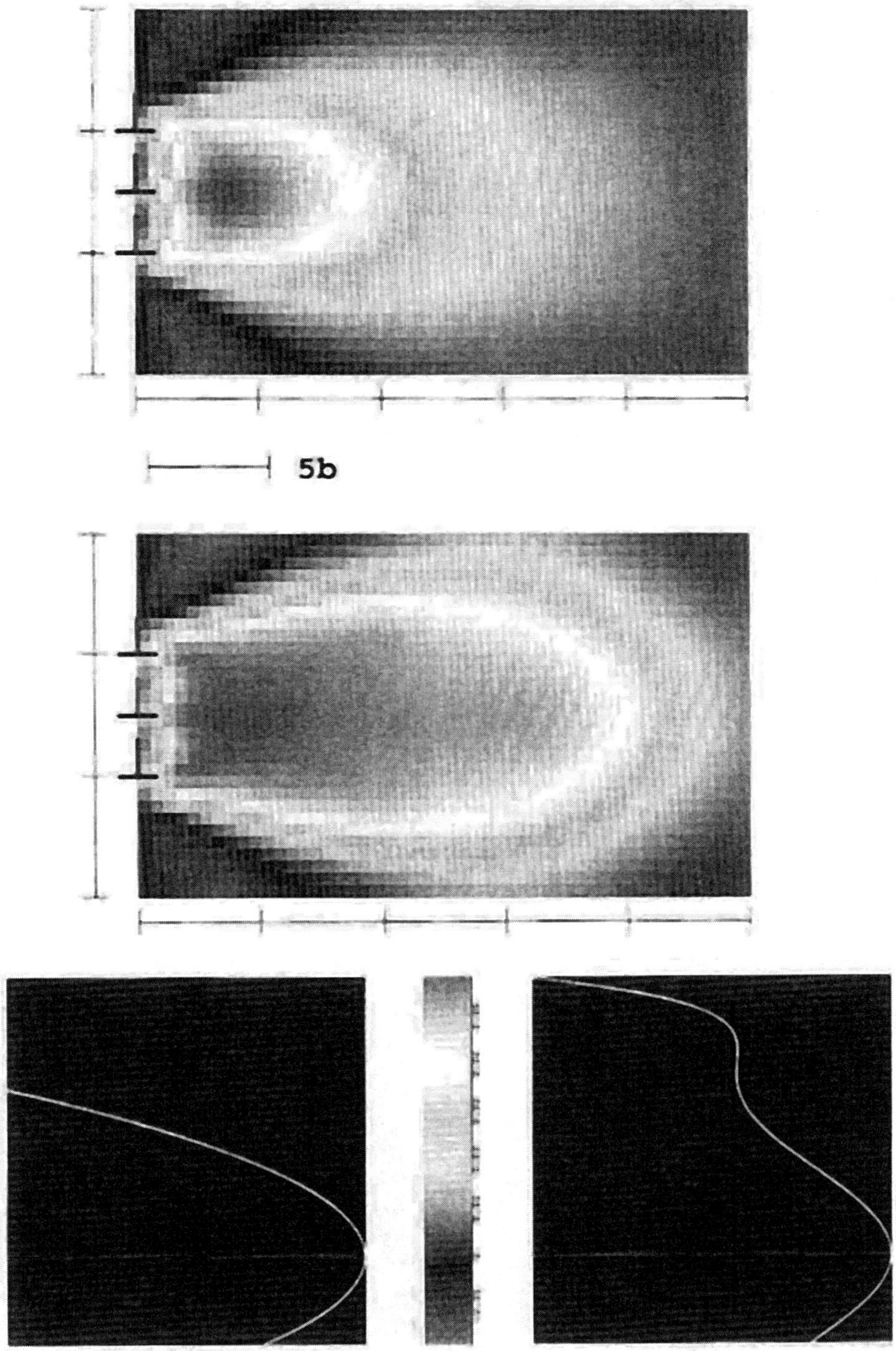

Fig. 4 Computed defect strain fields in linear clastic (upper) and nonlinear elastic (lower) materials, with corresponding energy-strain functions at left. Lower ease demonstrates martensitic embryo at conditions corresponding to instability point of martensitic phase. Source: Ref 35

shear transformation at a high temperature corresponding to the instability point of the martensite phase. At this extreme temperature, a substantial embryo of the fully formed martensite is already stabilized at the defect. The result is in sharp contrast to the alternative proposal of a "strain embryo" (Ref 32) or "localized soft model" (Ref 36) mechanism whereby a highly strained region collapses to the martensite structure at the M_s temperature. The rigorous calculation shows that the "soft mode" is already "frozen" into the product structure well above the M_s temperature, in a process closely resembling a defect dissociation. Additional calculations have revealed how the embryo can be further stabilized by interaction with a nearby free surface (Ref 37).

The nonlocal Landau-Ginzburg models aided by new numerical methods have provided long-sought quantitative answers regarding the two extremes of nonclassical homogeneous nucleation near a parent phase instability, and classical heterogeneous embryo formation near the martensitic phase instability. Further application to 3D problems such as FCC-BCT heterogeneous nucleation with varying slates of coherency promises further insights.

Growth

Addressing the second unit process of Fig. 1, Wang and Khachaturyan (Ref 38) employed the time-dependent Ginzburg-Landau (TDGL) equation, incorporating a linear-viscous dynamic response, in the 3D dynamic simulation of a cubic-tetragonal martensitic transformation with internal shuffles (improper ferroelastic) giving the sequence of structures depicted in Fig. 5. Starting from a supercritical embryo of arbitrary shape, the detailed simulation predicts the dynamic growth form of an

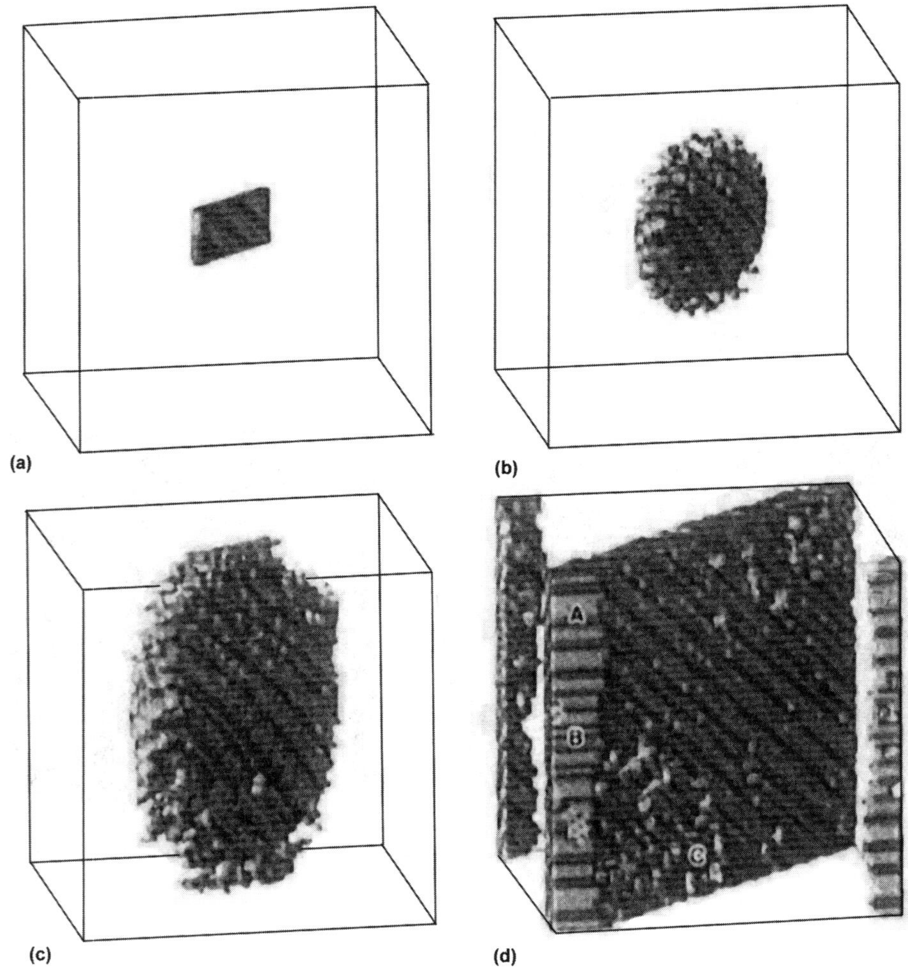

Fig. 5 Simulation of growth of supercritical semicoherent (internally twinned) martensitic embryo in nonlinear nonlocal continuum model. Source: Ref 38

internally twinned plate which is strikingly similar to the oblate spheroidal semicoherent embryo assumed by Kaufman and Cohen (Ref 34) and Knapp and Dehlinger (Ref 39) and predicted by the growth dynamic model of Raghavan and Cohen (Ref 40).

Relevant to the fourth process of multivariant growth in Fig. 1, the Wang-Khachaturyan simulations also treat multivariant microstructural evolution in the limiting case of semicoherent homogeneous nucleation. Detailed analysis of such simulations is a fertile area for further research on multivariant growth, and some analysis of the kinematic constraints in multivariant interactions has been initiated (Ref 41). For the case of mechanically induced transformation, Roytburd and Slutsker (Ref 42) extended a theory of polydomain microstructures incorporating the coherent thermodynamics of Larche and Cahn (Ref 43). With phase fraction and variant fractions as primary state variables, the evolution of multivariant microstructure under displacement control is predicted as a series of thermomechanical equilibrium states for well-catalyzed (easy nucleation) systems at high stresses.

Experimental characterization of multivariant microstructural evolution has centered on the evolution of mean plate volume with phase fraction $\bar{V}(f)$ as a primary input into kinetic theory (Ref 44). Extensive stereological study of nonthermoelastic plate martensites in ferrous alloys has shown a two-stage process in which initial grain-to-grain spreading of transformation characterized by a constant $\bar{V}$ is followed by an intragranular filling-in by finer plates characterized by a reciprocal power law. Preliminary measurements in a thermoelastic Fe-Pt alloy suggested a nonmonotonic behavior of $\bar{V}(f)$ (Ref 45), but a recent more thorough study in a coarse-grained thermoelastic TiNi alloy has given the same monotonic behavior as nonthermoelastic systems except for an overall finer plate size (Ref 46). It is surprising that, despite the high inteifacial mobility retained in thermoelastic systems, overall evolution of $\bar{V}(f)$ is similar to nonthermoelastic systems where growth units are frozen. This implies a dominance of universal geometric factors (Ref 47). A deeper understanding of observed $\bar{V}(f)$ behavior awaits more detailed analysis of multivariant growth simulations.

The model assumptions of the unit growth simulation of Fig. 5 correspond to a thermoelastic system. A preliminary model of unit growth dynamics in nonthermoelastic systems has demonstrated the important role of (size and rate dependent) plastic accommodation of the transformation shape strain in governing the technologically important lath/plate morphological transition in steels (Ref 33, 48). Working within a mesoscopic continuum nonequilibrium thermodynamic framework (Ref 49), Levitas (Ref 50) initiated more rigorous finite-element simulations of such unit growth in a plastic matrix. Contours of computed accumulated plastic strain for directional growth of a martensite lath are represented in Fig. 6. Calculating the incremental work of elemental growth steps offers insights into the evolution of morphology in a plastic environment. The behavior is found to be strongly path dependent, and nonlocal plasticity models are being adopted to more realistically treat size-dependent plastic flow at the 0.1 μm length scales of interest.

Defect Hierarchy—Distributed Systems

Consistent with Smith's vision of the importance of defect hierarchy in materials behavior and the current need for probabilistic structure/property theory to efficiently control variation in modern materials engineering as discussed in the section "Systems: Dynamic Hierarchy," martensitic transformations may represent the best developed example of quantitative analysis of defect hierarchy for prediction of macroscopic behavior. Spanning the hierarchy of Fig. 1, modeling and experiment have quantified the roles of fine scale defects on both thermal and atherrnal components of interfacial mobility (Ref 51, 52) and their dependence on alloy composition and plastic prestrain

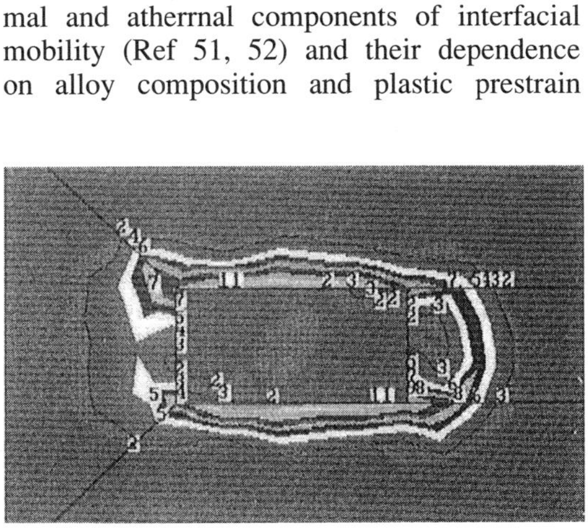

Fig. 6 Computed plastic strain contours for growth of lath martensite in ferrous alloy. Source: Ref 50

(Ref 53, 54), as well as the distributions of initial and autocatalytic nucleation sites which dominate the evolution of transformation under cooling or stress (55–58).

While it is the nature of states of perfection to be uniquely valued, it is an inherent characteristic of imperfection to be distributed in nature. The study of heterogeneous martensitic nucleation has naturally emphasized the form of defect potency distributions. Small particle experiments have defined the potency distribution of initial nucleation sites in terms of critical driving force $N_i(\Delta g)$ and analysis in term of heterogeneous nucleation theory has demonstrated that the distribution with respect to a defect size parameter $N_v(n)$ is of exponential form in both metals and ceramics (Ref 55, 57); extension to macroscopic polycrystals by sensitive acoustic emission experiments has measured the grain size dependence of the distribution amplitude (Ref 56). Fitting of stereological data for the evolution of plate number density in Fe-base alloys indicates that the number density of autocatalytic sites is of the form $N_a = pf$ (Ref 44), and more thorough analysis incorporating the exponential initial site distribution reveals that the autocatalytic potency distribution $p(n)$ is well described by a normal distribution (Ref 58). The relative potency of the two distributions determines smooth versus burst transformation modes, and probability functions defined by the initial potency distribution predict observed variation and sample-size dependence of the burst temperature (Ref 56). Preliminary modeling of the evolution of phase fraction f with temperature in ferrous alloys with lath martensite morphology (Ref 59) indicates that a normal potency distribution of autocatalytic sites $p(n)$ also applies if $\bar{V}(f)$ takes the same form as for plate martensites, but the direct measurement of $\bar{V}(f)$ in the very fine lath microstructures is difficult.

The recent extension of quantitative stereological studies to thermo elastic TiNi (Ref 46) indicates that the autocatalytic potency distribution $p(n)$ is also unimodal and reasonably approximated by a normal distribution as represented by the cumulative distribution curve of Fig. 7. That the stress field of a plate periphery should serve as a potent nucleation site with a characteristic potency defining a distribution mode is not surprising, but more quantitative insight into observed behavior awaits more detailed analysis of 3D computer simulation of multivariant transformation. Preliminary modeling of the evolution of phase fraction f with temperature in ferrous alloys with lath martensite morphology (Ref 59) indicates that a normal potency distribution of autocatalytic sites $p(n)$ also applies if $\bar{V}(f)$ takes the same form as for plate martensites, but the direct measurement of $\bar{V}(f)$ in the very fine lath microstructures is difficult.

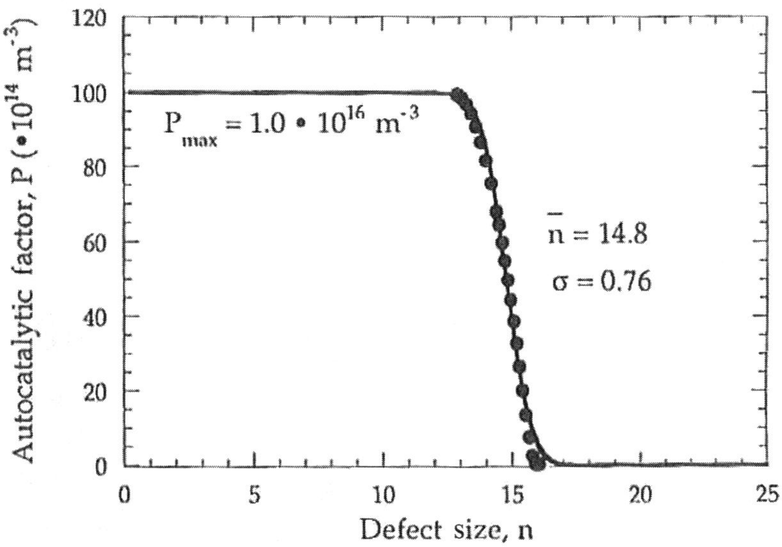

Fig. 7 Cumulative potency distribution of autoeatalytic nucleation sites determined from quantitative stercological measurements in thermoelastic TiNi alloy. Source: Ref 46

Some insight into the nature of autocatalytic defects can be gleaned from theoretical modeling of martensitic interfacial junctions (Ref 60), and the simulation of twin variant conversion in stressed martensitic microstructures (Ref 61, 62). For the latter case Killough and Kohn (Ref 61) used a TDGL-type model to generate a fully transformed multivariant microstructure via a continuous quasimartensitic instability, and then studied details of stress-induced twin nucleation within this microstructure for comparison with twin "tip splitting" observations conducted in Cu-Al-Ni (Ref 62). Such a simulated event is represented in Fig. 8. A twin band tapering from its K' plane intersects another band on the nearly orthogonal K', plane to spontaneously generate a defect with the same wedge-disclination dipole character as the nucleation site of Fig. 4. The stress-induced formation of a twin embryo in this environment as shown in Fig. 8 is strikingly similar to the martensitic embryo formation of Fig. 4. While Fig. 4 models a rare initial nucleation site inherited from the complex processing history of the parent phase, Fig. 8 represents an inherent defect generated by multivariant transformation. Further detailed analysis of simulated microstructures like Fig. 8 could generate theoretical nucleation site potency distributions for comparison with experiment, ultimately offering more predictive control of the distributions which pace the macroscopic evolution of transformation with increasing driving force.

Constrained Systems

Reaching to the higher levels of structure in Fig. 1, detailed considerations of the intergranular compatibility constraints in polycrystalline materials have employed a modified Taylor analysis to predict from lattice deformation symmetry the relative recoverable shape memory strains in a range of martensitic systems (Ref 63). A recent extension demonstrates that the relatively high recoverable strains in TiNi alloys compared to Cu-based memory alloys is associated with crystallographic texture, and desirable textures for improved properties arc predicted (Ref 64).

As denoted by the highest level of structure in Fig. 1, macroscopic component shape and surface boundary conditions can also constrain the evolution of multivariant microstructures, particularly the later stage processes of Fig. 1. Thin films with both free and constrained surfaces can constrain transformation at quite microscopic levels as well. Early experiments on ferrous martensitic transformations in thin films constrained by substrates demonstrated a wide range of crystal kinematic conditions compared to transformation in the bulk; the new conditions were shown to be consistent with invariant line strain (ILS) transformation with invariant lines defined by the intersection of the "Bain cone" of undistorted lines with the plane of the film (Ref 65). Roytburd et al. (Ref 66) applied a polydomain theory to multivanant microstructural evolution in constrained thin films, treating interactions with temperature dependent self stress associated with film-substrate thermal expansion mismatch. The model gives a series of local thermomechanical equilibrium polyphase polydomain states with a heating- cooling path dependence, reproducing essential features observed in TiNi films on Si. A kinematics-based theory incorporating elastic

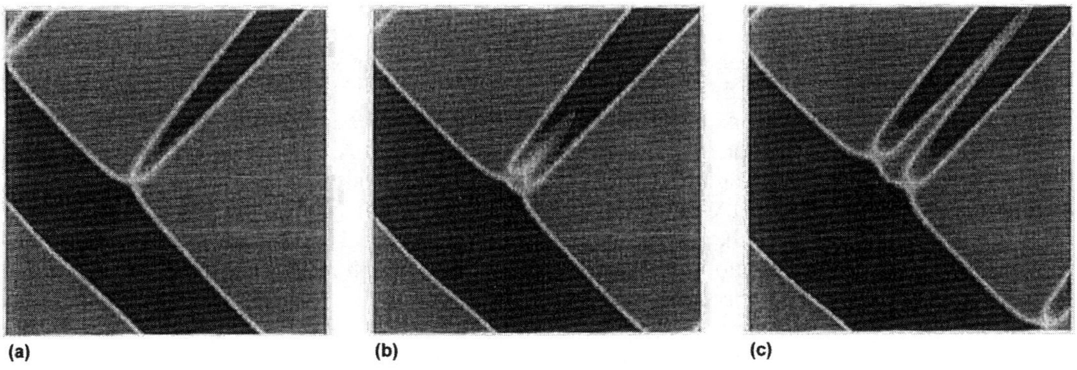

Fig. 8 Simulation of twin variant conversion in martensitic microstructure. Sequence shows twin "tip splitting" by heterogeneous nucleation analogous to Fig. 4. Source: Ref 60

and surface energies by Bhattacharya and James (Ref 67) predicts that even "released" thin films will adopt ITS single-variant growth units, and conditions for formation of "tunnel" and multivariant "tent" morphologies are specified as a basis for design of efficient microactuators.

Biological Systems—Molecular Martensite?

As first reported at ICOMAT-82 (Ref 68), biological protein cylindrical crystals undergo martensitic transformations to perform vital life functions in viruses and bacteria. In the case of the T4 bacteriophage, a virus that infects *Escherichia coli* bacteria, such transformation provides a tail-sheath contraction mechanism allowing the virus to inject its DNA into the bacterium. Analysis of the transformation kinematics accounting for constraints of 2D cylindrical symmetry identified a transformation shear strain of 120% coupled to a volume contraction of 40%. The transformation was shown to rigorously conform to the formal definition of martensitic transformations as proposed at ICOMAT-79 (Ref 69) and reassessed at ICOMAT-95 (Ref 70). Subsequent work has assessed evidence for similar transformations in higher organisms (Ref 33).

Recently the structure of the gp41 subunit of the envelope glycoprotein of the human immunodeficiency virus (HIV) has been determined to have the helical trimer structure represented in Fig. 9 (Ref 71). The subunit is known to be responsible for fusion of viral and cellular membranes, allowing release of viral contents into the host cell as an essential component of viral infection. A strong resemblance of the gp41 structure to that of the bistable hemagglutinin (HA) protein of influenza virus has led to the proposal that the gp41 fusion mechanism in HIV infection is the same as a displacive transformation "harpoon mechanism" established in the influenza case (Ref 72). The latter is represented schematically in the diagram of Fig. 10. In the "native" state of the HA trimer at the left, part of the protein is in the conformation of a nonhelical loop folded into the core of a "coiled coil." On transforming to the "fusogenic" state at the right, the entire trimer converts to the coiled coil configuration (as in Fig. 9 for the HIV case) extending the end by 100 Å during membrane fusion.

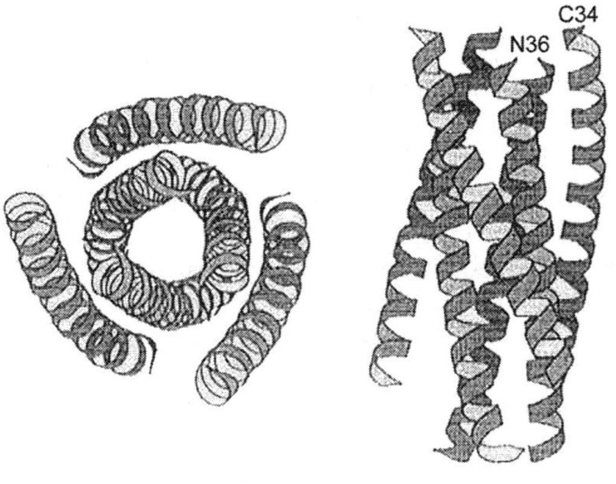

Fig. 9 Helical trimer structure of gp41 subunit of HIV envelope glycoprotein believed to undergo large-strain displacive transformation. Source: Ref 70

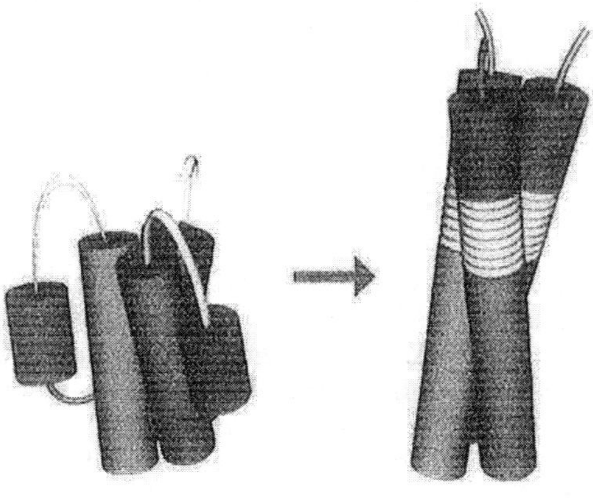

Fig. 10 Representation of two structural states of influenza hemagglutinin (HA) protein involved in 'harpoon mechanism' of cell infection. Source: Ref 71

The transformation is clearly displacive in character, involving a large distortion. Despite the 3-fold rotational symmetry, the trimer lacks the translational symmetry to be classed as a crystal, and in that sense the transformation cannot be formally classed as martensitic. Acknowledging the strong similarities, however, it is proposed that such transformations be called "molecular martensites" (by analogy with "liquid crystals") to encourage theoretical investigation of the applicability of martensitic concepts, Available information on the hydrophobic and hydrophilic components of the molecules offers insight into the origin of phase stability, and thermodynamic information exists in established phase boundaries with respect to pH and temperature for the influenza case

(Ref 72). Insights into nucleation path and dynamics may be fruitful in devising methods to prevent or disrupt the transformation. Consideration of detailed molecular architecture and phase stability has already led to speculation on small molecule inhibitors for an "austenite stabilization" approach to defeating HIV infection (Ref 71).

ACKNOWLEDGMENTS

Research on martensitic transformations at Northwestern University is sponsored by the National Science Foundation under Grant DMR-9500122. Many of the ideas presented here arose from discussions at a Martensite Theory Workshop held at Northwestern in June, 1998. Participants included G. Barsch (Penn State U.), K. Bhattacharya (Cal. Inst. Tech.), G. Ghosh (Northwestern), R. James (U. Minnesota), R. Killough and R. Kohn (New York U.), V. Levitas (U. Hannover, Germany), J. Morris (Iowa State U.), A. Reid (Northwestern) and A. Saxena (Los Alamos Natl. Lab.).

REFERENCES

1. C.S. Smith, *A Search for Structure*, MIT Press, 1981
2. M. Cohen. Mat. Sci. Eng. 25 (1976) 3
3. G.B. Olson. Science 277 (5330) (1997) 1237
4. T. Suzuki and H.M. Ledbetter, Phil. Mag. A 48 (1983) 83
5. Y.-Y. Ye, C.T. Chan, K.-M. Ho, and B.N. Harmon, Int. J. Supercomputer Appl. 4 (1990) 111
6. G.L. Krasko and G.B. Olson, Phys. Rev. B 40 (1989) 536
7. M. Schroter, E. Hoffmann, P. Entel and H. Akai, Proc. ICOMAT-95, J. Phys. (France) IV 5 (1995) C8–273
8. H.C. Herper, E. Hoffmann, P. Entel, and W. Weber, Proc. ICOMAT-95, J. Phys. (France) IV 5 (1995) C8–293
9. S.E. Kul'kova, K.A. Beketov, V.E. Egorushkin, and O.N. Muryzhnikova, Proc. ICOMAT-95, J. Phys. (France) IV 5 (1995) C8–539
10. Y-Y. Ye, C.T. Chan, and K.M. Ho, unpublished research
11. R.J. Gooding and J.A. Krumhansl, Phys. Rev. 38 (1990) 1695
12. D.M. Hatch, P. Hu, A. Saxena, and G.R. Barsch, Proc. ICOMAT-95, J. Phys. (France) IV 5 (1995) C8–131
13. G.R. Barsch, these proceedings
14. G.R. Barsch, private communication
15. J.W. Cahn and J.E. Hilliard. J. Chem. Phys. 31 (1959) 688
16. G.R. Barsch and J.A. Krumhansl, Phys. Rev. Lett. 53 (1984) 1069
17. A. Saxena and G.R. Barsch, Physica D 66 (1993) 195
18. P. Hong and G.B. Olson, Solid State Comm. 85 (1993) 681
19. F-R. Chen and G.B. Olson, Proceedings of the XIIth International Conference on Electron Microscopy, San Francisco Press. 1990
20. S. Kajiwara, K. Ogawa, and T. Kikuchi, Phil. Mag. Lett. 69 (1996) 405
21. J.R. Morris, Y-Y. Ye, K.M. Ho, C.T. Chan, and M.H. Yoo, Phil. Mag. Lett. 69 (1994) 189
22. R. Wu, A.J. Freeman, and G.B. Olson, Phys. Rev. B 50 (1994) 75
23. L. Zhong and A.J. Freeman, Northwestern University, unpublished research
24. L. Zhong, A.J. Freeman, and G.B. Olson, Northwestern University, unpublished research
25. K. Ullakko, J.K. Huang. C. Kantner, R. O'Haudley, and V.V. Kokorin, Appl. Phys. Lett. 69 (1996) 966
26. R. James, in: G.J. Dvorak (Ed.), *Research Trends in Solid Mechanics*, USNC/TAM, 1999.
27. T.B. Belytschko, Y.Y, Lu, and L. Gu, Intl. J. Numerical Methods in Eng. 37 (1994) 517
28. Y.A. Chu and B. Moran, Modeling and Simulation in Mats. Sci. Eng. 3 (1995) 455
29. Y.A. Chu, B. Moran. A.C.E. Reid, and G.B. Olson, Met, Trans. A (1999)
30. M. Lin, G.B. Olson, and M. Cohen, Acta Metall. Mater. 41 (1993) 253
31. T. Saburi and S. Nenno, Proc. ICOMAT-86, Jpn. Inst. Metals (1986) 671
32. M. Cohen, E.S. Machlin, and V.G. Paranjpe, Thermodynamics Physics Metallurgy, ASM, 1949. p 242
33. G.B. Olson and M. Cohen, in: F.R.N. Nabarro (Ed.), *Dislocation in Solids*, vol. 7, North-Holland, 1986, p 295
34. L. Kaufman and M. Cohen, Prog, Met. Phys. 1 (1958) 165
35. A.C.E. Reid, G.B. Olson, and B. Moran, Phase Transitions (1999)

36. P.C. Clapp, Phys. Stat. Sol. (B) 57 (1973) 561
37. A.C.E. Reid and G.B. Olson, Proc. ICOMAT-98 (1999)
38. Y. Wang and A.G. Khachaturyan, Acta Mater. 45 (1997) 759
39. H. Knapp and U. Dehlinger, Acta Met. 4 (1956) 289
40. V. Raghavan and M. Cohen, Acta Met. 20 (1972) 779
41. K. Bhattacharya, Acta Metall. Mater. 39 (1991) 2431
42. A.L. Roytburd and J. Slutsker, J. Appl Phys. 77 (1995) 2745
43. F. Larche and J.W. Cahn, Acta Metall. 21 (1973) 1051
44. V. Raghavan, in: G.B. Olson and W.S. Owen (Eds.), *Martensite*, ASM, 1992, p 197
45. G.B. Olson and H. Ohtsuka, Proceedings of the International Symposium on Shape Memory Materials, International Academic Publishers, Beijing, 1994, p 17
46. A. McGeorge, M.S. thesis, Northwestern University (1996)
47. E Hornbogen, Proceedings of the ICOMAT-89, Materials Science Forum, Vols. 56–58, Trans Tech, 1990, p 131
48. D.M. Haezebrouck. Sc.D. thesis. MIT (1987)
49. V.I. Levitas, J. Mech. Phys. Solids 45 (1997) 923–1203
50. V.I. Levitas, University of Hannover, Germany, unpublished research
51. M. Grujicic, G.B. Olson, and W.S. Owen, Metall. Trans. 16A (1985) 1713
52. M. Grujicic, G.B. Olson, and W.S. Owen, Metall. Trans. 16A (1985)
53. G. Ghosh and G.B. Olson, Acta Metall. Mater. 42 (1994) 3361
54. G. Ghosh and G.B. Olson. Acta Metall. Mater. 42 (1994) 3371
55. M. Cohen and G.B. Olson, *New Aspects of Martensitic Transformations*, Japan Institute of Metals, 1976, p 93
56. G.B. Olson, K. Tsuzaki, and M. Cohen, Proceedings of the Turnbull Symposium, MRS, 1987, p 129
57. I.-W. Chen, Y.-II. Chiao, and K. Tsuzaki, Acta Metall. 33 (1985) 1847
58. M. Lin, G.B. Olson, and M. Cohen, Metall. Trans. A 23A (1992) 2987
59. G. Ghosh, Northwestern University, research in progress
60. N.K. Simha and K. Bhattacharya, J. Mech. Phys. Solids 48 (1999) 2619
61. LR. Killough and R.V. Kohn, unpublished research
62. R. Abeyaratne, C. Chu, and R.D. James, Phil. Mag. 73 (1996) 457
63. K. Bhattacharya and R.V. Kohn, Proc. ICOMAT-95, J. Phys. (France) IV 5 (1995) C8–261
64. Y.C. Shu and K. Bhattacharya, Acta Mat. 46 (1998) 5457
65. M. Kato, S. Fukase, A. Sato, and T. Mori, Acta Metall. 34 (1986) 1179
66. A.L. Roytburd, T.S Kim, Q. Su, J. Slutsker, and M. Wuttig, Acta Mater. 46 (1998) 5095
67. K. Bhattacharya and R.D. James, J. Mech. Phys. Solids 47 (1999) 531
68. G.B. Olson and H. Hartman, Proc. ICOMAT-82, J. Phys. (France) 43 (1982) C4–855
69. M. Cohen, G.B. Olson, and P.C. Clapp, Proc. ICOMAT-79, MIT, 1979, p 1
70. J.W. Christian, G.B. Olson, and M. Cohen, Proc. ICOMAT-95, J. Phys. (France) IV 5 (1995) C8–13
71. D.C. Chan, D. Fass, J.M. Berger, and P.S. Kim, Cell 89 (1997) 263
72. C.M. Carr and P.S. Kim, Cell 73 (1993) 823

Transformations—Selected Works of G.B. Olson on Materials, Microstructure, and Design
C.E. Campbell, M.V. Manuel, and W. Xiong, editors

CHAPTER 6

Distributed-Activation Kinetics of Heterogeneous Martensitic Nucleation*

Minfa Lin, Bethlehem Steel Corporation
Gregory B. Olson, Northwestern University
Morris Cohen, Massachusetts Institute of Technology

Abstract

Martensitic Nucleation under normal circumstances is a heterogeneous process, and the heterogeneity can be explained by the potency distribution of defects that provide the nucleation sites, including both preexisting and autocatalytically generated defects. The potency distribution of the preexisting defects has been shown earlier by small-particle studies to follow an exponential function, whereas that of the autocatalytic defects is found in the present work to obey a Gaussian function. A major distinction between these two distributions is that the number density of the *preexisting* defects increases monotonically with decreasing defect potency, while the number density of the *autocatalytic* defects is distributed about a mode, giving a saturation level of its cumulative distribution. As a result of these potency distributions, the nucleating sites of both origins exhibit distributions in their activation energies for nucleation, and a distributed-activation kinetic model is now proposed to take these variations into account for martensitic transformations. The features of this model are tested with considerable success in experiments on an Fe-32.3Ni alloy, leading to calculations of the entire course of martensitic transformation curves (including the athermal, anisothermal, and isothermal contributions), the changing shapes of time-temperature-transformation (TTT) diagrams for a range of Fe-Ni compositions, and the grain-size dependence of "bursting" behavior.

Introduction

Kinetic phenomena in crystalline solids of relatively high perfection can often be adequately described with the approximation of a single activation energy. In contrast, it is well established that in highly imperfect solids, such as glasses (Ref 1, 2), a potency distribution of structural defects gives rise to a distribution of activation energies which is essential to quantitative description of kinetic behavior. When behavior of a crystalline solid is dominated by strong defect interactions, as in the case of heterogeneous nucleation of a phase transformation, it can similarly be anticipated that distributed-activation kinetics will prevail.

*M. Lin, G. B. Olson and M. Cohen, "Distributed-Activation Kinetics of Heterogeneous Martensitic Nucleation," *Metall. Trans.* 23A (1992) p 2987–98

The relative ease of direct nucleation-rate measurements allowed by rapid growth to macroscopic dimensions makes martensitic transformations an ideal case for quantitative study of heterogeneous nucleation kinetics.

While it has long been recognized that a defect potency distribution likely exerts a strong influence on both the athermal (temperature-dependent) (Ref 3) and isothermal (time-dependent) (Ref 4) aspects of heterogeneous martensitic nucleation kinetics, the most highly developed theory of martensitic kinetics has thus far employed the approximation of a single activation energy (Ref 5). A comprehensive review of martensitic kinetics (Ref 6) has emphasized common features of athermal and isothermal modes of transformation, offering hope of a universal kinetic model. We here explore such a unified framework for martensitic nucleation kinetics, employing a distributed-activation model to incorporate variations in defect potencies and their associated nucleation activation energies. Previous work upon which the model is based will be briefly reviewed first.

Preexisting Defect Size Distribution

The heterogeneity of martensitic nucleation is well demonstrated by the small-particle experiments of Cech and Turnbull (Ref 7) on an Fe-30.2Ni (weight percent) alloy, which showed that the martensite start temperature on cooling is particle size dependent. In these studies, the particles were found to transform over a wide range of temperatures and particle sizes; in fact, some of the smallest particles did not transform at all on cooling to the lowest temperature. These findings suggest two features of the transformation: (a) larger particles are more probable to contain nucleation sites or defects and (b) these defects have different potencies. The implication of a potency distribution for nucleation defects arising from these results was analyzed by Olson and Cohen (Ref 8) in terms of a statistical heterogeneous nucleation model, as follows. If a number density N_i of preexisting nucleation sites is randomly distributed throughout the volume, the probability p that a particle of volume V_p will contain at least one such site is:

$$p = 1 - \exp\left(-N_i V_p\right) \qquad \text{(Eq 1)}$$

Since martensitic transformations are nucleation controlled, p is then the transformation probability of the individual particle itself. It should be noted that in these small-particle experiments, the initial nucleation event in a given particle causes an easily detectable amount of transformation in the particle due to autocatalytic nucleation, but such autocatalysis is not transmitted to other particles because of their separation. If sites occur only on particle surfaces, a similar function can be obtained with V_p replaced by the surface area of the particle. In general, the V_p term can be represented as d_0^m, where d_0 is the particle diameter, $m = 2$ for surface nucleation, $m = 3$ for volume nucleation, and N_i is the number density of preexisting sites for surface or volume nucleation, respectively. Using Eq 1 for the Cech-Turnbull experimental data, Olson and Cohen (Ref 8) found that volume nucleation gives a better fit at the larger particle sizes, and that a cumulative defect-potency distribution, $N_i(n)$, that is, the cumulative density of nucleating defects with their potency equal or greater than n, can be expressed as:

$$N_i(n) = N_v^0 \exp\left(-\alpha \cdot n\right) \qquad \text{(Eq 2)}$$

where n is the thickness of critical nucleus stabilized by interaction with the defect (as quantified here in the section "Critical Defect Size for Martensitic Nucleation"), α is a factor that represents the distribution shape (0.84 for the Cech-Turnbull results), and N_v^0 is the total number density of preexisting nucleation sites of all potencies. This exponential distribution indicates that the preexisting nucleation defects become more and more numerous with decreasing potency (smaller n). Recent observations of nucleation at low-potency defects in small particles of ZrO_2 ceramics (Ref 9) have also confirmed the validity of this distribution function. (Some confusion has arisen in the interpretation of a recent Fe-Ni small-particle experiment due to the formation of ferrite during solidification of small droplets; Ref 10, 11. Ferrite formation has been analyzed by Ref 12.) The above preexisting defect distribution based on small-particle experiments has been extended to bulk polycrystal materials by Olson *et al* (Ref 13) In this case, even though the course of the overall transformation is determined by autocatalytically generated defects, the initial nucleation events are still governed by the preexisting defects, just as discussed above for small particles. In other words, the initial nucleation events are triggered by the highest potency defects existing in the parent phase. Olson *et al.* (Ref 13) measured this high-potency tail of the defect distribution by means of very sensitive acoustic-emission measurements to detect the initial nucleation events (prior to autocatalysis) in macroscopic polycrystalline specimens of an Fe-32.3Ni-0.004C (weight percent) alloy. The true M_s (temperature for initial nucleation) was found to decrease with increasing grain size, D. Assuming the same form of defect-potency distribution as derived from the Cech-Turnbull data and using the statistical heterogeneous nucleation model to fit the grain-size dependence of the true M_s temperature for initial nucleation events, they found an analytical form describing the grain-size depencence of N_v^0, the total number density of preexisting defects with all potencies:

$$N_v^0 = k_1 + k_2 D^{-m_1} \quad \text{(Eq 3)}$$

where (for the above Fe-32.3Ni alloy) $m_1 = 2.6$, $k_1 = 1.73 \times 10^{12}$ m^{-3}, and $k_2 = 1.4 \times 10^6$ (D in meters). Accordingly, the cumulative density of nucleating defects, $N_i(n)$, in bulk polycrystalline parent phases can be written as:

$$N_i(n) = (k_1 + k_2 D^{-m_1}) \exp(-\alpha \cdot n) \quad \text{(Eq 4)}$$

This equation shows how the preexisting nucleation defects are distributed with respect to their size n (potency).

Singly Activated Kinetic Model

By considering thermally activated nucleation in terms of plastic-deformation theory, Olson and Cohen (Ref 14) derived a relationship between activation energy Q and defect potency n, showing that Q is inversely related to defect potency. Therefore, the nucleation defects are also distributed with respect to their activation energies, and this should be taken into account in order to describe martensitic transformation kinetics involving both preexisting and autocatalytic nucleating sites. In contrast, the existing model (Ref 5) for martensitic nucleation kinetics at both preexisting and autocatalytic defects assumes that all nucleating defects have the same activation energy. With this simplification, the mathematical equation to describe the nucleation kinetics can be expressed as:

$$\frac{dN_v}{dt} = (N_i + Pf - N_v)(1 - f)\nu \exp\left(-\frac{Q}{kT}\right) \quad \text{(Eq 5)}$$

where N_i = number of preexisting nucleation sites in the parent phase
P = number of autocatalytic sites produced per unit volume of martensite formed
f = volume fraction of martensite formed
N_v = number of martensitic plates formed per unit volume of alloy
ν = nucleation attempt frequency
Q = thermal activation energy for the nucleation at temperature T
k = the Boltzmann constant

In this kinetic equation, variations in the defect potencies and their activation energies for nucleation are ignored as a first approximation. Magee (Ref 4), based on small-particle studies with atomized Fe-24.2Ni-3.6Mn and Fe-22Ni-0.49C alloys (the kinetics of which are due to the preexisting defects), demonstrated the inapplicability of singly activated kinetics to his experimental data.

In bulk polycrystal specimens where both preexisting defects and autocatalytically generated defects (the creation of new nucleation sites by the ongoing transformation) can come into play, the overall course of a martensitic transformation (volume fraction transformed versus temperature or time) is greatly influenced by autocatalytic nucleation. Despite its importance, the latter phenomenon has received very little quantitative study until now. The autocatalytic nucleation is believed to occur through any of the following three mechanisms (Ref 15): (a) stress-assisted nucleation: the activation by transformation-generated elastic stresses at less-potent preexisting defects that would not otherwise nucleate at the prevailing temperature; (b) strain-induced nucleation: the creation of new and relatively potent nucleating defects by the generation and rearrangement of dislocations during plastic accommodation of the parent phase in response to the transformational shape change; and (c) interfacial autocatalysis: direct nucleation of new martensitic units by a dislocation-dissociation process from the existing martensitic interfacial structure.

The preexisting defect distribution has been found to follow an exponential function in line with Eqs 2 and 4, but the autocatalytic defects can be expected to have a quite different distribution form. The potency of new defects generated by martensitic growth events is expected to be sensitive to the intensity of the elastic field surrounding the plates, which is, in turn, dependent on the martensite plate size. Since the latter is distributed about a mean value, we can expect the autocatalytic defect-potency distribution to be likewise distributed about a characteristic mean or mode.

The purpose of this investigation is to find the autocatalytic defect-potency distribution through quantitative metallography and then to propose a kinetic model in which the activation-energy distribution of the nucleation defects is taken into account. Such a model should allow description of the entire course of a martensitic transformation during cooling and during isothermal holding.

Experimental

The alloy chosen for the present study is Fe-32.3Ni-0.004C (weight percent), which is the same as used by Olson et al. in the acoustic-emission experiments, (Ref 13), from which the preexisting defect distribution for martensitic nucleation has been established. To describe the overall transformation kinetics in this alloy system, it remains to focus on the autocatalytic defect distribution.

The alloy in the form of a 3 × 3 × 8-cm bar was homogenized at 1100 °C for 7 days in an evacuated quartz tube. Pieces cut from this bar were then cold rolled to 50 and 75 pct reductions and annealed for 1 hour at 850 °C to achieve austenitic grain sizes of 200 and 80 μm, respectively. A still coarser grain size of 800 μm was obtained by annealing the homogenized pieces at 1300 °C for 12 hours. All of the specimens were fully austenitic after air cooling to room temperature.

The austenitic grain size was measured by lineal analysis and expressed as $D = 1.65\bar{l}$, where D is the average grain size and $\bar{l}$ is the mean linear intercept.

The martensitic transformation was followed at subatmospheric temperatures by the relative resistance changes measured with a four-point Keithley milliohmmeter on 0.5 × 1.0 × 0.1-cm specimens. (The method was capable of detecting a 5×10^{-6} ohm change, corresponding to about 0.8 pct martensite.) The resistance changes were converted into volume fraction of martensite by calibration with quantitative metallography and assuming the usual linear proportionality.

The mean volume of martensitic plates was determined metallographically using the Fullman equation (Ref 16):

$$\bar{V} = \frac{\pi^2 f}{8 N_A \bar{E}} \quad \text{(Eq 6)}$$

where f is the volume fraction of martensite, $\bar{E}$ is the mean reciprocal of the plate lengths as intersected by a random section, and N_A is the number of martensitic plates per unit area of a random section.

Standard methods were employed for metallographic preparation, and the martensitic structures were revealed by etching with a solution of 25 ml HF + 200 ml H_2O_2 + 275 ml H_2O for 5 seconds, followed with a 35 g $Na_2S_2O_5$ + 56 ml H_2O solution for 30 seconds. The latter step served to tint the microstructure selectively.

Experimental Results

The course of the martensitic transformation as a function of time was measured at several different temperatures, and the results are plotted in Fig. 1 (shown by solid lines) for the three different austenitic grain sizes. The specimens were quenched directly from room temperature (austenitic state) to the indicated holding temperatures, and hence, the amount of athermal (and rapid anisothermal) martensite formed prior to the measurable isothermal transformation is greater the lower the holding temperature. The initial part of the transformation is very rapid, in contrast to the marked retardation at later stages. The isothermal portion (out to the maximum holding period of t = 9000 seconds) is of the order of 5 pct for all holding temperatures and grain sizes.

Machlin and Cohen (Ref 3) explained similar phenomena in their Fe-29.5Ni (weight pct) alloy by an embryo distribution. The most energetic embryos are used up first, so that as time increases, the potency of the remaining embryos drops to lower levels. This is equivalent to the nucleation-site potency distributions, to be considered here more quantitatively taking autocatalytic-defect generation into account.

A typical microstructure of the alloy in this study is shown in Fig. 2 for an austenitic grain size of 200 μm, and the mean volume ($\bar{V}$) of martensitic plates from metallographic measurements is shown in Fig. 3 as a function of volume fraction of martensite for the three different austenitic grain sizes. Generally, $\bar{V}$ decreases with increasing martensitic transformation in the alloy under study and can be

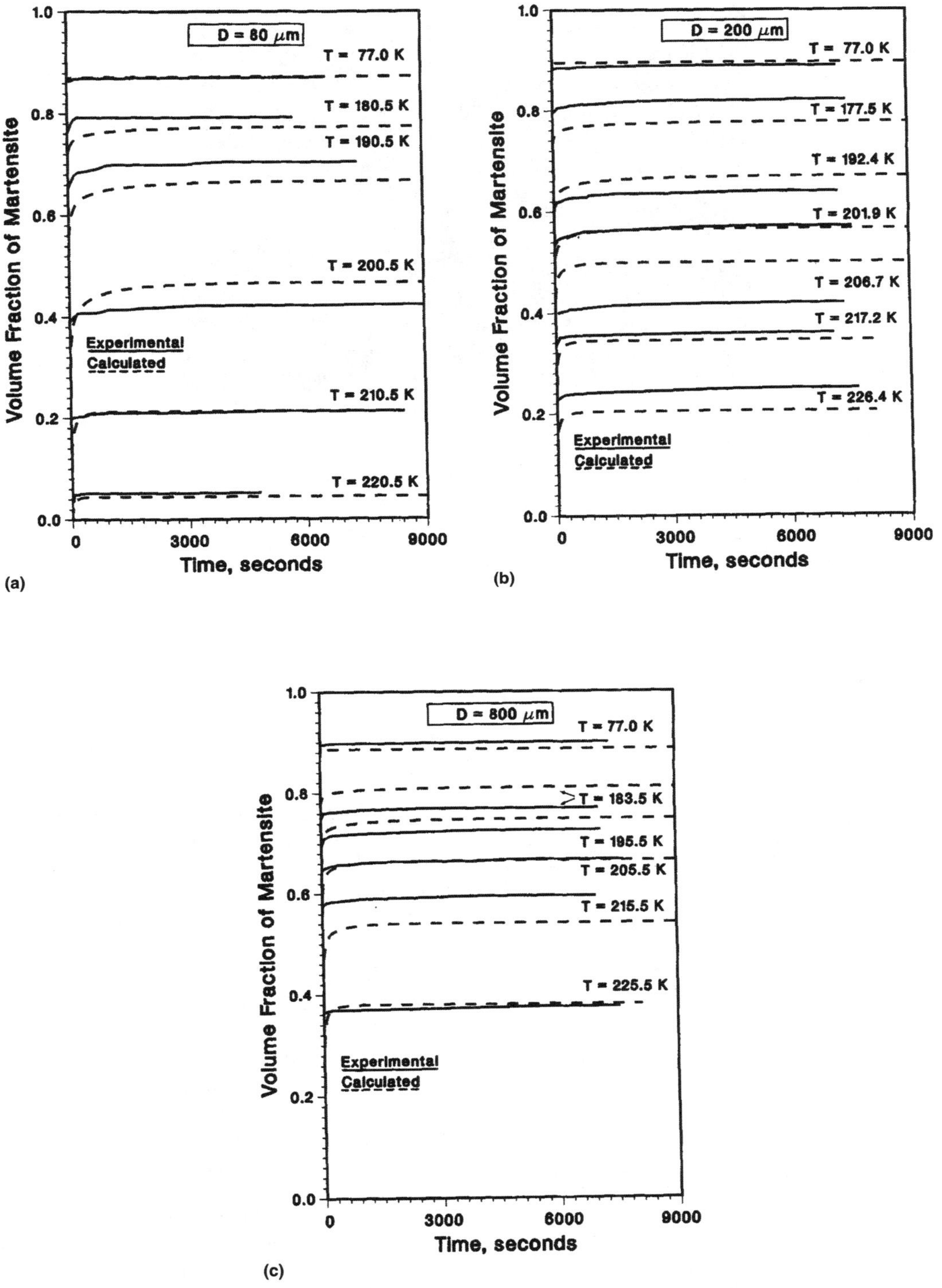

Fig. 1 Experimentally measured and model-calculated isothermal martensitic transformation kinetics after direct quenching to temperatures shown for an Fe-32.3Ni alloy with austenitic grain size: (a) $D = 80$ μm, (b) $D = 200$ μm, and (c) $D = 800$ μm.

Fig. 2 Optical micrograph of an Fe-32.3Ni alloy (with austentic grain size of 200 μm) quenched and held at 202 K for 150 min, showing lenticular plate morphology of martensite, with volume fraction of 0.6.

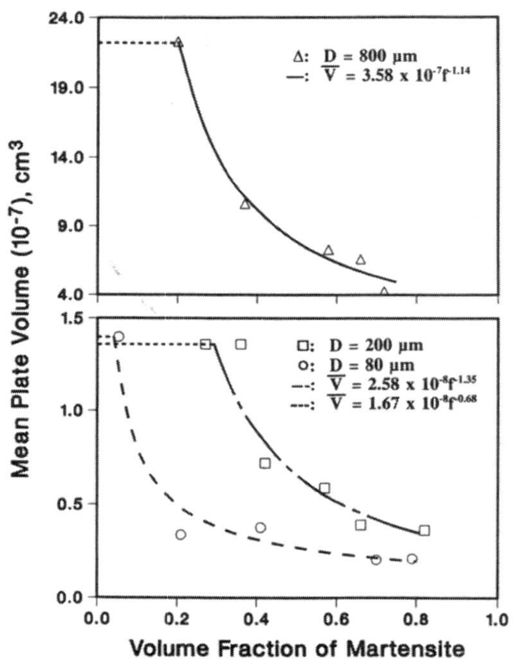

Fig. 3 Measured mean volume of martensitic plates versus volume fraction of martensite in an Fe-32.3Ni alloy with austenitic grain size D = 80 μm, D = 200 μm, and D = 800 μm.

approximated as a power function for all three grain sizes. The solid curves are the fitting results calculated from the power-function dependence of $\bar{V}$ versus f. These relationships will be used later in the kinetic model to be proposed. As shown in Fig. 3, when the volume fraction (f) is less than the experimentally measured smallest f, $\bar{V}$ is then treated as constant, which is justified by observations of previous studies (Ref 17).

Discussion

Critical Defect Size for Martensitic Nucleation

According to the Olson-Cohen dislocation-dissociation model for martensitic nucleation (Ref 14, 18, 19), the critical transformation volume free-energy change for barrierless nucleation from a defect of potency n can be written as:

$$\Delta G_{ch}(n) = -\left[\frac{2\gamma_s}{nd} + G_{el} + \frac{\mathbf{b}}{d}\tau_f\right] \qquad \text{(Eq 7)}$$

where n is the thickness of critical nucleus stabilized by the defect interaction, measured in units of the inter-planar spacing d of closest packed planes, G_{el} is the shape-insensitive component of strain energy per unit volume associated with distortions in the nucleus habit plane, γ_s is the interfacial energy per unit area, and τ_f is a frictional stress resisting the motion of interfacial-dislocations of Burgers vector **b**. The frictional stress τ_f has two components: athermal (τ_μ) and isothermal (τ_{th}) and is equal to $\tau_\mu + \tau_{th}$.

According to the thermal-activation theory of dislocation motion, the thermal component of τ_f can be written as:

$$\tau_{th} = \frac{Q_0 - Q}{V^*} \qquad \text{(Eq 8)}$$

where Q_0 is the total energy required to free a dislocation from a short-range obstacle, Q is the activation energy for dislocation motion under an applied shear stress (τ_f), and V^* is the "activation volume" defined by $V^* = -dQ/d\tau_f$. Furthermore, when the activation energy Q is greater than a certain value $Q_{\max}$, the nucleation rate will be too slow for experimental observation. In this study, the maximum isothermal holding time (t_m) is 9000 seconds, and therefore, the maximum activation energy $Q_{\max}$ can be estimated by:

$$t_m = \frac{1}{v \exp\left(-\frac{Q_{\max}}{kT}\right)} \qquad \text{(Eq 9)}$$

where v is the effective attempt frequency. Thus, Eq 8 becomes:

$$\tau_{th} = \frac{Q_{\max} - Q}{V^*} \qquad \text{(Eq 10)}$$

This thermal component of resistance to dislocation motion can be overcome by thermal fluctuations. By substituting τ_{th} and τ_μ for τ_f in Eq 7, the critical nucleating defect size (or potency) can be obtained:

$$n = -\frac{2\gamma_s}{d\left[\Delta G_{ch} + G_R + \frac{\mathbf{b}}{d}\frac{Q_{\max}-Q}{V^*}\right]} \qquad \text{(Eq 11)}$$

where $G_R = G_{el} + (\mathbf{b}/d)\tau_\mu$. The term G_R can be regarded as a net restraining force which includes the strain energy and the athermal resistance to interfacial-dislocation motion.

Equation 11 gives the sizes of defects that can nucleate martensite athermally (with $Q = 0$) or isothermally (with $0 < Q < Q_{\max}$) at a given temperature. The minimum size of defect ($n^{\min}$) that can nucleate martensite at a given temperature can be obtained by substituting $Q = Q_{\max}$ into Eq 11, leading to:

$$n^{\min} = -\frac{2\gamma_s}{d(\Delta G_{ch} + G_R)} \qquad \text{(Eq 12)}$$

It is seen that as the activation energy increases, the critical defect size n decreases and approaches $n^{\min}$ as the activation energy Q approaches $Q_{\max}$. This leads to the role of thermal fluctuations in the nucleation process and will be treated in the section "Distributed Activation Kinetics of Nucleation."

The interfacial energy γ_s and the restraining force G_R are assumed to scale with the shear modulus (μ) of the parent austenite. Using previous estimations (Ref 18) of G_R (= 420 J/mole = 5.9×10^7 J/m^3) and interfacial energy γ_s(= 150 mJ/m^2) for the Fe-30.2Ni (weight pct) alloy having a shear modulus μ of 6.0×10^{10} N/m^2, the proportionality constants for the restraining force G_R and interfacial energy γ_s are 9.9×10^4 and 2.5×10^{-13} m, respectively.

Autocatalytic-Defect Size Distribution

The number of nucleating defects available to operate at any given time and temperature per unit volume of alloy can be described as (Ref 5):

$$N_t = (N_i + Pf - N_v)(1 - f) \qquad \text{(Eq 13)}$$

The definition of each term in this equation is the same as in Eq 5. The $(1 - f)$ factor on the right-hand side of Eq 13 accounts for those potential nucleating sites that are swept up by the transformed martensite before they can nucleate. For the Pf autocatalytic term, it is assumed that the number of nucleating sites created by autocatalysis is proportional to the volume fraction of martensite formed.

When $N_t \rightarrow 0$ at a particular holding temperature, the transformation curve saturates. Since $(1 - f)$ is nonzero, the term $N_i + Pf - N_v$ must then be zero, and so, with $N_v = f/\bar{V}$, the autocatalytic parameter, P, can be determined from the following relationship:

$$P = \frac{1}{\bar{V}} - \frac{N_i}{f} \qquad \text{(Eq 14)}$$

where $\bar{V}$ and f are measured experimentally and N_i can be calculated from the preexisting defect distribution function of Eq 4 for the Fe-32.3Ni alloy. Thus, for any holding temperature, the autocatalytic factor P can be evaluated if the isothermal transformation reaches saturation at $f < 1$. The minimum activatable defect size, that is, when $Q = Q_{\max}$ in Eq 11, can also be calculated from Eq 12.

The above-obtained autocatalytic factor P is a cumulative quantity for a spectrum of defects with activation energies ranging from 0 to $Q_{\max}$. This is because the experimentally derived quantities (f and $\bar{V}$) and the number density N_i of preexisting nucleating defects, from which the autocatalytic factor P can be determined, are cumulative quantities.

The resulting cumulative P values are plotted against critical nucleating defect size n in Fig. 4 for three grain sizes of the Fe-32.3Ni alloy, with each P point corresponding to the minimum size of the critical nucleating defects at a given temperature. (The temperature dependence of the shear modulus μ required for the calculations of the minimum size of the critical nucleating defects was estimated from the temperature dependence of the elastic constants of Fe-Ni alloys measured by Hausch and Warlimont, Ref 20)

The curves in Fig. 4 demonstrate that a saturation level for the cumulative autocatalytic defect distribution is reached with decreasing defect size, in contrast to the exponential form of the potency distribution of the preexisting defects. The autocatalytic distribution is reasonably described by a Gaussian or normal distribution. The lines shown in the figure are the results of fitting, by a least-square method, three Gaussian distribution functions having different amplitudes, standard deviations, and means. The analytical forms for these distributions can be written as the integrated Gaussian distribution, described by the error function erf (x):

$$P = \frac{1}{2} P_{\max} \left[1 - \operatorname{erf}\left(\frac{n - \bar{n}}{\sqrt{2}\sigma} \right) \right] \qquad \text{(Eq 15)}$$

where $P_{\max}$, $\bar{n}$, and σ are the amplitude, mean, and standard deviation of the distribution, respectively. These parameters depend on the grain size, as illustrated in Fig. 5.

The distribution means of the autocatalytic factor increase slightly with grain size (Fig. 5a), implying that the potency of nucleating defects generated by the transformation increases with the martensitic plate size. The standard deviations defining the distribution width of the autocatalytic factor also increase with grain size (Fig. 5b) before approaching a saturation value of about 2.56. If the autocatalytic site-potency distribution is related to the distribution of martensitic plate sizes (as would

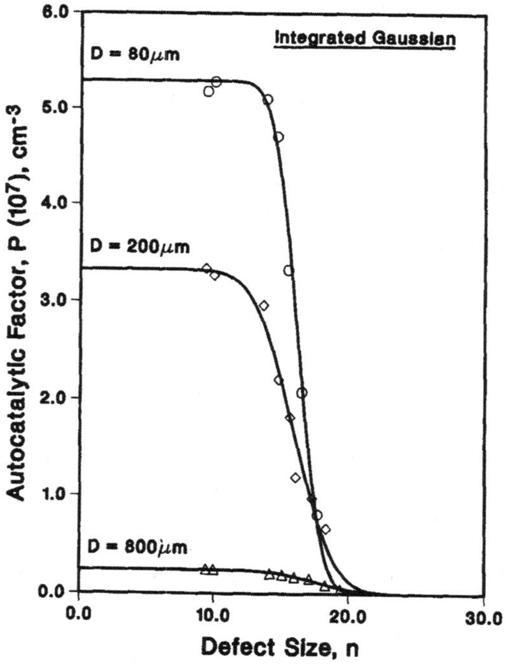

Fig. 4 Autocatalytic defect distributions generated by martensitic transformation in an Fe-32.3Ni alloy with different grain sizes. The defect size n is the critical nucleus thickness (in number of close-packed interplanar spacings) stabilized by the defect interaction.

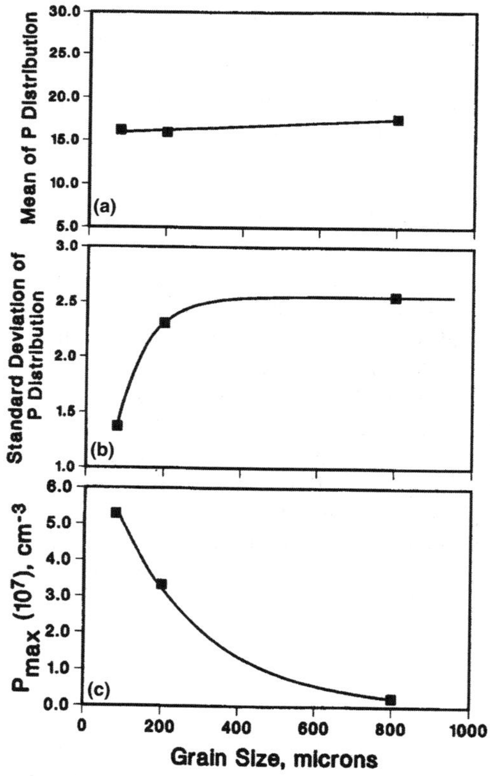

Fig. 5 (a) Mean ($\bar{n}$), (b) standard deviation (σ), and (c) amplitude ($P_{\max}$) of the autocatalytic-defect distribution *vs* austenitic grain size, generated by martensitic transformation in an Fe-32.3Ni alloy.

be the case for plate-tip stress concentrations), the existence of a constant distribution width at large grain size may indicate that the martensitic plates stop growth before hitting grain boundaries when the grain size is sufficiently large. This concept has been shown by Haezebrouck (Ref 19, 21, 22) in computer simulation of the nucleation and growth of a single martensitic plate. These studies found that rapid growth can be quickly arrested as soon as local plasticity sets in. The amplitudes of the P distribution decrease rapidly with grain size (Fig. 5c). This can be related to the assumed scaling of the autocatalytic number-density distribution with f in the definition of P. As the martensitic plates become larger with increasing grain size, fewer plates are then required to produce a given fraction of martensite, and the associated number densities of the autocatalytically generated plates are correspondingly lower. Thus, it may be appropriate to consider the autocatalytic distribution amplitude to scale more directly with N_v than with f; this would define a proportionality factor relatively insensitive to grain size.

The differential forms of the foregoing P distributions can be obtained by differentiating the cumulative functions represented by Eq 15. The results are plotted in Fig. 6 with respect to the defect size (n) for the three different grain sizes of the Fe-32.3Ni alloy studied here. A unique distribution function shape for the P parameter can be obtained by simply normalizing its individual values by the maximum amplitude and plotting against the normalized defect size n_{norm}, which is taken as $(n - \bar{n})/\sigma$. Equation 15 then becomes:

$$P_{\text{norm}} = \frac{1}{2}\left[1 - \operatorname{erf}\left(\frac{n_{\text{norm}}}{\sqrt{2}}\right)\right] \quad \text{(Eq 16)}$$

The normalized distributions (P_{norm}) for three grain sizes are shown in Fig. 7. The solid line is from Eq 16, and the points are individually normalized from the plotted points in Fig. 4.

In contrast to the exponential distribution form found for the preexisting nucleation sites, a saturation level for the autocatalytic parameter P is clearly evident. To verify the generality of this Gaussian distribution for other compositions in the Fe-Ni alloy system, two additional sets of P-n calculated data for an Fe-29.5Ni bulk specimen (Ref 3) and Fe-30.2Ni small particles (Ref 7) are also included in Fig. 7,

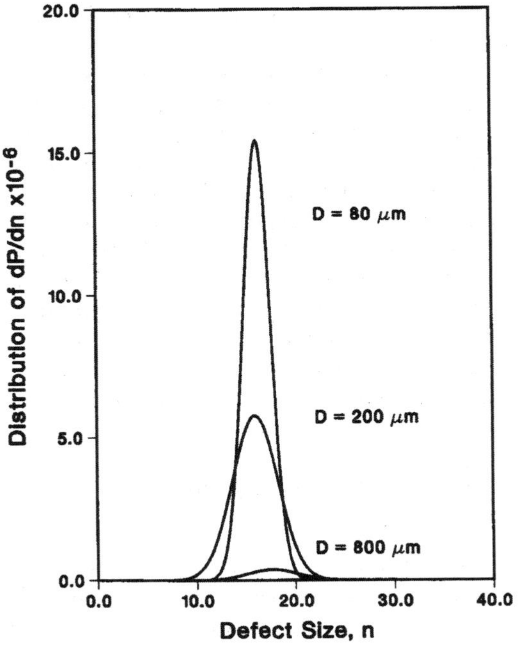

Fig. 6 Gaussian distributions of the autocatalytic defects generated by martensitic transformation in an Fe-32.3Ni alloy with different austenitic grain sizes.

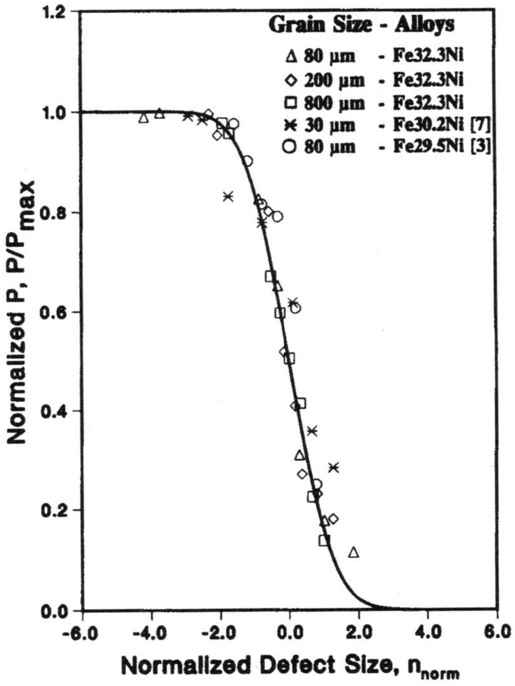

Fig. 7 Normalized cumulative form of the autocatalytic-defect distribution (P/P_{max}) versus normalized defect size ($(n - \bar{n})/\sigma$).

adopting the same forms of $\bar{V}$ dependence on volume fraction (f) as for the present Fe-32.3Ni. The latter assumption was necessary because $\bar{V}$-f data for the two additional alloys do not exist in the literature. It will be noted from Fig. 7 that all the P-n points fall on the same form of normalized distribution. Again, in contrast to the exponential preexisting-site distribution, the autocatalytic distribution in all these cases is consistent with a form that is distributed about a mode. The symmetric Gaussian distribution, for which the mode is equal to the mean value, appears to offer a valid approximation to the shape of the autocatalytic distribution deduced here.

Distributed Activation Kinetics of Nucleation

As discussed before, both preexisting and auto-catalytic nucleating defects are distributed with respect to the defect size (n), which, in turn, is related inversely to the nucleation activation energy (Q). Therefore, martensitic nucleation is a multiactivated process for both the preexisting and autocatalytically generated sites. The nucleation rate for this multiactivated process, generalizing on Eq 5, can be expressed as:

$$\frac{dN_v}{dt} = \int_0^Q \left[\left(\frac{dN_i}{dQ} + f\frac{dP}{dQ} - \frac{dN_v}{dQ}\right) \cdot (1-f)\nu \exp\left(-\frac{Q}{kT}\right)\right] dQ \qquad \text{(Eq 17)}$$

where dN_i/dQ is the activation-energy distribution of preexisting defects, dP/dQ is the activation-energy distribution of autocatalytic defects, and dN_v/dQ is the plate-number-density distribution as transformed from defects having an activation energy Q. For the purpose of numerical calculation, the integration is replaced by the summation:

$$\frac{dN_v}{dt} = \sum_0^Q [\delta N_i(Q) + f\delta P(Q) - N_v(Q)] \cdot (1-f)\nu \exp\left(-\frac{Q}{kT}\right) \qquad \text{(Eq 18)}$$

where $\delta N_i(Q) = (dN_i/dQ)\delta Q$ is the number density of the preexisting defects having activation energies between Q and $Q + \delta Q$, $\delta P(Q) = (dP/dQ)\delta Q$ is the number density of the autocatalytic defects having activation energies between Q and $Q + \delta Q$, and $N_v(Q) = (dN_v/dQ)\,\delta Q$ is the number density of the martensitic plates formed at time t from both the preexisting and autocatalytic defects with activation energies between Q and $Q + \delta Q$. The number density of martensitic plates $N_v(Q)$ is a function of both temperature and time, while $\delta N_i(Q)$ and $\delta P(Q)$ are temperature dependent only, and they can be calculated from the preexisting and the autocatalytic defect distributions which were presented earlier.

Since $dN_i/dQ = (dN_i/dn)\,(dn/dQ)$ and $dP/dQ = (dP/dn)\,(dn/dQ)$, $\delta N_i(Q) = (dN_i/dn)\,(dn/dQ)\delta Q$ and $\delta P(Q) = (dP/dn)\,(dn/dQ)\delta Q$. From the preexisting and auto-catalytic defect distribution functions, expressions for the derivatives dN_i/dn and dP/dn can be obtained; and from Eq 11, the relationship between n and Q, dn/dQ can be found. Therefore, the number density $\delta N_i(Q)$ of the preexisting defects having activation energies between Q and $Q + \delta Q$ and the number density $\delta P(Q)$ of the auto-catalytic defects having activation energies between Q and $Q + \delta Q$ can be written as:

$$\delta N_i(Q) = \frac{\alpha \mathbf{b} N_i n^2}{2\gamma_s V^*}\delta Q \qquad \text{(Eq 19)}$$

$$\delta P(Q) = \frac{\mathbf{b} n^2 P_{\max}}{2\sqrt{2\pi}\sigma\gamma_s V^*}\exp\left[-\left(\frac{n-\bar{n}}{\sqrt{2}\sigma}\right)^2\right]\delta Q \qquad \text{(Eq 20)}$$

Using parameter values described earlier, an activation volume V^* of 4.69×10^{-21} cm^3 is adopted, consistent with previous analysis of isothermal kinetics in Fe-Ni alloys (Ref 14). In the following calculations, the activation energy Q which ranges from 0 to Q_{max} was divided into 100 intervals, $\delta Q = Q_{max}/100$. Smaller intervals were also tested and gave the same results.

Equation 18 constitutes an initial-value problem with $f = 0$ at $t = 0$. This differential equation can be solved by the Euler method. Since $N_v = \sum_0^Q N_v(Q)$ and the volume fraction f can be assumed to be constant with respect to the activation energy Q at any time t because the contribution from each Q to the volume fraction is very small during the small time interval δt, we have:

$$\frac{dN_v(Q)}{dt} = [\delta N_i(Q) + f\delta P(Q) - N_v(Q)] \cdot (1-f)\nu \exp\left(-\frac{Q}{kT}\right) \qquad \text{(Eq 21)}$$

Rearranging and then integrating over the interval between t and $t + \delta t$, we obtain:

$$N_v(Q, t+\delta t) = \delta N_i(Q) + \delta p(Q)f - [\delta N_i(Q) + \delta p(Q)f - N_v(Q,t)] \cdot \exp[-F(Q)] \quad \text{(Eq 22)}$$

where $F(Q) = (1 - f)\delta t v \exp(-Q/kT)$ and the nucleation attempt frequency v is taken to be 1.0×10^{11}/s (Ref 4). Thus, the number density of plates transformed from defects having an activation energy between Q and $Q + \delta Q$ at any time can be calculated from the initial conditions.

The total number density of martensitic plates at time $t + \delta t$ is the sum of $N_v(Q, t + \delta t)$ with respect to all Q:

$$N_v(t+\delta t) = \sum_{0}^{Q_{max}} N_v(Q, t+\delta t) \quad \text{(Eq 23)}$$

and the volume fraction of martensite is:

$$f(t+\delta t) = N_v(t+\delta t)\bar{V} \quad \text{(Eq 24)}$$

where $\bar{V}$ is a function of f or N_v and has been determined experimentally. These calculated values of $N_v(Q, t + \delta t)$ and $f(t + \delta t)$ are then put back into Eq 22 to calculate $N_v(Q, t + 2\delta t)$ and $f(t + 2\delta t)$ for the next time interval. This process is reiterated from $t = 0$ to $t = 9000$ seconds. The integration time interval adopted for this calculation was 10^{-3} second, after finding that smaller (10^{-5}) and larger (10^{-2}) intervals yielded similar results.

The only variable to be chosen for the calculation is the maximum activation energy Q_{max}; this affects the rate but not the final saturation level. The value of Q_{max} was estimated from Eq 9. The saturation level is dominated by the cumulative autocatalytic defects (via the parameter P) which turn out to be much larger in number density than the preexisting defects N_i.

The calculated isothermal transformation curves, based on this multiactivated kinetics model, are plotted as dashed lines in Fig. 1, together with the experimental data (solid lines) for different temperatures and grain sizes. It can be seen from this figure that the calculated and experimental results are in rather good agreement, even extending from the initially rapid rate of transformation to the essentially flat isothermal tail.

Another feature of the distributed-activation energy model is illustrated in Fig. 8 for the course of martensitic transformation in the Fe-32.3Ni alloy on quenching to, and holding at, the liquid-nitrogen temperature. Here, the athermal contribution is derived from the very potent preexisting and autocatalytic defects having effective activation energies of $Q = 0$ at any given temperature. The anisothermal and isothermal parts are calculated from the less-potent defects having activation energies in the range of $0 < Q < Q_{max}$, as in Eq 11. Both the anisothermal and isothermal nucleation events are thermally activated, but the former take place too rapidly at any given temperature to be measurable as a function of time.

Range of Kinetic Behaviors

To further illustrate the full range of kinetic behaviors described by the distributed-activation model, the kinetic parameters evaluated for Fe-32.3Ni were employed to predict the kinetics of a range of Fe-Ni alloys, taking into account the composition dependence of the transformation thermodynamics as described by Kaufman (Ref 23). The calculated time-temperature-transformation (TTT) diagrams for 1 pct formation of martensite are shown in Fig. 9 for four different Fe-Ni alloys. As indicated in this figure, the transformation kinetics vary from a monotonically changing mode to suppressible C-curve behavior as the Ni content increases. The fully isothermal C-curve behavior in this alloy system appears when the Ni content is greater than 35.18 weight percent. The corresponding C-curve nose temperature is about 60 K. In contrast to Fe-Ni-Mn alloys which are characterized by a smaller activation volume V^* promoting fully isothermal kinetics over a wide range of conditions, measurable isothermal behavior at the start of transformation in the binary Fe-Ni alloys is predicted for a relatively narrow range of compositions. Such suppressible behavior in Fe-Ni alloys, traditionally considered to give athermal kinetics, has recently been demonstrated, showing "incubated" transformation in the range of 42 to 88 K (Ref 24).

Finally, the use of two potency distributions for the preexisting and autocatalytic defects can account for "burst" kinetics and its dependence on grain size, recognizing that the operation of at least one preexisting defect is a necessary condition for transformation. For

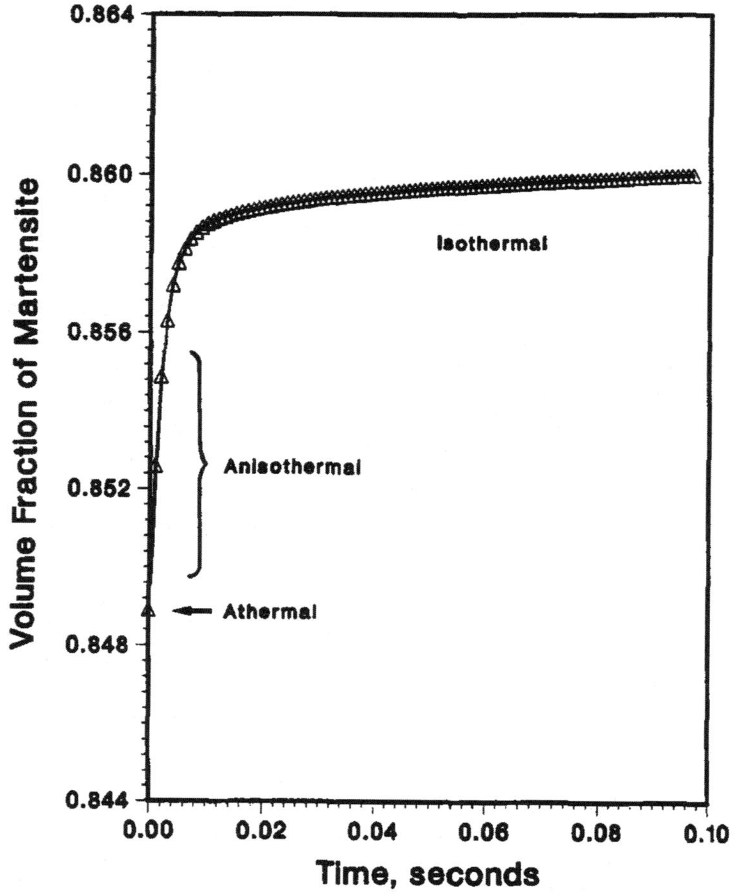

Fig. 8 Calculated short-time transformation curve on cooling to, and holding at, the liquid nitrogen temperature (77 K) for an Fe-32.3Ni alloy with austenitic grain size D = 80 μm, taking into account both the preexisting and autocatalytically generated nucleation defects and showing the contributions due to athermal, anisothermal, and isothermal components of transformation.

the 0.05 cm^3 specimen volume employed in these experiments, Fig. 10(a) shows (versus cooling temperature) the probability p that a specimen of the 80 μm grain-size material will contain a sufficiently potent initial defect according to Eqs 1 and 4. Also plotted is the transformed fraction that would result from the autocatalytic distribution once transformation is triggered. At this grain size, there is a high probability of an initial event prior to significant autocatalysis, and the fraction transformed increases smoothly with decreasing temperature. In contrast, Fig. 10(b) depicts the computed curves for the 800 μm grain-size material, where, as described by Eq 4, preexisting sites are less numerous and so the same nucleation probability requires cooling to a lower temperature; concurrently, the transformed fraction attainable from the autocatalytic distribution at a given temperature is higher, primarily through the increase in $\bar{V}$ due to the larger grain size. In this case, the M_s temperature for detectable transformation will show significant statistical scatter over a 45-K interval with a median value (p = 0.5) of 235 K, where the autocatalytic distribution gives an abrupt burst of transformation to f = 0.2. Here, the lower the transformation temperature, the larger the burst. Such behavior is consistent with the Cech-Turnbull small-particle experiments (Ref 7), where the transformation fraction measured in transformed particles was a unique function of cooling temperature, despite large variations in the onset temperature.

Fitting the measured $\bar{V}(f)$ and $P(n)$ parameters of Fig. 3 through 5 to smooth functions of grain size D, Fig. 11 summarizes the grain-size dependence of the characteristic potencies of the initial and autocatalytic defects at fixed

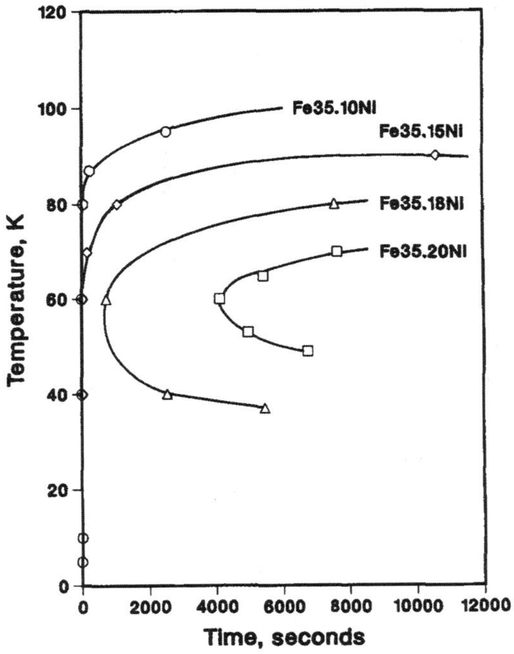

Fig. 9 Calculated TTT curves for 1 pct martensite in Fe-Ni alloys, showing the transition of transformation kinetics from monotonic to C-curve behavior beyond a critical Ni content.

specimen volume. Curves denoting initial nucleation probabilities of 0.05, 0.50, and 0.95 are seen to bracket the solid data points, representing detection by the high-sensitivity acoustic-emission measurements of Olson *et al.*, (Ref 13). Triangular points denote measurements of f = 20 pct in the current study, with curves calculated for transformation fractions of 10, 20, and 30 pct defined by the autocatalytic distribution (shown dotted at temperatures above the median M_s temperature). At grain sizes below 500 μm, relatively smooth transformation behavior is observed where acoustic emission detects initial nucleation events as high as 30 K above the M_s detection temperature by conventional electrical resistivity measurements (corresponding to autocatalysis). For grain sizes larger than 500 μm, the median potency of the most potent preexisting defect falls to a level comparable to the mean autocatalytic-defect potency, and the first event triggers substantial autocatalysis, with the resulting burst size tending to increase with further increase in grain size.

The two-distribution kinetic model is seen to account for a wide range of behaviors encompassing the dependence of transformation on temperature, time, grain size, and sample volume. This overall kinetic framework describes the full range of isothermal (including suppressible C-curve), anisothermal, athermal, and burst-athermal modes of transformation kinetics.

It should be noted in closing that the present analysis has focused on martensites with plate morphology. Recent experiments have revealed an isothermal "tail" giving kinetic curves similar to those of Fig. 1 in Fe-Ni alloys, forming lath martensites at temperature as high as 670 K (Ref 25); a dependence of lath packet size on cooling rate has been related to this isothermal component (Ref 26). Interpreting packet initiation in terms of preexisting site operation and parallel lath formation in terms of auto-catalysis, an adaptation of the two-distribution model and employing appropriate $\bar{V}(f)$ behavior for the lath morphology may provide a useful framework for treating these systems as well.

Conclusions

1. In order to quantify the nucleation behavior of a martensitic transformation as a function of temperature and time, the mean volume of martensitic plates ($\bar{V}$) was measured versus fraction transformation (f) in an Fe-32.3Ni alloy with three different grain sizes. Consistent with previous studies, $\bar{V}$ was found to decrease with increasing f for a given grain size of the parent phase and to increase with increasing grain size at a given f. The measured f and $\bar{V}$ were employed to define the number density of nucleated plates, N_v, and, hence, the number density of nucleation events.
2. Using a potency distribution of preexisting nucleation defects previously determined for the same alloy, an autocatalytic-defect distribution with respect to defect potency was calculated from the measured number density of nucleation events N_v. The latter distribution can be well described by a Gaussian function, whereas the preexisting-defect distribution is exponential in form. A major characteristic of the autocatalytic-defect distribution is that the cumulative number density of these defects saturates with decreasing defect size, while the number density of the preexisting defects continues to increase monotonically. The autocatalytic-defect mean potency

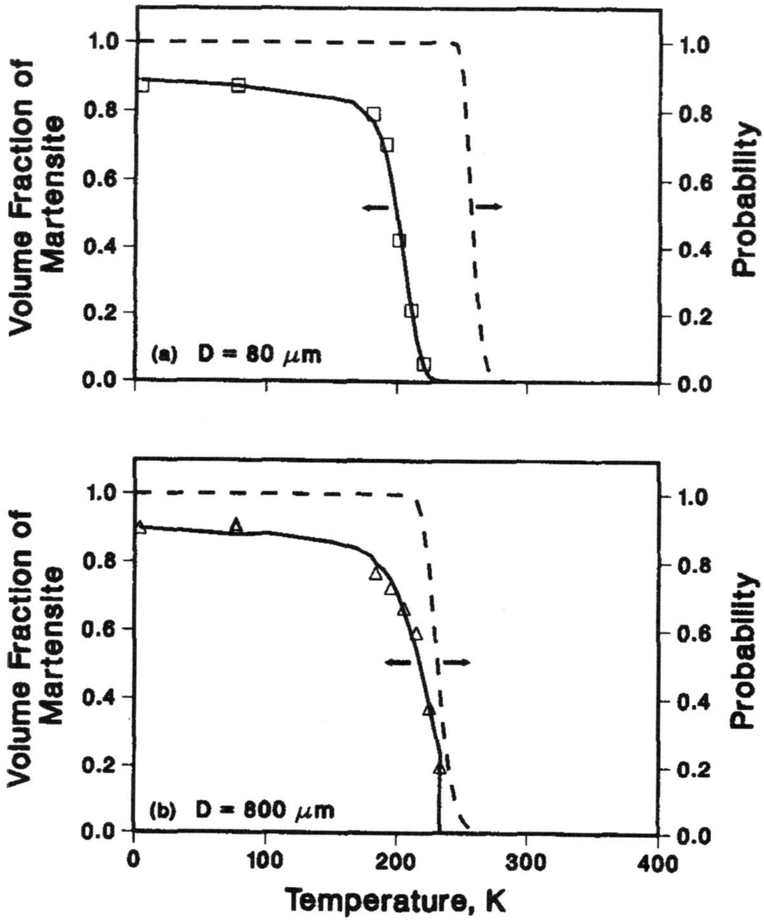

Fig. 10 Saturated volume fraction of martensite and probability of initial nucleation *vs* temperature for Fe-32.3Ni alloy with austenitic grain sizes (a) $D = 80$ μm and (b) $D = 800$ μm.

increases weakly with increasing grain size and is attributed to a greater effectiveness of larger martensite plates in generating autocatalytic defects. The distribution width also increases but appears to reach a limiting value at large grain size.

3. Employing the relation between defect size and activation energy for nucleation as given by the Olson-Cohen dislocation-dissociation model of martensitic nucleation, the distribution of nucleation defects (both preexisting and autocatalytic) with respect to the activation energy (Q) was also derived from the defect-size distribution. From the resulting activation-energy distribution, a distributed-activation model of nucleation kinetics was developed. The distributed-activation model accounts for the entire course of martensitic transformation curves (including the athermal, anisothermal, and isothermal contributions) and predicts the composition range of suppressible C-curve kinetic behavior as well as the grain-size dependence of burst behavior. The calculated results are found to be in good agreement with the experimentally observed transformation curves at different temperatures and grain sizes. The model appears capable of accounting for a wide range of observed kinetic behaviors in martensitic transformations.

ACKNOWLEDGMENTS

The authors would like to thank the National Science Foundation for the financial support of this research under Grant No. DMR 84-3802 at MIT and Grant No. DMR 88-20116

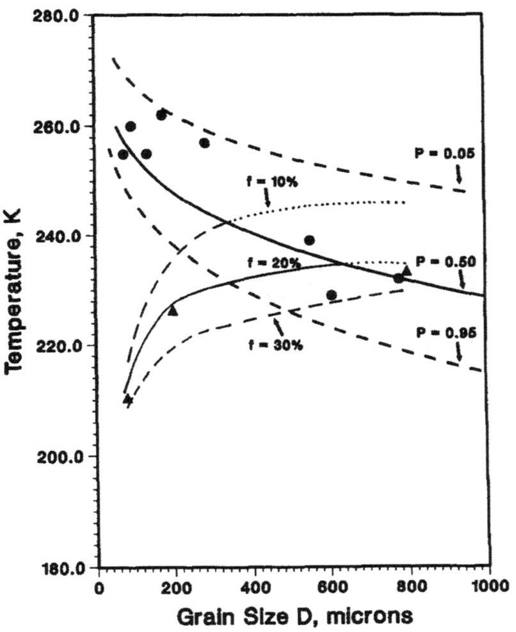

Fig. 11 Calculated transformation temperatures and initial nucleation probabilities as a function of grain size for Fe-32.3Ni alloy. The solid circles represent the very first plate-formation temperatures from high-sensitivity acoustic-emission measurements of Ref 13, and the triangular points are experimental data for 20 pct martensite.

at Northwestern University. Special thanks go to Dr. Dennis M. Haezebrouck for many helpful discussions.

REFERENCES

1. R.M. Kimmel and D.R. Uhlmann: *J. Appl. Phys.*, 1969, vol. 40, p 4254–60
2. A.S. Argon and H.Y. Kuo: *J. Non-Crystalline Solids*, 1980, vol. 37, p 241–66
3. E.S. Machlin and Morris Cohen: *Trans. AIME*, 1952, vol. 194, p 489–500
4. C.L. Magee: *Metall. Trans.*, 1971, vol. 2, p 2419–30
5. V. Raghavan and Morris Cohen: *Metall Trans.*, 1971, vol. 2, p 2409–18
6. V. Raghavan: in *Martensite*, G.B. Olson and W.S. Owen, eds., ASM, Metals Park, OH, 1992
7. R.E. Cech and D. Turnbull: *Trans. AIME*, 1956, vol. 206, p 124–32
8. G.B. Olson and Morris Cohen: *New Aspects of Martensitic Transformation*, Japan Institute of Metals, Kobe, Japan, 1977, p 93–98
9. I.-Wei Chen, Y.-H. Chiao, and K. Tsuzaki: *Acta Metall.*, 1985, vol. 33, p 1847–59
10. S. Kajiwara, S. Ohno, K. Honma, and M. Uda: *Proc. ICOMAT 86*, Japan Institute of Metals, Nara, Japan, 1986, p 359–64
11. S. Kajiwara, S. Ohno, and K. Honma: *Phil. Mag.* A, 1991, vol. 63, p 625–44
12. T.F. Kelly, M. Cohen, and J.B. Vander Sande: *Metall. Trans. A*, 1984, vol. 15A, p 819–33
13. G.B. Olson, K. Tsuzaki, and Morris Cohen: *David Turnbull Symp. on Phase Transitions in Condensed Systems*, G.S. Cargill, F. Spaepen, and K.N. Tu, eds., MRS, Pittsburgh, PA, 1987, vol. 57, p 127–48
14. G.B. Olson and Morris Cohen: *Metall. Trans. A*, 1976, vol. 7A, p 1915–23
15. G.B. Olson and Morris Cohen: *Ann. Rev. Mater. Sci.*, 1981, vol. 11, p 1–30
16. R.L. Fullman: *Trans. AIME*, 1953, vol. 197, p 447–52
17. D.G. McMurtrie and C.L. Magee: *Metall. Trans.*, 1970, vol. 1, p 3185–91
18. G.B. Olson and Morris Cohen: *Metall. Trans. A*, 1976, vol. 7A, p 1897–1904
19. G.B. Olson and Morris Cohen: in *Dislocations in Solids, Vol. 7*, F.R.N. Nabarro, ed., North-Holland, Amsterdam, 1986, p 295–407
20. G. Hausch and H. Warlimont: *Acta Metall.*, 1973, vol. 21, p 401–14
21. D.M. Haezebrouck: Sc.D. Thesis, Massachusetts Institute of Technology, Cambridge, MA, 1987
22. M. Grujicic, H.C. Ling, D.M. Haezebrouck, and W.S. Owen: in *Martensite*, G.B. Olson and W.S. Owen, eds., ASM, Metals Park, OH, 1992
23. L. Kaufman: Manlabs-NPL Thermochemical Database, 1977
24. K. Ullakko, M. Nieminen, and J. Pietikainen: *Martensitic Transformations*, Proc. ICOMAT 89, Sydney, Australia, B.C. Muddle, ed., *Materials Science Forum Vols. 56–58*, Trans Tech Publications, Aedermannsdorf, Switzerland, 1990, p 225–28
25. K. Tsuzaki, T. Fukiage, T. Maki, and I. Tamura: *Martensitic Transformations*, Proc. ICOMAT 89, Sydney, Australia, B.C. Muddle, ed., *Materials Science Forum Vols. 56–58*, Trans Tech Publications, Aedermannsdorf, Switzerland, 1990, p 229–34
26. K. Tsuzaki and T. Maki: *J. Japan Inst. Met.*, 1981, vol. 45, p 126–34

Transformations—Selected Works of G.B. Olson on Materials, Microstructure, and Design
C.E. Campbell, M.V. Manuel, and W. Xiong, editors

CHAPTER 7

Coupled Diffusional/Displacive Transformations (Part II)—Solute Trapping*

G.B. Olson, Northwestern University
H.K.D.H. Bhadeshia, University of Cambridge
M. Cohen, Massachusetts Institute of Technology

Abstract

An earlier theory for the nonequilibrium transformation of austenite in Fe-C alloys to partially supersaturated plates of ferrite is extended to include a recent solute trapping model by Aziz. The previous model yielded a relationship between interface velocity and supersaturation for a specified transformation temperature. However, a unique growth velocity was determined from a velocity maximization criterion that could not be justified. By including the Aziz solute trapping function as a third interface response function (in addition to functions describing the diffusion field velocity and interface mobility), it has been possible to obtain a more physically based unique solution to the growth velocity for a specified temperature. The new calculations on an Fe-C alloy indicate a smoothly increasing supersaturation during both nucleation and growth, as the transformation temperature is reduced. Both the nucleation and growth processes tend to become completely diffusionless only below the martensite start temperature.

Introduction

Martensitic transformation of austenite in steels has long been recognized to involve the diffusionless nucleation and growth of supersaturated plates. There is a large difference in the diffusivities of interstitial carbon and iron (or substitutional solutes). The possibility then arises that the substitutional lattice can be transformed by a displacive mechanism, while the carbon atoms redistribute between the parent and product phases (Ref 1). Circumstances could arise whereby the transformation interface then advances sufficiently slowly to enable the equilibrium or paraequilibrium partitioning of carbon. Alternatively, only some of the carbon may redistribute during transformation, the rest being trapped by the advancing interface.

Since any diffusion field is, by its very nature, subject to random perturbations, opportunities must arise for the field to adjust toward lower free energy states. It follows that there will be a tendency for the growth process to collapse toward equilibrium. Hence, growth involving partial supersaturation has to be stabilized by some process which occurs in series with the diffusion of solute (Ref 2, 3). One such stabilizing process is that of structural change across the interface (Ref 3). The rate at which the interface moves depends both on its intrinsic mobility (related to the process of structural change across the interface) and on the ease with which the

*G.B. Olson, H.K.D.H. Bhadeshia, and M. Cohen, "Coupled Diffusional/Displacive Transformations: Part II. Solute Trapping," *Metall. Trans.* 21 A, 1990, p 805–9

partitioned solute can diffuse ahead of the moving interface. Since the two processes are in series, the interfacial velocity as calculated from the interface mobility must always equal that computed from the diffusion of solute ahead of the interface.

By considering the simultaneous solution of the interface mobility and diffusion field velocity functions, we modeled the displacive transformation of Fe-C austenite to plates of ferrite which were partially supersaturated with carbon (Ref 3). The interface mobility law characteristic of martensitic transformations (Ref 3, 4–6) was utilized as one of the velocity solutions (or "interface response functions"), the other being the Ivantsov solution (Ref 3, 7) for the diffusion field velocity. Since the number of unknowns to solve for is three (the supersaturation, the composition in the austenite at the interface, and the growth rate) and since only two velocity laws were solved simultaneously, the method permitted the calculation of the *relationship* between plate growth rate and carbon supersaturation for a given transformation temperature. To assign a particular supersaturation and velocity to a given temperature, we assumed that the actual growth velocity adopted would be the maximum rate associated with that temperature (Ref 3). This heuristic approach was necessary since only two response functions were utilized in the model. The purpose of the study reported here was to attempt a more complete analysis which avoids the maximum rate hypothesis by utilizing an additional interface response function derived by Aziz (Ref 8 to 10) for the case of solute trapping during rapid solidification processes.

The Interface Response Functions

As mentioned earlier, the process of structural change across the interface and the diffusion of solute ahead of the interface both dissipate a proportion of the net free energy, ΔG, that is available for interfacial motion. The laws relating interface velocity to a particular process leading to the dissipation of free energy are the interface response functions, and we begin by introducing the three functions utilized in this study.

The Interface Mobility

The present work deals with a transformation product that is identical to martensite in structure and interface mobility but for which some partitioning of interstitial solute at the interface increases the available free energy change, allowing transformation to happen above the martensite start (M_s) temperature. Interfacial and elastic energy quantities are taken to be identical to previous estimates for martensitic transformation (Ref 3).

The theory for thermally activated dislocation motion can be adapted to represent the motion of the α/γ interface, using dislocation models of interfacial structure (Ref 4–6). The velocity, V_i, for the motion of the interface is then given by:

$$V_i = V_0 \exp\left\{-\frac{Q^*}{kT}\right\} \quad \text{(Eq 1)}$$

where Q^* is a free energy of activation and the pre-exponential factor, V_0, is taken to be 30 ms^{-1} based on experimental data from single-interface martensite transformations (Ref 5); k is the Boltzmann constant, and T the absolute temperature. The activation energy, Q^*, can be expressed as a function of the net interfacial driving force, G_{id}, through the relation (Ref 11):

$$Q^* = \int_{G_{id}}^{G'_{id}} v^* dG \quad \text{(Eq 2)}$$

where G'_{id} is the maximum glide resistance presented by obstacles to interfacial motion and v^* is the activation volume swept by the interface during a thermally activated event. For a wide range of obstacle interactions, the function $Q^*\{G_{id}\}$ can be represented by:

$$Q^* = Q_0\left\{1 - \left(\frac{G_{id}}{G'_{id}}\right)^y\right\}^z \quad \text{(Eq 3)}$$

where Q_0 is the total activation free energy necessary to overcome the resistance to dislocation motion without the aid of an interfacial driving force. The constants y and z define the shape of the force-distance function, and for solid solution interactions in the Labusch limit, we assume that $y = 0.5$ and $z = 1$ (Ref 12).

Analysis of kinetic data for the interface-controlled barrierless nucleation of martensite gives:

$$Q_0 = 0.31\,\mu\Omega$$

where μ is the shear modulus of the matrix and Ω is the volume per atom. Based on the behavior of Fe-Ni-C alloys (Ref 13), G'_{id} is taken to be:

$$G'_{id} = 1.22 \times 10^{-3}\mu$$

The Diffusion Field Velocity

This is calculated assuming a ferrite particle which is in the form of a disc with a curved edge of radius equal to the semithickness of the disc. The velocity, V_d, for steady-state growth of ferrite of constant composition, x_α, in a steel of composition $\bar{x}$, transformed at a temperature T and as a function of the diffusion of carbon in the austenite ahead of the interface, can be approximated by the Ivantsov solution for a parabolic cylinder (Ref 7):

$$\frac{\bar{x} - x_I}{x_\alpha - x_I} = (\pi p)^{0.5} \exp\{p\} \operatorname{erfc}\{p^{0.5}\} \qquad \text{(Eq 4)}$$

where x_I is the carbon concentration in the austenite at the interface and p is the Peclet number (a dimensionless velocity) given by:

$$p = \frac{V_d \rho}{2D} \qquad \text{(Eq 5)}$$

with ρ the effective plate tip radius consistent with the parabolic approximation of the plate shape.

As the diffusion coefficient, D, of carbon in austenite is a sensitive function of its concentration, a weighted average diffusivity, $\bar{D}$, is used instead (Ref 3, 14–16):

$$\bar{D} = \int_{x_I}^{\bar{x}} \frac{D\{x, T\}dx}{\bar{x} - x_I} \qquad \text{(Eq 6)}$$

The method for calculating the interface compositions as a function of the amount, G_{dd}, of free energy dissipated in the diffusion of solute ahead of the interface has been discussed by Ref 3 along with details of the thermodynamic theory and data necessary for the computations. It is, however, important to note that the concentrations x_α and x_I are not independent (see Fig. 1 of Ref 3) and that the choice of either fixes the value of the other.

Solute Trapping Law

It has long been established that under suitable transformation conditions, the transformation interface may move at a rate which is higher than would be consistent with the equilibrium partitioning of solute. The martensitic transformation in steels is a classic example of this effect, since during the nucleation and growth of martensite neither the interstitial nor substitutional atoms can diffuse. A similar effect arises during paraequilibrium transformation (Ref 17) in substitutionally alloyed steels, when interstitial elements may redistribute, but the substitutional atoms are configurationally frozen. In modern terminology, the atoms which are forced into the product lattice, that is, whose chemical potential is raised on transfer across the interface, are said to be trapped (Ref 18). When it is the solute atoms whose chemical potential is raised on transfer across the interface, the phenomenon is called *solute trapping*, as opposed to *solvent trapping*.

Aziz (Ref 8, 9) has developed a model for solute trapping during rapid solidification processing; a transition from diffusion-controlled to diffusionless solidification is predicted as the interface speed exceeds the maximum speed with which solute atoms can diffuse across the interface and are thus pinned down in the product phase. The model relates interfacial velocity to the partitioning coefficient, k_p, which is the ratio of the solute concentration in the (homogeneous) product phase to the solute concentration in the parent phase at the interface. In the present context,

$$k_p = \frac{x_\alpha}{x_I} \qquad \text{(Eq 7)}$$

and $k_p = k_e$ when the two concentration terms represent the respective equilibrium concentrations of the phases concerned; k_e is thus the equilibrium partitioning coefficient.

The model is based on chemical reaction rate theory, one of the assumptions of which is that the "reaction" consists of the repetition of unit steps involving the interaction of a small number of atoms. This approach seems justified for a solidification process, but as pointed out by Christian (Ref 19), the assumptions of chemical rate theory are unlikely to be applicable to displacive transformations involving the coordinated motion of large

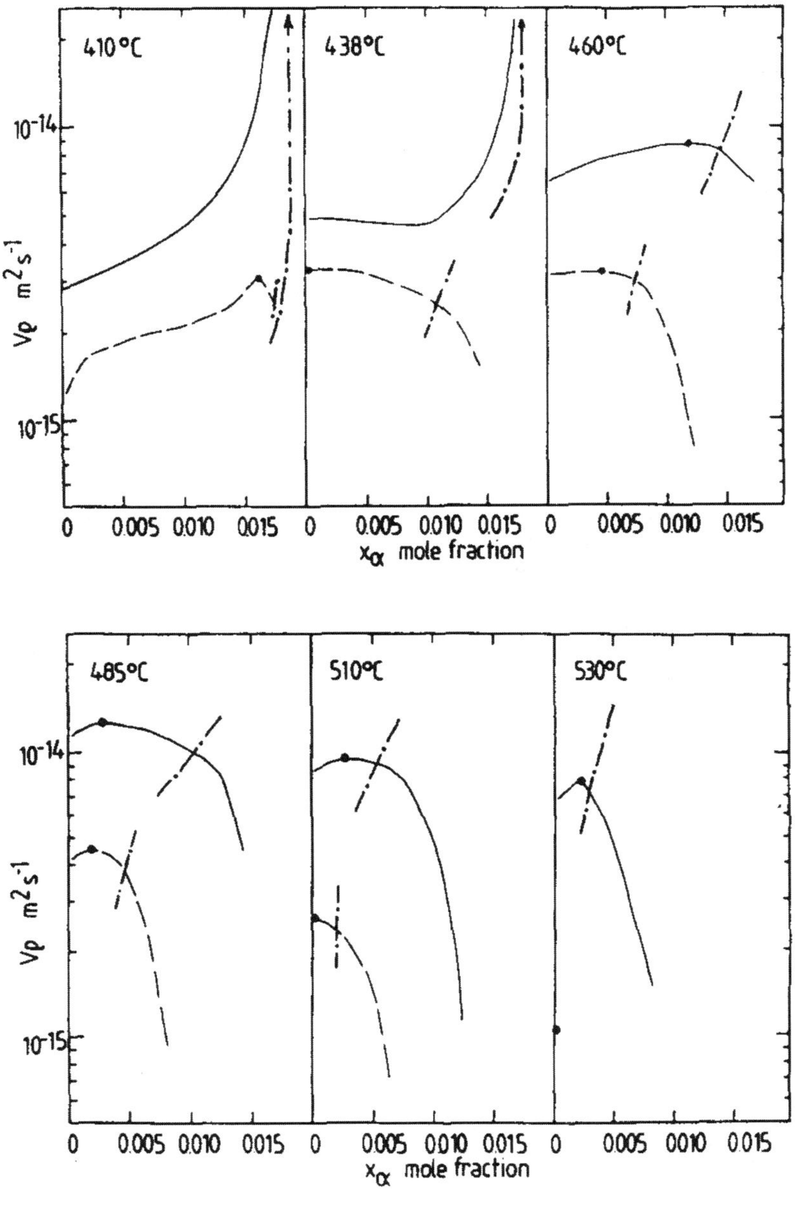

Fig. 1 Plots of $V\rho$ *vs* x_α computed for coupled transformation in Fe-0.4C wt pct steel ($\bar{x}$ = 0.0186). The dashed curves represent conditions at nucleation, while the solid curves depict growth. The points (dots) represent the locations of maxima in the velocity, when only the interfacial mobility and diffusion field velocity functions are solved simultaneously (Ref 3). The chain-link curves represent the solute trapping laws, and their intersections with the other curves depict the simultaneous solutions to all three interface response functions (V_d, V_i, and V_k).

numbers of atoms. Although the present work deals with displacive transformations, it is the substitutional atoms which participate in such coordinated atomic displacements. The interstitial carbon is, in most cases, sufficiently mobile to be treated as a species which is *diffusing* across the interface, that is, the carbon atoms undergo an uncoordinated transfer across the boundary. Consequently, the Aziz model, which relies on diffusive fluxes across the interface, may be utilized as a solute trapping function for the redistribution of carbon during the displacive transformation of austenite.

There are two basic mechanisms of interface displacement, one involving propagation by the displacement of steps and the other by the displacement of all elements of the boundary (this latter mechanism is called "continuous" motion). Aziz has derived slightly different trapping models for the two cases (Ref 8, 9). The step model only permits transformation below the T_0 temperature of the austenite in the vicinity of the interface, where it is thermodynamically possible to get composition invariant transformation in those regions. This would be too restrictive for the present study, where transformation is allowed above the T_0' temperature. (Strain energy is a dominant factor in displacive transformations, and it is below the T_0' temperature that composition invariant displacive reactions become thermodynamically possible.) Goldman and Aziz (Ref 10) have proposed another model for stepped growth (the "aperiodic step model"), in which the steps are assumed to pass at random intervals and which also restricts transformation to below the T_0 temperature of the parent phase at the interface. The interesting point is that, for that case, the trapping law they found turns out to be the same as for the continuous growth model.

For the present work, we assume the trapping law which happens to be common to the continuous growth and the aperiodic step model but emphasize that transformation is permitted above the T_0' temperature of austenite of composition x_I. The trapping model then gives a velocity function of the form (Ref 8–10):

$$V_k = \frac{D\{x_I\}}{\lambda}\left(\frac{k_p - k_e}{1 - k_p}\right) \quad \text{(Eq 8)}$$

where λ is the intersite distance of $\simeq$0.25 nm and $D\{x_I\}$ is the carbon diffusivity in austenite of composition x_I. The form $D\{x_I\}/\lambda$ amounts to the diffusion velocity of carbon, and trapping becomes significant as the actual interface velocity approaches this value. Note that the carbon atoms execute jumps across a glissile semi-coherent interface, so that it is appropriate to approximate the diffusion coefficient in the Aziz equation by the volume diffusivity (calculated for the composition of the austenite at the interface).

Equation 8 gives a third velocity relation which is a smooth function of x_α and x_I. As x_α approaches $\bar{x}$, the diffusion field velocity of Eq 4 diverges and the interfacial dissipation then imposes the condition that $x_I = \bar{x}$ (Fig. 1, Ref 3) such that V_k also smoothly diverges in the full trapping limit.

Results and Discussion

Calculations using the three interface response functions were carried out for a Fe-0.4C wt pct steel ($\bar{x}$ = 0.0186) for the transformation temperature range of 530 °C to 410 °C for both the nucleation and growth processes. The results are presented in Fig. 1. A value of 700 J mol^{-1} was taken as a reasonable representation of the elastic strain energy and surface energy terms (G_{el} + G_{surf}) during the growth of martensite; this term, in effect, represents the stored energy of martensite and is assumed to be independent of temperature. During barrierless nucleation, the stored energy term exceeds 700 J mol^{-1} and is taken to be 2×10^{-3} μ (corresponding to ρ = 1.5 nm), based on previous work on the nucleation of martensite (Ref 3). It is these differences in the stored energy that lead to different nucleation and growth kinetics.

In our previous work (Ref 3), the simultaneous solution of the interface mobility and the diffusion field velocity functions for a specified temperature led to the curves of velocity versus supersaturation reproduced in Fig. 1. Superimposed on those curves are the data obtained using the solute trapping model. Each intersection between the solute trapping function and one of the curves representing the simultaneous solution of the mobility and diffusion functions gives a unique combination of interface compositions, velocity, and supersaturation, which is compatible with all three interface response functions.

Examination of Fig. 1, 2, and 3 reveals that when compared with the earlier maximum velocity hypothesis, the new solutions, in general, lead to higher carbon concentrations in the ferrite during both nucleation and growth. Otherwise, the general results of the earlier model are verified; that is, at elevated temperatures, both nucleation and growth involve some degree of carbon partitioning. As the temperature is reduced, it becomes possible to sustain diffusionless growth even though nucleation requires diffusion. At higher undercoolings, true martensitic transformation becomes feasible with diffusionless nucleation and growth. The calculated time-temperature-

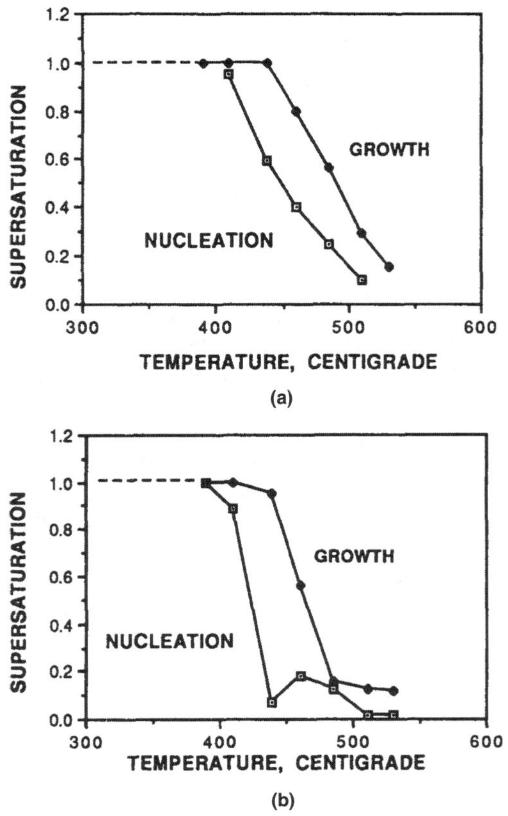

Fig. 2 (a) Plot of calculated normalized supersaturation ($x_\alpha/\bar{x}$) of carbon in ferrite versus the isothermal transformation temperature for a Fe-0.4C wt pct alloy, with the data obtained by the simultaneous solution of the interfacial mobility, diffusion field velocity, and trapping velocity functions. (b) The data obtained in our earlier analysis (Ref 3), in which the final solutions depended on a velocity maximization criterion.

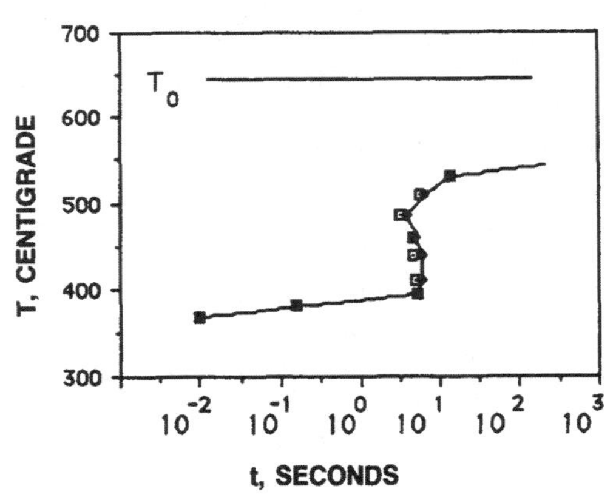

Fig. 3 Calculated isothermal transformation diagram for a Fe-0.4C wt pct alloy. The continuous curve is the time for 1 pet transformation for the present calculations, whereas the open points which do not lie on that curve are from the earlier analysis. Source: Ref 3

transformation diagram for the initiation of transformation is not greatly modified by the new analysis.

Conclusions

A more rigorous model for the coupled diffusional/displacive growth of plates of ferrite with varying levels of carbon partitioning has been established by incorporating a solute trapping function due to Aziz. The model relies on the simultaneous solution of three interface response functions, one describing the mobility of a glissile boundary, another the diffusion field velocity, and the third the velocity consistent with a specified degree of solute trapping. The solution of these functions is found to lead to the prediction of plate velocity and carbon supersaturation at any given temperature, without the need for a maximum velocity criterion.

The model retains many of the essential features of our earlier analysis. It predicts an increasing interfacial velocity and supersaturation during growth with decreasing transformation temperature. The nucleation velocity, however, goes through a maximum, thereby giving *C*-curve kinetics. The supersaturation during nucleation increases smoothly with undercooling, the nucleus becoming fully supersaturated at the martensite start temperature.

Nomenclature and Abbreviations

D	diffusivity of carbon in austenite
$\bar{D}$	weighted average diffusivity of carbon in austenite
ΔG	magnitude of Gibbs free energy change per unit volume
Q_0	activation free energy necessary to overcome the resistance to interfacial motion in the absence of an interfacial driving force
Q^*	activation free energy for interfacial motion
G_{dd}	Gibbs free energy per unit volume, dissipated in the diffusion of solute ahead of the transformation interface
G_{id}	Gibbs free energy per unit volume, dissipated in the process of interfacial motion
G_{el}	stored free energy per unit volume due to elastic strains

G_{surf} stored free energy per unit volume, due to interface
k Boltzmann constant
k_e equilibrium partition coefficient
k_p actual partition coefficient
p Peclet number
ρ plate tip radius
T absolute temperature
V velocity for the continuous motion of a planar interface
V_d velocity as calculated using the diffusion field velocity law
V_k velocity as calculated using the Aziz solute trapping law
V_i velocity as calculated using the interfacial mobility law
V_0 preexponential velocity factor for thermally activated interface motion
x carbon concentration, mole fraction
$\bar{x}$ average carbon concentration in alloy, mole fraction
x_I carbon concentration in γ at the α/γ interface
x_α carbon concentration in α at the α/γ interface
$x^{\alpha\gamma}$ equilibrium carbon concentration in ferrite
$x^{\gamma\alpha}$ equilibrium carbon concentration in austenite
α ferrite
γ austenite
μ shear modulus of austenite
G'_{id} maximum glide resistance presented by obstacles to dislocation motion
Ω volume per atom

ACKNOWLEDGMENTS

We are grateful to Professor M.J. Aziz for some valuable discussions on the subject of solute trapping. Research on this work, at Northwestern University and at MIT, is sponsored by the National Science Foundation under Grant No. DMR 88-20116.

This paper is based on a presentation made in the symposium "International Conference on Bainite" presented at the 1988 World Materials Congress in Chicago, IL, on September 26 and 27, 1988, under the auspices of the ASM International Phase Transformations Committee and the TMS Ferrous Metallurgy Committee.

REFERENCES

1. J.W. Christian: Iron and Steel Institute Special Report 93, Iron and Steel Institute, London, 1965, p 1
2. J.W. Christian and D.V. Edmonds: in *Phase Transformations in Ferrous Alloys*, A.R. Marder and J.I. Goldstein, eds., TMS-AIME, Warrendale, PA, 1983, p 293
3. G.B. Olson, H.K.D.H. Bhadeshia, and M. Cohen: *Acta Metall.*, 1989, vol. 37, p 381–90
4. M. Grujicic, G.B. Olson, and W.S. Owen: *Metall. Trans. A*, 1985, vol. 16A, p 1713
5. M. Grujicic, G.B. Olson, and W.S. Owen: *Metall. Trans. A*, 1985, vol. 16A, p 1723
6. M. Grujicic, G.B. Olson, and W.S. Owen: *Metall. Trans. A*, 1985, vol. 16A, p 1735
7. G.P. Ivantsov: *Dokl. Akad. Nauk. SSSR*, 1947, vol. 58, p 567
8. M.J. Aziz: *J. Appl. Phys.*, 1982, vol. 53, p 1158–68
9. M.J. Aziz: *Appl. Phys. Lett.*, 1983, vol. 43, p 552–54
10. L.M. Goldman and M.J. Aziz: *J. Mater. Res.*, 1987, vol. 2, p 524–27
11. U.F. Kocks, A.S. Argon, and M.F. Ashby: *Proc. Mater. Sci.*, 1975, vol. 19, p 1
12. F.R.N. Nabarro: *Proc. Roy. Soc.*, 1982, vol. A381, p 285
13. G.B. Olson: Northwestern University, Evanston, IL, unpublished research, 1984
14. R. Trivedi and G.M. Pound: *J. Appl. Phys.*, 1967, vol. 38, p 3569
15. R.H. Siller and R.B. McLellan: *Metall. Trans.*, 1970, vol. 1, p 985
16. H.K.D.H. Bhadeshia: *Met. Sci.*, 1981, vol. 15, p 477
17. A. Hultgren: *Jernkontorets Ann.*, 1951, vol. 135, p 403
18. J.C. Baker and J.W. Cahn: *Solidification*, ASM, Metals Park, OH, 1971, p 23
19. J.W. Christian: *Theory of Transformations in Metals and Alloys*, 2nd ed., Pergamon Press, Oxford, 1975, part I, p 77–78

Transformations—Selected Works of G.B. Olson on Materials, Microstructure, and Design
C.E. Campbell, M.V. Manuel, and W. Xiong, editors

CHAPTER 8

Kinetics of Martensite Transformations in Steels*

G.B. Olson and Z.D. Feinberg
Northwestern University

Abstract

Dislocation theory of the mechanism and kinetics of martensitic transformation has provided a foundation for the prediction of constitutive relations for transformation plasiticity. Application to the control of stress-state dependent shaping of stress-strain behavior has demonstrated substantial improvements in uniform ductility and ductile fracture toughness. The advance of such predictive science continues to play a central role in the new enterprise of science-based materials design.

Introduction

A martensitic phase transformation can be considered as the spontaneous plastic deformation of a crystalline solid in response to internal chemical forces. It is no surprise, then, that the theory of dislocations, after incorporating the generalized forces of chemical thermodynamics and special constraints of crystalline interfaces, has provided the most useful basis for understanding the mechanism of these transformations. Martensitic transformations have been defined as a subset of diffusionless/displacive solid-state transformations in which the strain energy arising from a shear-dominant lattice distortion controls the kinetics and product morphology during transformation (Christian *et al.*, Ref 1). Such transformations are first-order and heterogeneous in nature, proceeding by the propagation of relatively sharp interfaces. These characteristics make the approximations of traditional linear-elastic dislocation theory particularly appropriate, and it is for this reason that the application of the theory to martensitic transformations (Olson and Cohen, Ref 2) has been so successful.

These same characteristics give rise to the high mobility of martensitic interfaces underlying the quench-hardening of steel and the precisely prescribable transformation kinematics predicting their interaction with applied stress exploited in transformation plasticity applications. The latter applications span not only modern TRIP (transformation-induced plasticity) steels, but the still emerging class of shape memory alloys.

Mechanism and Kinetics of Martensitic Transformation

The application of classical nucleation theory to martensitic transformations, using

*G.B. Olson and Z.D. Feinberg, "Kinetics of Martensite Transformations in Steels," in *Phase Transformations in Steels*, Woodhead Publishing, 2012, vol. 2, p 59–82, doi:10.1533/9780857096111.1.59

both continuum-elastic and discrete-dislocation descriptions of a nucleus, has been extensively reviewed (Olson and Cohen, Ref 2 to 4). The classical approach treats the energetics of formation of a nucleus along a "path" of fixed structure and increasing size. The nucleus crystal structure is taken to be that of the fully formed transformation product, with relatively sharp interface; the specific interfacial energy is customarily assumed to be size independent.

The specific interfacial energy γ of a fully coherent particle will consist of a "chemical" core energy, while the interfacial energy in the semi-coherent case will be dominated by a short-range elastic-energy contribution. If the nucleus transformation strain (IPS) is an invariant-plane strain, the strain energy per particle volume for a thin oblate spheroid of radius r and semi-thickness c is of the form Kc/r, where K is an elastic constant. When the transformation strain deviates from an exact invariant-plane strain, an additional shape-insensitive strain energy contribution g_p^{str} arises from distortions in the particle habit plane. Following Kaufman and Cohen (Ref 5), the total free energy of an isolated martensitic particle can thus be described by:

$$\Delta G(r,c) = \frac{4}{3}\pi r^2 c\left(\Delta g^{ch} + g_p^{str}\right) + \frac{4}{3}\pi r^2 K + 2\pi r^2 \gamma \qquad \text{(Eq 1)}$$

Under the thermodynamic conditions for which martensitic transformations are known to occur, the critical energy barrier is far too large to be thermally surmountable, and it can be safely concluded that classical homogeneous nucleation is not possible.

In an imperfect crystal, defect interactions can substantially alter the energy of a classical nucleus such that the free-energy barrier for heterogeneous nucleation is greatly diminished relative to the homogeneous case. Consistent with the strain-energy dominant character of martensitic transformations, it is generally assumed that the most important interaction in heterogeneous martensitic nucleation is with the defect stress field. It is now well established by quantitative calculations that classical nucleation via such interaction is indeed quite plausible at experimentally observed heterogeneous nucleation sites. While this interaction can be described by traditional continuum-elastic methods, the structurally equivalent discrete-dislocation description in terms of a defect-dissociation process leads most directly to the essential behavior, and is actually the method by which the first viable theory of heterogeneous martensitic nucleation was developed (Olson and Cohen, Ref 6 and 7).

Heterogeneous Nucleation by Defect Dissociation

Adopting the discrete-dislocation description of Fig. 1 for the interfacial structure of a classical martensitic embryo, the long-range stress field and volume strain energy of the embryo arise primarily from the array of transformation "coherency" dislocations lying on crystal planes at a shallow angle to the embryo habit as represented schematically in Fig. 1(a). Clearly the free-energy barrier for nucleation is greatly diminished if the latter dislocation array is derived from the dissociation of an existing defect. A convenient defect model for this purpose is the extrinsic boundary defect represented by the array of "excess" interfacial dislocation in Fig. 2(a), residing in an existing grain boundary or interphase boundary. The defect has a finite length and height, and a net Burgers vector content giving rise to a stress field of range comparable to its length, l. Embryo formation by dissociation of this defect is depicted in Figs 2

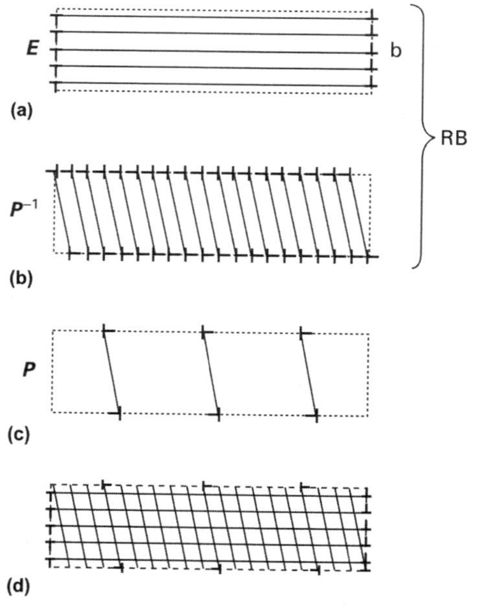

Fig. 1 Discrete-dislocation model of martensitic interface: (a) transformation "coherency" dislocations producing macroscopic shape strain ***E***; (b) coherency dislocations producing complementary shear $\boldsymbol{P}^{-1}$; (c) anticoherency dislocations producing lattice-invariant deformation ***P***; (d) total interfacial structure (Olson and Cohen, Ref 3).

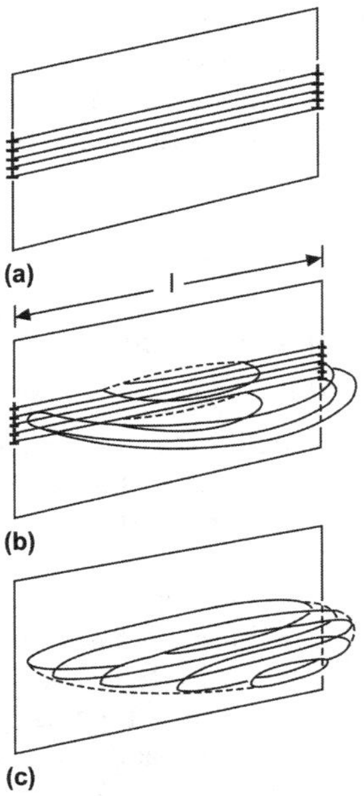

Fig. 2 Heterogeneous martensitic nucleation by defect dissociation: (a) nucleating defect; (b) dissociation to produce interfacial coherency dislocations of Fig. 1(a) in horizontal planes; (c) simultaneous generation of interfacial dislocation arrays of Figs 1(b) and (c) in nearly vertical planes (Olson and Cohen, Ref 3).

(b) and (c). The interfacial motion producing such a particle involves both the motion of the array of Fig. 1(a) as shown in Fig. 2(b), and the sequential formation of the discrete loops in Figs 1(b) and (c) under the acting thermodynamic driving force, represented by Fig. 2(c).

Following prediction of interfacial structure theory that the most significant coherency dislocations will be associated with planes of closest packing, it is appropriate for the construction of the dislocation arrays of Figs 1(a) and (b) to decompose the transformation lattice deformation into invariant-plane strains on close-packed planes. For face-centered cubic (fcc) to body-centered cubic (bcc) transformation as in ferrous alloys, such a decomposition was proposed by Bogers and Burgers (Ref 8) based on a hard-sphere deformation model. The resulting specific model for a martensitic nucleus provided the basis for the first detailed quantitative treatment of the energetics of classical heterogeneous nucleation. This model has predicted the essential characteristics of barrierless heterogeneous nucleation of martensitic transformations, with kinetic control by interfacial mobility. Combined with treatment of the composition dependence of solution hardening effects on the interfacial mobility, it has provided the framework for the most accurate prediction of the critical driving force for martensitic nucleation in complex alloys (Ghosh and Olson, Ref 9, 10). Using a distribution of total Burgers vector content for "superdislocation" defects of the type of Fig. 2(a) to describe potency distributions of both preexisting defects and new "autocatalytic" defects generated by martensite growth events, it has also provided the basis for a kinetic theory of the full evolution of martensitic transformations (Lin *et al.*, Ref 11). Incorporating both athermal and thermal components of the martensitic interfacial mobility, the general theory accounts for both systems exhibiting "athermal" (time-independent) behavior and the "isothermal" (time-dependent) mode of behavior, depending on the relative magnitude of the thermal terms in the interfacial mobility.

Mechanically Induced Transformations

Applied stress can assist the operation of the same nucleation sites (with the same total critical driving force) responsible for the transformation on cooling; such stress-assisted nucleation can be described for any stress state by the thermodynamic effect of stress as first derived by Patel and Cohen (Ref 12):

$$\Delta g^{\sigma} = \tau \gamma_T + \sigma_n \varepsilon_n \qquad \text{(Eq 2)}$$

where τ and σ_n are the resolved shear and normal stresses in the planes and directions of the transformation shear (γ_T) and normal strain (ϵ_n). However, if the applied stress exceeds the yield stress of the parent phase, strain-induced nucleation can occur on new potent nucleation sites created by the plastic strain. Under these circumstances, nucleation sites are plentiful, and this has allowed numerous detailed observations of strain-induced sites, consisting primarily of various types of shear-band intersections, and including intersections with grain boundaries (Venables, Ref 13; Lagneborg, Ref 14; Manganon and Thomas, Ref 15; Lecroise and Pineau, Ref 16; Suzuki *et al.*, Ref 17; Brooks *et al.*, Ref 18). The shear bands may be mechanical twins, hexagonal close-packed (hcp) ϵ-martensite, bundles of stacking faults, or slip bands. In each case, the intersection can be interpreted as producing a dislocation stacking of the type illustrated in

Fig. 2(a), and the relation of the active displacement to the orientation variant of the martensite being formed is consistent with the defect dissociation model (Lecroise and Pineau, Ref 16; Suzuki *et al.*, Ref 17; Brooks *et al.*, Ref 18). More potent defects can be generated in this way, compared to those existing in the annealed parent phase, and hence strain-induced nucleation can occur at lower driving forces.

The temperature regimes in which transformation initiates predominantly from either stress-assisted or strain-induced nucleation are depicted in the schematic stress-temperature diagram of Fig. 3. Spontaneous transformation triggered by existing nucleation sites occurs on cooling to the martensite start (M_s) temperature. Stress-assisted nucleation on the same sites will occur at the stress denoted by the solid line indicated. At a temperature designated M_s^σ, this stress reaches the yield stress σ_y for slip in the parent phase. Above M_s^σ, new potent nucleation sites introduced by plastic strain trigger strain-induced nucleation at the stress level depicted by the indicated solid curve. The temperature M_s^σ thus defines an approximate boundary between the temperature regimes where the two modes of nucleation dominate; near M_s^σ both modes will operate. Due to transformation plasticity, the observed yield stress follows the stress for

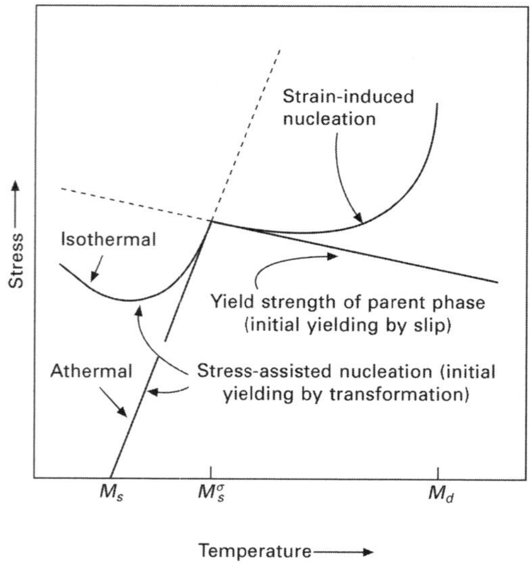

Fig. 3 Schematic representation of interrelationships between stress-assisted and strain-induced martensitic transformation (Olson and Cohen, Ref 19).

stress-assisted transformation below M_s^σ. A reversal of the temperature dependence of the yield stress thus provides a convenient determination of the M_s^σ temperature. Above the temperature M_d, no transformation occurs on deformation to fracture. Figure 3 represents the behavior for a particular stress state.

Mechanism of Stress-Assisted Transformations

Using the Patel-Cohen model (Patel and Cohen, Ref 12) of Eq 2 for the work of applied stress computed for the most favorably oriented martensitic plate, the calculated driving force contribution of applied stress in uniaxial tension ($\partial\Delta G/\partial\sigma$) is in good agreement with the observed stress dependence of the M_s temperature in alloys exhibiting nominally athermal kinetic behavior (Patel and Cohen, Ref 12; Bolling and Richman, Ref 20; Fisher, Ref 21). In addition to the numerous studies of the stress dependence of the M_s temperature, the stimulating influence of applied stress on the rate of the time-dependent isothermal transformation was first demonstrated by Machlin and Cohen (Ref 22).

The macroscopic transformation plasticity accompanying the transformation can arise in part from a biasing of the martensitic-plate variants which form under stress, but in ferrous alloys (particularly when a strong autocatalytic effect operates), much of the transformation plasticity is attributable to stress biasing of the extensive accommodation slip which occurs around the plates as they grow. When this deformation takes place at stresses below that for general (slip) yielding of the parent phase (i.e., the temperature regime where stress-assisted transformation is dominant), the kinetics of the transformation paces the macroscopic deformation. The simultaneous measurement of plastic flow and transformation in TRIP steels (Olson and Azrin, Ref 23) verifies the existence of a temperature regime in which plastic flow is controlled by stress-assisted martensitic transformation. Under these conditions, the volume fraction of martensite (f) and the plastic strain (ϵ) are confirmed to be linearly related (Olson and Azrin, Ref 23):

$$f = k\varepsilon \qquad \text{(Eq 3)}$$

This, in turn, defines a linear relationship between transformation rate and plastic strain rate. Here plastic flow will occur at the applied stress for which the combined (chemical and mechanical) driving force is sufficient to make the rate of transformation plasticity match the imposed strain rate; yielding in a constant strain-rate tensile test then corresponds to an imposed rate of stress-assisted isothermal transformation, with kinetic behavior predictable from transformation kinetic theory.

Mechanism of Strain-Induced Transformations

The kinetics of strain-induced martensitic nucleation at observed shear-band intersections can be adequately described by a simple relationship containing two temperature-dependent parameters and a fixed exponent (Olson and Cohen, Ref 24):

$$f^{\alpha'} = 1 - \exp\{-\beta[1 - \exp(-\alpha\varepsilon)]^n\} \qquad \text{(Eq 4)}$$

The first parameter (α) describes the rate of shear-band formation with respect to strain and is principally dependent on stacking-fault energy. The second parameter (β) is linearly related to the probability that a shear-band intersection will generate a martensitic embryo, and is governed by the chemical driving force for the martensitic reaction $\Delta G^{\gamma\rightarrow\alpha'}$, where γ refers to austenite and α' refers to the martensite. The transformation curve (volume fraction of martensite versus plastic strain) is sigmoidal in shape and approaches a saturation value determined by the β parameter. The transformation rate (relative to plastic strain) is controlled by both parameters. The temperature sensitivity of these transformation kinetics arises largely through $\Delta G^{\gamma\rightarrow\alpha'}$ and the stacking-fault energy, and may be minimized by a reduction of the entropy differences, $\Delta S^{\gamma\rightarrow\alpha'}$ and $\Delta S^{\gamma\rightarrow\epsilon}$.

Transformation Plasticity Constitutive Relations and Applications

The large transformation plasticity accompanying martensitic transformation allows substantial alteration of mechanical behavior under monotonic as well as cyclic loading conditions. A dramatic illustration of the potential of this is given by the original high-strength fully austenitic TRIP (transformation-induced plasticity) steels studied in the 1960s (Zackay *et al.*, Ref 25), which exhibit an extraordinary combination of strength, ductility, and toughness.

Transformation plasticity provides a significant departure from the classical view of structure/property relationships in materials. The traditional role of transformation kinetics is in the control of materials processing to achieve desired properties associated with the structure of the transformation product. Via transformation plasticity, one can design metastable materials to exploit the properties of the transformation itself, that is, properties of structural change. Under these circumstances, knowledge of transformation kinetics can be directly applied to the prediction of the constitutive flow relations underlying observed macroscopic mechanical behavior of materials during service as well as during processing.

Constitutive Flow Relations of Transformation Plasticity

Detailed comparison of stress vs. strain and fraction transformed martensite versus strain curves of TRIP steels reveals two major factors controlling the plastic-flow behavior during deformation-induced martensitic transformation (Olson and Azrin, Ref 23). In addition to the well-known static hardening contribution of the transformation product, a dynamic softening contribution arises from the operation of the transformation as a deformation mechanism. The latter effect is most amenable to quantitative treatment in the case of the stress-assisted transformation at sufficiently low temperatures (stresses) where plastic flow is entirely controlled by the transformation kinetics.

For barrierless martensitic nucleation controlled by interfacial mobility, time-dependent isothermal transformation can be predicted from the aforementioned thermal component of solution hardening, for which the activation energy Q can be approximated as a linear function of the acting total thermodynamic driving force,

$$Q = A + B\Delta G \qquad \text{(Eq 5)}$$

Following the basic kinetic theory of Pati and Cohen (Ref 26), the initial rate of transformation $\dot{f}$ can then be expressed as

$$\dot{f} = n_s V \nu \exp\left(-\frac{A + B\Delta G}{RT}\right) \qquad \text{(Eq 6)}$$

where n_s is the density of nucleation sites, V is the instantaneous mean martensitic volume, and v is the nucleation attempt frequency. The critical ΔG required to achieve a given $\dot{f}$ is then given by

$$\Delta G_{\text{crit}}(\dot{f}) = -\frac{1}{B}\left(A + RT\ln\frac{\dot{f}}{n_s V v}\right) \qquad \text{(Eq 7)}$$

which shows a linear dependence on temperature, T. For stress-assisted transformation, where yielding occurs by the rate of transformation plasticity matching the imposed strain rate, this corresponds to an imposed transformation rate of $\dot{f} = k\dot{\varepsilon}$. Calculation of the total thermodynamic driving force (chemical plus stress contributions) at the yield stress observed (Olson and Azrin, Ref 23) at temperatures below M_s^σ for high-strength TRIP steels verifies the linear ΔG_{crit}(T) relation of Eq 7 (Olson and Cohen, Ref 4). This linear relationship then predicts the critical stress for stress-assisted transformation σ_t determined by the solid curve of Fig. 4 where it is compared against measured stresses for 0.2% plastic strain and 1% transformation. In contrast to the linear behavior represented in Fig. 3, $\sigma_t(T)$ is in this case a curve which passes through a minimum at 158 K. The curvature arises from nonlinearity of $\Delta G_{ch}(T)$ at low temperatures, and the minimum is consistent with the behavior of relatively stable alloys which exhibit suppressible isothermal ("C-curve") transformation kinetics. In the stress-assisted regime, Eq 3 and 7 also predict a strain-rate sensitivity of flow stress of the form (Olson and Cohen, Ref 4):

$$\frac{\partial \sigma}{\partial \ln \dot{\varepsilon}} = -\frac{RT}{B\dfrac{\partial \Delta G}{\partial \sigma}} \qquad \text{(Eq 8)}$$

In line with the close relation between B and the activation volume for slip, the rate sensitivity of transformation plasticity is very similar to that for conventional plastic deformation.

Application of transformation plasticity to enhancement of uniform ductility and fracture toughness involves very different stress states, which must be taken into account in optimizing transformation stability for desired properties. A strong stress-state dependence of the Patel-Cohen work term (Patel and Cohen, Ref 12) arises from the transformation dilatation. Using the $\partial\Delta G/\partial\sigma$ values determined for uniaxial compression, uniaxial tension, and elastic crack tip, the corresponding σ_t curves calculated for the alloy of Fig. 4 are compared in Fig. 5. Also shown are 0.2% yield stress measurements for uniaxial compression. An unusually large strength-differential (S-D) effect, comparing

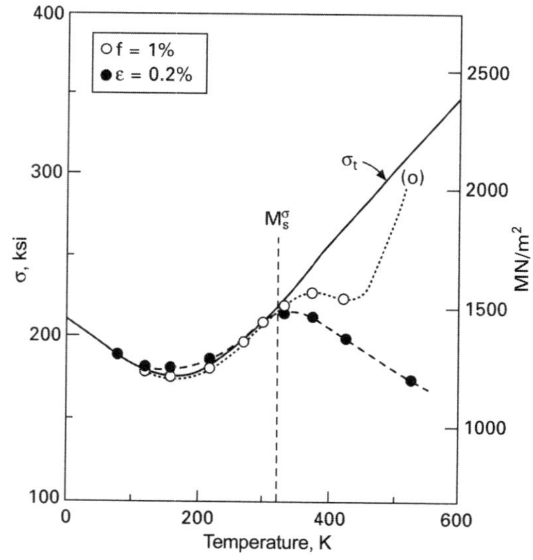

Fig. 4 Observed temperature dependence of 0.2% yield stress and stress at which 1% martensite is detected in high-strength TRIP steel (Olson and Azrin, Ref 23). Solid curve (σ_t) represents theoretical (tensile) stress for stress-assisted transformation (Olson and Cohen, Ref 4).

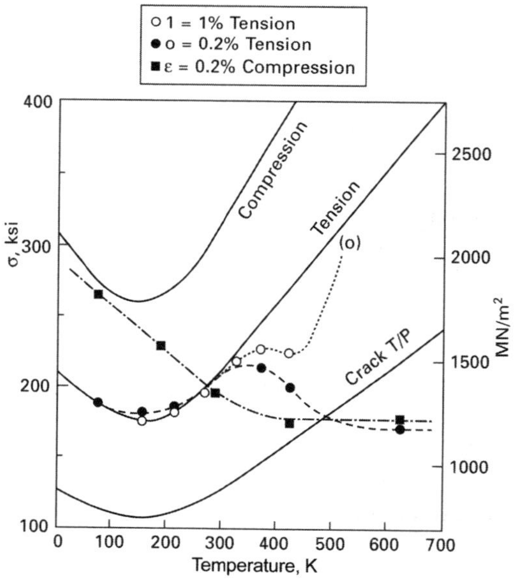

Fig. 5 Calculated transformation stresses and measured yield stresses for the TRIP steel of Fig. 4 under different stress states (Olson and Cohen, Ref 4).

flow stress in compression and tension, is expected in the stress-assisted regime. Although an anomalous reverse S-D effect is observed near 300 K, the expected effect is observed at low temperatures, suggesting that flow is controlled by transformation in tension and slip in compression. The prediction for the crack-tip case suggests that transformation will control flow over a wider temperature range and that the effective flow stress near 300 K is significantly reduced relative to uniaxial tension. As will be discussed further, this stress-state sensitivity is important to the achievement of improved strength/toughness combinations, allowing a high tensile strength material to benefit from the crack-tip plasticity of a softer material.

A knowledge of the kinetics of isothermal martensitic transformations can also be used to predict the shape of the σ-ϵ curve during stress-assisted transformation by taking into account the variation in the number of nucleation sites (n_s in Eq 7) with progress of the transformation (Olson and Cohen, Ref 4). The Pati-Cohen model, which satisfactorily accounts for the course of stress-free isothermal martensitic transformation, describes the nucleation-site density by an expression of the form (Pati and Cohen, Ref 26; Raghavan and Cohen, Ref 27):

$$n_s = (n_i + pf - N_v)(1 - f) \qquad \text{(Eq 9)}$$

Here n_i is the initial-site density, p is an "autocatalytic factor" accounting for new sites produced during transformation, and N_v is the number of martensitic plates per unit volume (accounting for sites which have already operated); the (1 − f) factor takes into account potential sites which have been "swept up" by transformation. While the p factor can cause n_s to increase in the initial stages of transformation, a decreasing average plate volume $\bar{V}$ due to "partitioning" can cause the $-N_v$ term to reduce n_s at later stages. A saturation level of transformation is reached when $n_s \rightarrow 0$. Substituting Eq 9 into Eq 7, the flow stress during stress-assisted transformation can be expressed as (Olson and Cohen, Ref 4):

$$\sigma_t(f, \dot{f}, T) = -\left(B\frac{\partial \Delta G}{\partial \sigma}\right)^{-1} \left[A + B\Delta G_{ch} + RT\ln\left(\frac{\dot{f}}{(n_i + pf - N_v)(1 - f)Vv}\right)\right] \qquad \text{(Eq 10)}$$

Substitution of Eq 3 then provides a complete constitutive relation for $\sigma_t(\varepsilon, \dot{\varepsilon}T)$. The shape of the σ-ϵ curve is determined by the behavior of the denominator in the last term of Eq 13. The yield stress is controlled by n_i, a stress drop is produced by the pf autocatalytic term, and approach to saturation causes rapid hardening as site depletion brings the denominator toward zero. Using the A and B parameters defined by the temperature dependence of σ_t in Fig. 4, and typical experimental values of the other kinetic parameters, a calculated σ-ϵ curve is compared with that measured at 158 K for the same TRIP steel in Fig. 6. The yield drop and flat portions of the curves are in good agreement, but the calculated final hardening (saturation) stage is too abrupt. This arises from the approximation of a singly activated process. As described earlier, the distribution of nucleation-site potencies is known to give a distribution of nucleation activation energies (Magee, Ref 11; Lin *et al.*, Ref 28), and this can account for a more gradual approach to saturation. It is important to note that the flow stresses depicted in Fig. 6 are all well below the stress for general yielding by slip. Relative to the "normal" slip-controlled flow behavior of this material, the phenomena thus far considered are entirely "softening" contributions. The high strain hardening associated with the saturation stage is best understood as the cessation of a softening phenomenon rather than the onset of a hardening mechanism. Once

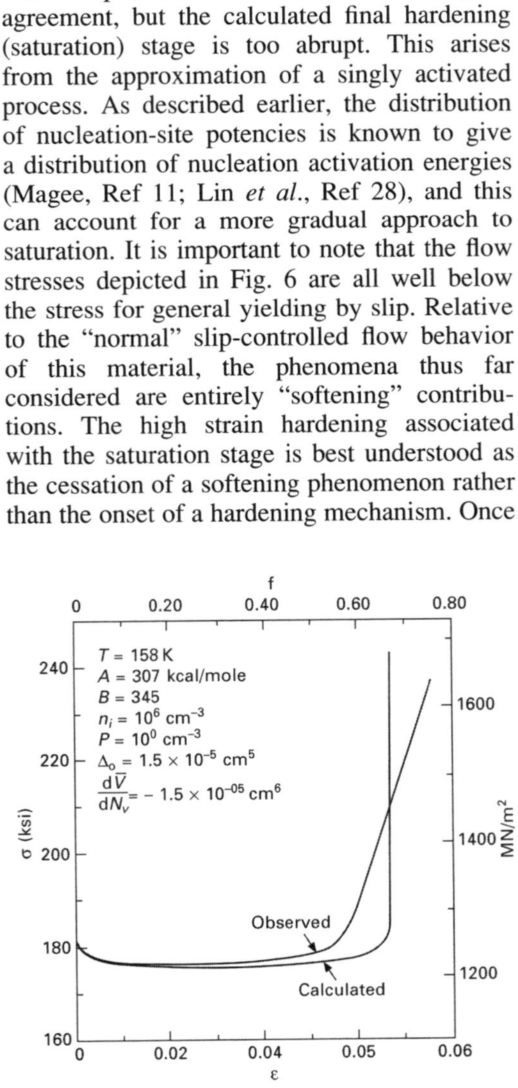

Fig. 6 Comparison of calculated and observed (Olson and Azrin, Ref 23) true σ-ϵ curves for the TRIP steel of Fig. 4 (Olson and Cohen, Ref 4).

the transformation plasticity phenomenon (dynamic softening) associated with stress-assisted transformation is essentially complete, the higher strength of the transformation product makes its "static hardening" contribution to the flow stress.

While control of plasticity by transformation kinetics allows rather precise prediction of flow behavior in the stress-assisted transformation regime, the simultaneous operation of general slip and transformation in the strain-induced regime makes the quantitative treatment of flow behavior more difficult. Although much attention has been given to the static-hardening aspects of the two-phase mixture formed during strain-induced transformation, careful comparison of σ-ϵ curves has demonstrated that dynamic softening is also an important contribution. A detailed study of the transformation and flow behavior of a metastable austenitic stainless steel, summarized in Fig. 7, has allowed a quantitative separation of the static hardening and dynamic softening contributions during strain-induced transformation (Narutani *et al.*, Ref 29). The measured σ-ϵ curve of the metastable steel is labeled σ_{exp}. The dashed $\sigma_{\alpha'}$ and σ_{γ} curves represent the corresponding σ–ϵ curves of the martensite and stable austenite measured separately on specially designed similar steel compositions. One estimate of the static hardening behavior of the two-phase mixture is obtained from a simple "rule of mixtures" using the $\sigma_{\alpha'}(\epsilon)$, $\sigma_{\gamma}(\epsilon)$ data and the measured $f(\epsilon)$ curve shown at the bottom of the figure; this estimate is depicted by the dashed curve labeled *RM*. Because the transformation plasticity contribution arising from biasing of the transformation shape strain does not contribute to the strain in either phase, the *RM* curve represents an

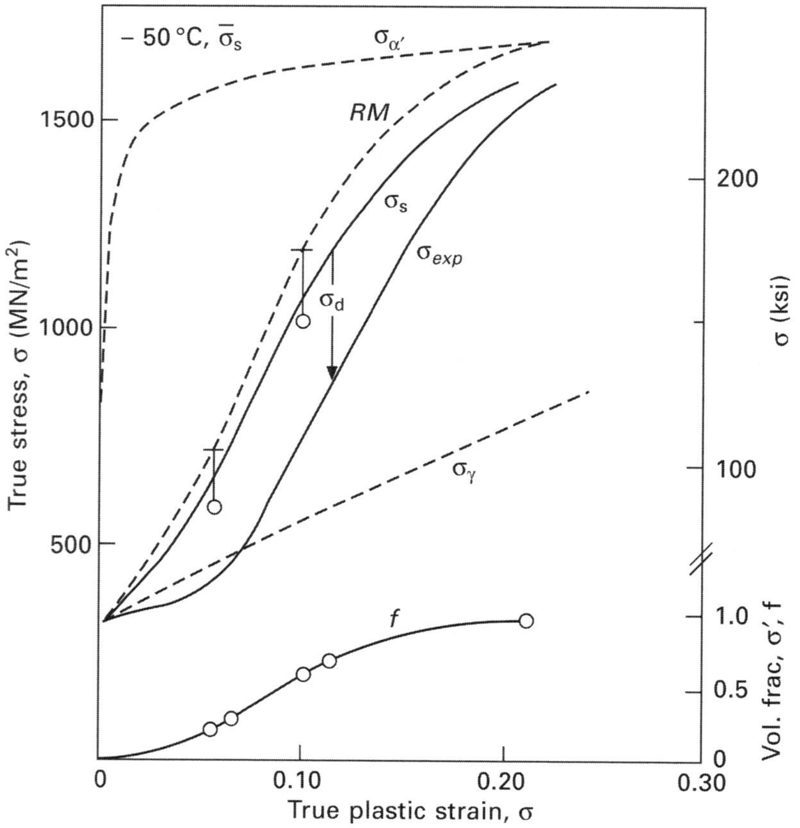

Fig. 7 Experimental flow stress, σ_{exp}, and volume fraction martensite, *f*, vs. plastic strain, ϵ, for metastable austenitic steel at −50 °C, $\dot{\epsilon}$, = 2.2 × 10^{-4}s^{-1}. Dashed curves represent the stable austenite flow stress, σ_γ, the martensite flow stress, σ_{α}', and the prediction of the rule of mixtures for two-phase hardening, *RM*. Solid curve, σ_s, is a prediction of strain-corrected rule-of-mixtures model (Narutani *et al.*, Ref 29).

upper limit to the static-hardening behavior. A strain-corrected rule of mixtures is obtained using $\sigma_{\alpha'}$ and σ_γ values at a strain of ϵ–αf, where αf corrects for the shape-strain contribution to the measured ϵ. An upper-limit estimate of the α coefficient is $\alpha = k^{-1}$, taking k from the measured stress-assisted transformation behavior at low temperatures. This gives the solid curve depicted just below the dashed *RM* curve. The open points represent static flow-stress estimates determined by prestrain experiments. After the transforming material had been deformed to a particular strain, the static flow stress of the resulting two-phase mixture (in the absence of dynamic transformation softening) was determined from its flow stress subsequently measured at a higher temperature where the austenite is stable, correcting for its temperature dependence. These estimates and the *RM* estimate bracket the strain-corrected rule of mixtures estimate which is therefore taken as a reasonable approximation of the flow stress arising from static hardening, σ_s. This is expressed as:

$$\sigma_s = [1 - f] \cdot \sigma_\gamma(\varepsilon - \alpha f) + f \cdot \sigma_{\alpha'}(\varepsilon - \alpha f) \qquad \text{(Eq 11)}$$

As indicated in Fig. 7, the dynamic softening increment $\Delta\sigma_d$ is then taken as the difference between σ_s and σ_{exp}. Correlation of the $\Delta\sigma_d$ increment thus obtained with f-ϵ behavior observed over a wide temperature range indicates that the fractional softening increment is proportional to $df/d\epsilon$:

$$\frac{\Delta\sigma_d}{\sigma_s} = \beta \cdot \frac{df}{d\varepsilon} \qquad \text{(Eq 12)}$$

where $\beta = 5.3 \times 10^{-2}$. This then gives a constitutive flow relation for strain-induced transformation of the form:

$$\sigma = \{[1 - f] \cdot \sigma_\gamma(\varepsilon - \alpha f) + f \cdot \sigma_{\alpha'}(\varepsilon - \alpha f)\} \cdot \left[1 - \beta \cdot \frac{df}{d\varepsilon}\right] \qquad \text{(Eq 13)}$$

Through Eq 13, $\sigma(\varepsilon, \dot{\varepsilon}, T)$ can be predicted from a knowledge of f, σ_γ and $\sigma_{\alpha'}$ as functions of ϵ, $\dot{\varepsilon}$, and T.

As in the case of stress-assisted transformation, dynamic softening is the dominant factor at low strains, causing the flow stress σ_{exp} of the transforming material in Fig. 7 to fall below that of the stable austenite, σ_γ. The static hardening becomes dominant at high strains. The combined effect of these two factors delays the maximum hardening rate $d\sigma/d\epsilon$ to a higher strain than that where $df/d\epsilon$ is maximum. The maximum $d\sigma/d\epsilon$ arises from both the static hardening (proportional to $df/d\epsilon$) and the diminution of dynamic softening (proportional to $-d^2f/d\epsilon^2$) as $df/d\epsilon$ decreases. Again, the cessation of a softening phenomenon provides a major contribution to the net rate of hardening. The generalization of this constitutive law to three-dimensional deformation has been developed by Stringfellow *et al.* (Ref 30).

Stability of Plastic Flow

With constitutive flow relations for transformation plasticity based on the transformation kinetics, we may now consider the associated macroscopic flow behavior as predicted by continuum plasticity theory. The ability of transformation plasticity to dramatically alter the shape of a σ-ϵ curve naturally raises the question of the optimum or "ideal" curve shape for stability of plastic flow. The minimum strain-hardening rate required to maintain stable flow during tensile deformation is expressed by (Backofen, Ref 31):

$$\frac{d\sigma}{d\varepsilon} = \sigma \qquad \text{(Eq 14)}$$

A consequence of strain hardening is that the hardening rate necessary to maintain stability is increased. Excessive hardening therefore makes it more difficult to maintain stability at higher strains. It follows that the most efficient use of a hardening increment $\Delta\sigma$ is to distribute it with respect to strain in such a way that $d\sigma/d\epsilon$ increases with ϵ. The ideal hardening curve, which provides the minimum hardening rate necessary to maintain stability at all strains, is given by the solution of the differential equation represented by Eq 15:

$$\sigma = \sigma_y \exp(\varepsilon) \qquad \text{(Eq 15)}$$

where σ_y corresponds to the yield stress. In contrast to the usual downward curvature associated with the structural hardening and recovery processes controlling deformation by slip, the ideal flow relation of Eq 15 possesses upward curvature (Kocks *et al.*, Ref 32). The "delayed hardening" behavior associated with the intersection of dynamic softening and static hardening phenomena in transformation plasticity provides an effective means of increasing $d\sigma/d\epsilon$ with ϵ to achieve this curvature.

The ideal hardening behavior of Eq 15 is contrasted with the conventional downward-curving behavior, approximated by a power law $\sigma = K\epsilon^n$, in Fig. 8. The open circle indicates the point where the power-law curve reaches the condition for flow localization by tensile necking, that is, where $d\sigma/d\epsilon$ falls below σ (Eq 14). From a local viewpoint, which considers the conditions at the necking strain, one would attribute the necking to insufficient hardening, and would propose to increase uniform ductility by increasing the hardening exponent, n. From a perspective which considers the σ-ϵ curve as a whole, the necking can equally well be ascribed to excessive hardening at low strains. In the region where the $K\epsilon^n$ curve lies above the $\sigma_0 \exp \epsilon$ curve, the superposition of a *softening* mechanism which lowers the flow stress to the $\sigma_0 \exp \epsilon$ curve would double the uniform ductility for the case depicted in Fig. 8. This illustrates the potential of the dynamic softening phenomenon in transformation plasticity for the enhancement of flow stability. At higher strains, where the $K\epsilon^n$ curve falls below the $\sigma_0 \exp \epsilon$ curve, superposition of a hardening mechanism, such as the high-strain static hardening contribution in transformation plasticity, is required to further maintain stable flow. If the ideal $\sigma_0 \exp \epsilon$ curve is followed, necking will never occur, and uniform ductility will be controlled by fracture rather than plastic instability. Figure 8 illustrates that the dynamic softening contribution to transformation plasticity at lower strains can be equally important as the high-strain static hardening contribution in promoting stability of plastic flow.

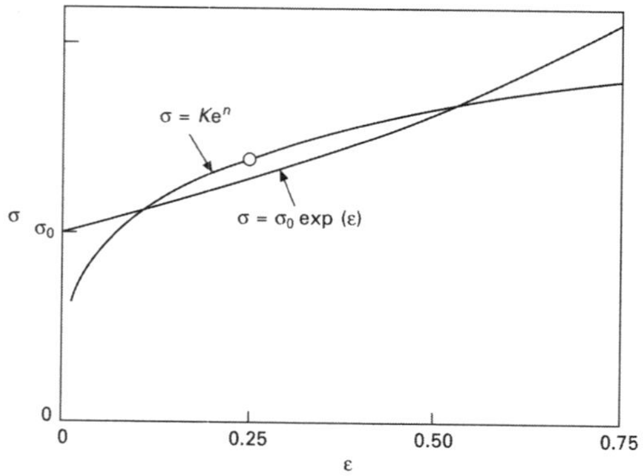

Fig. 8 Comparison of ideal exponential hardening behavior with power-law hardening. Open point indicates necking strain.

Applications of Transformation Plasticity to Ductility

The flow-stabilizing influence of the σ-ϵ curve-shaping effects associated with transformation plasticity allows dramatic enhancement of uniform ductility under conditions of optimum thermodynamic stability, and this phenomenon is the most well-established transformation plasticity application. Although temperature sensitivity and Lüders-band behavior limit the useful ductility of the highest strength TRIP steels, which depend almost entirely on their transformation behavior for stability of flow, the flow-stabilizing contribution of a moderate rate of transformation in lower strength metastable austenitic steels has been practically applied for many years. Control of transformation plasticity has led to the development of numerous commercial high-formability stainless steels.

The enhancement of flow stability in metastable austentic steels via transformation plasticity has been extensively investigated (Krivobok and Talbot, Ref 33; Angel, Ref 34; Fiedler *et al.*, Ref 35; Powell *et al.*, Ref 36; Bressanelli and Moskowitz, Ref 37; Zackay *et al.*, Ref 25; Tamura *et al.*, Ref 38; Olson and Azrin, Ref 23; Zackay *et al.*, Ref 39; Hecker *et al.*, Ref 40), and analyzed in terms of Eq 15 (Azrin *et al.*, Ref 41). A continuous local strain measurement technique allowed determination of true stress-true strain behavior of the original high-strength austenitic TRIP steels during both uniform and localized plastic flow under a variety of conditions (Azrin *et al.*, Ref 41; Olson and Azrin, Ref 23, unpublished research). Figure 9 summarizes such curves obtained for uniaxial tensile deformation of a fully austenitic TRIP steel of nominal composition Fe-9Cr-8Ni-4Mo-2Si-2Mn-0.3C strengthened by warm working to a 60% reduction at 450 °C (842 °F) (Olson and Azrin, Ref 23). Arrows indicate the necking strain or uniform strain, ϵ_u. The temperature (stability) dependence of the strength and ductility properties is summarized in Fig. 10. Consistent with the constitutive model predictions, the true σ-ϵ curves in Fig. 9 have a characteristic sigmoidal shape. The temperature dependence of the 0.2% yield strength σ_y in Fig. 10 is similar to

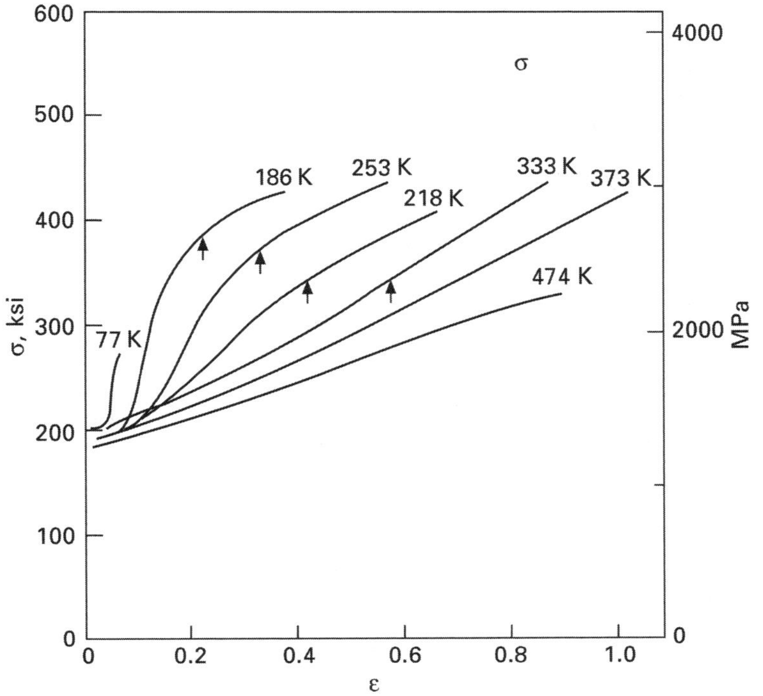

Fig. 9 True stress-strain curves of high-strength austenitic TRIP steel. Arrows indicate necking strain (Olson and Azrin, Ref 23).

that of Fig. 4, indicating an M_s^σ temperature near 280 K. The ultimate tensile strength (UTS) reaches a maximum in the stress-assisted transformation regime, limited at lower temperatures by fracture. The latter effect is indicated by the intersection of uniform strain ϵ_u and the fracture strain ϵ_f at the lowest temperature. The uniform ductility ϵ_u reaches a maximum above M_s^σ at a temperature near 350 K, reflecting an optimum overall rate of deformation-induced transformation. As indicated by the shape of the σ-ϵ curves, early completion of transformation at lower temperatures leads to early necking in the downward-curving portion of the σ-ϵ behavior that follows. Above 350 K the strain hardening associated with a reduced degree of transformation is insufficient to maintain stable flow in this high-strength material, and the uniform ductility is negligible. The existence of such an optimum stability for maximum uniform ductility in metastable austenitic steels in general is well established (Bressanelli and Moskowitz, Ref 37; Tamura *et al.*, Ref 38; Zackay *et al.*, Ref 39). The shape of the optimum stability 333 K curve in Fig. 9, giving a uniform ductility of 60% with a yield strength of 200 ksi (1.4 GPa), is quite close to the ideal exponential form.

Applications of Transformation Plasticity to Fracture Toughness

Transformation toughening in austenitic steels is complicated both by the greater complexity of the constitutive behavior of strain-induced as opposed to stress-assisted transformation, and the more complex processes of ductile fracture. The experimental study by Léal (Ref 43) of the interaction of transformation with ductility and fracture toughness is reviewed in Olson and Cohen (Ref 44). A series of six precipitation-hardenable austenitic steels based on A286 steel were designed to have a constant 1300 MPa yield strength, varying stability with respect to transformation, and a varying transformation volume change. The alloys also had a fairly strong temperature dependence of phase stability to allow variation of stability over wide limits. The fracture ductility and J_{IC} toughness were found to peak at the M_s^σ of their respective stress states.

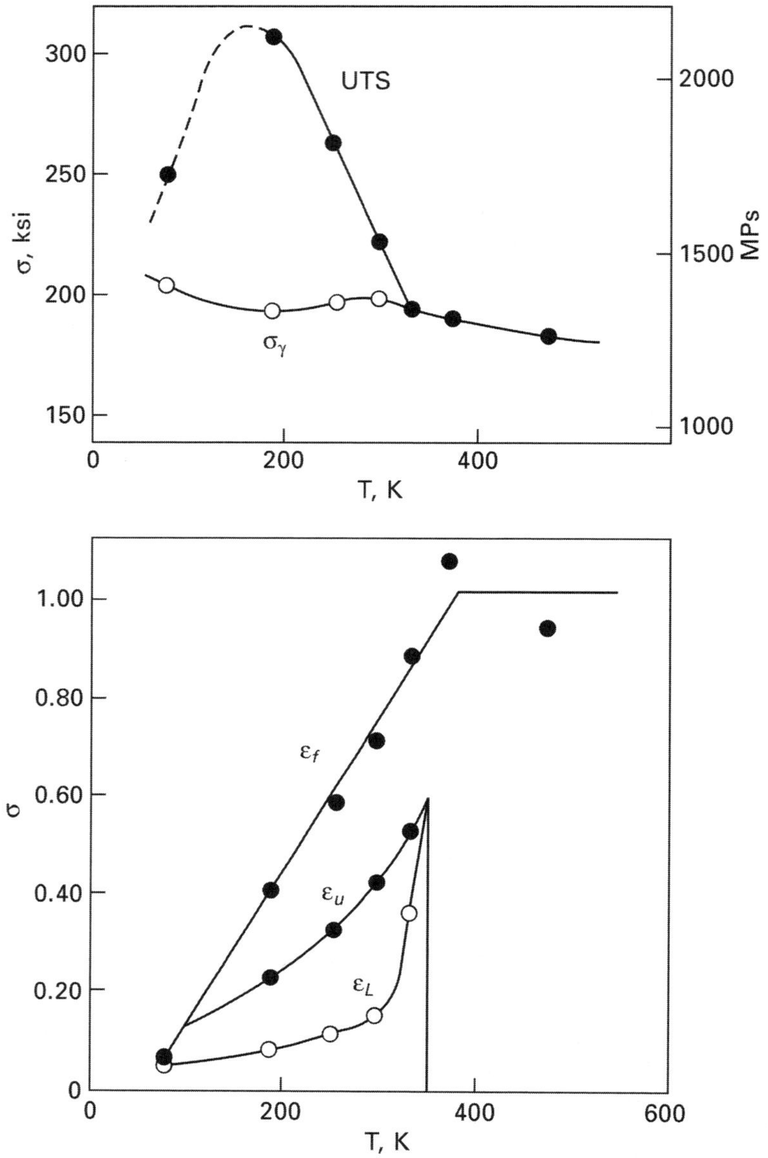

Fig. 10 Tensile properties of austenitic TRIP steel of Fig. 9, including Lüders strain ϵ_L, uniform (necking) strain ϵ_u, and fracture strain ϵ_f.

A subsequent study by Young (Ref 45), reviewed in Olson (Ref 46), involved a thorough evaluation of the transformation kinetics and constitutive relations of phosphocarbide-strengthened austenitic steels in which the transformation volume change and austenite-martensite hardness difference could both be varied. The measured enhancement of toughness and fracture ductility in these alloys was similar to that observed by Léal (Ref 43). In the regime of shear instability controlled ductile fracture, the increase in J_{IC} toughness varies linearly with the height of the transformation zone at the crack front. Correlation of the slope of this toughening response with transformation volume change $\delta V/V$, and austenite-martensite hardness differences δH_v (influencing strain hardening) shows that, in contrast to brittle solids, the toughening behavior is more dilatation sensitive, with the $\delta J/h$

scaling with the third power of $\delta V/V$. Higher δH_v also promotes more toughening.

The remarkable transformation toughening observed by Stavehaug (Ref 47) in a more recent series of γ'-strengthened austenites is summarized in Fig. 11, which plots the measured room-temperature J_{IC} toughness versus a thermodynamic stability parameter defined by the difference between the total acting thermodynamic driving force and its critical value at M_s^σ. Similar to the results of Léal (Ref 43) and Young (Ref 45), the toughening is maximum at the M_s^σ condition, but in these experiments the apparent toughness enhancement over the stable austenite toughness indicated by the dashed line is significantly higher.

Quantitative modeling of the mechanics of the transformation toughening has addressed mechanisms on two length scales. As an analysis of interactions of transformation hardening and void softening induced localization at the macroscopic level, Stringfellow and Parks (Ref 48) and Stringfellow (Ref 49) performed a series of model calculations incorporating a void softening model into the transformation constitutive model.

On a finer length scale, electron microscopy observations (Young, Ref 45) of the distribution of fine strain-induced martensite around 0.1 μm scale particles after straining to levels beyond the microvoid nucleation strain of nontransforming material indicate a more direct cross-interaction between transformation and the unit process of ductile fracture. Socrate and Parks (Socrate, Ref 50) have applied the transformation constitutive model in numerical simulations of microvoid nucleation, growth and coalescence employing a cohesive-zone model for interface debonding similar to that introduced by Needleman (Ref 51). Using interfacial strength estimates for the TiC particles observed in the experimental alloys, Fig. 12 shows the computed stress-strain curves for a cell model of the crack-tip process zone with a remote stress state of $\sigma_h\sqrt{\sigma} = 1.12$. Dotted, dashed, and solid curves represent material with (a) preexisting void, (b) a permanently bonded particle, and (c) a debonding particle, respectively. The curves labeled with a T depict the effect of strain-induced transformation at a stability near that of the crack-tip M_s^σ temperature. Comparing the transforming and nontransforming behaviors for material with either a preexisting void or a debonding particle, transformation greatly delays the onset of microvoid softening based negative strain hardening that drives shear localization fracture. Computed transformation fields prior to nucleation are in good agreement with the experimental observations. Simulations of interacting pairs of particles also demonstrate retardation of microvoid coalescence, but the largest toughening effect appears to be associated with delay of shear instability driven by void growth softening. The strain corresponding to the onset of terminal softening (stress maximum) in

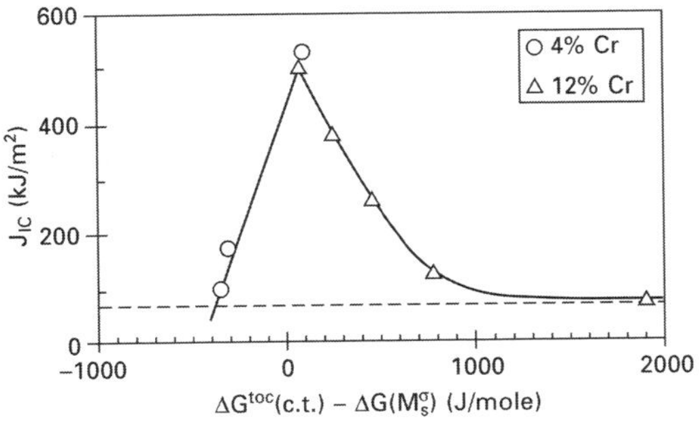

Fig. 11 Measured J_{IC} fracture toughness of 12Cr and 4Cr austenitic alloy series (1380 MPa yield strength) as a function of thermodynamic austenite stability parameter. (Stavehaug, Ref 47).

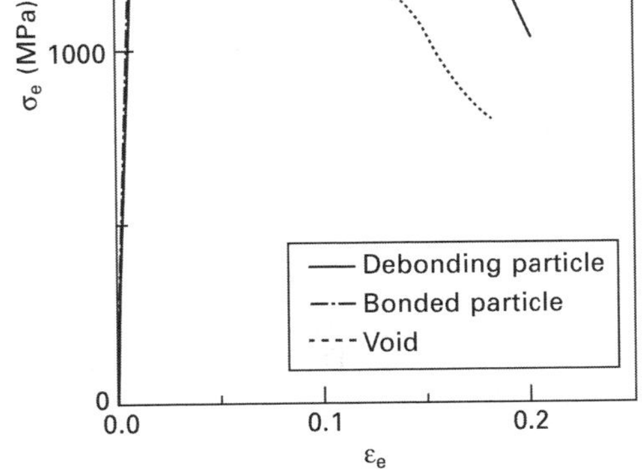

Fig. 12 Computed stress-strain curves for crack tip cell model; curves labeled T represent transforming material while others represent nontransforming material (Socrate, Ref 50).

Fig. 12 increases with transformation by a factor similar to the J_{IC} toughness enhancement of Fig. 11.

Conclusions

Building on dislocation-based theory of the mechanism and kinetics of heterogeneously nucleated martensitic transformations, quantitative models have been developed for the kinetics of stress-assisted and strain-induced martensitic transformations. Constitutive flow relations for transformation plasticity have been derived based on these kinetic models, incorporating the static hardening effect of the transformation product and the dynamic softening effect of the transformation as a deformation mechanism. The transformation softening and hardening effects can distort the stress-strain curve from the usual downward-curving shape to an upward-curving shape which approximates an ideal exponential hardening behavior promoting maximum uniform ductility. Numerical simulation of shear-instability-controlled ductile fracture demonstrates that the transformation plasticity behavior can also account for observed substantial enhancement of ductile fracture toughness.

The demonstrated behaviors in fully austenitic steels offer a foundation for predictive control of transformation plasticity in dispersed-phase systems. More broadly, the application of predictive theory of transformation plasticity to enhance mechanical properties of advanced steels remains a vital avenue of steel research. Predictive control of the complexities of martensitic behavior has served as the foundation of a systems approach to science-based materials engineering. In this way, ferrous metallurgy has played a leading role in the new technology of materials design (Olson, Ref 52).

REFERENCES

1. J.W. Christian, G.B. Olson, *et al.* (1995). Classification of Displacive Transformations: What is a Martensitic Transformation?, *Proc. ICOMAT 95*, supplement to *J. de Phys. III*, 5 Coll. C8: 3–10
2. G.B. Olson and M. Cohen (1986). Dislocation Theory of Martensitic Transformations. in *Dislocations in Solids Vol. 7*, ed. F. R. N. Nabarro. North-Holland, p 295–407
3. G.B. Olson and M. Cohen (1981). Dislocation Structure of Martensitic Interfaces. *International Conference of Solid-Solid Phase Transformations*, Carnegie Mellon University
4. G.B. Olson and M. Cohen (1982). Stress-Assisted Isothermal Martensitic Transformation – Application to Trip Steels. *Metallurgical Transactions A – Physical Metallurgy and Materials Science* 13(11): 1907–1914
5. L. Kaufman and M. Cohen (1958). Thermodynamics and Kinetics of Martensitic Transformations. *Prog. Metal Physics 1*: 165–246
6. G.B. Olson and M. Cohen (1976). A General Mechanism of Martensitic Nucleation: Part II: FCC-BCC and Other Martensitic Transformations. *Metall. Trans.* 7A: 1905–1914
7. G.B. Olson and M. Cohen (1976). Mechanism and Kinetics of Deformation-Induced Martensitic Transformations. *JOM – Journal of Metals* 28(12): A15
8. A.J. Bogers and W.G. Burgers (1964). Partial Dislocations on the (110) Planes in the bcc Lattice and the Transition of the fcc into the bcc Lattice. *Acta Metall.* 12: 225–261
9. G. Ghosh and G.B. Olson (1994a). Kinetics of Fcc→Bcc Heterogeneous Martensitic Nucleation. 1. The Critical Driving-Force for Athermal Nucleation. *Acta Metallurgica et Materialia* 42(10): 3361–3370
10. G. Ghosh and G.B. Olson (1994b). Kinetics of Fcc→Bcc Heterogeneous Martensitic Nucleation. 2. Thermal-Activation. *Acta Metallurgica et Materialia* 42(10): 3371–3379
11. M.F. Lin G.B. Olson, *et al.* (1992). Distributed-Activation Kinetics of Heterogeneous Martensitic Nucleation. *Metallurgical Transactions A – Physical Metallurgy and Materials Science* 23(11): 2987–2998
12. J.R. Patel and M. Cohen (1953). Criterion for the Action of Applied Stress in the Martensitic Transformation. *Acta Metallurgica* 1(5): 531–538
13. J.A. Venables (1962). Martensite Transformation in Stainless Steel. *Philosophical Magazine* 7(73): 35–44
14. R. Lagneborg (1964). Martensite Transformation in 18percent Cr-8percent Ni Steels. *Acta Metallurgica* 12(7): 823–843

15. D.L. Manganon and G. Thomas (1970). The Martensite Phase in 304 Stainless Steel. *Metall. Trans.* 1: 1577–1586
16. F. Lecroise and A. Pineau (1972). Martensitic Transformations Induced by Plastic-Deformation in Fe-Ni-Cr-C System. *Metallurgical Transactions* 3(2): 387–396
17. T. Suzuki, H. Kojima, *et al.* (1977). Experimental-Study of Martensite Nucleation and Growth in 18-8 Stainless-Steel. *Acta Metallurgica* 25(10): 1151–1162
18. J.W. Brooks, M.H. Loretto, *et al.* (1979). *In-situ* Observations of the Formation of Martensite in Stainless-Steel. *Acta Metallurgica* 27(12): 1829–1838
19. G.B. Olson and M. Cohen (1972). Mechanism for Strain-Induced Nucleation of Martensitic Transformations. *Journal of the Less-Common Metals* 28(1): 107–119
20. G.F. Bolling and R.H. Richman (1970). Influence of Stress of Martensite-Start Temperatures in Fe-Ni-C Alloys. *Scripta Metallurgica* 4(7): 539–544
21. P.J. Fisher (1974). The Effects of Stress on the Martensite Transformation. PhD thesis, University of New South Wales
22. E.S. Machlin and M. Cohen (1952). Isothermal Mode of the Martensitic Transformation. *Transactions of the American Institute of Mining and Metallurgical Engineers* 194(5): 489–500
23. G.B. Olson and M. Azrin (1978). Transformation Behavior of Trip Steels. *Metallurgical Transactions A – Physical Metallurgy and Materials Science* 9(5): 713–721
24. G.B. Olson and M. Cohen (1975). Kinetics of Strain-Induced Martensitic Nucleation. *Metallurgical Transactions A* 6(4): 791–795
25. V.F. Zackay, E.R. Parker, *et al.* (1967). Enhancement of Ductility in High-Strength Steels. *ASM Transactions Quarterly* 60(2): 252–259
26. S.R. Pati and M. Cohen (1969). Nucleation of Isothermal Martensitic Transformation. *Acta Metallurgica* 17(3): 189–199
27. V. Raghavan and M. Cohen (1971). Measurement and Interpretation of Isothermal Martensitic Kinetics. *Metallurgical Transactions* 2(9): 2409–2418
28. C.L. Magee (1971). Kinetics of Martensite Formation in Small Particles. *Metallurgical Transactions* 2(9): 2419–2430
29. T. Narutani, G.B. Olson, *et al.* (1982). Constitutive Flow Relations for Austenitic Steels during Strain-induced Martensitic Transformations. *J. de Physique* 43(C4): 429–434
30. R.G. Stringfellow, D.M. Parks, *et al.* (1992). A Constitutive Model for Transformation Plasticity Accompanying Strain-Induced Martensitic Transformations in Metastable Austenitic Steels. *Acta Metallurgica et Materialia* 40(7): 1703–1716
31. W.A. Backofen (1972). *Deformation Processing*. Addison-Wesley
32. U.F. Kocks, J.J. Jonas, *et al.* (1979). The Development of Strain-Rate Gradients. *Acta Metallurgica* 27: 419–432
33. V.N. Krivobok and A.M. Talbot (1950). Effect of Temperature on the Mechanical Properties, Characteristics, and Processing of Austenitic Stainless Steels. *Proceedings American Society for Testing and Materials* 50: 895–930
34. T. Angel (1954). Formation of Martensite in Austenitic Stainless Steels. *J. Iron Steel Inst.* 177: 165–174
35. H.C. Fiedler, B.L. Averbach, *et al.* (1955). The Effect of Deformation on the Martensitic Transformation. *Trans. ASM* 47: 267–290
36. G.W. Powell, E.R. Marshal, *et al.* (1958). Strain Hardening of Austenitic Stainless Steel. *Trans. ASM* 50: 478–479
37. J.P. Bressanelli and A. Moskowitz (1966). Effects of Strain Rate Temperature and Composition on Tensile Properties of Metastable Austenitic Stainless Steels. *ASM Transactions Quarterly* 59(2): 223–239
38. I. Tamura, T. Maki, *et al.* (1970). On Morphology of Strain-Induced Martensite and Transformation-Induced Plasticity in Fe-Ni and Fe-Cr-Ni Alloys. *Transactions of the Iron and Steel Institute of Japan* 10(3): 163–172
39. V.F. Zackay, M.D. Bhandarkar, *et al.* (1978). *Advances in Deformation Processing*. Plenum Press
40. S.S. Hecker, M.G. Stout, *et al.* (1982). Effects of Strain State and Strain Rate on Deformation-Induced Transformation in 304 Stainless-Steel. 1. Magnetic Measurements and Mechanical-Behavior. *Metallurgical Transactions A – Physical Metallurgy and Materials Science* 13(4): 619–626
41. M. Azrin, G.B. Olson, *et al.* (1976). Inhomogeneous Deformation and Strain-Rate Effects in High-Strength Trip Steels. *Materials Science and Engineering* 23(1): 33–41

42. G.B. Olson and M. Azrin (unpublished research). AMMRC
43. R. Léal (1984). Transformation Toughening of Metastable Austenitic Steels. PhD thesis, MIT
44. G.B. Olson and M. Cohen (1986). *Mechanical Properties and Phase Transformations in Engineering Materials*. TMS-AIME, p 367–390
45. C.C. Young (1988). Transformation Toughening in Phosphocarbide-Strengthened Austenitic Steels. PhD thesis, MIT
46. G.B. Olson (1990). *Science of Steel*. 34th Sagamore Army Materials Research Conference – Innovations in Ultrahigh-Strength Steel Technology
47. F. Stavehaug (1990). Transformation Toughening of γ' Strengthened Metastable Austenitic Steels. PhD thesis, MIT
48. R.G. Stringfellow and D.M. Parks (1991). A Self-Consistent Model of Isotropic Viscoplastic Behavior in Multiphase Materials. *International Journal of Plasticity* 7 (6): 529–547
49. R. Stringfellow (1990). Mechanics of Strain-Induced Transformation Toughening in Metastable Austenitic Steels. PhD thesis, MIT
50. S. Socrate (1995). Mechanics of Microvoid Nucleation and Growth in High-Strength Metastable Austenitic Steels. PhD thesis, MIT
51. A. Needleman (1987). A Continuum Model for Void Nucleation by Inclusion Debonding. *Journal of Applied Mechanics – Transactions of the ASME* 54(3): 525–531
52. G.B. Olson (2006). Advances in Theory: Martensite by Design. *Materials Science and Engineering A* 438: 48–54

Transformations—Selected Works of G.B. Olson on Materials, Microstructure, and Design
C.E. Campbell, M.V. Manuel, and W. Xiong, editors

CHAPTER 9

Spinodal Decomposition during Aging of Fe-Ni-C Martensites*

K.A. Taylor, Bethlehem Steel Corporation
L. Chang, Industrial Technology Research Institute
G.B. Olson, Northwestern University
G.D.W. Smith and M. Cohen, Massachusetts Institute of Technology
J.B. Vander Sande, University of Oxford

Abstract

A collaborative study of the aging of virgin Fe-Ni-C martensites has combined the techniques of transmission electron microscopy (TEM), atom-probe field-ion microscopy (APFIM), and electrical resistometry. Aging at room temperature leads to the rapid development of a fine-scale structural modulation along ⟨203⟩ lattice directions. Atom-probe analysis of Fe-15Ni-1C martensite reveals the formation of carbon-rich regions whose carbon concentration increases with time and approaches 11 at.% C on prolonged aging. The early stage kinetics of this process are composition-dependent and are consistent with carbon-diffusion control. The morphological features of the aging reaction are explained by elastic strain-energy considerations. In accordance with previous thermodynamic models, it is concluded that virgin Fe-C martensites are unstable and that phase separation occurs by a spinodal mechanism. The martensitic substructure does not appear to exert any substantial influence on this decomposition behavior.

Introduction

This paper deals explicitly with aging phenomena that occur in Fe-Ni-C martensites prior to any carbide precipitation. The latter process, constituting the first or T1 stage of tempering, is the subject of a separate paper (Ref 1) It has become increasingly clear that many types of change take place during the aging of ferrous martensites (Ref 2, 3), but there remains the need for a more complete understanding of the fundamental nature of these changes and their interrelationships.

Prior work had demonstrated that virgin martensites (i.e., those without prior autotempering or aging) undergo significant changes in hardness (Ref 4) and strength (Ref 5) on aging below, at, or somewhat above room temperature—well before the onset of any carbide precipitation. Transmission electron microscopy indicates that aging is accompanied by a redistribution of carbon atoms in the martensite, producing diffuse electron scattering along ⟨102⟩ lattice directions (Ref 6, 7). (Due to tetragonality of the martensites in this paper, it should be noted that the third Miller index (representing the *c*-axis) is not interchangeable with the other two indices when obtaining equivalent variants.)

Detailed observations of the positions, shapes, and integrated intensities of (200), (020), and (002) martensitic x-ray peaks in an Fe-18Ni-1.0C alloy, via a pseudo single-crystal technique (Ref 8), have led to the

*K.A. Taylor, L. Chang, G.B. Olson, G.D.W. Smith, M. Cohen, and J.B. Vander Sande, "Spinodal Decomposition during Aging of Fe-Ni-C Martensites," *Metall. Trans.* 20A (1989) 2717–2737

conclusion that, during aging, the carbon atoms undergo clustering but remain in a single interstitial sublattice, thus preserving a macroscopic tetragonal distortion. Aging studies by Mössbauer spectroscopy, which is capable of probing the immediate environments of iron atoms in martensites affected by neighboring interstitial carbon atoms, have been interpreted to indicate the formation of a metastable Fe_4C configuration within the martensite (Ref 9).

The kinetics of aging, particularly as a function of carbon concentration in Fe-Ni-C martensites, have been carefully explored by means of electrical resistivity measurements (Ref 10). An important feature of the aging process is the occurrence of a peak in the isothermal resistivity curves that increases in magnitude with carbon content. The attendant kinetics suggest that the underlying phenomena are carbon-diffusion controlled.

The formation of carbon-enriched regions during the aging of martensite has been shown by the fine-scale chemical analysis achievable with APFIM (Ref 11, 12). However, the measured carbon buildup of 8 at.% falls far short of the 20 at.% expected from the abovementioned Fe_4C.

In the foregoing research, there have not been appropriate correlations between the kinetics of the aging process and the fine-scale microstructural changes that have been observed. Moreover, many investigations of aging in the earlier literature did not pay adequate attention to the martensitic starting state because of inadvertent autotempering and aging sustained by martensites having M_s temperatures above room temperature. Another issue of considerable interest is the possible role of the martensitic substructure (dislocations and twin boundaries) in the basic aging process.

In the present work, special care was taken to start the aging experiments with martensites in the virgin state. Electrical resistivity measurements were conducted not only to evaluate the kinetics of aging but also to ascertain appropriate stages for detailed high-resolution TEM and APFIM. These new findings, some of which have appeared elsewhere in preliminary form (Ref 13, 14), now permit a fairly specific description of the underlying aging mechanism in Fe-Ni-C martensites.

Experimental Procedures

Materials and Processing

Two Fe-Ni-C alloys with subambient martensitic transformation temperatures were selected for the current research in order to minimize autotempering effects and thus obtain virgin martensite; in this way, aging could be monitored from its outset. The alloy compositions, nominally Fe-15Ni-1C and Fe-25Ni-0.4C, are given in Table 1. Austenitizing of these alloys was conducted at 950 °C for 1 hour in VYCOR (a trademark of Corning Glass Works) tubes backfilled with argon, followed by quenching in room-temperature water. Cast-iron chips were included in the capsules to minimize decarburization. The resulting austenitic grain sizes and martensite "burst" temperatures (M_b) are listed in Table 2. Several additional alloys from previous studies (included in Table 1) were also examined here in order to cover a broader carbon range. Austenitizing treatments and estimated M_b temperatures for these alloys are given in Table 2. The reader is referred to the original references for details concerning the processing of these alloys prior to the austenitizing treatments.

Table 1 Compositions of the experimental alloys, weight percent (atomic percent)

Alloy	C	Mn	Ni
Fe-25Ni-0.4C	0.39 (1.81)	...	24.9 (23.6)
Fe-15Ni-1C	0.88 (3.99)	...	14.8 (13.7)
Fe-33Ni-0.03C (Ref 15)	0.03 (0.14)	...	33.4 (32.3)
Fe-25Ni-0.2C (Ref 10, 16)	0.19 (0.89)	...	25.5 (24.4)
Fe-1.8Mn-1.8C (Ref 17)	1.80 (7.85)	1.80 (1.72)	...

Table 2 Austenitizing treatments and transformation temperatures for the experimental alloys

Alloy	Austenitizing treatment	Mean intercept grain size	M_b, °C	Volume fraction of martensite(a)
Fe-25Ni-0.4C	950 °C/1 h	61 μm	−48	0.85
Fe-15Ni-1C	950 °C/1 h	33 μm	−38	0.67
Fe-33Ni-0.03C	1200 °C/50 h	N.D.	−100(b)	N.D.
Fe-25Ni-0.2C	950 °C/1 h	N.D.	10(b)	N.D.
Fe-1.8Mn-1.8C	1140 °C/1 h	N.D.	−20(b)	0.59

(a) After cooling to −196 °C and warming back to room temperature. (b) Estimated. N.D. = not determined

Electron Microscopy

Discs for TEM studies were spark-cut from 3-mm-diameter austenitized rods and thinned mechanically to about 0.1 mm. Subambient quenching (to form martensite) and subsequent aging or tempering were done with the samples in this "bulk" state. Final thinning was conducted electrochemically in a twin-jet apparatus with a solution of 10 vol% perchloric acid/90 vol% methanol at about −55 °C and a potential of 20 V (corresponding to a current density of about 0.4 A/cm^2). Electron microscopy was performed with a JEOL JEM 200CX (operating at 200 kV) on naturally aged specimens; that is, specimens aged at room temperature (~18 °C to 20 °C).

Atom-Probe Field-Ion Microscopy

Atom-probe field-ion microscopy analyses were carried out primarily on the Fe-15Ni-1C alloy. Specimens were prepared from 0.8 × 0.8 × 12-mm blanks using a standard two-stage electropolishing technique, the first stage being carried out in 25 percent perchloric acid in acetic acid at 25 V and 20 °C, and the second stage in 2 percent perchloric acid in butoxy ethanol, also at 25 V /20 °C. For the examination of virgin martensite, specimens were electropolished at room temperature while still in the austenitic state and then inserted into the cryostat of the atom probe, which had been precooled to −173 °C, thus forming martensite in situ. Subsequent aging was conducted for various times at room temperature.

A Vacuum Generators FIM 100 time-of-flight atom probe was used in both the selected-area and random-area analysis modes. An imaging atom probe was also used in the time-gated mode in order to characterize the FIM image contrast. A pulse fraction of 0.15 of the DC voltage was used for probing. The "as-quenched" (virgin martensite) specimens were probed at −223 °C and −173 °C, whereas all aged specimens were probed at −173 °C. The probe size was in the range of 1.5 to 3 nm. Composition profiles were constructed by averaging over 200 or 100 ions; averaging over smaller numbers produced slightly higher peaks but greatly increased the background noise.

Since the alloys of this study contained appreciable levels of retained austenite, care was taken to ensure that APFIM observations were indeed carried out on the martensitic phase. Transformation twin contrast, as well as the FIM image characteristics of bcc crystals, was used to make certain that the apex regions of the specimens were martensitic.

Electrical Resistivity Measurements

Electrical resistivity specimens were typically 30-mm-long × 1.5-mm-square in cross section. Pure nickel lead wires were spot-welded to each end of the samples for the resistance measurements. Martensite was formed after spot welding by quenching the samples in liquid nitrogen (T = −196 °C) for 15 minutes. After quenching, the samples were held for 1 hour at −125 °C in order to allow the subzero isothermal transformation of the retained austenite to run its course (Ref 10). The samples were then isothermally aged or tempered and intermittent resistance measurements were performed at a reference temperature of −196 °C in liquid nitrogen (this temperature is sufficiently low that the thermal contribution to the resistance is small and no aging of the martensite occurs during the measurement itself). Aging temperatures were controlled within ~0.5 °C in suitable baths. The thermal history of a typical resistivity specimen is illustrated schematically in Fig. 1.

A Keithley Model 503 Milliohmmeter was used to measure resistance, with a rated accuracy of ~0.5 percent. Resistivities were calculated from the specimen geometry and were corrected for the presence of retained austenite (Table 2), as described by Sherman *et al.* (Ref 10). The accuracy of the absolute

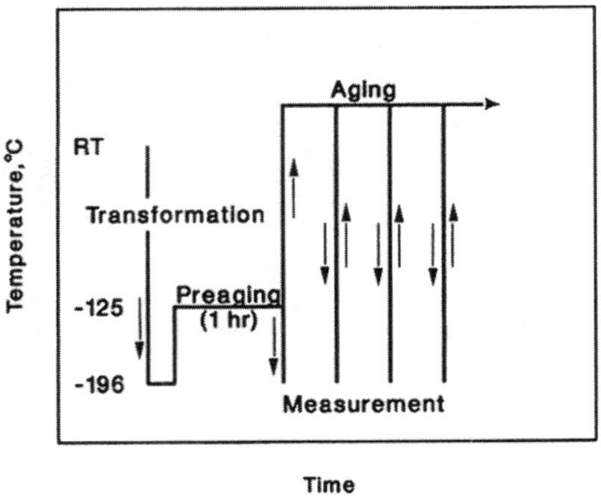

Fig. 1 Schematic representation of the thermal history of a typical electrical resistivity specimen.

resistivity values is estimated to be about 2.0 percent due to the combined error in dimensional and resistance measurements. However, the sensitivity to *relative* changes in resistivity was limited only by the accuracy of the milliohmmeter, which was ~0.5 percent. Smooth curves were drawn through the actual data points thus measured.

Results

Electron Microscopy

A typical microstructure associated with short aging times (several hours or less) is presented in Fig. 2. This bright-field/dark-field TEM image pair was obtained from the Fe-15Ni-1C alloy with the beam incident nearly along [100] (but tilted slightly to excite $\mathbf{g}_{020}$). These images exhibit a distinctive basketweave or "tweedlike" contrast consisting of two sets of fine striations, spaced about 1.5 nm apart on average. The inset diffraction pattern reveals two components of diffuse scattering (at the fundamental reflections) normal to each set of striations. The length of the diffuse streaks about $\mathbf{g}_{020}$, although difficult to measure experimentally, appears to be related reciprocally to the average striation spacing, that is, about $(1.5\ \text{nm})^{-1}$. In general, the details of the image contrast from the present aged martensites were found to be very sensitive to specimen orientation and were similar to the behavior reported by Sandvik and Wayman (Ref 18) for an Fe-8Cr-1.1C alloy.

The nature of the diffuse scattering is more clearly revealed in Fig. 3, which is a diffraction pattern obtained at exact [100] incidence. The two components described above are visible about each fundamental reflection, and the intensity distribution about each fundamental reflection appears to be similar. Additional diffraction patterns representing both high- and low-carbon alloys are shown in Fig. 4. Here, the beam direction is near [001], but tilted

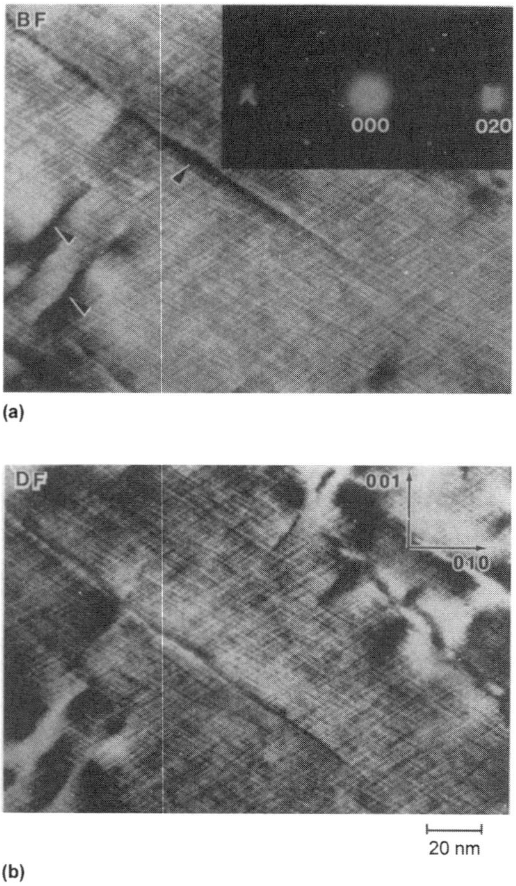

Fig. 2 Transmission electron micrographs ((a) bright-field/(b) dark-field pair, **g** = 020) of Fe-15Ni-1C martensite, aged 5.5 h at room temperature. Arrowheads indicate strain contrast at lattice dislocations.

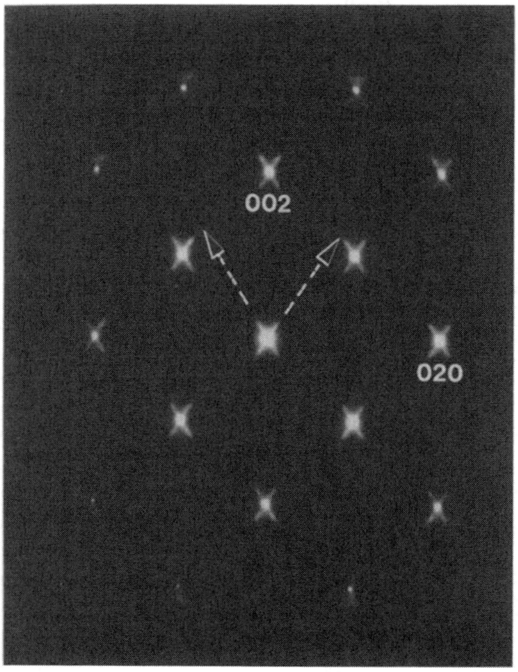

Fig. 3 [100] diffraction pattern obtained from Fe-15Ni-1C martensite after aging 26 min at room temperature ([023] and [0$\bar{2}$3] directions are indicated).

slightly about [100] in Fig. 4(a) and about [110] in Fig. 4(b). Four distinct spots or "satellites" are associated with the fundamental reflections, particularly the {110} reflections. Close examination of the higher-order fundamentals reveals that the arrangement of the satellites becomes asymmetric in moving away from the origin of reciprocal space. These satellite spots, as well as their geometric features, can be explained in terms of how the reflecting sphere intersects each of the four components of diffuse intensity (the two components in (100) and two other in the equivalent (010) plane) situated about each fundamental reflection. Figures 3 and 4 thus indicate that the diffuse scattering is roughly rodlike in shape, the habit line of which was determined by trace analysis of several diffraction patterns. The results, plotted in the standard cubic (001) stereogram of Fig. 5, show that the diffuse intensity lies roughly along each of the four equivalent (203) directions. (The relatively small tetragonality of the martensite lattice does not alter the representation significantly.) The diffuse scattering was also present about the transmitted beam, even when specimens were rocked about low-index orientations to change the diffracting conditions. However, the possibility that such diffuse intensity is produced by double diffraction cannot be ruled out completely.

The tetragonal symmetry of the martensite is demonstrated clearly by Fig. 3 and 4. Figure 3 does not possess the fourfold symmetry required for a cubic crystal, based on the fundamental reflections and the diffuse scattering. In general, regions free of plate boundaries and internal twinning yield single-crystal type diffraction patterns consistent with a tetragonal lattice symmetry. The axial ratio (c/a) measured directly from Fig. 3 is 1.058.

Microstructures (and/or the associated diffraction features) similar to those of the aged Fe-15Ni-1C alloy are found in all of the alloys studied in this investigation; Fig. 4(b) shows that the diffuse scattering is even present in the very low carbon Fe-33Ni-0.03C alloy. However, the "initial" structural wavelength (i.e., the average striation spacing after only a few hours of aging) depends on alloy carbon

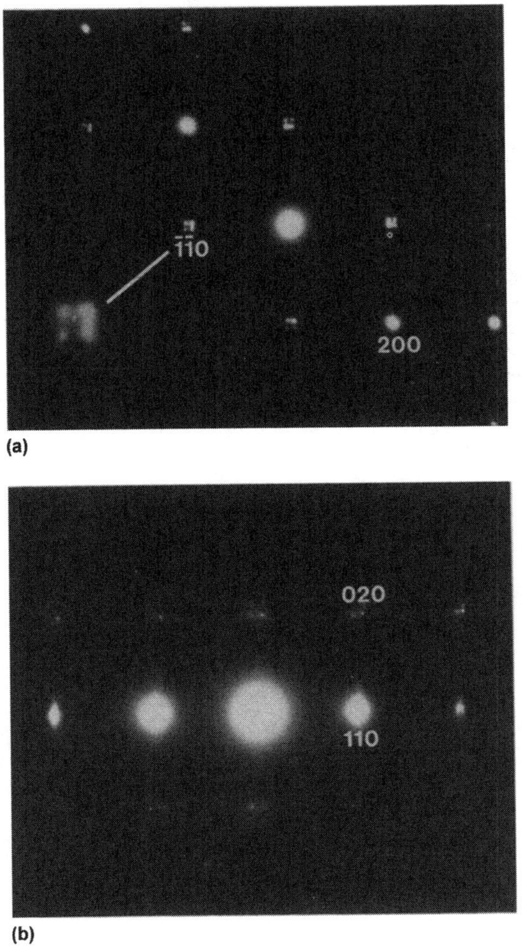

Fig. 4 Near-[001] diffraction patterns obtained from (a) Fe-15Ni-1C martensite (aged 6 d at room temperature) and (b) Fe-33Ni-0.03C martensite (aged 2 d at room temperature). The $(\bar{1}\bar{1}0)$ reflection is enlarged in the inset in (a).

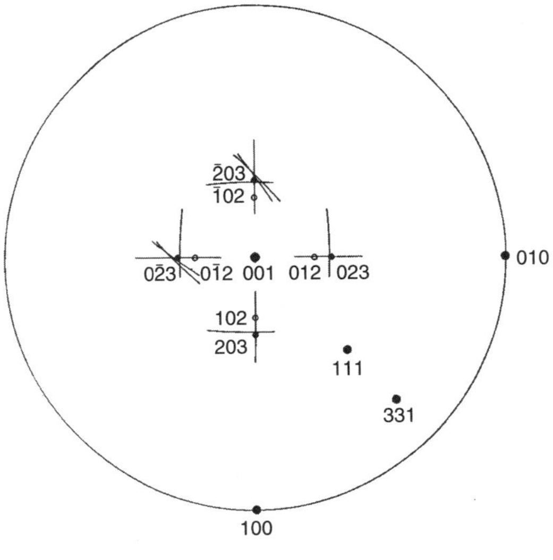

Fig. 5 Cubic (001) stereogram showing the projected directions of the diffuse scattering present in (100), (111), (331), and (001) diffraction patterns.

content. Wavelength measurements, taken directly from TEM micrographs, are plotted in Fig. 6 (the Fe-33Ni-0.03C alloy is not included, because micrographs at an appropriate orientation were not obtained). Included are the data reported by Kusunoki and Nagakura (Ref 7) for Fe-C alloys. The results, spanning a carbon range from about 0.2 to 2 wt% demonstrate that the wavelength generally decreases with increasing carbon content but becomes essentially independent of composition above about 1.4 wt% at a value of about 1 nm. This wavelength behavior is reflected in the diffraction patterns of Fig. 7 in that the diffuse scattering is spread over a larger angle as the carbon content increases. Another interesting observation is the tendency of the diffuse scattering to bend toward the ⟨011⟩ directions at the larger scattering angles (shorter wavelengths). This is most evident in the highest carbon alloy (arrows in Fig. 7).

Figures 8 and 9, combining TEM as well as APFIM data, indicate that the structural wavelength increases with aging time; for example, wavelengths of about 5 and 9 nm, respectively,

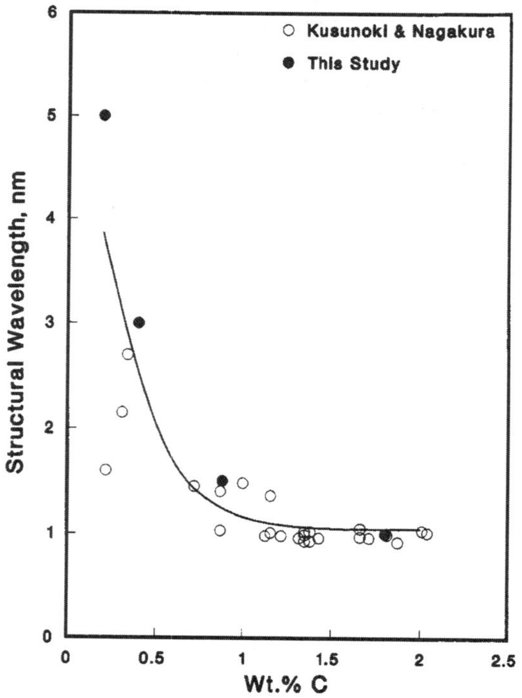

Fig. 6 Variation of initial modulation wavelength versus carbon content of martensite. Open circles are the data of Kusunoki and Nagakura (Ref 7) and solid circles are data from the present investigation (obtained after aging up to several hours at room temperature).

are attained when Fe-25Ni-0.4C and Fe-15Ni-1C martensites are aged for a month. During this microstructural coarsening, no significant change in morphology is detected, other than the increase in wavelength. However, a change of a more subtle nature—the development of asymmetry in the diffuse electron scattering—does appear after several weeks of aging. This phenomenon is related to carbide precipitation in the first stage of tempering (T1) and will be discussed elsewhere (Ref 1).

Atom-Probe Field-Ion Microscopy

The distribution of carbon atoms in as-quenched martensite was investigated via APFIM by varying the probe size and orientation. This had no appreciable effect on the averaged compositions of carbon (and nickel as well), indicating that carbon atoms were distributed uniformly in the virgin martensite, at least on the scale of the aperture size (1 to 3 nm).

Field-ion microscopy images of specimens aged for various times are shown in Fig. 10. Only a faint mottling effect is seen in the FIM images after short aging times, whereas a distinct banded structure emerges at 240 hours. Figure 11, which includes time-gated carbon images, shows that the dark bands are carbon-enriched. This suggests that the modulated structure detected by TEM consists of alternating carbon-rich and carbon-poor bands. The interface between adjacent dark and light bands appears diffuse, and the apparent continuity of atomic planes is consistent with the coherency demonstrated in previous high-resolution TEM studies (Ref 7). Not all FIM images exhibited the banded structure, however; this is probably due to a sensitivity of the image contrast to orientation, as was noted in the TEM observations.

Quantitative chemical analysis was performed by probing along martensite ⟨011⟩ directions, which are the closest low-index directions to the ⟨023⟩ structural modulations and, hence, are the best directions for chemical analysis. Composition profiles and their corresponding autocorrelation diagrams (which establish important parameters relating to second-phase composition and morphology) (Ref 19, 20) for Fe-15Ni-1C martensite in the virgin and aged conditions are presented in Fig. 12 and 13, respectively.

The composition profile for virgin martensite (Fig. 12a) exhibits no systematic variation in

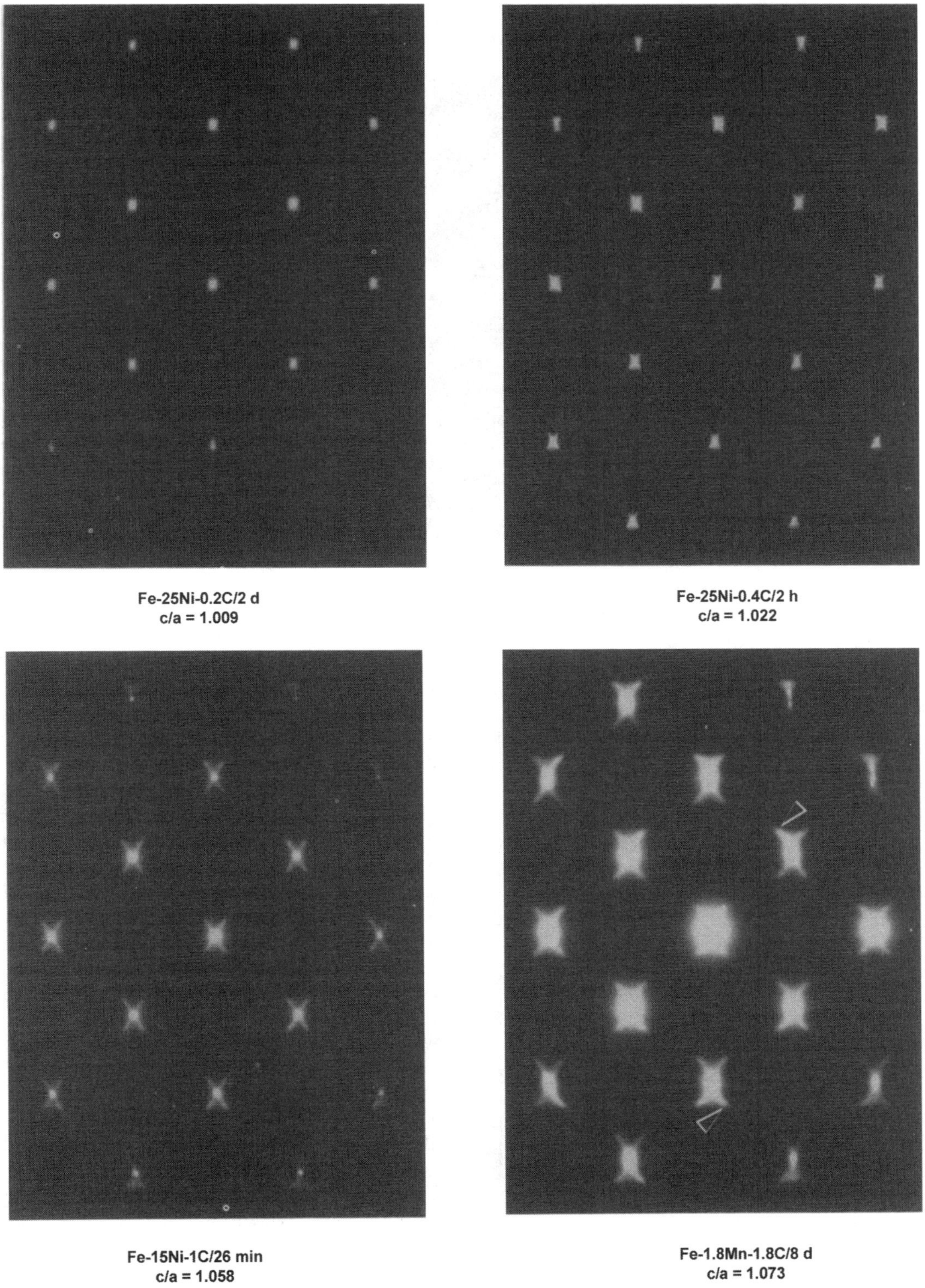

Fig. 7 [100] diffraction patterns from four different martensites aged for short times at room temperature.

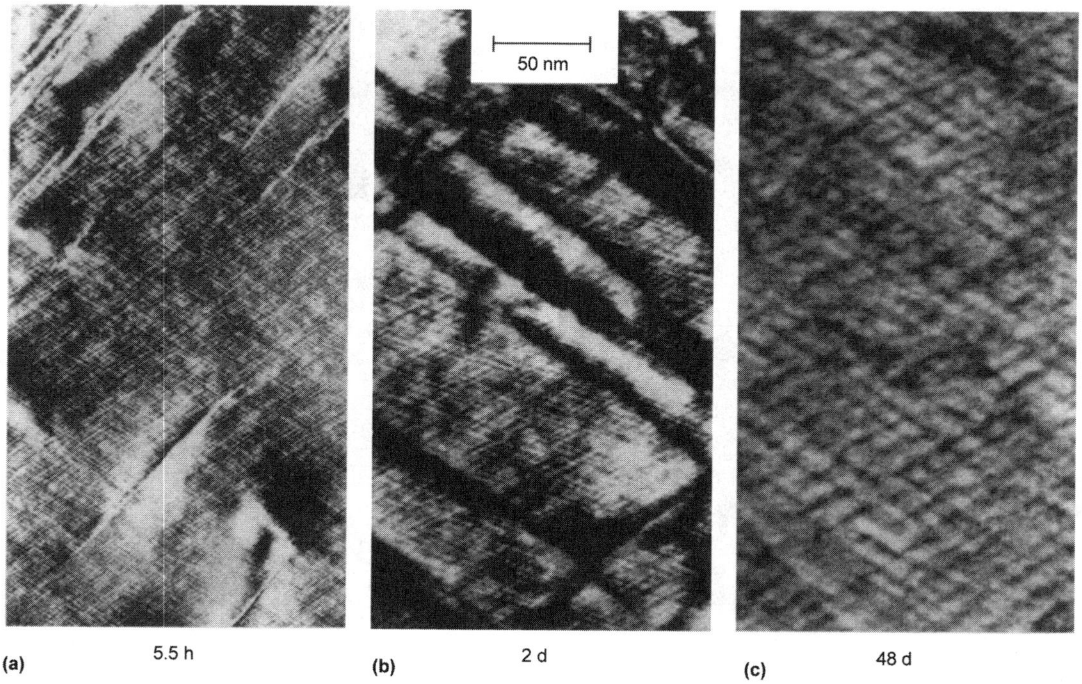

Fig. 8 Transmission electron micrographs of modulated structures in Fe-15Ni-1C martensite after aging (a) 5.5 h ([100] orientation), (b) 2 d ([331] orientation), and (c) 48 d ([100] orientation) at room temperature.

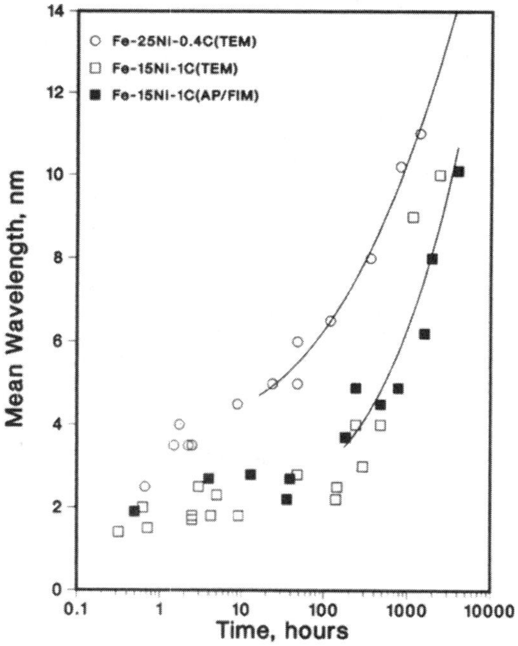

Fig. 9 Variation of modulation wavelength in Fe-25Ni-0.4C and Fe-15Ni-1C martensites *vs* aging time at room temperature. Solid curves represent fit of Eq 4 in the coarsening regime.

carbon concentration. This is reflected in the autocorrelation analysis (Fig. 13a), which is typical of a random solution. In contrast, the composition profile obtained after prolonged (1580 hours) aging (Fig. 12d) shows regular variations in carbon concentration, with a wave-length of about 30 atomic layers (for the aperture used, one layer ≅ 50 ions) and an average peak amplitude of about 11 at.% The carbon concentration in the low-carbon regions is estimated to be about 0.2 at.%. The periodic variation in composition is reflected in the autocorrelation diagram (Fig. 13d), in which regular and significant deviations from zero are apparent, this being typical for systems in which phase separation is occurring. The autocorrelation diagram indicates a "particle size" (k_0) of about 600 ions (2 to 3 nm) and a wavelength (k_1) of about 1300 ions (5 to 7 nm). The volume fraction of the high-carbon phase is given approximately by k_0/k_1 and is 0.44 in this case. These results are not corrected for stereological effects (i.e., the deviation of the probing direction from the true modulation direction).

Although the autocorrelation results for the intermediate aging times of 40 and 175 hours

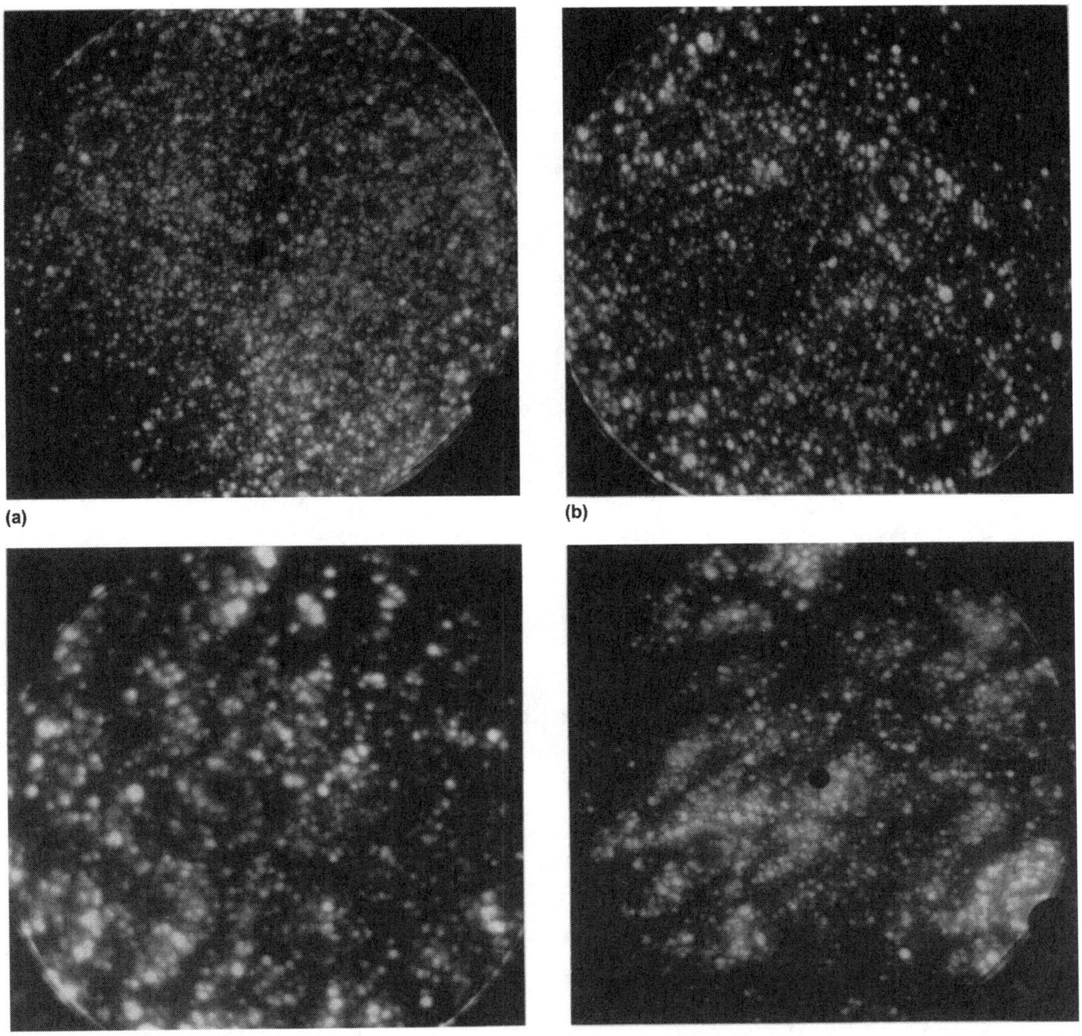

(a) (b) (c) (d)

Fig. 10 Field-ion micrographs of Fe-15Ni-1C martensite obtained in the (a) as-quenched condition and after aging (b) 40, (c) 240, and (d) 1870 h at room temperature. All micrographs were recorded using neon imaging gas at −173 °C (−279.4 °F). The field of view is about 70 nm in (a), (b), and (d) and about 50 nm in (c).

in Fig. 13 do not show obvious evidence of regular variations in carbon concentration, the TEM results already presented clearly show that modulations are well-developed by these times. However, atom-probe evidence of regular variations in carbon concentration for short aging times was detected during shorter data runs than those depicted in Fig. 12 and 13. In general, wavelength data (presented below) for any particular aging treatment are based on the most regular section of autocorrelation observed for that treatment.

The normalized frequency distribution of the composition of successively field-evaporated blocks of equal numbers of ions can reveal phase separation more readily than autocorrelation analysis, particularly during the early stages of aging (Ref 20). Figure 14 shows such results for a series of aging times. The frequency distribution for an ideal random solution (Fig. 14a) would follow a binomial distribution, with the peak frequency located at the average composition (about 4 at.% in this case). The frequency distribution for virgin martensite (Fig. 14b) shows considerable deviation from the binomial distribution, probably as the result of the formation of carbon molecular ions during field evaporation (Ref 11).

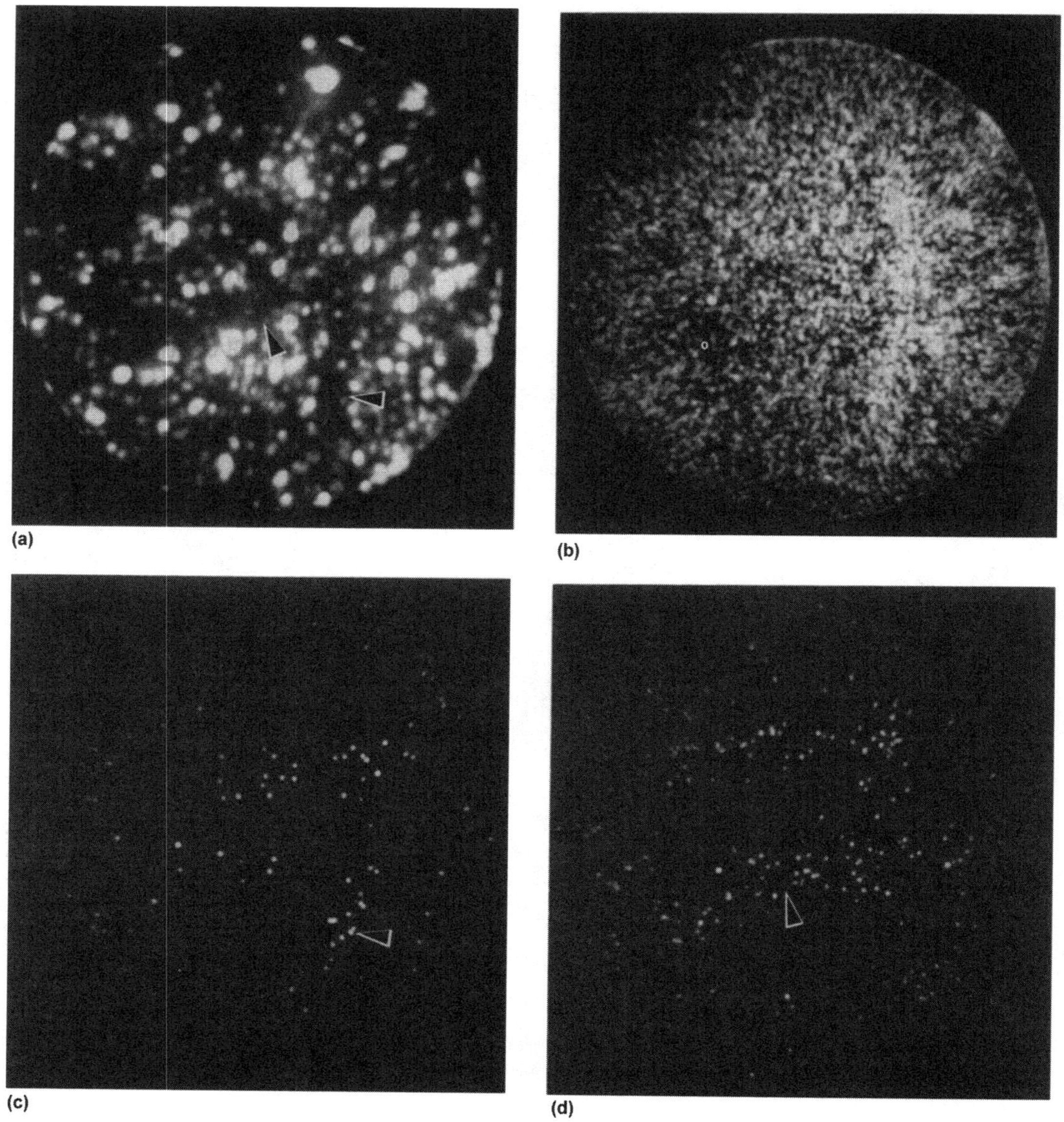

Fig. 11 A sequence of (a) a conventional FIM image, (b) a gated iron image, and (c) and (d) gated carbon images, obtained from Fe-15Ni-1C martensite after aging for 1680 h at room temperature. The field of view is about 66 nm in each case; (b) and (c) were obtained at 11.5 kV and (d) at 10.9 kV. Arrowheads show corresponding locations in (a), (c), and (d).

Although an accurate correction for the formation of carbon molecules could not be made, the frequency distribution for the virgin martensite nevertheless serves as a reference for the distributions of the aged specimens. As indicated in Fig. 14, aging tends to shift the peak frequency to lower carbon concentrations, with a general broadening of the frequency distribution. After aging for 1580 hours (Fig. 14d), a bimodal distribution appears at ~0 and 11 at.% C. The nature of the evolution of the frequency distribution clearly demonstrates that a phase-separation process is occurring.

The atom-probe results are summarized in Fig. 15 by plotting the maximum and minimum compositions (averaged over 100 ions) as a function of aging time. Each solid point is the averaged composition of either the high-carbon regions or the low-carbon regions. The bars on the data points represent the range of the extreme values. In practice, it was very difficult

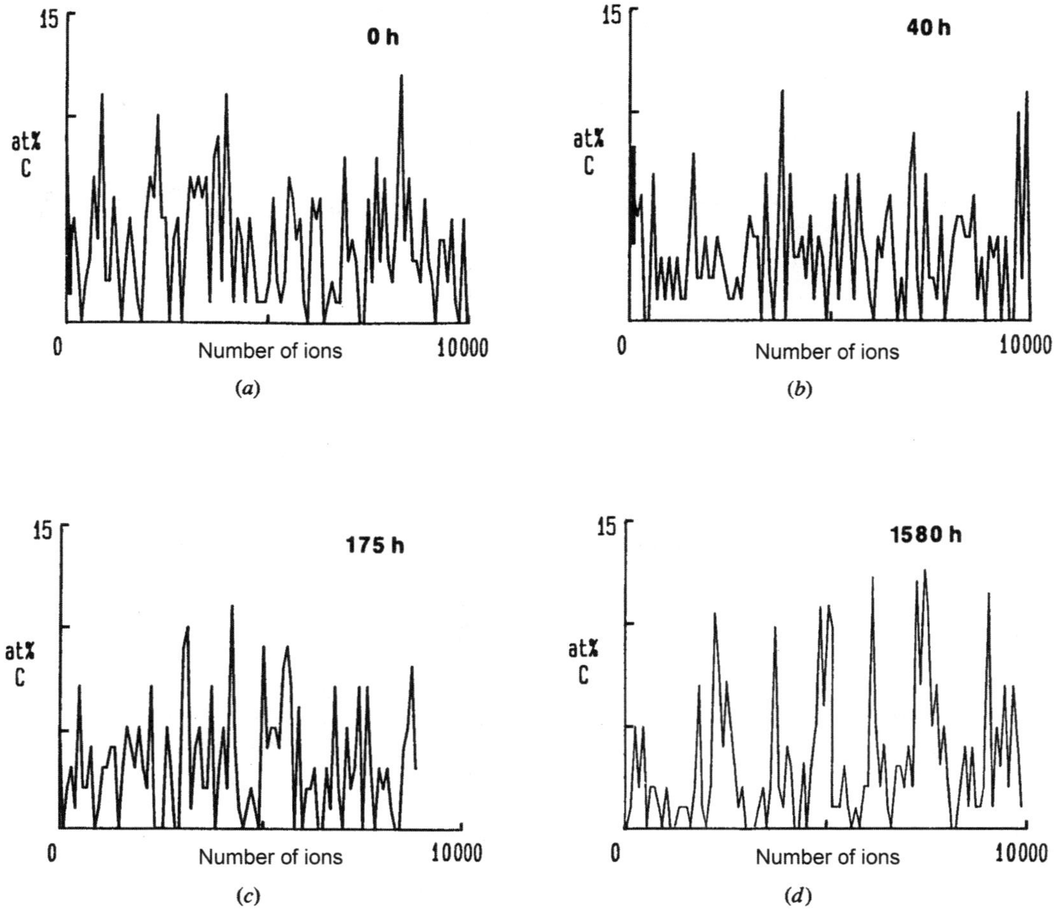

Fig. 12 Composition profiles from Fe-15Ni-1C martensite in (a) the virgin condition and after aging (b) 40, (c) 175, and (d) 1580 h at room temperature.

to specify these peak values at the shortest aging times, and, therefore, only rough ranges (dashed bars) are given. The composition of the high-carbon regions appears to level off at about 11 at.% C. After 1580 hours, the low-carbon regions are estimated to be at 0.2 at.% C, in agreement with earlier work by Miller *et al.* (Ref 12) on 0.4 wt% C alloys. For 40-hour aging, the composition frequency distribution indicates a composition amplitude significantly less than the final amplitude. This implies formation of the modulated microstructure via a continuous increase in composition amplitude, characteristic of a spinodal decomposition mechanism.

As mentioned earlier, compositional wavelengths were estimated from FIM images, composition profiles, and autocorrelation results; the data are included in Fig. 9. These values have been corrected for the deviation of the $\langle 011 \rangle$ probing direction from the true $\langle 023 \rangle$ modulation direction. The good agreement between the wavelengths measured by TEM and APFIM provides confidence that both techniques are indeed monitoring the same process.

In order to assess the temperature dependence of the decomposition behavior, a few specimens of the Fe-15Ni-1C alloy were examined after aging above room temperature. Similar banded FIM images were obtained, and selected-area atom-probe analyses of light- and dark-imaging areas sampled are summarized in Table 3. In addition to compositions near the 11 at.% C formed by room-temperature aging, some dark areas exhibited higher carbon contents near 25 at.% The relative

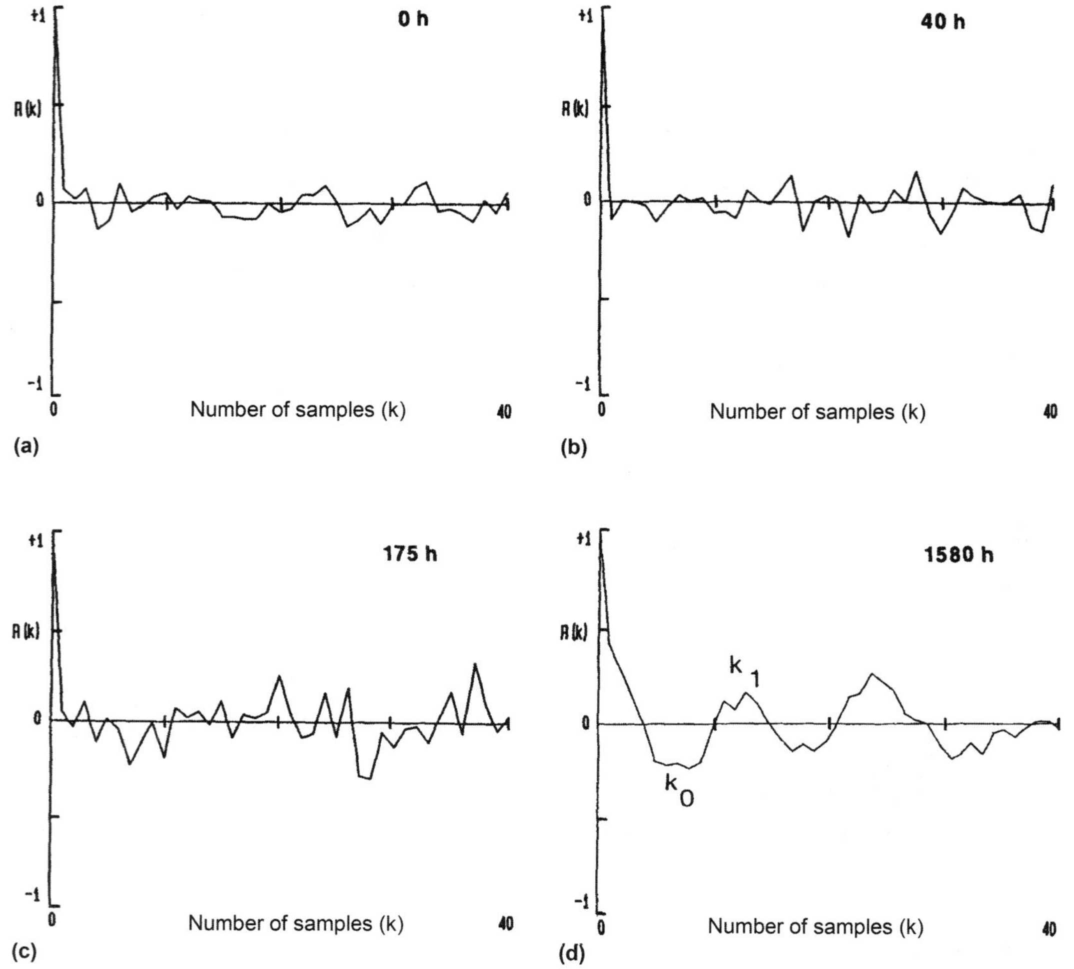

Fig. 13 Autocorrelograms (sample size: 100 ions) corresponding to the composition profiles of Fig. 12: (a) virgin martensite, and room-temperature aged martensites: (b) 40, (c) 175, and (d) 1580 h.

proportion of dark regions with compositions above 15 at.% increases as the temperature is raised. The conditions for the appearance of the higher carbon regions are consistent with the detection of ϵ-carbide by TEM, discussed in Ref 1. The observed coexistence of the two dark-imaging phases signifies that the decomposition process studied at room temperature precedes the precipitation of ϵ-carbide at temperatures at least as high as 150 °C. Further, the observations of Table 3 suggest that the decomposition-product compositions do not depart greatly from ~0.2 and 11 at.% over this temperature range.

An assessment of the dependence of compositional amplitude on carbon content was made by some measurements on two other alloys. An Fe-1.85 wt% C alloy was aged 14 days at 17 °C, similar to the composition and treatment employed by Choo and Kaplow in their Mössbauer study (Ref 9). Although dark-banded regions occasionally showed carbon contents as high as the 20 at.% of the Fe_4C phase originally inferred from the Mössbauer data, the observed composition distribution clearly indicated an average carbon content near 11 at.% in the high-carbon regions. This was also the case for 1-hour aging at 60 °C. To examine the behavior of a low-carbon alloy, an Fe-25Ni-0.2C alloy known to exhibit rapid aging kinetics (Ref 10) was aged 80 minutes at 17 °C Dark-imaging areas exhibited an average composition of 10.1 at.% C with a range of 6.9 to 12.1 at.% C.

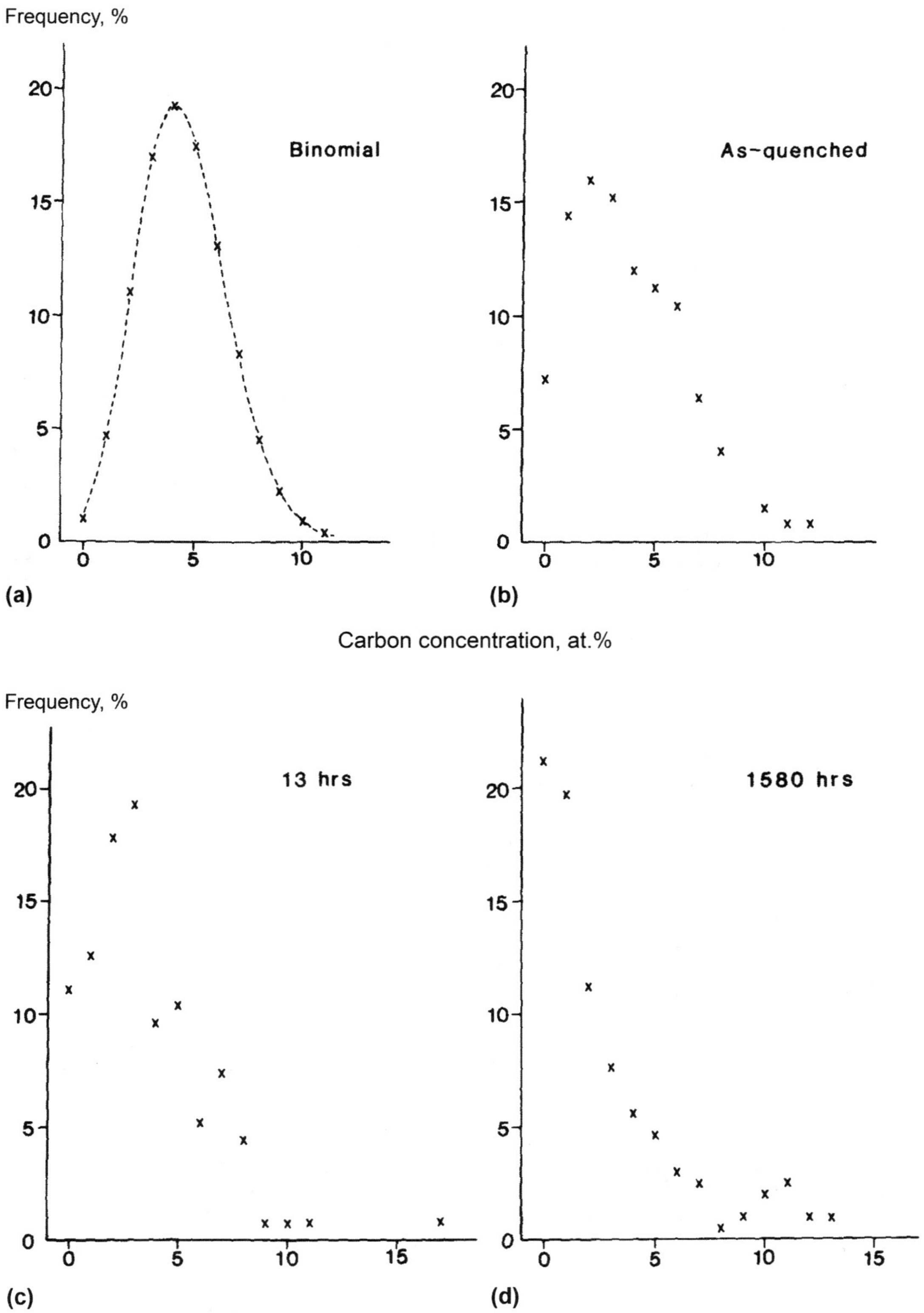

Fig. 14 Normalized frequency distributions present in composition profiles obtained by probing Fe-15Ni-1C martensite along ⟨011⟩ (averaged over 100 ions). The binomial distribution for an ideal random solution is plotted in (a). Frequency distributions corresponding to the virgin condition and after aging for 13 and 1580 h at room temperature are shown in (b) through (d).

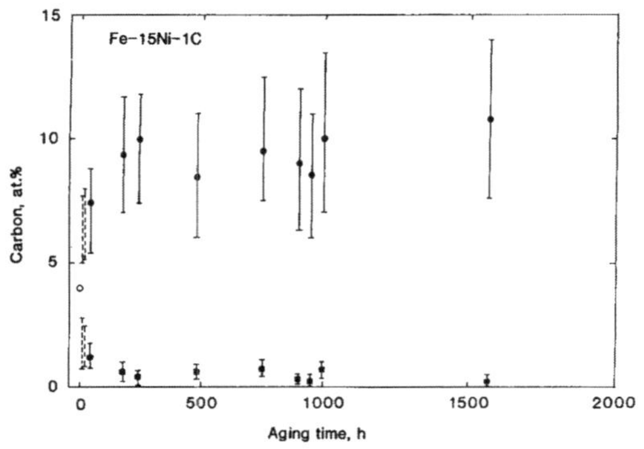

Fig. 15 Variation in the range of maximum and minimum carbon concentrations (averaged over 100 ions) in Fe-15Ni-1C martensite as a function of aging time.

Table 3 Selected-area atom-probe results for Fe-15Ni-1C martensite (atomic percent)

Treatment	Matrix % C	Dark areas minimum-maximum % C	Total number of samples	Samples <15 % C	Samples <15 % C
95 °C/1 h	0.17 to 0.62	10.1 to 15.6	11	10	1
100 °C/24 h	0.12 to 0.46	11.4 to 23.4	9	6	3
130 °C/1 h	0.12 to 0.17	9.5 to 25.3	48	19	29
150 °C/1 h	0.20 to 0.38	12.9 to 28.6	10	2	8

Electrical Resistivity

Plots of normalized electrical resistivity *vs* aging time for the Fe-25Ni-0.4C and Fe-15Ni-1C alloys are shown in Fig. 16. Both alloys undergo resistivity maxima in times that decrease with increasing temperature, which is indicative of a thermally activated process. Two trends emerge from the results. The magnitude of the resistivity peak is larger (1.067 versus 1.055), and the time required to reach the peak resistivity is longer in the higher carbon alloy. The resistivity behavior of the current investigation is very similar to that found in previous studies of Fe-C and Fe-Ni-C martensites over this range of carbon content (Ref 10, 21, 22).

The detailed structural information obtained in this study allows a correlation of positions on the resistivity curve with stages of the aging process. Based on the APFIM data for the 1C alloy, the full composition amplitude is reached beyond the resistivity peak at a time when the resistivity increase $\Delta\rho$ relative to ρ_0 has diminished to about 75 percent of its maximum value. This then denotes the boundary between the regimes of spinodal decomposition and

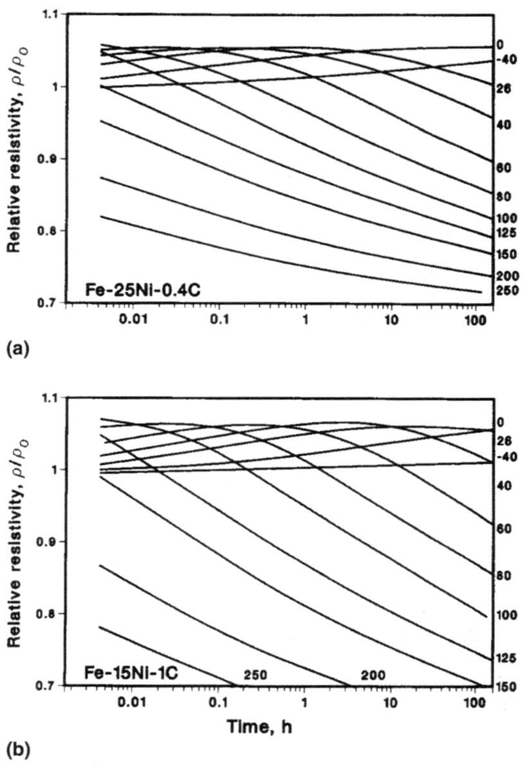

Fig. 16 Relative resistivity (at −196 °C) versus log (time) for (a) Fe-25Ni-0.4C and (b) Fe-15Ni-1C martensites (corrected for retained austenite) aged at the indicated temperatures (°C). The initial resistivities (ρ_0) were 25.3 and 36.3 $\mu\Omega$ cm, respectively.

coarsening, as indicated in the wavelength plots of Fig. 9. These regimes can be identified with the two substages of the aging process previously denoted as A1 and A2 (Ref 2). The point at which the composition amplitude reaches one-half of full amplitude appears to correspond to about the same resistivity value (~75 percent of maximum $\Delta\rho$) on the early side of the peak.

In contrast to the behavior observed for the Fe-Ni-C alloys of Fig. 16, Fig. 17 presents the relative resistivity changes measured in the Fe-1.8Mn-1.8C alloy. While a slope change near $\rho/\rho_0 \sim 0.9$ designates the onset of the first stage of tempering (T1), the smaller resistivity changes associated with the A1 and A2 aging stages near room temperature show no resistivity peak. Despite the TEM and APFIM evidence cited earlier for similar decomposition behavior, modulation formation at this extreme carbon content does not produce a resistivity peak. This is presumably a result of the higher initial resistivity of this higher carbon solid solution.

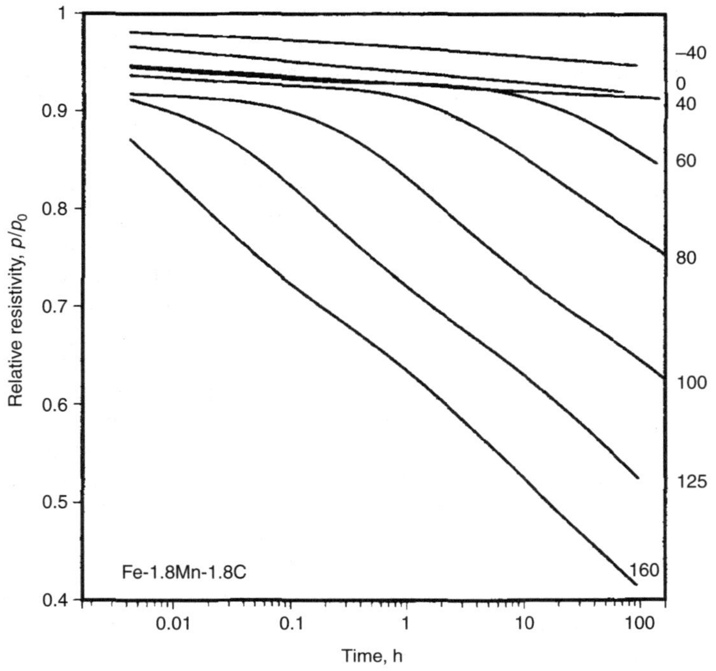

Fig. 17 Relative resistivity (at −196 °C *vs* log (time) for Fe-1.8Mn-1.8C martensite (corrected for retained austenite) aged at the indicated temperatures (°C). The initial resistivity (ρ_0) was 46.0 $\mu\Omega$ cm.

Discussion

Decomposition Morphology

X-ray and neutron diffraction (Ref 4, 8, 23, 24). and Mössbauer (Ref 25) investigations of martensite aging in Fe-Ni-C and Fe-C alloys have indicated that the octahedral interstitial sites are the preferred locations for carbon atoms in the body-centered iron structure. Furthermore, the tetragonal or orthorhombic distortion of the martensitic lattice is generally regarded as direct evidence of an unequal distribution of the carbon atoms over the three available octahedral interstitial sublattices. When carbon atoms occupy a single sublattice within an otherwise body-centered cubic iron lattice, the macroscopic lattice symmetry is reduced to tetragonal, and a macroscopic tetragonal distortion of the bcc iron lattice results. Zener (Ref 26) was the first to treat the thermodynamics of such carbon ordering, and tetragonal martensite is accordingly referred to as "Zener-ordered."

Selected-area electron diffraction patterns obtained in this investigation were generally consistent with the proposed tetragonal symmetry of fully ordered (in the sense that only one interstitial sublattice is preferentially occupied) martensite, both in terms of the arrangement of fundamental reflections as well as the geometry of the diffuse scattering. However, fine-scale {011} twinning was also frequently detected by TEM. The result of this twinning is to reduce the macroscopic lattice symmetry from tetragonal to orthorhombic, as has been detected through precise x-ray diffraction measurements on an Fe-18Ni-1.0C alloy (Ref 8). Details relating to {011} twinning in the alloys of the present study appear in Ref 27.

The apparent rodlike geometry of the diffuse electron scattering associated with the modulated structure is consistent with a planar morphology, regardless of whether one explains the diffuse scattering in terms of a shape effect or a modulated lattice strain (as have Kusunoki and Nagakura for Fe-C alloys; Ref 7). In general, a combination of both effects can be expected in a system where large coherency strains are encountered, as is likely the case in aged Fe-C martensites.

The observed directions of the structural modulation can be rationalized on the basis of elastic strain-energy considerations. Khachaturyan (Ref 28) has developed a theory which predicts the habit plane of a coherent platelike particle when it is derived from an infinite cubic matrix by a tetragonal lattice deformation. This theory is relevant to the present study inasmuch as the carbon occupancy of a single octahedral interstitial sublattice (referred to henceforth as the "Zener sublattice") results in a tetragonal distortion of the martensite lattice. For the case where both phases have the same elastic moduli, the habit plane of the second-phase particle is determined by the elastic anisotropy parameter, A:

$$(hkl)_{\text{habit}} = \begin{matrix} (h0l) & \text{if } A < 0 \\ (hhl) & \text{if } A > 0 \end{matrix} \qquad \text{(Eq 1)}$$

where A is given by $(c_{11} - c_{12} - 2c_{44})/c_{44}$. For pure iron, $c_{11} = 2.42 \times 10^5$ MPa, $c_{12} = 1.465 \times 10^5$, and $c_{44} = 1.12 \times 10^5$. The anisotropy parameter is thus less than zero for iron, implying a habit plane of the type $\{h0l\}$.

The exact habit plane is determined by the magnitude of the tetragonal lattice distortion associated with the second-phase particle. In general, the carbon dependence of the lattice constants of dilute Fe-X-C solutions can be expressed as (Ref 29):

$$\begin{aligned} a &= a_0(1 + u_{11}n_C) \\ c &= a_0(1 + u_{33}n_C) \end{aligned} \qquad \text{(Eq 2)}$$

where a_0 is the lattice constant of the carbon-free alloy, u_{ii} are elements of the lattice expansion tensor, and n_C is the fraction of occupied interstitial sites in the Zener sublattice. If the linear behavior given by Eq 2 is valid for the compositions measured here by APFIM, then the habit plane is a function only of the elastic constants and the "tetragonality factor," $t_1(=u_{11}/u_{33})$, for the interstitial solution (Ref 28):

$$(hkl)_{\text{habit}} = (\sin\theta_0, 0, \cos\theta_0) \qquad \text{(Eq 3)}$$

where:

$$\cos^2\theta_0 = 1 + \frac{c_{11} + 2c_{12}}{c_{11} + c_{12}} \frac{t_1}{1 - t_1}$$

The lattice expansion coefficients and the tetragonality factor applicable to the Fe-Ni-C alloys of this study are taken to be (Ref 27):

$$u_{11} = -0.275$$

$$u_{33} = 1.016$$

and

$$t_1 = -0.271$$

Combining these values with the elastic constants for iron yields a habit-plane normal of [0.542, 0, 0.841], which is less than 1 deg from [203] and is thus remarkably close to the observed directions. Although this calculation applies to a single second-phase particle in an infinite matrix of pure iron, the good agreement between the calculated and observed habit plane is consistent with the morphological development of the structural modulation being driven by a minimization of elastic strain energy. Strictly speaking, the t_1 for a stress-free lattice deformation should be used in Eq 3. However, the results are not altered if it is assumed that u_{11} and u_{33} change proportionately when elastic constraint exists.

It is noteworthy that this and numerous previous studies (Ref 4, 6, 7, 30) show that a tetragonal distortion/symmetry of the martensite lattice persists during aging, although the magnitude of (c/a) tends to decrease as aging proceeds. The reduction in (c/a) may be related to a decrease in the degree to which the high- and low-carbon regions scatter coherently as coarsening proceeds, in line with the suggestion of Hayakawa *et al.* (Ref 30), Nonetheless, the preservation of a tetragonal lattice distortion during aging indicates that carbon atoms remain within the Zener sub-lattice while undergoing a coherent redistribution.

Structure of the High-Carbon Phase

Previous Mössbauer (Ref 9, 31) and electron diffraction (Ref 6, 32) findings offer evidence that a secondary ordering of carbon interstitials may occur within the Zener sublattice during martensite aging, prior to the detection of the ϵ- or ϵ'-carbide. These results have been interpreted in terms of the formation of an Fe_4C phase with considerable deviation from the stoichiometric composition (Ref 31). The tetragonal Fe_4C configuration is depicted in Fig. 18. While the current Fe-1.8Mn-1.8C alloy did indeed exhibit the type of "superlattice" reflections previously attributed to Fe_4C (Fig. 19), none of the present Fe-Ni-C alloys displayed such superlattice reflections. However, the current as well as previous (Ref 12) APFIM results demonstrate that the compositions of the high-carbon regions in the latter alloys fall far short of the required 20 at.% C and are closer to the Fe_8C (or $Fe_{16}C_2$) stoichiometry. Interestingly, the structure of the well-established α''-$Fe_{16}N_2$ can be derived

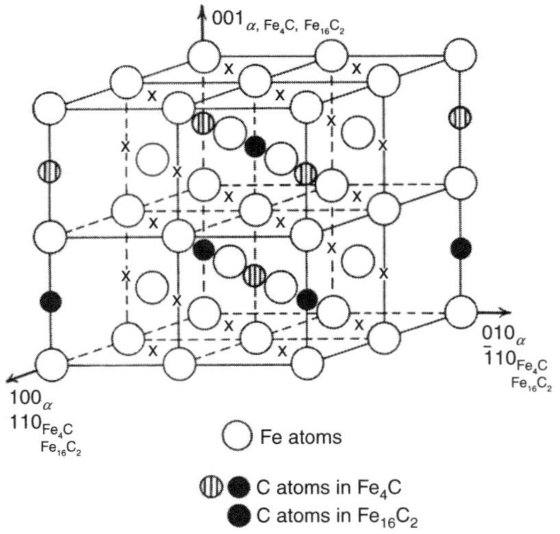

Fig. 18 Schematic representation of the Fe_4C structure. The lattice correspondence with the low-carbon ferrite suggested by electron diffraction patterns is also indicated. Large open circles are iron atoms; small solid and shaded circles are carbon atoms. The ×'s denote vacant "*c*-oriented" octahedral sites. The $Fe_{16}N_2$ structure is obtained by systematic removal of half of the interstitial atoms.

from the Fe_4C structure by systematic removal of one-half of the interstitial atoms (Fig. 18). This raises the possibility that a phase isomorphous with α'' exists in the Fe-C system.

Some further support for an α''-$Fe_{16}C_2$ phase can be found in the observation by Ferguson and Jack (Ref 33) that an α''-$Fe_{16}(C, N)_2$ carbonitride forms upon room-temperature aging of an Fe-C-N ferrite with comparable C and N contents (0.022 wt% C and 0.015 wt% N). They also found that the metastable phase decomposes at higher temperature (4 days at 200 °C) into ϵ-carbonitride and γ'-Fe_4N. Under metastable equilibrium, excluding the latter more stable phases, the α'' structure thus demonstrates significant C solubility.

As presented in the Appendix, a reassessment of Mössbauer data for aged Fe-C (Ref 9) and Fe-N (Ref 34) martensites, incorporating more recent measurements (Ref 35, 36), also supports the possibility of an α'' structure in both systems.

A recent theoretical study of the relative stability of interstitial phases based on body-centered iron has been published by Sato *et al.* (Ref 37). Using a molecular orbital-based discrete variational cluster method, the "eigenstrains" defining the local relaxations of Fe atoms around N and C atoms were computed by maximization of total bond order. Calculations of the elastic interactions between

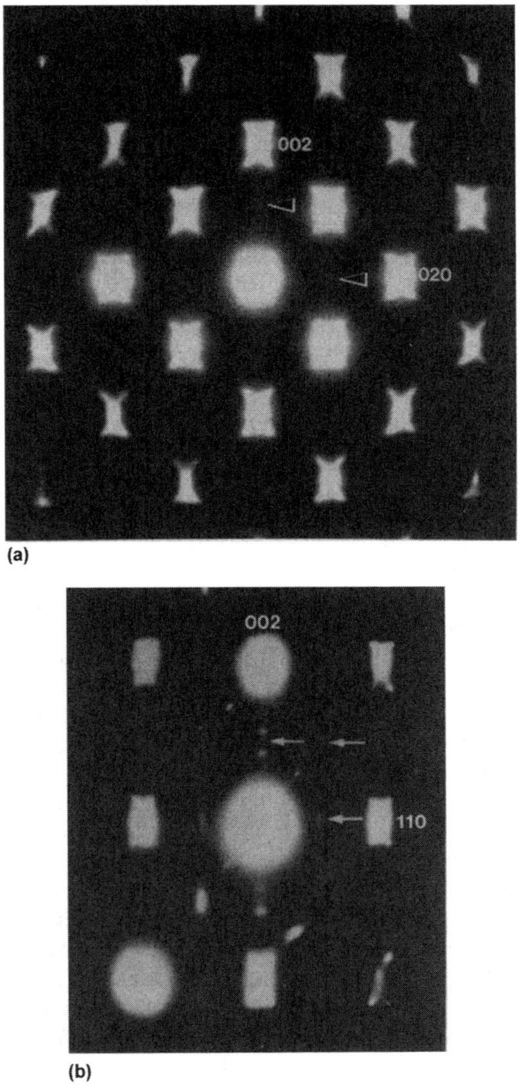

Fig. 19 "Superlattice" reflections (arrows) in the electron diffraction patterns of aged high-carbon martensites: (a) [100] pattern from Fe-1.8Mn-1.8C martensite (aged 8 d at room temperature) and (b) [$1\bar{1}0$] pattern from Fe-1.62C martensite. Source: Ref 63

interstitial atoms led to the prediction that the most stable ground-state structure for the body-centered solid solutions of both Fe-N and Fe-C systems is the α''-$Fe_{16}N_2$ structure. It thus seems quite reasonable that the metastable high-carbon phase produced by room-temperature decomposition of ferrous martensites is a form of this structure.

While further direct diffraction information would be of great value, the combined experimental evidence at this point favors the proposal that the high-carbon phase resulting from the spinodal decomposition studied here represents an imperfect form of the $Fe_{16}N_2$ structure. (An alternative structure proposed by Génin (Ref 38) is not consistent with the current composition results.)

We henceforth designate this phase as α''.

Thermodynamics of Decomposition

In addition to the aforementioned analysis by Sato *et al.*, (Ref 37) several notable treatments of the thermodynamics of carbon ordering have appeared in the literature. Zener (Ref 26) and Ren and Wang (Ref 39) have treated carbon ordering phenomenologically, while Khachaturyan (Ref 40) has taken a more fundamental approach based upon strain-induced interstitial interactions. These models have been formulated within the framework of linear elasticity theory, and they predict a composition-dependent critical temperature for carbon ordering. At room temperature, the Zener and Ren and Wang models predict that alloys containing less than about 0.25 wt% C should be disordered, whereas the Khachaturyan model places the critical composition at 0.03 wt% C. The apparent tetragonal symmetry of the present Fe-33Ni-0.03C alloy (as indicated by Fig. 4, and which we believe to be the first evidence of Zener order at such a low carbon level) is in line with the dilute critical composition corresponding to the Khachaturyan model.

The above models predict that the ordered solution is compositionally unstable; that is, phase separation via a "conditional spinodal" mechanism (Ref 41, 42) is indicated. Although the applicability of these models is limited to dilute solutions, there appears to be agreement between the predicted and the observed modes of phase separation; that is the experimental findings are indeed consistent with a spinodal mechanism. The most convincing evidence concerns the continuous increase in compositional amplitude with increasing aging time, as well as the modulated nature of the coherent two-phase mixture. An increase in compositional amplitude at constant wavelength is, according to linearized theory (Ref 41), a sufficient condition for defining a spinodal reaction. However, this condition must be relaxed in practice inasmuch as real systems, such as the alloys of this study, may begin coarsening before the metastable equilibrium phase compositions are attained (Ref 19).

The main shortcoming of the Zener and Khachaturyan models arises from the fact that the assumed equilibrium ground state (for all but the most dilute compositions) is a mixture of low-carbon bcc iron and the compound FeC. However, as shown in the present investigation, no experimental evidence exists to suggest the formation of such a carbon-rich compound.

A picture of phase relations in the Fe-C system consistent with the current experimental results is illustrated by the schematic free-energy curves in Fig. 20. The curve for the disordered (cubic) solution is taken to have positive curvature everywhere. The curve for the Zener-ordered solution, based on the behavior prescribed by the Khachaturyan model, has positive curvature initially but quickly develops negative curvature, reaching a minimum at the stoichiometric FeC composition. (A continuum of free-energy curves, corresponding to different values of an appropriate order parameter, can be constructed within the bounds of the "disordered" and "ordered" curves.) A more appropriate total free-energy function is illustrated by the solid curve in Fig. 20, reflecting a secondary ordering component which stabilizes the α'' structure.

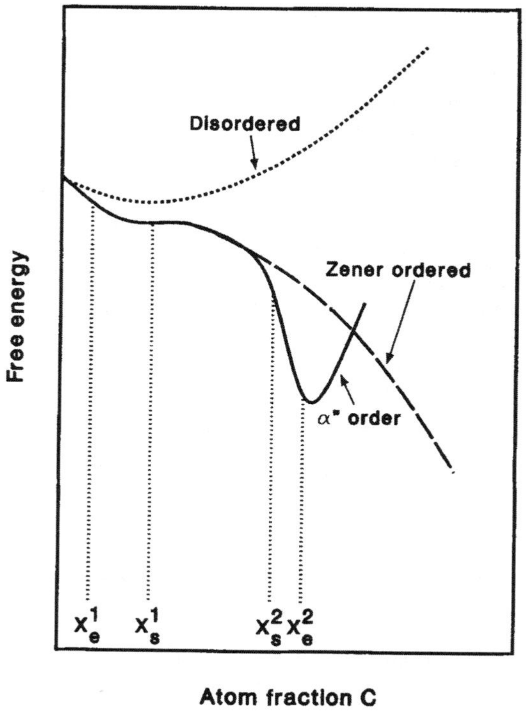

Fig. 20 Schematic free-energy curves for disordered, Zener-ordered, and α''-ordered body-centered Fe-C solutions.

The latter hypothetical ordered free-energy curve contains two inflections, denoted by X_s, that define the compositional limits within which spinodal decomposition can occur. Note that the reaction is predicted to proceed via a conditional spinodal mechanism, since Zener order is required to produce the initial compositional instability. The α'' composition of the high-carbon regions that ultimately develop is controlled by the secondary ordering imposed by longer range C-C repulsions. These two components of ordering would be expected to produce a metastable phase diagram similar in form to that of the Fe-Al system (Ref 42). The room-temperature phase boundaries, which the atom-probe results would place at 0.2 and 11 at.% C, are denoted by X_e. A qualitative description of the thermodynamics of this system could be obtained from a synthesis of the models of Khachaturyan (Ref 40) and Sato *et al.* (Ref 37). A full description of the coherent two-phase state would prescribe elastic energy contributions as treated by Cahn and Larché (Ref 43) and is beyond the scope of the current analysis. The relatively large coherency strains in this system, however, may offer a fertile area for experimental tests of interesting aspects of coherent equilibria.

Kinetics of Aging

Previous TEM studies of martensite aging have shown that the diffuse electron scattering associated with the modulated structure appears only after warming the virgin martensite from the liquid nitrogen temperature (Ref 7, 17), demonstrating that the modulated structure is not inherited from the parent austenite. The structural modulations develop very rapidly when initially virgin martensites are aged at room temperature. In this study, all alloys exhibited modulated structures after aging times as short as 5 minutes, the shortest time feasible for TEM without a specimen cold stage.

With the correlation previously established between microstructure and position on the resistivity curves of Fig. 16, the resistivity data can be used to quantify the kinetics of specific stages of the aging process. The apparent activation energies for the A1 aging stage corresponding to the points halfway up the electrical resistivity peak are 78 and 92 kJ/mole in the Fe-25Ni-0.4C and Fe-15Ni-1C alloys, respectively. The aging kinetics at this early stage of spinodal decomposition can be expected to be dominated by the rate of carbon diffusion, and these activation energies are similar to that for carbon diffusion in ferrite (82.8 kJ/mole; Ref 44). The current electrical resistivity results indicate that aging proceeds more rapidly in the Fe-25Ni-0.4C alloy, as compared with Fe-15Ni-1C; for example, the time at 40 °C required for the resistivity to pass through the peak and fall off to its initial value (~completion of the A1 stage) is about 25 hours for Fe-25Ni-0.4C, as compared with an estimated 175 hours for Fe-15Ni-1C. The generally higher apparent activation energy for the higher carbon alloy is consistent with Hillert's (Ref 45) hypothesis that the activation energy for a carbon-atom jump increases with the carbon content, that is, as the iron lattice tetragonal distortion increases. This mobility contribution is apparently strong enough to offset the kinetic influence of the shorter wavelength of spinodal decomposition (Fig. 6) in the higher carbon alloy.

For the A2 stage of coarsening, the Lifshitz and Slyozov (Ref 46) and Wagner (Ref 47) (LSW) theory of diffusion-controlled particle coarsening leads to the following functional relationship between particle diameter, or, in this case, structural wavelength, and aging time:

$$\lambda^n - \lambda_0^n = k(t - t_0) \qquad \text{(Eq 4)}$$

where λ is the wavelength at time t, λ_0 is the wavelength at time t_0, and n and k are constants, with n being equal to $^1/_3$. Equation 4 assumes that the product phase is of constant composition. The atom-probe data for Fe-15Ni-1C martensite show that the compositions of the high-carbon regions begin to level off (stage A2) after about 175 hours. This time was taken as t_0, and the data for longer times in Fig. 9 were then fitted to a function of the form of Eq 4 by a standard nonlinear least-squares regression method. The wavelength exponent, n, was found to be 0.39 (standard deviation = 0.17). No atom-probe data are available for Fe-25Ni-0.4C, and therefore, t_0 was estimated from that for the Fe-15Ni-1C alloy by assuming that the proportionality between the times required to reach the peak electrical resistivity in the two alloys holds for t_0 as well. On this basis, t_0 for the lower carbon alloy is about 7 hours. The coarsening data for times greater than 7 hours were then fitted to Eq 4, yielding a wavelength exponent of 0.235 (standard deviation = 0.032). The results of this fitting exercise are illustrated by the solid curves in Fig. 9. The deviation of the experimental *n*'s (in opposite directions) from the ideal value of 0.33 is not well understood, but other experimental (Ref 19) and theoretical (Ref 48, 49) investigations have also noted deviations from the LSW value in systems undergoing continuous transformations under combined phase separation and coarsening conditions. As noted in previous studies (Ref 10), the apparent activation energy was found to increase continuously with time in this investigation. This behavior may be related to an overlapping of phase separation and coarsening, perhaps with somewhat different rate-controlling processes.

Substructural Effects

Some previous studies of lath martensites have indicated that the martensitic morphology may exert a substantial influence on aging and tempering behavior (Ref 50, 51), presuming that dislocations and lath boundaries may act as potent traps for carbon atoms, accommodating more than 0.1 wt% C. The present results, however, indicate that the formation of a modulated structure is the dominant aging process in virgin martensite. The modulated structure emerges *uniformly* throughout the entire martensitic phase; it is detected in regions containing high densities of lattice dislocations as well as in regions containing fine transformation twins. Weak-beam TEM was performed in an effort to detect carbon segregation at dislocations, but no evidence of carbon clusters or precipitates on these defects was observed. Although the present results do not rule out the segregation of some carbon atoms to substructural defects, it appears that most of the carbon atoms are involved in a spinodal partitioning into high- and low-carbon modulations at the aging times investigated here, at least when starting with virgin martensites.

Properties of Aged Martensites

The anomalous increase in electrical resistivity in the early stages of martensite aging has been reported in a number of investigations, (Ref 4, 10, 21, 22) as well as in studies of other alloy systems where Guinier-Preston zone formation precedes precipitation of the equilibrium phase (Ref 52, 53). The origin of the resistivity increase in Fe-C martensites remains unsettled, but the most plausible explanations invoke atomic displacements or coherency strains (Ref 21, 54). The appearance of a resistivity peak seems to represent a situation in which the resistivity decrease due to the formation of low-carbon ferrite is more than offset by the resistivity increase due to coherency strains associated with second-phase formation in the spinodal process. As illustrated by the Fe-1.8Mn-1.8C alloy (Fig. 17), very high carbon martensites do not exhibit any increase in resistivity during aging. The opposing effects of coherency strain and "solute removal" appear to nearly balance each other in this instance.

The electrical resistivity peaks measured in the Fe-Ni-C alloys of the present study are associated with the development of a modulated structure. Since peaks can be detected on aging well above room temperature, it is evident that spinodal decomposition can occur over a wide range of temperature, as further supported by the APFIM data in Table 3. In principle, a composition-dependent temperature exists above which phase separation can no longer occur by the spinodal mechanism. Attempts to locate the coherent spinodal for the alloys of this study were unsuccessful, primarily because tempering proceeds so rapidly above about 175 °C that carbide precipitation sets in, even during the shortest holding times practicable. Future experiments will require especially rapid heating and quenching techniques to reach high temperatures without venturing into the first stage of tempering.

As mentioned earlier, significant changes in hardness and flow strength have been observed in association with the aging of virgin martensites (Ref 4, 5). A microhardness study has recently been performed on highly homogeneous, rapidly solidified samples of the Fe-Ni-C alloys investigated here (Ref 55). While a subtle hardness increase that might be attributable to local carbon-atom rearrangements near dislocations is detected at fairly short aging times, significant hardness peaks are only observed in association with completion of the A1 aging stage, and the onset of the T1 tempering stage. A detailed account of these observations will be reported elsewhere.

The present TEM results on virgin martensite aging in Fe-Ni-C alloys do not differ qualitatively from previous findings on Fe-C (Ref 7), Fe-Mn-C (Ref 56), and Fe-Cr-C (Ref 16, 18) alloys. The development of structural modulations by spinodal decomposition thus appears to be a general feature of the aging behavior of Fe-X-C martensites, and may also occur in freshly quenched, commercially important alloys. To test this point, thin foils were prepared of as-quenched 4340 steel, and the modulated structure shown in Fig. 21 was revealed. Furthermore, recent studies of platelike Fe-Ni-C and Fe-Ni-Cr-C alloys (Ref 58, 59) indicate that aging promotes the formation of microcracks in the martensite phase, particularly at points of plate impingement. This raises the possibility that the brittleness of nominally untempered martensite may be a consequence of the aging reaction and is not necessarily an inherent property of virgin martensite. Additional work needs to be directed toward a thorough characterization of the carbon-rich phase which forms during martensite aging and its influence on mechanical properties, as well as on the carbide precipitation phenomena during subsequent tempering.

Conclusions

The current results lead to the following conclusions regarding the aging behavior of virgin Fe-Ni-C martensites:

1. On aging, a fine-scale modulation of carbon concentration develops along the martensite ⟨203⟩ directions. The initial compositional wavelength depends on the carbon content; it is about 2.5 and 1.5 nm in Fe-25Ni-0.4C and Fe-15Ni-1C martensites, respectively. The aligned nature of the resulting modulated structure and the increase in compositional amplitude with time indicate that these virgin martensites decompose by a spinodal mechanism. The Zener-ordered state of the virgin martensite, as well as evidence for "secondary" ordering within the high-carbon regions, suggests that a coherent conditional spinodal reaction takes place in Fe-C martensites during aging.
2. The composition of the high-carbon regions levels off at about 11 at.% C with prolonged room-temperature aging in the Fe-Ni-C martensites of this study. Such results, combined with a reassessment of available Mössbauer data, suggest that the high-carbon phase which forms spinodally during the aging of Fe-C martensites may be isomorphous with the α''-$Fe_{16}N_2$ phase that precipitates during the tempering of Fe-N martensites.
3. The near-{203} habit of the spinodally evolved carbon-rich phase minimizes the elastic strain energy of the high- and low-carbon mixture.
4. The kinetics of the early stages of aging are controlled by carbon diffusion. The rate of spinodal decomposition is slower with increasing carbon content of the martensite and decreasing carbon mobility.
5. The modulated structure begins to coarsen before phase separation is complete. A wavelength of about 10 nm is attained before the onset of carbide precipitation corresponding to the first stage of tempering.
6. The martensitic substructure does not appear to exert any substantial influence on the overall aging behavior of virgin martensites.

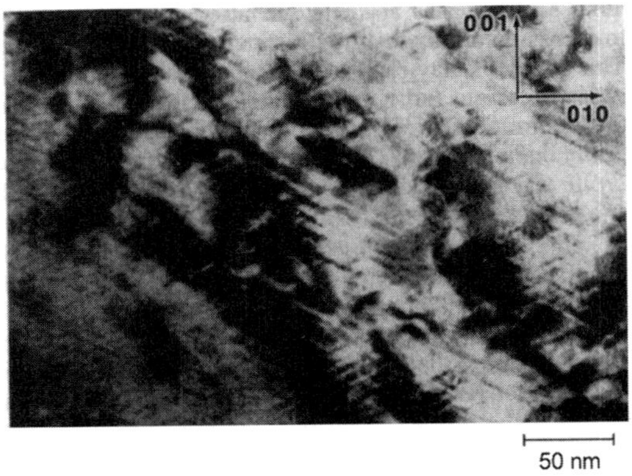

Fig. 21 Bright-field transmission electron micrograph of freshly quenched martensite in 4340 steel taken near the [100] orientation. The wavelength of the rather coarse modulated saucture is about 10 nm due to autotempering. Specimen furnished by Gore (Ref 27).

Appendix

Reassessment of Available Mössbauer Data on Fe-N and Fe-C Martensites

The Mössbauer effect, which provides information on the local environments of iron atoms through the perturbations created in their nuclear transitions by neighboring atoms, has been exploited by a number of investigators (Ref 9, 25, 31, 34–36, 60, 61) to deduce the arrangements of interstitial C and N atoms in virgin and aged martensites.

The most reliable Mössbauer parameter for which there is the greatest agreement among measured values is the hyperfine field, which provides information regarding the local magnetic moment of Fe atoms in their various environments. Due to the well-established structures of the phases in the Fe-N system, the interpretation of Mössbauer spectra of virgin and aged martensites is less ambiguous in this system. Table 4 summarizes the measured hyperfine fields for the various Fe environments in Fe-N martensites determined by DeCristofaro and Kaplow (Ref 34) and more recently by Rochegude and Foct. (Ref 36) In the notation of the latter reference based on the number of near-neighbor nitrogen atoms of a given Fe site, the sites I and I′ correspond to the first and second Fe neighbors of a N atom in a distorted octahedral interstice, while 0 designates Fe atom sites with no N near neighbors and II denotes Fe atoms with two N near neighbors. The phases in which the sites reside are indicated by subscripts in Table 4. Largely following the original interpretations of both studies, there is a clear correspondence among the known three, I, I′, and 0, environments of the established α'' structure produced on aging and the analogous three environments in the virgin martensite. (We have interpreted the fields as given in Table 3 of Ref 36 to increase in magnitude as follows: $0_{\alpha}<0_{\alpha'}<0_{\alpha''}$.) Also listed in Table 4 are 0_{α} sites corresponding to essentially nitrogen-free ferrite. Because of the similarity of the parameters for the α' virgin martensite and the α'' produced on aging, there is no need to invoke any nonoctahedral N interstitial occupancy in the virgin martensite. Formation of the γ'-Fe_4N phase generates two types of II sites (with 2 N neighbors) corresponding to positions on Fe_4N cube faces with normals parallel (||) and perpendicular ($\perp$) to the magnetization direction. In the perfect Fe_4N structure, only $II_{\gamma'}$ and $0_{\gamma'}$ sites exist, but sites corresponding to $I_{\gamma'}$ can be identified in the imperfect γ' phase. Compared to the 0_{α} site of ferrite, the relative hyperfine fields reflect a reduced moment at II sites with 2 N neighbors and a somewhat enhanced moment for 0 sites in the α', γ', and α'' structures (particularly for the α'') attributable to local lattice dilatation.

Table 4 Mössbauer hyperfine fields (in Tesla) for Fe-N martensites

	Site	Ref 34 12.0 at.% N	Ref 36 5.5 to 9.0 at.% N
Virgin martensite	$I_{\alpha'}$	31.6	30.8
	$I'_{\alpha'}$	34.6	33.3
	$0_{\alpha'}$	37.0	37.9
	(0_{α})	34.0	36.4
Aged martensite	$I_{\alpha''}$	28.8	30.0 to 30.3
	$I'_{\alpha''}$	31.4	32.8 to 32.5
	$0_{\alpha''}$	39.9	41.2 to 40.6
	(0_{α})	...	33.9 to 34.0
Fe_4N	$II_{\gamma'}(\parallel)$	21.9	23.5
	$II_{\gamma'}(\perp)$	21.6	23.6
	$I_{\gamma'}$	...	30.3
	$0_{\gamma'}$	34.1	37.0

A reinterpretation of Mössbauer hyperfine fields for Fe-C martensites, based on the behavior of the Fe-N system, is summarized in Table 5. The table incorporates the most recent precise measurements of Génin (Ref 35). In addition to the notation just employed in discussion of Fe-N, letter designations of spectral components previously employed in the literature (Ref 31) are listed for comparison. Following the clear correspondence of I, I′, and 0 sites in the α' and α'' structures of Fe-N, and in line with the original studies of Choo and Kaplow, (Ref 9) we interpret the Fe-C α' virgin martensite spectrum in terms of full octahedral C interstitial occupancy, defining the $I_{\alpha'}$, $I'_{\alpha'}$, and $0_{\alpha'}$ sites as in the Fe-N system. Compared to the Fe-N case, the relative hyperfine fields reflect a greater reduction of the magnetic moment at the I and I′ sites with a C near neighbor.

Table 5 Mössbauer hyperfine fields (in Tesla) for Fe-C martensites

	Site	Component (Ref 31)	Kaplow *et al.* (Ref 9, 34) 8.2 at.% C	Fujita *et al.* (Ref 60, 61) 7.9 at.% C	Ino *et al.* (Ref 31) 7.2 at.% C	Génin (Ref 35) 9.5 at.% C
Virgin martensite	$I_{\alpha'}$	C	27.9	27.5	...	28.3
	$I'_{\alpha'}$	B	33.1	32.0	33.0	33.1 to 33.4
	$0_{\alpha'}$	A	36.1	36.0	35.5	36.2
Aged martensite	$II_{\alpha''}$	F	...	...	18.0	16.5
	$I_{\alpha''}$	D	28.0	27.5	26.4	26.2
	$I'_{\alpha''}$	A′ (A″)	34.8	35.3	34.1	35.3
	$0_{\alpha''}$	E	38.6	36.0	37.1	37.6
	0_{α}	A″ (B)	...	32.0	31.5	32.4

Based on our APFIM measurements of a composition near (Fe, Ni)$_8$C for the high-carbon phase formed on aging, we have interpreted the hyperfine fields of the sites developed on aging in Table 5 in terms of an imperfect α''-$Fe_{16}C_2$ structure based on α''-$Fe_{16}N_2$. We thus regard the three sites identified by Choo and Kaplow as those of the α'' structure, recognizing that for the very high carbon alloy of these investigators, the volume fraction of the low-carbon regions would be small and may not have produced a separable contribution to their A′ spectrum, which we now attribute to $I'_{\alpha''}$. Compared to the Fe-N system, the relative hyperfine fields we ascribe to the α'' structure in Fe-C are consistent with stronger magnetic effects of C relative to N. The latter also implies that a $II_{\gamma'}$ site in Fe_4C would have a much lower magnetic moment than that proposed by Choo and Kaplow (Ref 9) for their γ'-Fe_4C model. This point has been borne out in preliminary electronic calculations on the Fe_4C structure by Krasko (Ref 62) using the Linear Muffin Tin Orbital-Atomic Spheres Approximation (LMTO-ASA) method. We propose instead that occasional $II_{\alpha''}$ sites with two C neighbors, resulting from imperfections of the α'' ordering, correspond to the low hyperfine field spectral component designated "F" in the most recent high-precision Mössbauer studies (Ref 31, 35). The relatively high moment enhancement of component "E," which we attribute to $0_{\alpha''}$, is also consistent with greater enhancement at 0 sites in the α'' compared to the γ' structure based on the Fe-N system. A lower perfection of the α'' structure formed in Fe-C compared to Fe-N would account for the larger number of spectral components in the aged Fe-C martensite as well as the relatively greater peak breadth of these components. This would also be consistent with the relatively sparse diffraction evidence for such C ordering, which may be of more short-range character. It is further suggested tentatively that a sometimes observed spectral component of the aged martensite with a slightly reduced hyperfine field, which we have designated 0_α in Table 5, may represent coherently strained regions of a low-carbon matrix with a decreased magnetic moment due to an elastic compression. This would then correspond to the regions of "negative tetragonality" associated with the "C peak" in the x-ray diffraction study of Chen and Winchell (Ref 8).

ACKNOWLEDGMENT

The authors gratefully acknowledge the financial support provided by the Office of Naval Research (United States) and the Science and Engineering Research Council (United Kingdom) throughout this investigation. K.A. Taylor and L. Chang also wish to thank the Rockwell International Corporation and the Republic of China (Taiwan) Ministry of Education, respectively, for their financial support. Helpful discussions relating to the interpretation of our results were held with Professor S.M. Allen and Drs. L. Kaufman, N. DeCristofaro, and J.R. Génin. Thanks are also due to Drs. G.T. Eldis, A.K. Sachdev, and S. Best for kindly supplying materials for this research.

REFERENCES

1. K.A. Taylor, G.B. Olson, M. Cohen, and J.B. Vander Sande: *Metall. Trans. A*, 1989, vol. 20A, p 2749–65
2. G.B. Olson and M. Cohen: *Metall. Trans. A*, 1983, vol. 14A, p 1057–65
3. G. Krauss: *Proc. Int. Conf. on Phase Transformations in Ferrous Alloys*, A.R. Marder and J.I. Goldstein, eds., TMS-AIME, 1984, p 101–23
4. P.G. Winchell and M. Cohen: *Trans. ASM*, 1962, vol. 55, p 347–61
5. G.T. Eldis and M. Cohen: *Metall. Trans. A*, 1983, vol. 14A, p 1007–12
6. V.I. Izotov and L.M. Utevskiy: *Phys. Met. Metallogr.*, 1968, vol. 25 (1), p 86–96
7. M. Kusunoki and S. Nagakura: *J. Appl. Crystallogr.*, 1981, vol. 14, p 329–36
8. P.C. Chen and P.G. Winchell: *Metall. Trans. A*, 1980, vol. 11A, p 1333–39
9. W.K. Choo and Roy Kaplow: *Acta Metall.*, 1973, vol. 21, p 725–32
10. A.M. Sherman, G.T. Eldis, and M. Cohen: *Metall. Trans. A*, 1983, vol. 14A, p 995–1005
11. M.K. Miller, P.A. Beaven, and G.D.W. Smith: *Metall. Trans. A*, 1981, vol. 12A, p 1197–1204
12. M.K. Miller, P.A. Beaven, S.S. Brenner, and G.D.W. Smith: *Metall. Trans. A*, 1983, vol. 14A, p 1021–24
13. L. Chang, A. Cerezo, G.D.W. Smith, M.K. Miller, M.G. Burke, S.S. Brenner, K.A. Taylor, T. Abe, and G.B. Olson: *J. Phys.*, 1984, vol. 45, p C9-409-C9-416

14. L. Chang, G.D.W. Smith, and G.B. Olson: *J. Phys.*, 1986, vol. 47, p C2-265-C2-275
15. S. Best: Ph.D. thesis, Massachusetts Institute of Technology, 1985
16. A.M. Sherman: Ph.D. thesis, Massachusetts Institute of Technology, 1972
17. A.K. Sachdev: Sc.D. thesis, Massachusetts Institute of Technology, 1977
18. B.P.J. Sandvik and C.M. Wayman: *Metallography*, 1983, vol. 16, p 429–47
19. P. Haasen: *Metall. Trans. A*, 1985, vol. 16A, p 1173–84
20. J. Piller and H. Wendt: *Proc. 20th Field Emission Symposium*, 1982, H.O. Andren and H. Norden, eds., Almqvist and Wiksell, 1983, p 265–74
21. H.W. King and S.G. Glover: *JISI*, 1959, vol. 193, p 123–32
22. V. Kerlins and C. Altstetter: *Trans. AIME*, 1963, vol. 227, p 94–98
23. G.V. Kurdjumov and A.G. Khachaturyan: *Acta Metall.*, 1975, vol. 23, p 1077–88
24. I.R. Entin, V.A. Somenkov, and S.Sh. Shil'shtein: *Sov. Phys. Dokl.*, 1973, vol. 17, p 1021–23
25. R. Kaplow, M. Ron, and N. DeCristofaro: *Metall. Trans. A*, 1983, vol. 14A, p 1135–45
26. C. Zener: *Trans. AIME*, 1946, vol. 167, p 550–83
27. K.A. Taylor, G.B. Olson, M. Cohen, and J.B. Vander Sande: *Metall. Trans. A*, 1989, vol. 20A, p 2739–47
28. A.G. Khachaturyan: *Theory of Structural Transformations in Solids*, John Wiley & Sons, 1983, p 244–49
29. A.G. Khachaturyan: *Theory of Structural Transformations in Solids*, John Wiley & Sons, 1983, p 191
30. M. Hayakawa, M. Tanigami, and M. Oka: *Metall. Trans. A*, 1985, vol. 16A, p 1745–50
31. H. Ino, T. Ito, S. Nasu, and U. Gonser: *Acta Metall.*, 1982, vol. 30, p 9–20
32. S. Nagakura and M. Toyoshima: *Trans. JIM*, 1979, vol. 20, p 100–10
33. P. Ferguson and K.H. Jack: *Tempering '81*, Metals Society, 1983, p 158–63
34. Nicholas DeCristofaro and Roy Kaplow: *Metall. Trans. A*, 1977, vol. 8A, p 35–44
35. J.-M.R. Génin: *Metall. Trans. A*, 1987, vol. 18A, p 1371–88
36. P. Rochegude and J. Foct: *Phys. Status Solidi A*, 1986, vol. 98, p 51–62
37. A. Sato, Y. Watanabe, and T. Mura: *J. Phys. Chem. Solids*, 1988, vol. 49, p 529–40
38. J.-M.R. Génin: *Metall. Trans. A*, 1988, vol. 19A, p 2901–9
39. S.B. Ren and S.T. Wang: *Metall. Trans. A*, 1988, vol. 19A, p 2427–32
40. A.G. Khachaturyan: *Theory of Structural Transformations in Solids*, John Wiley & Sons, 1983, p 530–41
41. J.W. Cahn: *Trans. AIME*, 1968, vol. 242, p 166–80
42. S.M. Allen and J.W. Cahn: *Acta Metall.*, 1976, vol. 24, p 425–37
43. J.W. Cahn and F. Larché: *Acta Metall.*, 1984, vol. 32, p 1915–23
44. R.P. Smith: *Trans. AIME*, 1962, vol. 224, p 105–11
45. M. Hillert: *Acta Metall.*, 1959, vol. 7, p 653–58
46. I. Lifshitz and V. Slyozov: *J. Phys. Chem. Solids*, 1961, vol. 19, p 35–50
47. C. Wagner: *Z. Elektrochem.*, 1961, vol. 65, p 581–91
48. J.S. Langer, M. Bar-on, and H.D. Miller: *Phys. Rev. A*, 1975, vol. 11, p 1417–29
49. J.S. Langer and A.J. Schwartz: *Phys. Rev. A*, 1980, vol. 21, p 948–58
50. G.R. Speich: *Trans. AIME*, 1969, vol. 245, p 2553–64
51. D. Kalish and M. Cohen: *Mater. Sci. Eng.*, 1970, vol. 6, p 156–66
52. J.M. Pelletier, G. Vigier, C. Mai, and R. Borrelly: *Acta Metall.*, 1983, vol. 31, p 1491–96
53. D. Turnbull, H.S. Rosenbaum, and H.N. Treaftis: *Acta Metall.*, 1960, vol. 8, p 277–95
54. D.W. Hoffman and M. Cohen: *Acta Metall.*, 1973, vol. 21, p 1215–23
55. S.E. Hartfield: S.M. thesis, Massachusetts Institute of Technology, 1988
56. Y. Tanaka and K. Shimizu: *Trans. JIM*, 1981, vol. 22, p 779–88
57. M. Gore: Ph.D. thesis research, Massachusetts Institute of Technology, 1985
58. T. Vilo and J. Pietikäinen: *Proc. Int. Conf. on Martensitic Transformations*, Cambridge, MA, 1979, p 721–26
59. J. Pietikäinen: *Trans. ISIJ*, 1985, vol. 25, p 340–44
60. F.E. Fujita, C. Shiga, T. Moriya, and H. Ino: *J. Jpn. Inst. Met.*, 1974, vol. 38, p 1030–37
61. F.E. Fujita: *Metall. Trans. A*, 1977, vol. 8A, p 1727–36
62. G. Krasko: Massachusetts Institute of Technology, unpublished research, 1986.
63. S. Nagakura, Y. Hirotsu, M. Kusunoki, T. Suzuki, and Y. Nakamura: *Metall. Trans. A*, 1983, vol. 14A, p 1025–31

PART II

MATERIALS DESIGN

Transformations—Selected Works of G.B. Olson on Materials, Microstructure, and Design
C.E. Campbell, M.V. Manuel, and W. Xiong, editors

Introduction to Part II—Materials Design

Charles Kuehmann

My first meeting with Professor Greg Olson was at Northwestern University in the fall of 1988. Greg was an energetic new addition to the department of Materials Science and Engineering with a vision to design new materials using emerging computational tools, akin to the computational revolutions that had fundamentally changed mechanical and aero engineering. At the time, I was a young aerospace engineer who had discovered that materials were a big opportunity to improve the systems I had been designing. I wanted to learn all I could about them, and Greg's approach was familiar and logical. For me, it was easy to underestimate the technical and cultural challenges that were barriers to realizing the vision of materials design, but as the body of work contained here demonstrates, Olson's dedication to this vision and his ability to enlist others to the cause overcame these challenges.

First among those challenges was to combine science and engineering, which had so long defined our field as separate endeavors, into a homogeneous whole. Using systems engineering to drive a goal-means paradigm of design within a hierarchical description of the process–structure–property–performance relationships of materials science, the performance-driven engineering of materials can be intimately combined with fundamental science that describes the individual connections in the process–structure and structure–property links. Chapters 10 through 12 in Part II of this work lay out this overall approach, and represent the first use of hierarchical flow-block "Olson" diagrams that follow the three-link chain paradigm for materials.

The second challenge for materials design was to derive science-based models that served the purpose of engineering rather than singularly scientific understanding. This meant that models needed to address necessary complexity yet be computationally efficient; focused on the goal to support design choices including component selection, composition, and process parameter optimization. Zhong, Freeman, Wu, and Olson (Chapter 13) demonstrate the ultimate level of fidelity in density functional theory calculations needed to address the complex mechanism for intergranular embrittlement. These calculations drove component selection for materials design in ultrahigh-strength secondary hardening steels as outlined in Chapter 14. Additional instances of design-driven science are presented in Chapter 15 and embedded in numerous materials design examples in Chapters 16 through 21. A corollary to the development of design-centered models was the development of experimental techniques targeting model calibration and validation (Chapters 22–24).

A critical hurdle for the wider adoption of materials design was demonstration through a variety of successful design examples. Exhibits are found in a compilation of a variety of steels (Chapters 16–19), shape-memory alloys (Chapter 20), and others (Chapter 21). Carrying these new materials through to end products and systems identified additional benefits of designed materials for qualification and validation. The

U.S. Defense Advanced Research Projects Agency (DARPA) Accelerated Insertion of Materials (AIM) program exploited the utility of design models, integrating new statistical methods to determine the variability of materials for design purposes significantly faster than could be accomplished by traditional testing alone (Chapters 22–25). Inherently stochastic properties of materials, such as fatigue and fracture, were addressed directly with mechanistic models to correlate microstructure distribution information with property variation (Chapters 25–26).

Based on these early successes, an initiative from the president's Office of Science and Technology Policy, called the Materials Genome, was established to embed computational materials into research across agencies and disciplines (Chapters 26–28). In the span of 30 years, materials design has developed from the vision of young researcher to a critical cornerstone of the field of materials science and engineering. Computationally designed materials are now in use in most aspects of our lives from transportation and energy to medicine and consumer products. More important, they are enabling many more advances in the future that will enhance our lives and even allow us to move beyond the confines of our planet.

The last hurdle to achieve these results is the education and training of a new generation of materials scientists and engineers with skills to fully integrate materials engineering into the design of all the products we make. Perhaps Greg's most enduring legacy will be his efforts to create the methods and principles to teach design to young materials engineers and scientists (Chapter 29).

Transformations—Selected Works of G.B. Olson on Materials, Microstructure, and Design
C.E. Campbell, M.V. Manuel, and W. Xiong, editors

CHAPTER 10

Computational Design of Hierarchically Structured Materials*

G.B. Olson, Northwestern University

Abstract

A systems approach that integrates processing, structure, property, and performance relations has been used in the conceptual design of multilevel-structured materials. For high-performance alloy steels, numerical implementation of materials science principles provides a hierarchy of computational models defining subsystem design parameters that are integrated, through computational thermodynamics, in the comprehensive design of materials as interactive systems. Designed properties combine strength, toughness, and resistance to impurity embrittlement. The methods have also been applied to nonferrous metals, ceramics, and polymers.

Introduction

For millennia, materials have been developed through the empirical correlation of processing and properties. The past century has seen the formation of a science of materials that has defined the structural basis of materials behavior, but its role has primarily been to explain the products of empiricism after their development. In the past decade, the numerical implementation of materials science principles and the integration of resulting computational capabilities within a systems engineering framework has given birth to a revolutionary approach (Ref 1) in the form of quantitative conceptual design of materials.

Materials as Systems

This approach to materials design is based on the philosophy of the late Cyril Stanley Smith (Ref 2, 3). Smith wrote extensively about interactive structural hierarchy in materials (and space-filling aggregates in all branches of science, including geology and biology). He envisioned a multilevel structure with strong interactions among levels, with an inevitable interplay of perfection and imperfection, and a duality of description in which structure can be equivalently regarded in terms of space-filling units or of the array of interfaces that bound them. This view of materials admits a necessary complexity.

In the modern form of materials science, Zener (Ref 4) has added a recognition of the dynamic nature of materials to Smith's spatial hierarchy. In association with a hierarchy of length scales, there is a spectrum of characteristic relaxation times (and therefore a spatio-temporal hierarchy), so that in any real structure, there is some level (such as the interatomic level or the grain boundary level) that has not had time to equilibrate. Thus, real structures are nonequilibrium and are therefore

*G.B. Olson, "Computational Design of Hierarchically Structured Materials," *Science*, Vol. 277, No. 5330 (1997) p 1237–1242. doi: 10.1126/science.277.5330.1237

path- (or history-) dependent, and recognizing this intrinsic dynamic nature further raises the essential complexity of materials.

Complexity sets natural limits on the degree of predictability. It is often argued that the complexity of materials makes them undesignable. But this would be so only if design required total predictability. Fortunately, another principle can be applied, one that Cohen (Ref 5) has termed "reciprocity," which can be illustrated by the example of structure-property relations. Although properties are typically regarded as being controlled by structure, Cohen argues that structure can be equally regarded as being controlled by properties, in that the perception of structure is governed by the properties that need to be understood. If a complex structure is examined from the viewpoint of specific properties, useful relations can be established. Much of materials science is the art of discriminating the essential from the nonessential (the latter being similar to the evolutionary vestiges of the biological world) as the products of empirical development are unraveled to control desired properties. Reciprocity then allows scientific analysis to provide the tools for materials design.

Although the powerful simplifying methods of scientific analysis provide the raw ingredients for design, these methods do not integrate results so that new complexity can be created and controlled. For this purpose, engineering has developed the systems approach. A concise summary of the approach, which is used in a materials design course at Northwestern University (Ref 6), is given in a review paper by Jenkins (Ref 7) of the Open University. System analysis begins with problem identification, organization of an appropriate interdisciplinary team, formulation of system design objectives from the function of the system in the wider system it serves (adopting a user-centered perspective), and identification of component subsystems and their interactions, typically represented in a flow-block diagram. Design synthesis starts with development and validation of appropriate (purposeful) models for subsystems and their interactions, assigning priorities and needed accuracy from the context of the design problem, followed by model integration to generate candidate design solutions offering satisfactory compromise among conflicting objectives. The implementation of prototypes then allows their experimental evaluation at the level of the models that created them, providing feedback for iterative reanalysis and design until objectives are met. Specifications can then be set for operation of the designed system.

In the context of materials, the prevalent practice of empirical development involves minimal up-front theoretical analysis and a large amount of parallel (and relatively superficial) evaluation of prototypes that leads to empirical correlations that produce materials with limited predictability of behavior. In an age of increasing cost of experiment and decreasing cost (and increasing power) of computation-based theory, a design approach making maximum use of science-based mechanistic models and the sequential, deeper evaluation of a small number of prototypes can not only reduce the time and cost of initial development but produce designed materials with more predictable behavior. This approach can also reduce the time and cost of process scale-up and material qualification.

Founded in 1985 with initial National Science Foundation (NSF) support, the Steel Research Group (SRG) (Ref 1), an international effort that includes several labs from industry, academia, and government, has adapted such a systems methodology to the science-based design of materials for the production of high-performance alloy steels. The research has integrated physical and process metallurgy, ceramics, applied mechanics, quantum physics, chemistry, mechanical engineering, and management science. Projects have investigated ultra-high-strength martensitic alloy steels, high-strength formable automotive sheet steels, and ferritic superalloys for power-generating turbine applications. The examples discussed here are drawn primarily from the largest project on martensitic alloy steels, which has more recently led to a project focused on high-performance gear and bearing steels. This class of steels undergoes a diffusionless (martensitic) structural transformation during quenching from high temperatures, to provide a fine microstructure offering the best combination of strength and toughness.

The materials property objectives motivating the research were developed with a property cross-plot approach, as generalized by Ashby (Ref 8) to quantify property-performance relations in a broad methodology for materials selection. The exercise defined combinations of strength (resistance to permanent deformation), toughness (fracture resistance), and resistance to environmental hydrogen (H) cracking that would allow a major advance in the usable strength level of structural steels, an advance that was recognized by industry participants as being beyond the reach of empirical development in the planned time frame.

Four primary elements are critical in materials science and engineering: processing, structure, properties, and performance (Ref 9, 10). There is no general agreement, however, on how these elements are interconnected. SRG research has found the linear structure shown in Fig. 1 to be crucial for systematic design. In the spirit of Smith's structural duality, Fig. 1 emphasizes that these elements form a three-link chain. Expanding Cohen's reciprocity, the structure offers a resonant bond between the science and engineering of materials, in which the deductive cause-and-effect logic of science flows to the right, while the inductive goal-means relations of engineering flow to the left. Further support for the utility of this paradigm is offered by its direct correspondence to the general axiomatic design approach developed by Suh and Albano (Ref 11, 12) to apply across all engineering disciplines.

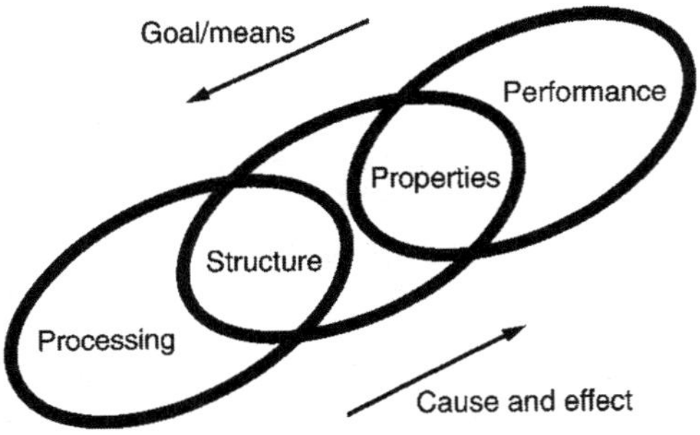

Fig. 1 Three-link chain model of the central paradigm of materials science and engineering.

Once a set of property objectives has been deduced from property-performance relations, the chain of Fig. 1 can serve as a backbone to which the addition of Smith's hierarchy can provide a first-order representation of a full system structure. The product of such an exercise as applied to the system structure of a high-performance alloy steel in SRG research (Ref 1) is represented in Fig. 2. The chart denotes the microstructural subsystems controlling the properties of interest, and the substages of processing (represented by a vertical process flow chart) governing the evolution of each subsystem. This representation of the full system was used to identify and prioritize the key structure-property and process-structure links to be quantified by the basic research of the SRG program.

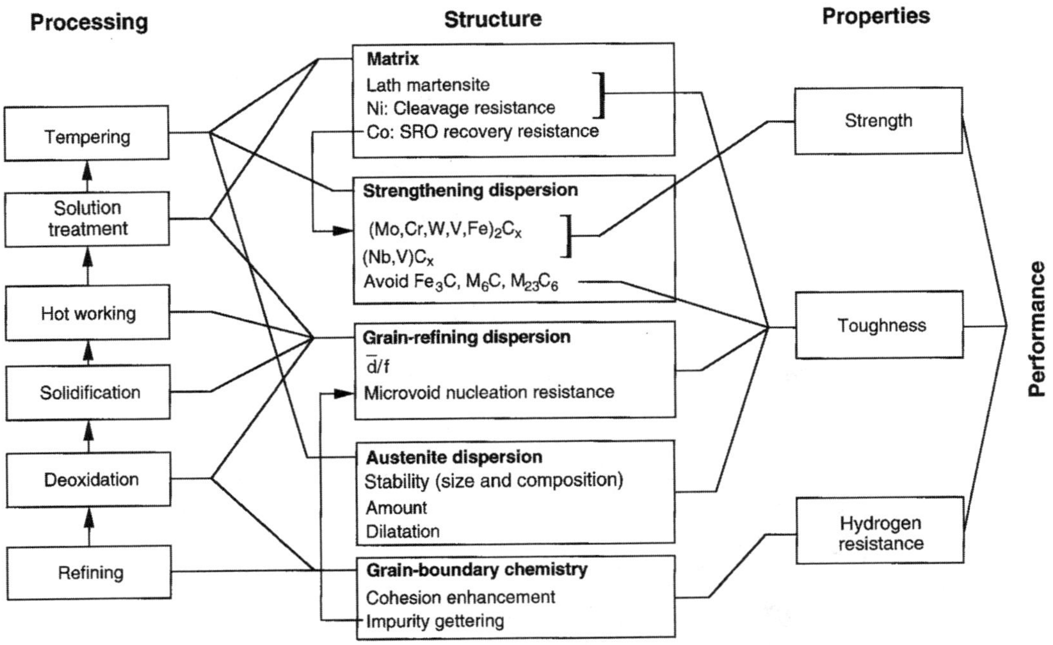

Fig. 2 Materials system chart for high-performance alloy steel. Source: Ref 1

Computational design of hierarchical structure requires a hierarchy of design models. Fig. 3 represents the computational models developed from research on the primary microstructural levels of Fig. 2. The experimental techniques used to create and validate these models are shown on the left; acronyms summarized on the right denote specific models and their software platforms.

The primary design tool used in this research for integrating the output of subsystem models was the THERMOCALC (TC) thermochemical database and software system (Ref 13) developed at the Royal Institute of Technology in Stockholm. Specifying subsystem requirements in terms of thermodynamic parameters, the flexible TC system is used to solve for complete alloy compositions that are capable of achieving desired microstructures under prescribed processing conditions. Recognizing the dynamic nonequilibrium nature of real microstructures, it should be emphasized that the thermodynamic parameters of interest rarely concern equilibrium states, but rather specify length scales and time scales of evolving metastable (or unstable) states. A remarkable degree of control of dynamic systems can be achieved through control of the thermodynamic forces that drive them.

Subsystem Modeling

The development of science-based computational subsystem models through focused basic research is reviewed briefly below.

Strength Subsystems. As denoted at the highest structural levels in Fig. 2 and 3, a primary consideration in strengthening is control of the structural transformation during quench hardening of an ultra-high-strength steel. The desired class of microstructure (denoted "lath martensite" in Fig. 2 and depicted at the top of Fig. 3) requires a diffusionless martensitic structural transformation at 200 °C or above. After refining the TC thermodynamic database, the development of a kinetic parameter database based on current transformation theory (Ref 14) provided the required computational model (MART) to predict transformation temperatures with required precision. Model predictions are validated by metallurgical quenching dilatometry (MQD), differential scanning calorimetry (DSC), light microscopy (LM), and transmission electron microscopy (TEM). Other transformation design codes shown in Fig. 3 (CASSIS and MAP) treat the kinetic competition with other structural transformations in the case of lower alloy steels.

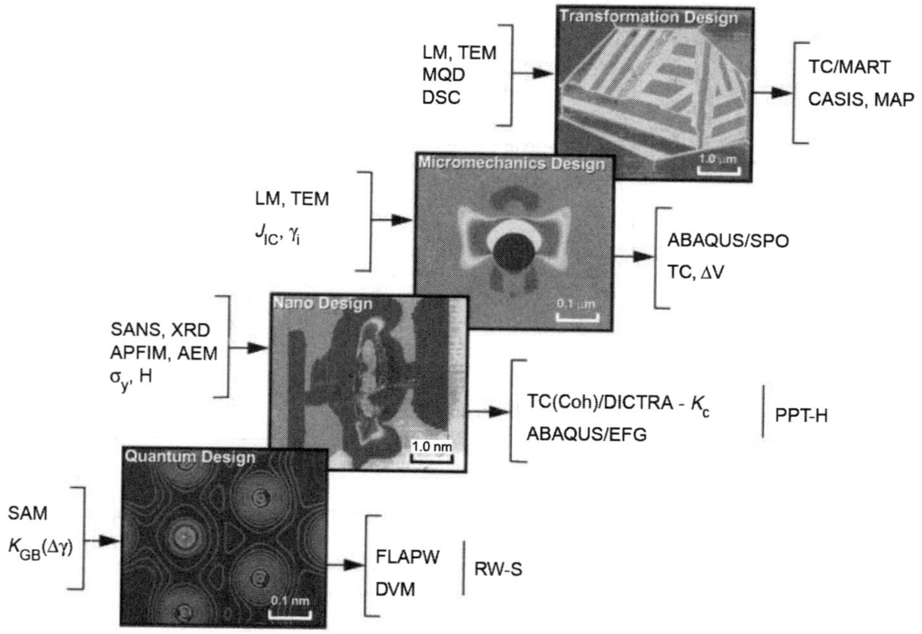

Fig. 3 Hierarchy of computational design models and the experimental tools used to create and validate them. Abbreviations are defined in the text.

The second structural level shown in Fig. 2 represents the final stage of strengthening by solid-state precipitation of alloy carbides during the last stage of heat treatment (Ref 15). The precipitation of ultrafine carbides corresponds to the "nano design" level in Fig. 3, represented by a model computation of the chemical composition field in the Fe-base matrix surrounding an ellipsoidal nanometer-scale carbide particle (Ref 16). At such small length scales, suppression of conventional structural relaxation processes promotes the continuity of crystal planes across the particle-matrix interface, causing extreme elastic distortion, and interfacial energy makes a dominant contribution to the thermodynamics governing particle size. Measurements of particle size by small-angle neutron scattering (Ref 17) (SANS), elastic distortion by x-ray diffraction (XRD), particle composition by atom-probe field-ion microscopy (Ref 18) (APFIM) and analytical electron microscopy (AEM), and calculations of elastic energies from continuum mechanics methods (Ref 16) (ABAQUS/EFG) are integrated with the TC thermodynamics (TC (Coh)) to deduce interfacial energies. Precipitation rate constants (K_c) are predicted (Ref 19) by use of the DICTRA diffusion software and database. These results are combined to generate a comprehensive model (PPT-H) for the evolution of alloy strength σ_y (controlled by particle size and spacing) during precipitation hardening. The resulting accuracy of particle size control has allowed the design of alloy compositions with efficient strengthening dispersions, achieving 50% greater strength at a given alloy carbon content.

Toughness Subsystems. A primary concern in fracture resistance is maintenance of a sufficiently fine polycrystalline grain size in order to inhibit competing brittle fracture mechanisms that allow fracture with less energy absorption. The grain size established during high-temperature heat treatments is determined by the geometry of a stable "grain-refining" particle dispersion, denoted by the third structural level of Fig. 2, which impedes the high-temperature grain boundary motion that is responsible for normal grain coarsening. Once sufficient grain refinement is achieved to promote fracture by higher toughness ductile-fracture processes, the actual level of fracture toughness becomes limited by the formation of microvoids, which occurs through interfacial separation at the grain-refining particles themselves (Ref 20). Simulation of such processes uses traditional continuum mechanics methods, as denoted by the micro-mechanics design level in Fig. 3. Such simulations (Ref 21, 22) have demonstrated the mechanism by which microvoids drive ductile fracture and have quantified the role of dispersion geometry (including average particle size d and volume fraction f) and particle interfacial properties as the basis for thermodynamic component selection and processing optimization in materials design (Ref 23). Experimental validation of models employs measurements of fracture energy absorption (J_{IC}) and critical plastic strain for strain localization (γ_i) in shear tests.

As denoted by the fourth structural level of Fig. 2, further resistance to microvoid fracture can be obtained through mechanical interaction with a fine dispersion of transformable particles (precipitated during final heat treatment), which can undergo a strain-induced structural transformation during ductile fracture (Ref 24). Using a matrix flow model (ABAQUS/SPO) (Ref 25) based on the kinetics of such transformations, the micromechanics simulation in Fig. 3 represents the computed contours of the fraction of the transformed phase produced during microvoid formation in such a material (Ref 26). The simulations demonstrate the role of pressure-sensitive transformation kinetics in stabilizing plastic flow during microvoid growth, and they define guidelines for optimizing transformation thermodynamic stability and dilatancy (ΔV). Based on TC thermodynamic modeling incorporating these guidelines, the dark bands denoted "TT" in Fig. 4 depict the toughness-hardness combinations achieved through multistep heat treatments controlling microstructure to exploit such transformation toughening. Properties achieved in the new commercial Aer-Met100 and an experimental alloy designated MTL1 (Ref 27) lie within the original SRG objectives box denoted by the dashed lines.

Embrittlement Resistance Subsystems. Denoted by the last structural level in Fig. 2, environmental cracking of ultra-high-strength steels occurs by an intergranular mechanism associated with the combined effects of environmental H and the prior segregation of embrittling impurities (Ref 28). Understanding its underlying mechanism has required the most fundamental and interdisciplinary research of the SRG effort, combining applied

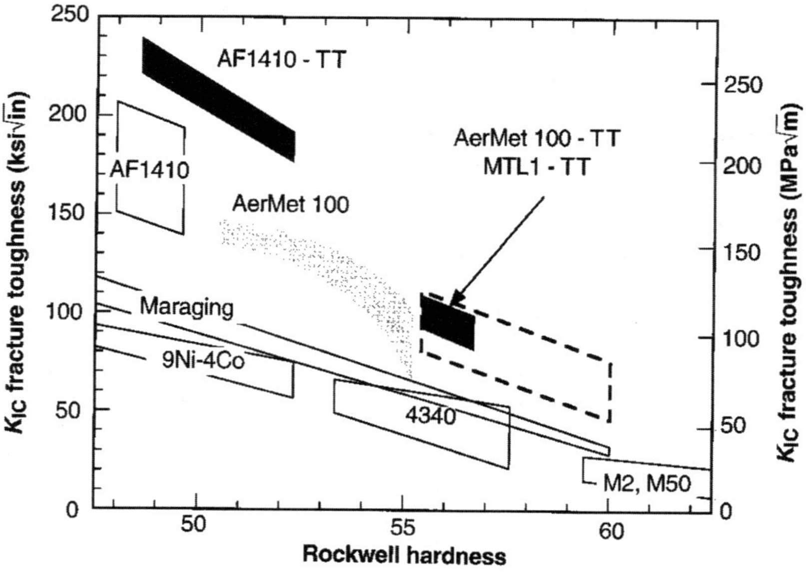

Fig. 4 Toughness-hardness plot of various steels. Dark bands for steels with TT designation show property improvement through multistep tempering to provide transformation toughening. K_{IC} fracture toughness (the critical stress-intensity for crack propagation) is given in both SI units (MPA$\sqrt{m}$) and English units (ksi$\sqrt{in}$, where ksi denotes kilopounds per square inch). Hardness (a measure of compressive strength) is measured on the Rockwell C (R_c) hardness scale.

mechanics, materials science, and quantum physics.

A crucial contribution has come from the work of Rice and Wang (Ref 29) in modeling the mechanics and thermodynamics of interfacial separation. A key prediction is that the embrittlement potency of a segregating solute should scale with the difference in its energies of segregation to the free surface (FS) versus a grain boundary (GB). Reported data for embrittlement potency in steels, measured as the shift in ductile-brittle transition temperature per amount of segregant (K/atomic%, with atomic% measured by scanning auger micro-analysis or SAM), shows a compelling correlation with this segregation energy difference, based on available thermodynamic data (Ref 29).

Based on this thermodynamic description of intergranular embrittlement, a series of electronic-level total energy calculations has used both the full-potential linear augmented plane wave (FLAPW) method (Ref 30) and the cluster discrete variation method (DVM), as represented by the valence charge-density contour plot denoted "quantum design" in Fig. 3. Although materials science has contributed the basic atomic structural models to support such calculations, a pivotal development has been the ability to compute interatomic forces within the FLAPW code to allow precise detailed atomic relaxations within the same method. The latest results (Ref 31, 32) are summarized in Fig. 5, which plots the experimental embrittlement potencies against the quantum mechanical theoretical predictions of the segregation energy difference ΔE, with and without the prior segregation of a monolayer of Mn, representing the most common alloying element in steels. The theoretical thermodynamics show a stronger correlation with embrittlement than the available experimental thermodynamics, particularly when Mn is taken into account. Correlation of measured grain boundary (GB) toughness (K_{GB}) with the ideal work of interfacial separation ($\Delta\gamma$) based on the FLAPW predictions has provided a modified Rice-Wang model (RW-S) for the design of grain boundary cohesion.

More important than the ability to determine these key energy differences is the ability to establish their underlying electronic basis. Computed charge-transfer plots, together with density-of-states curves and detailed electronic orbital plots, demonstrate that embrittling P and S undergo a nonhybridized electrostatic interaction with Fe, which is more adaptive to the FS environment, whereas the cohesion-enhancing B and C exhibit anisotropic hybridized bonding across the GB. Such calculations

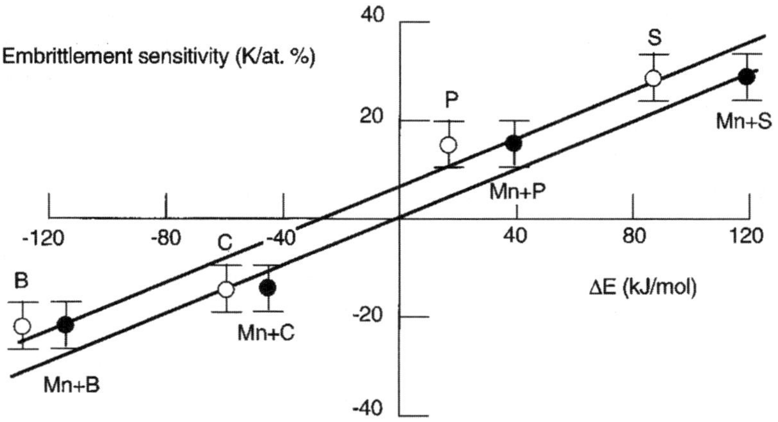

Fig. 5 Measured segregant embrittlement sensitivity in steels plotted versus the FLAPW prediction of difference (ΔE) in segregation energy to grain boundary and free surface environments, with and without prior monolayer segregation of Mn. Source: Ref 31, 32

further reveal that the embrittling effect of Mn is associated with promotion of in-plane bonding, which preferentially stabilizes the FS.

The most recent improvements in precision have now made possible a definitive calculation of the effect of H on Fe GB cohesion (Ref 33). In contrast to the hybridized versus electrostatic interactions displayed by the segregants shown in Fig. 5, H shows a third class of behavior that is more ionic in character. The underlying energetics of H embrittlement are associated with enhanced charge transfer from Fe to H in the FS environment. The intrinsic embrittlement potency per atom is comparable to that of P, which is consistent with experimental estimates (Ref 34), whereas it is well established that its effective potency is greatly amplified by its mobility (Ref 34). The small magnitude of this intrinsic potency, and the determination of its origin in charge transfer, offer the hope for its reversal or cancellation by predictive alloying.

The insights provided by these new predictive capabilities promise a new generation of "quantum steels" in which, analogous to semiconductors, boundaries are deliberately doped to attain desired electronic structures for enhanced intrinsic cohesion and altered interaction with impurities, including H.

Although the electronic-level approach offers the greatest potential for improvements, significant advances in environmental cracking resistance have already been demonstrated under SRG research through thermodynamics-based design of novel impurity-gettering phases such as lanthanum phosphate ($LaPO_4$), which, when accessed by appropriate processing, remove embrittling impurities from the grain boundaries (Ref 35, 36).

Design Examples

An effort to integrate these developing principles into the comprehensive design of a new alloy was undertaken by a team of students in the first materials design class at Northwestern University in 1989. The problem adopted was the design of a secondary-hardening stainless steel for use in bearings in the fuel and oxidizer turbopumps of the space shuttle main engine (SSME). The property objectives obtained from NASA and Rocketdyne engineers were Rockwell hardness R_c60 for wear and contact fatigue resistance, with a doubling of toughness and H resistance as compared with the current 440C alloy. Adopting the system structure of Fig. 2, composition was constrained for $LaPO_4$ impurity gettering for enhanced H resistance, while the matrix was constrained to contain a minimum of 12 weight % Cr for corrosion resistance. Estimating 0.30 weight % C as sufficient to maintain R_c60, based on the strengthening models, a line of Ni and Co compositions was computed to maintain a sufficiently high transformation temperature to yield a martensitic structure. Along this line, the computed thermodynamic stability and amount of transformation-toughening particles provided a unique Ni and Co content. With these composition variables fixed, Mo and V concentrations were optimized for the thermodynamics of precipitation strengthening, constrained by solubility limits at solution treatment temperatures. Under a grant from NASA, a prototype alloy was evaluated, and it demonstrated the desired doubling of fracture toughness at the required hardness level (Ref 37).

A related area of long-term interest has been the design of a new class of alloy steels for case-hardening gear and bearing applications. Research in this area was initiated with a property-performance analysis by a team of senior project students in mechanical engineering, who identified property objectives to allow a 50% reduction in gear weight. A team of students in materials design class then performed conceptual designs aimed at achieving these objectives by using recent strengthening models. Figure 6 depicts the increased hardness profile so far obtained in prototype gear steels (Ref 38). The fatigue properties are currently under evaluation, with promising preliminary results. A related custom gear steel is undergoing gear testing by the Newman-Haas Indycar racing team.

The range of design projects considered by the most recent materials design class is wide-ranging and includes not only alloy steel projects derived from SRG research but also nonferrous projects that test the generality of the design methodology and also explore the level of conceptual design that can be practiced without support from a major research project. This has included projects in ceramics and polymers. The ceramic project has assessed the feasibility of achieving processing and property requirements for dental fillings using Portland cement-based hydrate ceramic composites as a cost-effective replacement for mercury amalgams, and has so far included TC representation of metastable phase relations in the hydration of calcium silicates. The polymer project has explored the transfer of the gradient system technology of steel gears to case-hardening of plastics. Rubber-toughened epoxy systems have been identified as being most promising for achieving hardness gradients, controlled residual stress, and core toughness. The TC modeling of phase separation behavior during curing has aided process models for achieving desired phase fraction gradients.

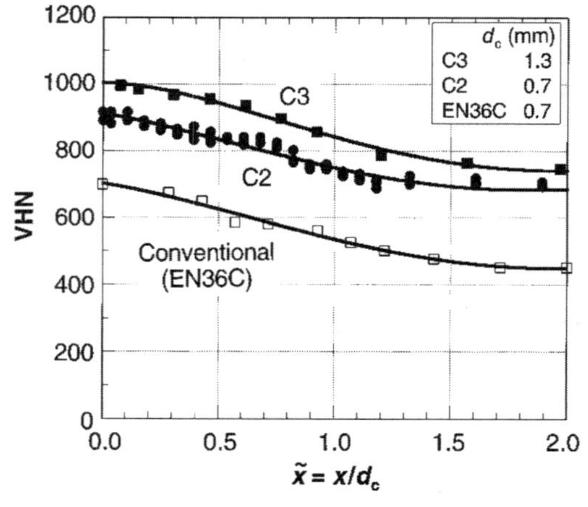

Fig. 6 Hardness profile for carburized prototype gear steels (C3 and C2) compared with conventional (EN36C) gear steel (Ref 38). Hardness is measured on the Vickers (VHN) scale. Distance x is normalized to the effective case depth d_c.

The most ambitious design project has been the "Terminator 3" self-healing, biomimetic, smart steel composite. A number of biomimetic concepts have been combined, starting with natural seashell, tooth, and bone architectures. These structures typically involve the reinforcement of a brittle ceramic by a rubbery polymeric component to provide "crack bridge" toughening in which rubbery ligaments stretch across cracks. More sophisticated adaptive behavior is exhibited by viruses and bacteria, which exploit martensitic structural transformations in cylindrical protein crystals to use reversible strain phenomena such as superelasticity and shape-memory effects (Ref 39).

A system integrating these concepts in an all-metallic composite consists of a precipitation-strengthened high-temperature superalloy reinforced by a thermodynamically compatible, precipitation-strengthened, shape memory alloy. The low-temperature brittleness of the superalloy is compensated for by crack bridging through martensitic transformation superelasticity that is analogous to the rubbery polymer reinforcement of the seashell system. After bridging of low-temperature damage, when the superalloy is returned to its operating temperature, the bridging ligaments contract by means of the shape memory effect to clamp the cracks closed. If the memory alloy has sufficiently greater strength, the clamping force can be maintained to promote diffusional rewelding of the damage, thus providing the ultimate biomimetic property of self healing.

Thermodynamic compatibility of the pair of two-phase alloys requires a four-phase equilibrium at operating temperatures and a two-phase equilibrium during solution treatment. A preliminary thermodynamic feasibility analysis, including assessment of memory alloy stability requirements, was performed by a team of juniors in materials design class. Continued evaluation (Ref 40) has included a test of mechanical concepts that uses a TiNi-reinforced Sn alloy composite prototype to demonstrate both macroscopic strain reversal and the desired crack-clamping behavior (Ref 41). Precise multicomponent phase relations for the Fe-based system have been evaluated with diffusion couple experiments, and prototype steel composites are being fabricated.

The success of these initial designs suggests that the integration of computational materials science within a systems engineering framework offers a powerful new approach for the creation of superior materials that have sophisticated control of a multi-level dynamic structure, combined with reduced time and cost of materials development. These first steps herald a new synergy of the science and engineering of materials.

ACKNOWLEDGMENTS

The Northwestern component of the SRG program has been sponsored by the Office of Naval Research, the Army Research Office, NSF, NASA, the U.S. Department of Energy, the Electric Power Research Institute, and the Air Force Office of Scientific Research, with industry gifts and fellowship support.

REFERENCES

1. G.B. Olson, M. Azrin, and E.S. Wright, eds., *Innovations in Ultrahigh-Strength Steel Technology* (GPO, 1990)
2. C.S. Smith, *A Search for Structure* (MIT Press, 1981)
3. C.S. Smith, in *Martensite*, G.B. Olson and W.S. Owen, eds. (ASM International, 1992), p 21–39
4. C. Zener, *Elasticity and Anelasticity of Metals* (Univ. of Chicago Press, 1948)
5. M. Cohen, *Mat. Sci. Eng.* (1976)
6. G.B. Olson, in *M.E. Fine Symposium*, P.K. Liaw, J.R. Weertman, H.L. Markus, and J.S. Santner, eds. (Minerals, Metals & Materials Society of AIME, 1991), p 41
7. G.M. Jenkins, in *Systems Behaviour*, J. Beishon and G. Peters, eds. (Harper and Row, for Open University Press, 1972), p 56–82
8. M.F. Ashby, *Materials Selection in Mechanical Design* (Pergamon, 1992)
9. COSMAT Summary Report, *Materials and Man's Needs* (National Academy of Sciences, National Academy Press, 1974)
10. *Materials Science and Engineering for the 1990s* (National Research Council, National Academy Press, 1989)
11. N.P. Suh, *The Principles of Design* (Oxford Univ. Press, 1990)
12. L.D. Albano and N.P. Suh, *Res. Eng. Des.* 4, 171 (1992)
13. B. Sundman, B. Jansson, and J.O. Andersson, *CALPHAD* 9, 153 (1985)
14. G. Ghosh and G. B. Olson, *Acta Metall. Mater.* 42, 3361 (1994)
15. G.R. Speich, in *Innovations in Ultrahigh-Strength Steel Technology*, G.B. Olson, M. Azrin, and E.S. Wright, eds. (GPO, 1990), p 89–111
16. K.C. King, P.W. Voorhees, and G.B. Olson, *Metall. Trans.* 22A, 2199 (1991)
17. A.J. Allen, D. Gavillet, and J.R. Weertman, *Acta Metall.* 41, 1869 (1993)
18. G.B. Olson, T.J. Kinkus, and J.S. Montgomery, *Surf. Sci.* 246, 238 (1991)
19. A. Umantsev and G.B. Olson, *Scr. Metall.* 29, 1135 (1993)
20. J.G. Cowie, M. Azrin, and G.B. Olson, *Metall. Trans. A*, 143 (1989)
21. A. Needleman, in *Innovations in Ultrahigh-Strength Steel Technology*, G.B. Olson, M. Azrin, and E.S. Wright, eds. (GPO, 1990), p 331–346
22. Y. Huang and J. W. Hutchinson, in *Modelling of Material Behavior & Design*, J.D. Embury, ed. (TMS-AIME, 1990), p 129–148
23. M.J. Gore, G.B. Olson, and M. Cohen, in *Innovations in Ultrahigh-Strength Steel Technology*, G.B. Olson, M. Azrin, and E.S. Wright, eds. (GPO, 1990), p 425–441
24. G.B. Olson, *J. Phys. V*, Colloque C1, 407 (1996)
25. R.G. Stringfellow, D.M. Parks, and G.B. Olson, *Acta Metall.* 1703 (1992)
26. S. Socrate, thesis, Massachusetts Institute of Technology (1996)
27. C.J. Kuehmann, J. Cho, T.A. Stephenson, and G.B. Olson, in *Metallic Materials for Lightweight Applications*, E.B. Kula and M.G.H. Wells, eds. (GPO, 1994), p 337–355

28. C.J. McMahon Jr., in *Innovations in Ultrahigh-Strength Steel Technology*, G.B. Olson, M. Azrin, and E.S. Wright, eds. (GPO, 1990), p 597–618
29. J.R. Rice and J.-S. Wang, *Mater. Sci. Eng. A*, 23 (1989)
30. R. Wu, A.J. Freeman, and G.B. Olson, *Science*, 376 (1994)
31. R. Wu, A.J. Freeman, G.B. Olson, *Phys. Rev. B*, 7504 (1996)
32. L. Zhong, R. Wu, A.J. Freeman, and G.B. Olson, *Phys. Rev. B*, 11 135 (1997)
33. L. Zhong, R. Wu, A.J. Freeman, G.B. Olson, unpublished results
34. P.M. Anderson, J.S. Wang, and J.R. Rice, in *Innovations in Ultrahigh-Strength Steel Technology*, G.B. Olson, M. Azrin, and E.S. Wright, eds. (GPO, 1990), p 619–649
35. J.F. Watton, G.B. Olson, and M. Cohen, in *Innovations in Ultrahigh-Strength Steel Technology*, G.B. Olson, M. Azrin, and E. S. Wright, eds. (GPO, 1990), p 705–737
36. G. Ghosh and G.B. Olson, in *Proceedings of 51st Annual Meeting of the Microscopy Society of America*, G.W. Bailey and C.L Rieder, eds. (San Francisco Press, 1993)
37. T.A. Stephenson, C.E. Campbell, and G.B. Olson, *Advanced Earth-to-Orbit Propulsion Technology 1992*, R.J. Richmond and S.T. Wu, eds., *NASA Conf. Publ. 3174* (1992), vol. 2, p 299–307
38. J. Wise, unpublished results
39. G.B. Olson and H. Hartman, *J. Physique* C4–855 (1982)
40. B. Files, "Design of a biomimetic self-healing superalloy composite," thesis, Northwestern University (1997)
41. G.B. Olson, K.C. Hsieh, and H.K.D.H. Bhadeshia, in *Microstructures LCS '94* (Iron and Steel Institute of Japan, Tokyo, 1994)

CHAPTER 11

Overview—Science of Steel*

G.B. Olson, Northwestern University

Introduction—Steels and Materials Design

Steel is the frontier of materials science. More is known about steels than any other class of materials. Paradoxically, despite this position in fundamental understanding, the actual process by which new steels are still developed amounts to the same empiricism which is applied to newer materials for which the alternative of science-based design is not an option. In the arena of empiricism the maturity of steels poses a severe disadvantage in that the "soft ground" has already been explored and empirical development has reached a point of diminishing returns. To compete with newer materials for which traditional empiricism can bring more rapid improvements, there is a pressing need for a better way to develop new steels. Moreover, the ultimate goal of materials science must certainly be the design of superior materials from basic understanding; our very limited ability to truly design materials calls into question the viability of this discipline. This circumstance makes the science of steel a most timely subject: no material is in a better position to be designed from scientific knowledge at the same time that no material is in more need of this approach.

The research reviewed here is based primarily on the Steel Research Group (SRG) program, a multi-institutional university/government/industry interdisciplinary effort directed at the scientific principles for the design of new classes of steels, motivated by specific property objectives of importance to industry. Two key classes of steels have been selected. The major effort reviewed here ("A" Program) is directed at ultrahigh-strength martensitic steels for advanced structural applications, while a smaller effort ("B" Program) is directed at high-strength ferrite-based microalloyed steels for high formability automotive sheet applications. Objectives have been deliberately chosen to represent a sufficiently ambitious advance to require a more fundamental approach to materials design than ever previously attempted.

Property Objectives

The performance of a material in service is controlled by a set of properties. Surprisingly, our ability to specify the long list of properties actually required for a given application can be the weakest link in the materials design process. In searching for a model system in which to develop a new approach to design, it is helpful to examine the most demanding applications for which the list of properties one can reasonably request is relatively short. Ultrahigh-strength steels for advanced structural applications are an excellent example where primary objectives can be expressed in terms of the three

*G.B. Olson, "Overview—Science of Steel," in G.B. Olson, M. Azrin, and E.S. Wright, eds., *Innovations in Ultrahigh-Strength Steel Technology*, 34th Sagamore Army Materials Research Conference Proceedings (1990) 3–66

properties of strength, toughness, and hydrogen resistance. If a sufficiently improved combination of these properties can be achieved, other properties such as cost and processability are secondary considerations.

Strength/toughness objectives adopted for the ultrahigh-strength steels are depicted in Fig. 1 which shows data compiled at MTL representing the typical K_{IC} fracture toughness levels of commercial steels currently in use in applications requiring hardness levels of R_c50 and above. Also indicated are the approximate ultimate tensile strength (UTS) corresponding to each hardness. Marginal fracture toughness and corresponding critical flaw size limit the useable hardness level in structural applications (involving significant tensile loads) to R_c55, with more conservative applications demanding greater flaw tolerance limited to R_c50. The dashed lines depict contours of constant critical flaw size at a design stress proportional to hardness, identifying the objectives shown to allow a 5 R_c point increase in usable hardness level. The latter corresponds to a substantial strength increment of 50 ksi or 400 MPa. The strength levels thus defined mark the toughness "frontier" of ultrahigh-strength steels; the objectives posed demanding K_{IC} improvements by a factor of 2 at R_c55 and 3 at R_c60.

While low alloy stage I tempered (ε - carbide strengthened) martensitic steels, exemplified by 4340, still show potential for toughness improvement and are the subject of further study under the SRG program, the new class of steels which has so far demonstrated the highest combinations of strength and toughness is the high alloy secondary-hardening (alloy carbide strengthened) martensitic steels, currently represented by the commercial HY180 and AF1410 steels. At R_c45, HY180 exhibits K_{IC} toughness levels in excess of $230\,\text{ksi}\sqrt{\text{in}}$, beyond the scale of Fig. 1. The higher-strength AF1410 reliably provides a K_{IC} toughness of $140\,\text{ksi}\sqrt{\text{in}}$ at R_c48 as depicted in Fig. 1, while exceptional commercial heats can show $190\,\text{ksi}\sqrt{\text{in}}$. Expanding this class of steel to the ultra-high-strength R_c55–60 range offers the greatest potential of meeting our toughness objectives. Moreover, in extending the useable strength of structural steels to R_c60, such secondary hardening steels would overlap with tool steels (represented by M2 and M50 in Fig. 1) of interest for bearing and gear applications, thus creating a class of steels with a wide range of potential applications.

Although fracture toughness is an important consideration, a more serious limitation on the

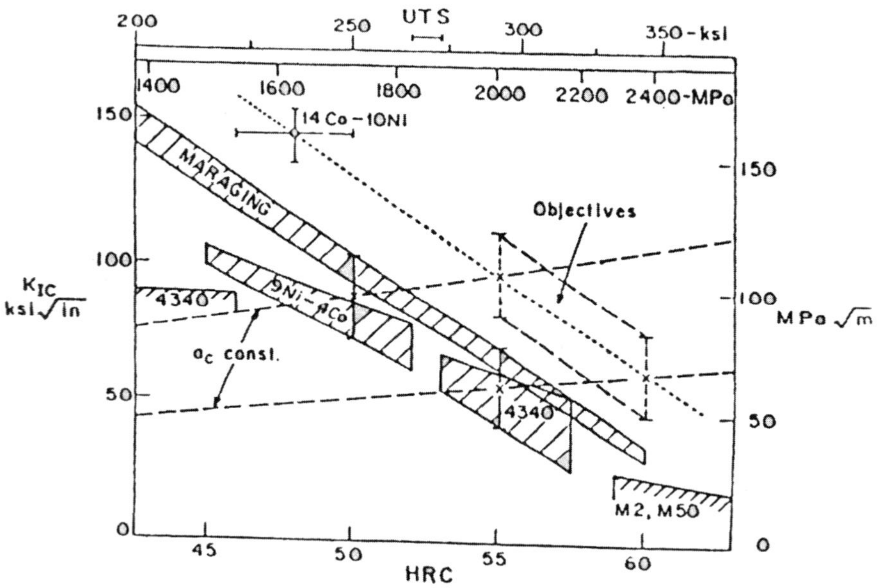

Fig. 1 Typical plane-strain fracture toughness (K_{IC}) versus Rockwell C hardness (HRC) for commercial ultra-high-strength steels currently employed in advanced structural applications. Upper scale shows approximate ultimate tensile strength (UTS) with indicated error limit. Strength/toughness objectives depicted allow an advance in useable strength corresponding to 5 points HRC

reliability of ultrahigh-strength steels is their low resistance to stress-corrosion cracking, which at these strength levels occurs by a hydrogen embrittlement mechanism associated with cathodic charging at the crack tip. At R_c55, the low hydrogen resistance limits the stress corrosion threshold K_{ISCC} to ~10 ksi$\sqrt{in}$. Raising usable strength to the levels depicted in Fig. 1 requires major improvements in hydrogen stress-corrosion resistance in order to maintain a K_{ISCC}/K_{IC} ratio of at least 0.5 out to R_c60. As traditional alloy development approaches have produced little change in this property of ultrahigh-strength steels in the past decade, the required K_{ISCC} improvements represent the most ambitious objective of the SRG effort, warranting the most interdisciplinary and fundamental approach.

Materials as Systems—An Approach to Design

The central paradigm of materials science is the sequential interrelation of processing, structure, properties, and performance. Although empirical development attempts to bypass elements of this sequence, for example, correlating processing and performance without regard to structure/property relations, scientific understanding demands that each link of the sequence be known. Cause-and-effect logic gives a natural direction to this sequence of the form: processing→structure→properties→performance, and from the philosophy of scientific analysis by which we distill simple laws from natural complexity, it is tempting to imagine a materials design process in which predictive theory specifies a composition and processing producing the necessary structure providing the required properties for desired performance. In practice each of these elements and their interrelation is of infinite complexity in real materials, severely compromising our predictive abilities. To usefully synthesize available understanding, we must combine with the scientist's methods of simplifying analysis the engineer's methods for dealing with real complexity, formalized in the "systems" approach (Ref 1).

Cyril Smith (Ref 2) has expounded on the notion of materials as systems, characterized by a hierarchical structure of interacting microstructural subsystems. Processing and properties can each be likewise described as hierarchies of interactive subsystems, each such subsystem involving important cross-interactions with microstructural subsystems. Morris Cohen (Ref 3) has observed that within this interlocking network there is a "reciprocity" such that equally valid to the notion that structure controls properties is the alternative view that properties control structure: our perception of structure is a function of the properties we are attempting to explain. This yields the important principle that knowledge of the relation of a particular component (subsystem) of structure to a property can be used to improve that property without ever requiring the impossible task of characterizing the entire structure. An important aspect of the engineering science of materials is then the discrimination between microstructural components which are actively linked to properties of interest, and inert ingredients such as nonessential evolutionary vestiges which are a common feature of complex systems. As emphasized by Cohen (Ref 3, 4) this will always involve a blend of predictive theory and empiricism. It is a reasonable expectation that by a combination of predicting the predictable and measuring the measurable we can identify sufficient subsystem interrelations to pose process/structure solutions to property/performance problems, tempered by a certain amount of discouragement from "unknowables" which are beyond our current abilities to predict or measure.

The flow-block diagram of Fig. 2 describes the structure of Ni-Co secondary hardening martensitic steels as a hierarchy of microstructural subsystems defined from the viewpoint of combined strength, toughness, and hydrogen resistance, and identifying the stages of processing affecting each. The multicomponent matrix of the steels consists of a lath martensitic microstructure (representing the best class of microstructure for strength/toughness) with a high Ni content for cleavage resistance. Based on the seminal research of Speich (Ref 5, 6) the Co is used to retard dislocation recovery (which we will later associate with short-range order), which has an important interaction with the strengthening subsystems via heterogeneous nucleation of the M_2C (and MC) carbide strengthening dispersion on the dislocations during tempering. Speich's study of HY180 (Ref 6), which formed the basis of the AF1410 development, further showed that an optimum strength/toughness combination is achieved by bringing the alloy carbide

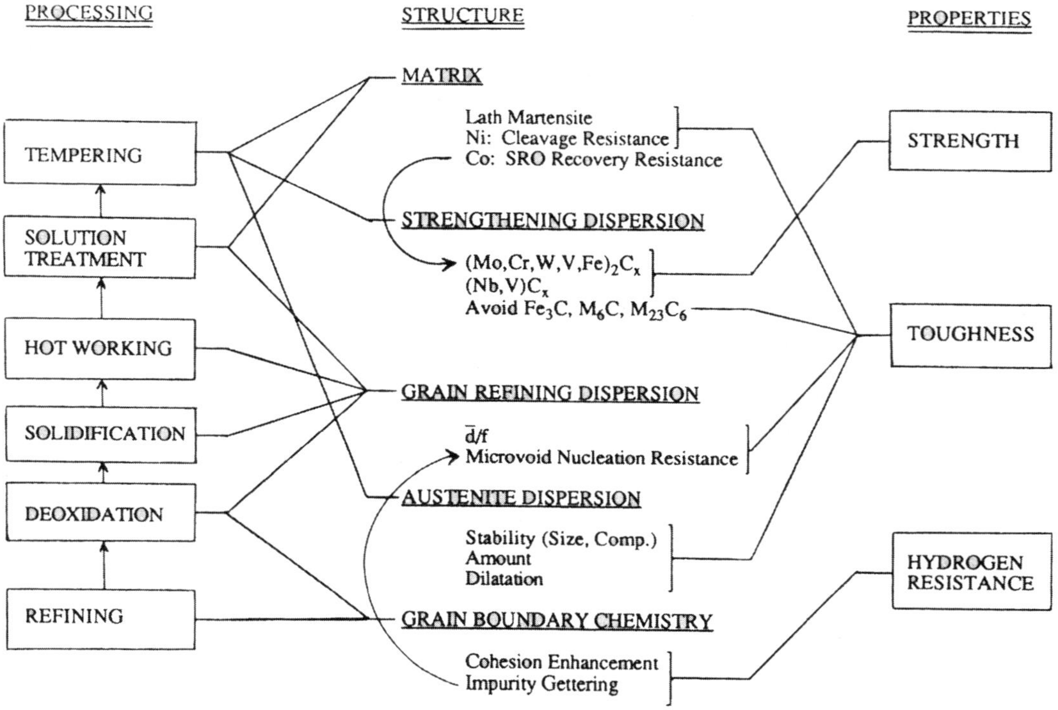

Fig. 2 Flow-block diagram illustrating the interactive hierarchical structure of a Co-Ni secondary hardening martensitic steel from the viewpoint of desired properties

precipitation reaction to completion in order to dissolve the relatively coarse transient θ-cementite (Fe_3C) which precipitates earlier and acts as a microvoid nucleation site limiting ductile fracture resistance. The M_2C carbide primarily responsible for secondary hardening is metastable, and high fracture toughness also demands that this metastable reaction reach completion without the nucleation of more stable carbides (M_6C, $M_{23}C_6$) which can precipitate incoherently on interfaces and degrade toughness.

Microvoid nucleation during plastic deformation accelerates ductile fracture through a process of shear localization. Once the precipitation kinetics of competing phases are controlled to achieve a fine strengthening dispersion without unnecessary coarser particles, resistance to microvoid nucleation is ultimately governed by the grain refining dispersion necessary to maintain a sufficiently fine grain size during solution treatment to inhibit competing brittle fracture modes. The grain-refining ability of this dispersion is set by the ratio of average particle diameter d to the volume fraction f. Subject to this constraint, toughness enhancement can be sought from an optimization of the geometry and composition of the dispersion (set by earlier processing stages), in order to increase microvoid nucleation resistance.

Further toughening can be achieved through modification of matrix constitutive behavior, the most dramatic example being the incorporation of transformation plasticity from a dispersion of metastable austenite particles precipitated during tempering. The transformation toughening is sensitive to the particle transformation kinetic stability (controlled by particle size and composition), the amount of dispersed phase, and the magnitude of its transformation dilatation.

Hydrogen stress corrosion cracking in ultrahigh-strength martensitic steels occurs by a brittle intergranular cleavage mechanism invariably associated with boundary segregation of embrittling metalloid impurities. The embrittlement appears to be controlled by the chemistry of intrinsic boundary cohesion. Under the SRG program, an interdisciplinary effort is directed at the ultimate application of an electronic-level understanding of intergranular cohesion to the design of grain boundary composition for both enhanced intrinsic cohesion and reduced interaction with hydrogen.

A parallel effort explores the use of novel processing techniques to getter embrittling impurities into highly stable compounds, potentially offering the additional toughening benefit of cleaning the interphase boundaries controlling microvoid nucleation.

As is a common feature of systems, the separate optimization of each of the subsystems of Fig. 2 poses conflicting objectives. Optimization of the total system demands compromise between subsystems such that local optimizations are highly constrained by the requirements of other subsystems. After reviewing progress in our understanding of the functioning of each of these subsystems, we will examine its application in conceptual designs which effect this compromise.

Strengthening

The strength of overaged precipitation-hardened structures is controlled by the Orowan bypass mechanism for which strength is inversely related to the particle mean free path. For a given amount of precipitating phase, maintaining the highest possible strength at completion of precipitation requires minimization of the final particle size. This is controllable in principle through application of transformation kinetic theory, once theory is properly calibrated by appropriate experiments in the system of interest. It is a fairly recent development that available instrumentation has attained the capabilities required to adequately characterize the fine-scale strengthening dispersions of ultrahigh-strength steels.

Results of an ongoing comprehensive study of M_2C carbide precipitation in AF1410 steel are summarized in Fig. 3. The figure shows the M_2C carbide particle size, shape, number density, volume fraction, lattice parameters, and composition, as well as the steel hardness, all as a function of tempering time at the standard temperature of 510 °C, following 1 hr solution treatment at the standard austenitizing temperature of 830 °C. The alloy composition studied is given in Table 1. The average particle size of the rod-shaped carbides is expressed by the diameter, d_s, of a sphere of equivalent volume. While there is considerable uncertainty in the size estimates from field-ion microscopy (FIM) (Ref 7) and transmission electron microscopy (TEM) (Ref 8), the greater sampling statistics of small-angle neutron scattering (SANS) (Ref 9) give a well-defined size distribution defining an average particle size with 90% confidence limits comparable to the size of the squares plotted in Fig. 3. Analysis of the SANS data incorporated the particle aspect ratio, β, determined from TEM of extracted particles, smoothly varying from near unity at initial nucleation to 4 at (near) equilibrium. The time dependence of average size suggests a smooth transition from nucleation to coarsening with suppressed growth characteristic of precipitation at high supersaturations as predicted by recent theories (Ref 10–12) based on the model of Langer and Schwartz (Ref 13). Nucleation in the ~0.1–0.5 h regime is followed by a short growth regime between 0.5 and 1.0 h, increasing average particle size from 36 to 56 Å. This is followed between 1 and 2 h (bracketing the reaction half-time $t_{1/2}$ indicated by the volume fraction f_v curve) by a surprising second stage of nucleation which reduces the average particle size through an increased number of finer particles, bringing the total particle number density N_v (determined by SANS) to a maximum. This may represent an extreme form of a phenomenon encountered in the numerical modeling of precipitation at high supersaturation by Kampmann and Wagner (Ref 11) in which oscillations in nucleation rate are associated with crossovers between the average and critical particle sizes. Overall, the observations are consistent with the general features of high supersaturation behavior in which the major portion of solute depletion is associated with nucleation rather than growth, and the average particle size never differs greatly from the critical size. This constitutes a highly useful phenomenon for maintaining the finest possible particle size at completion of precipitation.

The volume fraction curve, determined from SANS, shows that precipitation is virtually complete after 10 h tempering and very nearly complete after the standard 5 to 8 h temper employed in practice. This is consistent with nearly complete dissolution of transient cementite for optimal toughness as emphasized by Speich (Ref 5, 6).

Information on the state of interfacial coherency of the carbide particles can be deduced from the lattice parameters of extracted carbides measured by electron diffraction (Ref 8), also shown in Fig. 3. In a coherent state the carbides will tend to reduce elastic energy by adopting a modified composition

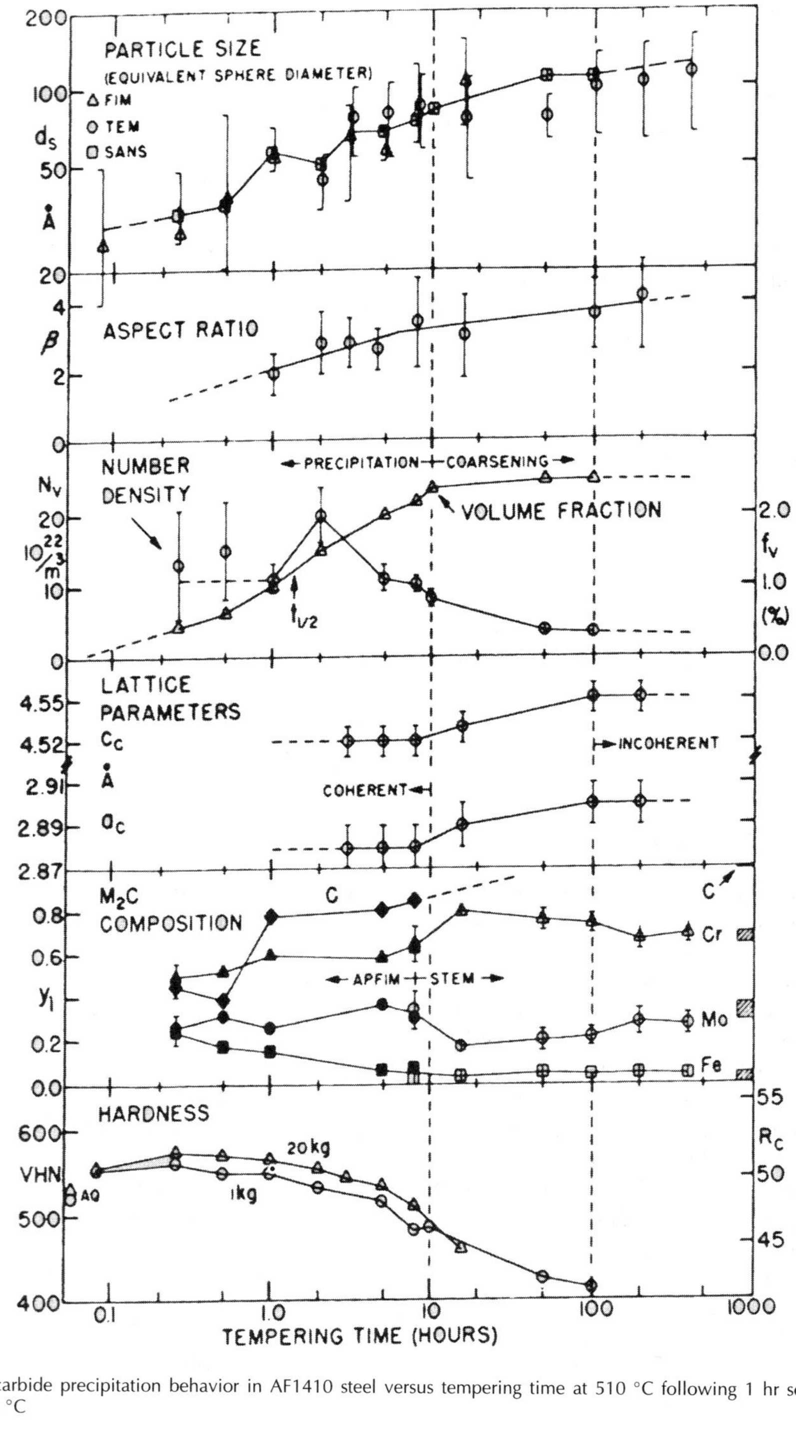

Fig. 3 M_2C carbide precipitation behavior in AF1410 steel versus tempering time at 510 °C following 1 hr solution treatment at 830 °C

which reduces the overall magnitude of the orthorhombic lattice deformation relating the coherent HCP carbide to the BCC matrix. The smaller lattice parameters at shorter tempering times, observed here and in similar alloys (Ref 8), are consistent with such a reduction in the large positive BCC→HCP transformation strains in this system. The observed lattice

Table 1 Alloy steel compositions, wt%

Commercial	C	Co	Ni	Mo	Cr	Other
HY 180	0.11	8.0	10.2	1.0	2.0	0.18Mn
AF1410	0.16	14.25	10.15	1.05	2.10	...
Experimental						
(1) 1410-4Mo	0.23	14.17	10.24	3.96	0.06	...
(2) 1605-4Mo	0.24	15.99	4.96	4.03	0.02	...
(3) 1605-MoCr	0.24	16.08	4.97	2.82	0.71	...
(4) 1605-CrMo	0.24	16.06	4.98	1.52	1.40	...
(5) 1605-4Mo (Nb)	0.24	16.02	4.97	4.03	0.02	0.029Nb
(6) 1605-4Mo (NbV)	0.25	16.10	4.97	4.03	0.02	0.013Nb 0.006V
(7) 1605-4Mo (Au)	0.23	15.92	5.03	4.10	0.02	0.07Au

parameter shifts suggest that the particles are coherent below 10 hr tempering and reach a fully incoherent state beyond 100 hr.

Measurement of the composition trajectories of the M_2C carbides presented in Fig. 3 is made possible by the atom-probe field-ion microscope (APFIM) providing the necessary microanalytical resolution at the atomic scale (Ref 14–16); microanalysis of the coarser particles obtained beyond 10 hr tempering (Ref 17, 18) is performed by scanning transmission electron microscopy (STEM). Compositions are represented as M_2C site fractions, with the bars at the right depicting the range of equilibrium compositions predicted by available thermodynamic databases (Ref 17). Although the carbide composition appears to ultimately reach equilibrium in the incoherent state, major departures from equilibrium exist at early stages of precipitation. Based on available data for the composition dependence of the M_2C lattice parameters (Ref 8), the depletion of C and enrichment in Fe are consistent with the observed lattice parameter shifts. In addition to elastic coherency effects, particle composition at small sizes will also be strongly influenced by capillarity. If the interfacial energy is not strongly composition dependent, the first particles will tend to adopt a composition corresponding to the maximum precipitation driving force; thermodynamic calculations of this composition suggest a reduced Cr/Mo ratio consistent with observation (Ref 19). The lattice parameter shifts and volume fraction measurements further suggest that the carbides achieve completion of precipitation in the coherent state, and are therefore close to a state of coherent equilibrium after 8 hr tempering of this steel.

The hardness curve in Fig. 3 demonstrates that the completion of M_2C precipitation (and elimination of cementite) at 8–10 hrs tempering corresponds to an overaged condition. As mentioned earlier, strength in this regime is controlled by the Orowan bypass mechanism for which the strength achieved from a given volume fraction of precipitate can be increased by refining the particle size. Employing the information obtained from Fig. 3, improved alloy compositions for more efficient strengthening can be sought from precipitation kinetic theory through control of appropriate scaling factors. In current theory of precipitation at high supersaturations (Ref 10–13), the principal scaling factors are (a) a diffusion time factor, which can be deduced from particle coarsening behavior; and (b) the initial critical nucleus size, inversely related to the precipitation thermodynamic driving force.

A rate constant for multicomponent particle coarsening, defining a diffusion time factor for multicomponent precipitation, has been derived by an extension (Ref 17, 20) of the model of Bjorklund, Donaghey, and Hillert (Ref 21). The model has been tested (Ref 17, 20) by comparison of the relative $(Cr,Mo)_2C$ carbide coarsening rates in the experimental alloys 1 to 4 listed in Table 1. This series of alloys has a total Cr and Mo content balanced to 0.25 w/o C in the stoichiometry of the M_2C carbide, representing the minimum alloy content for complete replacement of cementite by M_2C. Alloy 1 (1410-4Mo) has the 14Co-10Ni matrix composition of AF1410, while alloys 2 through 4 (1605 series) have a 16Co-5Ni matrix which eliminates interference from austenite precipitation at long tempering times (Ref 17). Figure 4 plots the computed rate constant K defined in Ref 20, as well as the relative value (k/k_1) normalized to alloy 1. Despite the higher diffusivity of Cr relative to Mo and the lower thermodynamic stability of

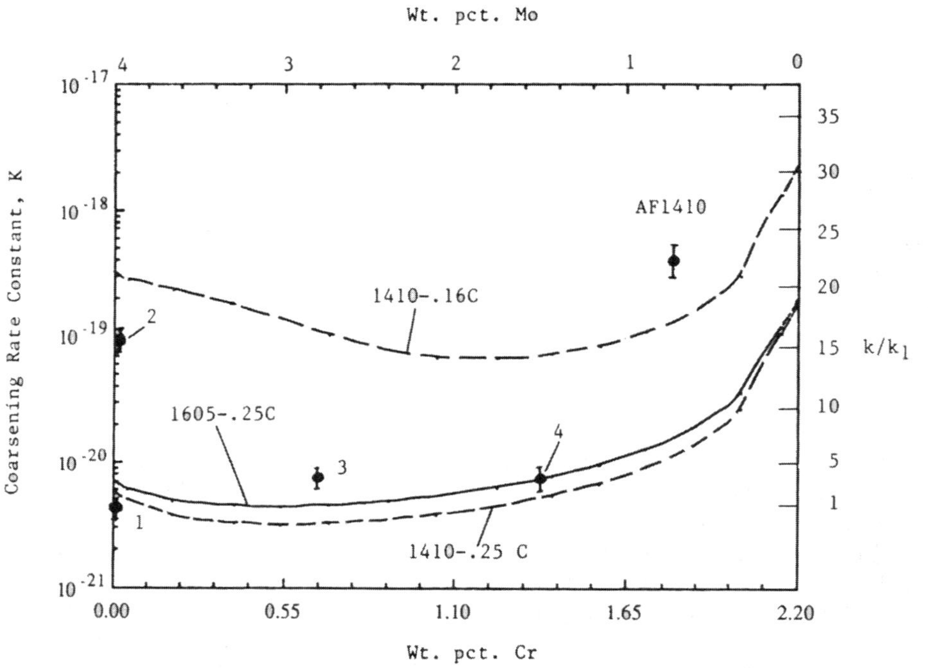

Fig. 4 Rate constants for $(Cr,Mo)_2C$ carbide coarsening in Ni-Co alloy steels with Cr and Mo contents stoichiometrically balanced to 0.24 w/o C

Cr_2C compared to Mo_2C, the model predicts a relatively low coarsening rate for the mixed $(Cr,Mo)_2C$ carbides (in part due to the stabilizing contribution of the entropy of mixing in the M_2C phase), as indicated in the lower curves for the 1605-.25C and 1410-.25C series. These stoichiometric compositions are predicted to have generally low coarsening rates (due to reduced Cr and Mo in matrix solution) as shown by comparison with the upper dashed curve representing 1410-.16 alloys with the lower carbon content of AF1410. Experimentally observed relative coarsening rates, k/k_1 as defined by the time dependence of average particle volume in 10 to 100 hrs tempering at 510 °C (Ref 17) are also plotted for alloys 1 to 4 and AF1410 in Fig. 4. Apart from an anomalously high rate for alloy 2, the comparison supports the predicted lower coarsening rate of the stoichiometric alloys (by a factor of 20) compared to the nonstoichiometric AF1410, and supports the predicted multicomponent effects giving the mixed $(Cr,Mo)_2C$ carbides of alloys 3 and 4 coarsening rates comparable to the (more stable) Mo_2C carbide of alloy 1. These data must be viewed with some caution, however, as the alloys experienced some degree of decarburization during heat treatment. Overall, the coarsening model appears reasonable and could be applied to design of alloys with lower coarsening rates for thermal softening resistance (as in tool steels); in structural steels for room-temperature applications it is more desirable to enhance the diffusional rate factor to accelerate secondary hardening at lower tempering temperatures.

The most important scaling factor governing the size scale of the precipitate dispersion (as well as the rate of nucleation) is the initial critical nucleus size. This in turn scales directly with the precipitate interfacial energy, and inversely with the precipitation thermodynamic driving force. The alloy composition dependence of the latter can be predicted from solution thermodynamics as conveniently treated by available computer software such as the THERMOCALC thermochemical database and software system (Ref 22). In order to treat coherent precipitation, however, it is necessary to modify the solution thermodynamics to incorporate the effects of elastic energy. The composition dependence of the M_2C carbide coherency strains (as described by the BCC→HCP lattice deformation accompanying coherent precipitation) can be prescribed from the composition dependence of the carbide lattice parameters (Ref 8). Using a simple isotropic homogeneous spherical inclusion model, it has been shown that the coherency elastic energy (in the limit of small carbide volume fractions) can be expressed in a form compatible with the solution free-energy functions employed in the THERMOCALC system allowing a simple modification of the database to treat the thermodynamics of coherent precipitation (Ref 19, 23). Newer elastic calculations for anisotropic inhomogeneous ellipsoidal inclusions (using observed aspect ratios as represented in Fig. 3) have been further refined to incorporate the effects of the redistribution of matrix substitutional solutes in the stress field of the coherent particle (Ref 24). This has been performed by an extension to multicomponent systems of the "open system" elastic constants method of Larché and Cahn (Ref 25) in which effective matrix elastic constants (incorporating solute redistribution) are defined from the solution chemical thermodynamics and the composition dependence of the matrix lattice parameters. Application to ferrite-M_2C coherent equilibrium in AF1410 is represented by Fig. 5, depicting computed contours of constant hydrostatic stress and composition in the matrix surrounding a coherent M_2C carbide viewed along the matrix $[100]_\alpha$ direction corresponding to the habit direction of the rod-shaped particle (Ref 24). Dashed contours depict compressive stress and solute depletion, while solid contours represent tensile stress and solute enrichment. The predicted solute distribution (particularly for Ni and Co) should be measurable by high-

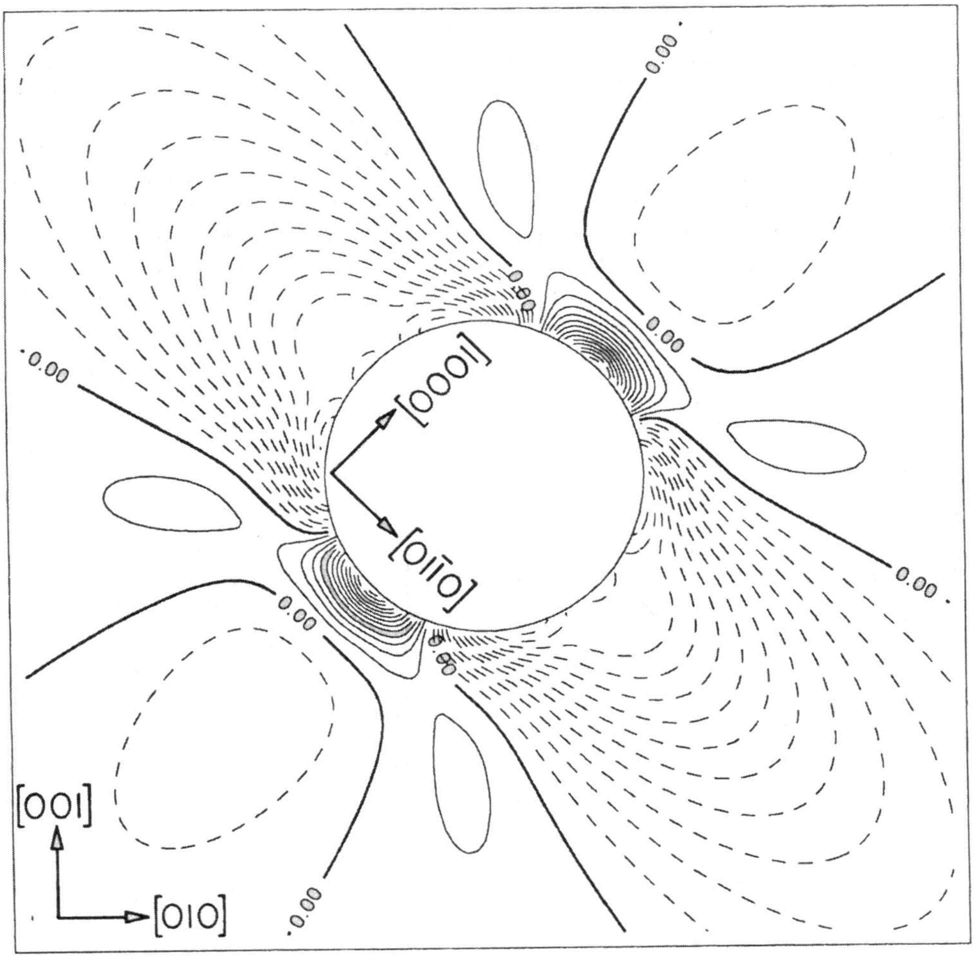

Fig. 5 Computed contours of constant hydrostatic stress and composition surrounding M_2C carbide in coherent equilibrium in AF1410 steel. Carbide viewed along $[100]_\alpha$ rod axis. Crystal axes of BCC α matrix and HCP M_2C carbide are indicated. Source: Ref 24

resolution microanalysis for the coherent particle size achieved after 8 hr tempering. Elastic energies computed from such linear elastic calculations represent an upper bound estimate. Assuming a composition-independent interfacial energy, reasonable agreement is obtained with measured M_2C composition trajectories in AF1410 (Fig. 3) and alloys 1–4 of Table 1 (Ref 17, 18) when a correction factor of 0.25 is applied to the composition dependent elastic energy function defined by these calculations (Ref 26). This factor likely compensates for neglect of particle interaction with matrix dislocations to be discussed later.

A quantitative model for the thermodynamics of coherent M_2C precipitation allows prediction of alloy compositions capable of more efficient strengthening through refinement of the particle size. As mentioned in discussion of Fig. 2, however, optimization of the metastable M_2C dispersion is constrained by kinetic phase competition with equilibrium phases such as M_6C and $M_{23}C_6$ which can precipitate incoherently on interfaces degrading toughness. This competition should also be amenable to control through alloy thermodynamics. Figure 6 shows a precipitation driving force map, constructed via THERMOCALC computations, showing contours of constant driving force (in kJ/mole) at 510 °C for precipitation of incoherent M_6C (Fig. 6a) and coherent M_2C (Fig. 6b) as a function of Cr and Mo content in alloys containing 14Co-10Ni-0.25C (Ref 26). The dashed straight lines represent stoichiometrically balanced compositions (as for the experimental alloys of Table 1) again corresponding to the minimum alloy content for complete elimination of cementite by M_2C precipitation. Intergranular embrittlement during 510 °C tempering of the experimental alloys (Ref 27), tentatively ascribed to M_6C precipitation, suggests a threshold driving force for M_6C nucleation under the tempering conditions of interest of 15 kJ/mole. The contour corresponding to this limit is superimposed as a dashed curve on the coherent M_2C driving force plot of Fig. 6(b). Through the competing influences of the chemical and mechanical thermodynamic contributions, the coherent M_2C driving force is nonmonotonic with Cr content, but increases with Mo. Maximization of the coherent M_2C driving force subject to the M_6C driving force constraint gives a unique optimum composition represented by the open point in Fig. 6(b). In parallel with these studies, an empirical alloy development program at Carpenter Steel (Ref 28) has arrived at a very similar composition demonstrating an excellent strength/toughness combination at a significantly higher strength than AF1410, supporting

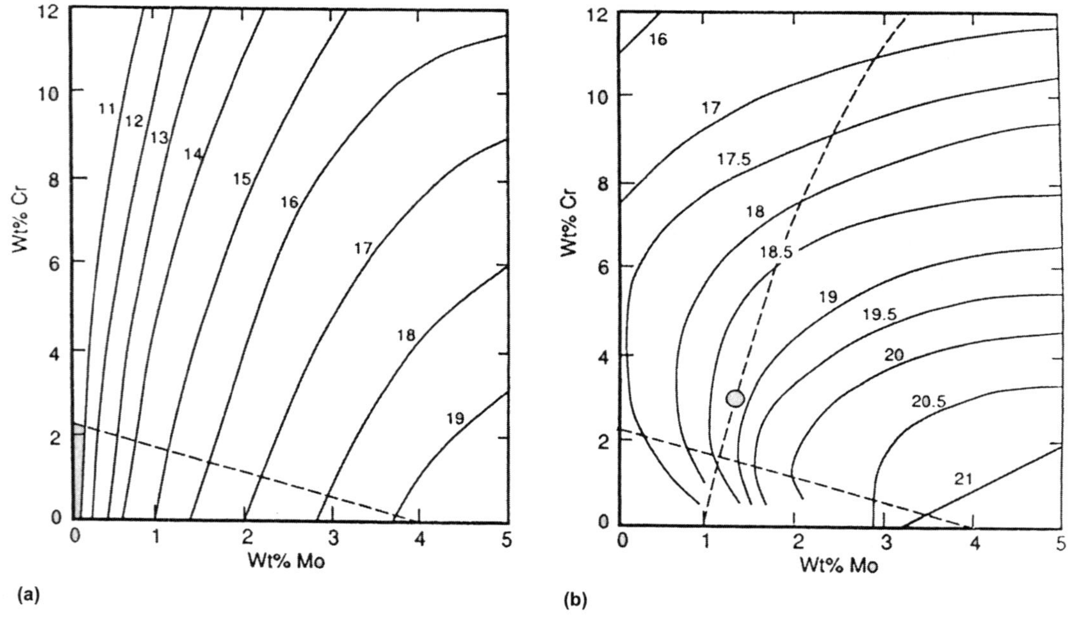

Fig. 6 Computed precipitation driving force (kJ/mole) at 510 °C for alloys containing 14Co-10Ni-0.25C: (a) incoherent M_6C, (b) coherent M_6C (26)

plausibility of the suggested thermodynamic design approach. It should be noted that the reference state for the driving force calculations of Fig. 6 was fully supersaturated ferrite, while alloy carbide precipitation is in practice preceded by cementite precipitation which lowers the carbon potential. The effect can be estimated by adopting an initial state of ferrite-cementite paraequilibrium in which the cementite is constrained to inherit the substitutional composition of the alloy (Ref 29). The lower carbon potential reduces the M_6C driving force by ~0.7 kJ/mole and the coherent M_2C driving force by ~2.7 kJ/mole with a slight distortion of the contour shapes of Fig. 6 (Ref 26).

The influence of the high Co recovery-resistant matrix on M_2C secondary hardening behavior is shown in Fig. 7 comparing 1 hr isochronal tempering hardness curves for a simple 4Mo-.23C steel (Ref 30) and for alloy 1 of Table 1, with the same Mo and C content but containing 14Co-10Ni (Ref 17). In the latter steel the secondary hardening peak temperature is moved down 100 °C with a $10R_c$ point increase in peak hardness. Comparison of the isothermal hardening behaviors at 500 °C (Ref 31, 32) shows an acceleration of secondary hardening in the Co-Ni steel by 2 to 3 orders of magnitude. Such a strong acceleration allowing a substantial reduction in secondary hardening temperature (with an attendant increase in precipitation driving force) is consistent with a switch to heterogeneous nucleation on dislocations in the recovery-resistant Co-Ni steels as proposed by Speich (Ref 5, 6).

The observed acceleration and enhancement of secondary hardening is similar to the nucleation effects of small additions of MC carbide formers (e.g., Nb, V) and GP-zone formers (e.g., Au) to simple Mo-C steels studied by Honeycombe and coworkers (Ref 31, 33). To investigate the possibility of superimposing the two types of nucleation phenomena, the effect of Nb, V, and Au additions on the secondary hardening response of the Co-Ni alloy 2 of Table 1 was studied in the series of alloys represented by alloys 5 to 7 (Ref 15). Observed hardening enhancement was limited to the level of ~1 R_c point, and extensive atom-probe microanalysis indicated preferential acceleration of M_2C nucleation by the high dislocation density such that there was little interaction with either MC precipitation or GP zone formation which required a longer time scale. Although some Au clustering was observed both in the matrix and M_2C suggestive of some degree of M_2C nucleation on Au zones, the Nb and V were found only in the M_2C phase with no indication of separate MC formation in near-peak-hardness microstructures. The

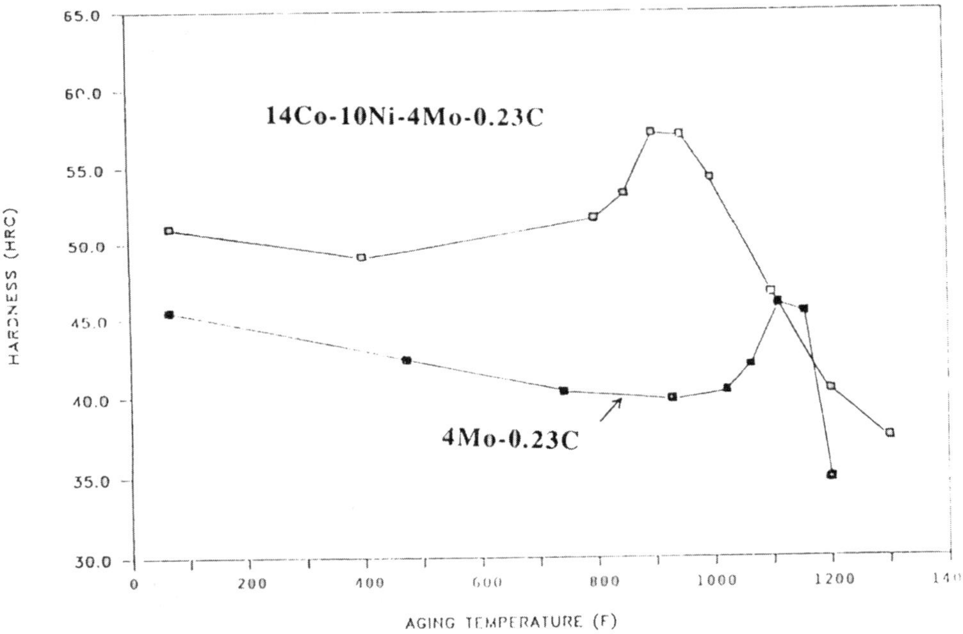

Fig. 7 Isochronal tempering hardness curves (1 h) for 4Mo-0.23C steels with and without 14Co-10Ni

measured compositions of M_2C carbides in alloys 5 to 7 are listed in Table 2. Following this demonstration of Nb and V solubility in M_2C, subsequent coherent thermodynamic calculations have indicated that V can be a very effective alloy addition to enhance the precipitation driving force for coherent M_2C (Ref 26).

The strong effect of heterogeneous nucleation at dislocations on the M_2C carbide precipitation behavior emphasizes the important role of Co in retarding dislocation recovery so that a high dislocation density can be retained at secondary hardening temperatures. Neutron diffraction has revealed the presence of short-range order (SRO) in precipitation-hardening high Co martensitic steels (Ref 34), and field-ion microscopy has indicated SRO in AF1410 (Ref 35). Based on the hypothesis that the recovery resistance is associated with SRO, recovery behavior has been investigated by TEM (Ref 36) in the simple ferritic binary Fe-Co and Fe-Al alloy systems where previous studies have demonstrated SRO by neutron diffraction (Ref 37) and TEM (Ref 38), respectively. The recovery behavior in the temperature range of 400–600 °C was also compared with that of Fe-Ni and Fe-Ni-Co martensitic alloys (Ref 36). Figure 8 shows the retarded softening behavior at 527 °C for cold-worked Fe-14Co and Fe-6Al compared to Fe as expressed by the hardness decrease ΔH relative to the decrease for full recrystallization ΔH_{REX}. As indicated in the figure, a preliminary neutron diffraction measurement (Ref 39) confirmed the existence of SRO in the Fe-14Co alloy after 1 hr at 527 °C. The heavily cold-worked initial structure of these alloys showed a deformation microband substructure very similar in character and size scale (~0.25 μm width) to the lath structure of the martensitic alloys, and showed similar stages of recovery consisting of (a) reduced general dislocation density, (b) sharpening of dislocation walls, and (c) polygonization and subgrain coarsening. The measured parabolic rate constant for subgrain coarsening at 527 °C showed an order of magnitude reduction for Fe-14Co (2.2×10^{-7} μm^2/s) compared to Fe (2.8×10^{-6}μm^2/s). While the Fe-Ni martensites showed monotonic softening behavior, the Fe-Ni-Co martensites showed hardening similar to previous observations interpreted as SRO strengthening (Ref 40). After full recrystallization at 800–830 °C, the ferritic Fe-Co and Fe-Al alloys also showed some hardening with aging at 420 °C to 575 °C. Although some preliminary Mossbauer spectroscopy on the Fe-Co and x-ray diffractometry on the Fe-Al did

Table 2 M_2C compositions in trace-addition alloys (510 °C/100 hr temper)

Alloy		Carbide Formula
#5	1605-4Mo (Nb)	$(Mo_{0.94\pm0.04}Fe_{0.04\pm0.01}Nb_{0.02\pm0.01})_2C_{0.88\pm0.02}$
#6	1605-4Mo (NbV)	$(Mo_{0.95\pm0.04}Fe_{0.03\pm0.01}Nb_{0.01\pm0.01}V_{0.01\pm0.01})_2C_{0.92\pm0.03}$
#7	1605-4Mo (Au)	$(Mo_{0.98\pm0.03}Fe_{0.018\pm0.01}Au_{0.002\pm0.001})_2C_{0.90\pm0.02}$

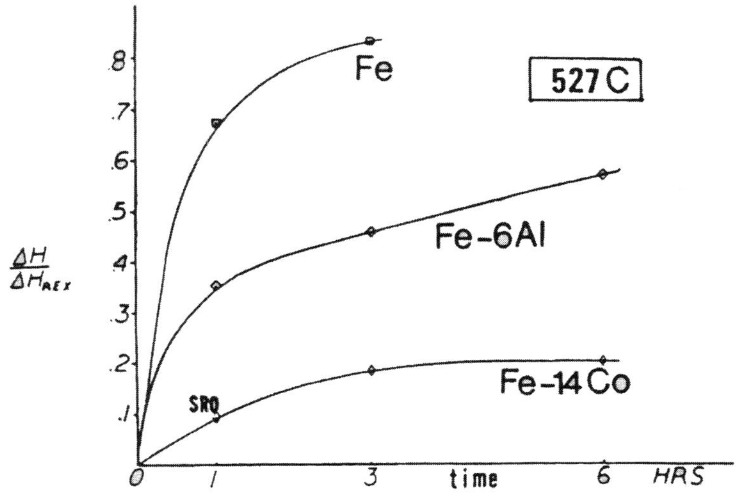

Fig. 8 Relative hardness decrease ΔH, normalized to full hardness decrease with recrytallization ΔH_{REX}, during 527 °C isothermal recovery of heavily cold-worked ferritic Fe, Fe-6Al, and Fe-14Co, showing retarded softening with alloying. Source: Ref 36

not yield supporting evidence for SRO (Ref 36), the bulk of the literature observations (Ref 34, 37, 38) and the neutron diffraction evidence (Ref 39) suggest that SRO is the most plausible basis for both the hardening and recovery resistance. In addition to increased dislocation glide resistance indicated by the observed hardening, dislocation climb processes important to recovery may be inhibited through a reduction of diffusivity by SRO as has been demonstrated in Cu-Zn (Ref 41). A more quantitative understanding of the mechanism of recovery resistance could allow both the more effective use of Co and some degree of substitution of other ordering elements such as Al in secondary hardening steels.

A longer-range issue regarding the achievement of maximum carbide strengthening for a given alloy carbon content concerns the properties of the carbide phase itself. In overaged structures characteristic of the completion of precipitation, size refinement will increase strength until a critical size is reached below which strength is controlled by particle shearing rather than the Orowan bypass mechanism. As discussed by Hornbogen (Ref 42) the critical particle size for this transition, and alloy strength in this regime, is controlled by the mechanical behavior of the dispersed phase; the high elastic modulus of alloy carbides compared to intermetallic phases makes them efficient obstacles capable of Orowan strengthening at smaller particle sizes. The ultimate design of optimal carbide strengthening dispersions will require information on the mechanical properties of the M_2C phase which can be obtained by synthesis of the multicomponent M_2C carbides of interest in bulk form. Such an investigation is currently under way (Ref 43).

Toughening and Shear Localization

The role of primary void formation at inclusion particles in ductile fracture is well established. Major improvements in the fracture toughness of high-strength steels in the past two decades have accompanied improved processing technology providing "clean" steels with reduced inclusion volume fractions; more recent work has demonstrated the importance of inclusion spacing in ductile fracture (Ref 44). Less well understood is the role of finer ~0.1 μm scale secondary particle dispersions which nucleate microvoids promoting shear localization, accelerating the process of primary void coalescence (Ref 45). In clean steels, the shear localization can dominate ductile fracture causing a mode I-loaded crack to locally adopt a mixed mode I and II "zig-zag" form of crack propagation. In addition to its microscopic role in mode I fracture toughness, shear localization also represents an important macroscopic failure mode limiting ballistic performance of ultrahigh-strength steels in armor and ballistically tolerant structural applications (Ref 46). Not only are the structure/property relations governing shear localization currently poorly defined, but due to experimental difficulties associated with the finer particle size relative to the well-studied ~1 μm scale primary inclusions, the microvoid nucleating particle dispersions are generally poorly characterized, constituting one of the least controlled components of steel microstructures. Control of shear localization represents the current frontier in the enhancement of ductile fracture resistance of ultrahigh-strength steels. As depicted in Fig. 2, efforts to control shear localization behavior focus on (a) the role of secondary particle dispersions (primarily grain refining dispersions) in microvoid nucleation and associated softening, and (b) the role of matrix constitutive behavior, as most effectively controlled through application of transformation plasticity provided by finely dispersed metastable austenite.

Shear Instability and Particle Dispersions

Shear localization behavior in ultrahigh-strength martensitic steels has been effectively investigated using shear tests conducted with and without compressive loading superimposed normal to the shear plane (Ref 47–49). Not only do the shear test specimen geometries allow relatively large amounts of material to be uniformly deformed to the point of shear instability, but the operating stress states provide no driving force for primary void growth, forcing the materials to fail entirely by the macroscopic shear localization mode. The experiments thus selectively study behavior associated with the microstructural component of interest: the microvoid nucleating particle dispersions. Tests are conducted using both thin-wall torsion and a linear-shear geometry allowing superimposed compressive loading. The critical shear instability strain, γ_i, is measured as the imposed uniform strain beyond which deformation localizes in shear bands, and corresponds to a maximum stress in the shear stress-strain (τ-γ) curve. The γ_i values measured in an air-melted 4340 steel with and without an applied normal compressive stress of 1/3 σ_y are plotted versus R_c hardness level in Fig. 9 (Ref 47, 48). Hardness is varied by 1 hr tempering at temperatures corresponding to Stage I (ε-carbide strengthening, R_c56), Stage II (retained austenite decomposition, R_c49–51), and Stage III (θ-cementite strengthening, R_c30–40) tempering. The instability strain generally decreases with increasing hardness level with an anomalous dip in Stage II tempering corresponding to "tempered-martensite embrittlement" associated with interlath cementite precipitation via austenite decomposition. The large increase in γ_i with compressive stress is consistent with a pressure-sensitive microvoid softening mechanism of shear instability. The role of microvoid formation in driving the instability is supported by TEM and SEM observations of microvoids in material deformation just to γ_i (Ref 47, 58) but such observations of fine voids associated with ~0.1 μm scale particles are not easily reproduced. Some preliminary experiments performed on the same material suggest that small-angle x-ray scattering measurements may allow detection of early stage microvoid formation in thin sections of deformed shear-test specimen gage sections (Ref 50).

Comparison of the instability strain measured in high-hardness 4340 under isothermal conditions at a strain rate of $10^{-3}s^{-1}$ and under adiabatic conditions at 10^2s^{-1} gives the surprising result that γ_i is unchanged (Ref 47, 48). While thermal softening under adiabatic deformation conditions may well contribute to shear localization in the highly compressive stress states associated with ballistic performance (Ref 46), it does not appear to significantly contribute to localization under pure shear stress. This would presumably also apply in tensile stress states such that improved resistance to microvoid formation should improve localization-controlled fracture toughness in ultrahigh-strength steels under both quasistatic and dynamic conditions.

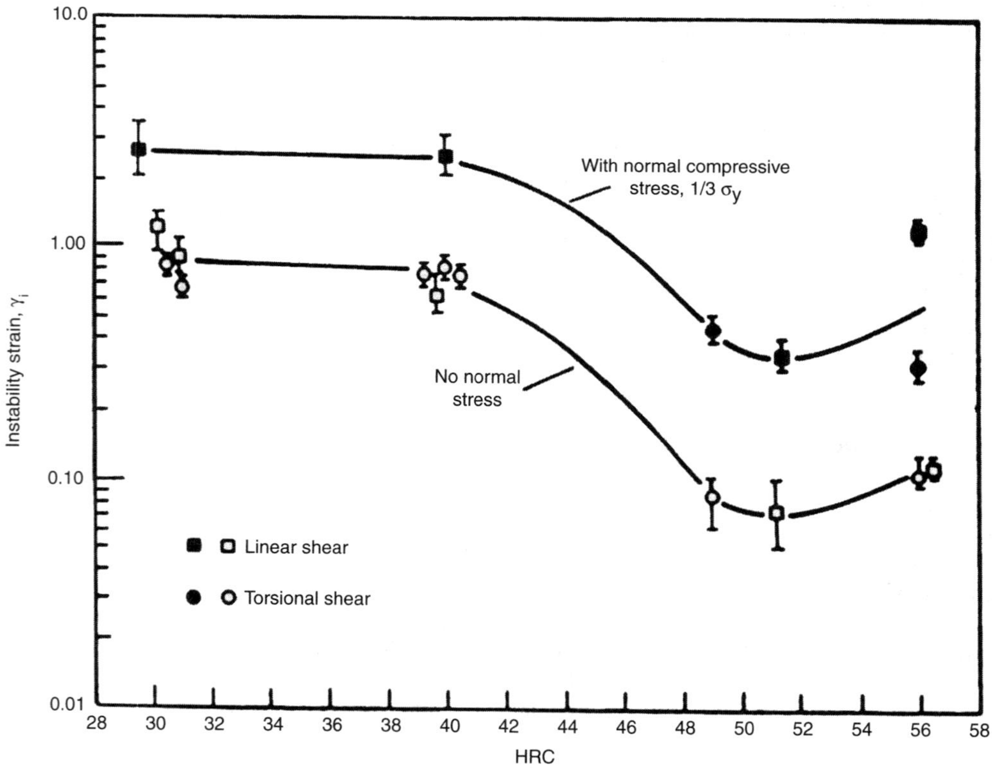

Fig. 9 Measured shear instability strain γ_i for 4340 steel as a function of hardness with and without normal compressive stress of 1/3 σ_y. Source: Ref 48

An abrupt nature of shear localization observed in steels at ultrahigh strength levels together with the small degrees of microvoiding observed just prior to its onset implies that the microvoid softening is nucleation controlled rather than growth controlled at these strength levels. A crucial tool for the theoretical investigation of structure/property relations governing microvoid nucleation and its associated softening has been the finite-element cohesive zone void nucleation model of Needleman (Ref 51, 52). Prescribing a local force-distance law for interfacial separation defining a characteristic length scale, the complete process of interfacial decohesion of a rigid spherical particle in an isotropically hardening elastic viscoplastic matrix is modeled as a function of particle diameter, d, particle volume fraction, f, and interfacial cohesive characteristics (separation work $\Delta\gamma$ and interaction range δ). The predicted nature of nucleation is depicted in Fig. 10 showing a quadrant of a cell with a particle centered at its lower left corner; equivalent strain contours in the matrix and interfacial configurations during separation are depicted for four applied remote strain levels $\varepsilon_1 < \varepsilon_2 < \varepsilon_3 < \varepsilon_4$ in axisymmetric tensile deformation. Abrupt decohesion at the top of the particle is followed by spreading along the interface. The macroscopic stress-strain response associated with these processes in a material with a periodic array of such cells is summarized in Fig. 11 comparing materials with three particle sizes $d_1 < d_2 < d_3$. The upper dashed curve represents the rigidly bonded case while the lower dashed curves correspond to replacement of the particles with voids of the same volume fraction, f. The largest particle size (d_3) shows the earliest nucleation with severe (brittle) softening behavior. Reduced particle size not only delays nucleation, but reduces softening via a more gradual (ductile) process of interfacial decohesion. Identical effects are observed by reducing f at fixed d. Fitting the model to measured microvoid nucleation conditions in steels for which d and f data are available allows an estimate of interfacial cohesive properties (Ref 52), with which the model can then predict quantitative benefits of variations in d and f. Stress states of the shear test experiments have been simulated by superimposing hydrostatic compressive stress, while retaining the axisymmetric deformation approximation (Ref 53).

Analytical modeling by Hutchinson and Tvergaard (Ref 54, 55) has elucidated both the nature of void/particle interactions promoting void growth in shear deformation, and the strong strain softening influence of void nucleation. Matrix plastic flow associated with load shedding from the decohering interface is shown to give stronger strain softening than had previously been expected. Further analyses (Ref 56) have gone beyond isolated particle and periodic cell models to explore the influence of nonuniformity of particle spatial distributions, demonstrating the manner in which clustering can further accelerate nucleation. As the basis of a more detailed accounting of the directional particle pair interactions underlying local cluster behavior, Tracey and Perrone (Ref 57, 58) have performed three-dimensional finite-element calculations of the deformation fields between pairs of spherical voids and particles in a plastic matrix undergoing the plane-strain shear deformation imposed in the shear test experiments, adopting constitutive relations appropriate to carbide particles in high hardness 4340 steel. Pairs of voids lying in the shear plane are found to exhibit the strongest interaction in a direction normal to the shear direction, while particle pairs exhibit strongest interaction when aligned parallel to the shear direction. Such analyses should ultimately allow control of anisotropic shear localization behavior in wrought alloys with anisotropic particle distributions.

To experimentally examine the influence of particle dispersion variations on shear localization behavior, the same shear tests performed on the air-melted (AM) 4340 of Fig. 9 have been extended to 4340 steels processed by vacuum-arc remelting (VAR) and rapid solidification processing (RSP) by centrifugal atomization and hot consolidation (Ref 48, 49, 59) and the particle dispersions quantitatively characterized by TEM and STEM analysis of thin foils and extraction replicas (Ref 49, 59, 60). In the Stage III tempered condition, shear localization is found to be dominated by the cementite dispersion precipitated on tempering and is relatively insensitive to earlier stages of processing. In contrast, the much finer ε-carbide strengthening dispersion precipitated in Stage I tempering does not participate in microvoid nucleation, and shear localization is controlled by the grain refining particle dispersion which remains undissolved during austenitizing, and which is sensitive to solution

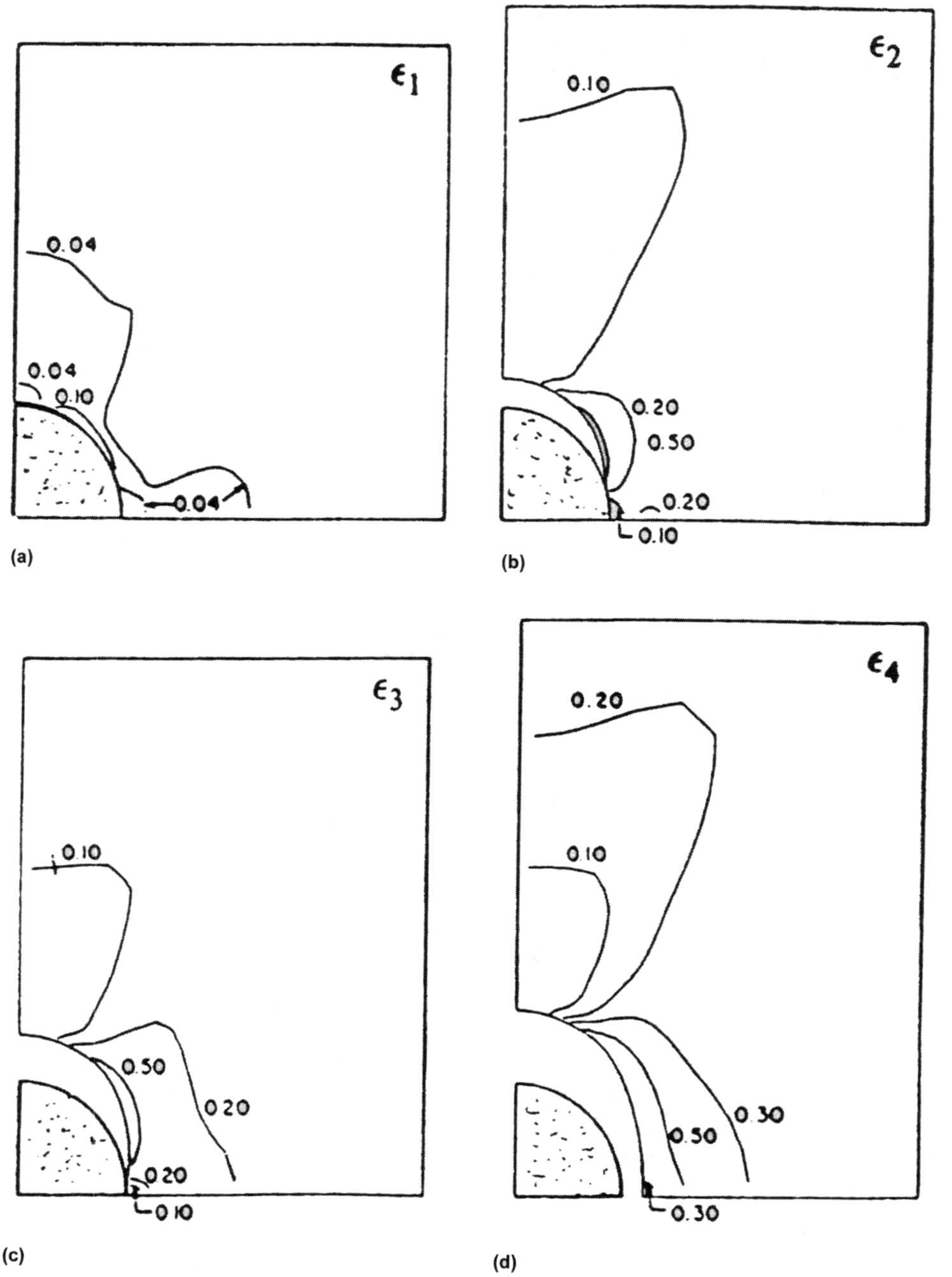

Fig. 10 Quadrant of cell model showing matrix strain contours and interface geometry during particle interfacial debonding at increasing applied strain levels. Source: Ref 51

treatment and melt practice. After conventional 1 hr solution treatment at 840 °C, the AM4340 contains a large number of 50 nm AlN particles and the distribution of MnS particles (predominantly 10–20 μm stringers) includes a small component of spherical 0.16 μm particles (Ref 60). Raising the solution temperature to 1100 °C dissolves the small sulfides and reduces the apparent number density of AlN particles by an order of magnitude (Ref 60). This is accompanied by an increase in the shear instability strain at R_c55 hardness from ~0.15 to 0.23 (Ref 47). In contrast to the AM4340, a more detailed study of the effect of solution

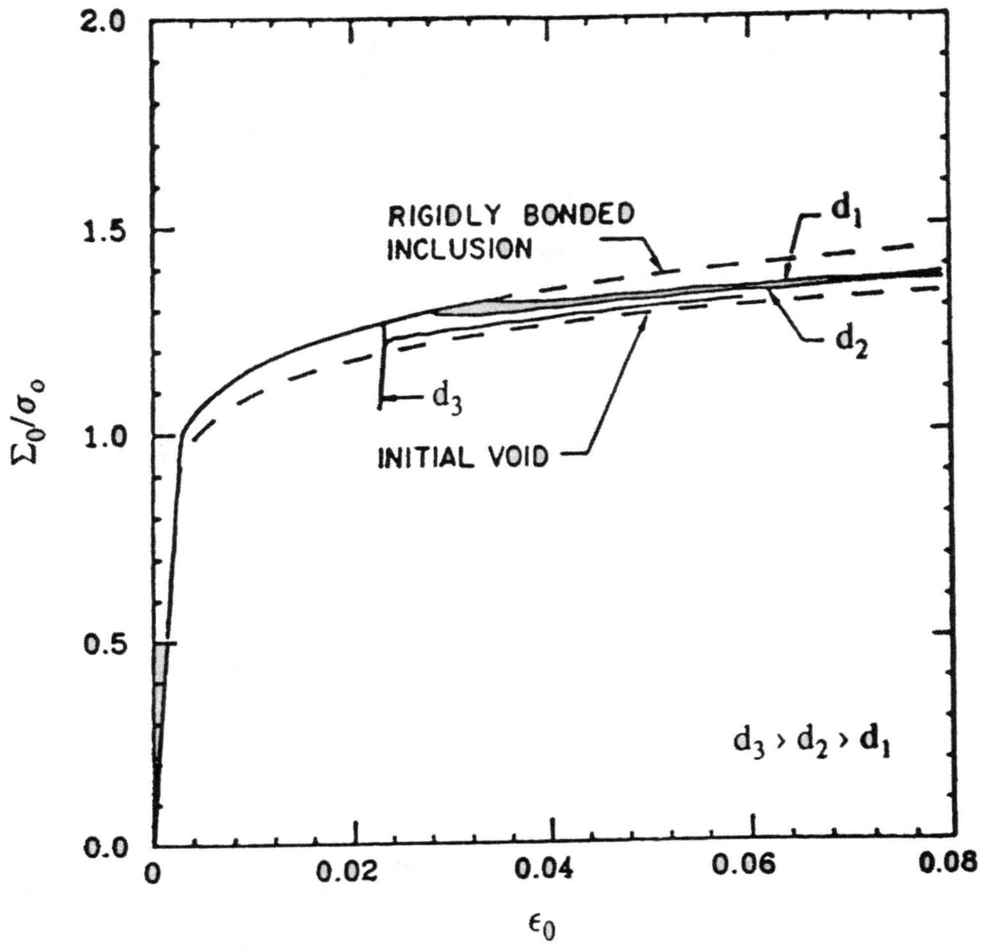

Fig. 11 Computed macroscopic stress-strain response during void nucleation at different size particles. Source: Ref 51

temperature on the VAR4340 in the range 870–1200 °C shows a monotonic decrease of γ_i with increasing solution temperature (Ref 49). In this case the dominant component of the grain refining dispersion at 870 °C is 0.17 μm diameter $M_{23}C_6$ carbides of composition $(Fe_{.74}Cr_{.25}Mo_{.01})_{23}C_6$ determined by STEM (Ref 60), with a total dispersion volume fraction of f = 0.44% estimated from SEM metallography (Ref 49). After 1200 °C treatment the estimated volume fraction of fine particles is reduced to f = .05%, consisting mainly of 80 nm Ti (C,N) and a small amount of 50 nm Al_2O_3 (Ref 49). Compared to the AM4340, the VAR4340 shows greater high-temperature grain coarsening, with a grain diameter of 17 μm at 870 °C coarsening to 280 μm at 1200 °C. In Stage III tempered material where cementite is the predominant microvoid nucleation site, γ_i decreases by the same factor on increasing solution temperature from 870 to 1200 °C as for the Stage I tempered material, suggesting the γ_i decrease is a feature of grain coarsening rather than the changing particle distribution. The RSP 4340 exhibits a highly stable grain refining dispersion of 0.27 mm Al_2O_3 with f = .26%; the dispersion, grain size, and γ_i are not significantly changed on increasing the solution temperature from 870 °C to 1200 °C (Ref 49).

Comparison of shear tests, slow-bend Charpy tests, and K_{IC} tests in the Stage I tempered (R_c53) VAR4340 (Ref 49) revealed a correlation between γ_i and Charpy energy consistent with the demonstration by metallographic sectioning of bend specimens that failure in the blunt notch initiates by shear localization for both coarse and fine-grained microstructures. Sectioning of K_{IC} specimens loaded to 90% K_{IC} showed that sharp crack propagation in the fine-grained 870 °C treated material occurs by "zig-zag" shear localization with a 10 μm wavelength, while an anomalous increase in K_{IC} in the 1200 °C treated coarse-grained material is associated with intergranular microcracking which increases the fracture process zone by more than an order of magnitude. This peculiar effect of grain coarsening does not appear to represent a practical toughening mechanism in view of the associated degradation in shear localization resistance revealed in other test geometries. Accelerated shear localization in the coarse structures may be related to texture softening effects associated with crystal slip behavior (Ref 61).

Comparative shear, Charpy, and K_{IC} testing of fine-grained VAR processed AF1410 steel shows strong correlations indicative of a dominant role of microvoid nucleation and shear localization in all three geometries. With a standard 830 °C solution treatment, comparison of underaged and overaged materials of the same hardness shows the accelerated microvoid nucleation effect of undissolved cementite in the underaged material as revealed in a reduction of both γ_i and K_{IC}. The results indicate a direct scaling between γ_i and the critical crack opening displacement (COD) for propagation (Ref 47). In the standard 510 °C 5 h tempered condition (slightly overaged), a modest increase in solution temperature from 830 to 885 °C is found to produce a significant improvement in strength/Charpy toughness combination without excessive grain coarsening (Ref 62). Examination of the particle dispersions present at these two temperatures demonstrates that the temperature increase greatly simplifies the dispersion from a list of six phases to a single (Ti,Mo) (C,N) phase as summarized in Table 3 (Ref 49). Although the strength is slightly reduced by the higher solution treatment, a lower temperature post-age treatment for enhanced Co ordering is found to restore a normal strength level while maintaining a 40% enhancement in Charpy toughness (Ref 49, 62). These experiments demonstrate how, for a fixed melt practice, fine-tuning of solution treatment to eliminate unnecessary particles can provide an optimal grain refining dispersion maintaining a sufficiently fine grain size while enhancing toughness through improved microvoid nucleation resistance.

Table 3 Particles in AF1410

Solution Temp.	Particle Type	Size Range, nm
830 °C (1525 °F)	$(Ti,Mo)_xC,N$	10–180
	$(Cr,Fe,Mo)_{23}C_6$	50–125
	$(Fe,Cr,Mo)_{23}C_6$	75–105
	$(Mo,Cr)_xC$	30–65
	$(Mo,Fe)_xC$	40–50
	Mo_xC	5–10
885 °C (1625 °F)	$(Ti,Mo)_xC,N$	20–35

Information on interfacial cohesion for phase selection for grain refining dispersions can be derived from nucleation experiments on well-characterized microstructures. Interpreting the measured shear localization strains as microvoid nucleation strains in the fine-grained 870 °C treated VAR4340 and RSP4340, a fit of the Needleman numerical nucleation model using the measured $\bar{d}$ and f for the $M_{23}C_6$ and Al_2O_3 dispersions has provided an estimate of the interfacial strength (or work of separation) indicating superior interfacial cohesion for $M_{23}C_6$ compared to Al_2O_3 (Ref 53). The intrinsic work of interfacial separation per unit area, $\Delta\gamma$, can be expressed by:

$$\Delta\gamma = \gamma_P + \gamma_M - \gamma_{IPB}$$

where γ_P and γ_M are the particle and matrix free surface energies and γ_{IPB} is the particle/matrix interphase boundary energy. While γ_{IPB} is difficult to anticipate, phases with high cohesive energy (which also promotes high stability and coarsening resistance to maintain fine dispersions) will tend to have higher free surface energy γ_P promoting higher $\Delta\gamma$. Such simple considerations imply Al_2O_3 should be superior to $M_{23}C_6$, contrary to experiment. In practice, interfacial cohesion will likely be strongly influenced by impurity segregation through mechanisms to be addressed in our discussion of grain boundary cohesion. High-resolution microanalysis of these interfaces will likely play an important role in the full characterization of the structural basis of microvoid nucleation resistance. In the meantime, quantitative estimates allowed by currently available nucleation experimental data indicate that TiC possesses the highest interfacial cohesion in steels (Ref 63). The observation of the closely related (Ti,Mo) (C,N) phase in an optimally solution-treated (Table 3) commercial AF1410 composition which is the product of an empirically optimized melt practice, further supports the merit of this compound for intrinsic nucleation resistance. Experiments to quantify the interfacial cohesion of this phase in AF1410 are under way (Ref 64).

While nucleation models can prescribe the dependence of microvoid nucleation resistance on average particle size, $\bar{d}$, and volume fraction, f, optimization of a grain refining dispersion requires a description of the dependence of grain refinement on these same variables. Based on an extensive quantitative study of grain refinement in VAR and RSP matrix tool steels, Hsu (Ref 65) has synthesized boundary pinning models of Gladman (Ref 66) and Hellman-Hillert (Ref 67) to define the average grain diameter D stabilized by the boundary pinning interaction of a dispersion of stable (static) particles of distributed size:

$$\bar{D} = \frac{4}{3\alpha}\left(\frac{3}{2} - \frac{2}{Z}\right)\left(\Sigma\frac{f_i}{d_i}\right)^{-1}$$

Here, f_i is the volume fraction of particles of diameter d_i, Z is a grain size distribution parameter defined by the ratio of the largest grain to the average grain size, and α represents the range of boundary/particle interaction normalized to the particle size. In the approximation of monodispersed particles of diameter, $\bar{d}$, and total volume fraction, f, the expression simplifies to:

$$\bar{D} = \frac{4}{3\alpha}\left(\frac{3}{2} - \frac{2}{Z}\right)\frac{\bar{d}}{f}$$

Figure 12 summarizes the observed correlation (with $\alpha = 1$) between $\bar{D}$ and $(\Sigma f_i/d_i)^{-1}$ for VAR

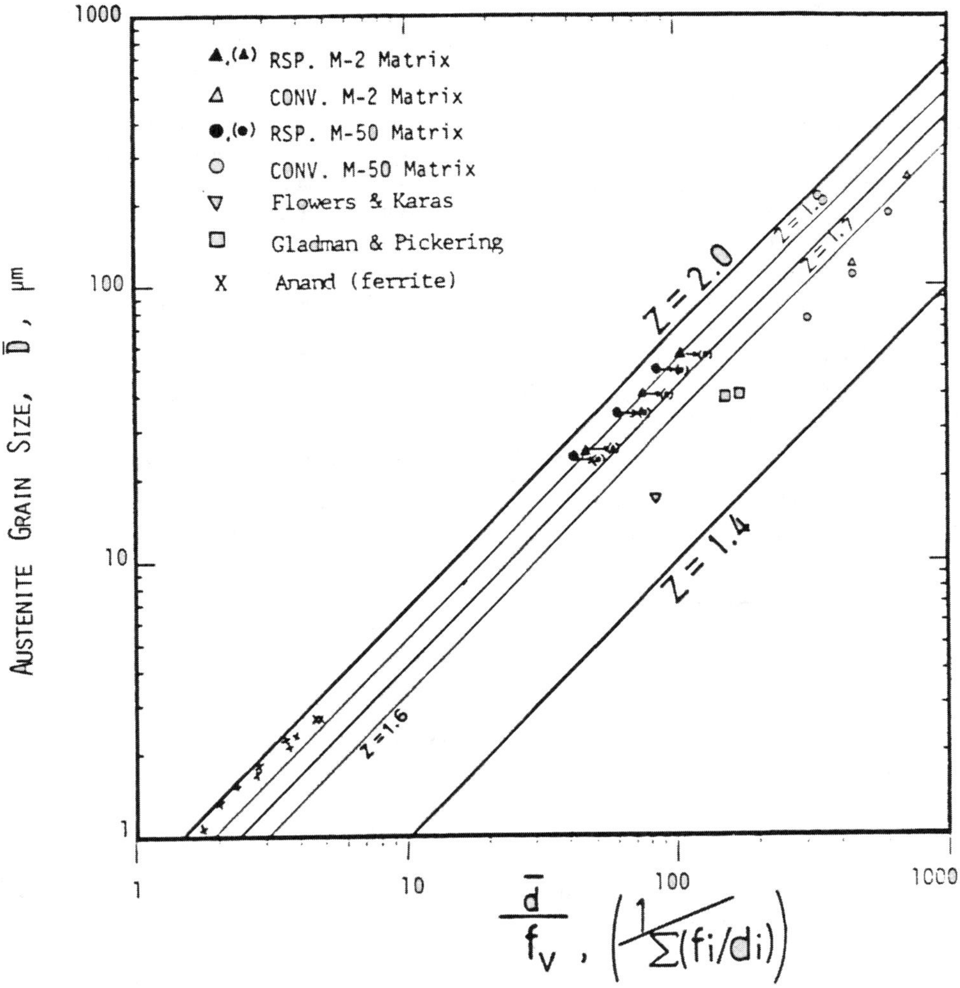

Fig. 12 Comparison of grain size data for steels with average particle size and volume fraction. Arrows show shift associated with summation over particle size distribution. Straight lines depict model predictions for various values of grain size distribution parameter, Z. Source: Ref 65

and RSP processed M-2 and M-50 matrix tool steels over a wide range of solution temperatures, and includes available literature data (Ref 68–70) relating D and $\bar{d}$/f in other steels. The beneficial grain refining effect of fine particle sizes in hot-consolidated RSP material is partly offset by a broader grain size distribution represented by a higher Z. The ability of RSP processing to disperse highly stable, coarsening resistant compounds such as oxides and oxysulfides on a fine size scale for efficient boundary pinning has allowed achievement of remarkable high-temperature grain coarsening resistance in a wide range of steels (Ref 71); the phenomenon offers the potential for practical exploitation of high solution treatment temperatures in the high toughness secondary hardening steels of concern here.

Since maintaining a desired grain size requires a fixed ratio of d to f, and nucleation resistance (as in Fig. 11) depends monotonically on both variables, geometric optimization of the grain refining dispersion then requires minimization of both d and f at fixed d/f. However, it can be anticipated that at sufficiently small particle size the Gladman type of static particle/boundary interaction just discussed will no longer apply as thermally activated boundary unpinning reduces the net interaction force exerted by the particles. Modeling such unpinning behavior by an adaptation of thermally activated deformation theory gives the effective pinning behavior described in Fig. 13 (Ref 72). Figure 13(a) depicts the required f to maintain a 10 μm grain size in a steel after 1 hr at 870 °C and 1200 °C as a function of particle size d. The straight line corresponds to the Gladman-type static interaction. Below a critical particle size, higher f is required for boundary pinning to compensate the effect of thermal activation. Adopting the f-d curve for 1200 °C solution treatment as inputs to the Needleman nucleation model, preliminary calculations of microvoid nucleation strain versus particle size (with f constrained to maintain constant D) for R_c53 VAR 4340 with an $M_{23}C_6$ dispersion give the behavior depicted in Fig. 13(b) (Ref 53). Although thermal activation ultimately limits improvements achievable at very small particle sizes, the model predicts that modest size refinement below the current level of d = 0.17 μm can yield substantial improvement in microvoid nucleation resistance. Maintaining the desired finer particle sizes during solution treatment will require more stable coarsening resistant phases such as those accessible through rapid solidification processing.

While experiments have already demonstrated substantial toughness enhancement in AF1410 through improved solution treatment to eliminate unnecessary dispersed particles, evolving theoretical guidelines and preliminary quantitative estimates indicate that even greater improvements in microvoid nucleation resistance await the design of optimal grain refining dispersions via combined phase selection and process control to achieve prescribable dispersion geometries.

Transformation Toughening

As emphasized by Krauss (Ref 45), equally important to shear localization as the role of

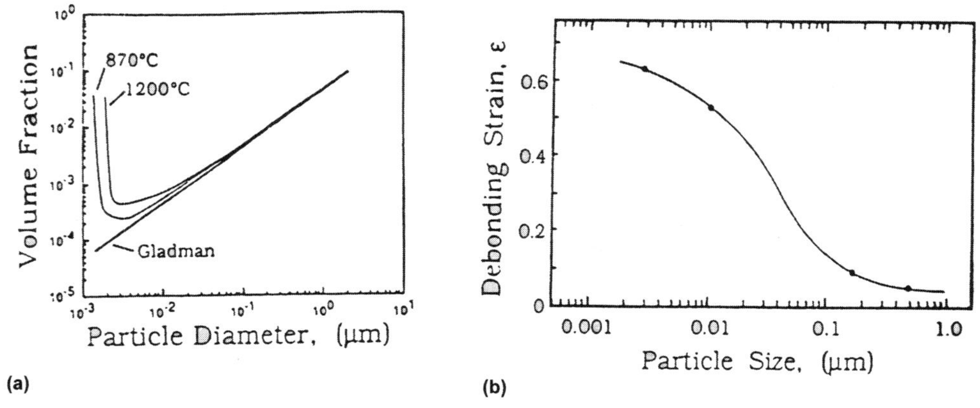

Fig. 13 Void nucleation on grain refining dispersion. (a) Necessary particle volume fraction versus particle diameter to maintain 10 μm grain size after 1 h at 870 °C and 1200 °C accounting for thermal activation, (b) particle debonding strain versus particle diameter employing volume fractions of (a) in Needleman model

particle dispersions is the role of the constitutive behavior of the matrix in which the particles reside. The most dramatic means of altering matrix constitutive behavior to delay shear localization and ductile fracture is the transformation plasticity that can be obtained from a dispersion of metastable austenite, as also represented in Fig. 2. Transformation plasticity resulting from martensitic transformation during deformation has been most thoroughly investigated in fully austenitic steels (Ref 73), identifying two basic modes of transformation. Stress-assisted transformation involves the same nucleation sites as the spontaneous transformation on cooling, but assisted by the thermodynamic effect of applied stress, while strain-induced transformation is controlled by the production of new nucleation sites by plastic deformation. Different transformation product morphologies tend to be associated with these two modes: stress-assisted transformation giving relatively coarse plates, while the strain-induced transformation develops as fine laths typically formed at intersections of shear microbands. Kinetic models and quantitative constitutive relations for associated transformation plasticity have been developed for both the stress-assisted (Ref 74) and strain-induced (Ref 75, 76) modes. Both modes promote sigmoidal stress-strain (σ-ε) curves with an initial upward curvature deriving from the combined influence of a softening contribution of the transformation as a deformation mechanism and a hardening contribution of the transformation product. Stress-assisted transformation gives plastic flow at stresses lower than the normal slip-controlled yield stress, and produces a reverse (positive) temperature dependence of the yield stress. The temperature where the transformation yield stress and slip yield intersect, characterized by a reversal in the yield-stress temperature dependence, is designated M_s^σ, and its value measured in uniaxial tension serves as a convenient measure of overall austenite stability with respect to mechanically induced transformation. As the transformation volume change interacts with the hydrostatic component of stress, the austenite stability and M_s^σ temperature vary significantly with stress state.

The upward σ-ε curvature associated with transformation plasticity imparts high stability to plastic flow behavior which is found to retard not only tensile necking but shear localization both within tensile necks and at crack tips. Figure 14 summarizes the measured effect of austenite stability (controlled by test temperature) on the relative change in (a) uniform ductility, ε_u, (b) tensile fracture ductility, ε_f, and (c) J_{IC} fracture toughness, in high-strength (1300 MPa σ_y) γ'-strengthened austenitic steels (Ref 77, 78). The stability

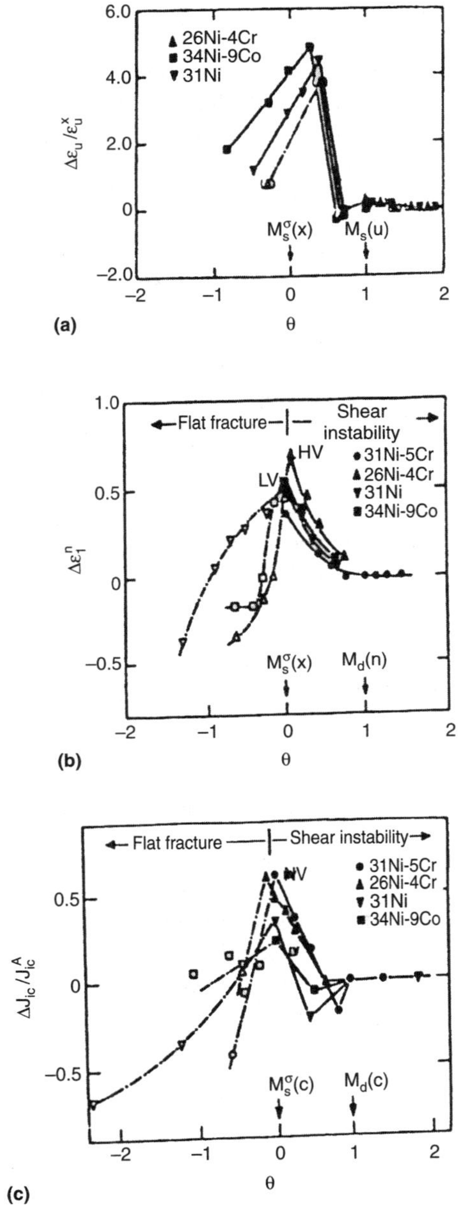

Fig. 14 Relative mechanical property enhancement versus transformation stability represented by normalized temperature, θ, for precipitation-strengthened metastable austenite steels. (a) Increase in uniform ductility $\Delta\varepsilon_u$ relative to stable austenite value ε_u^A, (b) increase in true fracture strain ε_f^n in axisymmetric tension, (c) increase in fracture toughness ΔJ_{IC} relative to stable austenite value J_{IC}^A, with LV and HV denoting low and high volume-change alloys, respectively. Source: Ref 77, 78

variable θ represents a normalized temperature defined by $\theta = \left(T - M_s^\sigma\right)/\left(M_d - M_s^\sigma\right)$ such that $\theta = 0$ corresponds to M_s^σ, and $\theta = 1$ corresponds to $T = M_d$, the highest temperature for which strain-induced transformation occurs. Here the M_s^σ temperature corresponds to that for the particular stress state controlling the property of interest. In addition to the well-known effect of transformation on uniform tensile ductility, the results of Fig. 14 demonstrate strong enhancement of ε_f and J_{IC} in materials where both failure initiation in the tensile neck and sharp-crack propagation are controlled by shear localization. Property enhancement is extremely sensitive to stability, with optimum properties achieved near the M_s^σ temperature. Hence, efficient use of transformation plasticity to improve a property at a given temperature requires placement of M_s^σ (for the stress state governing that property) at the temperature of interest. The precise mechanism by which transformation plasticity enhances J_{IC} toughness is being investigated by detailed metallographic observations of crack tip shear localization and transformation behavior (Ref 79) coupled with finite-element modelling of crack-tip transformation plasticity based on the transformation-kinetics-based constitutive models (Ref 80).

Composite systems with metastable austenite dispersions can exhibit essentially the same transformation plasticity features as fully austenitic alloys. The stability and transformation plasticity behavior of dispersed austenite in high strength steels has been most thoroughly investigated in the case of retained austenite in Stage I tempered 4340 steel (Ref 32, 81–83). For both 870 °C and 1200 °C austenitizing temperatures, controlled experiments varying austenite content with high magnetic field cryogenic treatments prior to 200 °C tempering have separated the role of the austenite in mechanical behavior. Microyielding experiments have quantified the retained austenite stability by measuring its M_s^σ temperature in uniaxial tension and compression, and revealed microyield σ-ε behavior consistent with transformation plasticity model predictions (Ref 82). The measured stress-state dependence of M_s^σ revealed an M_s^σ near room temperature for pure shear, and room temperature shear tests confirmed a significant increase in the shear instability strain, γ_i, from transformation plasticity interaction. The measured stability also indicated too low a stability level for optimum toughening interaction at a mode I crack tip, and controlled experiments confirmed that there was no significant effect of the retained austenite on K_{IC} toughness (Ref 81). The high stress levels operating in such high-strength steels contribute a large mechanical driving force for transformation. Achieving sufficiently high stability against transformation to obtain optimal toughening interaction is likely beyond the capabilities of the simple carbon-partitioning-based stabilization operative in the conventional retained austenite of Stage I tempered martensitic steels.

Martensitic nucleation theory (Ref 84) prescribes the manner in which particle size and composition control the stability of dispersed austenite. Size influences stability through the statistical behavior of heterogeneous nucleation sites. In metals and ceramics, a universal form of nucleation site potency distribution (Ref 85, 86) defines the manner in which characteristic defect potency diminishes with decreasing particle size. Composition affects stability in the same manner as for bulk materials, through the transformation chemical driving force and through a solid solution hardening effect on interfacial mobility (Ref 85). The high levels of austenite stability necessary for optimal transformation toughening in ultrahigh-strength steels can thus be attained by a combination of particle size refinement and enrichment in stabilizing solutes. The precipitation of austenite during tempering of secondary hardening steels offers the means to achieve this microstructural objective.

Multistep tempering treatments have demonstrated the ability to effect the desired microstructural control. Austenite stabilization also forms the central concept of "triple phase" ferrite/martensite/austenite low-alloy sheet steels designed to achieve sufficient plane-strain uniform ductility (20–25%) for adequate stretch formability at high strength levels (100 ksi 3% flow strength) desired for automotive applications (Ref 87). Short-time intercritical tempering after a low temperature temper which enriches cementite in Mn can produce a paraequilibrium carbide→austenite conversion, trapping a higher-than-equilibrium Mn content in the austenite (Ref 88). While this austenite normally forms at martensite lath boundaries where rapid growth ensues, thermodynamic calculations employing the paraequilibrium constraint can predict optimal cementite Mn contents and intercritical tempering temperatures to promote intragranular carbide conversion in recrystallized structures to achieve finer dispersions with higher stabilizing Mn contents (Ref 87). Similarly, in the high alloy AF1410 secondary hardening steel, solute enrichment of cementite at 420 °C prior to 510C tempering has been shown to provide a finer, more solute-enriched (intralath) austenite compared with the interlath austenite which precipitates during conventional single-step 510 °C tempering (Ref 83).

While optimal austenite stabilization in the low alloy steels requires trapping of Mn levels higher than equilibrium, the interlath austenite that precipitates during conventional tempering of high alloy steels like AF1410 involves incomplete partitioning such that the austenite solute content is lower than the equilibrium level, and further stabilization can be achieved by moving closer to equilibrium. In this case, more effective than cementite nucleation of austenite is the use of a short high temperature treatment prior to 510 °C tempering, which is found to nucleate a fine intralath dispersion of near-equilibrium austenite (Ref 83). The open symbols in Fig. 15 show Charpy toughness/hardness combinations achieved by various temperature (°C)/time(minutes) treatments prior to 510 °C/8 h tempering in AF1410; also shown are the properties (solid symbols) for conventional 510 °C tempering for 5 and 8 h. The curves represent contours of a quality parameter reflecting the normal hardness dependence of Charpy toughness. At R_c48 hardness, the 600 °C/15 min treatment provides a 24% toughness increase, while 570 °C treatments offer even better hardness/toughness combinations. Figure 16 shows a transmission electron micrograph of the microstructure produced using the 600 °C/15 min treatment (Ref 83). While relatively coarse interlath austenite (similar to that formed with single-step 510 °C tempering) is present, a fine intralath dispersion of 10–20 nm austenite particles is also evident throughout the structure. Austenite compositions determined by STEM microanalysis of this structure are compared with those for single-step 510 °C (5 and 8 h) in Table 4, together with associated mechanical properties. In addition to the substantial size refinement, it is evident that the Ni content is much increased in the fine austenite nucleated at 600 °C and is close to the predicted equilibrium composition (Ref 82). Whether compared to single-step 510 °C tempered materials on the basis of constant hardness (510 °C/5 h) or yield strength (510 °C/8 h), it is clear from the properties listed in Table 4 that the achievement of an

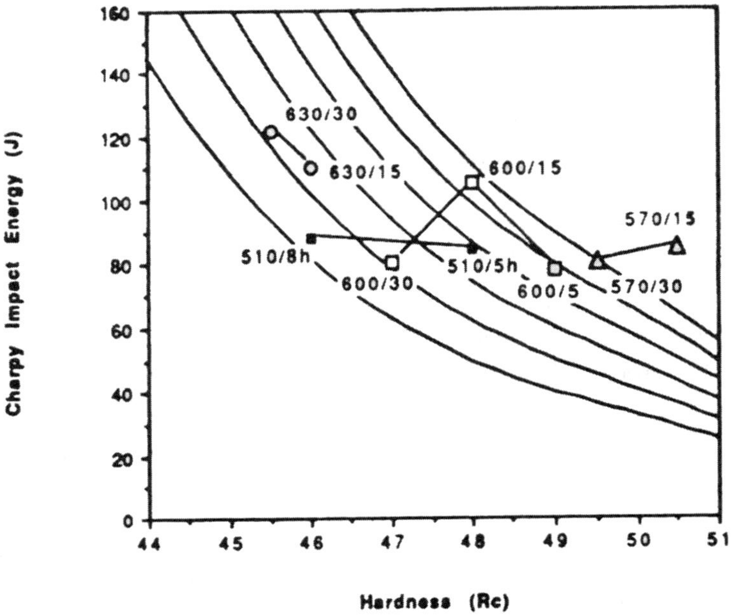

Fig. 15 Charpy V-notch impact toughness versus hardness achieved by multistep heat treatment to form high-stability austenite in AF1410 steel. Open symbols denote properties for indicated temperature (°C)/time (minutes) treatments prior to 510 °C/8 h tempering. Closed symbols denote properties for conventional 510 °C tempering for 5 and 8 h. Source: Ref 83

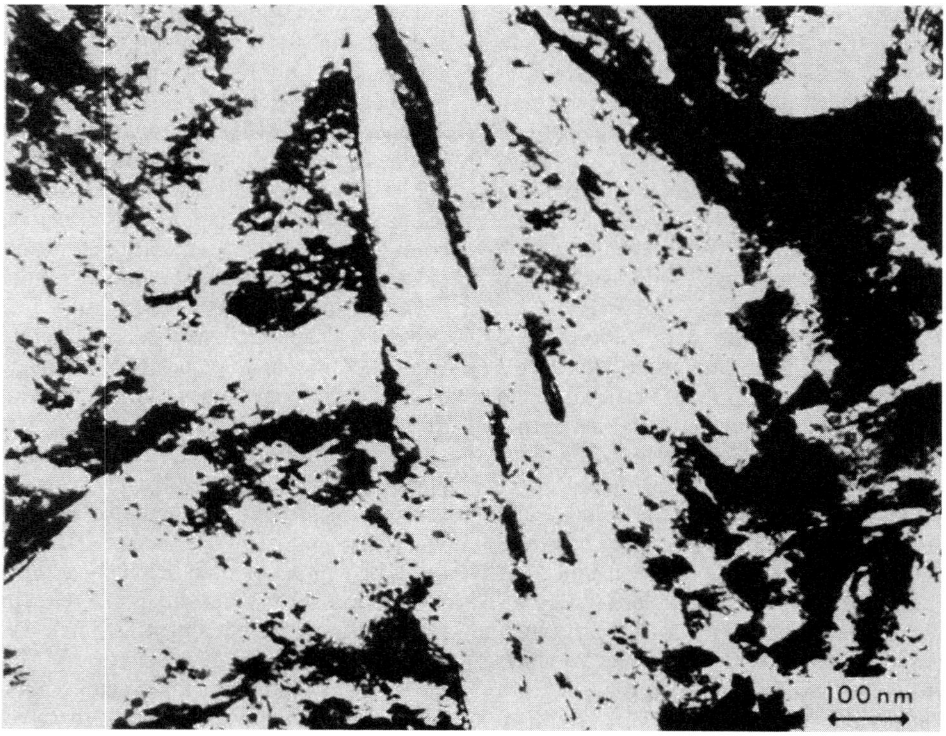

Fig. 16 TEM micrograph showing fine dispersion of intralath austenite formed by multistep heat treatment (600 °C/15 min + 510 °C/8 h); larger interlath austenite is also evident. Source: Ref 83

Table 4 AF1410 precipitated-austenite transformation toughening

	510 °C/5 h	510 °C/8 h	600 °C/15 min + 510 °C/8 h
γ Microstructure			
Composition			
interlath: (0.1–0.2 mm)		14Ni-13Co-3Cr	17Ni-15Co-4Cr
intralath: (100–200 Å)		...	29Ni-10.5Co-4Cr
f^γ	0.03	0.06	0.15
Properties			
Hardness	R_C48	R_C46	R_C48
Yield strength	1550 MPa	1360 MPa	1370 MPa
J_{IC} toughness	178 kJ/m^2	201 kJ/m^2	250 kJ/m^2

optimal austenite stability level has produced a substantial improvement in J_{IC} fracture toughness; equivalent to a K_{IC} of $230\,ksi\sqrt{in}$, this corresponds to a record toughness for a steel of R_c48 hardness.

Experiments in two series of model high-strength (R_c45) austenitic alloys strengthened by γ' precipitation and phosphocarbide precipitation have allowed a separation of the relative roles of transformation strain hardening and transformation dilatation in transformation toughening under condition of shear-localization-controlled ductile fracture (Ref 89). In the phosphocarbide-strengthened material, comparison of underaged and overaged material of the same hardness allows variation of transformation strain hardening behavior by altering the austenite/martensite hardness difference. Due to preferential martensite hardening by dissolved carbon, the underaged material has an austenite/martensite Vickers hardness difference of $\Delta H_V = 88 \pm 2$ while the overaged material has $\Delta H_V = 47 \pm 2$ identical to the γ'-strengthened alloys. Variations in Ni and Cr content provide variations in transformation dilatation. Figure 17 shows the transformation toughening efficiency determined in these alloys as the ΔJ_{IC} toughening increment per amount of crack-tip transformation measured as

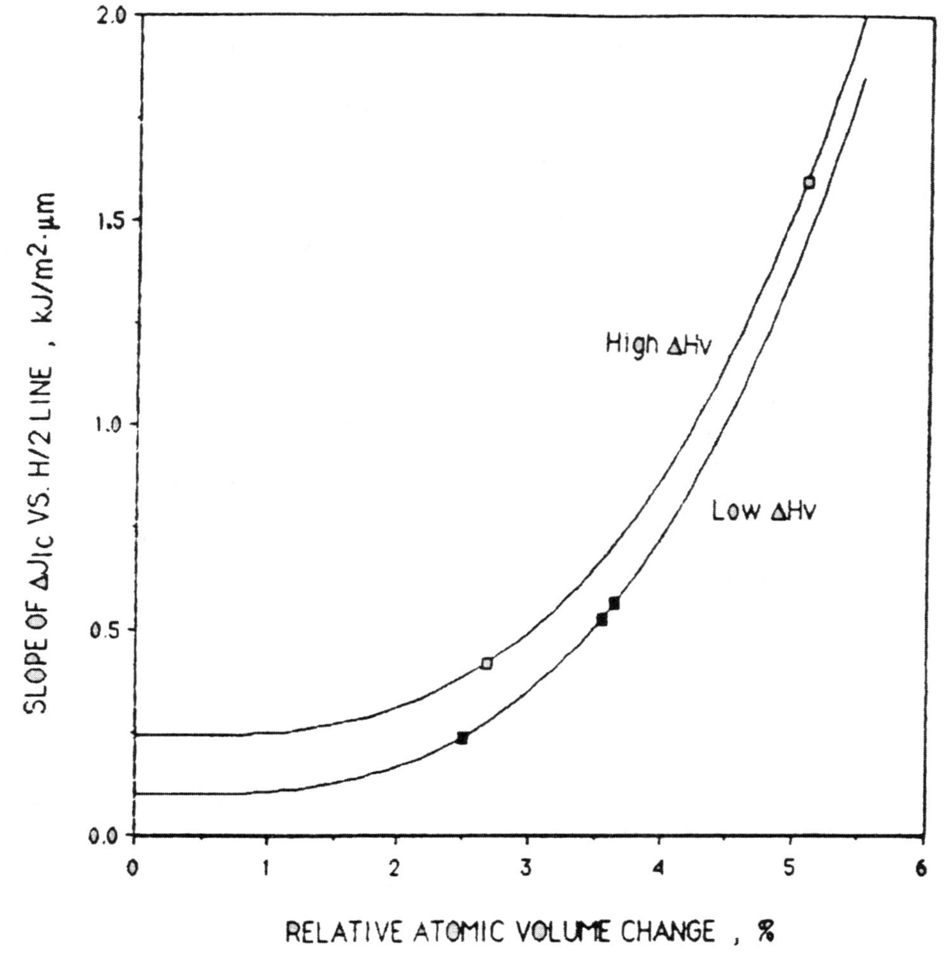

Fig. 17 Transformation toughening efficiency (slope of ΔJ_{IC} versus transformation zone half-height, H/2) versus transformation volume change for precipitation-hardened metastable austenitic steels with two levels of austenite-martensite hardness difference, ΔHV. Source: Ref 89

the transformation half-zone height, H/2. The curves indicated for the two ΔH_V levels suggest that higher transformation strain hardening enhances toughening, but a more dramatic effect is that toughening increases as the third power of the transformation dilatation; this behavior of transformation toughening in ductile solids contrasts with a second power dependence of crack-tip shielding-based transformation toughening in brittle solids (Ref 90). The strong dilatation dependence suggests that much more toughening can be achieved from precipitated austenite in secondary hardening martensitic steels by designing the austenite for maximum dilatation. Although the composition dependence of the transformation dilatation is highly nonlinear due to magnetic anomalies in austenite, magnetic models have been applied to its prediction (Ref 91). Such models are being calibrated by precision lattice parameter measurements in key compositions, to be applied in the design of optimal precipitated austenite for transformation toughening (Ref 92).

Hydrogen Resistance and Interfacial Cohesion

The stress corrosion cracking of ultrahigh-strength martensitic steels in aqueous environments, as quantified in standard K_{ISCC} tests, is known to occur by a hydrogen embrittlement mechanism associated with cathodic charging at the crack tip. Typical of hydrogen-assisted cracking at extreme strength levels, the fracture generally follows an intergranular path and is commonly associated with prior intergranular segregation of embrittling impurities. As discussed by McMahon (Ref 93), these are the same impurities responsible for the phenomenon of intergranular temper embrittlement, and extensive studies have revealed the relative embrittling potencies of impurities (e.g., P, S, Sn, Sb) and embrittling alloying elements (e.g., Mn, Si), as well as separating the role of cosegregating alloying elements which can alter the degree of impurity segregation. Thus far, the embrittling effect of hydrogen and prior segregants appear to be additive, and a common mechanistic basis can be sought for interfacial embrittlement by chemical segregants including hydrogen.

A general theoretical framework for treating the chemical embrittlement of interfaces has been developed by Rice and coworkers (Ref 94, 95) who have modelled the competition between brittle interfacial cleavage separation and crack-tip blunting by dislocation emission. Considering a flat crack in a bicrystal interface, the conditions for ductile versus brittle crack advance are determined by comparison of (a) the critical crack extension force (strain energy release rate) G_{disl} for crack-tip dislocation emission, as governed by crystal plasticity; and (b) the Griffith work of interfacial cleavage, $2\gamma_{int}$, as influenced by chemical segregation. Brittle propagation prevails when $G_{disl} > 2\gamma_{int}$, and ductile when $G_{disl} < 2\gamma_{int}$. For fixed $2\gamma_{int}$, model calculations predict that G_{disl} will depend on the direction of crack advance, such that brittle fracture can occur for one direction while ductile behavior occurs for the opposite direction. The prediction has been confirmed in critical bicrystal experiments on Cu embrittled by Bi (Ref 95).

While the precise manner in which chemical segregants might modify G_{disl} for dislocation emission remains unclear, a definite role of segregation in determining the Griffith work of interfacial cleavage $2\gamma_{int}$ has emerged from application of surface thermodynamics. Further, arguments have also been developed that $2\gamma_{int}$ is not only of crucial importance in governing the transition between ductile and brittle response, but even under conditions of substantial crack-tip plasticity the critical crack-extension force G_{IC} for intergranular fracture can be a monotonic function of $2\gamma_{int}$ (Ref 94). Control of $2\gamma_{int}$ would thus appear the most effective means of enhancing interfacial fracture resistance. The thermodynamic analysis (Ref 94, 95) reveals that for interfacial separation at fixed boundary solute coverage, Γ, nonlinear entropy contributions can be neglected such that $2\gamma_{int}$ changes linearly with Γ according to the relation:

$$2\gamma_{int} = (2\gamma_{int})_o - \left(\Delta g_b^o - \Delta g_s^o\right)\Gamma$$

where $(2\gamma_{int})_o$ is the work of separation of the clean interface, and Δg_b^o and Δg_s^o are the free energies of segregation of the solute to the boundary and free surface, respectively, evaluated at the temperature of fracture, taken as 300 K. The analysis clearly predicts that a segregating solute with a greater (more negative) segregation energy at a free surface compared to an internal boundary will embrittle (reduced $2\gamma_{int}$), while one with a lower energy at a boundary compared to the free surface will enhance interfacial cohesion (increased $2\gamma_{int}$). The embrittling potency of a segregant scales with the quantity $\Delta g_b^o - \Delta g_s^o$. The model further predicts that an embrittling solute which is mobile during fracture such that it can maintain a constant potential μ during separation will be far more embrittling than for the case of separation at fixed Γ (immobile solute).

The relative wealth of surface thermodynamic data (determined primarily from Auger electron spectroscopy studies) available for iron and steels compared to other materials make iron-based alloys the best case to test thermodynamic theory of interfacial cohesion. Unfortunately, however, both the grain boundary and free surface quantities are anisotropic, and insufficient data is available to either compare Δg_b^o and Δg_s^o at a fixed orientation or to apply consistent directional averaging procedures. Nevertheless, chemical trends are evident when $\Delta g_b^o - \Delta g_s^o$ is estimated from the available data for embrittling impurities. The best current measure of embrittling potency is the shift of the intergranular ductile-brittle transition temperature per amount of segregated solute (expected to depend linearly on $2\gamma_{int}$) and available measurements of this quantity are plotted versus $\Delta g_b^o - \Delta g_s^o$ in Fig. 18 (Ref 94). Despite considerable uncertainty associated with the available mix of polycrystalline and single crystal data, the predicted trend represented by the solid line is supported, particularly for the more well-defined cases of P and S. More recent thermodynamic data for B also correlate with its measured cohesion-enhancing effect (Ref 96). Estimates of $\Delta g_b^o - \Delta g_s^o$ for H

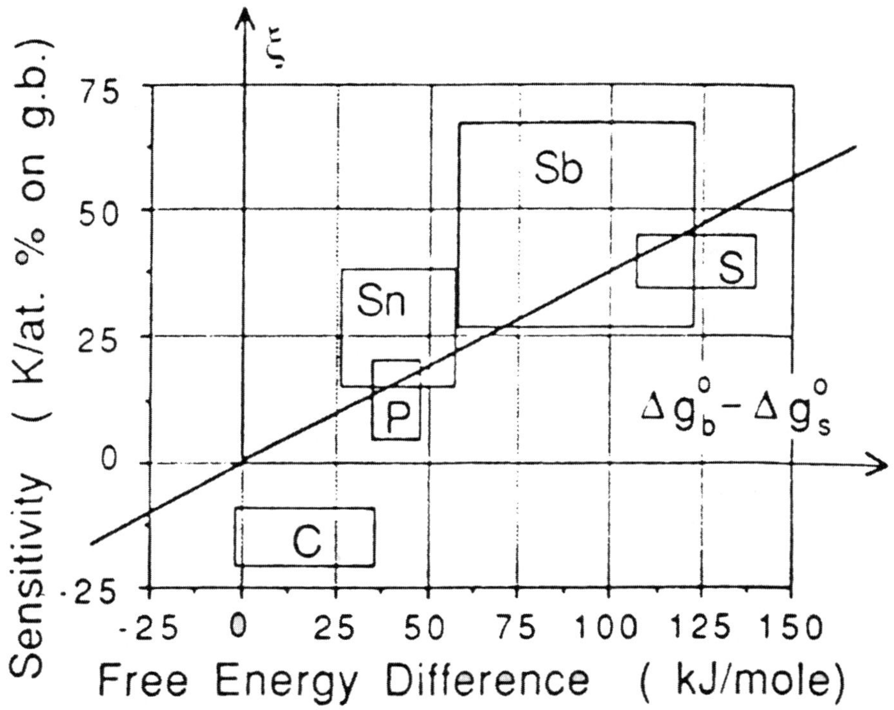

Fig. 18 Embrittling potency of impurity segregants in iron and steel (shift of transition temperature per amount of segregant) versus difference in segregation free energies to grain boundaries Δg_b^o and free surfaces Δg_s^o. Source: Ref 94

indicate an intrinsic embrittling potency less than or equal to that of P (Ref 95). The high mobility of H can allow separation at constant μ, however, and calculations for this case (Ref 95) predict substantial reduction of $2\gamma_{int}$ even for equilibrium matrix concentrations of $X_H \approx 10^{-9}$; the model is thus consistent with the extreme effectiveness of H as an embrittler (Ref 93).

A plausible thermodynamic description of the chemical embrittlement of interfaces opens the way to a more fundamental approach whereby first-principles quantum-mechanical calculations of the electronic state of impurities in grain boundary and corresponding fracture surface environments can provide insight into the origins of $\Delta g_b^o - \Delta g_s^o$ in chemical bond behavior, offering the potential for design of boundary composition for both enhanced intrinsic cohesion and reduced hydrogen interaction. The relatively low intrinsic embrittling potency represented by $\Delta g_b^o - \Delta g_s^o$ for P and H, which are the most common embrittlers of the highest strength steels, offers hope of neutralizing their embrittlement or even reversing their role to cohesion enhancement through appropriate boundary doping to achieve desired boundary electronic structure. While the most advanced techniques such as the full-potential linearized augmented plane wave (FLAPW) method (Ref 97) now appear capable of computing the relatively small energy differences governing metallurgical phenomena, the crucial first step in applying such techniques is to obtain realistic atomic positions for the structures of interest.

As reviewed by Harrison, et al., (Ref 98), simple hard-sphere models of grain boundary structure have revealed a striking similarity between the special polyhedral packing units present in grain boundary cores and the local coordination environments adopted by the light metalloid elements (e.g., B, C, P, S) in metal-rich transition metal compounds such as Fe_3P, as well as their coordination environments in metallic glasses. Specifically, these elements adopt the ninefold coordination of a capped trigonal prism which is a common structural unit in tilt boundaries. Harrison, et al., (Ref 98) have simulated the core structures of several high-coincidence tilt boundaries chosen to allow relatively short-period supercells appropriate for band-theory-based electronic calculations. To examine local atomic relaxations, a simple interatomic potential for Fe was derived based on the embedded atom method (EAM) and tested against observed free surface relaxations in Fe. (Although the ability to predict observed oscillitory variations in interplanar spacings at free surfaces is often used as an assessment of the validity of these potentials, they found such variations to be relatively insensitive to details of the potential, and found no better agreement with the observed surface relaxations of Fe than could be predicted from a simple model based on electrostatic effects neglected in the EAM approximations.) The model was initially applied to a Σ5 [100] tilt boundary on (031). Full occupancy of the trigonal prism sites of this boundary by P, however, would involve a minimum P-P spacing of 2.87 Å which is shorter than the minimum P-P distance of 3.48 Å exhibited in Fe_3P.

Furthermore, chemisorption of P on a (100) Fe surface, where the minimum site spacing is 2.87 Å, is observed to involve a c(2×2) surface reconstruction maintaining a minimum P-P spacing of 4.05 Å. A survey of possible boundaries allowing a minimum period supercell with full P site occupancy led to selection of the Σ3[1$\bar{1}$0] "incoherent" twin boundary on (111); the P site spacing of 4.05 Å is greater than the minimum P-P distance in Fe_3P, and P chemisorption on the corresponding (111) Fe free surface involves a simple p(1×1) structure with full site occupancy. The relaxed structure for this boundary computed with the Fe EAM potential is shown in the supercell of Fig. 19. Viewed along the [1$\bar{1}$0] tilt axis, the two sizes of open circles represent the two stacking positions of (1$\bar{1}$0) planes. The solid rectangles outline the projected positions of BCC units cells in the two grain orientations, and the capped trigonal prism 9-fold coordination sites are outlined in solid and dashed lines with solid circles depicting schematically the position of P sites. Repetition of supercells such as that outlined in Fig. 19 provides the necessary periodic structure for band-theory-based electronic calculations.

As a first assessment of the ability of band calculations to treat the 'chemical aspects of brittle fracture in transition metals, Krasko (Ref 99) has applied the linear muffin tin orbitals-atomic spheres approximation (LMTO-ASA) method to the classical problem of the effect of Ni on the resistance of Fe alloys to (100) cleavage fracture, using a simple periodic

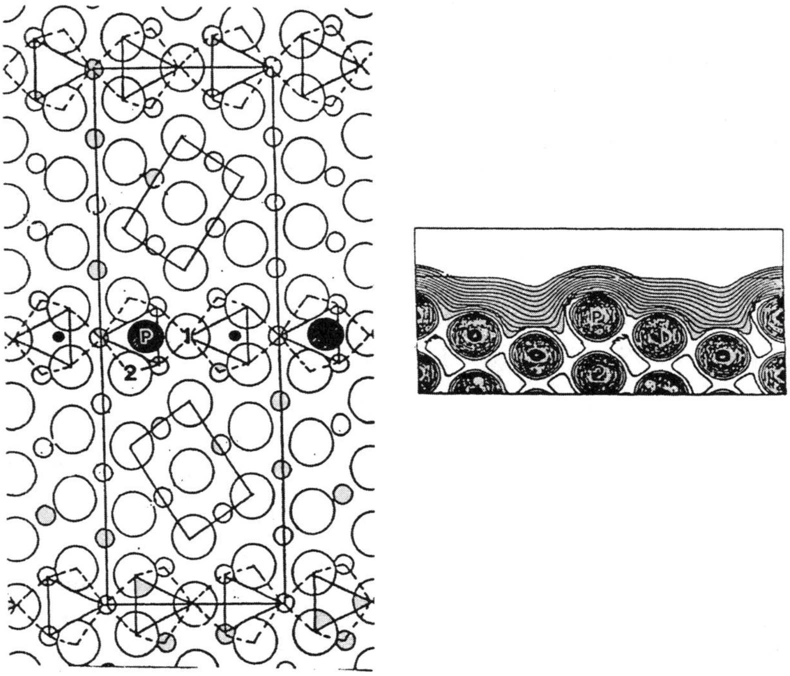

Fig. 19 Configurations employed in electronic calculations for P in Fe at grain boundary (left) and corresponding free surface (right) environments; free surface includes computed charge density contours. Source: Ref 100, 103

supercell in which vacuum is represented by empty atomic spheres to simulate (100) free surfaces. Using unrelaxed surface structures in which bulk interplanar spacings are rigidly maintained, the calculations overestimate the magnitude of the free surface energy, but clearly demonstrate that Ni at positions both on the free surface and one plane below substantially raise the (100) surface energy and therefore increase the Griffith work of separation. A decomposition of the surface energy change into magnetic and chemical contributions shows that while the strong effect of Ni below the surface is chemical in nature, the effect of Ni on the surface plane is entirely a magnetic effect arising from suppression of the normally enhanced magnetic moment of Fe atoms on a (100) surface which significantly reduces the surface energy and (100) cleavage resistance of pure Fe.

An initial application by Krasko (Ref 100) of the LMTO-ASA method to the Fe grain boundary of Fig. 19 has indicated that the near neighbor Fe atom separations across the boundary predicted by the EAM potential are unrealistically small. A modified EAM potential incorporating an explicit magnetic energy function is being developed (Ref 101) using results of LMTO-ASA calculations for large strain deformation of Fe between BCC and FCC structures (Ref 102). Preliminary LMTO-ASA calculations have been performed for the $\Sigma3[1\bar{1}0]$ (111) Fe tilt boundary containing P using a smaller six-layer supercell in which the interplanar spacing of the first two (111) planes at the boundary core are adjusted to maintain the Fe-P bond distances of the Fe_3P structure (100). The energy and electronic structure have been compared with and without P, holding the Fe atom configurations fixed. Comparison with the state of P on the corresponding (111) fracture surface is made possible by a preliminary FLAPW calculation performed for this case by Li and Freeman (Ref 103). Upon fracturing the boundary, the P atom can relax out of the plane of the grain boundary central Fe atoms (1 in Fig. 19) to a position like that represented in the computed charge density plot at the right of Fig. 19. In this preliminary calculation, the Fe-P distance was arbitrarily set equal to the Fe-Fe near neighbor distance.

Consistent with the expected electronegativity difference, the FLAPW calculation indicates charge transfer from Fe to P on the (111) free surface. In contrast, the LMTO-ASA calculation for P in the confined environment of the grain boundary indicates that the P atom (as defined by its Wigner-Seitz sphere volume) is charge deficient, and integration over this volume of the charge density difference relative to the superimposed charge densities of the clean Fe boundary and a pure P monolayer indicates a net transfer of ~0.1 electron from P. Such reverse charge transfer is consistent with the findings of other calculations for metalloid impurities in solid metallic environments and runs counter to the proposal of Briant and Messmer (Ref 104) based on simple cluster calculations that impurity-induced embrittlement is associated with weakened Fe-Fe bonding through charge transfer from Fe to the impurity. As is typical of free surfaces, the FLAPW calculations indicate the outer Fe atoms on the clean free surface are deficient 0.24 electron. Adsorption of a P monolayer restores 0.2 electron to these Fe atoms, and increases the Fe work function from 4.7 to 5.5 eV indicative of strong Fe-P bonding. The reversal of Fe-P charge transfer to a more natural state on the free surface and the restoration of surface Fe charge through P interaction may contribute to a lower energy of P on the free surface compared to the grain boundary. Energy comparisons will require relaxations of the surface P configuration, and a meaningful assessment of the role of charge transfer will require a more detailed analysis of the spatial redistribution of charge.

In analogy to the findings for the effect of Ni on Fe (100) cleavage resistance, an assessment of possible magnetic contributions to P embrittlement can be sought from the calculated Fe magnetic moments in the grain boundary and free surface environments. In both environments, Fe atom 1 of Fig. 19 in the absence of P shows an enhanced magnetic moment of ~$2.8\mu_B$ compared to $2.2\mu_B$ for bulk BCC Fe. Segregated P lowers the moment of this Fe atom in both environments, but more so in the grain boundary (~$1.1\mu_B$) than on the free surface (~$2.2\mu_B$). A lower magnetic free energy might thus contribute to the lower energy of P on the free surface versus the grain boundary.

Holding the Fe atom configuration the same as that adopted for the Fe-P case, a preliminary comparison of the states of B, C, P, and S in the grain boundary environment has been conducted with the LMTO-ASA method (100). The reverse charge transfer and strongly reduced Fe magnetic moment are similar for P and S, while weaker effects are found for B and C. Projected density of states (DOS) curves for the Fe atom adjacent to the boundary in Fig. 19 are directly compared with reflected (negative) DOS plots for the impurity atoms in Fig. 20. Overlap between the B and C 2p bands and the Fe 3d band in the region ~0.3 R_Y below the Fermi level, E_F, allows for strong Fe-impurity bonding, while such overlap with the 3p bands of P and S is virtually absent. These features, together with the relative magnetic moments and charge transfers, are in line with the greater boundary segregation energies for B and C compared to P and S. The relatively poorer bonding of P and S to the grain boundary offers the potential for a more profound change of bonding character on moving to the less constrained free surface environment.

An application of the simpler layer Korringa-Kohn-Rostoker (LKKR) technique to the Fe-P $\Sigma3[1\bar{1}0]$ (111) boundary reproduces qualitative features of the more rigorous LMTO-ASA results, including the relative reduction of the boundary Fe magnetic moment by P (Ref 105). Some preliminary calculations (Ref 106) using the discrete variational-X_α (DV-X_α) embedded cluster technique (Ref 107) have allowed direct comparison of the state of P in the Fe grain boundary and free surface configurations using the same atomic positions employed in the LMTO-ASA and FLAPW calculations. While the 3p component of the P DOS shows a similar shape to that in Fig. 20 for the grain boundary case, the shape distorts in the free surface case favoring energies near E_F which may reflect enhanced Fe-P interaction. Once essential electronic features of embrittlement are determined using the more rigorous LMTO-ASA and FLAPW methods, these simpler techniques may allow a quicker investigation of alloying effects as the basis of alloy design for enhanced interfacial cohesion.

While an electronic level understanding of interfacial embrittlement offers the greatest potential for property improvements in the long term, more immediate improvements based on empirical knowledge have been achieved through novel techniques for gettering of known embrittling impurities, and through constraint of alloy composition to exclude known embrittling alloying elements. Studies of rapid solidification phenomena in steels mentioned earlier (Ref 65, 71) have revealed that in addition to a substantial refinement in the particle size of "inclusion" phases, the normal sequential formations of oxides and sulfides is replaced by complex oxysulfide formation at the high melt undercoolings achievable by rapid solidification. Compared to sulfides, these phases getter sulfur in a much more stable form which resists dissolution at extreme temperatures. Phosphorus is normally very difficult to getter in conventionally processed steels due to the much lower stability of phosphides relative to sulfides. To explore the possibility of extending the rapid solidification effect to the formation of stable oxygen-phosphorus compounds, a thermodynamic survey was conducted of the potentially most stable phosphorus and sulfur gettering compounds in steels (Ref 108). Figure 21 summarizes the estimated free energy of formation of such compounds from dilute solution in Fe, plotted versus temperature in the range 900–1700 °C, identifying lanthanum as the most promising alloy addition to getter P as $LaPO_4$ and S as La_2O_2S. Thermodynamic and kinetic modeling further indicated feasibility of forming the metastable $LaPO_4$ phase at high melt undercoolings in compositions with the P and S contents of typical high quality steels, provided oxygen is kept sufficiently low (Ref 108). The high stability of such compounds should also provide the necessary coarsening resistance for an effective grain refining dispersion (once finely dispersed by rapid solidification) while their high cohesive energy should also in principle contribute interfacial cohesion (and microvoid nucleation resistance) through a high particle-phase free surface energy.

A simple stage I tempered martensitic steel composition was then designed as a substitute for 4340 steel capable of equivalent strength and toughness, but excluding Mn, Si, and Cr for enhanced interfacial cohesion. Carbon was maintained at 0.40C for strength, a comparable nickel content of 2.0Ni was held for cleavage resistance, and molybdenum (a possible cohesion enhancer) was increased to 1.5Mo for equivalent hardenability. Rapid solidification processing was performed by remelting a well-deoxidized billet of this composition, followed by a late addition of $LaNi_5$ (balanced to the melt P and S content with an empirical loss

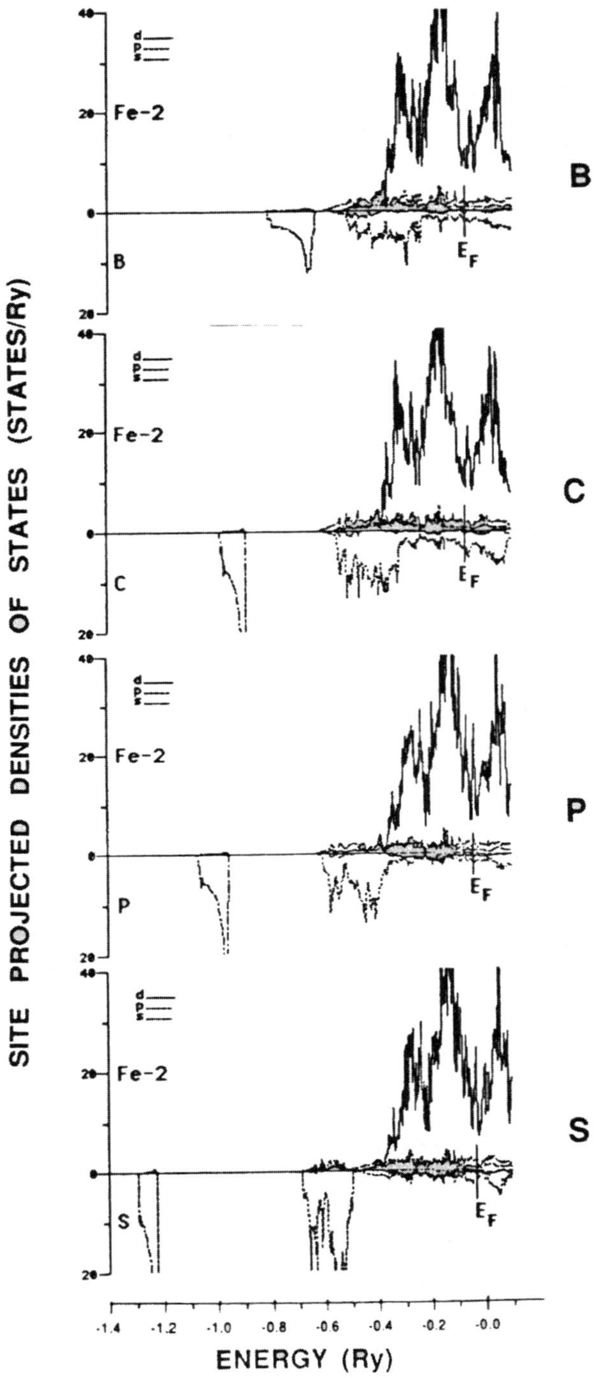

Fig. 20 Density of states (DOS) curves for the Fe atom adjacent to boundary of Fig. 19, directly compared to reflected (negative) DOS plots for B, C, P, and S in impurity site. Source: Ref 100

factor) just prior to rapid solidification via centrifugal atomization. A modest quantity of powder was produced and consolidated by hot extrusion at 1100 °C. STEM microanalysis employing a windowless x-ray detector revealed particles of the type shown in Fig. 22 with an x-ray fluorescence spectrum confirming the presence of La, O, and P; finer particles

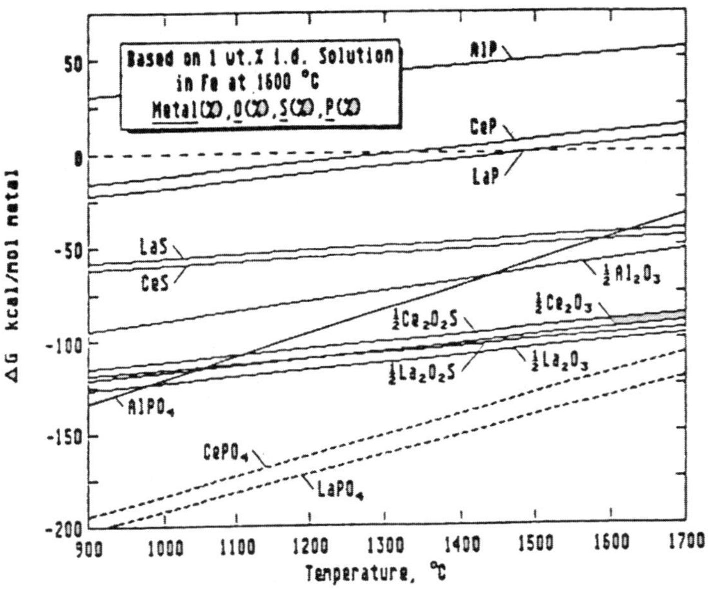

Fig. 21 Estimated free energy of formation (from solution in Fe) of potential impurity gettering phases. Source: Ref 108

containing La, O, and S and presumed to be La_2O_2S were also observed. The presence of the La-O-P phase after 1100 °C consolidation and its relatively fine ~0.1 μm size attests to a high stability of the compound. Observations on as-solidified powder by Hayzelden (Ref 60) confirms the presence of similar particles, but quantitative STEM microanalysis reveals a much higher O/P ratio than that of $LaPO_4$ suggesting an "oxyphosphide" phase rather than phosphate.

Despite serious powder cross-contamination, the experimental steels exhibited K_{IC} toughness superior to conventional 4340 and a reduced degree of quasicleavage fracture. Due to a higher-than-specified carbon content, a hardness of R_c57 was achieved with stage I tempering at 200 °C, and the grain coarsening resistance associated with the stable particle dispersions was also demonstrated. Although the small amount of material did not allow adequate K_{ISCC} determination, a rising-load K_{ISCC} screening test for which 4340 steel failed at $10\,ksi\sqrt{in}$ indicated a K_{ISCC} of $40\,ksi\sqrt{in}$, and a standard K_{ISCC} specimen survived 10^3 h at $20\,ksi\sqrt{in}$ without failure. The K_{ISCC} levels indicated by these preliminary experiments are compared against conventional ultrahigh-strength steels in Fig. 23, indicating a major property advance associated with achievement of the designed microstructure.

A larger heat of the experimental steel has been produced by gas atomization (Ref 109), but STEM microanalysis revealed the presence of lanthanum silicates attributed to a crucible reaction, with no evidence of the desired P and S gettering phases. Strength and toughness equivalent to 4340 was obtained, with good ballistic properties, but a measured 10^3 h K_{ISCC} of only $7\,ksi\sqrt{in}$ confirms the necessity of the impurity gettering to achieve the remarkable stress corrosion resistance (Ref 110). Another gas-atomized heat produced with appropriate processing modifications is currently being evaluated (Ref 110).

Materials Design—Stainless Bearing Steel

Our first effort to integrate these developments into the design of one alloy was undertaken by a team of students in the Materials Design Course at Northwestern University (Ref 111), who performed a conceptual design of a stainless bearing steel for the Space Shuttle Main Engine turbopumps, a demanding application requiring substantial improvements in toughness and stress-corrosion resistance. Maintaining a R_c60 hardness level required for rolling contact fatigue resistance and wear resistance, property objectives consisted of a doubling of K_{IC} and K_{ISCC}

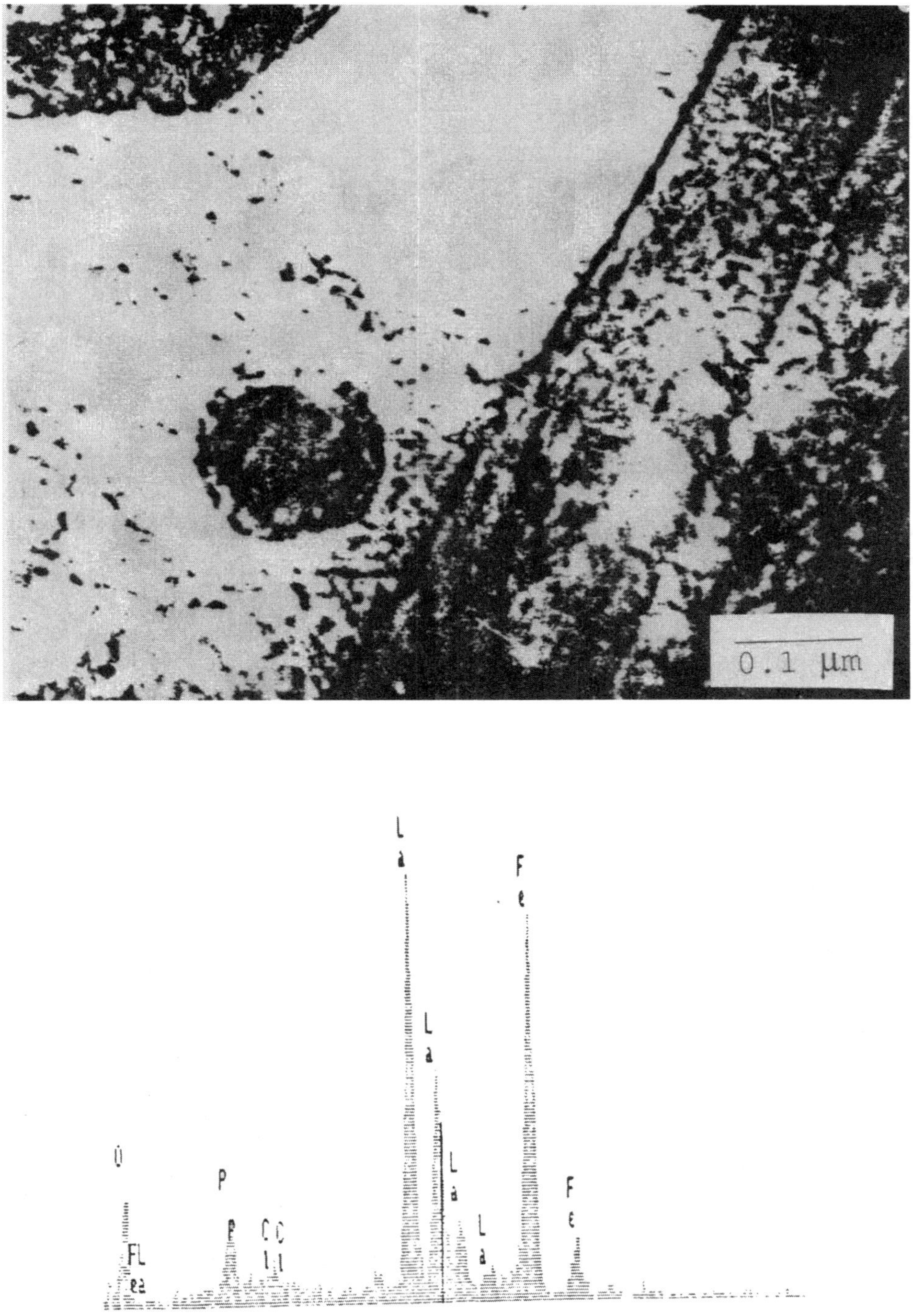

Fig. 22 STEM micrograph and associated x-ray fluorescence spectrum of La-O-P particle in rapidly solidified LaNiMo steel after consolidation at 1100 °C. Source: Ref 108

relative to the current alloy 440C employed in this application. Consistent with the system framework of Fig. 2, a sequence of calculations was devised to arrive at a unique alloy composition tailored to achieve quantitative microstructural objectives under specific processing conditions. For stress corrosion resistance, the overall composition was first constrained

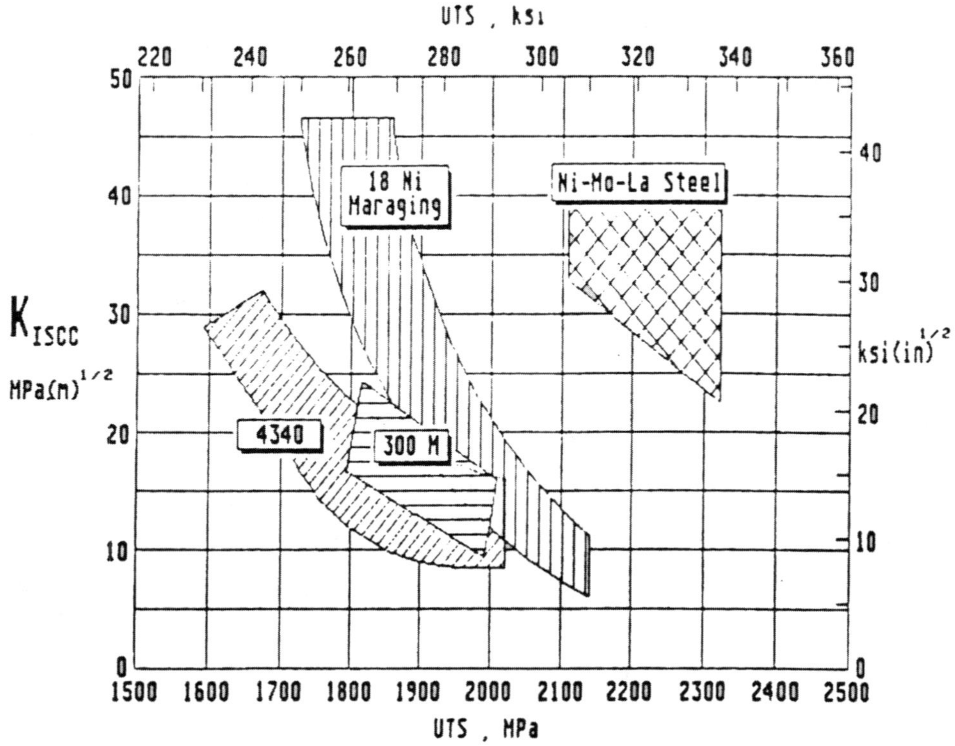

Fig. 23 Stress-corrosion resistance versus strength comparing rapidly solidified LaNiMo steel with conventional ultrahigh-strength steels. Source: Ref 108

for compatability with rapid solidification processing and La treatment, with exclusion of Mn and Si for interfacial cohesion. An Orowan strengthening analysis was applied to the available hardness and particle size data from the experimental steels of Table 1, defining the necessary particle spacing for R_c60 hardness, and a minimum carbon content of 0.30C was estimated to achieve this condition. Matrix composition was constrained to include 12Cr for stainless properties. THERMOCALC computations were then employed to compute a line of compositions in Ni and Co that maintained a sufficiently high M_s temperature to give the desired lath martensite structure. To incorporate transformation toughening, a unique Ni and Co composition was determined by the nature of the austenite that would precipitate during tempering near 500 °C, both in terms of its amount and its transformation stability as quantified by the martensitic transformation driving force at room temperature. To achieve maximum strengthening efficiency without embrittlement following the considerations of Fig. 6, V and Mo contents were then optimized to maximize the driving force for coherent M_2C at 500 °C while maintaining limits on the M_6C and $M_{23}C_6$ carbide driving forces, subject to a maximum allowed solution treatment temperature. A second iteration of this complete sequence was then followed to achieve self-consistency.

In this manner, a computer-aided thermodynamics-based design yielded a unique seven component alloy composition which is thermodynamically (and kinetically) capable of achiéving a desired multiphase microstructure under fairly well prescribed processing conditions. A small heat of this composition has been prepared and is currently undergoing evaluation. If this prototype alloy is successful, we will have arrived at a very different way of creating new materials.

Summary

Advances in theory and instrumentation focused on the principles underlying control of strength, toughness, and hydrogen resistance in ultrahigh-strength steels have achieved a level of fundamental understanding allowing true materials design. A comprehensive combined SANS, TEM, APFIM, STEM investigation of M_2C carbide precipitation has demonstrated important predicted features of precipitation at high supersaturation, and identified major effects of particle coherency. Recent theory of coherent thermodynamics and multicomponent diffusional kinetics calibrated by nanoscale experimental measurements has allowed application of the THERMOCALC thermochemical database and software system to predict the alloy composition dependence of basic scaling factors governing precipitation behavior. The thermodynamics-based control of competing precipitation reactions then allows design of optimal compositions for efficient strengthening without embrittlement.

Novel shear tests conducted in materials with well-characterized particle dispersions have quantified the pressure-dependent resistance to microvoid-nucleation-induced shear localization which accelerates ductile fracture. Continuum mechanical models show the role of dispersion geometry (size, fraction, spatial inhomogeneity) and interfacial properties on microvoid nucleation and softening behavior. The models predict improved geometries for grain refining dispersions and allow estimates of particle interfacial properties from measured nucleation behavior as a basis of phase selection for optimal grain refining dispersions with maximum resistance to microvoid nucleation. For a given alloy composition and melt practice, procedures for optimization of solution treatment to eliminate unnecessary particles have demonstrated substantial toughness enhancement.

Controlled experiments defining austenite stability and critically testing the role of austenite in mechanical behavior show that conventional retained austenite is too unstable to enhance fracture toughness in ultrahigh-strength steels. Transformation kinetic theory prescribing the particle size and composition dependence of dispersed austenite stability predicts that optimal stability for transformation toughening is achievable through austenite precipitation in high alloy steels. Guided by thermodynamic predictions, novel heat treatments and STEM microanalytical studies in AF1410 steel have demonstrated precipitation of optimally stable ultrafine intralath austenite of near-equilibrium composition, yielding a record level of fracture toughness in this steel. Critical experiments in model austenitic alloys have identified an important role of transformation dilatation in the transformation toughening of ductile solids, providing a basis for design of superior alloy compositions for maximum dispersed-phase transformation toughening.

Thermodynamic and continuum mechanical modeling of the competition between intergranular separation and crack-tip blunting predicts a thermodynamic basis for intergranular hydrogen stress corrosion with support from available surface thermodynamic data. Directed at the electronic basis of the governing thermodynamic quantities, band theoretic and cluster model calculations for embrittling impurities in grain boundary and corresponding fracture surface environments show charge transfer, magnetic moment, and band structural changes which correlate with embrittlement potency. The emerging understanding of the electronic basis of embrittlement will allow design of grain boundary composition for both enhanced intrinsic cohesion and reduced hydrogen interaction, offering the greatest potential for property improvements. Substantial advances in hydrogen stress corrosion resistance through cleaning of impurities from grain boundaries have already been achieved by design of novel metastable impurity-gettering compounds accessible through rapid solidification processing.

Integration of these concepts within a systems engineering framework has led to the first demonstration of computer-aided thermodynamics-based design of a complex multicomponent alloy directed at specific property and microstructural objectives. The frontier of materials design defined by this research constitutes both a new concept of materials science and the advent of true materials engineering.

ACKNOWLEDGMENTS

An initial phase of the Steel Research Group effort was conducted at MIT under NSF sponsorship. The expanded program at Northwestern University is funded by NSF, ONR, NASA, ARO, AFOSR, and DOE, supplemented by industrial fellowship support. Coordinated efforts at Harvard and Brown Universities are supported by NSF.

REFERENCES

1. G.M. Jenkins, "The Systems Approach," in *Systems Behavior*, ed. J. Beishon and G. Peters, Open University Press, 1972

2. C.S. Smith, *A Search for Structure*, MIT Press, 1981
3. M. Cohen, "Unknowables in the Essence of Materials Science and Engineering," *Mat. Sci. and Eng. 25* (1976) 3
4. M. Cohen, "Metallurgy and the Evolution of Materials Science and Engineering," in *Innovations in Ultrahigh-Strength Steel Technology*, ed. G.B. Olson, M. Azrin, and E.S. Wright, 34th Sagamore Army Materials Research Conference Proceedings (1990)
5. G.R. Speich, "Secondary Hardening Ultrahigh-Strength Steels," in *Innovations in Ultrahigh-Strength Steel Technology*, ed. G.B. Olson, M. Azrin, and E.S. Wright, 34th Sagamore Army Materials Research Conference Proceedings (1990)
6. G.R. Speich, D.S. Dabkowski, and L.F. Porter, *Metall Trans. 4* (1973) 303
7. J.A. Liddle, G.D.W. Smith, and G.B. Olson, "Alloy Carbide Precipitation in a High Cobalt-Nickel Secondary Hardening Steel," Proc. 33rd Int'l Field-Emission Symposium, *J. de Physique, 47* (1986) C7-223
8. J.S. Montgomery, and G.B. Olson, "Kinematics of M_2C Carbide Precipitation," in *Innovations in Ultrahigh-Strength Steel Technology*, ed. G.B. Olson, M. Azrin, and E.S. Wright, 34th Sagamore Army Materials Research Conference Proceedings (1990)
9. A.J. Allen, D. Gavillet, and J.R. Weertman, "SANS and TEM Studies of Isothermal M2C Carbide Precipitation in Ultrahigh-Strength AF1410 Steel," manuscript in preparation
10. H. Wendt and P. Haasen, *Acta Metall. 31* (1983) 1649
11. R. Kampmann and R. Wagner, in *Decomposition of Alloys*, ed. P. Haasen et al., Pergamon Press, NY (1984) p. 91.
12. R. Wagner and R. Kampmann, "Solid State Precipitation at High Supersaturations," in *Innovations in Ultrahigh-Strength Steel Technology*, ed. G.B. Olson, M. Azrin, and E.S. Wright, 34th Sagamore Army Materials Research Conference Proceedings (1990)
13. J.S. Langer and A.J. Schwartz, *Phys. Rev. A21* (1980) 948
14. G.M. Carinci, G.B. Olson, J.A. Liddle, L. Chang, and G.D.W. Smith, "APFIM Study of Multicomponent Precipitation," in *Innovations in Ultrahigh-Strength Steel Technology*, ed. G.B. Olson, M. Azrin, and E.S. Wright, 34th Sagamore Army Materials Research Conference Proceedings (1990)
15. G.M. Carinci, "Precipitation of Multicomponent Carbides in Co-Ni Martensitic Steels," Ph.D. thesis, MIT, 1989
16. T. Kinkus and G.B. Olson, Northwestern University, research in progress.
17. H.M. Lee, "Stability and Coarsening Resistance of M_2C Carbides in the Secondary Hardening Reaction," Ph.D. thesis, MIT, 1989
18. A.J. Garratt-Reed and H.M. Lee, MIT, unpublished research, 1988
19. M. Grujicic and G.B. Olson, *CALPHAD 12* (1988) 405.
20. H.M. Lee, S.M. Allen, and M. Grujicic, "Stability and Coarsening Resistance of M_2C Carbides in Secondary Hardening Steels," in *Innovations in Ultrahigh-Strength Steel Technology*, ed. G.B. Olson, M. Azrin, and E.S. Wright, 34th Sagamore Army Materials Research Conference Proceedings (1990)
21. S. Bjorklund, L.F. Donaghey, and M. Hillert, *Acta Met.* 20 (1972) 867
22. B. Sundman, B. Jansson, and J.O. Andersson, *CALPHAD, 9* (1985) 153
23. M. Grujicic, "Design of M_2C Carbides for Secondary Hardening," in *Innovations in Ultrahigh-Strength Steel Technology*, ed. G.B. Olson, M. Azrin, and E.S. Wright, 34th Sagamore Army Materials Research Conference Proceedings (1990)
24. K-U. King, P.W. Voorhees, T. Mura, and G.B. Olson, Northwestern University research in progress
25. F. Larché and J.W. Cahn, *Acta Metall. 21* (1973) 1051
26. G. Ghosh and G.B. Olson, Northwestern University, research in progress
27. M. Schmidt and R. Hemphill, "Characterization of 4% Molybdenum SRG Experimental Steel," in *Innovations in Ultrahigh-Strength Steel Technology*, ed. G.B. Olson, M. Azrin, and E.S. Wright, 34th Sagamore Army Materials Research Conference Proceedings (1990)
28. R. Hemphill, Carpenter Steel alloy AerMet100, patent pending
29. M. Grujicic and G.N. Haidemenopoulos, *CALPHAD 12* (1988) 219
30. J.L. Goss, "Austenitic Grain Growth and Precipitation Strengthening in Rapidly

Solidified Alloys," M.S. thesis, MIT, 1980
31. J.J. Irani and R.W.K. Honeycombe, *J. Iron and Steel Inst.*, 203 (1965) 826
32. G.N. Haidemenopoulos, "Dispersed-Phase Transformation Toughening in Ultrahigh-Strength Steels," Ph.D. thesis, MIT, 1988
33. A.T. Davenport and R.W.K. Honeycombe, *Metal Science, 9* (1975) 201
34. S. Spooner, H.J. Rack, and D. Kalish, *Metall. Trans. 2A* (1971) 2306
35. L. Chang, G.D.W. Smith, and G.B. Olson, *J. de Physique 47* (1986) C2-265
36. I. Aslanidis, "Softening and Recovery Resistance in Fe, Fe-Co, Fe-Al, Fe-Ni, and Fe-Ni-Co Alloys," Ph.D. thesis, MIT, 1989
37. V. Pierron-Bohnes, M.C. Cadeville, and G. Parette, *J. Phys. F: Met. Phys., 15* (1985) 1441
38. H. Warlimont and G. Thomas, *Metal Sci. 4* (1972) 47
39. S. Spooner, Oak Ridge National Laboratory, unpublished research
40. A.F. Yedneral, O.P. Zhukov, M.A. Kablukovskaya, B.M. Mogutnov, and M.D. Perkas, *Fiz. Metal. Metalloved, 36* (1974) 727
41. A.B. Kuper, D. Lazarus, J.R. Manning, and C.T. Tomizuka, *Phys. Rev., 104* (1956) 1536
42. E. Hornbogen, "Alloy Design for >3GPa Steels," in *Innovations in Ultrahigh-Strength Steel Technology*, ed. G.B. Olson, M. Azrin, and E.S. Wright, 34th Sagamore Army Materials Research Conference Proceedings (1990)
43. C. Knepfler, K.T. Faber, and J. Weertman, Northwestern University, research in progress
44. W.M. Garrison, Jr., and K.J. Handerhan, "Fracture Toughness: Particle-Dispersion Correlations," in *Innovations in Ultrahigh-Strength Steel Technology*, ed. G.B. Olson, M. Azrin, and E.S. Wright, 34th Sagamore Army Materials Research Conference Proceedings (1990)
45. G. Krauss, "Shear Fracture of Ultrahigh-Strength Low Alloy Steels," in *Innovations in Ultrahigh-Strength Steel Technology*, ed. G.B. Olson, M. Azrin, and E.S. Wright, 34th Sagamore Army Materials Research Conference Proceedings (1990)
46. J. Mescall and H. Rogers, "The Role of Shear Instability in Ballistic Penetration," in *Innovations in Ultrahigh-Strength Steel Technology*, ed. G.B. Olson, M. Azrin, and E.S. Wright, 34th Sagamore Army Materials Research Conference Proceedings (1990)
47. J.G. Cowie, M. Azrin, and G.B. Olson, "Microvoid Formation During Shear Deformation of Ultrahigh-Strength Steels," in in *Innovations in Ultrahigh-Strength Steel Technology*, ed. G.B. Olson, M. Azrin, and E.S. Wright, 34th Sagamore Army Materials Research Conference Proceedings (1990)
48. J.G. Cowie, M. Azrin, and G.B. Olson, *Metall. Trans. 20A* (1989) 143
49. M.J. Gore, "Grain Refining Dispersions and Mechanical Properties in Ultrahigh-Strength Steels," Ph.D. thesis, MIT, 1988
50. A. Benedetti and J.B. Cohen, Northwestern University, unpublished research, 1989
51. A. Needleman, *J. Applied Mechanics, 54* (1987) 525
52. A. Needleman, "A Numerical Study of Void Nucleation at Carbides," in *Innovations in Ultrahigh-Strength Steel Technology*, ed. G.B. Olson, M. Azrin, and E.S. Wright, 34th Sagamore Army Materials Research Conference Proceedings (1990)
53. M.J. Gore, A. Needleman, and G.B. Olson, unpublished research, 1988
54. J.W. Hutchinson and V. Tvergaard, "Effect of Particle-Void Interaction on Void Growth in Tension and Shear," in *Innovations in Ultrahigh-Strength Steel Technology*, ed. G.B. Olson, M. Azrin, and E.S. Wright, 34th Sagamore Army Materials Research Conference Proceedings (1990)
55. J.W. Hutchinson and V. Tvergaard, "Softening Due to Void Nucleation in Metals," in *Fracture Mechanics: Perspectives and Directions* ed. R.P. Wei and R.P Gangloff, ASTM STP 1020 (1989) p 61
56. Y. Huang and J.W. Hutchinson, "A Model Study of the Role of Nonuniform Defect Distribution on Plastic Shear Localization," in *Role of Modelling in Materials Design*, ed. J.D. Embury, TMS-AIME (1990)
57. D.M. Tracey and P.J. Perrone, "Modelling of Interactions in Void Nucleation

and Growth," in *Innovations in Ultrahigh-Strength Steel Technology*, ed. G.B. Olson, M. Azrin, and E.S. Wright, 34th Sagamore Army Materials Research Conference Proceedings (1990)
58. D.M. Tracey and P.J. Perrone, "Stress Distributions Near Microstructural Inhomogeneities,' *Trans. 6th Army Conf. Applied Math. and Computing* ARO Report #89-1 (1989)
59. M.J. Gore, G.B. Olson, and M. Cohen, "Grain-Refining Dispersions and Properties in Ultrahigh-Strength Steels," in *Innovations in Ultrahigh-Strength Steel Technology*, ed. G.B. Olson, M. Azrin, and E.S. Wright, 34th Sagamore Army Materials Research Conference Proceedings (1990)
60. C. Hayzelden, "Secondary Particle Dispersions and Impurity Gettering in Ultrahigh-Strength Steels," in G.B. Olson, M. Azrin, and E.S. Wright, eds., *Innovations in Ultrahigh-Strength Steel Technology*, 34th Sagamore Army Materials Research Conference Proceedings (1990)
61. R.J. Asaro and J.R. Rice, *J. Mech. Phys. Solids 25* (1977) 309
62. M. Schmidt and M.J. Gore, "Solution Treatment Effects in AF1410 Steel," in *Innovations in Ultrahigh-Strength Steel Technology*, ed. G.B. Olson, M. Azrin, and E.S. Wright, 34th Sagamore Army Materials Research Conference Proceedings (1990)
63. A.S. Argon and J. Im, *Metall Trans., 6A* (1975) 839
64. J.H. Beatty and M. Azrin, MTL, research in progress
65. C-Y. Hsu, "Grain-Growth Mechanisms in Rapidly Solidified Matrix Steels," Ph.D. thesis, MIT, 1984
66. T. Gladman, *Proc. R. Soc. A294* (1966) 298
67. P. Hellman and M. Hillert, *Scandinavian Journal of Metallurgy, 4* (1975) 211
68. J.W. Flowers, Jr., and S.P. Karas, *J. Appl. Phys. 38* (1967) 163
69. T. Gladman and F.B. Pickering, *JISI 205* (1967) 653
70. L. Anand and J. Gurland, *Metall Trans. 6A* (1975) 928
71. G.B. Olson and R.G. Bourdeau, "Rapidly Solidified Ferrous Alloys," in *Rapidly Solidified Alloys*, ed. S.K. Das, B.H. Kear, and C.M. Adam, TMS-AIME, (1985) p 185
72. M.J. Gore, M. Grujicic, G.B. Olson, and M. Cohen, *Acta Metall, 37* (1989) 2849
73. G.B. Olson, "Transformation Plasticity and the Stability of Plastic Flow," in *Deformation, Processing and Structure*, chpt. 9, ed. G. Krauss, ASM (1984) p 391
74. G.B. Olson and M. Cohen, *Metall. Trans. 13A* (1982) 1907
75. G.B. Olson and M. Cohen, *Metall. Trans. 6A* (1975) 791
76. T. Narutani, G.B. Olson, and M. Cohen, *J. de Physique 43* (1982) C4-429
77. R.H. Leal, "Transformation Toughening of Metastable Austenitic Steels," Ph.D. thesis, MIT, 1984
78. G.B. Olson and M. Cohen, "Martensitic Transformation as a Deformation Process," in *Mechanical Properties and Phase Transformations in Engineering Materials*," ed. S.D. Antalovich, R. O. Ritchie, and W.W. Gerberich, TMS-AIME (1986) p 367
79. F. Stavehaug, "Transformation Toughening of γ' - Strengthened Metastable Austenitic Steels," Ph.D. thesis, MIT, 1990
80. R.G. Stringfellow and D.M. Parks, MIT, research in progress
81. G.N. Haidemenopoulos, G.B. Olson, M. Cohen, and K. Tsuzaki, *Scripta Metall. 23* (1989) 207
82. G.N. Haidemenopoulos, M. Grujicic, G. B. Olson, and M. Cohen, *Acta Metall. 37* (1989) 1677
83. G.N. Haidemenopoulos, G.B. Olson, and M. Cohen, "Dispersed-Phase Transformation Toughening in Ultrahigh-Strength Steels," in *Innovations in Ultrahigh-Strength Steel Technology*, ed. G.B. Olson, M. Azrin, and E.S. Wright, 34th Sagamore Army Materials Research Conference Proceedings (1990)
84. G.B. Olson and M. Cohen, "Dislocation Theory of Martensitic Transformations," in *Dislocations in Solids, Vol. 7* ed. F.R. N. Nabarro, North-Holland, (1986) p 295
85. G.B. Olson, K. Tsuzaki, and M. Cohen, "Statistical Aspects of Martensitic Nucleation," *Turnbull Symposium: Phase Transitions in Condensed Systems*, ed. G.S. Cargill, F. Spaepen, and K.N. Tu, MRS (1987) p 129
86. I-W. Chen, Y-H. Chiao, and K. Tsuzaki, *Acta Metall. 33* (1985) 1847
87. M. Grujicic, M. Buonanno, S.M. Allen, G.B. Olson, and M. Cohen,

"Heterogeneous Precipitation of Austenite for Stabilization," in *Innovations in Ultrahigh-Strength Steel Technology*, ed. G.B. Olson, M. Azrin, and E.S. Wright, 34th Sagamore Army Materials Research Conference Proceedings (1990)
88. B.V.N. Rao, "Ferrite-Austenite Dual Phase Steel," U.S. Patent 4,544,422, October 1, 1985
89. C-C. Young, "Transformation Toughening in Phosphocarbide-Strengthened Austenitic Steels," Ph.D. thesis, MIT, 1988
90. B. Budiansky, J.W. Hutchinson, and J.C. Lambropoulos, *Intl J. Solids and Structures, 19* (1983) 337
91. G.N. Haidemenopoulos, M. Grujicic, G. B. Olson, and M. Cohen, *CALPHAD 13* (1989) 215
92. C. Kuehmann and G.B. Olson, Northwestern University, research in progress.
93. C.J. McMahon, Jr., "Hydrogen Embrittlement of High-Strength Steels," in *Innovations in Ultrahigh-Strength Steel Technology*, ed. G.B. Olson, M. Azrin, and E.S. Wright, 34th Sagamore Army Materials Research Conference Proceedings (1990)
94. J.R. Rice and J-S. Wang, *Mat. Sci. and Eng. A107* (1989) 23
95. P.M. Anderson, J-S. Wang, and J.R. Rice, "Thermodynamic and Mechanical Models of Interfacial Embrittlement," in *Innovations in Ultrahigh-Strength Steel Technology*, ed. G.B. Olson, M. Azrin, and E.S. Wright, 34th Sagamore Army Materials Research Conference Proceedings (1990)
96. J-S. Wang, private communication.
97. E. Wimmer, C.L. Fu, and A.J. Freeman, *Phys. Rev. Lett., 55* (1986) 2618
98. R.J. Harrison, F. Spaepen, A.F. Voter, and S-P. Chen, "Structure of Grain Boundaries in Iron," in *Innovations in Ultrahigh-Strength Steel Technology*, ed. G.B. Olson, M. Azrin, and E.S. Wright, 34th Sagamore Army Materials Research Conference Proceedings (1990)
99. G.L. Krasko and G.B. Olson, "Evaluation of Decohesion Mechanisms by LMTO-ASA-Stoner Cohesive Energy Calculations," in *Innovations in Ultrahigh-Strength Steel Technology*, ed. G.B. Olson, M. Azrin, and E.S. Wright, 34th Sagamore Army Materials Research Conference Proceedings (1990)
100. G.L. Krasko and G.B. Olson, "The Effect of Boron, Carbon, Phosphorus and Sulphur on Intergranular Cohesion in Iron," *Solid State Commun.* 76 (1990) 247–251
101. R.J. Harrison and G.L. Krasko, MTL, research in progress
102. G.L. Krasko and G.B. Olson, *Phys. Rev B 40* (1989) 11536
103. C. Li and A.J. Freeman, Northwestern University, research in progress.
104. C.L. Briant and R.P. Messmer, *Phil. Mag. B, 42* (1980) 569
105. M.E. Eberhart and J.M. MacLaren, "Mechanisms for Cleavage and Intergranular Embrittlement in Fe," in *Innovations in Ultrahigh-Strength Steel Technology*, ed. G.B. Olson, M. Azrin, and E.S. Wright, 34th Sagamore Army Materials Research Conference Proceedings (1990)
106. H. Eguchi, "Electronic Basis of Phosphorus Embrittlement of Iron Grain Boundaries," M.S. thesis, Northwestern University, 1990
107. D.E. Ellis, "Molecular Clusters," in *Handbook on the Physics and Chemistry of the Actinides*, ed. A.J. Freeman and G.H. Lander, North-Holland, (1985) p 1
108. J.F. Watton, G.B. Olson, and M. Cohen, "A Novel Hydrogen-Resistant UHS Steel," in *Innovations in Ultrahigh-Strength Steel Technology*, ed. G.B. Olson, M. Azrin, and E.S. Wright, 34th Sagamore Army Materials Research Conference Proceedings (1990)
109. G. Del Corso and G.B. Olson, unpublished research
110. J.H. Graves and A.A. Anctil, MTL, research in progress
111. A. Bilyk, T. Chen, N. Akaiwa, H. Eguchi, and G.B. Olson, Northwestern University 750-C95 Materials Design Class Project Report, June 1989

Transformations—Selected Works of G.B. Olson on Materials, Microstructure, and Design
C.E. Campbell, M.V. Manuel, and W. Xiong, editors

CHAPTER 12

Designing a New Material World*

G.B. Olson, Northwestern University

MATERIALS have paced the evolution of technology for millennia. Their importance in the advance of human civilization is apparent in the naming of historical epochs, from the Stone Age through the Bronze and Iron Ages and into the ongoing Silicon Age. The origin of diversity in the material world remains largely mysterious to the public, yet the specialists' ability to understand and manipulate the microstructures of materials has grown explosively over the past half-century (see the section, A Materials Science Timeline, in this chapter). As the new millennium unfolds, a confluence of natural philosophies—one that combines reductionist and synthetic thinking—is ushering in an Age of Design marked by new materials and ways of creating them that go beyond the dreams of the medieval alchemists.

Materials as Systems

The modern view of materials structure was best expressed by the late philosopher-scientist Cyril Stanley Smith (Ref 1). He described a universal multilevel nature of structure with strong interactions among levels and an inevitable interplay of perfection and imperfection at all levels. Smith argued that the materials scientist's distinct view of structure is defined by the desire to understand the structure and property relations underlying the technological and economic value of materials.

This view of matter integrates science and engineering and is built on a natural philosophy that is older than science. Smith identifies the origin of this philosophy within medieval alchemy, whose practitioners are mostly remembered for their attempts to "transmute" base metals into gold. They knew nothing about protons and electrons. Still, Smith asserts that their aspirations, motivations, and even concepts resembled the modern materials engineer's use of complex processing to "transmute" the multilevel microstructure of materials (from the atomic to macroscopic levels) to achieve the essential property of gold, namely, economic value.

The property-driven view of structure and processing for the creation of value, shared by modern materials science and alchemy, embraces an essential complexity of material structure. During the 17th-century birth of modern science, Smith identifies a short-lived Golden Age of materials science under the leadership of René Descartes, whose "corpuscular" philosophy inspired the development of a multilevel view of structure to account for the diverse properties of materials.

The prescient grasp of materials achieved by the corpuscularians is well represented by René Antoine Ferchault de Réaumur's 1722 sketch of the structure of quench-hardened steel in Fig. 1. He proposed that a single grain of steel (G), if enlarged, would reveal a set of "molecules" (M) and voids (v). Higher magnification would reveal a substructure of the molecules

*G.B. Olson, "Designing a New Material World," *Science*, 288 (May 12, 2000), p 993–998

(pp); and yet higher magnification would show an aperiodic arrangement of spheres. The finest scale Réaumur envisioned corresponds to a periodic packing of spheres, what we might think of as the nanoscale.

This elegant view contained many structural elements of modern materials science. There was no instrumentation with which to validate it, however, so the complex structural view was supplanted by more intellectually compelling but overly simplistic notions. One was Isaac Newton's continuum, which erased structural considerations entirely. The other was John Dalton's atomism, which held that there was only one level of structure that mattered. These conceptual idealizations were sufficiently compelling to divert the corpuscularian framework for two centuries.

We have had the past century to reinvent materials science. The atom and the continuum remain dominant philosophical forces; they are the foundation of existing theoretical tools. But armed with the growing reservoir of structural facts gleaned from instrumentation and computation, the challenge now is to adapt these theoretical tools to understand and control the complex structures of real materials.

The modern view of material structure differs from Réaumur's mainly in the detailed morphologies characteristic at the different length scales of a material's hierarchical structure. In the case of steel, the most significant difference is the over-estimation of porosity in the 18th-century depictions. And if Réaumur's voids are reinterpreted as "free volume," his sketch becomes a reasonable model of polymeric materials. It shows the remarkable ability of the human mind to infer necessary structure from the contemplation of properties alone.

Fig. 1 Materials savant. René de Réaumur and his 1722 sketch of steel's anatomy.

Smith made another important historical observation about materials development (Ref 2). Since prehistory, people have put newly discovered materials to practical use long before they understood much about them.

Consider the pattern-welded sword blades made and used by Merovingian Franks and Vikings (Fig. 2) as early as the 6th-century A.D. By hammer-welding steels of differing carbon content (a technical feature unknown to the early sword smiths), a laminate composite was fabricated by labor-intensive "hand lay-up" to produce a hybrid structure with a tough core supporting a hard cutting edge.

The actual mechanism of the quench hardening of steel, which is responsible for the hardened sword edge, was hotly debated even in the early 20th century. There were two factions: the "allotropists," who favored structural transformation, and the "carbonists," who pegged the hardening mechanism on dissolved carbon. A crucial observation was the discovery by Floris Osmond in 1893 of "martensite" in steel, the microstructural form that iron assumes during the quenching process. We now know that the combination of a martensitic transformation and the redistribution of trapped interstitial carbon (along with ambient aging during which additional microstructural evolution occurs) underlies the edge hardness of the 6th-century swords.

The ancient swords embody another illustration of how art and craft has traditionally preceded science. The sword smiths' use of chemical etching to bring out aesthetic metallic patterns set the foundation for modern metallographic observation of microstructure established by Henry Sorby in the 19th century. By etching metal samples with acid, Sorby revealed internal microstructures and correlated them with the properties and performance of the materials.

Fig. 2 Cutting edge. Etched and repolished tip region of a 6th-century sword.

Transmission electron microscopy (TEM), a modern analytical cohort of metallography that reveals finer structural levels (Fig. 3), suggests that the ancient sword-makers ever were accidental nanotechnologists. TEM analysis reveals nanometer-scale patterns of carbon in the hard edge of the sword. These patterns emerged from a process called "spinodal decomposition," in which a solid solution, such as the sword's high-carbon martensitic steel, becomes unstable and its constituents reorganize (Ref 3). In modern parlance, the swordmakers' products could be described as "self-assembled heterophase nanostructures."

The spatial dimension of materials is only part of what makes them tick. There is a spectrum of characteristic relaxation times associated with the various chemical and physical processes operating at the material's differing structural length scales. This adds the dimension of time.

The resulting dynamic spatiotemporal hierarchy means that any material at any time has structural features, such as grain sizes and dispersed particles, that have not yet reached equilibrium. That is why a material's structure and properties depend on how it was made and what conditions it endured in service.

Also fundamental to this dynamic view of materials is a realization that structural defects play vital roles at all length scales. Defects can make things fail in a thousand ways, but they're also often what makes materials so valuable. Perfect silicon crystal without dopant ions is not the semi-conductor that has changed society. Pure iron metal without the right spicing of carbon would never have become the steel backbone of the industrial revolution. Although "defect tolerance" remains a central tenet of modern materials science and is of incalculable commercial and safety importance, "defect engineering" is ascendant in the minds of many materials researchers. That's because defects on various hierarchical levels are a principal opportunity for controlling material behavior.

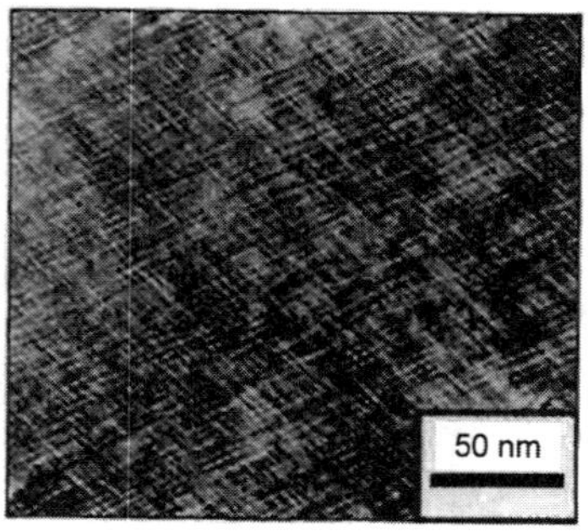

Fig. 3 Inside steel. TEM reveals nanoscale structure of a quench-hardened steel.

Because the personality of each material depends on all of these interacting spatial and dynamic attributes, it makes sense to approach materials as complex systems. Smith advocated such a systems view of materials decades ago. As more contemporary practitioners live by that insight, they are finding pathways to important new materials that can catalyze advances in manufacturing tools, computers, communications systems, and the myriad technologies whose very existence or improvement depends on more capable materials.

The Materials Discipline Comes of Age

Two principal branches of natural philosophy have evolved to form modern materials science. One is reductionist analysis, which takes nature apart to discern and understand its fundamental units. Reductionism has operated throughout the development of science. More sporadic has been the evolution of the synthetic systems view, which is better suited for understanding the connections holding nature together. In a new balance of these two philosophies, the systems view integrates the fruits of our investment in reductionism while replacing conventional discovery-based R&D with a far more effective and efficient design-based approach.

Although there were seeds of this turning point in materials science in early industrial laboratories at General Electric, Bell Labs, and elsewhere, the multifaceted field of materials research was deliberately synthesized as a single academic discipline in the late 1950s and 1960s with the founding of the first university materials departments (Ref 4). Specialists in the science and technology of metals, ceramics, polymers, and composites collaborated in pursuit of unifying principles for the creation of materials of all classes. This meeting of minds helped lay the sociological and cognitive groundwork for a systems approach to materials.

The development of operations research in World War II and large-scale national missions such as the Manhattan Project and the Apollo space program were also important. It is estimated that at least 70% of our unprecedented economic boom of the past decade derives from technology, which in turn derives in good part from radically improved productivity via new systems-based methods of product development.

After World War II, newly formed government agencies, including the Office of Naval Research, the Defense Advanced Research Projects Agency, the National Science Foundation (NSF), and others also helped midwife the multidisciplinary sensibility needed for developing complex materials. The same was true in academe. Starting with Smith's Institute for the Study of Metals at the University of Chicago, a series of university materials research laboratories established a national infrastructure for the interdisciplinary enhancement of materials science. The private sector added sociological foundations for the emerging discipline through a diverse mix of professional materials societies, as well as the establishment of numerous materials-centered journals and conferences.

The field now consists of many thousands of practitioners who share a growing sense of community, yet whose collective expertise runs a stunning gamut of materials categories. There are synthetic diamond makers, metal alloy designers, polymer scientists, optical fiber experts, thin film ceramic makers, developers of compound semiconductors, and "biomimetic" materials researchers who aim to emulate or adapt biology's unmatched brilliance in materials innovation. The list goes on and on, with a diversity akin to the living kingdom's millions of species.

Over the past four decades in materials R&D, however, there has been a consistent emphasis on "good science," as defined by reductionism, rather than "good materials," which emerge when engineering, manufacturing, and economic factors are included in the mix. This bias has limited the technological impact of the materials science community, particularly the academic portion. And materials engineering per se has been left primarily to industry, which has yet to fully benefit from the expanded base of materials science.

There is a twist in store. The science of materials has reached a level at which it now can radically change engineering practice. The possibilities are akin to what has come from the relationship between the life sciences and medicine. Until the 19th century, there was little or no science to guide medical technology and practice. Since then, however, the ever growing corpus of biomedical knowledge has been leading to an ever more amazing stream of health care innovations. Here, a genuine desire to meet societal needs has produced a healthy mix of reductionist and systems viewpoints, yielding a culture that naturally integrates scientific understanding into practical use. The materials research community is poised to emulate this model.

Materials by Design—Efficient Innovation

There is a general engineering design movement under way. It draws on the vast information pool generated by reductionist analysis, but adds the component of design, for which the systems approach is crucial.

Central to the materials design approach is a powerful logical structure connecting the "four elements" of materials science: processing, structure, properties, and technological performance (Fig. 4). By connecting adjacent pairs of these elements, a three-link chain representing a versatile materials paradigm emerges. The deductive, cause-effect logic of reductionist science flows from processing to performance. All along the way, science reveals the relevant structures and phenomena, often in astounding clarity and detail. The inductive logic of systems engineering flows the other way, from performance to processing, thereby enabling designers to arrive at specific procedures likely to yield materials with the desired sets of properties and performance.

Not often successful in terms of producing useful new materials, early efforts at materials design

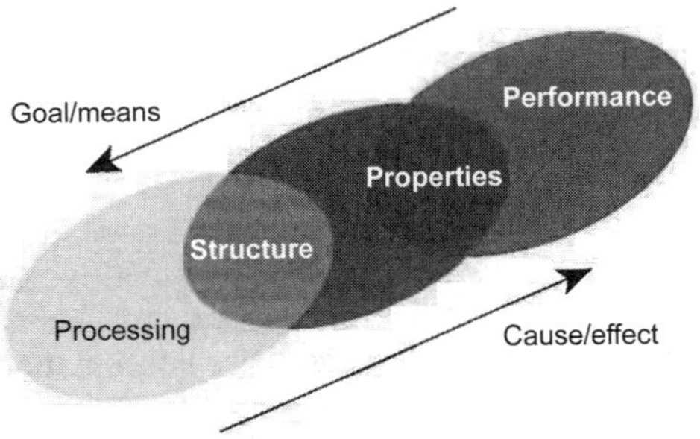

Fig. 4 Big four. Four-element paradigm of modern materials science and engineering.

nonetheless were harbingers of what materials development is to become. Generally, what has made or broken past efforts was whether they included the element of design. One notable success is the work of H.K.D.H. Bhadeshia and coworkers whose ambitious assault on the complex problem of weld metal design has spawned productive efforts at several national laboratories in the United States and Japan.

The initiative I know best is the Steel Research Group (SRG) at Northwestern University, which my colleagues and I have continuously developed since 1985 (Ref 5, 6). This university-industry-government program was organized within the context of systems engineering to explore general methods, tools, and databases for the design of materials, using high-performance steels as a test case (see the section, "Materials by Design," in this chapter). There is a faith underlying this framework: The scientific knowledge base is now robust enough to supplant the traditional, empirically driven development of materials with a more efficient theory-driven and computationally based approach (see the section, "Virtualizing Materials to Create Real Ones," in this chapter).

In the SRG, we begin by combining the perspectives of materials users, suppliers, modelers, and designers into a set of specific materials property objectives. Those specifications, in turn, help us define how to use, adapt, or expand science-based models and databases of material behavior. We then use these models and databases to zero in on compositions and processing protocols that can transform those compositions into alloys fitting our specifications. In the past decade, we have used this framework to develop new alloys with unprecedented sets of properties. Some are now under evaluation by industrial and government partners for use in airplanes, power generators, aircraft carriers, and other applications.

Our own projects for designing steel are just a beginning. A recent NSF-sponsored workshop on Materials Design Science and Engineering (Ref 7) has called for broadening this approach to the design of all classes of materials. And in recognizing materials as one of five critical technologies for U.S. competitiveness, the President's Office of Science and Technology Policy (Ref 8) has identified computational materials design as a principal opportunity.

Opportunities abound for application of the new systems-driven computational design approach. Successful examples from our efforts include a stainless steel bearing for a Space Shuttle application, high-strength, high-toughness steels for aircraft landing gear and armor applications, and a new class of ultrahard steels for advanced gear and bearing applications. The more general validity of the design approach has been explored with computational design projects focusing on nonferrous alloys, hydrate ceramics, case hardening polymers, and nanostructured thin film materials for microelectromechanical devices and hard coatings.

The new design capabilities will also help realize the dream of biomimetic materials, which emulate the complex adaptive microstructures of the living world that are beyond the reach of traditional empirical development. One successful demonstration is self-healing metallic composites. These incorporate "shape memory" alloys that exploit martensitic structural transformations to change their shape controllably. Integrating such components into thin film electronic devices to create efficient microactuators is leading to smart materials systems that unite the worlds of structural and electronic materials. Here the growing philosophy of predictive materials design is now combining the electrical engineer's realm of perfection-driven, artificially structured microcircuits with the materials traditions of self-assembly (Ref 9) and defect tolerance (Ref 10).

Following the same philosophy can reduce the cost of discovery. In contrast to computational materials design, the prevalent industrial methods of materials development are based on an intrinsically slow and expensive process of trial-and-error empiricism. Theoretical input has been qualitative at best. It typically takes tens of millions of dollars over two decades to fully develop and qualify a new material in a critical application (Ref 11). This sluggishness stands out in an era when engineers are expected to deliver new generations of products, such as automobiles, on an 18-month cycle. What's more, under these competitive pressures many industrial materials-development groups have been severely reduced or closed over the past decade. A recent National Research Council study (Ref 11) has concluded that the greatest challenge to the materials field today is the short time and cost constraints of the full materials development cycle.

Another reason these industrial materials development efforts have been downsized is anchored in the old R&D model in which new materials are discovered rather than designed.

A typical estimate is that $1 of discovery costs $10 of development. If eliminating that first dollar was the only advantage of computational materials design, its impact on the total discovery and development cost would be small. The real advantage of materials design is that good design in the first place requires much less development later. And rather than using materials design approaches simply to provide an initial prototype for subsequent empirical development, there's yet more to be gained by integrating predictive modeling throughout the full design and development process.

Consider two costly and time-consuming phases of the standard materials development cycle: process optimization and qualification testing. A major concern in process optimization is scale-up. Because processing phenomena such as heat transfer depend on product size, a prototype material investigated on a small scale is not likely to behave the same way when processed on a large scale. "Solidification design" can preempt this common showstopper. By using models that simulate how materials respond to processing conditions for designing a material at the ultimate process scale of interest, we can reduce the number of expensive large-scale experiments while shortening the development cycle.

Improving qualification testing is more challenging. To design with confidence, designers must know how properties vary within the materials they intend to use. Current practice requires much testing to define these variations statistically. But most structure-property theories provide mean values of the variables of interest; they do not tell a designer about location-to-location variations within materials.

Help is coming, however. Researchers are developing a probabilistic materials science whereby structural distributions are mapped into property distributions. These are the kinds of data designers can use to make better predictions about how different materials will affect their products. Models of this kind already exist for fracture properties and heterogeneous phase transformations, and more efforts are being planned for this vitally important area.

Structure of Education

Education offers the most leverage for moving materials R&D into the systems-based paradigm. In its post-Cartesian form, modern metallurgy began with an emphasis on the direct correlation of processing and properties. The advent of physical metallurgy a century ago opened the "black box" of structure and brought a revolution in the fundamental understanding of the mechanistic link between processing and properties. This understanding created the foundation for the more recent generalization to materials science, making possible the general materials design and systems engineering methodologies described in this essay. Given the demonstrated potential of materials design, institutionalizing it via education would result in many benefits to society.

Due to the historic dominance of reductionist philosophy, however, we now have in place an analysis-oriented education system. In the name of objectivity, we train students to shut down the more synthetic and emotional tools of thought, which are precisely the ones best suited for doing good systems engineering. The future of engineering education, therefore, ought to target these subjective reasoning powers.

An activity of our new materials engineering curriculum at Northwestern, the Dragonslayer Project, is designed to do this. The project explores the design of an "aesthetic" material through collaboration of freshman and upperclass design teams. The widespread presence of Western dragon combat in literature is used to integrate frontier steel technology with the history and legend of swordmaking to design an ultrahigh-performance sword that our market analysis shows would be of maximum value to sword collectors.

Much of the mystique of the legendary Samurai sword, which even five centuries ago achieved performance levels equivalent to those of a modern carburized blade, stemmed from its ability to cut through other swords of the day. The equivalent performance advance has been adopted as a benchmark for property requirements.

Fanciful as the project might seem, it's no different from the kind of materials design that goes on at the DuPonts and Cornings of the world. Searches of literature, albeit medieval literature, were used to determine technical specifications, including the need for flame resistance when fighting fire-breathing dragons. Historians of the Dark Ages helped to identify and select a historically accurate sword style—the patterned double-edge broadsword. And market analysis helped determine what the target buyers, sword collectors, would be willing to buy and how much they might pay.

Finite-element mechanics simulations of cutting through a modern carburized blade led to a conceptual design for radical surface-hardening technologies to achieve this ambitious sword-cutting objective. We also were guided by the unlikelihood that anyone would face supernatural evil armed only with technology. This led to an additional design parameter, which market surveys indicated would constitute an important attribute in a collectible dragonslaying sword: The sword should be made from material of heavenly origin, namely, meteoritic iron. This is a technical feature with historical and legendary precedent (including the Excalibur legend).

Our systems approach to materials design has established the feasibility of using tabletop-scale aqueous electrolytic refinement of available meteoritic iron to achieve the necessary purity for producing the unprecedented alloy steel performance needed for a sword fit to defeat the most evil of adversaries. The freshman design team's proposed market plan includes auctioning a single "technomystical" sword for publicity, followed by commercial marketing of a range of high-performance steel products, including the "Dragonslayer golf club."

In this millennium, a new architecture of synthetic thought will continue its symbiosis with modern computational capabilities. And an age of empirical exploration will continue to be superceded by an age of design. This will open up powerful pathways by which human creativity, fused with scientific knowledge, will bring new levels of control over the material world widely applicable both to society's problems and its ambitions. The manifestation of the design paradigm, whether in an undergraduate or corporate setting, corresponds to a form of transmutation beyond the alchemist's dreams: the creation of materials from thought.

MATERIALS BY DESIGN

To design a material is to try to meet a material user's need. A good place to start is with property cross-plots, like those of Michael Ashby, that graphically define property-performance relations (Ref 12). These help engineers select materials for their product designs. They're useful for defining a quantitative set of property objectives that will sum into the materials performance needed by users. These performance specifications are determined by the role of the potential material in the wider system it serves. They also help define economic parameters, such as the cost of raw materials and processing, for the overall material design task.

With objectives and economic constraints defined, the linear three-link framework of materials science and engineering (Fig. 4) serves to guide the design and development phases. The Steel Research Group (SRG) (see the section, "Materials by Design: Efficient Innovation," in this chapter) uses flow-block diagrams that represent key microstructural subsystems (such as crystal grain sizes), the primary links of these subsystems to the properties they control (such as strength and toughness), and the stages of processing (such as tempering or reheating) that govern their dynamic evolution (Fig. 5).

MATERIALS BY DESIGN (*continued*)

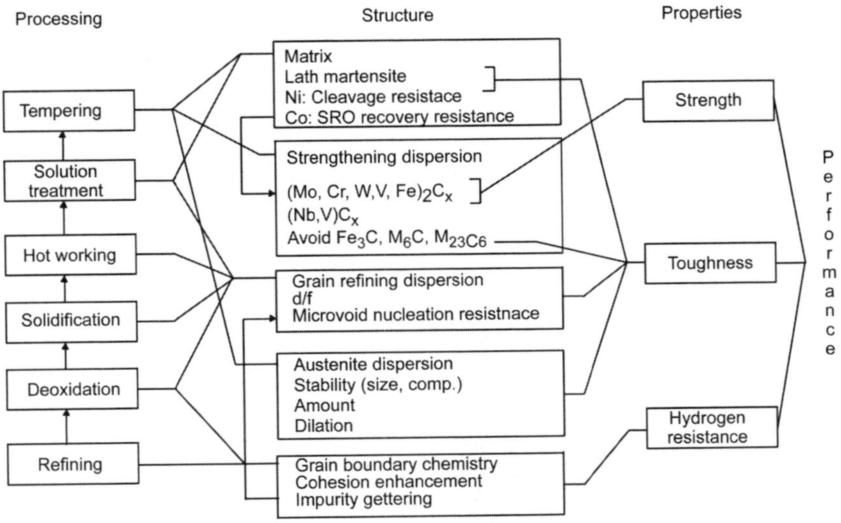

Fig. 5 Planning materials. Flow-block diagrams guide materials design.

With these in hand, systems analysis is then applied to identify and prioritize key structure-property and processing-structure relations. Often, part of this exercise involves some additional modeling or empirical data gathering to fill gaps in the knowledge required for making practical decisions about composition and processing details. The systems view operates here at the strategic level, but it is supported by traditional reductionist analysis at a tactical level. On balance, the procedure greatly reduces the amount of costly experimentation in materials creation. Instead of making tens of prototypes along the way to a useful new material, SRG designers reach their target metal using only a few actual melts to refine the computation-heavy design efforts.

VIRTUALIZING MATERIALS TO CREATE REAL ONES

Design of the hierarchical structure in materials requires a hierarchy of models based on materials science, applied mechanics, and even quantum physics (Fig. 6). Models to help design features on the coarsest structural level of solidification, such as the chemical banding visible in the patterned sword blades mentioned earlier, employ powerful thermodynamic codes such as THERMOCALC (Ref 13). These models enable designers to simulate the 10 μm scale of structure. In metal alloys, this is the level at which the chemical partitioning between liquid and solid phases evolves during solidification processing. Application of these models aids decision-making about thermal processing details, a practice we like to call "solidification design."

VIRTUALIZING MATERIALS TO CREATE REAL ONES (*continued*)

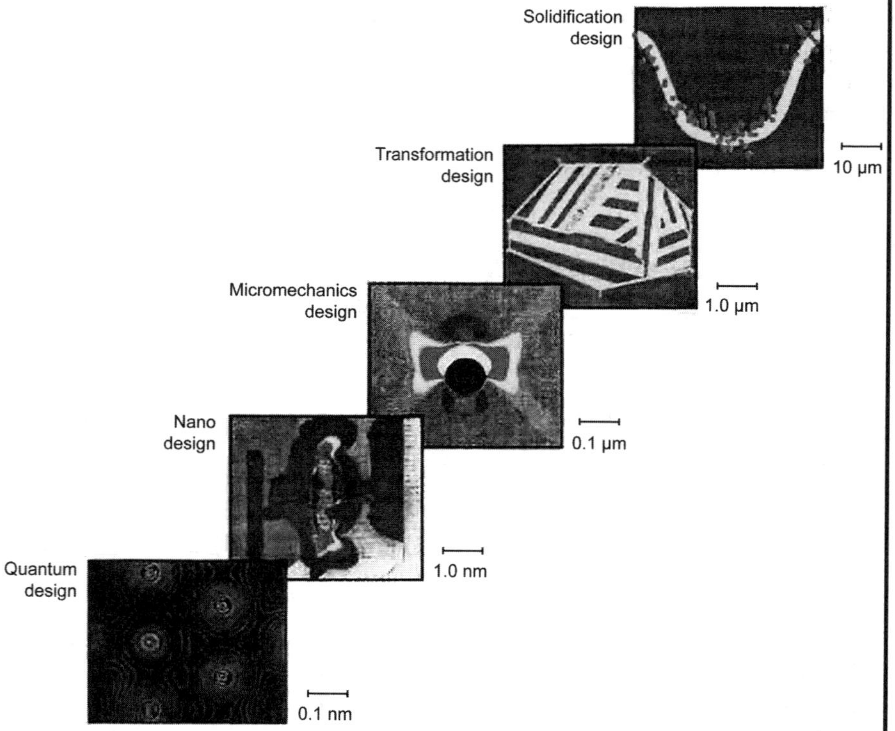

Fig. 6 Model relay. Computational design of materials requires a hierarchy of models.

At the 1.0 µm scale of structure, "transformation design" is the goal. This concerns the evolution of structural changes during quenching, whereby crystal grains present at high temperatures transform and subdivide as the hot alloy cools into a hierarchy of lower temperature crystalline units. The design goal here is to specify and control processing temperatures so that desirable microstructures will form, while hindering the formation of competing microstructures that are less beneficial.

The 0.1 µm scale represents the micromechanics design level. An example of the phenomena relevant at this scale is "grain refining," in which the large grains formed at high temperatures that can embrittle alloys are made smaller by more precise thermal or compositional control. There's a trade-off here, because more smaller particles can also catalyze ductile fracture (breaking, that is), as there are more interfaces that can separate from one another. Micromechanics models typically are based on continuum descriptions of mechanical phenomena that can simulate the evolution of microstructures during material deformation and fracture.

In recent years, an even finer structural level—the nanoscopic level—has become better understood and more controllable. The control of 1 nm scale particle dispersions in alloys created through solid state precipitation during "tempering" at intermediate temperatures, for

VIRTUALIZING MATERIALS TO CREATE REAL ONES (*continued*)

example, provides efficient obstacles for resisting plastic deformation. Said differently, this nanometer-scale structuring strengthens the metal.

The development of design models for these diminutive scales builds on a half-century evolution of theory for both precipitation and strengthening in metals. A major materials science breakthrough of the 1950s was the identification of dislocations as the key defects that enable the sliding of crystal planes, which shows up as plastic deformation of many materials. While most structure-property relations, such as the Hall-Petch relation for grain refinement strengthening, are based on empirical correlations, a major triumph of the 1960s was the Orowan particle strengthening equation. Derived directly from dislocation theory, it relates strength to the inverse spacing of dislocation obstacles (such as nanoscale precipitates). Such developments in theory, which mathematically codify materials behavior, are cornerstones for the claim that materials can be designed largely in silico.

The new accuracy of theoretical predictions of structure at the nanometer scale is only possible because of recent advances in high-resolution instrumentation allowing precise calibration and validation of theory in this regime. This includes such techniques as x-ray and neutron diffraction, various electron microscopies, and atom-probe microanalysis. The latter is represented by the three-dimensional atomic reconstruction of a 3 nm strengthening carbide particle in an ultrahigh-strength steel shown in Fig. 7. Such new capabilities in structural and chemical analysis down to the atomic scale open a new era of quantitative materials nanotechnology.

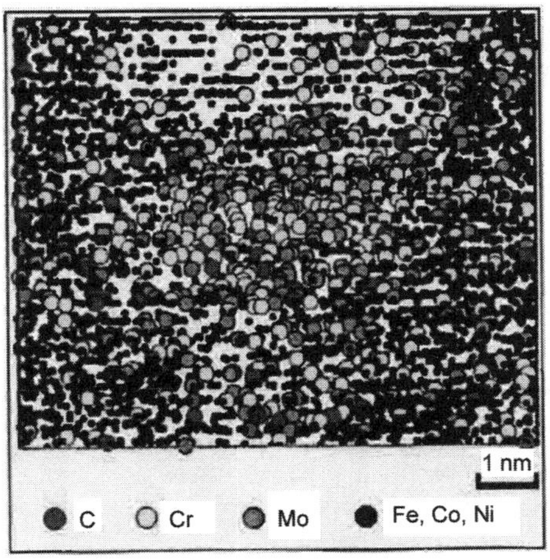

Fig. 7 Atomscapes. A view of a Cr-Mo carbide particle that strengthens steel alloy.

The electronic level is the finest level relevant to real materials. This is the realm of quantum design. As acknowledged by the 1998 Nobel Prize in Chemistry shared by John Pople and Walter Kohn, the development of computational quantum mechanics and its extension via density functional theory constitute a profound advance that already has had significant industrial impact. A collaboration of materials science, applied mechanics, and quantum physics has enabled some of us in the Steel Research Group (see the section,

VIRTUALIZING MATERIALS TO CREATE REAL ONES (*continued*)

"Materials by Design: Efficient Innovation," in this chapter) to apply computational quantum mechanics to engineer steel at the subatomic level.

Our approach was to recast models of the impurity-induced embrittlement of grain boundaries into thermodynamic terms (Ref 14) and to rely on precise models of the atomic structures at grain boundaries. As represented by the computed valence charge density contours for a phosphorus atom at the core of an iron grain boundary, total energy calculations are sufficiently precise and accurate to explain the known effects of interstitial components such as boron, carbon, phosphorus, sulfur, and hydrogen on the cohesion of iron grain boundaries. What's more, the calculations lead to new mechanistic insights at the electronic bonding level. Extension to elements that occupy substitutional Fe sites in the boundary has enabled us to predict new alloying elements for enhancing boundary cohesion in steels. This is creating a new generation of "quantum steels" that incorporate these properties derived directly from electronic-level predictions.

A Materials Science Timeline

1556	Georgius Agricola's *De re metallica*, a compendium of 16th-century mining, metallurgical, and general materials production, is published.
1664	Cartesian corpuscular philosophy recognizes material properties as emerging from a multilevel structure.
1665	Robert Hooke publishes *Micrographia*, which reveals levels of material microstructure never before seen.
1722	René de Réaumur publishes the first technical treatise on iron.
1782	Josiah Wedgwood develops an early form of process control with his invention of the pyrometer for measuring furnace temperatures.
1808	John Dalton publishes his *New System of Chemical Philosophy*, which establishes atomic theory.
1824	Joseph Aspdin invents portland cement, which remains one of the most used materials in the world.
1839	Charles Goodyear accidentally discovers vulcanization, which ultimately renders raw rubber latex into a widely useful material.
1856	Henry Bessemer patents a process for large-scale steel production.
1860s	Henry Sorby applies light microscopy to the study of the microstructure of metals and rocks.
1869	John Hyatt successfully commercializes celluloid, an artificial plastic material.
1869 and 1870	Dmitri Mendeleev and Julius Lothar Meyer publish versions of what will become known as the Periodic Table of the Chemical Elements.
1886	Charles Hall and Paul Héroult independently discover cost-effective methods for producing aluminum metal from ore.
1893	Floris Osmond discovers martensitic transformation.
1900	Max Planck formulates the idea of quanta, thereby setting the stage for the development of quantum mechanics.
1906	Alfred Wilm discovers age hardening in aluminum alloy, which is later used for making dirigibles and other aircraft.

1909	Leo Baekeland patents Bakelite, the first entirely synthetic plastic, and commercializes it widely.
1911	Heike Kamerlingh Onnes discovers superconductivity in mercury chilled to temperatures near absolute zero.
1911–12	The father-son team of William Henry and William Lawrence Bragg, along with Max von Laue, develops the basis of x-ray crystallography, one of the most important analytic techniques for studying material structure.
1921	A.A. Griffith postulates role of defects in fracture strength.
Late 1920s	Hermann Staudinger argues that polymers are made of small molecules that link to form chains.
1934	Wallace Hume Carothers invents nylon.
1940s	The wartime practice of organizing multidisciplinary research collaborations to achieve technological goals becomes a model for the subsequent organization of a field that later becomes known as materials science and engineering.
1947	John Bardeen, William Shockley, and Walter Brattain invent the transistor.
1950s to 1960s	Much of the theoretical foundation behind the formation and evolution of material microstructure is developed. Among them is the Hall-Petch relation for grain refinement strengthening and the theory of diffusion of solids.
1953	Karl Ziegler develops catalysts that make it easier and cheaper to polymerize ethylene into stronger, more capable polymers.
1955	A team of scientists at General Electric combine high temperatures and enormous pressures to create synthetic diamond.
1957	John Bardeen, Leon Cooper, and John Schrieffer provide theoretical basis for superconductivity, discovered in 1911.
1959	The U.S. government funds the first IDLs, or interdisciplinary laboratories, which mark a beginning of the modern academic model of materials science and engineering.
	J.W. Cahn and J.F. Hilliard develop theory of microstructural evolution in diffuse-interface systems.
1970	Researchers at Corning develop optical fibers transparent enough to make fiber optic communication practical.
1974	A study by an NAS committee, COSMAT, defines field of materials science and engineering, creating a community sensibility.
1980	Gerd Binnig and Heinrich Rohrer invent scanning tunneling microscopy, which has led to a family of imaging tools often capable of molecular- and atomic-scale resolution.
1985	First university "materials by design" initiatives attempt computational materials design.
1986	K. Alex Müller and J. Georg Bednorz discover high-temperature superconductivity in ceramic materials.
1990s	The field of materials science and engineering begins shifting into a more systems-based approach to materials innovation and toward materials design in which researchers can predict new materials they would like to have rather than having to discover them.

REFERENCES

1. C.S. Smith, *A Search for Structure* (MIT Press, 1981)
2. C.S. Smith, *A History of Metallography* (MIT Press, 1988)
3. K.A. Taylor *et al., Metall. Trans.* 20A, 2717 (1989)
4. I. Amato, *Stuff: The Materials the World Is Made Of* (Basic Books, 1997)
5. G.B. Olson, M. Azrin, E.S. Wright, eds., *Innovations in Ultrahigh Strength Steel Technology*, Proceedings of the 34th Sagamore Army Materials Research Conference, 1990
6. G.B. Olson, *Science* 277, 1237 (1997)
7. D.L. McDowell and T.L. Story, "New Directions in Materials Design Science and Engineering," National Science Foundation Workshop Report, Georgia

Institute of Technology Materials Council, 1999
8. "New Forces at Work: Industry Views Critical Technologies," 4th Office of Science and Technology Policy Report on Critical Technologies, February 2000
9. G.M. Whitesides, J.P. Mathias, C.T. Seto, *Science* 254, 1312 (1991)
10. S. Williams, *Technol. Rev.* 102, 92 (September/October 1999)
11. D. Stein, *MSE: Forging Stronger Links to Users* (National Research Council, 1999)
12. M.F. Ashby, *Materials Selection in Mechanical Design* (Pergamon, 1992)
13. L. Kaufman and H. Bernstein, *Computer Calculation of Phase Diagrams* (Academic Press, 1970)
14. J.R. Rice and J.-S. Wang, *Mater. Sci. Eng.* A107, 23 (1989)

Transformations—Selected Works of G.B. Olson on Materials, Microstructure, and Design
C.E. Campbell, M.V. Manuel, and W. Xiong, editors

CHAPTER 13

Charge Transfer Mechanism of Hydrogen-Induced Intergranular Embrittlement of Iron*

Lieping Zhong, Northwestern University and Seagate Technology
Ruqian Wu, California State University at Northridge
A.J. Freeman and G.B. Olson, Northwestern University

Abstract

Impurity-induced reduction of intergranular cohesion—a major factor limiting the usable strength level of ultrahigh-strength steels—is particularly severe when aggravated by mobile hydrogen through environmental interaction, as in the case of hydrogen stress corrosion cracking. As an aid in establishing an understanding on the electronic level, the influence of hydrogen on the cohesion of an iron grain boundary was determined using the full-potential linearized augmented plane wave (FLAPW) method with the generalized gradient approximation. Through precise calculations on both grain boundary and free surface environments, we found that hydrogen is a strong embrittler. Analysis of the results in terms of structural relaxation, bonding character, and magnetic interactions shows that the hydrogen-iron chemical bond is stronger on the free surface and a charge-transfer mechanism is found to play a dominant role for the hydrogen-induced reduction of cohesion across the iron grain boundary. These results provide a quantitative explanation from First principles for the technologically important phenomenon of hydrogen-induced intergranular embrittlement.

Introduction

H-induced embrittlement of steels remains a highly complex unsolved problem (Ref 1). Various mechanisms have been proposed (Ref 2) to explain the H-induced embrittlement (HIE), such as (a) precipitation of gaseous hydrogen, (b) formation of hydrides, (3) deformation localization, and (4) reduction of cohesion across the grain boundary. While the role of the first two factors has been confirmed in some systems, the relevance of the latter two is still controversial.

It has been well established that H atoms, like many other nonmetal or metalloid impurities in Fe, segregate to the Fe grain boundary (GB) (Ref 1, 2). In addition, H in Fe has a uniquely high mobility (compared to other metalloid impurities such as P and S, the mobility of H can be orders of magnitude higher). From a thermodynamic point of view, a normal segregant inevitably alters cohesion

*L. Zhong, A.J. Freeman, R. Wu, and G.B. Olson, "Charge Transfer Mechanism of Hydrogen-Induced Intergranular Embrittlement of Iron," *Phys. Rev. B*, 21 (2000) p 938–941

in a local environment. Based on a plausible thermodynamic description by Rice and Wang (Ref 3), the fracture mode of solids is determined by the competition between brittle interfacial cleavage separation and crack-tip blunting through dislocation emission. It has been proposed (Ref 3, 4) that for a fixed boundary solute (impurity) coverage, Γ, nonlinear entropy contributions can be neglected and thus the relation

$$2\gamma_{int} = (2\ \gamma_{int})_0 - (\Delta g_b^0 - \Delta g_s^0)\Gamma \qquad \text{(Eq 1)}$$

holds, where $(2\ \gamma_{int})_0$ is the work to separate the clean grain boundary, and Δg_b^0 and Δg_s^0 are the free energies of segregation of the solute to the grain boundary and free surface (FS), respectively. Hence, if the free energy of segregation for the solute atom at the FS is more negative than it is at the GB, this positive energy difference for $\Delta g_b^0 - \Delta g_s^0$, will lead to embrittlement.

Rice and Wang (Ref 3) also showed that intergranular fracture toughness is a monotonic function of the ideal work of Eq 1 and demonstrated a compelling correlation between measured embrittlement potencies of segregants and corresponding $\Delta g_b^0 - \Delta g_s^0$ values from available surface thermodynamic data. They further predicted the manner in which the embrittlement potency of a *mobile* solute is amplified beyond that of Eq 1 and used critical experiments in bictystals (including boundaries of the geometry considered in this paper) to verify the predicted temperature dependence of intergranular hydrogen embrittlement, which can be nonmonotonic in nature through a competition between thermodynamic and kinetic factors (Ref 2). Their analysis reveals that the more complex embrittlement behavior of a mobile segregant is still fundamentally driven by the thermodynamic quantity $\Delta g_b^0 - \Delta g_s^0$. Obviously, the value of $\Delta g_b^0 - \Delta g_s^0$ is intrinsically related to the interatomic chemical interaction and thus can be determined using present state-of-the-art *ab initio* quantum mechanical calculations (Ref 5) provided the method employed treats both the GB and FS at the same level of precision.

We report here the results of precise calculations of the electronic and magnetic properties of the H/Fe $\Sigma3[1\bar{1}0](111)$ GB (Ref 6) and the corresponding H/Fe(111) free surface. (Σ3 refers to a special orientation between crystals for which one in three lattice points are coincident, $[1\bar{1}0]$ denotes the crystal Miller indices of the axis of misorientation, and (111) denotes the habit plane of the boundary, both referred to the bcc crystal lattice. The boundary corresponds to the "incoherent" twin boundary.) The mechanism of the H-induced embrittlement is investigated through a comparison of total energies, chemical bonding behavior, and magnetic interactions at 0 K. As sketched in Fig. 1(a), a 23-layer slab is adopted to simulate the clean $Fe\Sigma3[1\bar{1}0]$ (111) GB; the H impurity is placed at the center of the trigonal prism formed by Fe host atoms. As shown in Fig. 1(b), the H/Fe(111) free surface is simulated by a 13-layer slab that contains a bcc Fe(111) substrate film covered by a pseudomorphic H monolayer on each side. The two-dimensional (2D) lattice constant and the unrelaxed Fe-Fe interatomic distance are chosen from experimental values for bulk bcc Fe, that is, d_{Fe-Fe} = 4.69 a.u., whereas the final equilibrium atomic geometries are optimized through the calculated atomic force (Ref 6–8) on each atom.

Both the GB and FS are treated by the highly precise all-electron full potential linearized augmented plane wave (FLAPW) method (Ref 5) with the generalized gradient approximation (GGA) (Ref 9) for the exchange and correlation interactions between electrons. No assumptions are made in the FLAPW calculations about the nature of bonding, or shape of the wave functions, charge density or potential, and are self-consistent fully converged (as also found in previous extensive tests on similar systems (Ref 10–13).

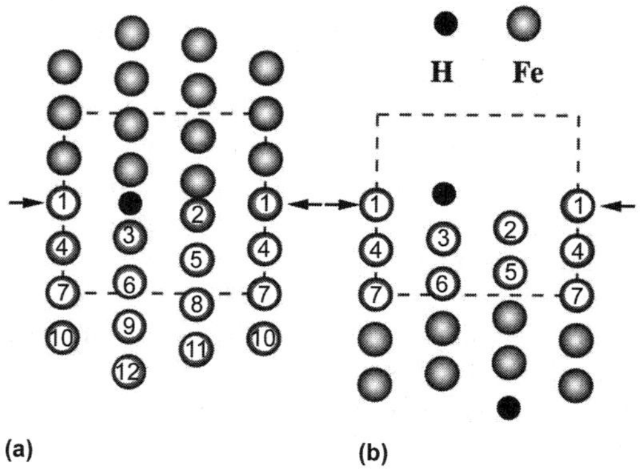

Fig. 1 Model and notation for the atomic structures of (a) $Fe\Sigma3[1\bar{1}0]$ (111) grain boundary, and (b) Fe(111) free surface. The dashed lines mark the region mapped in Fig. 2. The arrows denote grain boundary plane and corresponding fracture surface.

The equilibrium atomic structures were obtained by determining the vertical interplanar distances according to the calculated atomic forces for both H/Fe GB and FS as well as for the corresponding clean reference systems. The structure within the lateral (111) planes was kept unchanged in order to maintain the in-plane threefold rotational symmetry. We fixed the three outermost Fe layers while adjusting the GB environment. Due to its small size, an H atom in the GB and FS alters the atomic positions of the surroundings very slightly. Compared to the structure in the clean Fe GB, the Fe(2) and Fe(3) atoms are pushed away by 0.03 a.u. and 0.08 a.u. from the central plane (relative to the clean Fe GB). Compared with the metalloid impurities case (Ref 11, 12), where the impurity on the Fe(111) FS induces a strong multilayer relaxation in the substrate, the H-induced relaxation is only within 0.11 a.u.

Strikingly, we found that the H-Fe bond length is much shorter in the FS environment than in the GB; the calculated $d_{H-Fe(3)}$ in H/Fe(111) is 3.12 a.u., which is about 8% shorter than that in the GB environment (3.40 a.u.), and is expected to significantly affect the chemical interactions in GB and FS. As presented in Fig. 2, the charge density differences between the self-consistent charge density of the H/Fe systems and the corresponding superpositions of the charge densities of the reference systems (i.e., a free H monolayer, the clean Fe GB and Fe(111) FS), indicate pronounced charge redistributions in both GB and FS environments. Significant charge accumulations around the H sites can be seen from the contours in Fig. 2, which show that the H in both GB and FS acts as an electron acceptor. Compared with the typical embrittling metalloid impurity P, where the P-Fe bonding undergoes a nonhybridized embeddedlike electrostatic interaction (Ref 11, 12), the contour profile in Fig. 2 suggests that the H-Fe(3) bonding is more ioniclike, which is consistent with an earlier GB calculation (Ref 14, 15). Within the muffin-tin spheres (r_{MT} = 1.0 a.u.), the integral charges

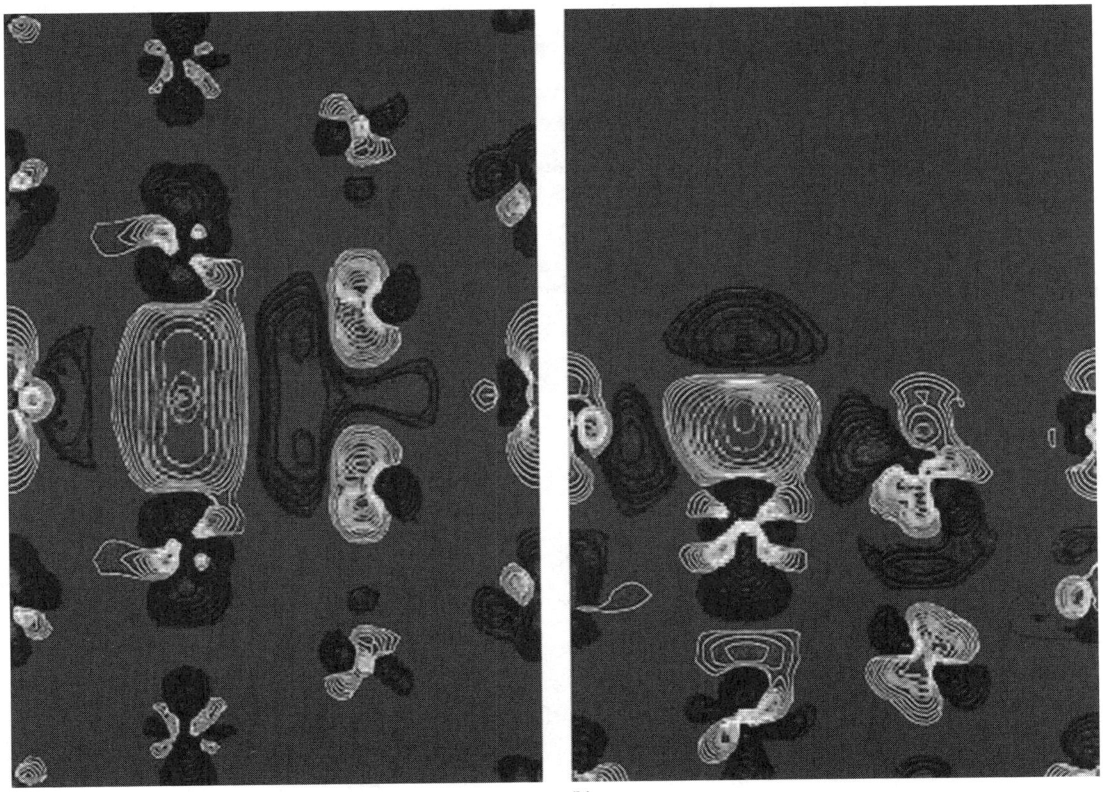

(a) (b)

Fig. 2 The valence charge density difference calculated by FLAPW with GGA for (a) H/Fe GB and (b) H/Fe FS. Contours start from $\pm 5\times10^{-4}$ e/a.u.^{3} and increase successively by a factor of $\sqrt{2}$; white color indicates charge accumulation, black color indicates charge depletionn.

obtained from Fig. 2 are 0.07 and 0.11 electrons for H in the GB and FS, respectively. In the GB case, the presence of H leads to a decrease of charge density in the region between Fe(2) atoms across the GB core, as also seen for other metalloid impurities (Ref 13, 16). With a shorter H-Fe(3) distance, the chemical interaction between H and Fe(3) atoms in the FS is much stronger than their counterpart in the GB, as also evidenced by the integral charges in the muffin-tin spheres. Strikingly, even in the vacuum in Fig. 2(b), the presence of H results in charge removal in the region above the H adatom, which indicates a much stronger H-Fe chemical interaction in FS.

This ioniclike bonding behavior can be understood from examining more fundamental electronic and magnetic properties. In fact, the basic atomic characteristics that control the impurity-host interaction are the radial extension and the relative energy position of the impurity valence states: the former is important in the covalent bond-formation ability of the impurity; the latter determines the relative amounts of ionic and covalent character in the bonding. In the case of H, the 1 *s* valence orbitals are too extended to form stable covalent hybrids with the Fe host in both GB and FS. In contrast, since the energy location of H-1 *s* is much lower than the *d* bands of the Fe host, the bonding behavior between H and Fe is expected to be more ioniclike than covalent.

The effect of H on the magnetization of the Fe host is similar to that of C in the Fe GB and FS (Ref 11, 12), namely, a reduction on Fe(1) and Fe(3) and an enhancement on Fe(2) atoms. Referring to the clean Fe GB, the H reduces the magnetic moment of Fe(3) and Fe(1) by $0.23\mu_B$ and $0.07\mu_B$, respectively. By contrast, the H-induced magnetic moment enhancement is remarkable for Fe(2) (by $0.22\mu_B$), Fe(4) (by $0.13\mu_B$), and even for Fe(6) (by $0.05\mu_B$). At the FS, the reduction of the Fe(3) magnetic moment is as large as $0.27\mu_B$. In both environments, a small induced magnetic moment ($-0.01\mu_B$) is found in the H muffin-tin sphere.

The binding energies for a H impurity in the Fe GB and on the Fe(111) surface are calculated and their differences are given in Table 1. The calculated results using the FLAPW method with the local spin density approximation (LSDA) for the exchange and correlation interactions between electrons are also listed

Table 1 Binding energies (in eV) and their differences for H in Fe GB and FS and the decomposition into chemical, mechanical contributions calculated with LSDA and GGA

		ΔE_b	ΔE_s	$\Delta E_b - \Delta E_s$
Unrelaxed (chemical)	LSDA	−3.64	−3.85	+0.21
	GGA	−3.09	−3.42	+0.33
Relaxation (mechanical)	LSDA	+0.23	+0.11	+0.12
	GGA	+0.08	+0.15	−0.07
Relaxed (total)	LSDA	−3.41	−3.74	+0.33
	GGA	−3.01	−3.27	+0.26

for comparison. Results are listed for the "chemical" interaction contribution (defined as the result obtained without structural relaxation, called "unrelaxed" in Table 1), the "mechanical" energy released during the structural relaxation of the clean Fe GB and FS after impurity removal (called "relaxation" in Table 1), and the sum of these contributions that defines the total interaction energy. Although the impurity segregation energies used in the Rice-Wang model are expressed relative to an impurity in dilute solution in crystalline Fe, the binding energies, ΔE_b and ΔE_s, computed here are referred to the calculated binding energy of an isolated 2D monolayer with the same structure as the segregated monolayer as shown in Fig. 1. The segregation energy differences, which governing the effect of the impurity on the Fe GB cohesion, will be the same as the energy difference $\Delta E_b - \Delta E_s$. By including the H-induced structural relaxation, the binding energy ΔE_b and ΔE_s, when calculated with GGA, are −3.01 eV/adatom and −3.27 eV/adatom for H in the Fe GB and FS environments, respectively. As a result, the calculated binding-energy difference ($\Delta E_b - \Delta E_s$) is + 0.26 eV/adatom (or 25.21 kJ/mol), which agrees well with the 25 ± 19 kJ/mol estimated from experimental data (Ref 4). According to the Rice-Wang thermodynamic theory (Ref 3), H is thus a strong embrittler for the cohesion across the Fe GB.

From Table 1, we see that the atomic relaxation in both the clean GB and FS environments calculated with GGA has only a minimal contribution to ΔE_b and ΔE_s (0.08 and 0.15 eV/adatom respectively) and even a negative difference (−0.07 eV/adatom). By contrast, the chemical energy contributes a positive value of $\Delta E_b - \Delta E_s$ (+ 0.33 eV/adatom); it is larger in the FS than in the GB, despite the fact that one out of two vertical bonds is broken. As discussed, this is because of the shortened H-Fe bond-length associated with the much stronger H-Fe interaction in the FS environment, which favors greater FS stability and hence an embrittling effect. This result further supports the argument that charge-transfer from Fe to H is the key mechanism for HIE in the Fe GB.

Before discussing our results further, it is important to make comparison for the results calculated by GGA with those by LSDA. It is well known that the most widely used LSDA overestimates bulk moduli and underestimates volumes of 3*d* transition metal bulk solids. Of its most widely studied shortcomings is the incorrect prediction of an anti-ferromagnetic fcc ground state for Fe bulk solid, rather than the ferromagnetic bcc structure (Ref 16). It is well documented that the GGA is a significant improvement over LSDA on the properties governed by a realistic description of bond formation (Ref 9) especially in the case of Fe. Hence, we expect that in these systems, the GGA approach is more preferred. Indeed, for the GB and FS systems considered here, we can see from Table 1 that the calculations with GGA and with LSDA give different ΔE_b and ΔE_s values. For the energy differences, which is crucial for determining the cohesion properties of H in GB, remarkably improvement is made in the mechanical (relaxation) part. The $\Delta E_b - \Delta E_s$ calculated with LSDA is + 0.12 eV, which means a contribution to embrittlement; the counterpart calculated with GGA gives −0.07 eV, which indicates even a slight enhancement for Fe GB cohesion. From the calculation with GGA, we can see more clearly that only the chemical part contributes the HIE in the Fe GB.

It is interesting to compare the HIE in the GB with the effects of other embrittling metalloid impurities such as P, where the electrostatic embeddedlike P-Fe bond is found to tune the embrittlement behavior (Ref 11, 12). For P, because of the large spatial extension of its 3*p* wave functions, the valence states can be easily affected by the surrounding Fe atoms; indeed the charge density in the inner region of the P atom is significantly decreased. Meanwhile the Coulomb and Pauli repulsions push the P 3*p* wave function away from the high charge density region around the Fe atoms. The net action from these two effects is to squeeze the P 3*p* electrons into the region intermediate between P and Fe. Thus, the P-Fe bond shows the nature of an embeddedlike electrostatic interaction in which the P-Fe(1) bonding is almost as strong as the P-Fe(3) bonding. With the spatial isotropy of embedded- like P-Fe bonds, the energy loss due to fracture is small and can be easily compensated through release of the lattice stress (mechanical relaxation) since only one out of five bonds is cut from the GB to FS.

In sharp contrast with this physical picture, the main mechanism we find for HIE is charge transfer with the H impurity acting as an electron acceptor in both the GB and FS. Although the number of neighbors of H changes from five Fe atoms (two Fe(3) and three Fe(1) atoms) in the GB to four Fe atoms (one Fe(3) and three Fe(1) atoms), in the FS the energy loss due to the removal of one Fe(3) after the fracture is compensated by enhancement of the H-Fe chemical interaction. We note that a well-known early theory of hydrogen embrittlement by Troiano (Ref 17) also invoked a charge-transfer mechanism, but the proposed direction of transfer was opposite to that demonstrated by the rigorous calculations presented here. In a general sense, the information obtained here may also prove to be important for investigations of H-induced effects in other alloy systems, and further electronic studies should identify new directions in alloy composition design for improved hydrogen resistance.

ACKNOWLEDGMENTS

This work was supported by the Office of Naval Research (Grant No. N00014-94-1-0188) and a grant of computer time at the Pittsburgh Supercomputing Center from the NSF Division of Advanced Scientific Computing and the Arctic Region Supercomputing Center. We thank J. S. Wang of Harvard University for helpful discussions.

REFERENCES

1. C.J. McMahon, Jr., in *Innovations in Ultrahigh-Strength Steel Technology*, ed, G.B. Olson, M. Azrin, and E.S. Wright, Proceedings of the 34th Sagamore Army Materials Research Conference (Government Printing Office, 1990), p 597
2. J.S. Wang, in *Hydrogen Effects in Materials*, ed. A.W. Thompson and N.R. Moody (TMS, 1966), p 61
3. J.R. Rice and J.S. Wang, *Mater. Sci. Eng.*, A 107, 23 (1989)

4. P.M. Anderson, J.S. Wang, and J.R. Rice, in *Innovations in Ultrahigh-Strength Steel Technology*, ed. G.B. Olson, M. Azrin, and E.S. Wright, Proceedings of the 34th Sagamore Army Materials Research Conference (Government Printing Office, 1990), p 619
5. E. Wimmer, H. Krakauer, M. Weinert, and A.J. Freeman, *Phys. Rev.* B 24, 864 (1981)
6. J.M. Soler and A.R. Williams, *Phys. Rev.* B 40, 1560 (1989)
7. R. Yu, D. Singh, and H. Krakauer, *Phys. Rev.* B 43, 6411 (1992)
8. R. Wu and A.J. Freeman (unpublished)
9. J.P. Perdew, K. Burke, and M. Ernzerhof, *Phys. Rev.* Lett. 77, 3865 (1996)
10. A.J. Freeman and R. Wu, *J. Magn. Magn. Mater.* 100, 497 (1992)
11. R. Wu, A.J. Freeman, and G.B. Olson, Science 265, 376 (1994)
12. R. Wu, A.J. Freeman, and G.B. Olson, *Phys. Rev.* B 53, 7504 (1996)
13. L. Zhong, R. Wu, A.J. Freeman, and G.B. Olson, *Phys. Rev.* B 55, 11 133 (1996)
14. G.L. Krasko and G.B. Olson, Solid State Commun. 76, 247 (1990)
15. G.L. Krasko and G.B. Olson, Solid State Commun. 79, 113 (1991)
16. C.S. Wang, B.M. Klein, and H. Krakauer, *Phys. Rev.* Lett. 54, 1852 (1985)
17. A.R. Troiano, *Trans. Am. Soc. Met.* 52, 54 (1960)

Transformations—Selected Works of G.B. Olson on Materials, Microstructure, and Design
C.E. Campbell, M.V. Manuel, and W. Xiong, editors

CHAPTER 14

Influence of Alloying Additions on Grain Boundary Cohesion of Transition Metals—First-Principles Determination and Its Phenomenological Extension*

W.T. Geng, A.J. Freeman, and G.B. Olson
Northwestern University

Abstract

The toughness and ductility of ultrahigh-strength alloys is often limited by intergranular embrittlement, particularly under conditions of unfavorable environmental interactions such as hydrogen embrittlement and stress corrosion cracking. Here we investigated the mechanism by which the segregated substitutional additions cause intergranular embrittlement. An electronic-level phenomenological theory is proposed to predict unambiguously the effect of a substitutional alloying addition on grain boundary cohesion of metallic alloys, based on first-principles full-potential linearized augmented plane-wave method (FLAPW) calculations on the strengthening and embrittling effects of the metals Mo and Pd on the Fe grain boundary cohesion. With the bulk properties of substitutional alloying addition A and the matrix element M as inputs, the strengthening or embrittling effect of A at the grain boundary of M can be predicted without carrying out first-principles calculations once the atomic structure of the corresponding clean grain boundary is determined. Predictions of the embrittlement potency of a large number of metals, including the $3d$, $4d$, and $5d$ transition metals, are presented for the Fe $\Sigma3$ (111) and the Ni $\Sigma5$ (210) grain boundaries. Rigorous FLAPW calculations on the effect of Co. Ru. W. and Re on the Fe $\Sigma3$ (111) grain boundary and Ca on the Ni $\Sigma5$ (210) grain boundary cohesion confirm the predictions of our model. This model is expected to be applicable to other high-angle boundaries in general and instructive in the quantum design of ultrahigh-strength alloys with resistance to intergranular fracture.

Introduction

Grain boundary cohesion is often the controlling factor limiting the ductility of

*W.T. Geng, A.J. Freeman, and G.B. Olson, "Influence of Alloying Additions on Grain Boundary Cohesion of Transition Metals—First-Principles Determination and Its Phenomenological Extension," *Phys. Rev. B* 63 (2001) 165415

high-strength metallic alloys (Ref 1). Thus, understanding the influence of transition metal alloying additions is of great importance in predicting and controlling grain boundary embrittlement (GBE) since the complexity of GBE behavior is often associated with the presence of substitutional alloying elements. Attempts at qualitatively explaining GBE on the electron-atom level go back to the 1960s (Ref 2). Although predictions based on simple thermodynamic (Ref 3–5) and pair bonding models (Ref 6, 7) have been proposed, the role of alloying elements on the cohesion of grain boundaries in alloys remains controversial. A thermodynamic theory by Rice and Wang (Ref 8) demonstrated that previous thermodynamic models are incorrect and identified the key surface-thermodynamic energy difference governing the quantitative role of a boundary segregant. In addition to the simple pair-bonding approaches (Ref 6, 7) which, like thermodynamic theories, employ energetics to describe the effect of segregants, electronic structure theorists have applied both empirical (Ref 9, 10) and first-principles quantum-mechanical methods (Ref 11, 12) in attempts to directly deduce the effect of segregants on grain boundary (GB) fracture resistance. In contrast to these studies, our recent line of research (Ref 13–18) has applied rigorous first-principles methods to evaluate the surface-thermodynamic energy difference identified by Rice and Wang (Ref 8) as directly connected to the ideal work of GB fracture.

In spite of the complexity of the mechanical behavior and GB atomic structures, general trends in certain mechanical properties can be correlated with specific features of the electronic structure (Ref 12, 19). First-principles computation has proved to be an accurate and powerful tool in attacking the problem of the mechanical properties of real materials such as GBE. An inherent advantage of the first-principles electronic theory is that it is independent of any adjustable parameters, and so, numerical results provide a solid basis or starting point for phenomenological theories. Without such a theory starting from first principles, one has always to repeat the full procedure of calculations to predict quantitatively the effect of any substitutional element even on the same GB (Ref 16, 18). Mn this sense, the exact mechanism by which segregation elements cause embrittlement remains unclear.

Very recently (Ref 18), the effect of Mo and Pd substitutional segregation on the cohesion of the Fe $\Sigma3$ (111) GB was investigated using the first-principles full-potential linearized augmented plane wave (Ref 20, 21) (FLAPW) total energy/atomic force method with the generalized gradient approximation (GGA) (Ref 22). Based on the Rice-Wang model, (Ref 8) our total energy calculations show that Mo has a significant beneficial effect on the Fe GB cohesion, while Pd behaves as a weak embrittler. An analysis of the geometry optimization indicates that both Mo and Pd have a moderate atomic size to fit well in the GB substitutional site. The elastic energy associated with the Mo and Pd segregation was estimated with a rigid environment approximation. It was found that both Mo and Pd introduce a beneficial volume effect. Studies of the electronic structures show that the strong bonding capability associated with its half-filled *d* bands makes Mo a cohesion enhancer (−0.90 eV) for the Fe $\Sigma3$ (111) GB. By comparison, the weak bonding capability associated with its filled *d* bands leads Pd to be a weak embrittler (+0.08 eV). These first-principles quantum-mechanical results support, in general, the main idea of the simple pair-bonding models (Ref 6, 7) in that the elemental cohesive energy difference between the alloying element and the host element plays an important role in determining the effect of this alloying element on the host GB cohesion. They differ, however, quite significantly: for example, for Mo and Pd in the Fe GB, the embrittlement potency given by Seah's pair-bonding model (Ref 6) is −1.5 and +0.9 eV, respectively. This significant difference with the values of −0.90 and +0.08 eV from our rigorous calculations indicates that a quantum-mechanical treatment is necessary in an accurate study of the GBE. Also, the numerical results for Pd, which has a similar elemental cohesive energy to that of Fe, point to the importance of the role played by the GB free volume, which can be determined with high accuracy only by quantum-mechanical calculation.

Based on the physical insight obtained from the results of our rigorous calculations, here we develop a phenomenological model by which the embrittlement potency of a substitutional atom in a given GB can be predicted without carrying out full first-principles calculations. To be accurate and reliable, such a theory must use first-principles results, rather than adjustable parameters, as inputs. To be capable of prediction without solving the quantum-mechanical Schrödinger equation, such a theory must also make use of atomic, or bulk quantities as inputs. This means that this model should have an electronic, rather than an atomistic or thermodynamic basis, in order to yield a quantum description of the mechanical behavior of a substitutional element at the GB. The rest of the paper is organized as follows. In the section, "Phenomenological Model," the development of a phenomenological model starting from first-principles is presented. Confirmation of the model with rigorous FLAPW calculations is discussed in the section, "Confirmation of the Model,", and then we give a short summary.

Phenomenological Model

Volume Effect

Since a segregant can introduce a significant relocation of the atoms near the GB (Ref 17) the volume effect of a segregant is of great importance for understanding many of its physical and mechanical properties at the GB. In a highly precise first-principles treatment, one has to fully relax the atoms near the GB.

To address the problem of volume mismatch, the atomic size of the segregant and also the size of the GB site should be well defined. Unfortunately, the geometric size of an atom has no absolute meaning and its definition depends on the physical, chemical, or mechanical problem under consideration. The problem of defining the size of an atom in metallic solid solutions was discussed in detail by King (Ref 23) in the study of substitutional solid solutions. One of these definitions uses the atomic volume in the structure of the elemental crystals and was adopted by de Boer *et al*, (Ref 24) in the study of energy effects in transition metal alloys and solid solutions. In ourview, the substitutional alloying additions segregated to the GB core can be taken as a special kind of solid solution. Thus, in the current paper, we also employ this definition.

Volume expansion (or "free volume") is an important property in the atomic structure of the GB. Its contribution to the space available for the substitutional element has a significant influence on the physical and mechanical behavior of this element (Ref 18). The existence of GB volume expansion makes the problem of the volume effect in the GB different from those in the alloys and solid solutions, where the otherwise undisturbed atomic structure is perfect. The FLAPW method employed in our investigations (e.g., Ref 13, 16–18) is well known for both its high precision and computational effort. To simulate an Fe $\Sigma 3$ (111) or Ni $\Sigma 5$ (210) GB, more than 20 atoms need to be included in a calculational unit cell. And the transition metals, especially those with magnetism such as Fe and Ni, are known to have a poor convergence in solving the Kohn-Sham equation self-consistently. Therefore, we relaxed the atomic positions only in the direction normal to the GB plane and kept the lateral symmetry and also the mirror symmetry in the normal direction, which is thought to be a reasonable approximation as we achieved fairly good agreement with the experiment on the mechanical behavior of the impurities (Ref 13, 17) and additions (Ref 16, 18). Therefore, our measure for the GB expansion involves only the relative normal displacements. A local measure for the GB expansion is the relative normal displacement of the two nearest atoms, $M(2)$ across the boundary plane as represented in Fig. 1(a) and 2(a). This definition, however, is not appropriate for the investigation of the volume mismatch between the substitutional element and the GB site, as the relocation of the atoms near the GB involves $M(4)$, which, along with $M(2)$ and $M(3)$, determines the local environment of the GB core. The volume of the GB site, V^{GB}, can be taken as the displacements of $M(4)$. The calculated V^{GB} for the Fe $\Sigma 3$ (111) GB is 27.4 a.u.^{3} (34.5% of the bulk value 79.4 a.u.^{3}; Ref 25); for the Ni $\Sigma 5$ (210), it is 18.6 a.u.^{3} (25.2% of the bulk value 73.8 a.u.^{3}; Ref 25).

In recent work, Ochs *et al.*, (Ref 26, 27) studied the atomic structures for the $\Sigma 5(310)$ GB in bcc Nb, Mo, Ta, and W with the first-principles pseudopotential method. For Nb, Mo, and W, their calculations indicate noticeable lateral displacements of the grains. Based on their first-principles results, we might suppose that lateral atomic displacements in cases of 3*d* transition metals such as Fe and Ni can also possibly lower the total energy of the GB. If this is the case, a more precise measure of the GB expansion in Fe and Ni should also consider the lateral displacements of the GB atoms as well.

As we have pointed out previously (Ref 18) not all the expanded volume near the GB is *available* for the GB core atom. The GB core atom $A/M(1)$ can form bonds only with $M(2)$ and $M(4)$, but not with $M(3)$. This means that

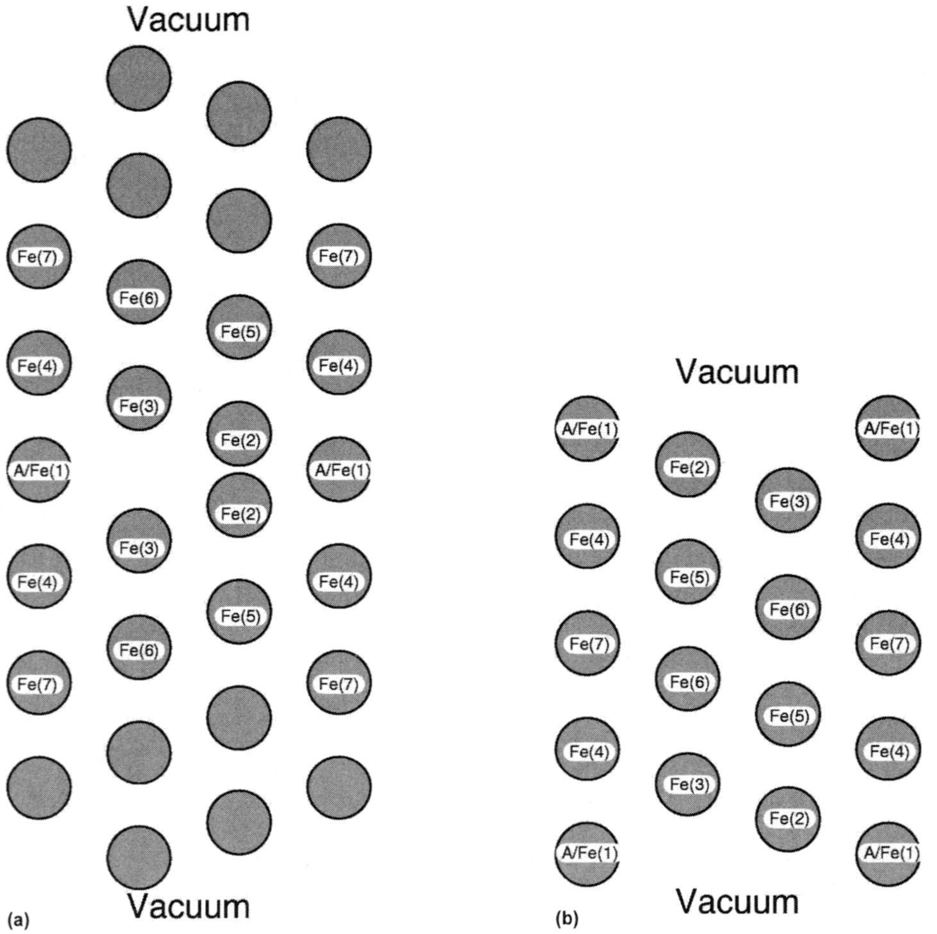

Fig. 1 Model and notation for the structure of (a) the Fe Σ3 (111) [001] grain boundary, and (b) the Fe (111) free surface.

only two-thirds of the GB free volume per site is available for $A/M(1)$. Thus, the volume mismatch ΔV^A between a segregant and the GB can be defined as:

$$\Delta V^A \equiv V^A - \left(V^M + \frac{2}{3}V^{GB}\right) \quad \text{(Eq 1)}$$

To address the volume effect of a substitutional addition A, one should compare the elastic energy of the clean and segregated GB. This elastic energy is associated with the volume mismatch between $A/M(1)$ and the GB site. Since the crystal lattice near the GB is not perfect even without the volume mismatch between $A/M(1)$ and the GB site, the bulk modulus and shear modulus (of M) near the GB no longer retain the perfect bulk values and are not well defined. As an approximation, these deviations are neglected in the present model. In our previous work (Ref 18) we employed a rough approximation to estimate the elastic energy associated with the volume mismatch between $A/M(1)$ and the GB site. We neglected the relaxation of the GB during segregation and took the effect of the volume mismatch between the GB core atom and the GB hole as only the compression or expansion of the segregant in a rigid environment. As an improvement, we now calculate the elastic energy in the framework of a classical elasticity theory with the *sphere and hole* model developed by Friedel and Eshelby (Ref 28, 29). In this model, a spherical hole with volume V^M (atomic cell of metal M) in the matrix is partly filled by a sphere of metal A with volume V^A (dissolved atomic cell). The remaining volume, $(V^M - V^A)$, is accommodated by elastic deformation of M and A. If this is a state of purely internal stress, the total volume is

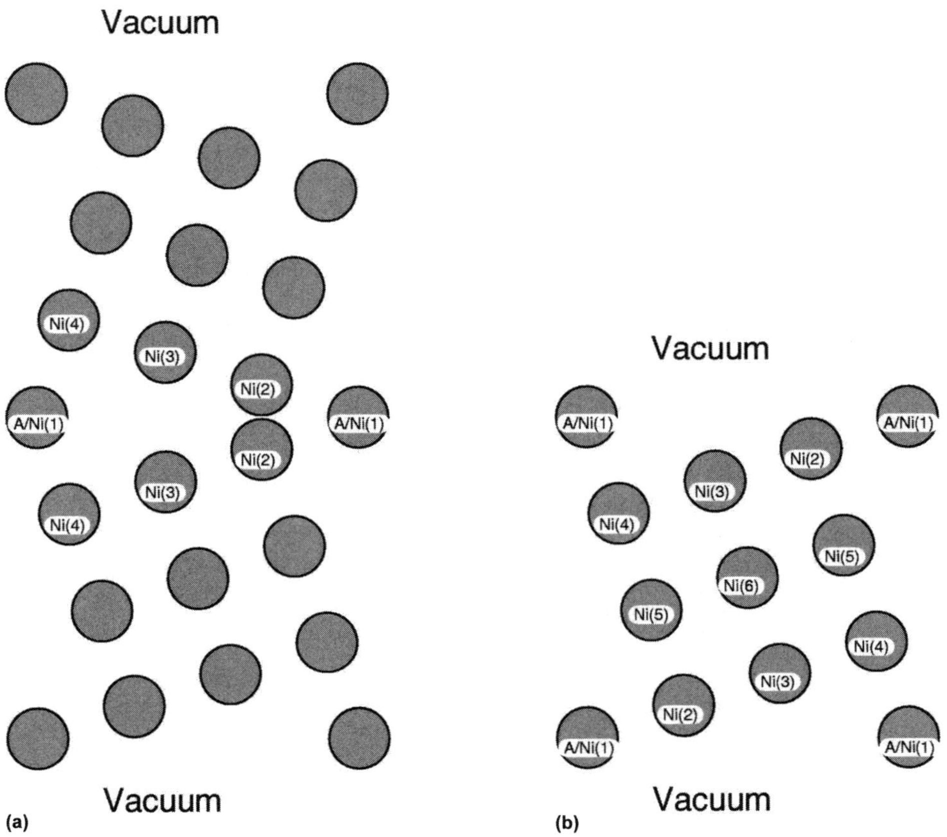

Fig. 2 Model and notation for the structure of (a) the Ni Σ5 (210) [001] grain boundary, and (b) the Ni(210) free surface.

unaffected. Both A and hole are then subjected to a uniform hydrostatic pressure. The stress on A is related to its bulk modulus K_A, while that on the matrix is related to an effective modulus equal to 4/3 times the shear modulus of the matrix (Ref 28, 29), G_M.

The elastic energy yields:

$$E_V^A = \frac{K_A(\Delta V^A)^2}{2V^A} + \frac{2G_M(\Delta V^M)^2}{3V^M} \qquad \text{(Eq 2)}$$

where ΔV^A and ΔV^M are the volume changes of sphere and hole due to the internal stress. The pressures are adjusted such that they are continuous across the interface between matrix and "inclusion," which leads to an expression for the elastic energy per mole of the solute metal:

$$E_V^A = \frac{2K_AG_M(V^M - V^A)^2}{3K_AV^M + 4G_MV^A} \qquad \text{(Eq 3)}$$

In the GB environment, the hole cut from the matrix has a volume V^M + 2/3 V^{GB} rather than V^M. Therefore, to describe the GB case, Eq 3 should take the form:

$$E_V^A = \frac{2K_AG_M(V^A - V^M - \frac{2}{3}V^{GB})^2}{3K_AV^M + 4G_MV^A} \qquad \text{(Eq 4)}$$

$A = M$ corresponds to the clean GB case:

$$E_V^M = \frac{8K_AG_M(V^{GB})^2}{27K_AV^M + 36G_MV^A} \qquad \text{(Eq 5)}$$

When $M(1)$ is replaced by A, the change of the elastic energy near the GB core, ΔE_V^A, is:

$$\Delta E_V^A = E_V^A - E_V^M \qquad \text{(Eq 6)}$$

which is the volume elastic effect of a substitutional addition A in the GB core site.

We calculated ΔE_V^A for a large number of metals, including the 3d, 4d, and 5d transition metals on the Fe Σ3 (111) and the Ni Σ5 (210) GB. The calculated ΔE_V^A value for each alloying addition is listed in column 8, Tables 1

Table 1 Model calculated (see Eq 15) embrittling effect ΔE^A_B (eV) for all of the transition elements on the Fe Σ3 grain boundary cohesion

Atom	$-E^A_{Coh}$	ΔE^A_{Coh}	ΔE^A_{Heat}	ΔE_{Stru}	V^A	K_A	$\delta\Delta E^A_V$	ΔE^A_B
Li	1.63	2.66	0.94	0.00	143.58	0.116	−0.04	1.16
Be	3.32	0.97	−0.20	0.03	55.77	1.003	0.44	0.71
Na	1.11	3.18	2.75	0.00	254.46	0.068	0.18	2.16
Mg	1.51	2.78	0.67	0.01	156.93	0.354	0.19	1.34
Al	3.39	0.90	−0.91	0.12	112.09	0.722	−0.07	−0.03
K	0.93	3.36	4.80	0.00	481.33	0.032	0.33	3.05
Ca	1.84	2.45	1.28	0.01	293.40	0.152	0.76	2.01
Sc	3.93	0.36	−0.52	0.04	158.04	0.435	0.26	0.22
Ti	4.86	−0.57	−0.74	0.02	119.22	1.051	−0.01	−0.44
V	5.30	−1.01	−0.29	0.00	93.46	1.619	−0.11	−0.54
Cr	4.10	0.19	−0.06	0.00	81.01	1.901	−0.02	0.02
Mn	2.98	1.31	0.01	...	82.49	0.596	−0.07	0.37
Fe	4.29	0.00	0.00	0.00	79.39	1.683	0.00	0.00
Co	4.39	−0.10	−0.02	0.20	75.23	1.914	0.07	0.10
Ni	4.44	−0.15	−0.06	0.06	73.83	1.86	0.09	0.04
Cu	3.50	0.79	0.50	0.01	79.86	1.37	−0.01	0.42
Zn	1.35	2.94	−0.14	0.07	103.02	0.598	−0.11	0.85
Rb	0.85	3.44	4.75	0.00	587.83	0.031	0.47	3.20
Sr	1.72	2.57	1.90	0.00	379.12	0.116	0.97	2.46
Y	4.39	−0.10	−0.06	0.09	223.45	0.366	0.92	0.90
Zr	6.32	−2.03	−1.17	0.01	157.30	0.833	0.47	−0.59
Nb	7.47	−3.18	−0.70	0.00	121.37	1.702	0.05	−1.24
Mo	6.81	−2.52	−0.09	0.00	105.11	2.725	−0.09	−0.96
Tc	6.85	−2.56	−0.13	0.19	95.85	2.97	−0.11	−0.94
Ru	6.62	−2.33	−0.20	0.53	91.68	3.208	−0.10	−0.77
Rh	5.75	−1.46	−0.23	0.36	92.95	2.704	−0.10	−0.54
Pd	3.94	0.35	−0.19	0.08	99.24	1.808	−0.11	−0.03
Ag	2.96	1.33	1.23	0.00	115.35	1.007	−0.04	0.81
Cd	1.16	3.13	0.42	0.04	145.43	0.467	0.15	1.35
Cs	0.83	3.46	5.21	0.00	745.67	0.020	0.40	3.29
Ba	1.86	2.43	2.12	0.00	421.77	0.103	1.04	2.56
La	4.49	−0.20	0.25	0.11	249.93	0.243	0.84	0.89
Hf	6.35	−2.06	−0.98	0.10	149.29	1.09	0.43	−0.55
Ta	8.09	−3.80	−0.67	0.00	148.31	2.00	0.61	−0.88
W	8.66	−4.37	0.00	0.00	107.11	3.232	−0.08	−1.54
Re	8.10	−3.81	−0.01	0.27	99.24	3.72	−0.11	−1.29
Os	8.10	−3.81	−0.17	0.85	94.51	4.18	−0.11	−1.15
Ir	6.93	−2.64	−0.38	0.64	95.58	3.55	−0.11	−0.91
Pt	5.85	−1.56	−0.59	0.16	101.93	2.783	−0.11	−0.77
Au	3.78	0.51	0.37	0.00	114.37	1.732	−0.03	0.26
Hg	0.69	3.60	0.69	...	158.41	0.382	0.22	1.65
Tl	1.87	2.42	1.06	0.00	192.80	0.359	0.55	1.71
Pb	2.04	2.25	0.95	0.00	204.49	0.430	0.81	1.88

Also listed are elemental cohesive energies E^A_{Coh} and ΔE^A_{Coh} (eV), heat of formation of solid solution *A*Fe ΔE^A_{Heat}, atomic volumes V^A, and volume mismatch ΔV^A (a.u.^{3}), bulk moduli K (10^{11} N/m^2), volume mismatch correction $\delta\Delta E_V$, and also the work needed to change the ground fcc (hcp is approximated by fcc) structure to bcc structure for an element. The shear modulus of Fe, G_{Fe}, is (Ref 32) 0.816×10^{11} N/m^2

and 2. A positive value represents an increase of the elastic energy near the GB, and so, this segregant is not favored energetically. It is seen that for both the Fe Σ3 (111) and the Ni Σ5 (210) GB, simple metals (except for Li and Al) introduce harmful volume effects, due to their large atomic size. Most transition metals have a moderately larger atomic size compared to Fe and Ni and introduce a beneficial volume effect to the GB cohesion. For instance, the volume effects of Mo and Pd in the Fe Σ3 (111) GB are −0.09 and −0.11 eV, respectively. If the GB hole gets to be, say, 10% smaller due to the lateral relaxation, these volume effects would be −0.06 and −0.09 eV, about 20–30% smaller. In Seah's pair-bonding model (Ref 6), volume expansion of the GB is not considered.

Bonding Character

The other factor, even more important in general, in determining the behavior of a segregant in the GB is its bonding character in both the GB and free surface (FS) environments. At the electronic level, a quantitative description of the chemical bonding will generally employ the concepts of charge transfer, or electronegativity. However, neither charge transfer nor electronegativity is well defined in a non- elemental crystal and therefore is not useful to build into a unified theory. The macroscopic

Table 2 Model calculated (see Eq 15) embrittling effect ΔE_B^A (eV) for all of the transition elements on the Ni $\Sigma 5$ grain boundary cohesion

Atom	$-E^A_{Coh}$	ΔE^A_{Coh}	ΔE^A_{Heart}	ΔE_{Stru}	V^A	K_A	$\delta\Delta E^A_V$	ΔE^A_B
Li	1.63	2.81	0.03	0.02	143.58	0.006	−0.04	0.99
Be	3.32	1.12	−0.22	0.00	55.77	1.003	0.27	0.57
Na	1.11	3.33	1.40	0.01	254.46	0.068	0.27	1.85
Mg	1.51	2.93	−0.25	0.00	156.93	0.354	0.36	1.25
Al	3.39	1.05	−1.39	0.00	112.09	0.722	0.06	−0.05
K	0.93	3.51	2.35	0.00	481.33	0.032	0.40	2.35
Ca	1.84	2.60	−0.37	0.00	293.40	0.152	0.91	1.65
Sc	3.93	0.51	−1.79	0.00	158.04	0.435	0.46	0.03
Ti	4.86	−0.42	−1.54	0.00	119.22	1.051	0.18	−0.47
V	5.30	−0.86	−0.75	0.17	93.46	1.619	−0.05	−0.53
Cr	4.10	0.34	−0.27	0.39	81.01	1.901	−0.05	0.10
Mn	2.98	1.46	−0.33	0.00	82.49	0.596	−0.06	0.32
Fe	4.29	0.15	−0.06	0.20	79.39	1.683	−0.05	0.05
Co	4.39	0.05	−0.01	0.00	75.23	1.914	−0.01	0.00
Ni	4.44	0.00	0.00	0.00	73.83	1.86	0.00	0.00
Cu	3.50	0.94	0.14	0.00	79.86	1.37	−0.05	0.31
Zn	1.35	3.09	−0.63	0.00	103.02	0.598	−0.02	0.80
Rb	0.85	3.59	2.59	0.01	587.83	0.031	0.54	2.60
Sr	1.72	2.72	−0.06	0.00	379.12	0.116	1.10	1.99
Y	4.39	0.05	−1.62	0.00	223.45	0.366	1.17	0.65
Zr	6.32	−1.88	−1.37	0.00	157.30	0.833	0.78	−0.30
Nb	7.47	−3.03	−1.36	0.36	121.37	1.702	0.29	−1.05
Mo	6.81	−2.37	−0.32	0.40	105.11	2.725	0.06	−0.70
Tc	6.85	−2.41	0.03	0.00	95.85	2.97	−0.04	−0.83
Ru	6.62	−2.18	0.02	0.00	91.68	3.208	−0.06	−0.78
Rh	5.75	−1.31	−0.04	0.00	92.95	2.704	−0.05	−0.50
Pd	3.94	0.50	0.00	0.00	99.24	1.808	−0.02	0.15
Ag	2.96	1.48	0.68	0.00	115.35	1.007	0.12	0.84
Cd	1.16	3.28	−0.24	0.00	145.43	0.467	0.33	1.34
Cs	0.83	3.61	2.84	0.01	745.67	0.020	0.46	2.61
Ba	1.86	2.58	0.01	0.00	421.77	0.103	1.16	2.02
La	4.49	−0.05	−1.46	0.00	249.93	0.243	1.03	0.53
Hf	6.35	−1.91	−2.04	0.00	149.29	1.09	0.76	−0.56
Ta	8.09	−3.65	−1.33	0.19	148.31	2.00	1.07	−0.53
W	8.66	−4.22	−0.14	0.52	107.11	3.232	0.10	−1.18
Re	8.10	−3.66	0.10	0.00	99.24	3.72	0.00	−1.19
Os	8.10	−3.66	0.06	0.00	94.51	4.18	−0.04	−1.24
Ir	6.93	−2.49	−0.07	0.00	95.58	3.55	−0.04	−0.89
Pt	5.85	−1.41	−0.22	0.00	101.93	2.783	0.02	−0.52
Au	3.78	0.66	0.33	0.00	114.37	1.732	0.17	0.50
Hg	0.69	3.75	0.04	0.00	158.41	0.382	0.41	1.67
Tl	1.87	2.57	0.14	0.00	192.80	0.359	0.77	1.67
Pb	2.04	2.40	0.08	0.00	204.49	0.430	1.08	1.91

Also listed are elemental cohesive energies E^A_{Coh} and ΔE^A_{Coh}(eV), heat of formation, heat of solid solution ANi ΔE^A_{Heart}, atomic volumes V^A, and volume mismatch ΔV^A (a.u.^{3}), bulk moduli K (10^{11} N/m^2), volume mismatch correction $\delta\Delta E_V$, and also the work needed to change the ground bcc structure to fcc (hcp is approximated by fcc) structure for an element. The shear modulus of Ni, G_{Ni}, is (Ref 30) 0.839×10^{11} N/m^2

quantity that can be a measure of the bonding capability of an element, as adopted in the simple thermodynamic (Ref 5) or pair-bonding theory (Ref 6) of GBE, is the elemental cohesive energy. It is also employed in the present model.

According to the Rice-Wang thermodynamic theory, the potency of a segregation impurity in reducing the "Griffith work" of a brittle boundary separation is a linear function of the difference in *binding energies* for that impurity at the GB and the FS. For a substitutional addition, the above *binding energies* should be binding energy differences between the segregant and the host (GB core) atom. As discovered in our previous work (cf. Table III in Ref 18), the first-principles FS and GB (Fe $\Sigma 3$) chemical energies, defined as the work needed to remove the segregant while not permitting the host (Fe) atoms to relax, have a simple relation:

$$E^{FS}_{Chem}(A) \simeq \frac{2}{3} E^{GB}_{Chem}(A) \qquad \text{(Eq 7)}$$

Thus, the difference in *binding energies* for A at the GB and the FS, E^A_{Chem}, is then:

$$E^A_{Chem} \simeq \frac{1}{3} E^{GB}_{Chem}(A) \qquad \text{(Eq 8)}$$

From Table III in Ref 18, it is also found that:

$$E^{GB}_{Chem}(A) - E^{GB}_{Chem}(M) \simeq E^A_{Coh} - E^M_{Coh} \qquad \text{(Eq 9)}$$

Combining Eq 7 and 9, we have:

$$\Delta E^A_{Chem} \equiv E^A_{Chem} - E^M_{Chem} \simeq \frac{1}{3}(E^A_{Coh} - E^M_{Coh}) \quad \text{(Eq 10)}$$

This means that the embrittlement potency of a substitutional atom in the GB of Fe is about 1/3 of the cohesive energy difference between that element and the host Fe atom if the volume effect is not significant. We note that in Seah's pair-bonding model, this factor is 1/2. This difference indicates that a quantum mechanical treatment is necessary in an accurate study of the GBE. The factor 1/3 can be understood by the $\sqrt{Z}$ theory (Ref 31, 32). The simplest expression of band character is in the second-moment approximation to the tightbinding model, in which the cohesive energy per atom varies as $\sqrt{Z}$, where Z is the atomic coordination that can range from 1 (diatomic molecule) to 12 (fcc crystal). For the segregant in the Fe Σ3 (111) GB, Z = 8, and for that on the Fe (111) FS, $Z = 4$. Hence, by applying the $\sqrt{Z}$ rule, one will get:

$$E^A_{Chem} \simeq 0.293 \times E^{GB}_{Chem}(A). \quad \text{(Eq 11)}$$

The relation is treated as approximate rather than exact because in both FS and GB systems, the bond lengths of $M(1) - M(n)$ (n = 2,3,4) differ from the bulk values. Taking the contributions from the volume effect (Eq 6) and the bonding character (Eq 10) together, the embrittling effect, ΔE^A_B, of an alloying addition A is now:

$$\Delta E^A_B = \frac{1}{3}(E^A_{Coh} - E^M_{Coh}) + \Delta E^A_V. \quad \text{(Eq 12)}$$

Although the relation in Eq 8 is drawn from only a few cases, it can be expected to hold in general for other alloying additions in this high-angle boundary system. The difference between 1/3, obtained from our first-principles case studies, and 0.293, given by the $\sqrt{Z}$ theory, is only 0.04. If we take this difference as the error of the relation in Eq 8, the relative error in the factor 1/3 is only about 12%.

Heat of Formation for Solid Solutions—Macroscopic Atom Model

Since the unit cell we employed for computation is periodic and the lattice type of the matrix is kept as its bulk crystal, the systems under our investigation, strictly speaking, are solid solutions. The heat of formation for solid solutions can be viewed as a chemical shift of the bonding capability of the solute atoms. One difference between solid solutions and alloys is that in the former system, atoms with different sizes have to occupy equivalent lattice positions. This gives rise to an additional positive contribution, in the form of elastic energy, to the alloying enthalpy, due to the lattice deformation (or "atomic size mismatch"). This elastic energy is discussed in the "Volume Effect," in this chapter. Since our interest is not limited to a few specific additions (in Fe or Ni), we need comprehensive thermodynamic data for the heat of formation of all the A elements in M (M = Fe and Ni) alloys. The existing experimental data is far from complete and the first-principles determination of these quantities is beyond the computational effort we can afford in the present paper. As an alternative, we employ the *macroscopic atom* model (Ref 24) to estimate the heat of an alloy with a specific concentration that is determined by our slab model.

In the *macroscopic atom* picture, the heat of formation of an alloy A in M with a concentration c_A is:

$$\Delta E^A_{Sol} = (1 - C_A)[1 + 8C_A^2 \times (1 - C_A)^2] \times \Delta E^A_{Sol}(0) \quad \text{(Eq 13)}$$

where C_A is:

$$C_A = \frac{c_A(V^A)^{2/3}}{c_A(V^A)^{2/3} + (1 - c_A) \times (V^M)^{2/3}} \quad \text{(Eq 14)}$$

and $\Delta E^A_{Sol}(0)$ is the heat of an alloy of A in M in infinite dilution (Ref 24). In our first-principles calculation, c_A is 1/23 for A in Fe, and 1/21 for A in Ni.

Another difference between solid solutions and alloys is that in the former case there is a structure-dependent enthalpy related the preference for metallic elements to crystalize in one of the main crystallographic structures, namely, bcc, fcc, and hcp, depending on the number of their valence electrons (Ref 24). We use ΔE^A_{Stru} to denote the total energy difference of elemental crystal A between its ground-state structure and that of the host. Skriver (Ref 33) studied systematically the crystal structure of all the transition metals by the first-principles linear muffin-tin orbital method in the local-density approximation, and showed that with the experimental atomic volume, the energy difference between bcc and fee is one order of magnitude larger than that between fcc and hcp. To our knowledge, there are no systematic first principles investigations with GGA, so far. To make all (or, as many as possible,) contributions in our model to be founded on the same basis, we prefer not to cite the above LDA results but to carry out full GGA calculations for all the elements under consideration. In doing so, we approximate the hcp by the fcc structure in order to save computational effort. This approximation will introduce an error to ΔE^{A}_{Stru}, which is about 10%.

Taking the previous two corrections into account, Eq 12 becomes:

$$\Delta E^{A}_{B} = \frac{1}{3}(E^{A}_{Coh} - E^{M}_{Coh} + \Delta E^{A}_{Sol} + \Delta E^{A}_{Stru}) + \Delta E^{A}_{V} \qquad \text{(Eq 15)}$$

For Mo and Pd in the Fe Σ3 (111) GB, the model calculated values for ΔE^{A}_{B} are −0.96 and −0.03 eV, respectively. As mentioned above, our first-principles results are −0.90 and +0.08 eV, respectively. Note that for Mo and Pd in the Fe GB, the embrittlement potency given by Seah's pair-bonding model (Ref 6) is −1.5 and +0.9 eV, respectively, which are significantly greater than our results.

The model predicted embrittlement potencies of all the simple and transition metal elements as substitutional alloying additions to the Fe Σ3 (111) and Ni Σ5 (210) GB are listed in column 9, in Tables 1, and 2, To be more illustrative and more readily compared, these values are plotted in Fig. 3 (for Fe) and Fig. 4 (for Ni), respectively.

Confirmation of the Model

In order to verify our theory, we carried out first-principles calculations on the effects of 3*d* transition metal (Co), 4*d* transition metal (Ru), and 5*d* transition metals (W and Re) segregation on the cohesion of the Fe Σ3 (111) GB and a simple metal (Ca) on the Ni Σ5 (210) GB by using the same FLAPW method. As sketched in Fig. 1 for the Fe Σ3 (111) GB case and Fig. 2 for the Ni Σ5 (210) GB case, both the FS and GB were simulated by a slab model (Ref 34, 35). To obtain reliable values, the FS and GB systems were treated on an equal footing with high accuracy and the atomic structures of the FS and GB were also optimized in the normal direction for the cases with and without segregated atoms. Bearing this in mind, we used the same set of numerical parameters (*k* points, plane-wave cutoff, muffin-tin radius, etc.) in the FLAPW calculations for both the FS and GB.

In the FLAPW method (Ref 20, 21), no shape approximations are made to the charge densities, potentials, and matrix elements. For both host and alloying additions, the core states are treated fully relativistically and the valence states are treated semirelativistically (i.e., without spin-orbit coupling). The GGA formulas for the exchange-correlation potential are from Perdew *et al.*, (Ref 22).

For Co, Ru, W, and Re on the Fe Σ3 GB, our first-principles results are +0.05, −0.65, −1.31, and −1.31 eV, respectively. The values calculated with the model are +0.10, −0.77, −1.54, and −1.29 eV, respectively. The largest discrepancy between the first-principles and the model results is about 0.2 eV in the case of W, whereas for Ca in the Ni Σ5 (210) GB, our model gives +1.65 eV and the first-principles results show an embrittlement potency of +1.4(± 0.2) eV. In general, the agreement between our phenomenological model and first-principles is quite good.

Summary

Starting from first principles, we formulated and developed an electronic-level phenomenological theory to quantitatively predict the mechanical behavior of a substitutional metallic element in the grain boundary without carrying out full first-principles calculations, once the atomic structure of the clean grain boundary is determined. From our results, it is concluded that the strongest cohesion enhancer in the Fe Σ3 (111) GB is W, followed by Re, Nb, and Os. The strongest cohesion enhancer in the Ni Σ5 (210) GB is Os, followed by Re, W, and Nb. This model was tested and verified by detailed rigorous first-principles calculations on Fe- and Ni-based alloys. Although we have focused on two specific GB types, that is, bcc Fe Σ3 (111) and fee Ni Σ5 (210), we expect our theory to be applicable to other high boundaries in general and instructive in the quantum design of ultrahigh-strength alloys through selection of alloying elements for GB cohesion enhancement. Further calculations

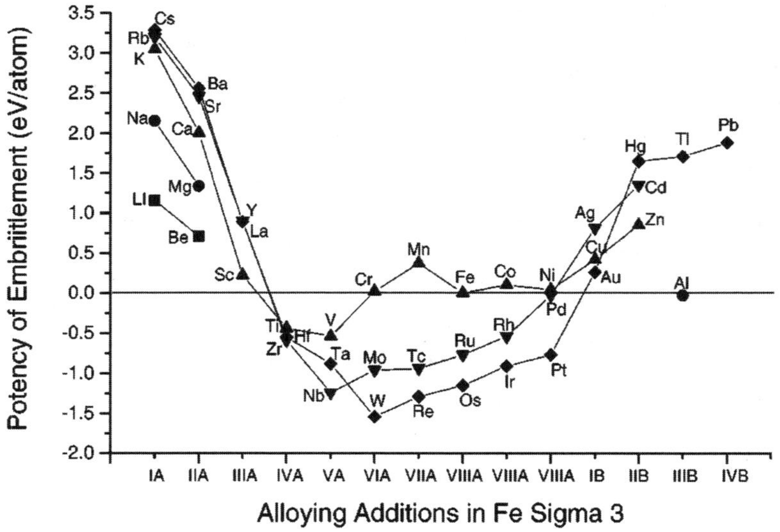

Fig. 3 Embrittlement potency of substitutional additions on the Fe Σ3(111) GB.

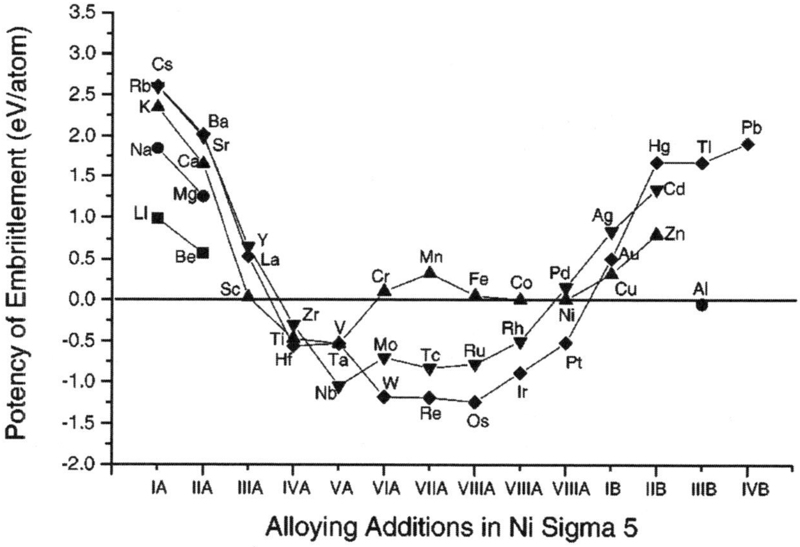

Fig. 4 Embrittlement potency of substitutional additions on the Ni 25 (210) GB.

and modelings of the segregation energy from crystalline solution to GB are under way to next control the segregation of desired components to the boundary.

ACKNOWLEDGMENTS

Work supported by the Office of Naval Research (Grant No. N00014-94-1-0188) and a grant of Cray-T90 computer time at the NSF-supported San Diego Supercomputing Center and Cray-J90 computer time at the Arctic Region Supercomputing Center, supported by the DOD.

REFERENCES

1. C.L. Briant and S.K. Banerji, in *Embrittlement of Engineering Alloys*, ed. C.L. Briant and S.K. Banerji (Academic Press, 1983), p 21
2. A.R. Troiano, Trans. Am. Soc. Met. 52. 54 (1960)

3. R.H. Doremus, J. Appl. Phys. 47. 1833 (1976)
4. J.P. Stark and H.L. Marcus. Metall. Mater. Trans. A 8A, 1423 (1977)
5. D.Y. Lee, E.V. Barrera. J.P. Stark, and H.L. Marcus. Metall. Mater. Trans. A 15A, 1415 (1984)
6. M.P. Seah, Acta Metall. 28, 955 (1979)
7. P. Wynblatt and R.C. Ku. Surf. Sei. 65, 511 (1977)
8. J.R. Rice and J.S. Wang. Mater. Sei. Eng., A 107, 23 (1989)
9. M.S. Daw and M.I. Baskes, Phys. Rev. Lett. 50, 1285 (1983)
10. T. McMullan, M.J. Scott, and E. Zaremba, Phys. Rev. B 35, 1076 (1987)
11. G.S. Painter and F.W. Averill, Phys. Rev. Lett. 58, 234 (1987)
12. M.E. Eberhart and D.D. Vvedensky, Phys. Rev. Lett. 58, 61 (1987)
13. R. Wu, A.J. Freeman, and G.B. Olson, Science 265, 376 (1994)
14. L.P. Sagert, G.B. Olson, and D.E. Ellis, Philos. Mag. B 77, 871 (1997)
15. L.P. Sagert, Ph.D. thesis. Northwestern University, 1995
16. L.P. Zhong, R. Wu, A.J. Freeman, and G.B. Olson, Phys. Rev. B 55, 11 133 (1997)
17. W.T. Geng, A.J. Freeman. R. Wu, C.B. Geller, and J.E. Raynolds, Phys. Rev. B 60, 7149 (1999)
18. W.T. Geng, A.J. Freeman, R. Wu, and G.B. Olson, Phys. Rev. B 62, 6208 (2000)
19. G.B. Olson, Science 277, 1237 (1997)
20. E. Wimmer, H. Krakauer, M. Weinert, and A.J. Freeman, Phys. Rev. B 24, 864 (1981)
21. M. Weinert, E. Wimmer, and A.J. Freeman, Phys. Rev. B 26, 4571 (1982)
22. J.P. Perdew. K. Burke, and M. Ernzerhof, Phys. Rev. Lett. 77, 3865 (1996)
23. H.W. King, *Alloying Behavior and Effects in Concentrated Solid Solutions* (Gordon and Breach, 1966)
24. F.R. de Boer, R. Boom, W.C.M. Mattens, A.R. Miedema, and A.K. Niessen, *Cohesion in Metals* (North-Holland, 1988), p 44
25. C. Kittel, *Introduction to Solid State Physics*, 4th ed. (Wiley, 1971), p 39
26. T. Ochs, O. Beck, C. Elsässer, and B. Meyer, Philos. Mag. A 80, 351 (2000)
27. T. Ochs, O. Beck, C. Elsässer, M. Mrovec, V. Vitek, J. Belak, and J.A. Moriarty, Philos. Mag. A 80, 2405 (2000)
28. J. Friedel, Adv. Phys. 3, 446 (1954)
29. D.J. Eshelby, J. Appl. Phys. 25, 255 (1954); Solid State Phys. 3, 19 (1956)
30. G. Bradfield, *Use in Industry of Elasticity Measurments in Metals with the Help of Mechanical Vibrations* (HMSO, 1964), p 5
31. G. Allan, in *Handbook of Surfaces and Interfaces*, ed. L. Dobrzynski (Garland SPTM, 1978)
32. D. Spanjaard and M.C. Desjonqueres, Phys. Rev. B 30, 4822 (1984)
33. H.L. Skriver, Phys. Rev. B 31, 1909 (1985)
34. R. Wu, A.J. Freeman, and G.B. Olson, J. Mater. Res. 7, 2403 (1992)
35. G.L. Krasko and G.B. Olson, Solid State Commun. 76, 247 (1990)

CHAPTER 15

Strong Interface Adhesion in Fe/TiC*

J.H. Lee, A.J. Freeman and G.B. Olson, Northwestern University
T. Shishidou, Hiroshima University
Y.J. Zhao, NanoStellar Inc.

Abstract

As an aid to understanding the superior toughness of Ti-modified steels provided by fine Ti(C, N) particles, first-principles full-potential linearized augmented plane wave (FLAPW) density functional calculations were performed on the Fe matrix/TiC particle interface. It was found that at equilibrium a strong covalent bonding between Fe-C is formed at the interface, and the magnetic moment of the interface Fe (1.98 μ_B) is reduced from that of the tetragonally strained structure (2.51 μ_B). We then calculated with a rigid separation model the separation energy curve and the force separation law for the Fe-C debonding process at the interface, which predicts 2.45 J m^{-2} for the work of separation and 30.66 GPa for the force maximum. We also found that the strong Fe-C bond provides an interfacial fracture strength equal to that of the pure bcc Fe matrix. A clear picture is given for the microscopic origin of this strong metal/ceramic adhesion based on density of states (DOS) considerations. For a more realistic understanding of the Fe-C bonding, structural optimization calculations were performed at each separation distance. The effect of relaxation was found to be larger at short separation distances than in the large separation region, which leads to a crossover behavior in the separation energy curve from the elastically deformed to the clearly separated regime at a critical distance (~1.75Å), and to a discontinuity in the force separation law. Despite this large relaxation effect, the work of separation, 2.52 J m^{-2}, is not changed much from that of rigid separation.

Introduction

The ultimate goal of steel research is to design new classes of high-performance steels (Ref 1) with sufficient advances in fracture toughness and strength to visibly increase component life with minimal increase in cost. To this end, the control of dispersed inclusion phases within the Fe-based matrix plays a crucial role. It has been demonstrated (Ref 1) that a marked increase of fracture toughness can be achieved in Ti-modified steels with controlled dispersions of Ti nitrides (primary inclusions with size ~1 μm) and carbides (secondary particles ≤ 0.1 μm). Bulk Ti(C, N) in the NaCl(B1) structure is a well-known hard material widely used for cutting tools and wear-resistant coatings. Its electronic structure (Ref 2) is characterized by strong hybridization between Ti 3p and C (or N) 2s2p orbitals, and clear bonding and antibonding peaks appear in the density of states (DOS) that are well separated by a pseudogap. Its valence-electron number is close to the *optimal* value: almost all (none) of the bonding

*J.-H. Lee, T. Shishidou, Y.-J. *Zhao*, A.J. Freeman, and G.B. Olson, "Strong Interface Adhesion in Fe/TiC *Philosophical Magazine*, Vol. 85, No. 31 (2005) 3683–3697

(antibonding) states are occupied. Furthermore, it was found that the pda bonding state possesses extremely high resistance to shear strain (Ref 2). In spite of these understandings of bulk Ti(C, N), a fundamental understanding of the role of Ti(C, N) particles in the Fe matrix is, however, not yet clearly established.

The problem of metal/ceramic adhesion has attracted considerable and growing interest in recent years (Ref 3–16) for its broad significance for materials technology. From the theoretical side, much effort has been devoted to the ab initio determination of the adhesion strength. It is widely believed that the adhesion strength is best described by the *ideal work of separation* (Ref 8), which is the reversible work to separate the interface into two free surfaces (separation energy in the infinite separation limit, W_∞), and W_∞ has been calculated for several metal/ceramic interfaces. (Ref 8, 11–13). Most of these studies, however, addressed ceramic oxides, while research on interfaces between metal and conductive-ceramics [such as Ti(C, N)] is still very limited.

Recently, Dudiy *et al*., (Ref 17–20) investigated the Co/Ti(C, N) interface to gain insights into metal/ceramic adhesion. Using a plane-wave pseudopotential method, they discussed the energetics and electronic structure of Co/Ti(C, N)(001) and found that strong bonding between Co and (C, N) is formed at the interface. These authors also examined layer- and orbital-projected DOS (PDOS) and compared it with bulk fcc Co(C, N). From this analysis, they argued that the antibonding states clearly seen in bulk Co(C, N) are substantially broadened in energy and space when the Co-(C, N) bond comes to the interface, due to their coupling with the delocalized metallic states of Co and that this is the most important and fundamental mechanism for the unexpectedly strong Co-C bonding at the interface. Using:

$$W_\infty = (E_{sl1} + E_{sl2} - E_{int})/2A \qquad \text{(Eq 1)}$$

(where E_{int} is the total energy of the supercell with the interface, E_{sl1} and E_{sl2} are the total energies for the same supercell when one of the slabs is kept and the other is replaced by vacuum, and A is the interfacial area within one supercell), they calculated W_∞ for several interface geometries. For all the complex interfaces of Co/TiC, their calculated W_∞'s are close (within 10%) to the work of adhesion from wetting experiments, 3.64 J m^{-2}. For Co/TiN, their calculated values are 0.6–1.0 J m^{-2} lower than those for Co/TiC. Like others, these authors used the ideal work of separation as measures of the interface energetics. Also, Arya and Carter (Ref 21, 22) studied bulk TiC, bulk Fe, their low index surfaces, as well as one Fe/TiC interface using a plane-wave pseudopotential method. With a careful study of stability on the TiC and bcc Fe surfaces, they selected Fe(110)/TiC(100) interface for the calculations, which shows a lattice misfit of ~2.1%. Using Eq 1 with A instead of $2A$ (they have only one interface), these authors calculated the work of adhesion by changing the number of TiC layers up to three. Their calculated work of separation is larger for 1 TiC layer (3.99 J m^{-2}) than for the others (2.57 J m^{-2} for 2 TiC layers, and 2.55 J m^{-2} for 3 layers).

In this work, to obtain some understanding of the role of the fine particles of TiC in the Fe matrix, we have performed first-principles calculations on the Fe/TiC interface, the same interface that Arya and Carter (Ref 21, 22) studied, but with a different orientation. Using a supercell modeled with $(Fe)_4/(TiC)_4/(Fe)_4$ layers, we determined the electronic and magnetic structure at equilibrium. For the interface energetics, we took a different approach; instead of calculating W_∞ only, we inserted rigid separations at the interface with different separation distances and calculated the separation energies as a function of a separation distance. W_∞ is obtained as the separation energy in the infinite separation limit without using Eq 1. This series of calculations serves to elucidate how the metal/ceramic interface actually behaves under separation and provides a more detailed insight about the adhesion strength than W_∞ itself. Also the force separation law can be obtained from these calculations, which is, together with the separation energy curve, a useful input for multiscale continua modeling in steel research. Finally, full structural optimizations were performed at each separation to see how the separation energy and the force separation law change with atomic relaxation.

This paper is organized as follows. The section "Computationl Details" describes the methodology and the model system used in our calculations. The results of the calculations and discussion about the results are provided in the section "Results and Discussion." Finally, we summarize our work.

Computational Details

All calculations are performed with the highly precise all-electron full-potential linearized augmented plane wave (FLAPW) method (Ref 23, 24). In the FLAPW method, many-body interactions between electrons are treated within either the local (spin) density approximation [L(S)DA] (Ref 25, 26) or the generalized gradient approximation (GGA) (Ref 27) in the framework of density functional theory. In the present work, we adopt GGA because it well describes many-body interactions between electrons for the bulk ground-state properties of transition metals and titanium carbides/nitrides (Ref 28, 29). Core states were treated fully relativistically and valence states scalar relativistically. Fe 3d4s, Ti 3p3d4s, and C 2s2p electrons were treated as valence electrons, and muffin-tin (MT) sphere radii were chosen to be 2.1 (Fe), 2.2 (Ti), and 1.35 (C) atomic units (a.u.). Spherical harmonics with $l \leq 8$ were used inside the MT spheres to represent the charge density, potential, and augmented wave function. A cutoff of the plane-wave basis, K_{max} (in a.u.), was set so that the relation $K_{max}R_{ave} \sim 8$ is satisfied, where R_{ave} is the averaged MT sphere radius. In the interstitial region, a star-function is used to represent the charge density and potential with a cutoff, G_{max}, of $2K_{max} + 2$. The Monkhorst-Pack (MP) scheme (Ref 30), or the improved tetrahedron (IT) method (Ref 31), was applied to k-point summations, and convergence in terms of the number of irreducible k-points, n_k, was carefully examined. Specific values for K_{max} and n_k will be given later for each calculation.

Results and Discussion

Bulk Properties of Fe and TiC, Interface Orientation, and Site Preference

First, we determined the bulk properties of bcc Fe and B1 TiC with K_{max} = 4.0 for Fe and 4.3 for TiC. For k-point summations, the IT method was used with (15, 15, 15) mesh points which gives n_k = 120 for both cases. To find the equilibrium lattice constant, we used a third-order polynomial to fit the total energy as a function of lattice constant, and the bulk modulus, B, was obtained from its definition, $B = -V\partial P/\partial V = V\partial^2 E/\partial V^2$, where V is the volume of the unit cell, P pressure, and E the total energy. The results are summarized in Table 1. Our results agree well with other calculations (Ref 18, 28, 32, 33).

Next, for the interface orientation, we adopted the Baker-Nutting relation (Ref 36). which is often observed between bcc Fe and B1 carbides or nitrides. Since we focus on the Fe(001)/TiC(001) interface with the orientation $\langle 100\rangle_{bcc} \| \langle 110\rangle_{B1}$, we used the smallest size of the in-plane cell with a fully coherent interface. Since this restriction gives a rather large lattice misfit (using the experimental lattice constants from Table 1), $(\sqrt{2}a_{Fe} - a_{TiC})/a_{TiC} = -0.064$ (the theoretical lattice constants give a misfit of −0.087), we determined a reasonable in-plane lattice constant by calculating the elastic energy cost for both Fe and TiC under tetragonal strain: for $\sqrt{2}a_{Fe} = a_{TiC} = 4.260$ Å, both Fe and TiC pay the same energy cost (65 meV), where the c lattice constant is 2.684 Å (Fe) and 4.437 Å (TiC). (The Poisson effect was fully taken into account. At each a (in-plane lattice constant), the c constant was found by minimizing the total energy.) The strained coherent interface can be taken to represent the bonding behavior of the coherent patches that reside between primary interfacial dislocations in a structurally relaxed interface.

Regarding the interface geometry, we determined the *site preference* of Fe atoms at the TiC(001) interface by considering three types of interface positions for Fe: (a) on-top of the C site; (b) on-top of the Ti site; and (c) in the middle of the TiC square (cave site). Our two-dimensional (2D) FLAPW code (Ref 23, 24) was applied to a film containing two layers of Fe and three layers of TiC (Ref 37). Each Fe and TiC slab held its bulk-like structure (bct and fct, respectively), while all atoms were relaxed along the c direction until the calculated atomic forces (Ref 38) were negligible (<0.05 eV/Å). For these calculations, K_{max} = 4.31 and G_{max} = 10.62 were used, and MP scheme was applied to k-point summations

Table 1 Calculated bulk properties of bcc Fe and B1 TiC

	Source	a_0 (Å)	B (Mbar)	M (μ_B)
Fe	Present work	2.831	1.92	2.17
	Ref 32	2.88	1.82	2.127
	Ref 33	2.84	1.86	2.17
	Ref 34	2.86	1.72	2.12
TiC	Present work	4.383	2.38	...
	Ref 18	4.33	2.52	...
	Ref 28	4.315	2.2	...
	Ref 35	4.317	2.4	...

Lattice constants a_0, bulk moduli B, and magnetic moment M are given and compared with experiments with the mesh (12, 12) in the 2D Brillouin zone, which gives n_k = 21 for the structures (a) and (b), and n_k = 36 for (c). From total energy comparisons, it was found that structures (a) — Fe on top of C — is by far the most stable with the relative total energies being (a) 0, (b) 2.00, and (c) 1.03 eV. For the structure (a), a strong covalent bonding was formed between Fe and C with a bond length 1.89 Å, and no clear bonding features were seen for (b) and (c).

Interface Properties of Fe/TiC

To determine the interface properties more precisely and to calculate the separation energy curve, the number of superlattice layers was increased to $(Fe)_4/(TiC)_4/(Fe)_4$ (Fig. 1), where the interface geometry (a) was adopted and the 3D periodicity was continued with repeating slabs. The unit cell contained 16 atoms; since it has inversion symmetry, there are 8 types of atoms: 4 of Fe, 2 of Ti, and 2 of C. A K_{max} = 4.13 and n_k = 21 with the mesh (12, 12, 2), and the MP scheme was applied for summations over Brillouin zone. In addition to the relaxation of atomic z positions, we determined the equilibrium c length of the superlattice unit cell as 19.610 Å by examining the total energy of seven different c lengths.

Equilibrium Properties. Using the supercell structure determined in the section "Bulk Properties of Fe and TiC, Interface Orientation, and Site Preference," in this chapter, we investigated the equilibrium properties of the Fe/TiC interface. Plotted in Fig. 2 is the valence charge density and spin density at the equilibrium optimized structure. Given in Table 2 are the calculated interlayer distances and atomic magnetic moments. It is clear from Fig. 2 that the interface Fe atom forms a strong covalent bond with the C atom. The strong metal/ceramic adhesion found by Dudiy *et al.*, (Ref 17) for the Co/TiC interface also occurs here at the Fe/TiC interface.

The interface bond length determined from the atomic relaxation, 1.88 Å, is very short but close to that of a separate $(Fe)_2/(TiC)_3$ film calculation, 1.89 Å. This is to be compared with cementite Fe_3C, the most stable Fe carbide in steels: the current Fe-C interface bond length is even shorter than the shortest Fe-C bond length in Fe_3C, 1.96 Å (Ref 39), or shorter than our calculated Fe-C bond length of unstable B1 FeC, 2.00 Å. And the current

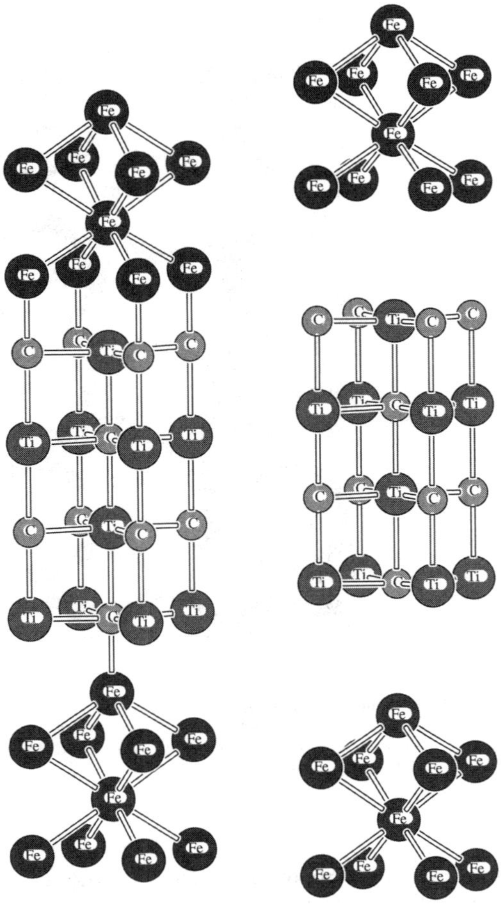

Fig. 1 (Left) Supercell structure of $(Fe)_4/(TiC)_4/(Fe)_4$ at equilibrium. Shown is the conventional unit cell. The primitive cell contains 16 atoms. (Right) Supercell structure with separation at the interface.

Fe-C bond length is shorter than that of the other work on Fe/TiC (Ref 21, 22) where, unlike the present study, the C atom is not on top of the Fe atom. We note that there is a rather extensive work on C dissolution and diffusion in Fe (Ref 40), in which the authors calculated Fe-C bond length for various C positions. Since these authors obtained the Fe-C bond lengths ranging from 1.75 to 2.11 Å, our result shows good agreement with those calculations. Another interesting observation is the buckling bewteen C and Ti atoms at the interface: C atoms go up relative to Ti atoms with a distance of 0.067 Å. The charge density in the middle of the interface Fe-C bond is 0.12 electron/(a.u.)^3, which is considerably higher than that of the Ti-C bond (0.07) and the Fe-Fe bond (0.06), and even higher than that of Fe_3C (0.10). Due to this strong

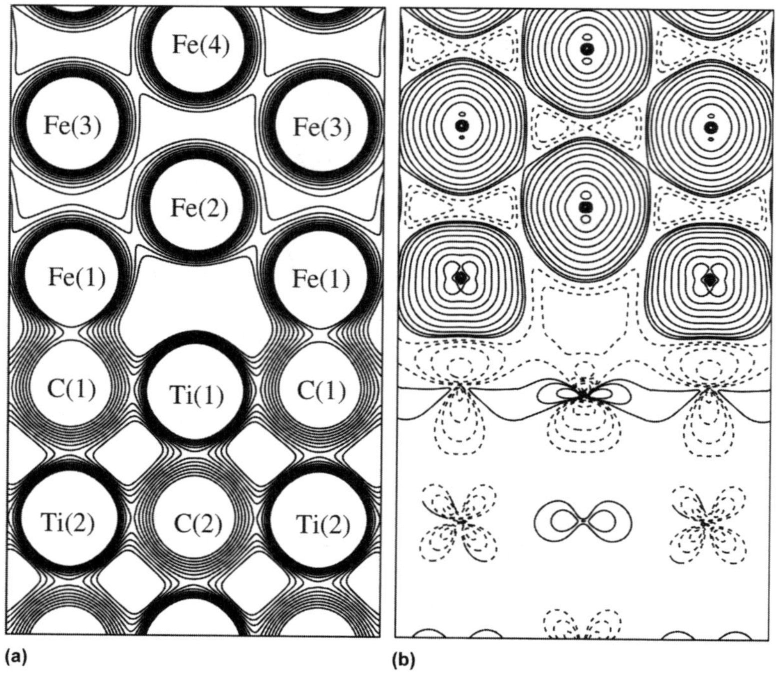

Fig. 2 (a) Valence charge density and (b) spin density contour plot on a B1 (100) section. The number in the parentheses represents the symmetry type. For charge density, the low density resulting from the metallic bonding of Fe and the high density around the nuclei are out of range for this plot: contours start from 0.047 and end at 0.14 electron/(a.u.)^3. For spin density, solid and dashed lines represent positive and negative contours, respectively: contours start from ±0.001 and end at ±0.77 spin/(a.u.)^3.

Table 2 Summary of calculated quantities at the fully optimized geometry of $(Fe)_4/(TiC)_4/(Fe)_4$

	z-distance (Å)	M (μ_B)
Fe(4)–Fe(1)	1.36 (+1.3%)	Fe(4) = 2.48 (−1.3%)
Fe(3)–Fe(4)	1.32 (−1.8%)	Fe(3) = 2.45 (−2.6%)
Fe(2)–Fe(3)	1.38 (+2.5%)	Fe(2) = 2.44 (−3.1%)
Fe(1)–Fe(2)	1.21 (−9.8%)	Fe(1) = 1.98 (−21.3%)
C(1)–Fe(1)	1.88	C(1) = −0.02
C(1)–Ti(1)	0.067	Ti(1) = 0.01
C(1)–Ti(2)	2.25 (+1.3%)	Ti(2) = −0.02
Ti(1)–C(2)	2.15 (−2.9%)	C(2) = 0.01
bct Fe–Fe	1.34	2.51
fct Ti–C	2.22	...

Interlayer distances and atomic magnetic moments are given with atomic types referred in Fig. 2. Values in parentheses indicate the deviation from the tetragonally deformed bulk values listed in the lower panel

bonding, the interface Fe atom shows a considerably reduced magnetic moment (2.00 μ_B), which is a 20% reduction from the bulk value of this tetragonally strained structure (2.51 μ_B), while the second Fe layer (2.46 μ_B) almost recovers its bulk bct value. The second layer of Fe relaxes down toward the interface and causes an enhancement of the Fe(1)–Fe(2) bonding. The interlayer distance in the Fe block recovers its bulk value in the middle of the Fe slab.

Rigid Separation. Now, let us focus on the character of the interface Fe-C bond. To see exactly how strong this interface Fe-C bonding is, we inserted a rigid separation at the interface, as demonstrated in Fig. 1, and increased the separation distance (which corresponds to normal cleavage) and obtained the separation energy curve $W(x)$ shown in Fig. 3 as a function of separation distance $x = (c_s - c_e)/2$, where c_e and c_s are the unit cell length in z-direction at equilibrium and with separation, respectively. We examined not only the interfacial separation but also two different types of Fe-Fe separation for comparison: separation inserted (a) between the interfacial and the second Fe layers and (b) between the second and third Fe layers. Since we have two equivalent interfaces in the unit cell, the separation energy, W_s, is defined as half of the total energy difference between the equilibrium, E_e, and separated geometry, E_s:

$$W_s = (E_s - E_e)/2 \quad \text{(Eq 2)}$$

The calculations were done at six different separation distances and fitted to the universal binding energy curve (Ref 41), $W(x) = W_\infty - W_\infty(1 + bx)\exp(-bx)$, whose obtained parameters are listed in Table 3. From Fig. 3, the actual separation is seen to occur at the Fe(1)–C(1) interface. The calculated work of separation, 2.17 eV (3.83 J m^{-2}), is smaller than the Griffith work of Fe(1)–Fe(2) or Fe(2)–Fe(3) debonding (see Table 3), and is smaller than the calculated (001) Griffith work of pure bcc Fe of 2.7 eV (5.5 J m^{-2}). (The calculated (001) separation of pure bcc Fe shows very similar behavior to the Fe(2)–Fe(3) debonding process.) However, this does not mean that the interface Fe-C bonding is weak; as is apparent from Fig. 3, *it does resist normal separation in the same way as does the Fe-Fe bond* in the small separation region. Consequently, the force maximum and its position, which are of importance in characterizing mechanical properties, also show similar values among the three types of debonding (Fig. 3b and Table 3), giving a comparable critical normal stress of 30 GPa. (The force separation law calculated here from first principles has been integrated with the gradient-plasticity and multiscale-continua models (Ref 42) with which to simulate the ductile fracture dynamics.)

We remark that the calculated work of separation in the present work can be compared to those of Arya and Carter (Ref 21 and 22) as mentioned in the introduction. Although the current work of separation is close to that of one TiC layer case in their work, it is important to note that the present interface structure is under a large in-plane strain to make a coherent interface. Since 65 meV per layer is required to make Fe and TiC match at the interface, this energy should be released at the infinite separation, which would result in the reduction of the calculated work of separation from 2.17 eV to 2.17 − 0.065 × 12 (12 layers in the present study) =1.39 eV. And, this corresponds to 2.45 J m^{-2}, which is very close to the 2 and 3 TiC layer cases of Arya and Carter.

The difference among the three debonding processes appears in the large separation region, and results in different W_∞ values. In addition to the covalent directional bond

Table 3 The parameters obtained for the universal binding energy curve

	W_∞ (eV) (J m^{-2})	$1/b$ (Å)	σ_{max} (eV/Å) (GPa)
C(1)–Fe(1)	2.17 (3.83)	0.46	1.74 (30.66)
Fe(1)–Fe(2)	3.28 (5.78)	0.68	1.77 (31.26)
Fe(2)–Fe(3)	2.70 (4.76)	0.56	1.77 (31.28)

$1/b$ gives the force-maximum separation distance, and σ_{max} is the maximum value of the force

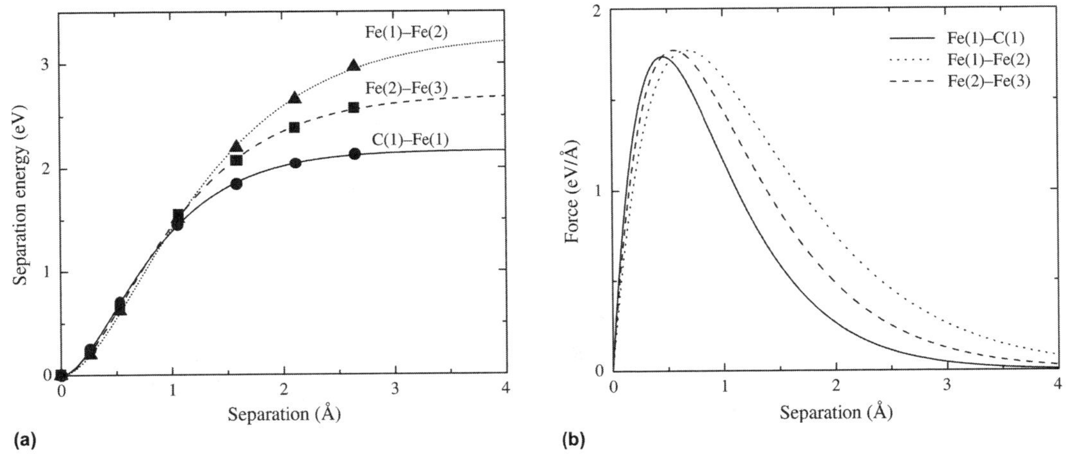

Fig. 3 (a) Separation energy per unit cell for the (001) plane cleavage. (b) Corresponding force separation law; the C(1)–Fe(1) debonding (solid line), Fe(1)–Fe(2) (dotted line), and Fe(2)–Fe(3) (dashed line). The lines were obtained by fitting with the universal binding energy curve (see text for details).

through 3d–3d hybridization, the Fe-Fe bond is governed by the metallic bond mediated via the sp conduction electrons, which brings features of long-range interactions. In the case of the interface Fe-C bond, this metallic-bonding feature is weaker than in the Fe-Fe bond, leading to the lack of interaction in the large separation region: the interface Fe and C atoms are bonded by the strong and short-range covalent interactions, resulting in the smaller W_∞ value than those of the other Fe-Fe bonds.

Another important factor for structural properties is the shear strength. From the charge density plot of Fig. 2, we can easily expect that this system would have high resistance to shear deformation or slip in the (001) plane compared to pure bulk Fe. Now, TiC has strong and highly directional bonds, which are oriented 90° with respect to each other and strongly resist shear strain. The interface Fe is pinned to the TiC surface by this directional bond along the *c* axis, and thus high resistance to (001) shear is clearly expected. This aspect will be addressed in future work through first-principles calculations on the same footing as the present one.

Now, the question is: *why is the interface Fe-C bonding so strong?* To answer this, let us examine the projected DOS (PDOS) onto the interface atoms depicted in Fig. 4. Despite the strong covalent Fe-C bonding and strong metal/ceramic adhesion realized at the interface, each PDOS resembles remarkably well the bulk one — meaning that the interface electronic structure can retain its original bulk features to a considerable extent. This retention of the bulk features at the interface, which deters embrittlement, arises from the special role of the interface Fe $3d_{3z^2-r^2}$ state in the current interface structure. In Fig. 5, the interface PDOS is further decomposed into the specific Fe $3d_{3z^2-r^2}$ and C $2p_z$ orbitals, which are relevant to the pda hybridization. We can see clear evidence of strong hybridization with several resonance peaks in their PDOS. The $3d_{3z^2-r^2}$ PDOS is significantly different from the bulk one, while the other 3d orbitals (not shown) and even the C $2p_z$ orbital retain more or less their bulk features. (Loosely hybridizing with Ti(1) atoms, Fe(1) can keep a bcc-like environment, while strongly hybridizing with the Fe(1) $3d_{3z^2-r^2}$ orbital C(1) keeps a B1-like environment.) The e_g orbitals of Fe are less important in promoting bcc cohesion than the t_{2g} orbitals since in the stretching direction of the e_g orbitals there are no bcc nearest neighbor atoms. In this sense, the $3d_{3z^2-r^2}$ orbital can change its features from the bulk rather

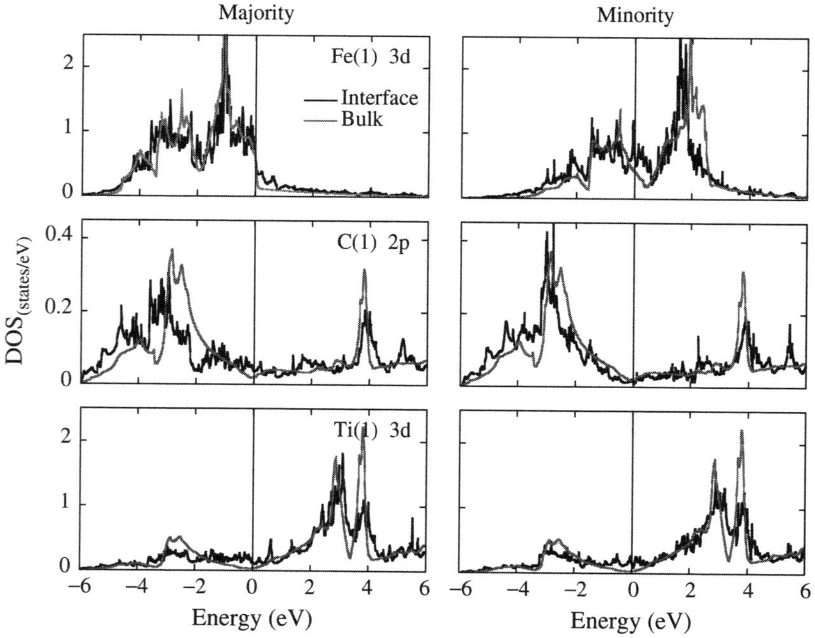

Fig. 4 Projected DOS onto the interface atoms. Left panels are for majority spin, while right minority. The corresponding bulk PDOS (bct Fe or fct TiC) is also shown by gray line. Fermi energy is set at 0 eV.

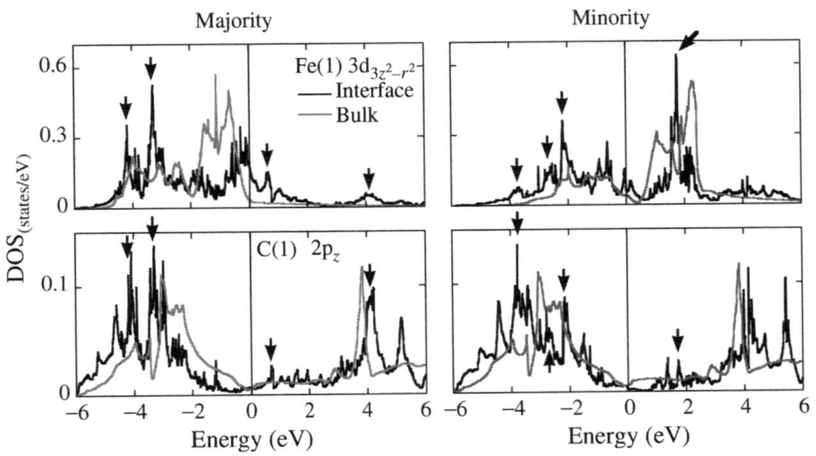

Fig. 5 Projected DOS onto the Fe(1) $3d_{3z^2-r^2}$ and C(1) $2p_z$ orbitals. Some of the clear resonance peaks due to the pdσ hybridization are indicated by arrows.

freely, and provide an optimal situation for strong pdσ hybridization which appears in the bulk Ti-C, without much affecting the bcc cohesive energy—except for the magnetic energy cost (the occupied minority states in the $3d_{3z^2-r^2}$ orbital are increased and the majority states decreased as a result of the pdσ hybridization). Thus, the C(1) $2p_z$ state need not greatly change its PDOS from the bulk in forming the pdσ bond with Fe(1), and is still able to keep optimal features for the strong Ti-C bonds.

Separation with Relaxation. In order to gain more realistic and physical insights into the Fe/TiC interfacial bonding, we performed the full structural optimization at each separation distance to see how the separation energy changes from that of the rigid separation case. Figure 6 shows the separation energies for all the three types of separation with relaxation. In the case of rigid separation, the separation was seen to occur at the Fe(1)–C(1) interface; the relaxed results also show that the actual separation does occur there. In the small separation region, all the three processes behave exactly the same, but at large separation distances, the separation at the Fe(1)–C(1) has the lowest energy, implying that it is the ground state at a given separation distance among the three separations. Thus, we restrict ourselves henceforth to the separation at the Fe(1)–C(1) interface only.

Plotted in Fig. 7 are the separation energy and the force separation law with atomic relaxation as a function of separation distance. For better fitting accuracy, we inserted three more separations, 1.75, 1.85, and 1.96 Å. As is clear from the figure, the separation energies change remarkably upon relaxation. The effect of relaxation is larger at small separation distances (<~1.75 Å) than at large separation distances (>~1.75 Å). At small separations, the separation energy is well fitted with a parabola, $W(x) = ax^2$ with $a = 0.57$ (eV/Å^2), whereas at large separation distances it follows a universal binding energy curve (Ref 41) as in the rigid separation case, but with a different σ_{max} (= 2.00 (eV/Å) = 35.3 (GPa)). The idea of rigid and relaxed separation was applied to other systems such as α-Al_2O_3, Al, and Si (Ref 43, 44), and the same behavior in the separation energy was observed.

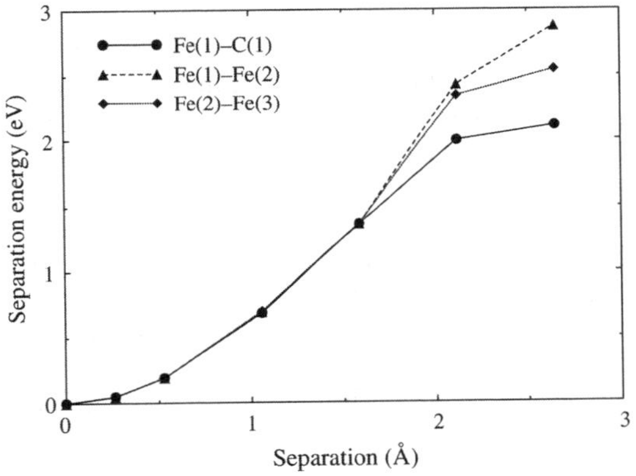

Fig. 6 The separation energy for three types of separation after the full structural optimization. Solid line with circles denotes the separation at the Fe(1)–C(1) interface, dashed line with triangles at the Fe(1)–Fe(2) interface, and dotted line with diamonds at the Fe(2)–Fe(3) interface.

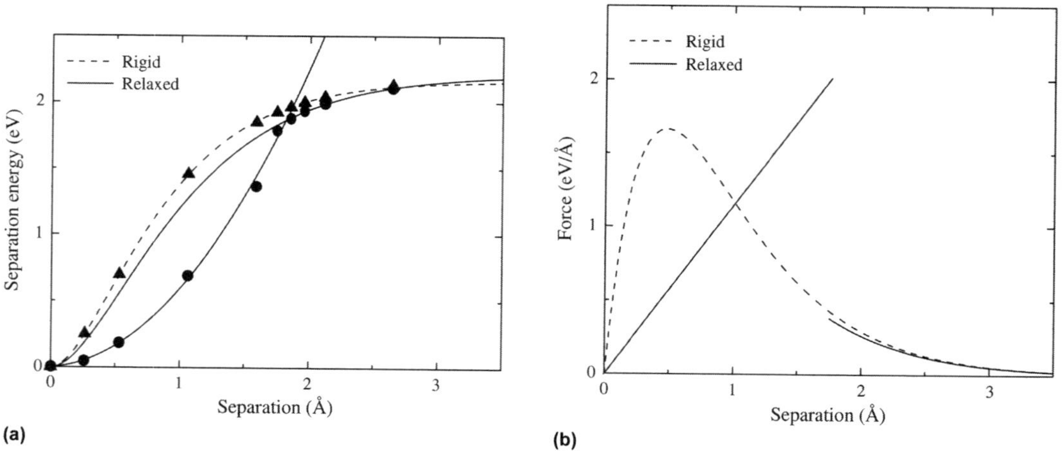

Fig. 7 (a) The separation energy curve and (b) force separation law after the structural optimization. Solid lines are for relaxed results, and dashed ones for rigid separations.

This crossover behavior can be easily understood: for separation distances less than a critical separation (1.75 Å), the quantum mechanical potential energy barrier between the surfaces is low, and the interfacial atoms can tunnel through the barrier and move toward each other to restore bonding upon relaxation, which results in an elastically elongated system without a clear separation. Beyond the critical separation, however, the potential barrier between the two surfaces is too high for the atoms to restore bonding, and Fe and TiC remain separated with a well-defined separation distance. Thus it may be inferred that the actual breaking of the Fe-C bond occurs near 1.75 Å. This is confirmed in Fig. 8, in which the normalized interlayer distances between Fe(1)–C(1), Fe(1)–Fe(2), and Fe(2)–Fe(3) are given. As is clear from the figure, all the distances quickly restore their equilibrium values after a critical separation except that of Fe(1)–C(1), which grows larger. And, due to this crossover behavior of the separation energy, the force separation law has a discontinuity at a critical separation, where the force has a maximum value, σ_{max} (Fig. 7b). The position of σ_{max}, 1.75 Å, is significantly shifted from that of rigid separation case in which σ_{max} appears at 0.46 Å. In addition, σ_{max} is increased from its rigid separation value by 4.64 GPa. We note that despite this large relaxation effect, the work of separation remains almost the same even under relaxation (W_{∞} = 2.21 [eV] = 3.9 [J m^{-2}], which is also reduced to 2.52 J m^{-2} due to lattice misfit). This is because the work of separation is obtained at the infinitely separated limit, and the relaxation effect is very small at large separation distances.

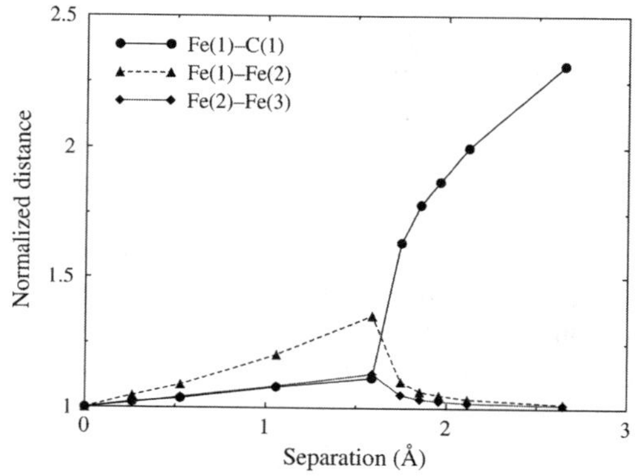

Fig. 8 Interlayer distances under atomic relaxation, normalized with respect to values at equilibrium. Solid line with circles denotes distances for Fe(1)–C(1), dashed line with triangles for Fe(1)–Fe(2), and dotted line with diamonds for Fe(2)–Fe(3).

Summary

We performed extensive first-principles FLAPW calculations on the Fe(001)/TiC(001) interface, and determined the equilibrium and separation properties. At equilibrium, a strong covalent bond is formed between the interface Fe and C. Due to this strong covalency, the magnetic moment of interface Fe is significantly reduced from its bct value. In order to see how strong the Fe-C interfacial bonding is, we performed rigid separation calculations, and obtained the separation energy curve and the first-principles force separation law. These calculations, along with a detailed investigation of the PDOS, revealed: (a) The Fe-C bond is governed by short-range interaction coming from pdσ hybridization. (b) At small separations, the Fe-C bond resists normal separation in the same way as does the Fe-Fe bond. The difference appears in the large separation region resulting in different W_∞ values due to the different metallicity in the bonds. (c) The actual separation occurs at the Fe-C interface with $W_\infty = 3.83$ J m^{-2}, which is significantly reduced to 2.45 J m^{-2} due to lattice misfit. The force separation law was also obtained and has been integrated into gradient plasticity and multiscale continua models. (d) The Fe $3d_{3z^2-r^2}$ orbital at the interface plays a dominant role in this strong metal/ceramic adhesion because it is rather free to change from the bulk and enables other interface states to retain their original bulk features. This allows the interfacial C $2p_z$ to keep its bulk character and the optimal features of the strong Ti-C bonds in forming the pdσ bond with the interface Fe.

To gain a more realistic understanding of the Fe-C interfacial bonding, a full structural optimization was performed at each rigid separation, and the results also show that the separation occurs at the Fe-C interface. The detailed study of the Fe-C interfacial separation revealed that the relaxation effect is more vivid in the small separation regime than at large separation distances, which leads to a crossover behavior in the separation energy curve: the separation energy curve is well fitted by a parabola at small separation distance (<1.75 Å), and follows a universal binding energy relation in the large separation region (>1.75 Å). This crossover behavior was reflected in the force separation law, which has a discontinuity at the critical separation, 1.75 Å, at which the actual breaking of the Fe-C bond occurs. In spite of this large relaxation effect, W_∞ does not change much from that of rigid separation because the relaxation effect is small at large separation distances, and W_∞ is obtained in the infinite separation limit.

ACKNOWLEDGMENTS

The authors are grateful to Y.N. Gornostyrev, O.Y. Kontsevoi, and C.F. Woodward for their fruitful discussions. This work was supported by the ONR (grant No. N00014-01-1-0953) under the Naval Materials by Design Grand Challenge and by grants of computer time at the NAVO and ERDC Supercomputing Centers.

REFERENCES

1. G.B. Olson, Science 277 1237 (1997)
2. S.-H. Jhi, J. Ihm, S.G. Louie, *et al.*, Nature 399 132 (1999)
3. M.J. MacLachlan, M. Ginzburg, N. Coombs, *et al.*, Science 287 1460 (2000)
4. H.L. Meyerheim, R. Popescu, J. Kirschner, *et al.*, Phys. Rev. Lett. 87 076102 (2001)
5. J.S. Moodera, J. Nowak and R.J.M. van de Veerdonk, Phys. Rev. Lett. 80 2941 (1998)
6. D.A. Muller, D.A. Shashkov, R. Benedek, *et al.*, Phys. Rev. Lett. 80 4741 (1998)
7. D.A. Shashkov and D.N. Seidman, Phys. Rev. Lett. 75 268 (1995)
8. M.W. Finnis, J. Phys. Condens. Matter 8 5811 (1996)
9. A. Bogicevic and D.R. Jennison, Phys. Rev. Lett. 82 4050 (1999)
10. C. Verdozzi, D.R. Jennison, P.A. Schultz, *et al.*, Phys. Rev. Lett. 82 799 (1999)
11. R. Benedek, A. Alavi, D.N. Seidman, *et al.*, Phys. Rev. Lett. 84 3362 (2000)
12. I.G. Batirev, A. Alavi, M.W. Finnis, *et al.*, Phys. Rev. Lett. 82 1510 (1999)
13. J. Hartford, Phys. Rev. B 61 2221 (2000)
14. F. Rao, R. Wu and A.J. Freeman, Phys. Rev. B 51 10052 (1995)
15. C. Li, R. Wu, A.J. Freeman, *et al*, Phys. Rev. B 48 8317 (1993)
16. C. Li and A.J. Freeman, Phys. Rev. B 43 780 (1991)
17. S.V. Dudiy, J. Hartford and B.I. Lundqvist, Phys. Rev. Lett. 85 1898 (2000)
18. S.V. Dudiy and B.I. Lundqvist, Phys. Rev. B 64 045403 (2001)
19. M. Christensen, S.V. Dudiy and G. Wahnstrom, Phys. Rev. B 65 045408 (2001)
20. S.V. Dudiy and B.I. Lundqvist, Phys. Rev. B 69 125421 (2004)
21. A. Arya and E.A. Carter, J. Chem. Phys. 118 8982 (2004)
22. A. Arya and E.A. Carter, J. Chem. Phys. 120 1142 (2004)
23. E. Wimmer, H. Krakauer, M. Weinert, *et al.*, Phys. Rev. B 24 864 (1981)
24. M. Weinert, J. Math. Phys. 22 2433 (1981)
25. L. Hedin and B.I. Lundqvist, J. Phys. C 4 2064 (1971)
26. U. von Barh and L. Hedin, J. Phys. C 5 1629 (1972)
27. J.P. Perdew, K. Burke and M. Ernzerhof, Phys. Rev. Lett. 77 3865 (1996)
28. R. Ahuja, O. Eriksson, J.M. Wills, *et al.*, Phys. Rev. B 53 3072 (1996)

29. V. Ozoliņš and M. Körling, Phys. Rev. B 48 18304 (1993)
30. H.J. Monkhorst and J.D. Pack, Phys. Rev. B 13 5188 (1976)
31. P.E. Blöchl, O. Jepsen and O.K. Andersen, Phys. Rev. B 49 16223 (1994)
32. D.J. Singh, W.E. Pickett and H. Krakauer, Phys. Rev. B 43 11628 (1991)
33. G.Y. Guo and H.H. Wang, Chinese J. Phys. 38 949 (2000)
34. American Institute of Physics, *American Institute of Physics Handbook*, 3rd ed. (McGraw-Hill, 1970)
35. R.W.G. Wyckoff, *Crystal Structure*, Vol. 2, 3rd ed. (Wiley, 1963)
36. G. Baker and J. Nutting, Spec. Rep. 64 1 (1959)
37. T. Shishidou, J.-H. Lee, Y.-J. Zhao, *et al*, J. Appl. Phys. 93 6876 (2003)
38. R. Yu, D. Singh and H. Krakauer, Phys. Rev. B 43 6411 (1991)
39. F.H. Herbstein and J. Smuts, Acta Crystallogr. 17 1331 (1964)
40. D.E. Jiang and E.A. Carter, Phys. Rev. B 67 214103 (2003)
41. A. Banerjea and J.R. Smith, Phys. Rev. B 37 6632 (1988), and references therein
42. S. Hao, B. Moran, W.K. Liu, *et al*, J. Comp.-Aid. Mat. Des. 10 99 (2003)
43. E.A.A. Jarvis, R.L. Hayes and E.A. Carter, Chem. Phys. Chem. 1 55 (2001)
44. R.L. Hayes, M. Ortiz and E.A. Carter, Phys. Rev. B 69 172104 (2004)

Transformations—Selected Works of G.B. Olson on Materials, Microstructure, and Design
C.E. Campbell, M.V. Manuel, and W. Xiong, editors

CHAPTER 16

Systems Design of High-Performance Stainless Steels—I. Conceptual and Computational Design*

C.E. Campbell and G.B. Olson
Northwestern University

Abstract

Application of a systems approach to the computational materials design led to the development of a high-performance stainless steel. The systems approach highlighted the integration of processing/structure/property/performance relations with mechanistic models to achieve desired quantitative property objectives. The mechanistic models applied to the martensitic transformation behavior included the Olson-Cohen model for heterogeneous nucleation and the Ghosh-Olson solid-solution strengthening model for interfacial mobility. Strengthening theory employed modeling of the coherent M_2C precipitation in a BCC matrix, which is initially in a paraequilibrium with cementite condition. The calibration of the M_2C coherency used available small-angle neutron scattering (SANS) data to determine a composition-dependent strain energy and a composition-independent interfacial energy. Multi-component pH-potential diagrams provided an effective tool for evaluating oxide stability. Constrained equilibrium calculations correlated oxide stability to Cr enrichment in the metastable spinel film, allowing more efficient use of alloy Cr content. The composition constraints acquired from multicomponent solidification simulations improved castability. Then integration of the models, using multicomponent thermodynamic and diffusion software programs, enabled the design of a carburizable, secondary-hardening martensitic stainless steel for advanced bearing applications.

Introduction

As the cost of experimentation increases and the demand for application specific materials rises, the need to reduce the amount of experimentation in the alloy development process grows. The recent number of efforts to model the composition dependence of materials properties (Ref 1–8) reflects this trend. The traditional alloy development process consists of producing numerous heats with systematic composition variations and testing the compositions to discover the alloy with the optimal properties. A computational systems design approach combines mechanistic models of the known processing/structure/property/performance interactions to predict a single alloy

*C.E. Campbell and G.B. Olson, "Systems Design of High Performance Stainless Steels: I. Conceptual Design," *J. Computer-Aided Mat. Des.* 7 (2001) p 145–170

composition, which is evaluated for its ability to meet the design goals. With this approach the design goals should be achieved through the iteration of two or three prototypes, thus significantly reducing the experimental cost of development. However, to use a systems approach, the relationships between the processing/structure/property/performance interactions must be organized in a manner that recognizes the hierarchical structure of materials (Ref 9). That is, a substructure is present within a given microstructural element, such as the martensitic laths within a packet formed within a prior austenitic grain.

To organize this hierarchical system of interactions, systems design (Ref 10) emphasizes the interaction of the subsystems and the function of the component within the larger system. The four primary steps of systems design are systems analysis, synthesis, implementation, and operation. These steps are outlined in Fig. 1. The first step, systems analysis, requires defining the application and the material property/performance objectives. A flow block diagram is used to clarify the processing/structure/property/performance relations. The first stage of synthesis prioritizes the property objectives and the interactions. After prioritizing, the types of models needed to characterize the interactions and the related subsystems are determined. The required model depends on the priority of the interaction being described. A highly complex model should not be developed to represent an interaction of low priority.

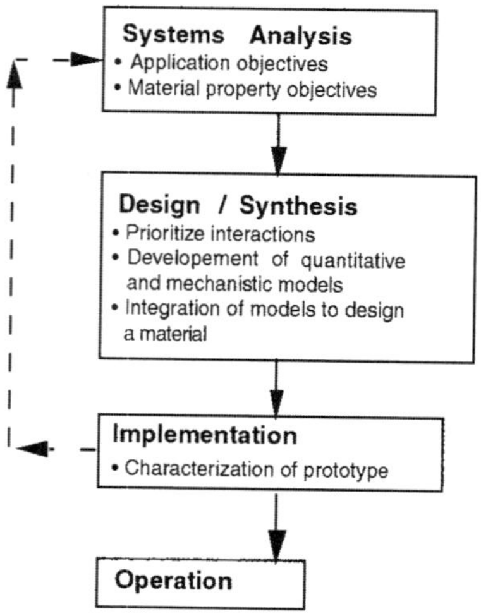

Fig. 1 Stages of systems design as applied to materials design.

Mechanistic-based models are preferred for describing interactions of high priority. Once the appropriate models are developed, the models are integrated to design a new material for a specific application. During this design process, it is important not to optimize a subsystem at the expense of the total system. In addition, the design composition should be a robust one: a composition from which small deviations will not result in dramatically different properties. Once the optimal composition is determined, the prototype alloy is made and characterized. Evaluation of the prototype determines whether further modeling of the system is needed or whether continuation to the final step of the design process may be effected: putting the material to use in the intended application.

Application of a systems approach to the design of stainless steels is ideal: there is a wealth of information describing the thermodynamic, mechanical, and corrosion properties of stainless steels and there is a need to double the toughness of the existing highest-strength stainless steels. Industrial bearing applications requiring increased toughness and aqueous corrosion resistance range from automotive to nuclear. In addition, the bearing components of the mainshaft of subsonic and supersonic aircraft gas turbines must operate at increasing temperatures, speeds, and loads (Ref 11). The increasing operating temperatures require greater dimensional stability and the increasing engine speeds make raising the fracture-toughness more critical. The low toughness present in current through-hardening alloys makes bearings susceptible to fast-fracture mechanisms that can result in potentially serious secondary damage to the engine. Furthermore, as the desire to improve engine efficiency increases, bearings increasingly operate in more severe environments, thus requiring improved corrosion resistance. This need to improve corrosion resistance is emphasized by the loss of 50% of the aerospace bearings during storage due to moisture exposure causing corrosion (Ref 12).

The research presented here follows a systems approach to the computational design of a high-performance stainless steel. This process includes the determination of the desirable property objectives for a new class of stainless steels, the development of the necessary models, the integrated design of a prototype composition, and the characterization of the design prototype. It should be noted that the mechanistic models developed are not limited to use in the design applications of high-performance stainless steels, but can be applied to design applications of other ferrous alloys.

Conceptual Design

Property Objectives

Analysis of prototype alloys, bearing steels, secondary-hardening steels, and martensitic stainless steels and of projected industrial needs determined the property objectives for a new class of high-performance stainless steels, specifically for aerospace bearings (Ref 13). Achieving the desired high surface hardness and core toughness property objectives, without the use of thermal mechanical processing, requires the use of a case/core system that has two sets of property objectives: one set for the case material and one set for the core material. The development of a case/core system also demands that case/core compatibility always be maintained and that the case and core properties not be optimized at the expense of one another. The properties defining the case include surface hardness, aqueous corrosion resistance, rolling-contact fatigue (RCF) resistance, and wear resistance. The essential core material properties are the toughness and the strength.

The microstructure characteristics controlling the case properties include the surface hardness, the case depth, and the formation of a protective passive film. The surface hardness and case depth determine the fatigue and wear resistance of the case. Cyclic loading of contacting surfaces at low sliding ratios, RCF, results in the initiation of microcracks that form from dislocation accumulation about surface or subsurface defects. RCF failure occurs when the microcracks grow and the stress concentration exceeds the fracture strength of the material. At a fixed applied stress, increases in the Hertzian contact fatigue strength extend the RCF life as the initiation of microcraks is suppressed. The Hertzian contact fatigue strength is proportional to the more easily measured material parameter, material hardness, as shown in Fig. 2. Thus for the currently available aerospace bearing materials, to increase the contact fatigue strength by 15% requires a surface hardness of R_C 65. With this increased surface hardness and a corresponding increase in toughness, an exponential increase in the fatigue life is expected, based on the empirical formulations (Ref 14, 15) and experimental work (Ref 16, 17). In addition to maintaining a high surface hardness at room temperature to extend fatigue life, for some applications hot hardness is required due to higher operation temperatures. For such cases, the desired temperature at which the alloy retains a R_C 58 hardness is 400 °C.

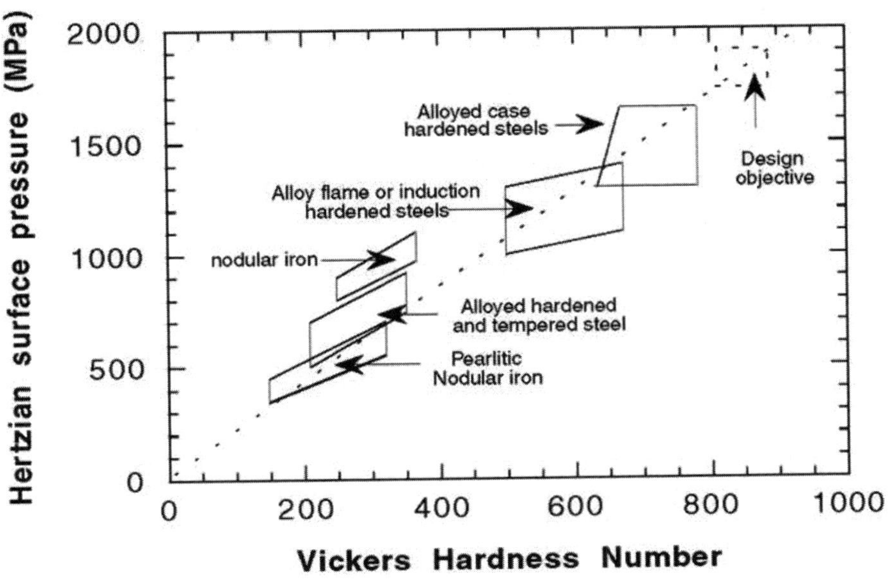

Fig. 2 For various ferrous alloys, the contact fatigue strength represented by the Hertzian surface pressure as a function of material hardness, as previously shown in the unpublished work by Breen at the Gear Institute, Evanston, IL, 1992.

To maintain compatibility between the case and the core, the hardness is gradually decreased from a high surface value to a lower core value. The distance over which this decreasing hardness occurs is the case depth. A hardness gradient, which provides critical material strength to exceed the applied stress during RCF, determines the desired case depth. The von Mises stress criterion defines the position of the maximum applied stress within the case.

Wear is the loss or displacement of material due to cyclic loading of contacting surfaces at high sliding ratios. Wear resistance improves with increased dimensional stability and high surface hardness. In comparing the fatigue life and wear resistance of the through-hardening 52100 and the carburizing 8620 bearing alloys, Hengerer et al., (Ref 18) found an exponential decrease in weight loss with increasing hardness. (Tables 1 and 2 list the compositions and properties of the alloys described.) Optimal fatigue life and wear resistance were achieved with surface hardnesses between R_C 62–64. As wear resistance at high temperatures is also important design objectives include a wear resistance equal to that of the high temperature equivalents of the 52100 and 8620 alloys, through-hardening M50 alloy and the carburizing M50NiL alloy.

The aqueous corrosion resistance should be superior to that of the 440C monolithic stainless bearing alloy, the current standard alloy for high-strength applications requiring mild corrosion resistance. Anodic polarization measures the applied potential at which the design alloy and the 440C alloy exhibit a decreased current indicating a transition from the active to passive regions. The decreased current associated with the passive region, due to the increasing oxidizing conditions, results in a lowered corrosion rate. A second evaluation method of the corrosion behavior is salt-spray testing. Despite the reproducibility and correlation limitations of the salt-spray testing, testing in a 5% salt environment provides a simple comparison of material behavior using an industrial accepted standard accelerated corrosion test. It is noted that a corrosion rate of less than 0.051 mm yr^{-1} provides excellent corrosion resistance, according to Fontana (Ref 19).

The core toughness and strength levels are critical to preventing catastrophic failure. Based on the material selection methods of Ashby (Ref 20), Fig. 3 depicts a cross-plot of the yield strength and K_{IC} fracture toughness properties of existing stainless and bearing steels and reveals the trade-off between the two properties. The squared ratio of the K_{IC} fracture toughness over the yield strength is proportional to the critical flaw size, which defines the minimum toughness needed for a given hardness level. For the core material the desired ratio of $(K_{IC}/\sigma_y)^2$, equivalent to the nonstainless carburizing bearing steel M50NiL, is 0.185 cm, corresponding to a $K_{IC} = 65\,\text{MPa}\sqrt{\text{m}}\,(60\,\text{ksi}\sqrt{\text{in}})$ at a hardness of R_C 50 ($\sigma_y \approx$ 250 ksi (1.72 MPa)). In Fig. 3, the dashed line represents this ratio of $(K_{IC}/\sigma_y)^2$.

Design Approach

These property objectives are achieved by a structure of hierarchical interacting microstructural subsystems. The flow block diagram in Fig. 4 illustrates the interactions among the desired hierarchical microstructural subsystems required for the processing and the property objectives. Of the two major microstructural

Table 1 Compositions of commercial and prototype alloys (Fe is the balance)

Alloy	Weight (%)								
	C	Co	Cr	Mn	Mo	Ni	Si	V	W
52100	0.98–1.10	...	1.30–1.60	0.25–0.45	...	...	0.15–0.30	...	...
M50	0.80–0.85	0.00–0.25	4.00–4.25	0.15–0.35	4.00–4.50	0.00–0.15	0.00–0.25	0.90–1.00	0.00–0.25
M50NiL	0.11–0.15	0.00–0.25	4.00–4.25	0.15–0.35	4.00–4.50	3.20–3.60	0.10–0.25	1.13–1.33	0.00–0.15
HY180	0.11	8	2	0.18	1	10.2	...	...	...
AF1410	0.16	14.25	2.1	...	1.05	10.5	...	...	...
AerMet100	0.24	13.4	3	...	1.18	11	...	...	...
440C	0.95–1.20	...	16–18	1.00	0.75	...	1.0	...	...
Pyrowear 675	0.07	5.4	13.0	0.65	1.80	2.6	0.4	0.6	...
NASA1	0.30	22.5	12.0	...	0.30	8.5	...	0.25	...
NASA2	0.30	25.0	12.0	...	0.30	2.5	...	0.25	...
1605-8A	0.253	15.97	8.97	...	...	4.97	...	...	...
1605-2D	0.25	16	4.1	5					

subsystems, the case and core subsystems, the case microstructure is the more complex subsystem, having to provide strength, corrosion resistance, and fatigue resistance. The primary function of the core subsystem is to provide fracture-toughness.

Case Subsystem The desired high strength and aqueous corrosion resistance case properties require the carburization of a lath martensitic matrix microstructure. The lath microstructure produced after quenching from the solution treatment temperature has a high-dislocation density that provides a direct strengthening contribution, as well as nucleation sites for a fine-scale precipitate dispersion. The carburization of the case material within the limits of the carbon solubility in the FCC phase field increases the strength of the martensitic matrix by increasing the strain dilation in the matrix with the carbon occupation of interstitial sites. If the carburization is done in the single phase FCC field, the gradient carbon concentration results in a beneficial compressive stress profile. If carburization occurs outside the single FCC phase field, primary carbides may form interfacially and act as sites for crack initiation. In addition the formation of primary carbides removes Cr from the matrix decreasing the aqueous corrosion resistance. Preliminary results of notch-fatigue testing by Kuehmann (Ref 21) and rolling contact fatigue testing by Wise (Ref 22) indicate the effectiveness of reducing the primary carbide fraction to increase the fatigue life.

Precipitation of fine-scale M_2C (HCP, M= Cr, Mo, V, Fe) carbides using secondary hard ening further reduces the interparticle spacing and thus increases the strength. During secondary hardening (stage IV tempering), M_2C carbides precipitate heterogeneously on dislocations and produce a fine-scale dispersion (Ref 23). To maintain the initially nucleated small particle size requires precipitation in a highly supersaturated solution, which fosters

Table 2 Mechanical properties of commercial and prototype alloys

Alloy	Type	Hardness, R_C Case/ Core	Yield strength (MPa)	Fract. Tough. K_{IC}(MPa$\sqrt{m}$)
52100	Bearing-M	62–65 (core)	345	29
M50	Bearing-M	62	2200	23
M50NiL	Bearing-C	(60–64)/ (40–46)	1210	51
HY180	UHS	45	1344	250
AF1410	UHS	48	1650	154
AerMet100	UHS	54.0	1724	126
AerMet100	UHS-Cast	54	1692	125
440C	Stainless-M	57	1896	20
Pyrowear675	Stainless-C	63/40	986	163
NASA1	Stainless-N	70/58	. . .	47
NASA2	Stainless	40	. . .	32

M = Martensitic, C = Carburizing, N = Nitriding, UHS = Ultrahigh-strength.

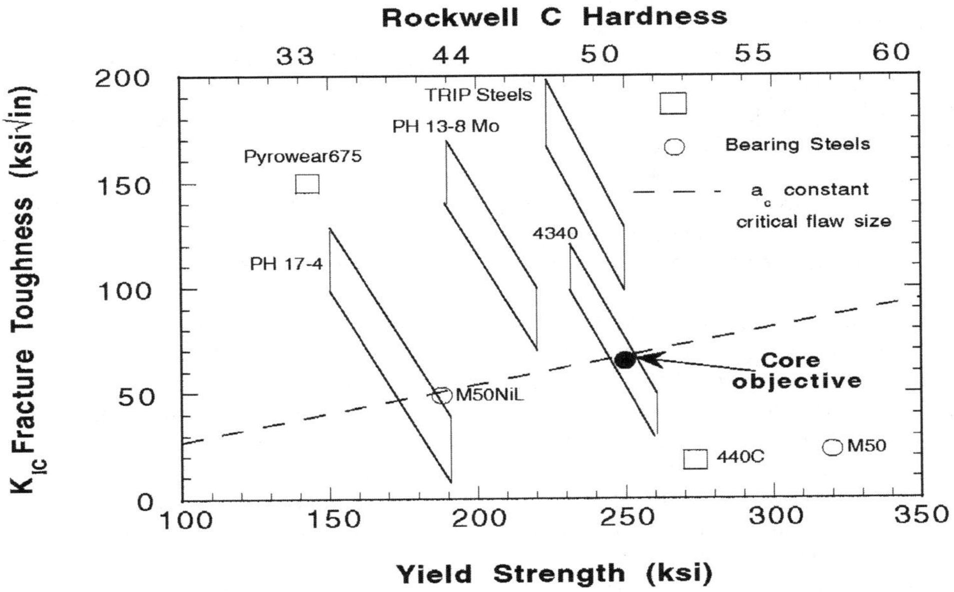

Fig. 3 K_{IC} fracture toughness versus yield strength of stainless steels and reference bearing steels. The dashed line represents the desired critical flaw size.

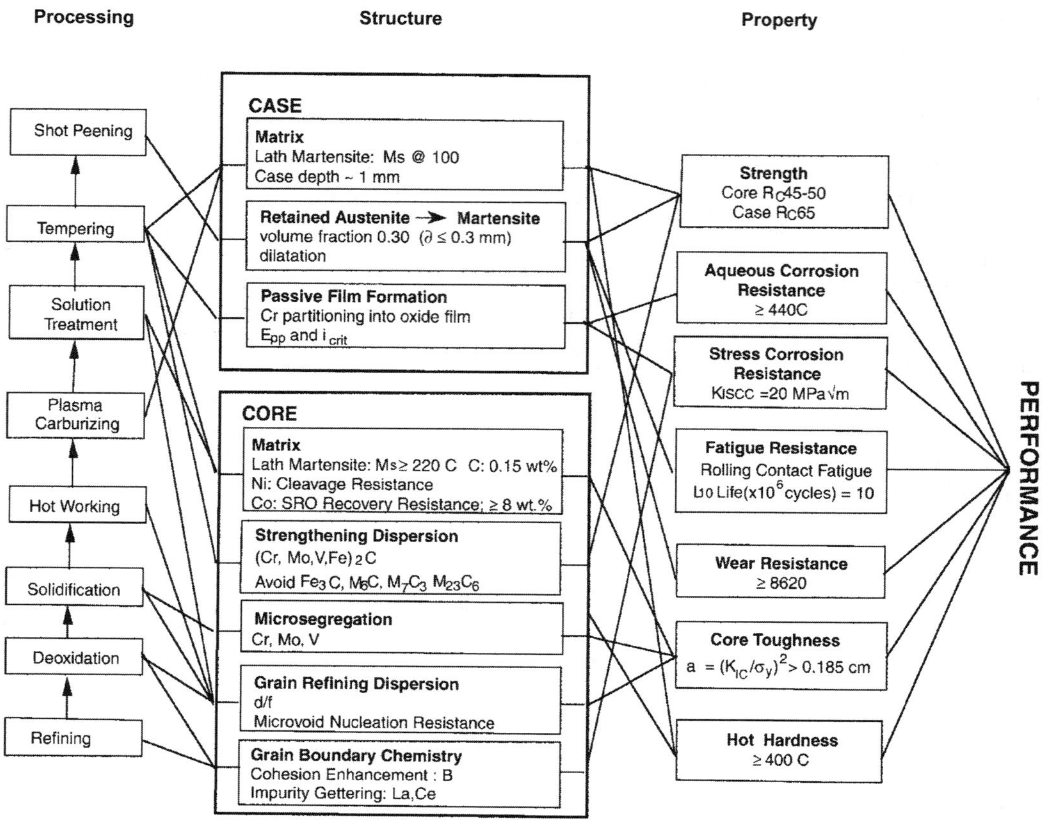

Fig. 4 Flow-block diagram of the processing/structure/property/performance interactions for a carburizing, secondary-hardening, martensitic stainless steel.

nucleation and coarsening without growth. Langer and Scharwtz (Ref 24) and Kampmann and Wagner (Ref 25) demonstrated the effectiveness of this type precipitation in maintaining the initial particle size. An additional mechanism to improve the strength dictates maintaining the dislocation substructure of the martensitic microstructure at the secondary hardening temperatures. Speich (Ref 26) determined that this can be achieved with a matrix composition containing at least 8 wt% Co.

Formation of a passive film on the surface of the case material supplies the needed aqueous corrosion resistance for the alloy. This passivating film is primarily composed of Me_2O_3 (Me = Cr, Fe, Ni) and Me_3O_4 oxides; therefore, the amount of Cr partitioning to the film controls the formation and stability of the film. Experiments by Kirchhiem and co-workers (Ref 27, 28) in Fe-Cr alloys determined the amount of Cr in the oxide film is logarithmically dependent on the matrix Cr concentration. Their work supports the conclusion that the corrosion resistance for an alloy with a fixed Cr content can be improved by increasing the activity of Cr in the matrix, and thus, promoting Cr partitioning to the oxide film. Therefore, the design parameters to improve the aqueous corrosion resistance focus on decreasing the fraction of precipitated carbides, which decrease the matrix Cr concentration, and on increasing the activity Cr in the matrix to enhanced the partitioning of Cr to the oxide film.

Core Subsystem The following constituents control the fracture toughness: the matrix Ni composition, the prior austenite grain size, the martensitic microstructure, the reduction of microvoid nucleation sites, and the grain boundary cohesion. Optimization of these constituents yields ductile fracture at room temperature, that is, shifting the ductile-to-brittle transition (DBT) temperature below room temperature. Early work (Ref 29) showed additions of 1 wt% Ni to ferrous alloys have effectively reduced the DBT temperature below room temperature and the same behavior was observed in low strength binary Fe-Ni alloys by Jolley (Ref 30) and Leslie (Ref 31). Sufficiently decreasing the prior austenitic grain size also aids in shifting the DBT to below room temperature. This desired reduction of the grain size is achieved by using Ti carbide (Ti,Mo)C,N grain refiners (with diameters between 10 nm and 180 nm) (Ref 32) and by optimizing the solution temperature to achieve high hardness with the minimum solution treatment time. The beneficial effects of refining the solution temperature have been demonstrated for the AF1410 (Ref 33) and AerMet100 (Ref 34) alloys.

The lath martensitic microstructure formed within the prior austenite grains provides an essential microstructural element for obtaining both high strength and toughness. The fine-scale lath microstructure produced after quenching is dependent on the martensitic start, M_s, temperature, which is composition dependent. Experimental observations showed that an M_s temperature above 150 °C is required to obtain a lath microstructure upon quenching from the solution treatment temperature. As the M_s temperature decreases the fraction of retained austenite increases and lath microstructure coarsens to a plate-like microstructure. Thus, the first step in controlling the type of martensitic microstructure formed is to determine the composition dependence of the M_s temperature.

In corrosive environments where chemical embrittlement can occur, grain boundary cohesion also becomes important factor in maintaining toughness. The segregation of impurities to grain boundaries results in brittle-fracture as impurities lower the interfacial cohesion. For a grain boundary with a given impurity, Rice and coworkers (Ref 35) modeled the competition between crack-tip blunting by dislocation emission and brittle interfacial cleavage. The model predicts that ductile fracture will prevail if the energy for dislocation emission is greater than the Griffith work for interfacial cleavage controlled by interfacial thermodynamics. The results of literature thermodynamic data (Ref 36) confirmed the detrimental effect of phosphorus, tin, antimony and sulfur. Boron was shown to be a superior cohesive enhancer, as supported by first-principles electronic calculations (Ref 1). Detrimental impurities can be removed from the grain boundaries using gettering techniques (Ref 37–40). Watton et al. (Ref 41) demonstrated that lanthanum and cesium can be used to effectively getter both P and S impurities.

For the integrated design of a case/core material, the system of interactions between the desired properties, structure, and processing is depicted in Fig. 4. Following a systems approach, the next step is defining the priorities of the processing/structure/property interactions and identifying the conflicting objectives of the various subsystems. The case properties are the primary property objectives, while the core properties are secondary objectives. These case property objectives correspond to the following microstructural and processing objectives: a quench-hardenable lath martensitic matrix microstructure, the formation of a passivating film at the surface, and the avoidance of primary carbides. Secondary microstructural objectives include the efficient use of secondary hardening to optimize the strength and toughness and the reduction of microsegregation during solidification.

Within this system, conflicting objectives exist. The M_s temperature is controlled primarily by the C, Co, Cr, and Ni levels. Co additions increase the M_s temperature, while increasing the C, Cr, and Ni compositions strongly decreases the M_s temperature. Maintaining a constant M_s temperature by limiting C conflicts with the desired strength objectives, which require high carbon. Aqueous corrosion resistance depends on increasing the Cr that partitions to the oxide film; however, increasing Cr lowers the M_s temperature and promotes undesirable carbides.

Within the interactions of the secondary objectives are other conflicting objectives. Vanadium increases the driving force for precipitation of the coherent M_2C carbide; however, V solubility at the solution treatment temperature, T_s, is limited. Increasing the amounts of carbide formers, Cr, Mo, and V increases the tendency for solidification microsegregation. The tempering process increases the strength via secondary hardening, but can promote the relaxation of desired residual compressive stresses.

Model Development

Based on the priorities of the property objectives, the models developed emphasize the martensitic transformation behavior, the precipitation of coherent M_2C carbides, the formation and stability of passivating oxide films, and the microsegregation occurring during solidification processing. These models correlated the desired microstructural features and processing conditions with thermodynamic quantities and diffusional reactions using the Thermo-Calc (Ref 42) and DICTRA (Ref 43) software platforms. The application of these models to the current design application is now described. However, for details regarding the specific models, the reader is encouraged to consult the literature.

Martensitic Transformation Behavior

The FCC→BCT martensitic transformation, which occurs upon quenching from the solution temperature to room temperature, exhibits a high-strength matrix. This displacive diffusionless transformation results in a large distortion strain and a high dislocation density. The strain energy associated with the large lattice deformation dominates the kinetics and morphology of the transition. To achieve the desired lath morphology requires a M_s temperature of approximately 150 °C and thus the modeling objective is to determine the composition dependence of the M_s temperature.

Using Thermo-Calc, the M_s temperature for a multicomponent ferrous-based alloy was predicted implementing the Olson-Cohen model (Ref 44, 45) for heterogeneous nucleation and the Ghosh-Olson solid-solution strengthening model (Ref 46, 47) for the description of the frictional work associated with interfacial motion. Recognizing the considerable uncertainty in the thermodynamics of carbon ordering in the BCT martensite, the chemical thermodynamics of the system were modeled using the descriptions of the FCC and BCC solid solutions based on the SGTE Solution database (Ref 48). Modifications to the SGTE description were made by Ghosh (unpublished thermodynamic assessment work done by G. Ghosh at Northwestern University, Evanston, IL, 1992) and by Campbell (Ref 13), which incorporate Fe-Co-Cr ternary interaction parameters based on the evaluation of model ternary alloys. For a wide range of temperatures and compositions, the predicted M_s temperatures correlate well with the measured M_s temperatures, Fig. 5. The standard deviation of the difference between the measured and calculated M_s is ± 35 °C. Several prototype and commercial alloys were used to test the validity of the model. All of the commercial alloys show good agreement with the model. In particular, the deviation for the commercial carburizing stainless steel Pyrowear

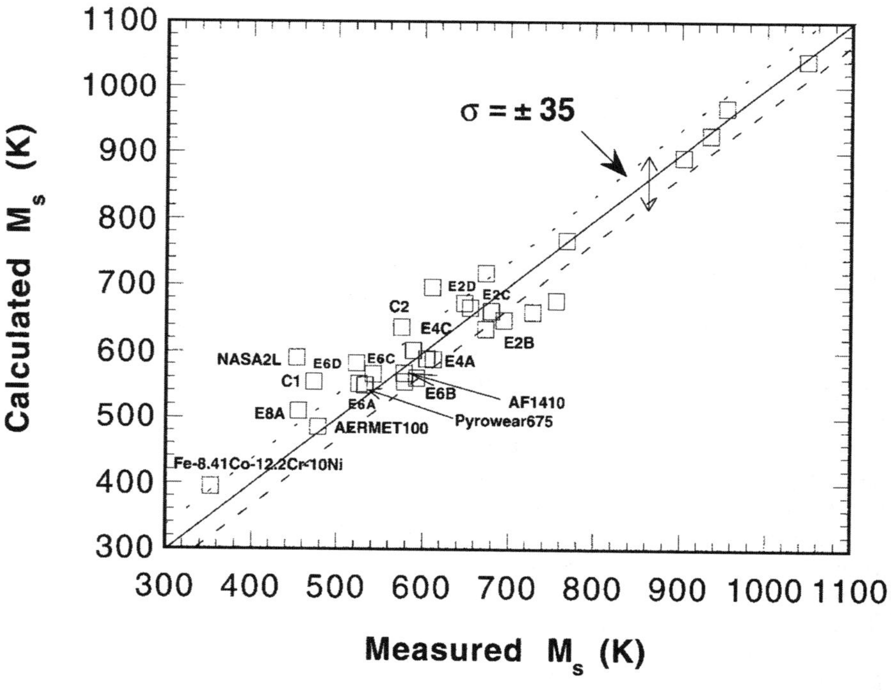

Fig. 5 Correlation of the calculated and measured M_s temperatures for model and commercial alloys.

675 and a stainless prototype (E8A) is ± 25 °C. These alloys demonstrate the success of the model to predict the martensitic behavior in high Co-Cr ferrous alloys.

Precipitation of Coherent M_2C Carbides

The efficiency of the strengthening dispersion is determined by the critical particle size and the interparticle spacing. Characterization of the AF1410 alloy (Ref 49–51) demonstrated the effectiveness of the precipitation of coherent M_2C carbides in a highly supersaturated BCC matrix. This precipitation in a highly supersaturated BCC matrix resulted in the suppression of the growth regime leading to precipitation by a coarsening process, as predicted by the Langer-Schwartz model (Ref 24, 25, 52), and thus, the mean carbide size remained near the critical particle size, r^*, which smoothly increased as the supersaturation dropped. Modeling of the critical particle size as a function of the fraction of the precipitation reaction completed uses the Gibbs-Thompson Eq 1 and small angle neutron scattering (SANS) carbide size data for the AF1410 (Ref 50) and model alloys (unpublished research by P. Jemian and J.R. Weertman, Northwestern University, Evanston, IL, 1992):

$$r^* = \frac{2\gamma}{\Delta G} \qquad \text{(Eq 1)}$$

where γ is the interfacial energy (J m^{-2}) and $-\Delta G$ is the precipitation driving force (J m^{-3}) associated with the supersaturation level. The modeling efforts require that the initial state from which precipitation initiates be defined and the free energy of the coherent M_2C carbide phase be described.

In the tempering temperature range of 400 to 550 °C the carbon mobility is much greater than that of the substitutional elements and results in the initial precipitation of paraequilibrium cementite, which only requires carbon diffusion (Ref 53–56). With additional tempering time, diffusion of metallic substitutional elements occurs and M_2C carbides precipitate, dissolving the cementite. The BCC-cementite paraequilibrium conditions assume equal carbon chemical potentials and no redistribution of the substitutional species:

$$\mu_C^{CEM} = \mu_C^{BCC} \sum_i y_i \left(\mu_i^{CEM} - \mu_i^{BCC}\right) = 0 \text{ where } i \neq C \qquad \text{(Eq 2)}$$

where μ_C^{CEM} is the carbon chemical potential in cementite and μ_C^{BCC} is the carbon chemical potential in BCC. For the current thermodynamic description, the substitutional species (Fe, Cr, Co, Mo, Ni, and V) and the interstitial species (C and vacancies) are modeled on separate sublattices and thus, y_i is the site fraction of i on its corresponding sublattice. As there is no substitutional diffusion, $y_i^{CEM} = y_i^{BCC}$ where i does not equal carbon. The chemical potentials of the substitutional species in the BCC and cementite phases are given by μ_C^{BCC} and μ_i^{CEM}, respectively. Experiments confirmed that this paraequilibrium condition is the initial state from which M_2C carbide precipitation begins (Ref 57).

The free energy of the coherent M_2C carbides is the summation of the chemical free energy, the elastic energy, and the interfacial energy.

$$G_m^{M_2C} = G_{\text{chemical}} + G_{\text{elastic}} + G_{\text{interfacial}} \qquad \text{(Eq 3)}$$

The HCP phase description of the SGTE solution database (Ref 48) is used to describe the chemical thermodynamics of the M_2C carbide. The linear-elastic model for an inhomogeneous inclusion treated by Liarng (Ref 58) and King et al., (Ref 59) represents an upper bound on the composition-dependent elastic energy. It has been assumed that the upper bound estimate of the elastic self-energy scales by a fixed correction factor and is independent of carbide phase fraction at the low fractions of interest. The interfacial energy is assumed to be composition and size independent for a given state of coherency. The model of Johnson and Cahn (Ref 60) describes the shape dependence of the capillary energy employing the experimental observations of the M_2C carbides size and shape. The fraction of linear-elastic energy and interfacial energy were calibrated using SANS data from the AF1410 alloy (Ref 51) and other model alloys.

Calibration with the SANS data determined an elastic energy correction factor of 0.3 for the fully coherent state (Ref 13). A three coherency state model was necessary to fit the data, with full coherency below a 30 Å particle size diameter, a semicoherent state between 30 Å and 50 Å, and an incoherent state above 50 Å. The interfacial energy associated with the coherent particles is 120 mJ m^{-2}, with semicoherent particles is 404 mJ m^{-2}, and with incoherent particles is 520 mJ m^{-2}. With these calibrations, the Gibbs-Thompson equation predicts the mean particle size as a function of the precipitation reaction completed (Ref 13). Figure 6 shows the correlation of the calibrated model predictions of the evolution of the shape-corrected effective particle size with the experimental measurements. Thus, the M_2C strengthening is optimized by maximizing the driving force for coherent M_2C carbides precipitating in the BCC matrix in paraequilibrium with cementite. Maximizing the driving force decreases the particle size and increases the strength.

Aqueous Corrosion Resistance

The pH-potential diagrams for multicomponent alloys and constrained equilibrium calculations determining the degree of Cr partitioning to the oxide film were used to optimize the aqueous corrosion resistance. The pH-potential diagrams, as demonstrated by Pourbaix (Ref 61), represent the equilibrium phase boundaries within dimensions of pH and applied potential for the given solid phases and the aqueous solution. Originally Pourbaix (Ref 61) assumed dilute solution concentrations with the range of 1.0e-4 molal to 1.0e-6 molal. The pH-potential diagram calculations for multicomponent alloys assume a total ionic concentration less than 1.0e-4 molal in the aqueous solution and that the ionic concentration of the solution scales with base alloy composition. The latter assumption is based on the work of Kirchheim (Ref 28); study of the passivation of Fe-Cr alloys at long times (steady-state conditions) revealed that the relative Cr concentration of the electrolyte solution was approximately equal to the Cr composition of the base alloy. The pH assumed in the pH-potential diagram calculation is the pH at the metal/film interface. However, when comparing to experiments typically the referenced pH is the bulk pH, even though the bulk pH may differ substantially from the interface pH. This inconsistency occurs due to the many difficulties associated with measuring the interface pH (Ref 62).

Figure 7 is the calculation of the pH-potential diagram for a Fe-12Cr (wt%) alloy in a water-based aqueous solution assuming nonstoichiometric multicomponent oxide phases. The addition of Cr shifts the passive boundaries to lower pH levels, thus increasing the region over which the passivation occurs. The pH-potential diagrams for multicomponent alloys can be used to indicate whether the desired oxide passivation is thermodynamically favored.

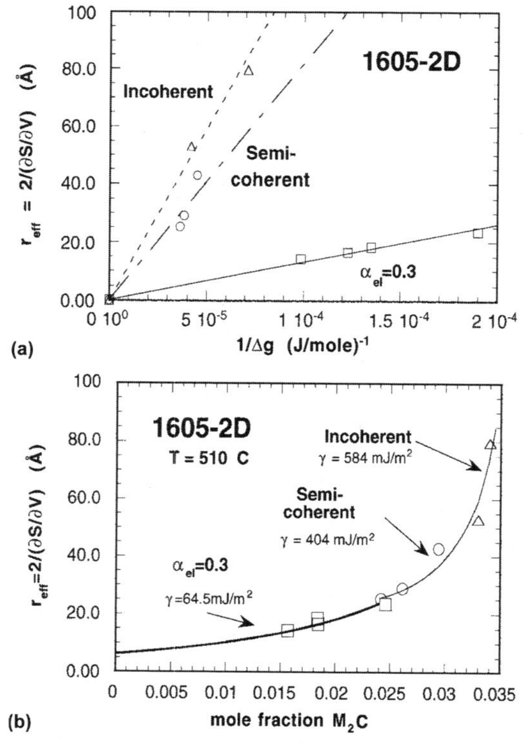

Fig. 6 For alloy 1605-2D at 510 °C (a) r_{eff} versus $1/\Delta G$ for coherent α_{el} = 0.3(□), semi-coherent M_2C with C in paraequilibrium with cementite (O), and incoherent M_2C with C in BCC-M_2C equilibrium (Δ). (b) r_{eff} as a function of mole fraction of M_2C. The data symbols in (b) correspond to the fit used in (a).

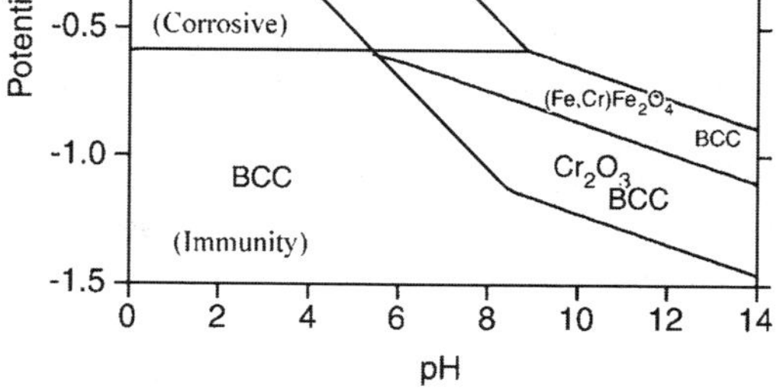

Fig. 7 pH-potential diagram for Fe-12Cr considering nonstoichiometric oxides.

To determine if the desired passivation occurs requires kinetic modeling. A constrained metastable equilibrium calculation is used as a first order approximation of the dynamic state. A strained-layer model with composition dependent strain energy terms based on the coherent M_2C model (Ref 58, 59) was developed to calculate the constrained equilibrium between the metal alloy and the initial forming spinel-type oxide, $(Fe,Cr)_3O_4$. The model strain energy terms were calibrated to the observed amount of Cr partitioning to the oxide film. Figure 8 shows the calibration of the Cr partitioning to the data of Kirchheim (Ref 28) and the application of the model to high-order systems, which fosters more efficient partitioning.

Solidification Microsegregation

As a large amount of solidification microsegregation was associated with an earlier stainless prototype (NASA2) and the AerMet100 alloy was characterized to have acceptable levels of microsegregation, a method to predict an degree of microsegregation in an alloy was developed using the DICTRA software and calibrating the calculations with the experimental measurements from the AerMet100 alloy (Ref 63). The DICTRA simulation of solidification assumes equilibrium is maintained at the liquid/solid interface and diffusion occurs in both the solid and liquid phases. The simulation also assumes that initial liquid composition is homogeneous and the liquid/solid interface is planar. For each time increment, the flux equations are numerically solved along each point of the geometric grid subdividing the assumed symmetrical dendrite.

DICTRA simulations of the AerMet100 solidification process show good agreement with the measured microsegregation after solidification (Ref 63). Comparison of the composition profiles, Fig. 9, shows that the thermodynamic and kinetic parameters used produced quantitative results, which require no further calibration. Thus, solidification simulations for secondary hardening ferrous alloys can be used to determine compositions that minimize microsegregation.

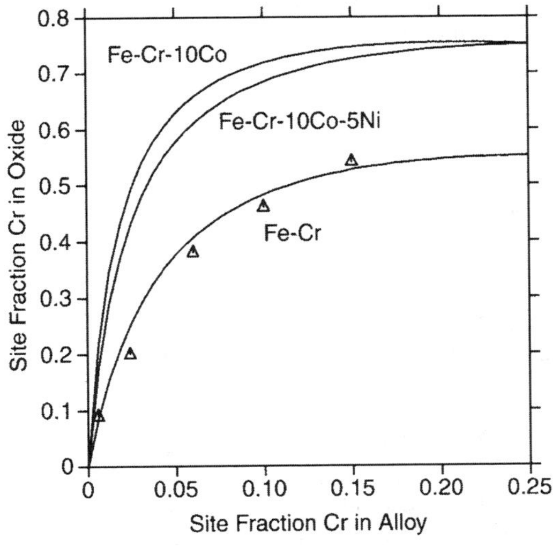

Fig. 8 Calculation of Cr partitioning in the metastable spinel phase (Me_3O_4). A strain term of 50 kJ/mole is added to the terms describing the Fe^{+3}-Cr^{+3} exchange on the octahedral sublattice. The Δ symbols represent the experimental work of Kirchhiem (Ref 28).

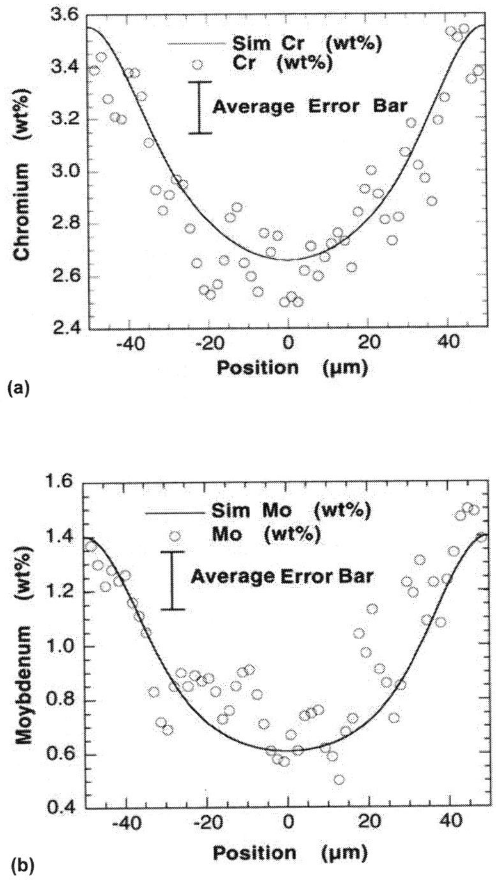

Fig. 9 For the as-cast AerMet100, experimental and DICTRA simulated composition profiles across a 100 μm secondary dendrite arm for (a) Cr and (b) Mo.

Computational Design Integration

To complete the synthesis step of systems design (Fig. 1), the design models were integrated to predict an optimal composition to meet the previously established objectives. Figure 10 illustrates the thought process guiding the integration of the models.

Matrix Optimization

The first step in the design of a carburizing, secondary-hardening, martensitic stainless steel is to estimate the carbon content needed to achieve the desired core and case strengths. The strengthening predictions (Ref 22) give some guidance to the strength achieved for a given carbon level. The predictions assume that a given alloy composition dictates a specific carbide volume fraction and considers carbides of fixed particle diameter. Based on these assumptions, a modified version of Orowan (Ref 64) and Ashby (Ref 65) theory predicts the strength:

$$\Delta\tau_{\text{Total}} = \Delta\tau_{\text{bypass}} + \Delta\tau_{\text{BCC}}$$

$$\Delta\tau_{\text{bypass}} = 1.88\frac{Gb}{2\pi(1-\nu)^{1/2}L}\ln\left(\frac{2R}{b}\right)$$

$$\text{where } L = R\left(\frac{2}{f}\right)^{1/2} \qquad \text{(Eq 4)}$$

and where $\Delta\tau_{\text{BCC}}$ is the estimated martensite matrix strength, 155 VHN, G is the BCC shear modulus, υ is Poison's ratio for the matrix, R is the particle radius, b is the Burger's vector, and f is the carbide volume fraction. Figure 11 depicts the predicted hardness as a function of carbon for given carbide particle sizes and provides an estimate of the desired carbon composition. The positions of the hardness values for the AF1410 and AerMet100 alloys are consistent with the SANS measurements of the particle size (Ref 50). The higher driving force associated with the NASA1 prototype predicts a smaller particle size; the measured hardness is consistent with a particle size of 40 Å. Assuming a precipitation driving force equivalent to NASA1 can be obtained to achieve the same particle refinement, for a R_c 65 case hardness and a R_c 50 case hardness the estimated carbon compositions are 0.5 wt% C in the case and 0.2 wt% C in the core.

With these carbon content estimates, the matrix composition in both the case and the core is optimized to achieve the correct M_s temperature gradient, aqueous corrosion resistance, and sufficient cleavage resistance in the core material. As previously indicated the desired M_s temperature needed to achieve a lath martensitic microstructure is 150 °C; thus, the desired minimum case M_s temperature is 150 °C. The desired core M_s temperature is chosen to be 100 deg higher than the case M_s temperature to establish a reasonable M_s temperature gradient, which will correspond to the desired carbon gradient and will provide a compressive stress profile at the surface. However, the core M_s temperature is not a primary objective and the temperature gradient may be altered to accommodate objectives with higher priority. In addition, to compensate for error in the M_s model for high Co-Cr compositions the design M_s values are raised approximately

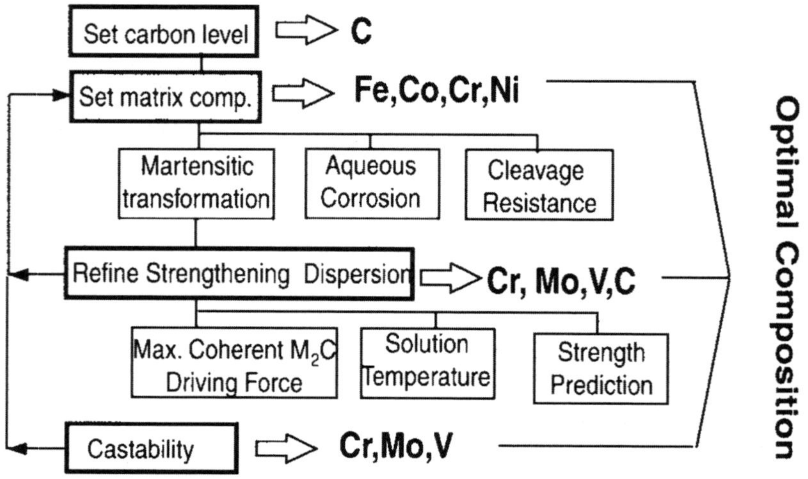

Fig. 10 Schematic of the design optimization procedure.

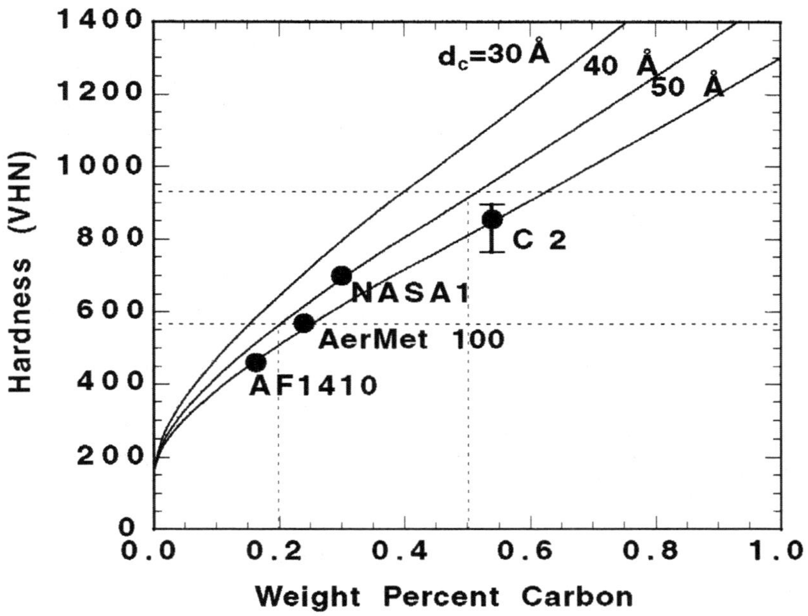

Fig. 11 Hardness as a function of alloy carbon content (Ref 22). The dashed lines represent the desired case and core carbon and the corresponding hardness levels at 40 Å particle size.

25 deg. Thus, the matrix core composition is optimized with respect to constant M_s contours at 550 K (277 °C) as functions of Co and Ni compositions at various Cr levels, as shown in Fig. 12. The contours have a parabolic shape for which the maximum Ni content shifts to lower Co levels as the Cr content increases. However, the benefits of short-range order dislocation recovery resistance require at least 8 wt% Co (Ref 26). Ni contents greater than 1 wt% Ni shift the DBT temperature below room temperature improving the fracture toughness resistance. Cost constraints suggest limiting the Co content to less than 25 wt%. The constraints on the Ni and Co compositions, represented by the dashed lines in Fig. 12, limit the Cr composition to 10 wt% Cr or less. The optimization of the aqueous corrosion resistance also influences the matrix Cr composition, and this will be discussed further in the re-optimization of the composition. The high Co contents of interest (15–20 wt%) promote the partitioning of Cr to the oxide film increasing the oxide stability, as seen in Fig. 8.

The carbon gradient resulting from carburizing and the corresponding M_s temperature gradient limit the optimum Ni content. For various Ni concentrations, Fig. 13 represents the M_s gradient as the C increases from the core to the case. The compositions of Ni and C that

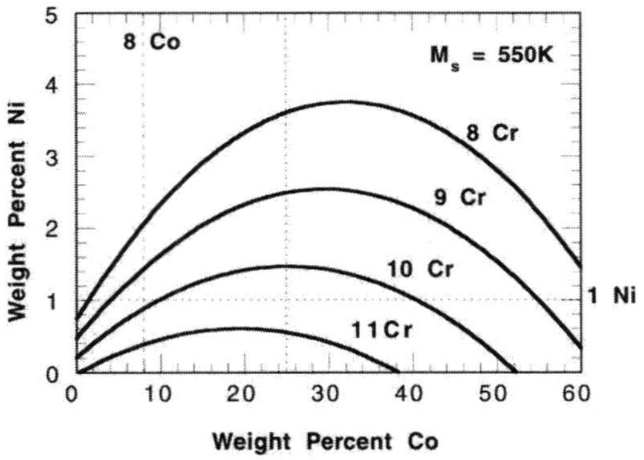

Fig. 12 Constant M_s contours at 550 K defining the Co and Ni content of the matrix. The dashed lines represent the system constraints.

produce a M_s of 450 K (177 °C) in the case are chosen. Finally, the Ni and Co contents were refined once more using the 9Cr M_s contour at 450 K, Fig. 14. The broad maximum suggests there is little benefit derived from increasing the Co from 15 to 20 wt%. Thus matrix composition is fixed at Fe-15Co-1.5Ni-9Cr-0.2C (0.5C case) wt%.

Strengthening Dispersion Optimization

With an initial matrix composition set, refinement of the secondary hardening dispersion maximizes the strength. At the assumed

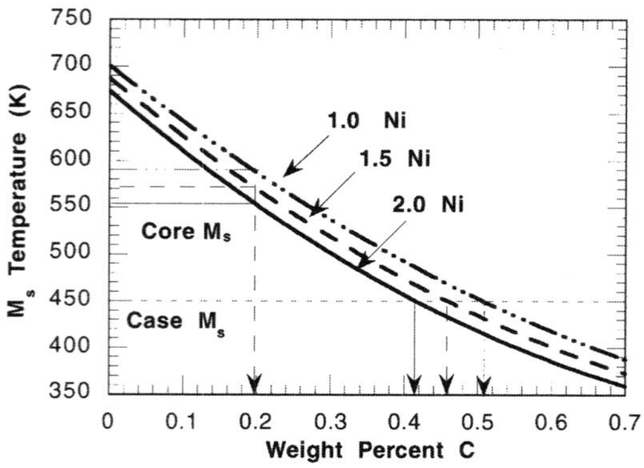

Fig. 13 M_s temperature as a function of matrix carbon content for Fe-15Co-9Cr-Ni (wt%) alloys. The dashed lines represent the case and core M_s temperatures corresponding to the various Ni and C compositions. The core C composition is fixed at 0.2 wt% C and the case M_s temperature is fixed at 450 K.

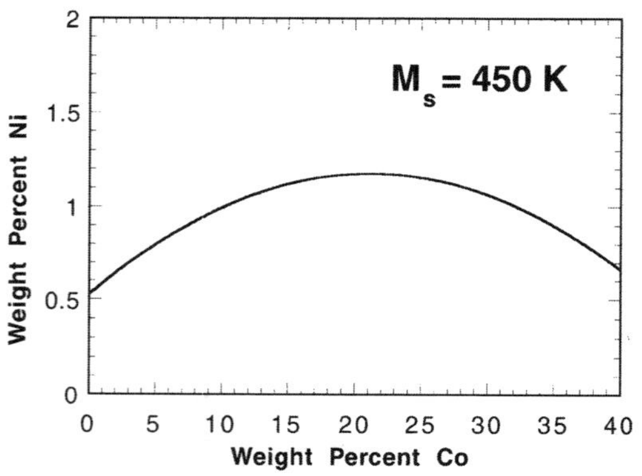

Fig. 14 M_s contour at 450 K defining the case matrix composition for the Fe-Ni-Co-9Cr-0.5C (wt%) system.

tempering temperature of 482 °C, the coherent M_2C driving force is maximized, within the composition constraints of the solubility at the solution temperature (<1100 °C). In addition to these composition constraints, the paraequilibrium conditions between BCC and cementite restrict the amount of carbon in the matrix decreasing the M_2C driving force. For both the core and case compositions, the design calculates the coherent M_2C driving force under paraequilbrium conditions using the coherent M_2C model. The M_2C coherency model assumes a carbide aspect ratio of 1, appropriate to initial nucleation. To insure sufficient driving force to obtain the desired strengthening, the driving force for the new alloy should equal the driving force of a prototype alloy with sufficient hardness, NASA1, which has a driving force of 15.9 kJ/mole at the tempering temperature of 455 °C. Figure 15 explores the composition dependence of the coherent M_2C driving force with respect to the Fe-15Co-9Cr-1.5Ni-0.48C matrix, assuming a tempering temperature of 482 °C. Both Mo and V increase the driving force, while increased Cr decreases the driving force.

The carbon-vanadium section at 1100 °C, seen in Fig. 16, shows the constraints imposed on the V composition. To avoid primary carbides, the material must be solution treated in the single FCC phase field. To achieve a solution treatment temperature of 1100 °C for both the core and case restricts the V composition to 0.2 wt% V and the case carbon level to 0.48 wt% C.

Aqueous Corrosion

Thermodynamics are needed to predict the formation of a passive film on the surface, the increase of Cr activity in the matrix, and the avoidance of primary carbides. For a 9 wt% Cr addition, the pH-potential diagram indicates that the passivating oxides Me_3O_4 and Me_2O_3 are thermodynamically stable. The additional Co in matrix increases the Cr activity in the matrix and correspondingly increases the partitioning ratio of Cr, as shown in Fig. 8. Based on the work of Kirchheim (Ref 28), this increase in Cr activity should improve the kinetics for the formation of the Me_3O_4 oxide film. Using the fine scale secondary-hardening M_2C carbides and following processing schedules that do not include phase fields in which primary carbides are thermodynamically stable prevents the depletion of matrix Cr needed for the formation of a passive film.

Castability

Efforts to improve castability constrain the molybdenum concentration. The castability of the alloy improves by decreasing the microsegregation and by improving the fluidity through a reduced freezing range (Ref 66). The fluidity, the ability to fill a mold, is greatest when freezing occurs at a constant temperature allowing the slow inward growth of solid metal. Large freezing ranges result in the liquid surrounding interlacing dendrites (i.e., the metal is in a mushy condition), which restricts the flow of the liquid. The dependence of fluidity on the freezing range was verified in a study of

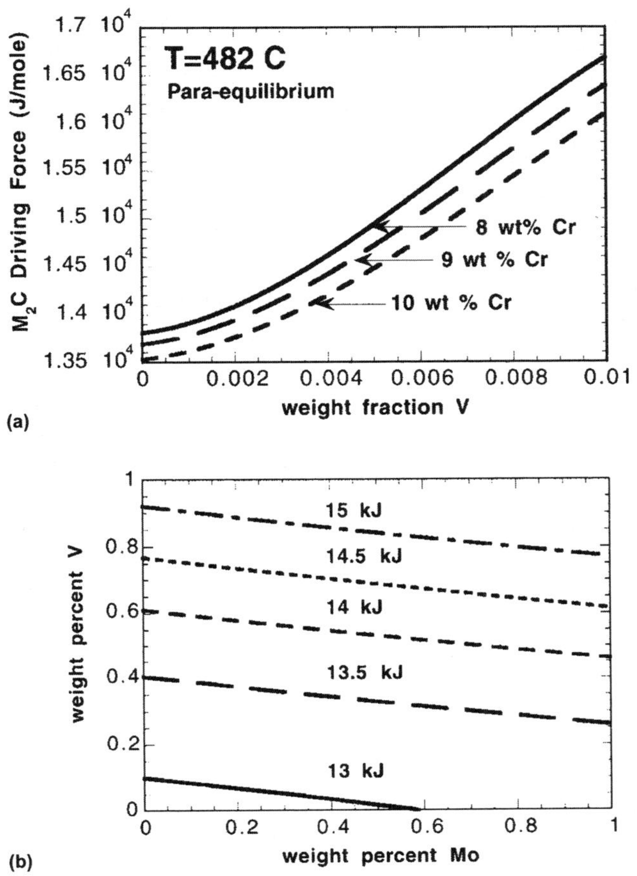

Fig. 15 Composition dependence of the coherent M_2C driving force, calculated assuming the initial conditions are BCC-cementite paraequilibrium at 482 °C (a) Fe-15Co-1.5Ni-Cr-0.5Mo-V-0.5C (wt%) (b) Fe-15Co-1.5Ni-9Cr-Mo-V-0.48C (wt%).

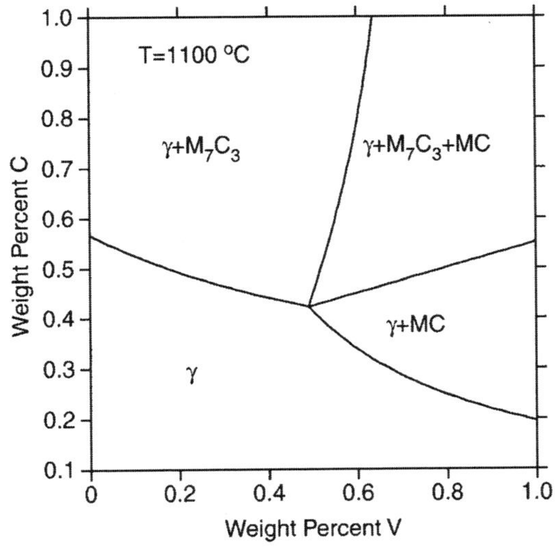

Fig. 16 Phase section at 1100 °C with respect to C-V in the Fe-15Co-1.5Ni-9Cr-0.5Mo-V-C (wt%) system, calculated using the SGTE database.

aluminum alloys with similar solidification characteristics (Ref 67).

The segregation ratios, $\delta = C_{max}/C_{min}(C =$ concentration) characterizes the microsegregation after solidification. Diffusion simulations of the solidification process allow the microsegregation occurring in the composition space Fe-9Cr-(0-1)Mo-(0–1)V-0.2C (wt%) to be investigated. As the segregation ratios of Co and Ni are expected to be low based on the cast AerMet100 work (Ref 63), these constituents are excluded from the composition space to reduce the simulation time. The solidification simulations assume a cooling rate of 0.5 °C s^{-1}, a total solidification time of 1000 s, a secondary dendrite size of 100 μm and no carbide precipitation. These assumptions are based on the results of the AerMet100 solidification study (Ref 63). Tables 3–5 contain the calculated segregation ratios of the composition matrix studied. Figure 17 illustrates the effect of composition on the freezing range. Analysis of the predicted segregation ratio shows Mo to be the largest segregant. The addition of only Mo or V decrease Cr segregation. The additions of both Mo and V at high levels promote Cr segregation. V segregation increases with Mo additions. The freezing range increases with both Mo and V additions. Due to the large segregation ratio for Mo and the demonstrated ability to achieve a high M_2C driving force without Mo, Mo is removed from the design composition consideration. The first iteration design composition is then Fe-15Co-1.5Ni-9Cr-0.2V-0.2C (0.5C case) (wt%).

Table 3 Matrix of the predicted δ_{Cr} for Fe-9Cr-Mo-V-0.2C (wt%) alloys

	0.0	0.5	1.0
0.0	1.56	1.50	1.51
0.2	1.49	1.51	...
0.5	1.50	1.52	1.54
1.0	...	1.55	...

Table 4 Matrix of the predicted δ_{Mo} for Fe-9Cr-Mo-V-0.2C (wt%) alloys

	0.0	0.5	1.0
0.0	...	2.51	2.56
0.2	...	2.55	...
0.5	...	2.61	2.67
1.0	...	2.72	...

Table 5 Matrix of the predicted δ_V for Fe-9Cr-Mo-V-0.2C (wt%) alloys

	0.0	0.5	1.0
0.0	...	...	...
0.2	1.75	1.79	...
0.5	1.79	1.84	1.88
1.0	...	1.94	...

Reoptimization

Next, the first iteration composition is reexamined to insure self-consistency. Table 6 summarizes the predicted properties achieved by the first iteration composition. The strength is predicted for 50% completion of the M_2C precipitation reaction using the coherency model to determine the particle size input for the strengthening model (Ref 65). The particle size is calculated assuming coherent carbides with an aspect ratio of 3, where this is the observed state just before coherency loss, corresponding typically to the peak hardness condition. The mole fraction of M_2C is set at 50% of the equilibrium value. An average coherent interfacial energy of 120 mJ m^{-2} is assumed. The strengthening mechanism is assumed to be dislocation bypass and the modified Orowan theory was applied, Eq 4. The initial driving force for M_2C precipitation is slightly lower than the desired NASA1 driving force of 15.9 kJ/mole. Solution temperature constraints on the V solubility and Mo segregation during

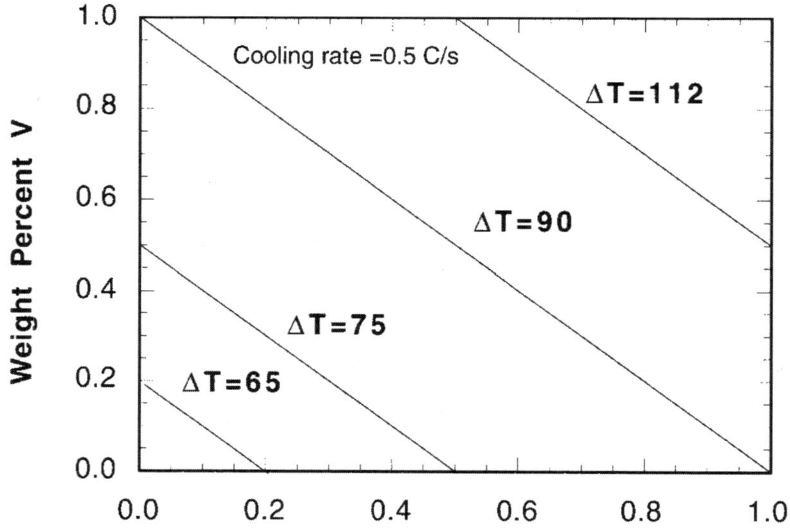

Fig. 17 Contours of constant freezing ranges in the Mo-V composition space for Fe-9Cr-Mo-V-0.2C (wt%) alloys. A cooling rate of 0.5 °C s^{-1} and no carbide precipitation were assumed.

solidification restrict increasing the driving force of the alloy composition.

The Cr partitioning in the oxide film exceeds that of 440C, which has a value of $y_{Cr} = 0.58$. The pH-potential diagram for the alloy composition is shown in Fig. 18 and indicates that at the pH and electrode applied potential of interest (pH = 7 and potential range of (−0.1 to 0.1 V, SHE) passivating oxides (Me_3O_4 and Me_2O_3) form. Note the alloy will enter a corrosive region if the pH is reduced to a pH less than 6 for potentials greater than −0.47 V, SHE.

The initial composition possesses almost all of the desired properties. The calculated M_2C driving force for precipitation is 90% of the desired NASA1 value. This lower driving force is accepted as increasing the driving force would be done at the expense of other objectives. This completes the optimization of the design composition. To the design composition late boron and cesium additions are made at conventional levels during the vacuum-melt processing to improve grain boundary cohesion and getter impurities.

Table 6 Predicted properties of design prototype

Property	Value
Case properties	
M_s	446 K (172 °C)
Hardness at 50% completion	825 VHN
Final mole fraction M_2C	0.0668
Coherent M_2C driving force at 482 °C	14.2 kJ/mole
Partitioning of Cr in oxide film, y_{Cr}	0.75
Solution Temperature, T_s	1100 °C
Core properties	
M_s	570 K (297 °C)
Hardness at 50% completion	575 VHN
final mole fraction M_2C	0.0282

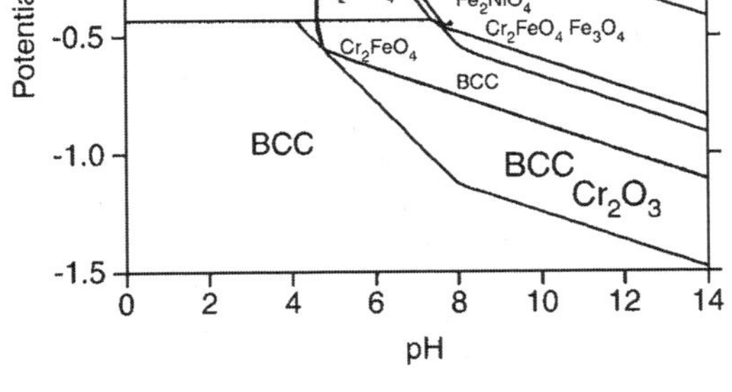

Fig. 18 pH-potential diagram for Fe-15.5Co-1.5Ni-9Cr considering stoichiometric oxides.

Discussion

The presented computational design calculations represent the most recent work in the development of a martensitic secondary-hardening stainless steel. Previous design iterations for monolithic stainless steels and carburizable secondary-hardening gear steels directed the model development. The first iteration of this computational design process resulted in the design of the NASA1 prototype (Ref 68, 69). While NASA1, a monolithic stainless bearing composition, was designed with somewhat limited design models, it achieved many of the desired design objectives and illustrated the feasibility of systems design. The weaknesses of the NASA1 composition highlighted the processing/structure/property interactions that required further model development. The greatest need was for further development of the modeling of the martensitic transformation behavior. With kinetic factors (Ref 46, 47) added to the modeling of the martensitic transformation behavior, the NASA2 prototype was designed. Characterization of NASA2 prototype showed improvement in obtaining the desired martensitic microstructure; however, the measured M_s temperature was still 150 deg below the desired M_s temperature, indicating further model development was needed. Evaluation of the NASA2 prototype also showed that significant microsegregation had occurred with the high alloy concentration and implied that modeling of the solidification process would be useful. The C2 prototype (Ref 21, 70), a carburizing secondary-hardening alloy, was designed with approximate elastic strain energies added to the chemical thermodynamics of the M_2C carbide phase to account for coherency effects. This approximation underestimated the strengthening efficiency of the coherent M_2C; thus, the current modeling incorporated SANS data on the particle size evolution to better calibrate the coherency strain energy and addressed the initial state of M_2C precipitation.

The current computational design of the carburizable secondary-hardening martensitic stainless steel incorporates refined models of the martensitic transformation behavior and the kinetics of the secondary-hardening process and adds models for the stability of the passivating film and the microsegregation resulting from solidification processing. Using these models, the present computational design predicts that an alloy composition with 9 wt% Cr will provide more aqueous corrosion resistance than the 440C composition, which has 11 wt% Cr in the matrix. The higher corrosion resistance for the lower Cr alloy is achieved by increasing the Cr activity in the matrix composition, and thus, increasing the Cr partitioning to the metastable oxide film. The design composition has a higher matrix Cr activity because of the large Co addition (15 wt%) to obtain the desired martensitic microstructure. Avoiding the precipitation of primary carbides reduces the amount of Cr taken from the matrix composition.

Conclusions

A computational systems design approach determined property objectives for a new class of high-performance stainless steel, guided the development of mechanistic models, and determined a specific composition to meet the desired property objectives. To achieve the desired high fatigue strength, toughness, and aqueous corrosion resistance, a carburizing secondary-hardening, martensitic stainless steel requiring only thermal processing was conceived. The design was completed using the Thermo-Calc and DICTRA multicomponent thermodynamic and kinetic software packages to integrate models of the martensitic transformation behavior, coherent M_2C carbide precipitation, the formation and stability of passivating oxide films, and microsegregation resulting from solidification processing.

To complete the systems design process, the prototype must be characterized to determine if the desired microstructures are achieved and property objectives are met. The prototype characterization will also serve to determine strength and weaknesses of the models used. Evaluation of the prototype composition is discussed in part II (Ref 71).

ACKNOWLEDGMENTS

This research was supported by a NSF Graduate Fellowship for CEC and was conducted as part of the multi-institutional Steel Research Group program. The authors thank G. Ghosh for assistance with the modeling of the martensitic transformation behavior. Pyrowear 675 was provided by Carpenter Technology. Alloy 1605-8A was produced by NKK Corporation, Japan.

REFERENCES

1. G.B. Olson, Science, 277 (1997) 1237
2. C.J. Small, and N. Saunders, MRS Bull., 24 (1999) 22
3. B. Sundman, and J. Ågren, MRS Bull., 24 (1999) 32
4. K.E. Spear, T.M. Besmann, and E.C. Beahm, MRS Bull., 24 (1999) 37
5. C. Bernard, M. Pons, E. Blanquet, and R. Madar, MRS Bull., 24 (1999) 27
6. P. Villars, N. Onodera, and S. Iwata, J. Alloys Compd., 279 (1998) 1
7. A.A. Howe, and D.C.J. Farrugia, Mater. Sci. Technol., 15 (1999) 15
8. P.J. Spencer, Thermochim. Acta, 314 (1998) 1
9. G.B. Olson, Science, 288 (2000) 993
10. G.M. Jenkins, in J. Beishon, and G. Peters, (Eds.), *Systems Behavior*, Open University Press, 1972, p 56–79
11. E.N. Bamberger, in W.R. Loomis, (Eds.), *New Directions in Lubrication, Materials, Wear, and Surface Interactions: Tribilogy in the 80's*, Noyes Publications, 1983, p 736–756
12. T.J. McCaffrey, and D.E. Wert, in J.J.C. Hoo, (Ed.), *Creative Use of Bearing Steels*, ASTM STP 1195, ASTM, 1993, p 137–148
13. C.E. Campbell, Doctoral dissertation, Northwestern University, 1997
14. J. Morrow, *Internal Friction, Damping, and Cyclic Plasticity*, ASTM, 1964, p 45–84
15. O.H. Basquin, *Proceedings*, Am. Soc. Testing Mats., 1910, p 625–640
16. E. Hornborgen, and C. Verpoort, In D. Francois, (Eds.), Fifth International Conference on Fracture (1980), Pergamon Press Ltd., 1982, p 315–322
17. M.R. Stoudt, R.C. Cammarata, and R.E. Ricker, Scri Mater, 43 (2000) 491
18. F. Hengerer, U. Brockmüller, and P.O. Sörström, In J.J.C. Hoo, (Eds.), Creative Use of Bearing Steels, ASTM STP 1195, ASTM, 1993, p 21–33
19. M.G. Fontana, Corrosion Engineering, McGraw-Hill, 1986

20. M.F. Ashby, Acta Metall., 37 (1989) 1273
21. C.J. Kuehmann, and G.B. Olson, Adv Mater Processes, 153 (1998) 40
22. J.P. Wise, Doctoratal dissertation, Northwestern University, 1998
23. J.S. Montgomery, and G.B. Olson, in G.B. Olson, M. Azrin, and E.S. Wright, (Eds.), *Innovations in Ultrahigh-Strength Steel in Technology* (34th Sagamore Army Materials Research Conference), U.S. Government Printing Office, 1987, p 147–178
24. J.S. Langer, and A.J. Schwartz, Phys. Rev. A, 21 (1980) 948
25. R. Kampmann, and R. Wagner, in P. Haasen, V. Gerold, R. Wagner, and M.F. Ashby, (Eds.), *Decomposition of Alloys: the early stages*, 2nd Acta-Scripta Metallurgica Conference, Pergamon Press, 1983, p 91–103
26. G.R. Speich, D.S. Dabkowski, and L.F. Porter, Metall. Trans., 4 (1973) 303
27. R. Kirchheim, Electrochimia Acta, 32 (1987) 1619
28. R. Kirchheim, B. Heine, H. Fischmeister, S. Hofmann, H. Knote, and U. Stolz, Corr. Sci., 29 (1989) 899
29. A.M. Hall, Nickel, in *Iron and Steel*, John Wiley & Sons, 1954
30. W. Jolley, J. Iron Steel Inst., 206 (1968) 170
31. W.C. Leslie, R.J. Sober, S.G. Babcock, and S.J. Green, Trans. AMS, 62 (1969) 690
32. G.B. Olson, in G.B. Olson, M. Azrin, and E.S. Wright, (Eds.), *Innovations in Ultrahigh-Strength Steel in Technology* (34th Sagamore Army Materials Research Conference), U.S. Government Printing Office, 1987, p 3–66
33. M. Schmidt, and M.J. Gore, in G.B. Olson, M. Azrin, and E.S. Wright, (Eds.), *Innovations in Ultrahigh-Strength Steel in Technology* (34th Sagamore Army Materials Research Conference), U.S. Government Printing Office, 1987, p 407–424
34. C.J. Kuehmann, Doctoral dissertation, Northwestern University, 1994
35. J.R. Rice, and J.-S. Wang, Mater. Sci. Eng., A107 (1989) 23
36. P.M. Anderson, J.S. Wang, and J.R. Rice, in G.B. Olson, M. Azrin, and E.S. Wright, (Eds.), *Innovations in Ultrahigh-Strength Steel Technology*, Sagamore Army Materials Research Coference Proceedings, 1990, p 619–649
37. M. Guttman, Surf. Sci., 53 (1975) 213
38. P.E. Waudby, Int. Met. Rev., 2 (1978) 74
39. M.P. Seah, P.J. Spencer, and E.D. Hondros, Met. Sci. (1979) 307
40. C.I. Garcia, G.A. Ratz, M.G. Burke, and A.J. DeArdo, JOM (1985) 22
41. J.F. Watton, G.B. Olson, and M. Cohen, In G.B. Olson, M. Azrin, and E.S. Wright, (Eds.), *Innovations in Ultrahigh-Strength Steel in Technology* (34th Sagamore Army Materials Research Conference), U.S. Government Printing Office, 1987, p 705–737
42. B. Sundman, B. Jansson, and J.O. Andersson, CALPHAD, 9 (1985) 153
43. A. Borgenstam, A. Engström, L. Höglund, and J. Ågren, J. Phase Equil., 21 (2000) 269
44. G.B. Olson, and M. Cohen, Metall. Trans., 7A (1976) 1905
45. G.B. Olson, and M. Cohen, Metall. Trans., 7A (1976) 1915
46. G. Ghosh, and G.B. Olson, Acta Metall. Mater., 42 (1994) 3361
47. G. Ghosh, and G.B. Olson, Acta Metall. Mater. 42 (1994) 3371
48. SGTE, Scientific Group Thermochemica Europe, Solution Database, 1994
49. G.B. Olson, T.J. Kinkus, and J.S. Montgomery, Surf. Sci., 246 (1991) 238
50. A.J. Allen, D. Gavillet, and J.R. Weertman, Acta Metall. Mater., 41 (1992) 1869
51. J.S. Montgomery, and G.B. Olson, in G. Krauss, and P.E. Repas, (Eds.), R. Gilbert, Speich Symp. Proc.: *Fundamentals of Aging and Tempering in Bainitic and Martensitic Steel Products*, 1992, p 177–214
52. R. Wagner, and R. Kampmann, in G.B. Olson, M. Azrin and E.S. Wright, (Eds.), *Innovations in Ultrahigh-Strength Steel in Technology* (34th Sagamore Army Materials Research Conference), U.S. Government Printing Office, 1987, p 209–221
53. A. Hultgren, ASM, 39 (1947) 915
54. M. Hillert, Acta Metall., 3 (1955) 34
55. M. Hillert, in L.H. Bennett, T.B. Massalski, and B.C. Giessen, (Eds.), *Alloy Phase Diagrams: Materials Reseearch Society Symposia Proceedings*, Elsevier, 1983, p 295–309
56. M. Hillert, and J. Ågren, in J.D. Embury, and G.R. Purdy, (Eds.), *Advances in Phase Transitions: International Symposium*, Pergmon Press, 1987, p 1–19
57. G. Ghosh, C.E. Campbell, and G.B. Olson, Metall. Mater. Trans, 30A (1999) 501

58. R.-H. Liarng, Doctoral dissertation, Northwestern University, 1996
59. K.C. King, P.W. Voorhees, G.B. Olson, and T. Mura, Metall. Trans., 22A (1991) 2199
60. W.C. Johnson, and J.W. Cahn, Acta Metall., 32 (1984) 1925
61. M. Pourbaix, *Atlas of Equilibria in Aqueous Solution*, Pergamon, 1966
62. B.D. Craig, *Fundamental Aspects of Corrosion Films in Corrosion Science*, Plenum Press, 1991
63. H.E. Lippard, C.E. Campbell, T. Björklind, U. Borggren, P. Kellgren, V.P. Dravid, and G.B. Olson, Metall. Mater. Transa., 29B (1998) 205
64. E. Orowan, *Internal Stresses in Materials Alloys*, Institute of Metals, 1948
65. M.F. Ashby, Phil. Mag., 14 (1966) 1157
66. R.W. Heine, R. Carl, J. Loper, and P.C. Rosenthal, *Principles of Metal Casting*, McGraw-Hill, 1967
67. J.P. Dennsion, and E.V. Tull, J. Inst. Metals, 85 (1956) 1712
68. A. Bilyk, class report, Northwestern University, 1989
69. G.B. Olson, U.S. Patent, 5,221,372, 1993
70. D.J. Cook, and G.A. Lewis, Class report, Northwestern University, 1994
71. C.E. Campbell, and G.B. Olson, J Comput-Aided Mater Des, 2001

Transformations—Selected Works of G.B. Olson on Materials, Microstructure, and Design
C.E. Campbell, M.V. Manuel, and W. Xiong, editors

CHAPTER 17

Systems Design of High-Performance Stainless Steels—II. Prototype Characterization*

C.E. Campbell and G.B. Olson
Northwestern University

Abstract

Within the framework of a systems approach, the design of a high-performance stainless steel integrated processing/structure/property/performance relations with mechanistic computational models. Using multicomponent thermodynamic and diffusion software platforms, the models were integrated to design a carburizable, secondary-hardening, martensitic stainless steel for advanced gear and bearing applications. Prototype evaluation confirmed the predicted martensitic transformation temperature and the desired carburizing and tempering responses, achieving a case hardness of R_c 64 in the secondary-hardened condition without case primary carbides. Comparison with a commercial carburizing stainless steel demonstrated the advantage of avoiding primary carbides to resist quench cracking associated with a martensitic start temperature gradient reversal. Based on anodic polarization measurements and salt-spray testing, the prototype composition exhibited superior corrosion resistance in comparison to the 440C stainless bearing steel, which has a significantly higher alloy Cr concentration.

Introduction

A systems approach (Ref 1) to computational materials design was used to design a prototype high-performance stainless bearing steel. The first three steps of system design — establishment of the property objectives, the model development, and integration and optimization of the design models — were described in Part I (Ref 2; chapter 16 in this volume). Figure 1 illustrates the interconnections between the desired property objectives, structure, and processing for the design of a carburizing secondary-hardening, martensitic stainless steel. The next step of the systems design approach is the characterization of the design composition to determine whether the desired property objectives were obtained. Experimental evaluation of the design prototype also identifies the strengths and weaknesses of the design models applied. That is, if the prototype does not meet the required property objectives, then either the desired property objectives must be reevaluated or the design models must be improved.

This work describes the characterization of the prototype carburizing, secondary-hardening, martensitic stainless steel designed in Part I (Ref 2) and verifies whether the established property objectives were obtained. These objectives include a lath martensitic microstructure obtained with a M_s temperature greater than 150 °C, a carburized case that has a surface hardness exceeding R_c 62 and that does not include the formation of primary carbides, an aqueous corrosion resistance

*C.E. Campbell and G.B. Olson, "Systems Design of High Performance Stainless Steels: II. Prototype Characterization," *J. Computer-Aided Mat. Des.* 7 (2001) p 171–194

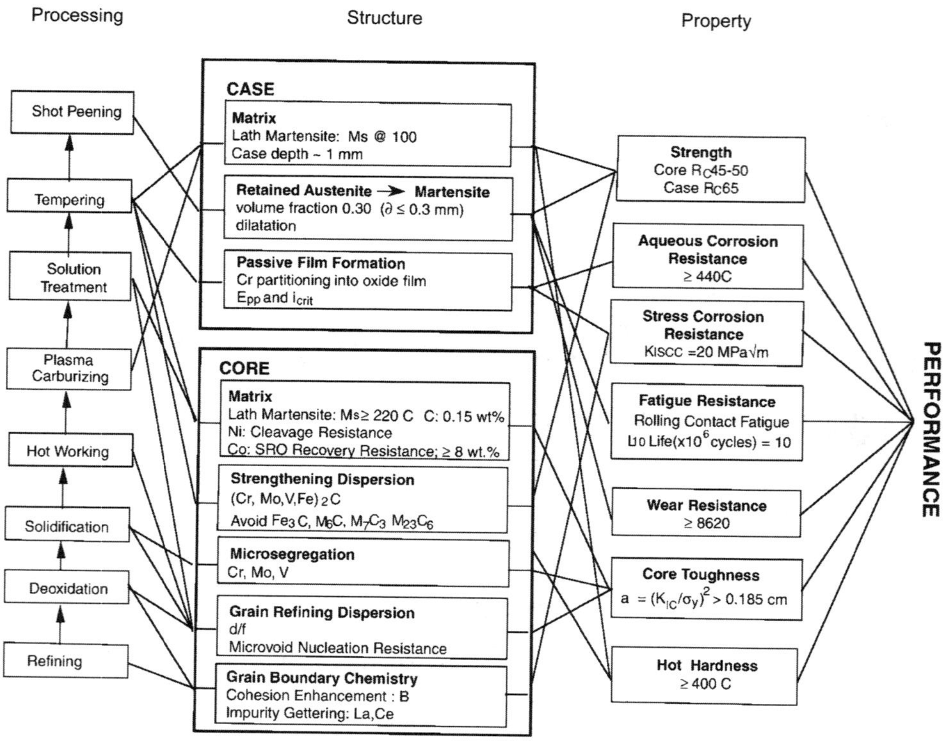

Fig. 1 Flow-block diagram of the processing/property/structure/performance interactions for a carburizing secondary-hardening stainless steel.

exceeding that of 440C, a rolling contact fatigue resistance equilvalent to high temperature bearing alloys M50 and M50NiL, and a core toughness twice that of 440C. The characterization of this prototype includes comparisons to commercial stainless steels and various experimental alloys.

Materials and Methods

Materials

An experimental, 22.68 kg (50 lb), vacuum induction melted heat of the design composition, CS62 (Ref 3), was produced by Republic Steel using standard Al and Ti deoxidation practices. Boron additions were made to improve grain boundary cohesion and cesium additions were made to getter sulfur and phosphorous impurities. After solidification, the ingot was reheated to 1232 °C (2250 °F) and forged to a 0.0584 m (2.3 in) square bar. The forged bar was immediately reheated to 1232 °C and reduced to a 0.033 m (1.3 in) square bar. The chemical analysis of the studied experimental heat of CS62 is given in Table 1. In general, the chemical analysis of the experimental composition corresponds to desired composition for the prototype. However, the oxide content at 0.005 wt% O and the aluminum content at 0.008 wt% are higher than expected in comparison to other commercial alloys. These compositions deviations may have an adverse effect on the material properties, especially the fracture toughness.

In characterizing the design alloy, the performance of the prototype was compared to two other alloys: the model secondary-hardening steel 1605-8B (Ref 4) and the commercial carburizing martensitic stainless steel, Pyrowear 675 (Ref 5). The 1605-8B alloy was produced by NKK Corporation using vacuum induction processing. Carpenter Technology provided rod material from an experimental heat of Pyrowear 675.

Experimental Methods

Dilatometry was used to study the martensitic transformation using a computer-controlled dilatometer. Cylindrical samples, 3 mm in

Table 1 Alloy compositions (wt%).

Element	Design Prototype (CS62)	AerMet100	AF1410	1605-8B	NASA1	Pyrowear 675
Fe	balance	balance	balance	balance	balance	balance
Co	15.51	13.10	14.25	15.8	22.53	5.4
Ni	1.56	11	10.5	4.88	8.48	2.6
Cr	9.13	3.0	2.1	7.87	11.75	13.0
Mo	0.002	1.18	1.05	1.5	0.3	1.8
V	0.21	NA	NA	NA	0.25	0.6
Si	<0.0002	0.04	NA	NA	0.01	0.4
Mn	0.02	0.01	NA	NA	<0.01	0.65
C	0.21 (core)	0.24	0.16	0.248	0.29	0.07 (core)
Al	0.008	0.001	0.005	...	...	...
B	0.0023	...	...	...	...	...
Ce	0.015	...	...	...	<0.001	...
La		...	...	...	<0.001	...
N	0.002	...	...	...	...	...
O	0.005	...	...	...	...	...
P	0.001	0.005	...	...	0.005	...
S	0.002	0.006	...	...	0.003	...
Ti	0.011	0.02	...	...	...	...

diameter and 10 mm in length, were heated in an induction furnace and were quenched using helium gas. The temperature was monitored by a Pt-Pt10%Rh thermocouple spot welded directly to the sample surface. The dilatometry experiments were conducted under a vacuum of $< 10^{-2}$ Pa. The thermal expansion and contraction were recorded as functions of temperature as the samples were heated at 500 °C/ minute to a temperature above the austenitic finish temperature, A_f, then held at this temperature for 15 min, and finally quenched to room temperature. The austenitic and martensitic start (A_s, M_s) temperatures were identified by the discontinuous length changes.

Bulk and surface heat treatments were performed to produce the desired microstructures. The typical sample size for such treatments was a 0.0125 m cube. The solution treatments were performed with the material either encapsulated in quartz tubing or in an furnace backfilled with argon. Tempering to precipitate the strengthening dispersion was also performed in an inert environment. Using a hydrogen furnace, decarburizing experiments were conducted using polished samples of dimensions 2.5 mm × 5 mm × 20 mm.

Macro-hardness measurements were made using a Rockwell C tester on samples polished to a 600 grit finish. A Vickers indenter with a 100 g load was used to measure the microhardness on samples polished to a 1 μm finish.

Plane-strain fracture toughness testing was conducted following the ASTM standard E399 (Ref 6) for K_{IC} testing. The machined compact tension samples were oriented longitudinally with respect to the forging axis and had a thickness of 1.27 cm (0.5 in). These samples were polished to a 6 μm finish so that crack propagation could be observed during precracking. Testing was performed using a servo-hydraulic load-frame machince with a digital controller. The load line displacement was obtained using a clip gauge, which has a gauge length of 0.254 cm (0.1 inch) and a range of 0.381 cm (0.15 inch). The compact-tension samples were fatigue precracked at 20 Hz frequency with a stress ratio (*R* ratio) of 0.1. The peak load was established at 4.89 kN (1.1 kips) and maintained until crack initiation was detected. Following precracking, side grooves of a depth of 20% of the total width, were added to the samples. The side-grooved samples were then loaded at a cross-head speed of 0.0013 cm s^{-1} (0.0005 in s^{-1}) until fracture occurred. Two samples were tested for each overaged heat treatment condition.

Additional three-point bend measurements were made on subsize samples using a screw-driven machine. The dimensions of the sub-sized three-point bend samples were determined by the maximum decarburizing sample size. The corresponding dimensions of the samples were 20 mm in length, 5 mm in width, and 2.5 mm in thickness. Electronic discharge machining (EDM) was used to cut a starter notch, with a width of 0.025 mm and a length of 1.125 mm, at the center of the 20 mm × 2.5 mm face. These samples were fatigue precracked at 10 Hz frequency at a stress ratio (*R* ratio) of 0.1. The samples were tested using a cross-head speed of 0.00083 cm s^{-1}.

Optical and scanning electron microscopy (SEM) were employed to characterize the microstructures. Optical microscopy determined the general microstructure and the grain size, as well as variations in microstructures produced by composition gradients. The samples were polished a 1 μm surface finish. A sodium-metabisulfite etchant (167 ml H_2O, 33 ml HCl, 0.5 g $Na_2S_2O_5$) was used to revealed the martensitic structure within the prior-austenitic grains. A 10% Nital etchant was used to investigate the carbide structures. The fracture surfaces were characterized using a cold field emission scanning electron microscope (cFEG-SEM) with a energy dispersive spectroscopy (EDS) detector.

The corrosion behavior was evaluated using anodic polarization measurements. An electrode was soldered to the sample before mounting in epoxy. The exposed surface area was polished to a 1 μm and measured. A computer-controlled potentiostat was used to measure the anodic polarization curves in derated solutions. For consistency with the previously obtained polarization data (Ref 7), the materials were tested in aqueous solutions of 1% sucrose (for conductivity) and 3.5% NaCl with pH = 7. The free corrosion potential was established by allowing the system to reach a steady-state conditions after sitting for 30 min. The anodic potential was scanned at a sweep rate of 2 mV s^{-1} continuously recording potential and current.

Salt-spray corrosion testing was conducted following the ASTM standard B117 (Ref 8). Before testing, the samples were weighed to the nearest 0.0001 g and the exposed surface area was measured. The samples were polished to a 600 grit finish and cleaned with acetone. Then the samples were suspended in a fog chamber for a total of 48 h and were exposed to cycles of alternating environmental conditions of 2 h at 20 °C with a 5% salt spray solution and then 1 h dry heat at 30 ± 1 °C (86 ± 1.8 °F). After exposure, the samples were again weighed to determine a corrosion rate and the surface corrosion products were noted. The corrosion rate, R, is calculated using the expression:

$$R = \frac{8.7610^4 w}{\text{mm yr}^{-1}} \quad \text{(Eq 1)}$$

where w is the measured weight loss in g, ρ is the density of the specimen in g cm^{-3}, a is the surface area exposed in cm, and t is the exposure time in hours. The presented corrosion rates serve as a preliminary analysis of the corrosion behavior of the materials as only two data points were collected for each sample.

Characterization of Prototype (Results)

The first step in evaluating the new alloy (CS62) was the characterization of the core material to determine the correct processing conditions of the forged material. While the case material properties are more important than the core properties, the correct processing of the core material must be established before carburizing the material to evaluate the case material properties and processing conditions. To assist in the characterization of the prototype, the Thermo-Calc (Ref 9) and DICTRA (Ref 10) software programs helped correlate observed microstructures to processing and material properties and aided in process optimization.

Core Material

The measurement of the M_s temperature is the first step in the core material evaluation. Dilatometry determined a M_s temperature of 291 °C ± 15 °C, and an A_s temperature of 761 °C ± 10 °C. The temperature at which the free energy of the FCC phase equals the free energy of the BCC phase, the T_0 temperature, was calculated 16 degrees lower than the experimental T_0 of 521 °C. Figure 2 presents the recorded dilatometry curves, along with the optical microscopy of the quenched microstructure revealing the desired martensitic structure. In addtion, the observed microstructure showed no evidence of large segregation ratios, which would have resulted in bands of the martensitic laths corresponding to the composition variations.

Previous work (Ref 11) showed that optimizing the solution treatment temperature (T_s) maximizes the fracture toughness of an alloy. The optimal 1 h T_s is determined by the processing that produces the maximum hardness for the smallest grain size. When all of the carbon is in the FCC solid solution the hardness, measured after quenching, will level off with increasing T_s, but the grain size will continue to increase. The solution treatments were followed by an oil and liquid N_2 quenching to produce a martensitic transformation. The material was then tempered at 200 °C for 1 h (Stage I) to reduce quenching variations effects. Figure 3(a) shows the hardness dependence on temperature and Fig. 3(b) depicts the corresponding grain size dependence. After reaching a peak hardness of R_c 51.3 for a T_s of 1050 °C, the hardness leveled off with increasing T_s, suggesting the full dissolution of the carbides. At the optimal T_s of 1050 °C, the grain size is 30 μm, using the mean linear intercept method. For comparison, the commercial carburizing stainless steel, Pyrowear 675, has a solution treatment of 1050 °C for 1.5 h and the prototype 1605-8B alloy has a solution treatment of 1050 °C for 1 h. The optimized solution treatment for CS62 is also similar to that of an earlier stainless prototype, NASA1, which had an optimized solution treatment of 1150 °C for 1 h. The higher optimal solution temperature for the NASA1 alloy is attributed to its higher carbon (0.30 wt% C) and vanadium contents (0.25 wt% V) (Ref 7).

The secondary-hardening tempering response was investigated at three different temperatures: 510 °C, 482 °C, and 450 °C. At 510 °C only monotonic softening was observed, as demonstrated in Fig. 4. A broad peak hardness level at R_c 53.5 was observed at 482 °C. At 450 °C, peak hardness was not established after 24 h of tempering. Consequently, 482 °C, the assumed tempering temperature in the design calculations (Ref 2), was confirmed as an acceptable tempering temperature.

The fracture toughness, determined by K_{IC} testing, was characterized as a function of tempering time at 482 °C. Two slightly overaged conditions were chosen: 16 h and 22 h. Both conditions failed to maintain stable crack growth during precracking: unstable crack growth was initiated after the fatigue precrack had grown to 75% of its desired length. Although a valid K_{IC} could not be ascertained

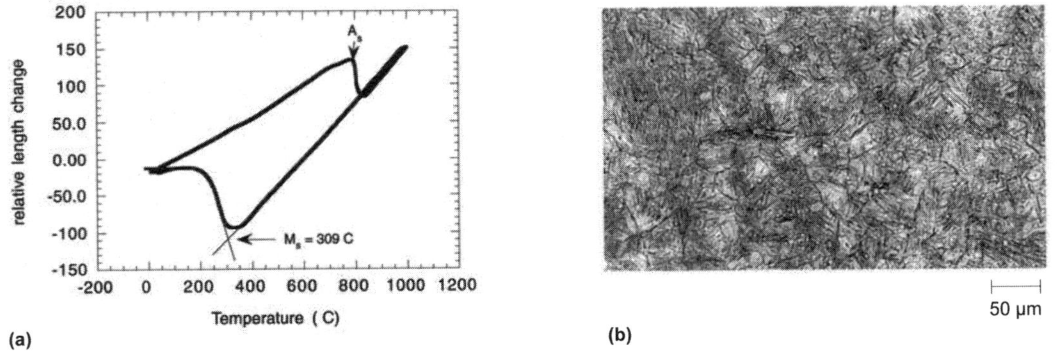

Fig. 2 (a) Dilatometry curve representing the martensitc transformation. (b) Optical micrograph of the lath martensitic microstructure produced upon quenching from a T_s of 1050 °C.

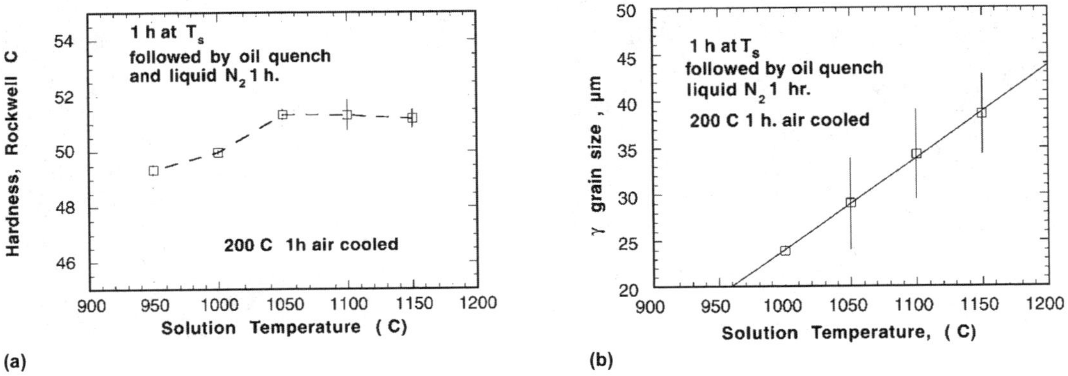

Fig. 3 (a) Hardness as a function of solution temperature. (b) Grain size as a function of solution temperature.

for either the 16 h or 22 h tempered conditions, the fracture surfaces were evaluated to assess the fracture mechanisms and to estimate the toughness levels. The validity of the fracture surface evaluation to determine the active fracture mechanism in plane strain is based on the previous work developing the Stretch Zone Width (SZW) technique (Ref 12–14). It was observed that the fracture surface just ahead of the SZW is indicative of the initial fracture mechanism under plane strain constraints. SEM examination of the fracture surfaces, just beyond the fatigue precrack, reveal quasi-cleavage surfaces with signs of intergranular embrittlement, Fig. 5. The intergranular facets have a diameter of approximately 30 μm, which corresponded to the average grain size achieved after the 1050 °C solution treatment.

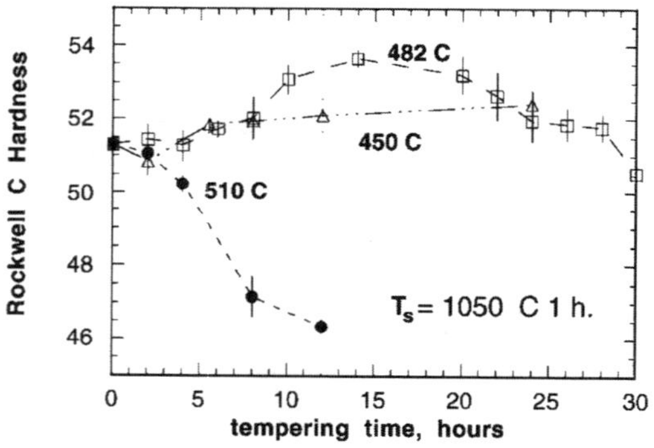

Fig. 4 Tempering response at 510 °C, 482 °C, and 450 °C.

The underlying brittle failure modes achieved using secondary-hardening conditions did not meet the design objectives for the ductile fracture criteria. This was not completely unexpected. As noted earlier the processing of the alloy was not as clean as desired, and the core composition was designed that the carbon content could be adjusted to achieve the desired toughness without significantly impacting the case microstructure design. Further investigation of the fracture mechanisms as a function of heat treatment and composition was worthy of study to determine how future modeling efforts could be improved.

Evaluation of the fracture mechanisms as a function of heat treatment include investigation of the stage I tempering conditions. After tempering for 1 h at 200 °C, a K_{IC} of $68 \pm 1.5\,\mathrm{MPa}\sqrt{\mathrm{m}}\left(62 \pm 1.5\,\mathrm{ksi}\sqrt{\mathrm{in}}\right)$ at R_c 52 was measured and a ductile fracture surface was observed, as seen in Fig. 6. This result at R_c 52, which is only 1 R_c point below the secondary-hardening conditions examined, indicated that with a fine ε-carbide strengthening dispersion the alloy has sufficient Ni content to resist cleavage fracture at room temperature. It should be noted in Fig. 6 that there are some large voids, approximately 25 μm in diameter, which can be attributed to oxide inclusions resulting from the higher than desired oxygen content in the experimental heat.

The reduction in fracture-toughness associated with secondary hardening is most likely the result of the loss of grain boundary cohesion. Spaulding (Ref 15) evaluated the P and

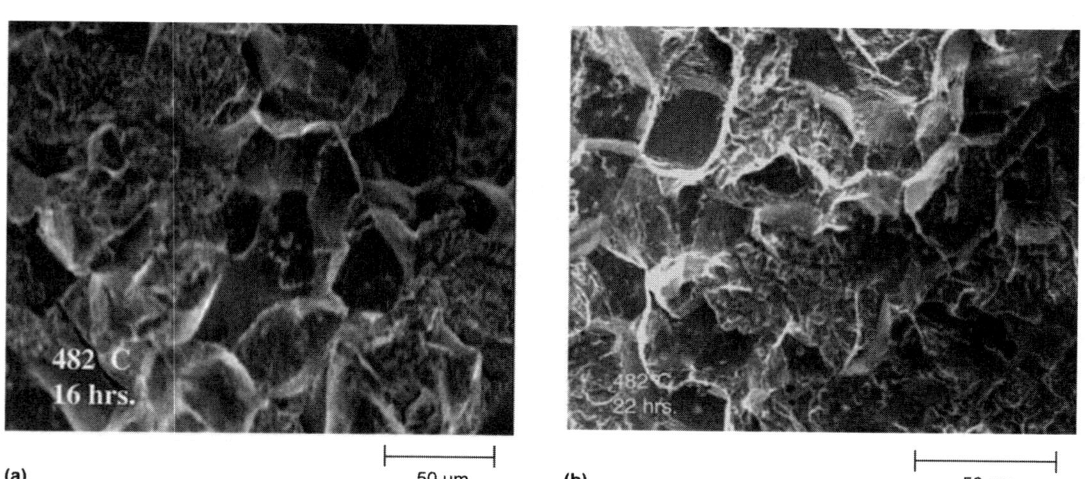

(a) (b)

Fig. 5 SEM micrographs of the fracture surfaces observed after (a) 16 h and (b) 22 h of tempering at 482 °C.

S segregation and decarburization embrittlement during secondary hardening of the 1605 model alloys and a prototype gear steel. The study concluded that the critical Griffith work for interfacial separation depended on the hardness level and demonstrated that ductile fracture modes could be obtained with reduced carbon levels. Thus, to determine the necessary reduction in alloy hardness to change the fracture mechanism in the secondary-hardened condition, decarburization studies were conducted.

Based on DICTRA simulations (Ref 16), decarburization experiments were performed on CS62 and 1605-8B in wet hydrogen for 8 h, 10 h, and 20 h. Following decarburization, the final vacuum homogenization treatments were also guided by DICTRA simulations. For the 8 h decarburized samples, homogenization was completed at 1100 °C for 10 h. The 10 h and 20 h decarburized samples were homogenized for 20 h at 1050 °C. Following solution treatment and tempering (see Table 2 for specific conditions), the subsized three-point bend samples were precracked and fractured.

The results of the fracture testing were evaluated with the understanding that the fixed subsized specimen geometry could result in varying crack-tip stress states with decreasing hardness levels, making fracture analysis difficult. In addition to the constraints of the subsize specimen geometry, the fracture-toughness analysis was limited by the higher than desired oxide content. The following analysis is intended to provide a qualitative description of the fracture mechanisms present in the lower carbon version of the alloy. It is accepted that for the fixed sample geometry, as the crack grows in a ductile material the plastic zone may exceed the sample width and plane strain conditions will not be maintained. However, the initial crack will form under plane strain conditions, so that fractography at the beginning of stable crack growth will be representative of the initial acting fracture mechanism in plane strain.

After 8 h of decarburization, both CS62 (480 Vickers Hardness Number (VHN)) and 1605-8B (373 VHN) had quasi-cleavage fracture surfaces. Ten and 20 h of decarburization reduced the carbon content of 1605-8B enough to produce a ductile fracture at less than 290 VHN. Unfortunately, 20 h of decarburization was not sufficient to produce a ductile fracture in CS62 with a 437 VHN. However, the fracture surface in Fig. 7 reveals some evidence of a more ductile fracture mode with little evidence of intergranular fracture. The microhardness profiles across the decarburized samples were measured and the strengthening model (Ref 17) was used to estimate the carbon level. The relatively flat hardness profiles indicated that the homogenization treatments were successful. The hardness levels, estimated carbon contents, and estimated K_{IC} values are given in Table 3.

Due to the small size of the samples, the estimated K_{IC} values require scaling to better represent values obtained from standard size samples. Using the work of Bergstrom (Ref 18) and calibrating the K_{IC} measured on the original CS62 material with a standard thickness (K_{IC} standard = 68 MPa$\sqrt{m}$: K_{IC} subsize = 132

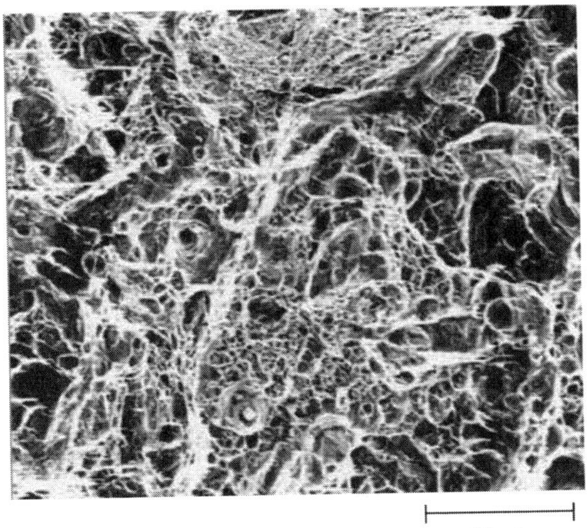

Fig. 6 SEM micrograph of the ductile fracture surface produced after tempering at 200 °C for 1 h.

Table 2 Heat treating conditions for the decarburized samples.

Alloy	Homogenization	Solution treatment	Tempering
CS62 – 0 h	NA	1050 °C 1 h	200 °C 1 h
CS62 8 h	1100 °C 10 h	1050 °C 1 h	510 °C 2 h 482 °C 6 h
CS62 – 10 h	1050 °C 20 h	1100 °C 0.5 h	510 °C 2 h 482 °C 6 h
CS62 – 20 h	1050 °C 20 h	1100 °C 0.5 h	510 °C 2 h 482 °C 6 h
1605-8B 0 h	NA	1050 °C 1 h	200 °C 1 h
1605-8B 8 h	1100 °C 10 h	1050 °C 1 h	510 °C 2 h 482 °C 6 h
1605-8B 10 h	1050 °C 20 h	1100 °C 0.5 h	510 °C 1 h
1605-8B 20 h	1050 °C 20 h	1100 °C 0.5 h	510 °C 1 h

MPa$\sqrt{m}$), the experimental K_Q values were scaled using a factor of 0.5 to predict a more representative K_{IC} estimate for screening purposes. Figure 8(a) shows the measured hardness decreasing as a function of decarburization time. From Fig. 8, it is estimated that 45 h of decarburizing at 1100 °C is required to reduce the carbon level of the design composition to 0.05 wt% C, and thus produce a ductile fracture condition after secondary hardening.

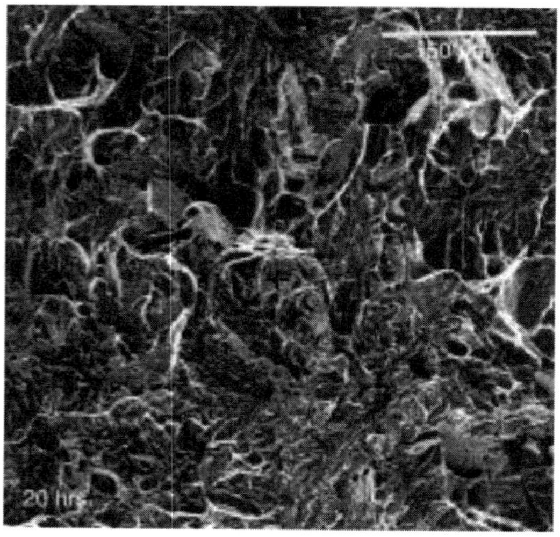

Fig. 7 Fracture surface of CS1 decarburized for 20 h at 1100 °C (0.14 wt% C).

Table 3 Results of decarburization experiments.

Alloy and condition (h of decarb.)	Hardness (VHN)	Estimated C, wt%	Measured K_Q	Estimated K_{IC} MPa$\sqrt{m}$
CS62 – 0 h	590	0.21	132	66
CS62 – 8 h	480	0.17	NA	NA
CS62 – 10 h	473	0.17	NA	NA
CS62 20 h	437	0.14	55.5	27.7
1605-8B – 0 h	585	0.24	124	62.1
1605-8B 8 h	373	0.10	87.4	43.6
1605-8B – 10 h	290	0.04	140	70(a)
1605-8B – 20 h	287	0.04	122	61(a)

(a) Estimated value; unable to break samples using Sintech.

Combining the results of the various investigations of the fracture mechanisms resulting from secondary-hardening treatments, a mechanism map of the dominating fracture modes defined by the carbon content and hardness was constructed, as seen in Fig. 9. The data represented on the map included the decarburization experiments on CS62 and 1605-8B and the overaging experiments of CS62. The fracture map is divided into three regions. Above a hardness level of 525 VHN intergranular fracture dominates. Intergranular fracture also occurs at low hardness levels, to the right of the line:

$$\text{VHN} = 1527 \cdot (\text{wt\% C}) + 175 \qquad \text{(Eq 2)}$$

To the left of the intergranular boundary and between the hardness levels of 375 VHN and 575 VHN is a region of quasi-cleavage. At hardness levels below 375 VHN and carbon contents to the left of the intergranular boundary (≈< 0.10 wt% C), ductile fracture mechanisms prevail.

The two regions of intergranular fracture indicate two different mechanisms. Intergranular fracture occurring at high carbon and hardness is represented by standard overaged tempering treatments of 1605-8B and CS62. Spaulding's work (Ref 15) showed that

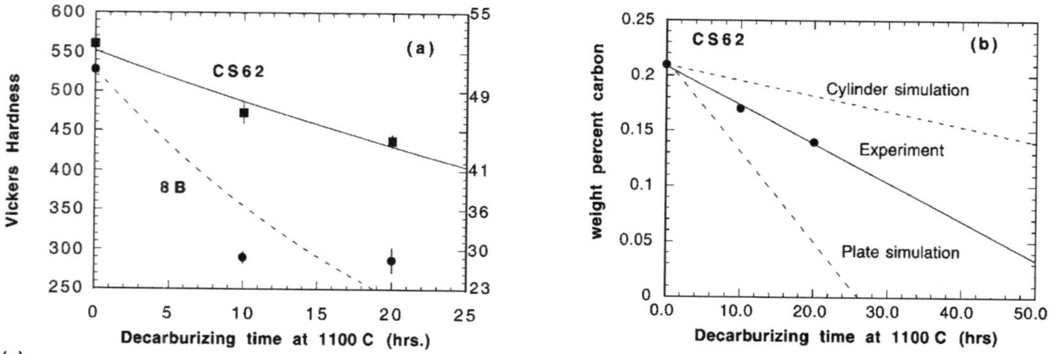

(a) (b)

Fig. 8 (a) Hardness as a function of decarburizing time a 1100 °C for alloys 1605-8B and CS62. (b) Alloy carbon content as a function of decarburizing time for CS62. Experimental carbon concentration estimates are compared to the predictions of the DICTRA simulations for the plate and cylinder geometries.

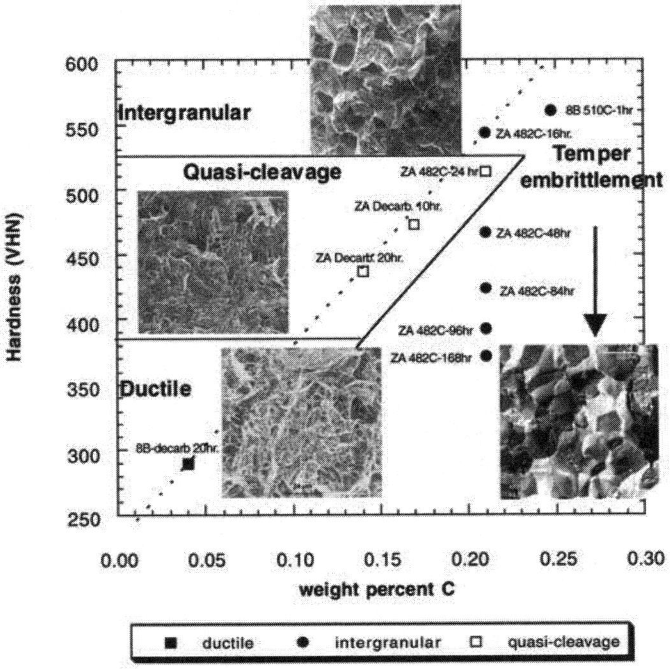

Fig. 9 Fracture mechanism map defining the region of intergranular, quasi-cleavage, and ductile fracture by the carbon content and hardness. The fracture surfaces are representative of the different fracture modes.

depletion of cohesion-enhancing C at the grain boundaries can cause for intergranular fracture at hardness levels greater than 450 VHN. The lower region of intergranular fracture is characterized by the long time overaging experiments on CS62, where the segregation of P and S impurities is likely to be the cause of the lost grain boundary cohesion.

The fracture mechanism map of Fig. 9 suggests that to achieve a ductile fracture mode using secondary hardening, the carbon content of the prototype composition must be decreased to approximately 0.05 wt% C. Such a lower carbon content alloy is currently under evaluation. The upper region of intergranular fracture indicates that fracture surface analysis of the grain boundaries is needed to determine if the boron additions, which were added to improve grain-boundary cohesion, are present, or if boron was oxidized during processing.

Case Material

Optimal carburizing would produce 0.48 wt% surface C with a 1 mm case depth. Using industrial gas-carburizing practices performed by Caterpillar, preoxidation at 900 °C was followed by gas-carburizing for 5 h at 927 °C at nominal 1.0 surface carbon potential to produce a 1.6 wt% surface carbon with 1.5 mm case depth, as shown Fig. 10. The higher than nominal surface carbon content is predicted for the high Cr level. To produce a case without primary carbides the surface carbon was reduced. The computed equilibrium carbon isopleth indicates that at 927 °C M_7C_3 carbides are present. The predicted carbon distributions in the FCC matrix at the carburizing and solution treatment temperatures are shown in Fig. 11 and the corresponding phase fields are shown in Fig. 12.

Diffusion treatments were used to reduce the surface carbon and to dissolve the primary carbides. DICTRA simulations were again used as a guide in determining the best treatment. At 1175 °C, the carbon solubility in the FCC phase is 0.7 wt% C. Using the measured carbon profile after carburizing and assuming an equilibrium spheroidal M_7C_3 carbide distribution, Fig. 13 presents the DICTRA simulations of the austenite carbon diffusion process at 1175 °C in the two-phase field. The simulation predicts that after 2 h a 0.65 wt% surface carbon level should be achieved and the primary

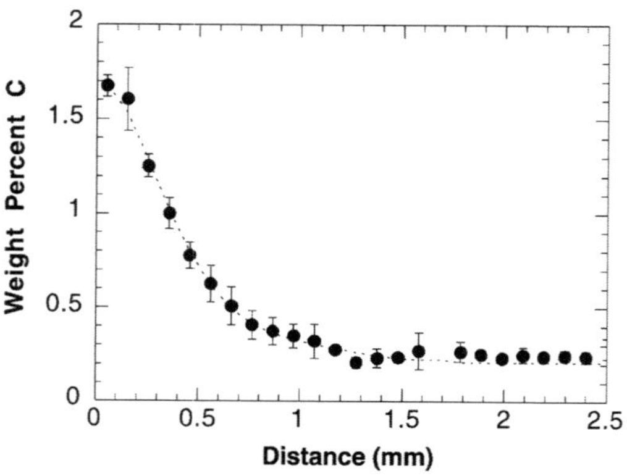

Fig. 10 Measured carbon profile after initial carburizing at 927 °C for 5 h at nominal 1.0 surface carbon potential.

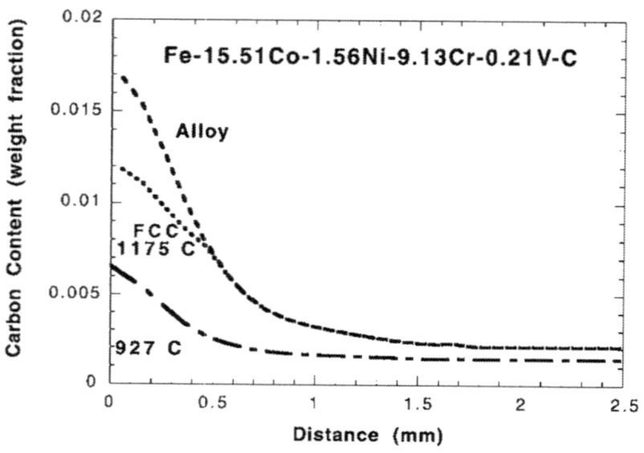

Fig. 11 Computed carbon distribution in matrix at 927 °C and 1175 °C assuming equilibrium with the M_7C_3 carbide.

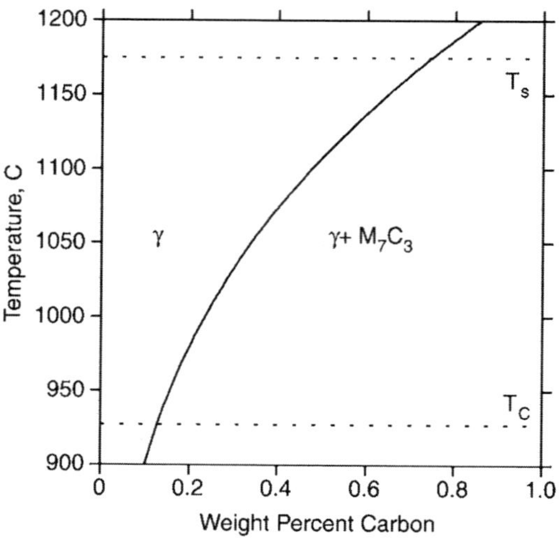

Fig. 12 Carbon isopleth for CS62. Dashed lines denote the carburizing (T_c) and solution (T_s) temperatures.

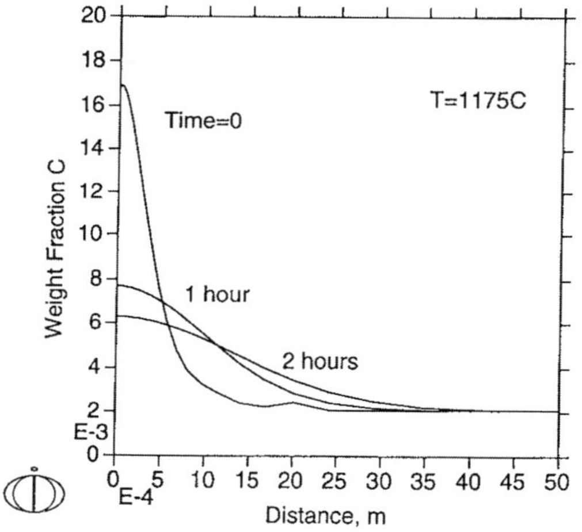

Fig. 13 DICTRA simulation at 1175 °C for the diffusion of the carbon distribution produced after carburizing at 927 °C.

carbides would be dissolved at equilibrium. This diffusion step was performed followed by oil and liquid N_2 quenching and a 200 °C 1 h temper. The resulting hardness profile with a surface hardness of 790 VHN is shown in Fig. 14(a). Figure 14(b) show the hardness profile after secondary hardening at 482 °C for 18 h and indicates a measured surface hardness of 816 VHN. Secondary hardening produced slightly higher case and core hardnesses. Both profiles show a large case depth (≈3 mm) due to the long carbon diffusion step.

Optical microscopy of the case microstructure confirms a martensitic matrix, as seen in Fig. 15. The amount of retained austenite and plate martensite increases toward the surface. Microscopy also shows the presence of some primary carbides (M_7C_3) in the first 250 μm of the case, indicating that equilibrium at 1175 °C was not fully established.

Although the measured case hardness met the original property objective, detailed analysis of the hardness and carbon profiles shows quantitative discrepancies with the design model predictions. A maximum hardness of 816 VHN is obtained at a case depth of 200 μm with the presence of some primary carbides. The maximum hardness achieved without primary carbides is 800 VHN at 300 μm. The DICTRA carburization simulation predicted a 0.625 wt% C composition at a case depth of 200 μm and 0.62 wt% C composition at 300 μm. To evaluate the design strengthening model, the DICTRA simulations were used to determine the

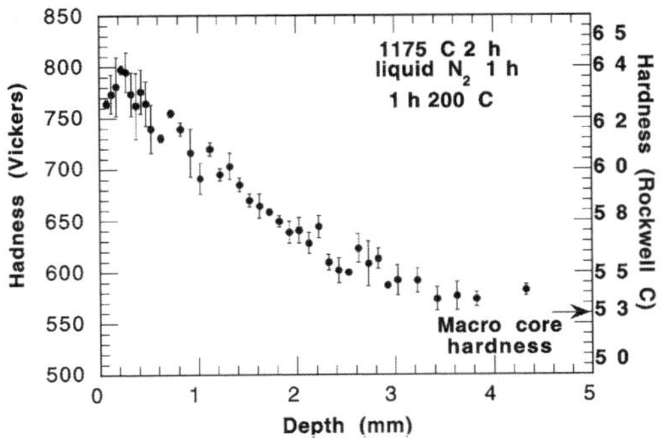

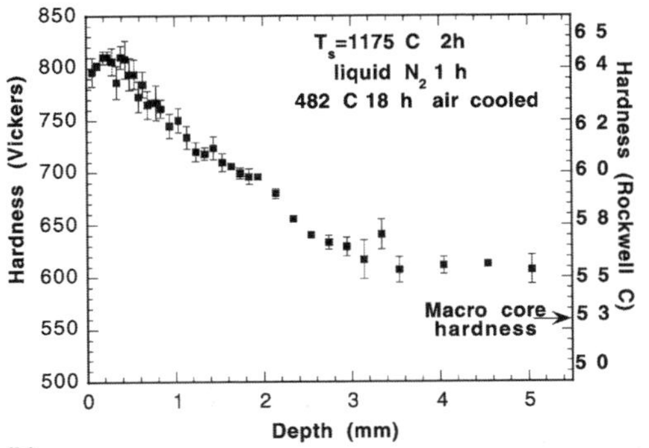

Fig. 14 Hardness profiles after (a) after Stage I tempering at 200 °C for 1 h and (b) after secondary-hardening at 482 °C for 18 h.

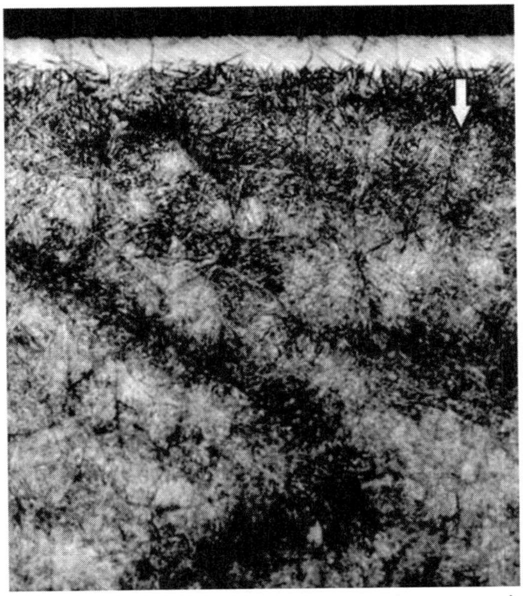

Fig. 15 Optical micrograph of the case microstructure after 2 h at 1175 °C followed by 18 h at 482 °C. Arrow denotes direction of carbon diffusion.

case depth at which the design carbon content (0.48 wt%) was obtained. The hardness at this case depth of 1.25 mm is 775 VHN, 50 points below the design value of 825 VHN. The slightly lower than desired hardness can be attributed to the M_2C driving force being lower than the original target value, producing a larger M_2C particle size with less efficient strengthening.

The fine size scale associated with the alloy M_2C carbides precipitated during secondary hardening should provide higher resistance to fatigue and wear corrosion compared to alloys that use primary carbides as a strengthening dispersion. For comparison with such an alloy, the commercial Pyrowear 675 alloy was studied. Initial carburizing of the primary carbide strengthened alloy using pack-carburizing highlights a large difference in alloy processability. Despite using various slow cooling methods from the recommended carburizing temperature of 871 °C a crack-free surface could not be obtained, Fig. 16. Modeling indicates that the surface cracks were generated from the residual tensile stresses produced by a M_s profile that reverses directions with increasing carbon, Fig. 17(a). For comparison, the desirable montonic M_s temperature profile for CS62 is shown in Fig. 17(b). For the Pyrowear 675 alloy, the combination of low-matrix carbon and high-matrix Cr results in an increasing M_s temperature as the carbon content increases. The additional alloy C from carburizing removes Cr from the matrix to form M_7C_3 carbides. This decrease of matrix Cr increases the M_s temperature. Figure 18 depicts the C isopleth of this primary carbide strengthened alloy and shows that both the carburizing and solutionizing temperatures correspond to multiphase regions.

Case cracking of the primary carbide strengthened alloy was avoided when a more controlled carburizing method was utilized. Gas-carburizing for 24 h at 871 °C at a nominal 0.7 surface carbon potential, by MRC Bearing Corp., produced a crack-free case. Both carburizing processes were followed by the standard solutionizing (1050 °C for 1.5 h) and tempering (510 °C for 4 h) procedures for the alloy. This heat treatment and carburizing process produced a hardness of R_c 60 at case depth of 0.11 cm and a R_c 50 at 0.19 cm.

Anodic polarization measurements in 1% sucrose and 3.5% NaCl aqueous solutions were made on the core material of Pyrowear 675 and CS62 and the carburized CS62. The measurements were compared to previous measurements made on 440C and the NASA1 prototype stainless bearing steel (Ref 7). Figure 19(a) compares the anodic polarization resistance of the various monolithic alloys and Fig. 19(b) compares the core and case material of CS62. The core material of CS62 behaved similarly to the core material of Pyrowear 675. Pyrowear 675, NASA1, and the case and core material of CS62 all surpass the anodic polarization curve of 440C. Figure 19 (b) shows the remarkable result that carburization of CS62 causes only a slight reduction in the corrosion resistance compared to the core material. This suggests that the fine-scale strengthening dispersion does not remove a signifcant concentration of Cr from the matrix and thus, the Cr is still available to contribute to passive film formation.

Fig. 16 Case-cracking observed after pack-carburizing Pyrowear 675.

With testing in the 3.5% NaCl solution, Fig. 20, NASA1 exhibits the superior resistance. The 440C alloy showed no distinguishable evidence of passivation, as evident by the shallow slope of the potential versus current line, in contrast to the behavior of Pyrowear 675 and CS62 alloys. At high potentials where pitting occurs, the Mo-bearing Pyrowear 675 has slightly better performance than CS62. However, both Pyrowear 675 and CS62 show only very small passivating regions and further experimental work is needed to clarify the electrochemical reactions. Carburized Pyrowear 675 was not tested as a crack-free surface was not available at the time of testing. Due to the high content of primary carbides in the carburized condition, significant degradation of the corrosion resistance is expected.

Salt-spray testing illustrated the superior performance of the prototype composition as seen in the measured corrosion rates, Table 4. The guidelines given by Fontana (Ref 19) for corrosion rates in ferrous and nickel-based alloys hold that corrosion rates <0.0254 mm yr^{-1} are "outstanding" and rates between 0.0254 mm yr^{-1} and 0.127 mm yr^{-1} are "excellent." The carburized sample of Pyrowear 675 test is affected by case cracking. It was also noted that the surface of carburized Pyrowear 675 was covered with black oxide

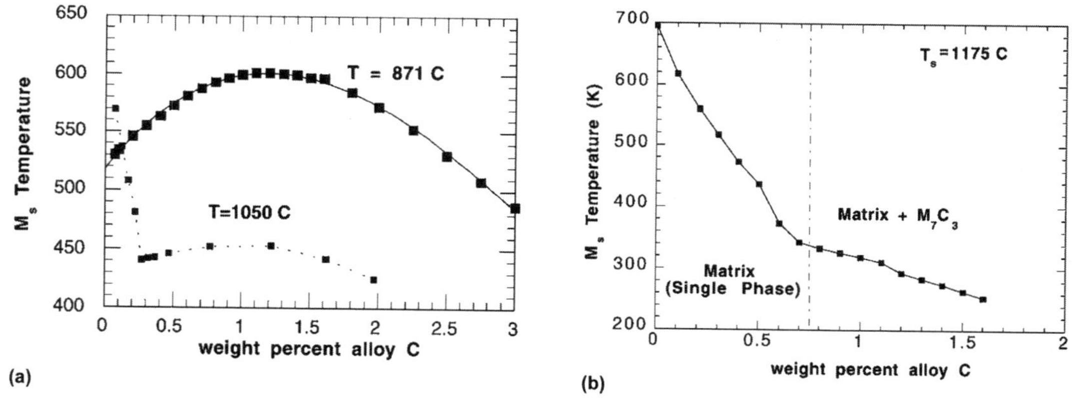

Fig. 17 (a) Predicted M_s temperature distribution corresponding to the Pyrowear 675 matrix carbon content at the carburizing and solutionizing temperatures. (b) Predicted M_s temperature distribution for CS62 at the solution temperature of 1175 °C.

scale after testing. No such scale was formed on the carburized CS62.

Initial rolling contact fatigue (RCF) testing using a three ball-on-rod method indicates that at a contact stress of 5.42 GPa (786 ksi) carburzied CS62 has a fatigue life equivalent to M50, a commerical high-temperature monolithic bearing steel. Wells et al., (Ref 20) performed similar RCF testing using the same contact stress of 5.42 GPa and compared the M50 alloy to carburised Pyrowear 675. With an optimal heat treatment, the B50 lifetime (Weibull distribution for 50% accumulated failure) of Pyrowear 675 was five times greater than M50.

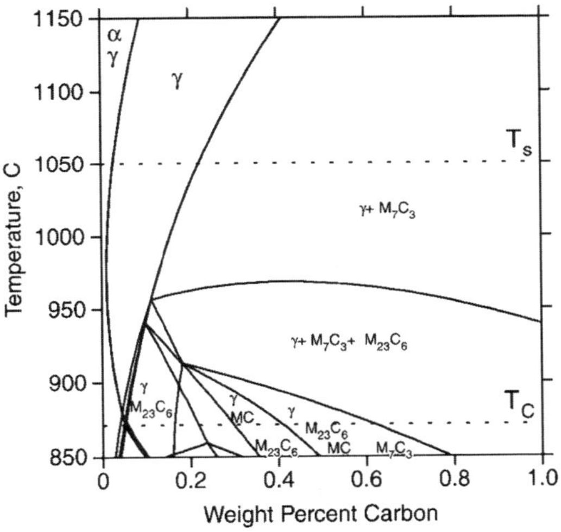

Fig. 18 Computed carbon-isopleth for Pyrowear 675. Dashed lines represent the carburizing (T_c) and solutionizing (T_s) temperatures.

Discussion

Using a systems approach to computational materials design, a carburizable, secondary-hardening, martensitic stainless steel was designed, which meets all of the case material property objectives, requires only thermal processing, and appears capable of meeting all of its property objectives. Establishing the desired property objectives, by analysis of existing bearing, secondary-hardening, and stainless steels, a quantitative conceptual design was embodied by developing and integrating design models for the martensitic transformation behavior, the coherent M_2C carbide precipitation strengthening, the formation and stability of passivating oxide films, and the microsegregation occurring during solidification processing.

The design properties are compared to the measured properties in Table 5. A case surface hardness of R_c 64.5 was obtained without the use of extensive primary carbides. The benefit of this microstructure was illustrated by the difficulty in processing the Pyrowear 675 alloy, which uses primary carbides to achieve its R_c 64 surface hardness. When pack-carburizing, the formation of primary carbides resulted in a tensile stress at the surface as the M_s temperature distribution was reversed. The martensitic case microstructure free of primary carbides in CS62 improved the RCF life as supported by preliminary studies showing the RCF life of carburized CS62 equilivalent to the M50 bearing steel.

The case diffusion temperature was higher than desired due to limitations of standard

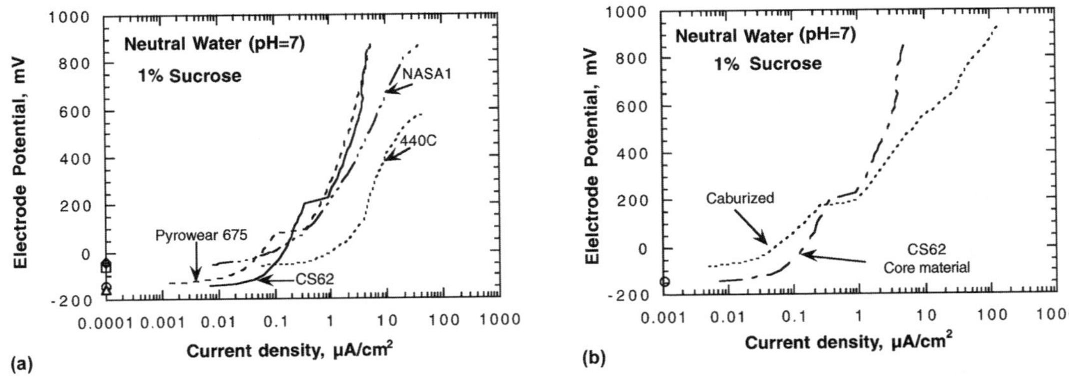

Fig. 19 (a) Anodic polarization measurements in 1% sucrose solution for monolithic alloys and (b) comparison of the behavior of the case and core material of CS62. The E_{corr} values are represented by the open symbols where ○ = CS62, □ = 440C, Δ = Pyrowear 675, and ◇ = NASAI.

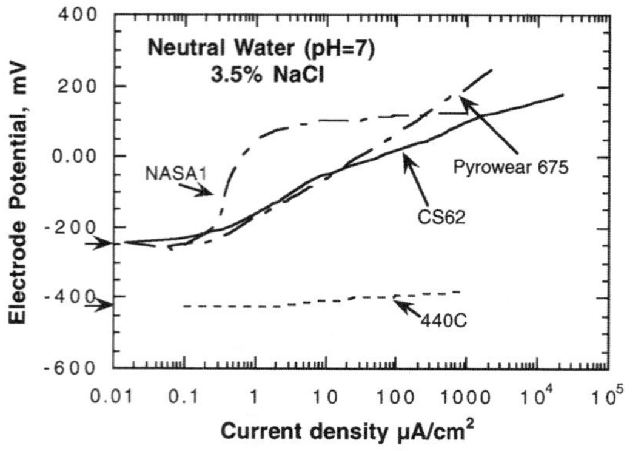

Fig. 20 Anodic polarization measurements in 3.5% NaCl solution for monolithic alloys.

Table 4 Corrosion rates after salt-spray testing.

Alloy	Corrosion rate, mm yr^{-1}
CS62	0.00256
CS62-carburized	0.0234
Pyrowear 675	0.00663
Pyrowear 675-carburized	0.0747

Table 5 Predicted and measured properties of the prototype composition

Property	Design	Measured
Core properties		
M_s temperature, °C	300	291
Hardness at 50% M_2C completion	575 VHN (R_c ~ 54)	R_c 53.5
Hardness stage I	...	R_c 52
Solution temperature, °C	...	1050
grain size at T_s for 1 h		30 μm
K_{IC} MPa$\sqrt{m}$	50	68 (Stage I) 29 (Stage IV)
Corrosion rate, mm yr^{-1}	...	0.0256
Case properties		
Matrix structure	martensitic	martensitic
Hardness at 50% M_2C completion	825 VHN	816 VHN R_c 64.5
Hardness stage I	...	800 VHN R_c 64
Solution temperature, °C	1100	1175
Corrosion rate, mm yr^{-1}	<0.051	0.0234

industrial carburizing conditions. With better control of the carburizing conditions the diffusion temperature could be decreased to the design solution temperature while achieving a reasonable case depth.

Consistent with model predictions of alloying effects on Cr partitioning to the passive film, anodic polarization measurements and salt-spray testing demonstrated the success of the 9Cr (CS62) alloy to achieve similar corrosion properties as traditional 12–13Cr steels. Anodic polarization measurements in 1% sucrose and 3.5% NaCl show similar resistance for both the core material of CS62 and the 13Cr alloy, Pyrowear 675. Both alloys exceed the corrosion resistance of 440C stainless bearing alloy. In the 3.5% NaCl solution the NASA1 alloy (12Cr–25Co) has the best resistance. However, further anodic polarization measurements are needed to clarify electrochemical reactions.

Measured corrosion rates from salt-spray testing also illustrate the excellent performance of CS62. Quantitative comparison of the carburized CS62 and carburized Pyrowear 675 material was hampered by case cracking of the latter. However, carburized CS62 should have superior corrosion resistance to carburized Pyrowear 675, as the case CS62 does not have a large fraction of primary carbides depleting the matrix of Cr. This is supported by the demonstration that carburizing causes only a slight reduction in the corrosion resistance of CS62.

The evaluation of the core material included the measurement of the M_s temperature and the fracture toughness. For the received prototype composition, the measured M_s temperature is within 10 degrees of the predicted temperature. The initial core hardness objective was achieved, with R_c 51.3 as-quenched hardness. The core fracture toughness objective was surpassed using Stage I tempering (200 °C, 1 h), $K_{IC} = 68\,\text{MPa}\sqrt{m}$ at R_c52. In the secondary-hardened condition, the fracture toughness at R_c 53 was similar to current bearing steels, $K_{IC} = 29\,\text{MPa}\sqrt{m}$. Higher K_{IC} values are expected for cleaner alloy heats (< 0.008 wt% O). The limited fracture toughness achieved with secondary hardening is primarily attributed to grain boundary embrittlement. Retaining the originally desired high core hardness of R_c 53 at 0.20 wt% C in the secondary-hardening condition will require enhancement of grain boundary cohesion. Decarburization experiments indicate that a ductile core can be obtained by lowering the carbon content to 0.05 wt% C corresponding to a core hardness of R_c 30.

Table 6 compares the properties of CS62, Pyrowear 675, and 440C. If used in the Stage I tempering conditions, CS62 surpasses in all properties. Although Pyrowear 675 has a much higher core fracture toughness, its core strength is much lower. Using CS62 as a secondary-hardening alloy, it has a fracture-toughness similar to 440C; however, it has a

Table 6. Properties of CS62, Pyrowear 675 and 440C

Property	CS62	Pyrowear 675	440C (monolithic alloy)
Matrix Cr (wt%)	9 core/ 9 case	13 core/ 5 case	11 matrix / 17 alloy
Core hardness, R_c	52	40	60
Case hardness, R_c	64	64	60
Core fracture Toughness $K_{IC}(MPa\sqrt{m})$	68 Stage I 29 Stage IV	150	20
Predicted y_{Cr} in oxide film	0.75	0.64	0.54
Corrosion rate, mm yr^{-1}	0.00256 (core) 0.0234 (carb.)	0.00663 (core) 0.0747 (carb)	NA

higher surface hardness, and should have greater dimensional stability as it has no primary carbides. CS62 should also have fewer solidification segregation problems than Pyrowear 675, which has 1.8 wt% Mo and higher Cr. Microsegregation calculations show that the Mo has the greatest potential for segregation.

Reducing the core carbon level to achieve high-toughness with secondary-hardening in CS62 will require minimal change in the system of interactions depicted in Fig. 1. The case structure and composition met the desired properties objectives and is actually the more complex subsystem to design in the case/core system. Lowering the core carbon level will not alter the case composition, and will also increase the core M_s temperature, resulting in a desirable steeper M_s gradient in the case. The processing of the alloy should remain the same, as the case composition is unaltered.

While the models used were sufficient to design the current prototype, further modeling work will improve the accuracy of the models and increase variety of design applications to which the models can be applied. Current modeling of the martensitic transformation behavior emphasizes the prediction of the M_s temperature; however, future applications may require control of the martensitic morphology as well. For example, controlling the fraction of plate and lath martensite in the case of a carburized steel. Characterization of the design prototype highlighted the need for improved modeling capabilities describing fracture toughness. These improvements are most likely to come from quantam mechanics modeling of impurity-induced embrittlment, which provide insights into the electronic bonding of various impurities at the grain boundaries, and the implementation of these results into a design process. While current simplistic modeling of the aqueous corrosion resistance was sufficient for this application, a better understanding of the kinetic process is needed. Ideally, the aqueous corrosion modeling should include the diffusion simulation of the growing passive film and indicate the composition gradient in the passive film. As all of the models developed rely correlating material properties and processing to thermodynamic quantities and diffusion reactions, all the modeling efforts are enhanced with improved thermodynamic and diffusion mobility assessments.

Conclusions

By designing a case material which avoids the use of primary carbides and optimize the partitioning of Cr into the passivating oxide films for corrosion resistance, a carburizing, secondary-hardening, martensitic stainless steel was developed that maintains the strength levels of the 440C composition while increasing the aqueous corrosion resistance and fracture toughness. Although the development of CS62 was a three-year process based on two previous prototypes, an important output was the development of transferable process kinetic and thermodynamic models. These models, defining the martensitic transformation behavior, the precipitation of the coherent M_2C strengthening dispersion, the microsegregation resulting from solidification processing, and the passivation behavior of oxide films formed, can now be used to design high performance stainless steels to meet the specific property requirements of a given application, as outlined in Part I (Ref 2; Chapter 16 in this volume) in the design objectives. The models developed are also useful in the design of other high-performance ferrous alloys, such as high-performance aerospace gear steels. The same systems design strategy can also be applied to other materials, as already demonstrated by the design of case-hardenable polymer gears (Ref 21). Thus, a systems design approach provides an efficient and organized methodology for computational materials design requiring far fewer experimental prototypes. Once the required design models are developed for a given system, the time and cost for alloy development can be greatly reduced.

ACKNOWLEDGMENTS

This research was supported by a NSF Graduate Fellowship for CEC and was conducted as part of the multi-institutional Steel Research Group program. The authors thank C.J. Kuehmann for assistance with the toughness and fatigue testing. Pyrowear 675 was provided by Carpenter Technology. Alloy 1605-8B was produced by NKK Corporation, Japan. Gas-carburizing of the alloys was performed by Caterpillar Inc. and MRC Bearings Corp.

REFERENCES

1. G.M. Jenkins, in J. Beishon, and G. Peters, (Eds.), *Systems Behavior*, Open University Press, 1972, p 56–79
2. C.E. Campbell, and G.B. Olson, J. Comput.-Aided Mat. Des. (2001)
3. C.J. Kuehmann, G.B. Olson, J.P. Wise, and C. Campbell, U.S. Patent, 6, 176.946, 2001
4. C.J. Kuehmann, and G.B. Olson, in E.B. Hawbolt, (Ed.), *International Symposium on Phase Transformations During the Thermal Mechanical Processing of Steel - Honouring Professor Jack Kirkaldy*, Metallurgical Society of the Canadian Institute of Mining Metallurgy and Petroleum, 1995, p 345–356
5. CarTech, U.S. Patent 5,002,729, 1993
6. *Annual Book of ASTM Standards*, Vol. 03.01, ASTM, 1997, p 413–443
7. T.A. Stephenson, C.E. Campbell, and G.B. Olson, in R.J. Richmond, and S.T. Wu, (Eds.), *Advanced Earth-to-Orbit Propulsion Technology*, NASA Conference Publication 3174, 1992, p 299–307
8. *Annual Book of ASTM Standards*, Vol. 02.05, ASTM, 1996, p 1–8
9. B. Sundman, B. Jansson, and J.O. Andersson, CALPHAD, 9 (1985) 153
10. A. Borgenstam, A. Engström, L. Höglund, and J. Ågren, J. Phase Equilibria, 21 (2000) 269
11. C.J. Kuehmann, Doctoral dissertation, Northwestern University, 1994
12. K.F. Amouzouvi, and M.N. Bassim, Mater. Sci. Eng., 55 (1982) 257
13. D.T. Hoelzer, J.A. Ali, and F. Ebrahimi, Scr. Metall., 20 (1986) 1575
14. M. Srinivas, S.V. Kamat, and P.R. Rao, J. Testing and Evaluation, 22 (1994) 302
15. D. Spaulding, "Grain boundary cohesion and segregation in ultrahigh strength alloy steels," Doctoral dissertation, Northwestern University, 1995
16. C.E. Campbell, "Systems design of advanced gear steels," Doctoral dissertation, Northwestern University, 1997
17. J.P. Wise, "Transformation toughening in ultrahigh strength austenitic steels," Doctoral dissertation, Northwestern University, 1998
18. D. Bergstrom, Doctoral dissertation, Northwestern University, 1996
19. M.G. Fontana, *Corrosion Engineering*, McGraw-Hill, 1986
20. M.G.H. Well, J.C. Beck, R.M. Middleton, P.J. Huang, and D.E. Wert, Surf. Engineer., 15 (1999) 321
21. R. D'Oyen, "Systems design of case hardened polymers for gears," Doctoral dissertation, Northwestern University, 1997

Transformations—Selected Works of G.B. Olson on Materials, Microstructure, and Design
C.E. Campbell, M.V. Manuel, and W. Xiong, editors

CHAPTER 18

Computer-Aided Design of Transformation Toughened Blast-Resistant Naval Hull Steels—Part I*

A. Saha and G.B. Olson
Northwestern University

Abstract

A systematic approach to computer-aided materials design has formulated a new class of ultratough, weldable secondary-hardened plate steels combining new levels of strength and toughness while meeting processability requirements. A theoretical design concept integrated the mechanism of precipitated nickel-stabilized dispersed austenite for transformation toughening in an alloy strengthened by combined precipitation of M_2C carbides and BCC copper both at an optimal ~3 nm particle size for efficient strengthening. This concept was adapted to plate steel design by employing a mixed bainitic/martensitic matrix microstructure produced by air-cooling after solution treatment and constraining the composition to low carbon content for weldability. With optimized levels of copper and M_2C carbide formers based on a quantitative strength model, a required alloy nickel content of 6.5 wt% was predicted for optimal austenite stability for transformation toughening at the desired strength level of 160 ksi (1100 MPa) yield strength. A relatively high Cu level of 3.65 wt% was employed to allow a carbon limit of 0.05 wt% for good weldability, without causing excessive solidification microsegregation.

Introduction

With scientific advances in the past century, the property-driven view of structure and processing for the creation of value has motivated the development of a systematic strategy in computer-aided design of materials (Ref 1, 2). The approach combines materials science, quantum physics, and continuum mechanics in the integration of process/structure/property/performance relations for predictive design of high-performance steels as multilevel dynamic structures. Throughout the history of materials development, there has been an ever-increasing need for stronger, tougher, more fracture-resistant, and easily weldable plate steels for structural applications at minimal cost. Unfortunately, however, any increase in strength is rarely achieved without concomitant decreases in toughness and ductility, which limits the utility of most ultrahigh-strength steels. The best combinations of strength and toughness have usually been obtained from

*A. Saha and G.B. Olson, "Computer-Aided Design of Transformation Toughened Blast Resistant Naval Hull Steels: Part I," *J Computer-Aided Mater Des*, 14, 2 (2007) 177–200

martensitic microstructures as shown in Fig. 1. High-strength bainitic steels have not been as successful in practice because of the coarse cementite particles in bainite that are detrimental to toughness. The primary benefit motivating the research of air-hardened steels containing bainite/martensite mixtures is the ease of processing, which finally leads to a product with good performance at a relatively lower cost. There is then the possibility of improving the strength and toughness simultaneously using fine-grained bainitic ferrite plates and enhancing the toughness by transformation toughening effects. Further improvements of strength can be achieved with coprecipitation of alloy carbides and BCC copper for easily weldable low-carbon steels (Ref 3), where alloy carbide precipitation dissolves the deleterious cementite.

It is now well established that the interaction of deformation-induced martensitic transformation of dispersed austenite with fracture-controlling processes such as microvoid induced shear localization results in substantial improvements in fracture toughness called Dispersed Phase Transformation Toughening (DPTT). This is the toughening mechanism modeled and investigated in this work. Transformation toughening is attributed to modification of the constitutive behavior of the matrix through pressure-sensitive strain hardening associated with the transformation volume change (Ref 4, 5). The transformation behavior and the toughening effects are controlled by the stability of the austenite dispersion. For transformation toughening at high strength levels, the required stability of the austenite dispersion is quite high and can be achieved only by size refinement and

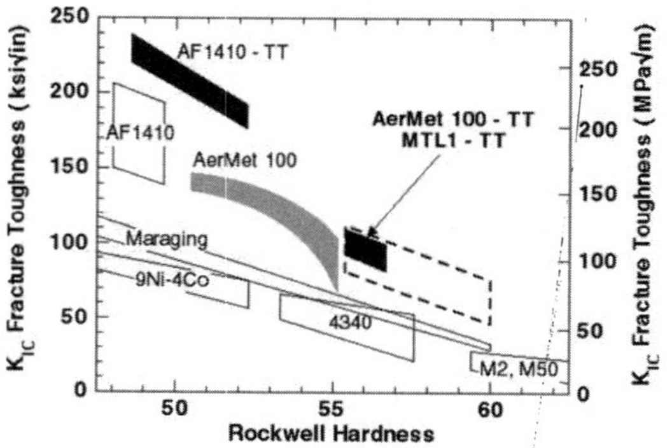

Fig. 1 K_{IC} toughness versus R_C hardness cross-plot for ultrahigh-strength martensitic steels. Source: Ref 1

compositional enrichment of the austenite particles. The size influences the characteristic potency of nucleation sites in the particles while the composition influences the chemical driving force and interfacial friction for the martensitic transformation. The size refinement and the compositional enrichment of the austenite can be controlled with heat treatments such as multistep tempering (Ref 6).

Combining new levels of strength and toughness while meeting processability requirements, the design addressed here will focus on a new class of ultratough, weldable secondary hardened bainitic/martensitic plate steels for blast-resistant naval hull applications.

Design Objectives

Looking ahead to the projected naval hull material requirements in the year 2020, the primary design objectives motivating this research are the achievement of extreme impact fracture toughness ($C_v > 85$ ft-lbs or 115 J corresponding to fracture toughness, $K_{Id} > 200$ ksi in$^{1/2}$ or 220 MPa m$^{1/2}$ and $K_{Ic} > 250$ ksi in$^{1/2}$ or 275 MPa m$^{1/2}$) at high strength levels of 150–180 ksi (1030–1240 MPa) yield strength in weldable, formable plate steels with high resistance to hydrogen stress corrosion cracking ($K_{ISCC}/K_{IC} > 0.5$) (Ref 7, 8). Because of difficulties in measurement of K_{Id} and K_{Ic} fracture toughness at such extreme levels, toughness of prototypes will be assessed by Charpy impact energy (C_V) absorption measurements; details of the $K_{Ic} - C_V$ and $K_{Id} - C_V$ toughness correlation will be discussed later. The primary design goals are marked by the star in the cross-plot of K_{Ic} fracture toughness and yield strength illustrated in Fig. 2 (Ref 9). This design aims to substantially expand the envelope marked as "steels" to the top right corner of the plot.

Design Approach

Based on Cyril Smith's (Ref 10) modern systems view of materials structure as "universal multilevel structure with strong interactions among different levels," this materials design approach breaks down the complex nature of the structural hierarchy to better understand the structure and property relations underlying

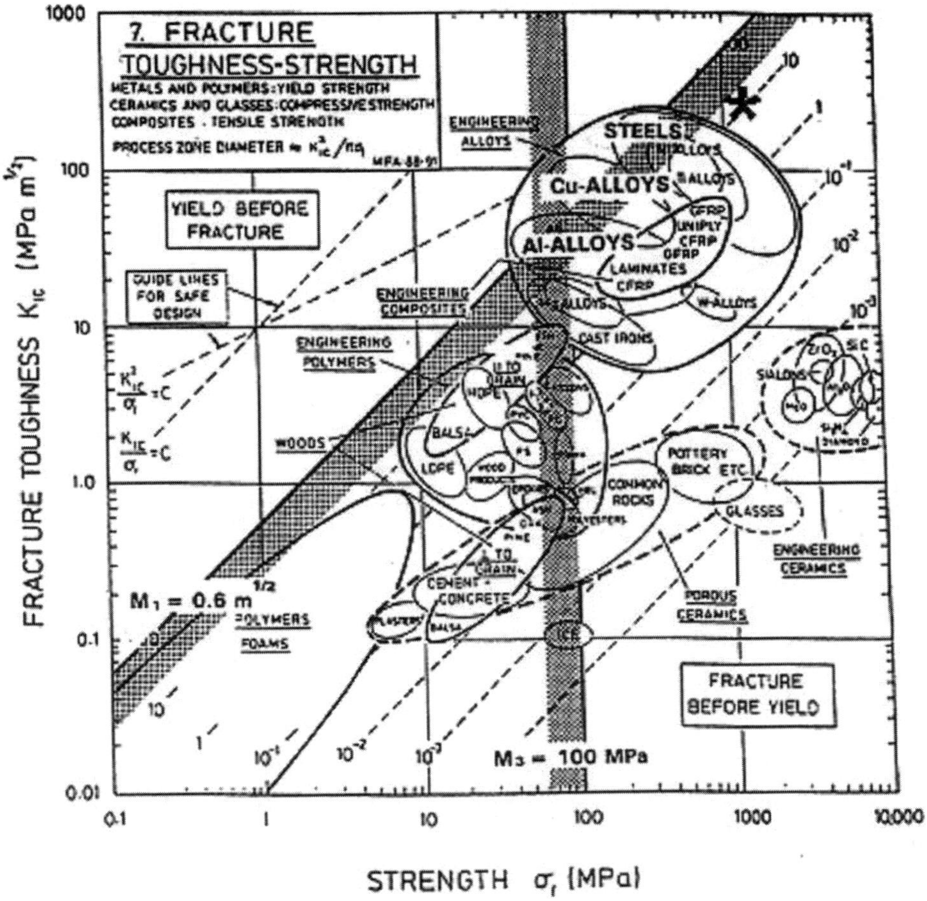

Fig. 2 K_{IC} toughness versus σ yield strength cross-plot for different classes of materials. Source: Ref 9

the technological and economic value of materials. Our approach follows the framework of systems engineering as outlined by Jenkins (Ref 11). Once the final goal has been set, the structure is described as subsystems with graphical representation of the interactions through a system flow block diagram. The flow block diagram represents the key microstructural subsystems, links them to the properties they control and then to the stages of processing that govern their dynamic evolution. Because of the complex nature of materials, it should be realized that the interaction between the subsystems is as important as the subsystems themselves. The systems analysis is then applied to identify and prioritize the key structure-property and processing-structure relations. Figure 3 describes the system structure adopted here to design a steel with the specified strength and toughness levels while meeting composition and processing constraints for weldability and hydrogen resistance.

Our system-level design integration follows a multiobjective "satisficing" approach (Ref 11), with aspects of single-objective optimization applied at the subsystem level based on previously established quantitative structure/property relations for strength and toughness. Quantitative design integration follows a parametric approach, expressing microstructural subsystem requirements in terms of thermodynamic parameters and employing computational thermodynamics to iteratively solve for a self-consistent alloy composition balancing the conflicting subsystem requirements. Alloy processability is addressed as constraints on overall composition and transformation temperatures.

With toughness being a major priority for this design, the matrix was chosen as a secondary-hardened bainite/martensite mixture, in which cementite particles should be eliminated by alloy carbide precipitation in "Stage IV" tempering, as they are detrimental

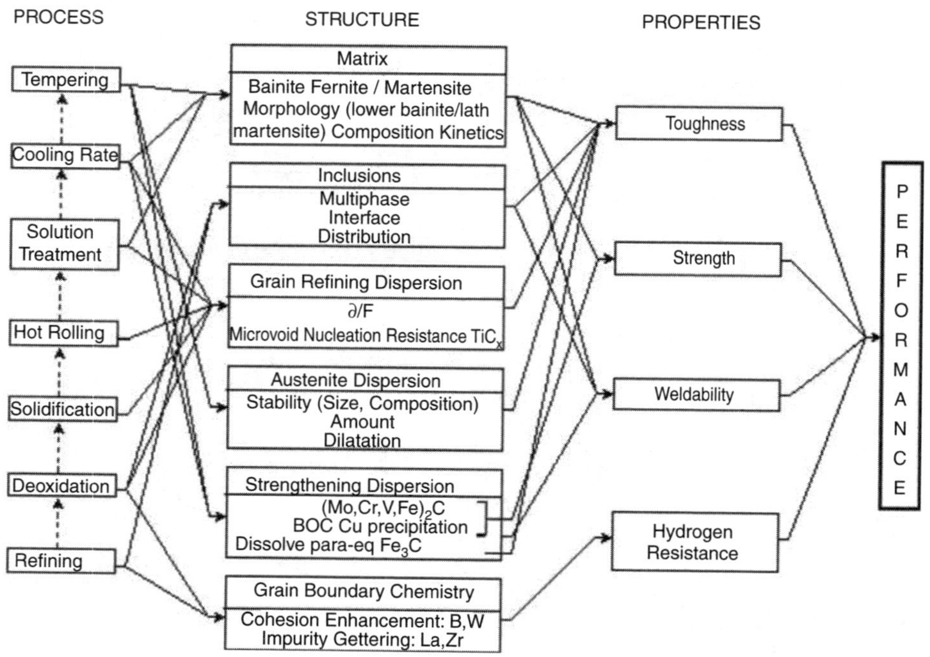

Fig. 3 Material system chart for design of blast resistant naval hull steels.

to the toughness of the steel. In addition, an optimum-stability austenite dispersion in the bainitic/martensitic matrix can also help in increasing the toughness of the steel. Our previous studies (Ref 12, 13) have demonstrated exceptionally large fracture toughness values in high-strength precipitation-hardened metastable austenitic steels. This remarkable increase in the fracture toughness is attributable to the process of transformation toughening which will be discussed in greater detail later. Recent studies (Ref 14) have also shown that selection of fine Ti(C, N) as a grain refining dispersion contributes to increasing the fracture resistance by delaying the coalescence of microvoids among the primary voids. Studies (Ref 15) have also suggested that the resistance to primary void formation and coalescence is proportional to inclusion spacing. It is thus desirable to reduce the volume fraction of primary inclusions or coarsen inclusions for a given volume fraction. This can be achieved by clean melt practices and tight composition control. However, engineering design fracture toughness parameters like K_{Ic} and K_{Id} are difficult and expensive to measure. Thus for preliminary design analyses, small-scale inexpensive fracture measurements like Charpy V-notch impact energy (C_V) values can be used to estimate K_{Ic} and K_{Id}. Studies of fracture toughness dependence on loading rate measured over a temperature range (Ref 16) have shown that K_{Ic} fracture toughness values under static and intermediate loading are about 20% higher than the K_{Id} measured under impact loading. While it is known that K_{Ic} and C_V cannot be universally correlated (Ref 17), Barsom and Rolfe (Ref 16) established an approximate relation between K_{Ic} and C_V test results for conventionally grain-refined steels:

$$K_{IC}^2 = AC_V \quad \text{(Eq 1)}$$

where A is a constant of proportionality. Fitting Eq 1 to results from high Ni steels shown in Fig. 4, an empirical correlation can be established for the class of steels explored in this study.

According to these relationships, the C_V impact toughness objective of 85 ft-lbs is equivalent to the K_{Ic} fracture toughness of 250 ksi in$^{1/2}$(275 MPa m$^{1/2}$) represented in Fig. 2, corresponding to a dynamic K_{Id} of 200 ksi in$^{1/2}$ (220 MPa m$^{1/2}$).

A fine carbide dispersion must be obtained in order to achieve the desired strength level. Coherent M_2C carbides have been used in

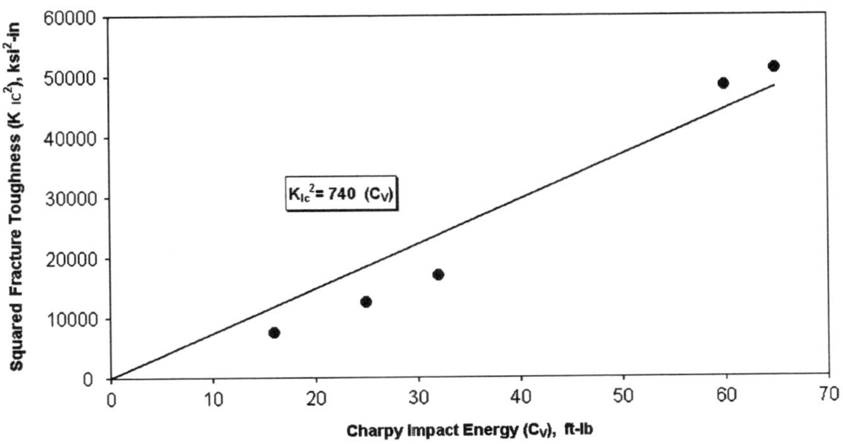

Fig. 4 Correlation between K_{Ic} and C_V test results (Ref 16) for high Ni steels.

secondary hardened steels that are currently in use (Ref 18). Previous work (Ref 19) to optimize the carbide particle size for strength has shown that 3 nm carbide precipitates corresponding to the transition from particle shear to Orowan bypass provide maximum strength. Thermodynamics and kinetics of carbide precipitation have to be controlled to obtain such a fine M_2C carbide dispersion. The driving force for M_2C nucleation should be maximized by proper control of the amount and ratio of carbide formers in the alloy to refine the M_2C particle size. Sufficient M_2C precipitation must be achieved to dissolve cementite in order to attain the desired toughness levels because coarse cementite particles are extremely deleterious as microvoid nucleation sites. Tempering times should also be minimized to prevent impurity segregation at grain boundaries.

Even if we maintain low alloy carbon levels, higher alloying content tends to reduce weldability by increasing hardenability. For any structural material, the heat-affected zone (HAZ) adjacent to the welded joints are considered to be the weakest links. Weldability of steels is controlled by both the matrix and the strengthening dispersion structures. As a rule of thumb, for adequate weldability of the steel C content of the alloy should be kept below 0.15 wt%. This in turn limits the C available for M_2C strengthening. For the bainitic matrix, if we modify the hardenability of the steel, we can obtain bainite with a much lower cooling rate. This becomes a trade-off since weldability can deteriorate as the hardenability increases.

Ultrahigh-strength steels are prone to a decrease of fracture toughness in aqueous environments due to hydrogen-assisted cracking. This reduction of toughness is caused by intergranular brittle fracture associated with impurity segregation to grain boundaries, which may reduce toughness of the steel by as much as 80% in a corrosive environment. The common impurities in steel are P and S, both of which are embrittlers since they have lower free energy on a surface than at a grain boundary. The most effective way of reducing them is by cleaner processing techniques or impurity gettering. Impurity gettering can tie up P and S as stable compounds formed during solidification. La and Zr have been found to be effective impurity gettering elements. Another approach to minimize impurity effects is by the design of grain boundary chemistry. Segregating elements like W and Re (Ref 7, 8) preferentially on the grain boundaries that enhance grain boundary cohesion is beneficial to the stress corrosion cracking resistance. Small amounts of dissolved B also help in grain boundary cohesion. As denoted at the lowest level of the system chart of Fig. 3, these issues are addressed by trace element composition constraints set at the earliest stage of material processing.

Based on the above considerations for the property → structure → processing relationships of the overall design system, Fig. 5 presents a schematic process flow of the iterative design integration procedure employed in this work to determine a theoretical design composition. Each of the steps of this process will next be discussed in turn.

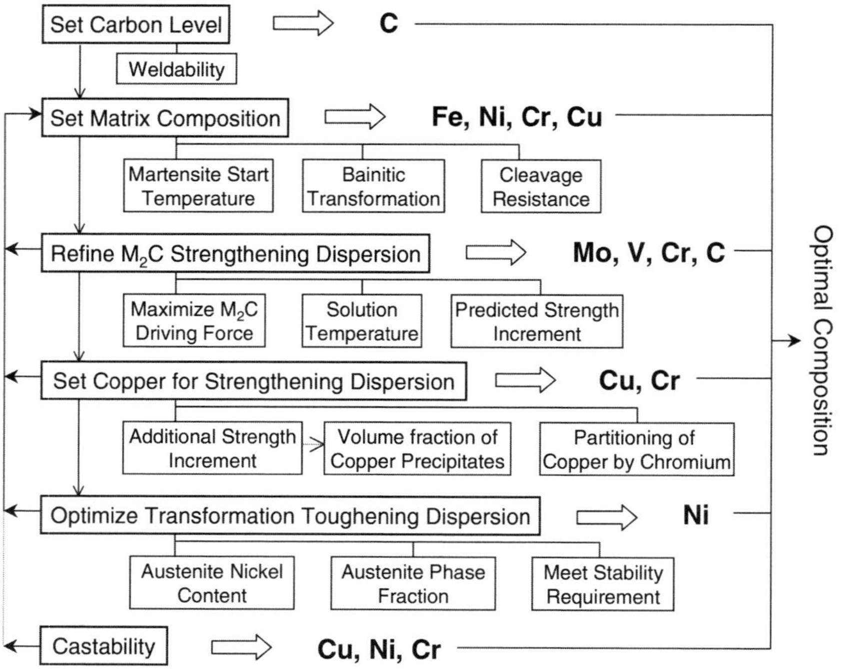

Fig. 5 Schematic of the design procedure.

Strength Design

An efficient approach to strengthen the steel while limiting carbon content for weldability is co-precipitating M_2C carbides and BCC copper. By trading off the interaction between particle size and the phase fraction of the precipitates, the goal of high strength can be achieved.

Quantitative Strengthening Contributions. As highlighted by the system design chart (Fig. 3), the strength of the alloy will be designed by using quantitative strengthening models to predict its dependence on the structure of the steel. To achieve a goal of 160 ksi (1100 MPa) yield strength, quantitative models will be employed to relate the contribution from dispersions of M_2C carbide precipitates (Ref 19) and BCC copper precipitates (Ref 20) in secondary-hardened steels. The levels of M_2C carbide formers and copper will be selected based on the strength contribution from each of these substructures. In this work, assessment of the yield strength of the material has been made directly from the hardness data because of the ease and convenience in measurement of the latter. Hardness of a material is a direct manifestation of its resistance to plastic flow, monotonically relating to yield stress. An empirical relationship has been developed between hardness and yield stress based on experimental data from previous research on related steels: HSLA100 data from Foley (Ref 21), AerMet100 data from Kuehmann (Ref 22) and SRG C2 experimental gear steel data from Spaulding (Ref 23). The best-fit curve in a log-log plot of hardness versus yield stress has been used to determine the relationship incorporating strain hardening effects. Figure 6 presents the experimentally measured hardness-yield stress data from the previous research superimposed with the best-fit power-law relationship and the theoretical straight-line relation describing the same for an ideal plastic material (Ref 24). The higher hardness of the empirical power-law relationship relative to the ideal-plastic case represents the effect of strain hardening, which is more pronounced at lower strength levels. The point at which the two curves meet represents the prediction limit of the relationship.

Thus, the hardness estimate for the target yield strength of 160 ksi (1100 MPa) from the power-law relationship is 389 VHN. The relationship obtained is:

$$VHN = 6.116 YS^{0.8184} \quad \text{(Eq 2)}$$

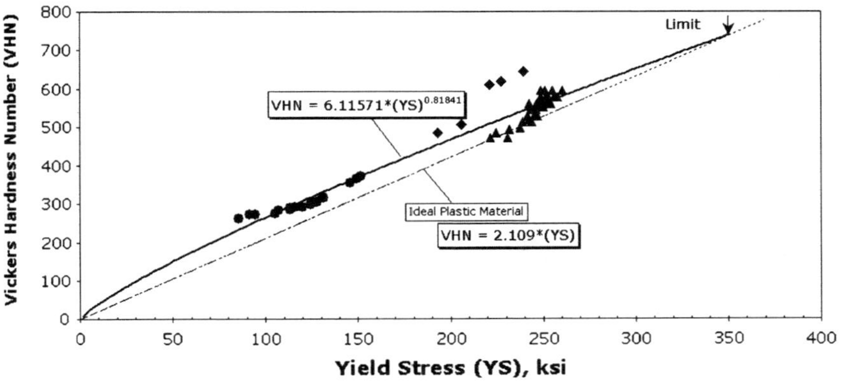

Fig. 6 Power-law relationship relating hardness of related steels to yield stress from experimental data from Foley (Ref 21) (circles), Kuehmann (Ref 22) (triangles), and Spaulding (Ref 23) (diamonds) shown in comparison to straight-line relationship for ideal plastic material.

where VHN (Vickers Hardness Number) is in kg/mm^2 and YS (Yield Strength) is in ksi.

The first step in the design involved setting the carbon content of the alloy to ensure good weldability. Figure 7 presents the Graville diagram summarizing effects of overall carbon content and alloy carbon equivalent. This shows that at 0.05 wt% C, the steel is not susceptible to hydrogen-induced cold cracking in the heat affected zone (HAZ) of weldments. Consistent with the carbon level in current Navy HSLA100 steel, the lower limit C content of 0.05 wt% C was set for the alloy, thus allowing relatively high carbon equivalent levels of alloy content.

Based on the effect of M_2C carbide precipitates in the Orowon bypass regime, Wise (Ref 19) developed a quantitative strength model to predict the strengthening achieved for a given carbon level. The predictions assume that a given carbon content of the alloy is precipitated to the full equilibrium carbide volume fraction and considers carbides of fixed particle diameter. Based on the predicted change in hardness-carbon content (wt%) plot shown in Fig. 8, at a C level of 0.05 wt% the hardness increment due to M_2C carbide precipitation is estimated to be 175 VHN provided a sufficient driving force is maintained to achieve the particle size range of ~4–5 nm in Fig. 8. The base strength of a lath martensitic substructure was estimated as 63 VHN by Wise (Ref 19). The additional strength increment of 151 VHN to achieve the strength goal of 389 VHN must then be attained through BCC copper precipitation strengthening, described later. The effect of solid solution strengthening is assumed to be negligible for steels having low carbon and low hardenability. Thus, the total strength of the alloy has been modeled by superimposing three contributions. The strength is described by the effects of M_2C carbide precipitates, τ_{M2C}; BCC copper precipitates, τ_{Cu} and matrix martensitic structure, $\tau_{\alpha'}$:

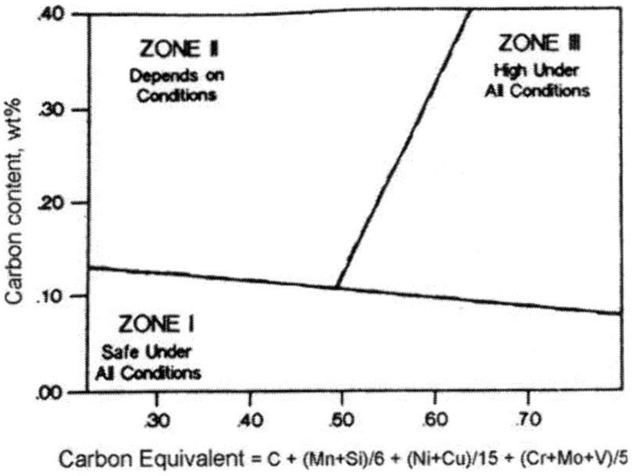

Fig. 7 Graville diagram for determining susceptibility to HAZ cracking in plate steels. Source: Ref 25

$$\tau = \Delta\tau_{M2C} + \Delta\tau_{Cu} + \Delta\tau_{\alpha'} \equiv 389\,\text{VHN} \qquad \text{(Eq 3)}$$

The contributions of the individual mechanisms to achieve the strength goal equivalent to 389 VHN are graphically presented in Fig. 9.

M_2C Carbide Strengthening. For the high-strength design, we want to ensure that all of the carbon is taken up by the M_2C carbide formers (Cr, Mo and V) in order to dissolve the cementite in the matrix. Cementite negatively affects strength and toughness. Therefore, we want the sum of the atomic concentrations of

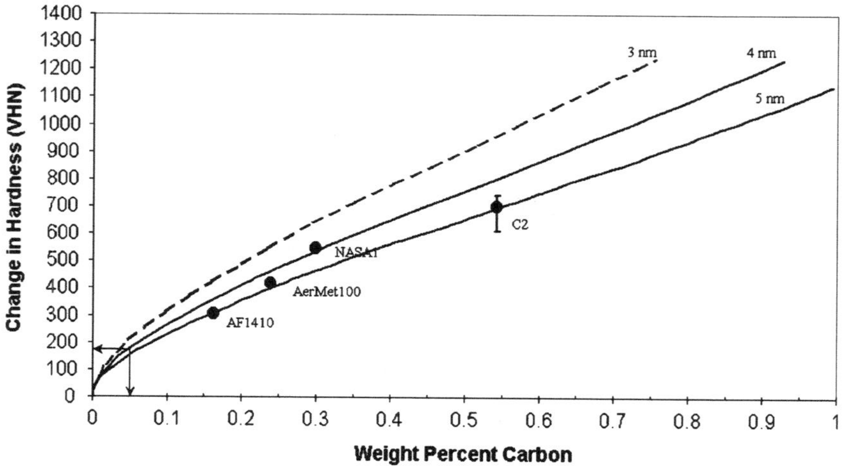

Fig. 8 Change in hardness as a function of alloy carbon content for M_2C carbide strengthening contribution (Ref 19). The arrows represent hardness increment of 175 VHN is achieved at C level of 0.05 wt% set for the alloy. Experimental results of other secondary hardening steels are shown.

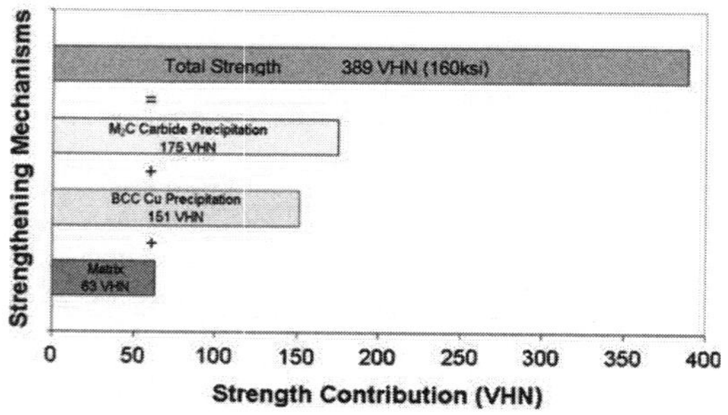

Fig. 9 Graphical representation for contributions of the individual mechanisms to achieve the strength goal equivalent to 389 VHN.

Cr, Mo, and V to be at least double the concentration of C for the M_2C stoichiometry.

A series of calculations were performed in order to design a steel that meets the strength requirements. Preliminary compositions were set using the guideline (consistent with the HSLA100 alloy) that carbon content should be limited to 0.05 wt% for weldability (Fig. 7); Cu should be at least 1.5 wt% for significant strengthening (Ref 21, 26), minimum Ni content should be at least half that of Cu to avoid hot shortness, and the relative amounts of carbide formers Cr, Mo, and V was initially set equal in atomic percent. A feasibility study was then performed to ensure that this strengthening concept, in conjunction with our approach for toughening (described later) is thermodynamically possible, that is, all of the phases needed for precipitation strengthening and nickel-stabilized austenite could coexist at least in metastable equilibrium at processing temperatures. A BCC Cu-rich precipitate is necessary for Cu precipitation strengthening, an M_2C carbide phase is necessary for carbide strengthening, and FCC austenite is critical for transformation toughening. Considering these constraints, thermodynamic feasibility of the preliminary alloy composition was verified by ThermoCalc (Ref 27) calculations at kinetically reasonable tempering temperatures of 400–500 °C using the SGTE/SSOL thermochemical database.

Before the driving forces were calculated, a Mo-Cr phase diagram section at a solution temperature of 900 °C was calculated in order to determine the relative solubility of the carbide formers in the austenite phase at a reasonable solution temperature. The solubility limit should not be exceeded in order to achieve full conversion to M_2C for maximum carbide strengthening. Recalling that the M_2C stoichiometry must be maintained, the stoichiometric constraint, as well as the solubility limit, can be superimposed onto a contour plot of driving force versus Mo and Cr concentration to optimize the M_2C driving force. The Mo-Cr phase diagram section is shown in Fig. 10. It is apparent from the diagram that solubility is not a limiting factor in the region of interest, due to the relatively limited C content.

The stoichiometric constraints of the M_2C carbide dictated that the total amount of carbide formers (Cr, Mo, V) needed to balance the carbon content would be 0.468 at.%. Using this constraint, initial plots were constructed of the driving force for M_2C nucleation versus at.%(Mo) and at.%(Cr), setting V at different levels. Figure 11 is a representative plot of driving force contours with varying at.%(Mo) and at.%(Cr) at a constrained composition of 0.05 at.% V at 500 °C. The stoichiometric constraint line has been drawn on the plot

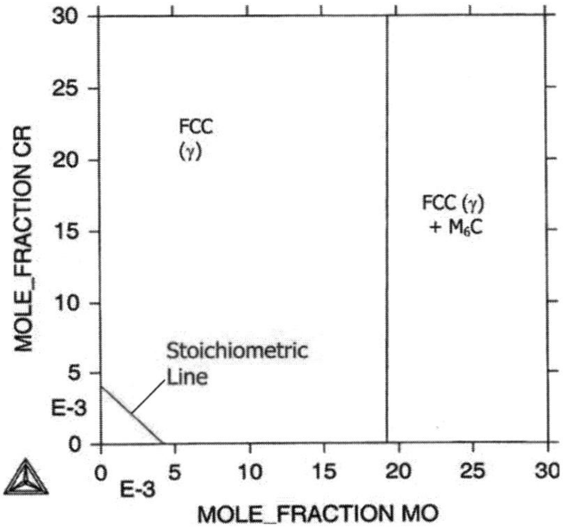

Fig. 10 Cr-Mo phase diagram section at 900 °C with alloy composition in atomic %: Fe-0.234C-1.32Cu-6.21Ni-0.055V. This diagram shows the phase fields of the FCC austenite and FCC + M_6C revealing that the M_2C stoichiometric line is well within the solubility limit.

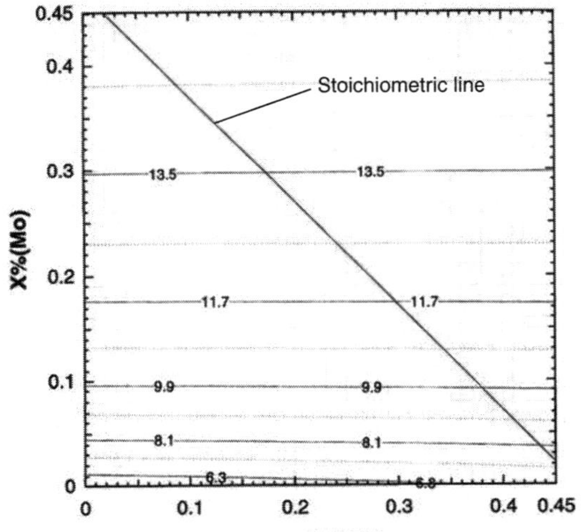

Fig. 11 Driving forces (in kJ/mole) for M_2C carbide nucleation contour plot varying at.%(Mo) and at.% (Cr) with superimposed M_2C stoichiometric (heavy) line at 500 °C at alloy compositions at.% Fe-0.234C-1.32Cu-6.21Ni-0.055V.

indicating the line of allowed compositions for M_2C. This study indicated that Cr has the least effect on driving force, especially at the higher contents of interest.

Based on this finding, another set of driving force plots were created varying at.%(Mo) and at.%(V) while setting the Cr level at fixed values. Due to the very small Cr dependence, all the plots were very similar and so only a representative graph (Fig. 12) is included here at 0 at.%(Cr). A similar M_2C stoichiometric line was drawn as before, constraining a maximum driving force at about 14.4 (kJ/mole). This plot revealed an almost equal effect on driving force for Mo and V, indicating that any allowed ratio of the two should give a maximum driving force value, so a series of calculations were done along the stoichiometric line (maximum driving force). We found a feasible alloy composition where all the desired phases as mentioned before coexisted, which is indicated by the point and arrow in Fig. 12. An initial feasible alloy composition in wt% was thus determined without any Cr: Fe-0.05C-1.5Cu-6.5Ni-0.6Mo-0.1V.

The V-Mo phase diagram section at a solution temperature of 900 °C with the feasible alloy composition was then calculated. Again, the solubility of these carbide formers is not a limiting factor in the region of interest. This plot is shown in Fig. 13.

Copper Precipitation Strengthening. In addition to M_2C carbide strengthening, BCC

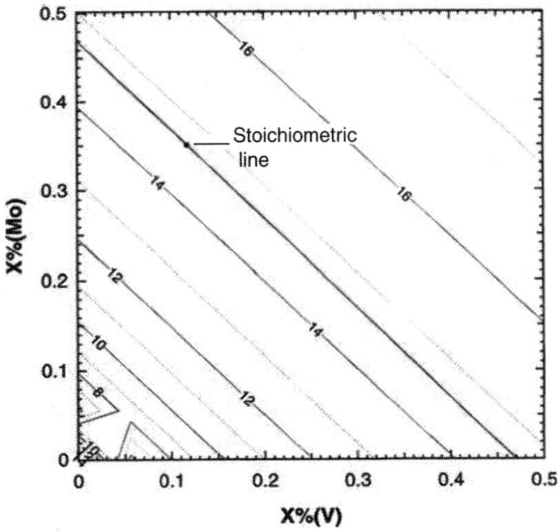

Fig. 12 Driving force (in kJ/mole) for M_2C carbide nucleation contour plot varying at.%(Mo) and at.% (V) with superimposed M_2C stoichiometric line at 500 °C at alloy compositions at.% Fe-0.234C-1.32Cu-6.2Ni.

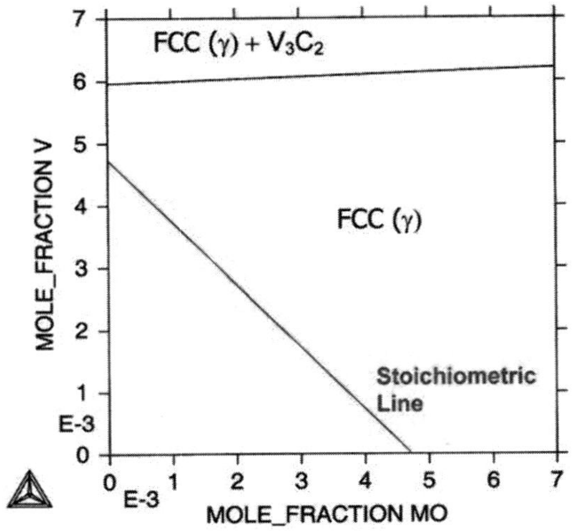

Fig. 13 Mo-V phase diagram section at 900 °C with alloy composition in atomic %: Fe-0.234C-1.32Cu-6.2Ni. This diagram shows the phase fields of the FCC austenite and FCC + V_3C_2 revealing that the M_2C stoichiometric line is well within the solubility limit.

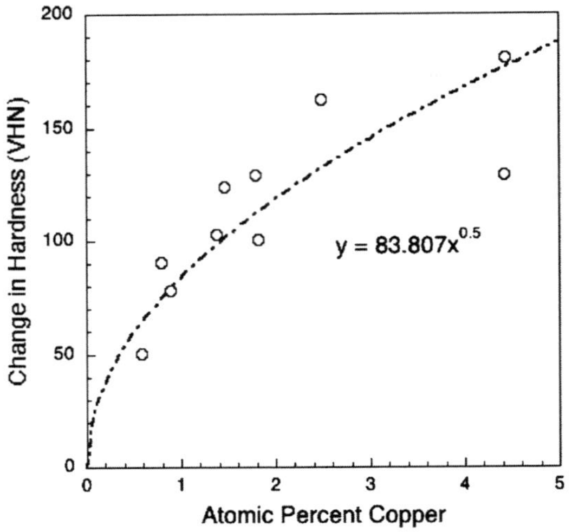

Fig. 14 Change in peak hardness as a function of alloy copper content for BCC copper strengthening contribution (Ref 20). Experimental results of other copper strengthened steels are shown. The dotted line represents the best-fit line for one-half power law given by Eq 5.

copper precipitation strengthening will be modeled to control the phase fraction of the precipitates through the alloy copper content to provide the additional increment of strength (≡151 VHN). The copper precipitates that contribute to strengthening in steels have a metastable BCC structure, which are fully coherent with the matrix having an average diameter of 1–5 nm (Ref 28–35). The strengthening mechanism has been described by the Russell-Brown model (Ref 20) based on the interaction between the matrix slip dislocation and the second phase copper-rich particle of lower shear modulus than the matrix. The shear stress has a maximum value, τ_{max}, given by Eq 4:

$$\tau_{max} = \frac{0.041 G b f^{1/2}}{r_0} \quad \text{(Eq 4)}$$

where G is the matrix shear modulus, b is the burgers vector, f is the volume fraction of precipitates, and r_0 is the core radius of the dislocation. The maximum strength that can be achieved (which empirically occurs at a particle diameter of 2–3 nm) is thus proportional to the square root of the volume fraction of the precipitate. Based on this volume fraction dependence of the precipitate on yield stress, the peak hardening increment from available data of copper precipitation strengthened steels (Ref 20) was plotted as shown in Fig. 14. The best-fit line described by a one-half power law defined the hardening increment dependence on the alloy content (at.%) of copper:

$$\Delta\tau(\text{VHN}) = 83.807 X_{Cu}^{1/2} \quad \text{(Eq 5)}$$

giving a strengthening coefficient within a factor of ~1.2 of the Russell-Brown estimate. Based on this relationship, the hardness increment of 151 VHN can be achieved by addition of 3.25 at.% Cu to the alloy composition, provided this level can be made soluble at the alloy austenizing temperature.

Transformation Toughening Design

For design of tough steels for such high-strength levels (160 ksi or 1100 MPa YS) we need to develop a fully secondary-hardened microstructure with high-stability austenite produced by precipitation. At high strength levels we need high stability of precipitated austenite since the mechanical driving force for transformation is very high. This design seeks to improve the toughness of higher strength steels by utilizing the beneficial properties of Ni-stabilized precipitated austenite. This form of austenite can precipitate during annealing or tempering at elevated temperatures above about 470 °C. The fact that this dispersed austenite forms by precipitation is significant because it allows greater overall control of the amount and stability of the austenite. Further processing and treatments can be used in the form of multistep tempering to first nucleate particles in a fine form at a higher tempering temperature and then complete Ni enrichment during completion of precipitation strengthening (cementite conversion to M_2C) at a lower final tempering temperature (Ref 4, 6).

Shear localization by microvoid nucleation is known to be the most dominant fracture mode in high toughness steels. Studies (Ref 36) have shown that fine particle dispersions with adherent interfaces are most useful for controlling microvoid nucleation. The most promising microstructure modification is achieved by nucleating an optimal stability austenite dispersion, which increases toughness by suppressing microvoid nucleation to higher strain levels. Thus, emphasis will be put on the design of intralath dispersions, as the greater stability associated with their finer size makes them the primary toughening form of austenite precipitates.

The austenite dispersion must have sufficient stability and proper formation kinetics to ensure maximum toughening enhancement. Other factors controlling the stability of austenite are particle size and stress state sensitivity, the latter being related to the transformational volume change. The Olson-Cohen classical heterogeneous martensitic nucleation model can be applied to describe dispersed austenite stability for transformation toughening (Ref 4). Stability of an austenite precipitate is defined by chemical and mechanical driving force terms. At the M_s^σ temperature (where transformation occurs at yield stress) for the crack-tip stress state, the total driving force equals the critical driving force for martensite nucleation, as represented by Eq 6:

$$\Delta G^{ch} + \sigma_y \frac{d\Delta G^\sigma}{d\sigma}\bigg|_{cracktip} = -\left[\frac{2\gamma}{nd} + G_0 + W_f\right] \quad \text{(Eq 6)}$$

Rearranging the terms and substituting a dependence of defect potency on particle volume V_p (Ref 37), we can define a convenient stability parameter:

$$\Delta G^{ch} + W_f + \frac{K}{\ln(V_p)} = -\left[\sigma_y \frac{d\Delta G^\sigma}{d\sigma}\bigg|_{cracktip} + G_0\right] \quad \text{(Eq 7)}$$

ΔG^{ch} is the transformation chemical free energy change, and W_f is the athermal frictional work term. ΔG^{ch} is temperature and composition dependent while W_f is only composition dependent. W_f will vary with tempering temperature due to the change in austenite composition. σ_y is the yield stress of the material, ΔG^σ is set by the stress state, and G_0 is a fixed stored energy term. K is a proportionality constant, γ is the nucleus specific interfacial energy, and d is the crystal interplanar spacing.

The austenite stability for a given set of conditions or service temperature for a given dispersion can be assessed by the parameter given by the left-hand side of Eq 7. If we assume an austenite particle size equivalent to that achieved in previous studies of AF1410 and AerMet100 steels, our austenite stability parameter becomes the sum of the chemical driving force for transformation of FCC austenite to BCC martensite at room temperature (300 K) and the frictional work term for martensitic interfacial motion: $\Delta G_{ch} + W_f$. ThermoCalc can be used with an appropriate thermochemical database (Ref 38) to predict the temperature and compositional dependence of the chemical energy term. Ghosh and Olson (Ref 38, 39) have modeled the composition dependence of the frictional work term as a power law with an exponent of 0.5 and a fit to experimental data. Appropriate superposition laws considering relative strengths of the solutes were applied for complex systems. The model is represented in Eq 8:

$$W_f = \sqrt{\sum_i \left(K^i X_i^{1/2}\right)} + \sqrt{\sum_j \left(K^j X_j^{1/2}\right)} + \sqrt{\sum_k \left(K^k X_k^{1/2}\right)} + K^{Co} X_{Co}^{1/2} \quad \text{(Eq 8)}$$

where the K's represent the coefficients used to fit the solid solution strengthening data and i = C, N; j = Cr, Mn, Mo, Nb, Si, Ti, V; and k = Al, Cu, Ni, W. Equation 7 further indicates that the stability parameter is a linear function of the yield strength of the material.

Thorough studies of transformation toughening (Ref 12, 13) have demonstrated that maximum toughening occurs at the M_S^σ temperature for the crack-tip stress state, thus quantifying an optimal austenite stability. The strength dependence of this optimal stability level at fixed V_P was determined from previous dispersed-phase transformation toughening optimization experiments on the AF1410 and AerMet100 steels (Ref 40). Figure 15 gives a plot of the measured optimal stability, $\Delta G_{ch} + W_f$, at room temperature against Vickers hardness of the alloy. The required room temperature stability of the austenite dispersion projected from the hardness (or strength) requirement of the design is marked by the shaded region in the figure and quantitatively expressed in Table 1. To achieve a goal of 160 ksi (1100 MPa) yield strength equivalent to Vickers hardness of 389 (Rc40 equivalent), the estimated value of $\Delta G_{ch} + W_f$ for the required stability is determined to be 2,837 J/mole.

The design of transformation-toughened austenite has been calibrated against this stability parameter to determine the optimal level of austenite-stabilizing nickel in the alloy. Plots of both the phase fraction of austenite and nickel content in the austenite phase versus alloy atomic fraction Ni were computed. Figure 16 was calculated at an estimated final tempering temperature of 500 °C for substitutional diffusion and revealed that a minimum of 3.5 at.% Ni is required to get austenite and a maximum content of nickel in the austenite of about 0.30 could be obtained. It also showed that at the 6.25 at.% Ni composition from initial feasibility studies, about a 0.10 phase fraction of austenite would be formed as shown by the arrows. This compares well to the phase fraction of austenite employed in previous transformation toughened steels. We thus set the alloy Ni level to 6.25 at.%, which also saturates the austenite Ni content to 30 at.%.

After conducting the feasibility study mentioned earlier and verifying that the corresponding austenite stability was within the design limits given in Fig. 15, a thermodynamically viable alloy composition was found using 6.25 at.% Ni. This Ni level was deemed acceptable as it fell well within the nickel content guidelines, while being low enough to limit the material cost.

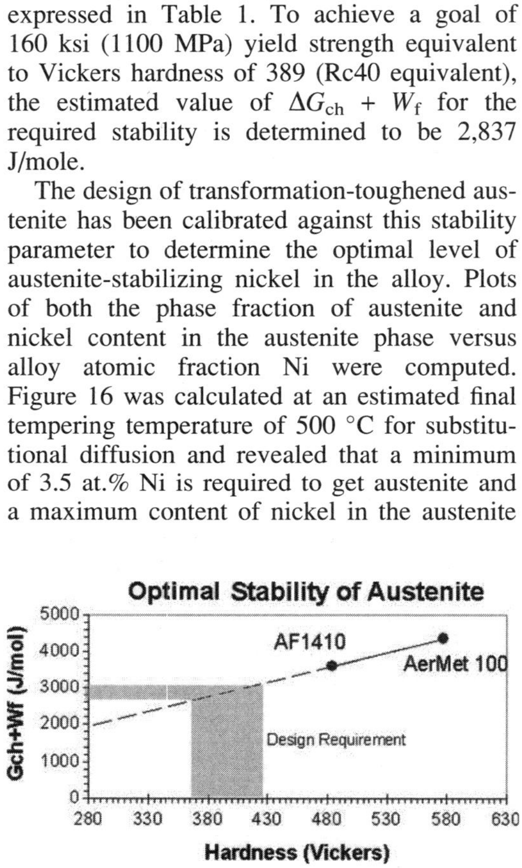

Fig. 15 Optimal room temperature (300 K) austenite stability plotted as a function of VHN. The shaded region shows our range of interest for austenite stability corresponding to a yield strength requirement of 150–180 ksi (1030–1240 MPa) after extrapolation of data from previous alloys, AF1410 and AerMet100.

Table 1 Target chemical driving force (ΔG_{ch}) + frictional work (W_f) value.

Alloy	Rockwell C Hardness, R_c	VHN (kg/mm^2)	$\Delta G_{ch} + W_f$ J/mol
AerMet 100	54	577	4350
AF1410	48	484	3600
Design	40	389	2837

Composition Integration

The overall composition was next optimized so that all of the phases necessary for strengthening and toughening are simultaneously present. The full sequence of iterative calculations followed the flow chart of Fig. 5. Since the maximum M_2C driving force at fixed total solute is obtained with no chromium, equilibrium phase calculations were done for the initial feasible composition. For this initial composition, it was found that the copper added for precipitation strengthening went instead into the austenite phase. It was found that Cr partitions Cu out of austenite and into the BCC precipitate phase effectively at 2 at.% and above, and the Cr content was thus set to this

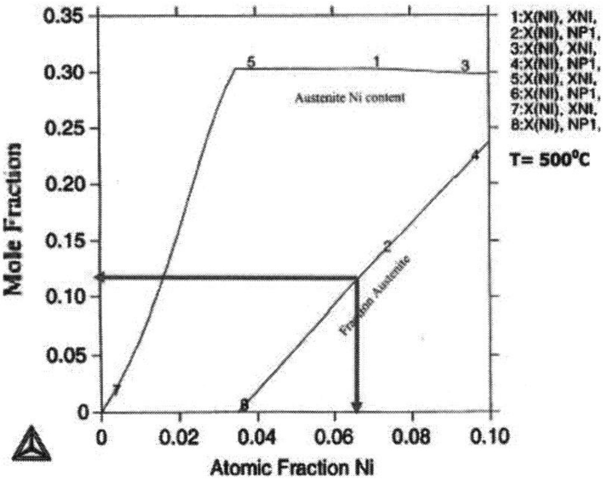

Fig. 16 Fraction of Ni in austenite and phase fraction of austenite in alloy versus mole fraction of Ni at 500 °C with alloy composition in weight%: Fe-0.05C-3.65Cu-1.85Cr-0.6Mo-0.1V.

minimum value. The iterative calculation sequence of Fig. 5 then yielded a self-consistent alloy composition meeting all phase relation requirements. The composition was then further evaluated from the viewpoint of processability requirements.

Processing Considerations

Solution Treatment Temperature and Allotropic Transformations

A solution treatment temperature of 900 °C was chosen based on previous optimization studies (Ref 22, 41) on grain size dependence of hardness and toughness based on 1-h solution-treated samples in secondary-hardened steels. With the increased levels of Cu and Cr it was confirmed that the alloy was solution treatable at 900 °C as shown by the ThermoCalc phase fraction plot in Fig. 17.

For this alloy composition, our martensite and bainite kinetic models predict an M_S temperature of 298 °C and a bainite start (B_S) temperature of 336 °C. These should be sufficiently high to allow formation of bainite/martensite mixtures with air-cooling.

Scheil Simulation for Microsegregation Behavior

Solidification of alloys generally occurs with segregation, which can have a strong effect on the alloy's final properties. Thus, it is important to model segregation to assess the processability of the designed alloy. This investigation uses thermodynamic modeling to predict microsegregation of the as-cast material. Macrosegregation effects can also be addressed by modeling liquid buoyancy associated with the microsegregation amplitude but are not addressed in this assessment.

Scheil simulation is a fast method of estimating microsegregation (Ref 42). The main approximations are infinite diffusion in the liquid but no diffusion in the solid phase. This has been coupled to the multicomponent SGTE thermodynamic database using ThermoCalc from which the solid/liquid equilibrium was calculated repeatedly during the simulation. Figure 18 presents the solidification simulation result as temperature versus fraction solid using the nonequilibrium Scheil simulation and compares it with the full equilibrium case. Figure 19 presents the composition profiles calculated by the Scheil simulation showing the degree of microsegregation in the solid after solidification. Here, the fraction of solid is equivalent to position relative to a dendrite arm center. Previous comparison with more rigorous calculations incorporating solid back diffusion indicate that the Scheil result at 95% solid is a reasonable estimate of the maximum microsegregation amplitude under typical ingot solidification conditions (Ref 43). The results presented in Table 2 predict that Mo has the

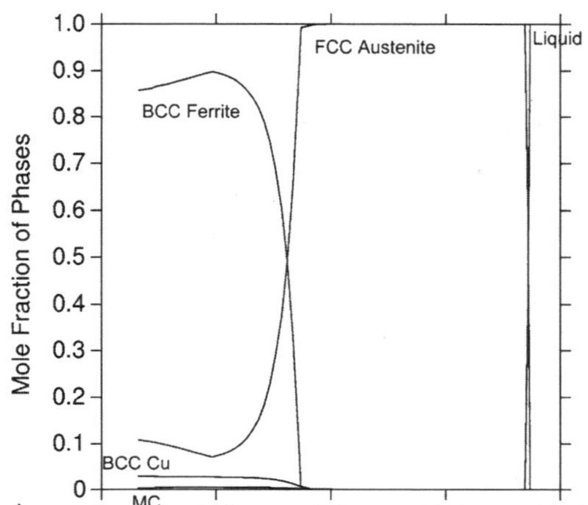

Fig. 17 Equilibrium mole fraction of phases in the alloy as a function of temperature, showing that the alloy is solution treatable at 900 °C

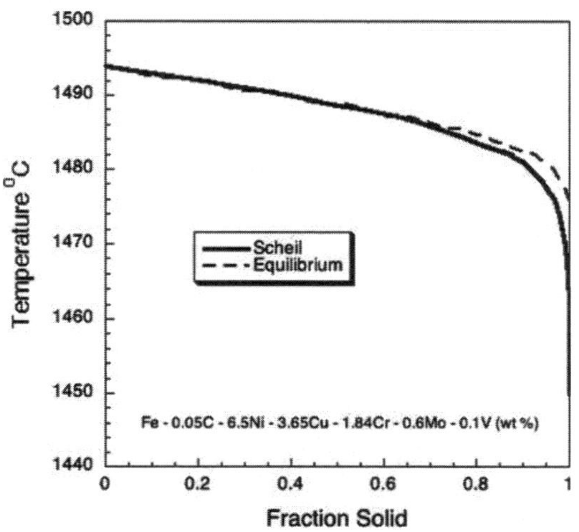

Fig. 18 Scheil simulation for evolution of the fraction solid with cooling for designed alloy Fe-0.05C-6.5Ni-3.65Cu-1.84Cr-0.6Mo-0.1V (wt%) in comparison with equilibrium solidification.

greatest potential for segregation. However, since the level of Mo in the alloy is relatively low, no serious Mo microsegregation problems are predicted for the designed composition. While the variation of Cu is significant, it was assessed to be acceptable for homogenization treatment at 1200 °C.

Selection of Tempering Temperature

Using multistep tempering to initially nucleate austenite at a high tempering temperature, the austenite stability for this transformation toughened alloy is dependent on the final tempering temperature. With the alloy composition fixed, the austenite stability was then calculated as a function of tempering temperature as shown in Fig. 20. It illustrates that the $\Delta G_{ch} + W_f$ value of 2836 J/mole desired for this alloy is achieved for a tempering temperature of 490 °C, very close to the originally assumed temperature of 500 °C.

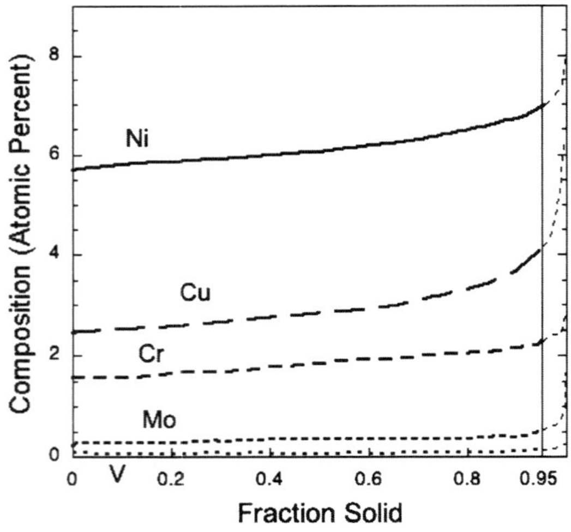

Fig. 19 Scheil simulation for composition profile of each alloying element after solidification for designed alloy Fe-0.05C-6.5Ni-3.65Cu-1.84Cr-0.6Mo-0.1V (wt%). Solid fraction corresponds to position relative to dendrite arm center.

Table 2 Amplitude of microsegregation with respect to each alloying element predicted by Scheil simulation at 95% solidification

Alloying elements	Ni	Cu	Cr	Mo	V
Nominal alloy composition—C_{alloy} (at.%)	6.38	3.31	2.04	0.36	0.11
Microsegregation amplitude $C_{0.95} - C_0$ (at.%)	1.29	1.67	0.72	0.34	0.05

Based on the design calculations, we derive a composition for the ultratough, high strength weldable plate steel (in wt%) with final tempering at 490 °C: Fe-0.05C-3.65Cu-6.5Ni-1.84Cr-0.6Mo-0.1V.

The composition should be solution treatable at 900 °C, with predicted M_S and B_S transformation temperatures of 298 °C and 336 °C, respectively. Based on previous transformation toughened steels, it is planned that initial tempering at a slightly elevated temperature will be employed to nucleate the austenite before tempering at 490 °C. to enrich the Ni content to the designed level. In our judgment, the highest risk assumption of this design is that fine austenite can be heterogeneously nucleated on Cu particles as effectively as it has been observed to nucleate on M_2C carbides (Ref 40). The observed effectiveness of Cu as a nucleant in numerous alloy systems (Ref 44, 45) supports this assumption.

Summary and Conclusions

A systematic strategy in computer-aided design of materials has been successfully applied to demonstrate feasibility of an ultratough high-strength weldable plate steel for naval hull applications. The approach integrated processing/structure/property/performance relations with mechanistic models to achieve the desired quantitative property objectives. Quantitative models were used to design the toughening and strengthening dispersions, which were the two major property requirements to be met under stringent processability constraints.

The overall strategy behind the thermodynamic modeling has been to map the mechanical properties objectives to structure and then to thermodynamic parameters to set goals for the design of the microstructural subsystems.

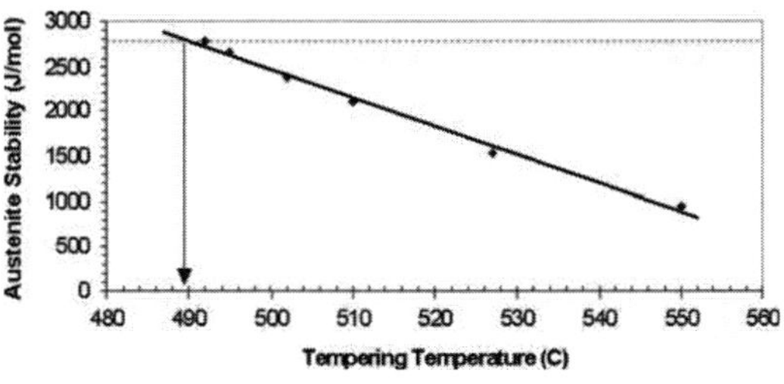

Fig. 20 Room temperature (300 K) stability of austenite as a function of tempering temperature. The required stability is predicted for 490 °C

The Olson-Cohen model for heterogeneous martensitic nucleation aided the determination of a thermodynamic stability parameter for the austenite. The optimal stability of the transformation-toughened austenite was calibrated against the parameter ($\Delta G_{ch} + W_f$) combining thermodynamic driving force and interfacial friction to obtain the required toughness, which was the top priority of this design. Then, based on the strength requirement projected to hardness values, the design space was identified. The alloy composition and the processing conditions were subsequently determined by conforming to these parametric design requirements.

The explored design concept is based on the mechanism of dispersed austenite stabilization for transformation toughening adapted to weldable high-strength steels. The concept employs mixed bainitic/martensitic microstructures produced by air-cooling of solution-treated plate, combined with copper and alloy carbide precipitation strengthening during secondary tempering, constrained by a low carbon content for weldability. A fine particle dispersion of optimal ~3–5 nm size for effective strengthening was designed by precipitation of M_2C carbides and BCC copper from a highly supersaturated BCC solution. The carbon content of the alloy was set at 0.05 wt% to meet weldability constraints. Based on the carbon level set, a quantitative carbide-strengthening model was used to determine the strength contribution from M_2C carbides, with the driving force for M_2C precipitation maximized at ~14 kJ/mole (while initially maintaining a stoichiometric balance between carbon and the carbide formers Mo, Cr, V) to obtain a fine 3–4 nm precipitate particle size. The additional strengthening required to meet the yield strength goal of 160 ksi (1100 MPa) was achieved by setting an optimal level of copper at 3.65 wt% based on a quantitative copper-strengthening model. This relatively high Cu level was necessary to allow the low carbon limit. A high Ni/Cu ratio (1.8) was also maintained in the multicomponent alloy design to prevent hot shortness problems during processing.

Transformation toughening arises from dispersed austenite precipitates, which undergo a martensitic transformation at the crack-tip stress state. This leads to inhibition of crack growth by delay of microvoid shear localization during ductile fracture. Thus to achieve high toughening by this mechanism, required stability of the austenite was designed by optimizing Ni as an FCC stabilizer. Thermodynamic calculations predicted an alloy Ni content of 6.5 wt% to enable the equilibrium nickel content of 30% in the austenite to meet the requirement for transformation toughening.

The design also revealed that although Cr did not have a strong effect on the driving force for carbide precipitation, it helped in partitioning Cu out of the austenite phase for effective copper precipitation strengthening; the alloy Cr level was set at 1.8 wt%. The processability conditions were then evaluated under stringent restrictions for the designed alloy (Fe-0.05C-3.65Cu-6.5Ni-1.84Cr-0.6Mo-0.1V; wt%) based on solution treatment condition, microsegregation behavior and tempering condition. A design solution treatment condition of 900 °C for 1 h was found sufficient to dissolve M_2C carbides without excessive austenite grain growth. No serious microsegregation problems were predicted for the designed composition based on results from Scheil simulation. A final tempering temperature of 490 °C. (after a higher temperature austenite nucleation step) was predicted to achieve sufficient austenite stability for transformation toughening. Thus, thermodynamic calculations demonstrated feasibility of combining copper and M_2C carbide precipitation for strengthening in combination with nickel-stabilized austenite for transformation toughening in a relatively low-cost weldable plate steel.

ACKNOWLEDGMENTS

This research was carried out under the financial support by the Office of Naval Research (ONR) under grant number N00014-01-1-0953 as a part of the ONR Grand Challenge in Naval Materials by Design. The authors are thankful to Gautam Ghosh, Ph.D. (Northwestern University), C.J. Kuehmann, Ph.D. and H.J. Jou, Ph.D. (QuesTek Innovations LLC) for their helpful discussions during the alloy design process.

REFERENCES

1. G.B. Olson, *Computational Design of Hierarchically Structured Materials*. Science 277, 1237–1242 (1997)
2. G.B. Olson, Pathways of Discovery: Designing a New Material World. Science 288 993–998 (2000)

3. M.S. Gagliano and M.E. Fine, Precipitation Kinetics of Niobium Carbide and Copper in a Low Carbon, Chromium-Free Steel. CALPHAD 25, 207–216 (2001)
4. G.N. Haidemenopoulos, G.B. Olson, and M. Cohen, Dispersed-Phase Transformation Toughening in Ultrahigh-Strength Steels. In G.B. Olson, M. Azrin, and E.S. Wright, (eds.) *Innovations in Ultrahigh-Strength Steel Technology* (34th Sagamore Army Materials Research Conference), p 549–593. US Government Printing Office (1990)
5. S. Socrate, Mechanics of Microvoid Nucleation and Growth in High-Strength Metastable Austenitic Steels. Doctoral dissertation, Department of Materials Science and Engineering, Massachusetts Institute of Technology, Cambridge, MA (1995)
6. G.B. Olson and C.J. Kuehmann, Transformation Toughening in Dispersed Phase Systems. In Austenite Formation and Decomposition Symposium in Mat Sci & Tech 2003 Proc (ISS & TMS), p 493–504 (2003)
7. W.T. Geng, A.J. Freeman, and G.B. Olson, Influence of Alloying Additions on Grain Boundary Cohesion of Transition Metals: First-principles Determination and its Phenomological Extension. Phys. Rev. B 63, 165415 (2000)
8. C.D. Kantner, Designing Strength, Toughness and Hydrogen Resistance: Quantum Steel. Doctoral dissertation, Department of Materials Science and Engineering, Northwestern University Evanston, IL (2002)
9. M.F. Ashby, Materials Selection in Conceptual Design. In Proc. Ashby Symp. On Materials Design, ASM World Materials Congress, Chicago, IL (1988)
10. C.S. Smith, *A Search for Structure*. MIT Press, Cambridge, MA (1981)
11. G.M. Jenkins, The Systems Approach. In J. Beishon, and G. Peters, (eds.) *Systems Behavior*, p 56–79. Open University Press, Birmingham, UK (1972)
12. F. Stavehaug, Transformation Toughening of γ'-Strengthened Metastable Austenitic Steels. Doctoral dissertation, Department of Materials Science and Engineering, Massachusetts Institute of Technology, Cambridge, MA (1990)
13. R.H. Leal, Transformation Toughening of Metastable Austenitic Steels. Doctoral dissertation, Department of Materials Science and Engineering, Massachusetts Institute of Technology, Cambridge, MA (1984)
14. K.C. Hsieh, Fracture Toughness of Ti-modified Air-melted UHS Steels, Doctoral dissertation proposal Department of Materials Science and Engineering, Northwestern University, Evanston, IL (2000)
15. W.M. Garrison and K.J. Handerhan, Fracture Toughness: Particle-Dispersion Correlations. In G.B. Olson, M. Azrin, and E.S. Wright, (eds.) *Innovations in Ultrahigh-Strength Steel Technology* (34th Sagamore Army Materials Research Conference), p 443–466. US Government Printing Office, (1990)
16. J.M. Barsom and S.T. Rolfe, *Fracture and Fatigue Control in Structures: Applications of Fracture Mechanics*. Prentice Hall p 159–188 (1987)
17. R.O. Ritchie, B. Francis, and W.L. Server, Evaluation of toughness in ALSL 4340 alloy steel austenitized at low and high temperatures. Metall. Trans. A 7A, 831–838 (1976)
18. G. Ghosh, C.E. Campbell, and G.B. Olson, An Analytical Electron Microscopy Study of Paraequilibrium Cementite Precipitation in Ultra-high Strength Steel. Metall. Mater. Trans. A 30A, 501–512 (1999)
19. J. Wise, Systems Design of Advanced Gear Steels. Doctoral dissertation, Department of Materials Science and Engineering, Northwestern University Evanston, IL (1998)
20. K.C. Russell and L.M. Brown, A Dispersion Strengthening Model Based on Differing Elastic Moduli Applied to the Iron-Copper System. Acta Metall. 20, 969–974 (1972)
21. R.P. Foley and M.E. Fine, Microstructure and Property Investigation of Quenched and Tempered HSLA-100 Steel. In A.J. DeArdo, (ed.) *ISS, Proceedings of the International Conference on Processing, Microstructure and Properties of Microalloyed and Other Modern High Strength Low Alloy Steels*, p 315–329 Warrendale, PA (1991)
22. C.J. Kuehmann, Thermal Processing Optimization of Nickel-Cobalt Ultrahigh-Strength Steels. Doctoral dissertation, Department of Materials Science and Engineering, Northwestern University Evanston, IL (1994)

23. D. Spaulding, Grain Boundary Cohesion and Segregation in Ultrahigh Strength Alloy Steels. Doctoral dissertation, Department of Materials Science and Engineering, Northwestern University, Evanston, IL (1995)
24. M.F. Ashby and D.R.H. Jones, *Engineering Materials: An Introduction to their Properties and Application.* Pergamon Press (1980)
25. B.R. Somers, Introduction to the Selection of Carbon and Low-Alloy Steels. In: *ASM Handbook: Welding, Brazing, and Soldering*, 6th edn., p 405–407. ASM International (1993)
26. M.S. Gagliano, Co-precipitation of Copper and Niobium Carbide in a Low Carbon Steel. Doctoral dissertation, Department of Materials Science and Engineering, Northwestern University Evanston, IL (2002)
27. B. Sundman, B. Jansson, and J. Andersson, The ThermoCalc databank system. CALPHAD 9, 153–190 (1985)
28. S.R. Goodman, S.S. Brenner, and J.R. Low, An FIM-Atom Probe Study of the Precipitation of Copper from Iron-1.4 At. Pct Copper. Part I: Field-Ion Microscopy. Metall. Trans. A 4, 2363–2369 (1973)
29. S.R. Goodman, S.S. Brenner, and J.R. Low, An FIM-Atom Probe Study of the Precipitation of Copper from Iron-1.4 At. Pct Copper. Part II: Atom Probe Analyses. Metall. Trans. A 4, 2371–2378 (1973)
30. W.J. Phythian, A.J.E. Foreman, C.A. English, J.T. Buswell, M. Hetherington, K. Roberts, and S. Pizzini, The Structure and Hardening Mechanism of Copper Precipitation in Thermally Aged or Irradiated Fe-Cu and Fe-Cu-Ni Model Alloys. In R.E. Stoller, A.S. Kumar, D.S. Gelles, (eds.) *Effects of Radiation on Materials: 15th International Symposium*, p 131–150. ASTM STP 1125, (1992)
31. A. Deschamps, M. Militzer, and W.J. Poole, Precipitation Kinetics and Strengthening of a Fe-0.8 wt% Cu alloy. ISIJ Int. 41, 196–205 (2001)
32. G.M. Worrall, J.T. Buswell, C.A. English, M.G. Hetherington, and G.D.W. Smith, A Study of the Precipitation of Copper Particles in a Ferrite Matrix. J. Nucl. Mater. 148, 107–114 (1987)
33. E. Hornbogen, In: G.R. Speich, and J.B. Clark (eds.) *Precipitation from Iron-Base Alloys*. Gordon and Breach, New York, NY (1965)
34. S. Pizzini, K.J. Roberts, W.J. Phythian, C.A. English, and G.N. Greaves, A Fluorescence EXAFS Study of the Structure of Copper-rich Precipitates in Fe-Cu and Fe-Cu-Ni Alloys. Phil. Mag. Lett. 61, 223–229 (1990)
35. W.J. Phythian, S. Dumbill, P. Brown, and R. Sinclair, Stability of Thermally Induced Copper Precipitates Under Neutron Irradiation. In: *Proceedings of the 6th International Symposium on Environmental Degradation in Nuclear Power Systems–Water Reactors*, p 729–737. TMS (1993)
36. M.J. Gore, G.B. Olson, and M. Cohen, Grain-Refining Dispersions and Properties in Ultrahigh Strength Steels. In: G.B. Olson, M. Azrin, and E.S. Wright, (eds.) *Innovations in Ultrahigh-Strength Steel Technology* (34th Sagamore Army Materials Research Conference), p 425–441. US Government Printing Office, (1990)
37. G.B. Olson, K. Tsuzaki, and M. Cohen, Statistical Aspects of Martensitic Nucleation. In: G.S. Cargill, F. Spaepen, and K.N. Tu, (eds.) *David Turnbull Symposium on Phase Transitions in Condensed Systems*, vol. 57, p 127–148. MRS, (1987)
38. G. Ghosh and G.B. Olson, Kinetics of FCCBCC Heterogeneous Martensitic Nucleation – I. The Critical Driving Force for Athermal Nucleation. Acta Mater. 42, 3361–3370 (1994)
39. G. Ghosh and G.B. Olson, Kinetics of FCCBCC Heterogeneous Martensitic Nucleation – II. Thermal Activation. Acta Mater. 42, 3371–3379 (1994)
40. H.E. Lippard, Microanalytical Investigations of Transformation Toughened Co–Ni Steels. Doctoral dissertation, Department of Materials Science and Engineering, Northwestern University Evanston, IL (1999)
41. C.E. Campbell and G.B. Olson, Systems design of high performance stainless steels I. Conceptual and computational design. J. Comput. Aided Mat. Design 7, 145–170 (2000)
42. E. Scheil, Bemerkungen zur Schichtkristallbildung. Z. Metallkd. 34, 70–72 (1942)
43. H.E. Lippard, C.E. Campbell, T. Bjorklind, U. Borggren, P. Kellgren, V.P. Dravid, and

G.B. Olson, Microsegregation behavior during solidification and homogenization of AerMet100 steel Metall. Mater. Trans. B 29, 205–210 (1998)
44. K. Hono, D.H. Ping, M. Ohnuma, and O. Onodera, Cu Clustering and Si Partitioning in the Early Crystallization Stage of an $Fe_{73.5}Si_{13.5}B_9Nb_3Cu_1$ Amorphous Alloy. Acta Mater. 47, 997–1006 (1999)
45. T. Ohkubo, H. Kai, D.H. Ping, K. Hono, Y. Hirotsu, Mechanism of Heterogeneous Nucleation of α-Fe Nanocrystals from $Fe_{89}Zr_7B_3Cu_1$ Amorphous Alloy. Scripta Mater. 44, 971–976 (2001)

Transformations—Selected Works of G.B. Olson on Materials, Microstructure, and Design
C.E. Campbell, M.V. Manuel, and W. Xiong, editors

CHAPTER 19

Prototype Evaluation of Transformation Toughened Blast-Resistant Naval Hull Steels—Part II*

A. Saha and G.B. Olson, Northwestern University
J. Jung, Quest Tek Innovations

Abstract

Application of a systems approach to computational materials design led to the theoretical design of a transformation toughened ultratough high-strength plate steel for blast-resistant naval hull applications. A first prototype alloy has achieved property goals motivated by projected naval hull applications requiring extreme fracture toughness ($C_v >$ 85 ft-lbs or 115 J corresponding to $K_{Id} \geq$ 200 ksi.in$^{1/2}$ or 220 MPa.m$^{1/2}$) at strength levels of 150–180 ksi (1030–1240 MPa) yield strength in weldable, formable plate steels. A continuous casting process was simulated by slab casting the prototype alloy as a 1.75 in. (4.45 cm) plate. Consistent with predictions, compositional banding in the plate was limited to an amplitude of 6–7.5 wt% Ni and 3.5–5 wt% Cu. Examination of the oxide scale showed no evidence of hot shortness in the alloy during hot working. Isothermal transformation kinetics measurements demonstrated achievement of 50% bainite in 4 min at 360 °C. Hardness and tensile tests confirmed predicted precipitation strengthening behavior in quench and tempered material. Multistep tempering conditions were employed to achieve the optimal austenite stability resulting in significant increase of impact toughness to 130 ft-lb (176 J) at a strength level of 160 ksi (1100 MPa). Comparison with the baseline toughness-strength combination determined by isochronal tempering studies indicates a transformation toughening increment of 65% in Charpy energy. Predicted Cu particle number densities and the heterogeneous nucleation of optimal stability high Ni 5 nm austenite on nanometer-scale copper precipitates in the multistep tempered samples was confirmed using three-dimensional atom probe microanalysis. Charpy impact tests and fractography demonstrate ductile fracture with $C_v >$ 80 ft-lbs (108 J) down to −40 °C, with a substantial toughness peak at 25 °C consistent with designed transformation toughening behavior. The properties demonstrated in this first prototype represent a substantial advance over existing naval hull steels. Achieving these improvements in a single design and prototyping iteration is a significant advance in computational materials design capability.

*A. Saha, J. Jung, and G.B. Olson, "Prototype Evaluation of Transformation Toughened Blast Resistant Naval Hull Steels: Part II," *J Computer-Aided Mater Des*, 14, 2 (2007) 201–233

Introduction

A systems approach (Ref 1; also Chapter 18 in this book) to computational materials design integrating processing/structure/property/performance relations with mechanistic models was applied to the design of high-toughness plate steels at high strength employing transformation toughening phenomena while constraining the alloy carbon content of the steel for weldability. The toughening concept explored is based on the mechanism of dispersed austenite stabilization for transformation toughening adapted to weldable bainitic/martensitic plate steels in an alloy strengthened by precipitation of M_2C carbides in combination with copper. The desired microstructure is thus a matrix containing a bainite-martensite mix, BCC copper, and M_2C carbide for strengthening with a fine dispersion of optimum stability austenite for transformation toughening. The bainite-martensite mix will be formed by air-cooling from a solution treatment temperature and subsequent multistep aging at secondary hardening temperatures will precipitate the toughening and strengthening dispersions. The strengthening approach is based on design concepts of the current Navy HSLA100 steel (Fe-0.06C-0.9Mn-0.4Si-3.5 Ni-1.6Cu-0.6Mo-0.03Nb; in wt%) with a quench and temper processing treatment. This is integrated with modeling of nickel-stabilized austenite produced by precipitation as demonstrated in transformation toughened AerMet100 and AF1410 steels with multistep tempering treatments (Ref 2–4). The proposed methods of toughening and precipitation strengthening have been modeled in this design to assess theoretical feasibility in order to minimize necessary experimentation.

Details of the computational design are presented in Ref 1. This paper reports the experimental evaluation of a single prototype of the designed alloy, to validate both the design and the models that created it.

Material and Experimental Procedures

Material

Special Metals Corporation in New Hartford, New York, produced the alloy as a 34-pound experimental heat by Vacuum Induction Melting (VIM) from 100% virgin raw materials and cast into a 9.5 × 8 × 1.75 in. (24.1 × 20.3 × 4.5 cm) slab ingot as a simulation of a continuous casting process. The as-cast ingot was subsequently homogenized at 2200 °F (1204 °C) for 8 h and then hot rolled to 0.45 in. (1.1 cm) thickness followed by air-cooling to room temperature by Huntington Alloys in Huntington, West Virginia. The final dimension of the plate measured roughly 33 × 10 × 0.45 in. (83.8 × 25.4 × 1.1 cm). The hot-rolled plate was annealed at 900 °F (482 °C) for 10 h to improve machinability. The designed and the actual compositions (in wt%) of the alloy are given in Table 1. The impurity level in the alloy was measured as 0.002 wt% S, 13 ppm O, and 2 ppm N. The measured composition is very close to the designed composition and the variations are within the tolerance limits of the design.

Experimental Procedures

Heat Treating. Mechanical test samples were evaluated in the martensitic condition. These samples were solution treated at 900 °C for 1 h and quenched in water followed by a liquid nitrogen cool for 30 min prior to every tempering treatment to ensure a fully martensitic starting microstructure with minimal retained austenite. Solution treatments were performed in an argon atmosphere to prevent oxidation of samples. To ensure rapid heating of the entire sample, the short-time nucleation stage heat treatments were conducted using a molten salt bath followed by water-quenching to room temperature. The salt used for the molten bath was Thermo-Quench Salt (300–1100 °F) produced by Heat Bath Corporation. The residue layer from the salt pot treatment was ground off before the second step aging treatment. The standard aging treatments for longer times (1–10 h) were performed in a box furnace with vacuum encapsulation to prevent oxidation and decarburization, and then air-cooled to room temperature. Vacuum encapsulation employed 0.75 in. diameter Pyrex tubes evacuated by a diffusion pump. After evacuation, the tubes were backfilled with argon three times before reaching a final vacuum of <5 mtorr, and then sealed.

Metallographic Sample Preparation. All samples were ground and polished directly to 1 µm finish using a Buehler Ecomet-4 variable speed automatic grinder/polisher. The samples prepared for measuring hardness were mounted in room temperature curing acrylic, while those prepared for microsegregation studies were hot

Table 1 Designed and Measured Composition (in wt%) of alloy

Alloy	Fe	C	Cu	Ni	Cr	Mo	V
Design	Bal.	0.05 ± 0.01	3.65 ± 0.05	6.5 ± 0.2	1.84 ± 0.05	0.6 ± 0.05	0.1 ± 0.01
Measured	Bal.	0.040	3.64	6.61	1.78	0.58	0.11

mounted with conductive phenolic resin using a Stuers LaboPress-1 after nickel-plating for edge retention of the oxide layer during grinding and polishing. Microsegregation samples were etched in a 2% nital (2% nitric acid in ethanol) solution for 10–30 s to reveal the compositional banding close to the metal-oxide interface associated with scale formation during hot working. Following etching, the samples were viewed with an optical microscope and analytical SEM.

Dilatometry. Dilatometric study of phase transformation kinetics employed a computer-controlled MMC Quenching Dilatometer. Specimens were prepared by EDM (Electro-Discharge Machining) wire cutting into cylindrical rods 10 mm long and 3 mm in diameter. The samples are heated by an induction furnace and cooled by jets of helium gas. They are mounted between two low-expansion quartz platens, which are lightly spring-loaded and connected to an LVDT transducer to record length. The temperature is monitored by a Pt-Pt 10%Rh thermocouple spot welded directly to the sample surface. The sample stage is enclosed in a vacuum chamber connected to a turbo-mechanical pump system capable of achieving a vacuum of 10^{-4} torr.

The dilatometry was used for determining the martensite start temperature (M_S) and for evaluating the bainite transformation kinetics. For estimating the experimental M_S temperature, samples were heated at a rate of 2–3 °C/s to 1050 °C, held for 5 min and then rapidly quenched (>100 °C/s) to room temperature. For studying the bainite kinetics, samples were held isothermally for 2 h at bainite transformation temperatures between 360 and 420 °C after quenching (Cooling rate from 800 to 500 °C, $T_{8/5}$ = 50 °C/s) from the austenizing temperature. All samples were austenized at 1050 °C for 5 min and then rapidly quenched prior to the actual runs of martensite and bainite transformation in order to ensure a uniform starting microstructure.

Microhardness Testing. Vickers hardness was measured for every aging condition as a measure of strength. Hardness measurements of materials in this study were performed using the Buehler Micromet II Micro Hardness Tester based on the method prescribed in ASTM standard E384. Prior to testing, all the heat-treated samples were mounted in acrylic molds and polished to 1 μm. The samples were at least 8 mm thick and ground to reveal opposite surfaces to avoid any errors due to anvil effects. At least 10 hardness measurements were recorded uniformly across the cross-section for every sample tested and the average was documented as the hardness value.

Impact Toughness Testing. The impact toughness properties for the different heat treatment conditions of the alloy were measured using a Tinius Olsen 260 ft-lb (352 J) impact-testing machine. Prior to testing, the samples were machined according to the ASTM standard Charpy V-notch dimensions (1996 ASTM E23) 10 × 10 × 55 mm (0.39 × 0.39 × 2.17 in.) with a 45° center notch of depth 2 mm and root radius of 0.25 mm. The specimens corresponded to the L-T orientation. The impact fracture energy was measured directly on an analog scale and the given impact energy data was typically based on a two-sample average. For the low-temperature impact fracture properties, the samples were held for 20 min at the test temperature in an Instron low-temperature furnace connected to a liquid nitrogen supply. Within 5 s of removal from the furnace, the samples were placed inside the machine and struck with the hammer.

Tensile Testing. Tensile test specimens were machined from blanks measuring approximately 10 × 10 × 70 mm (0.39 × 0.39 × 2.76 in.) from the original plate parallel to the longitudinal rolling direction. Prior to machining, the samples were solution-treated and aged as discussed in the section "Heat Treating." From each blank, subsized tensile specimens, scaled in accordance to ASTM standards (1996 ASTM E8M) were machined as shown schematically in Fig. 1. The final specimen had a gage diameter of 6 mm (0.24 in.) and a gage length of 30 mm (1.18 in.).

All tensile tests were performed at room temperature using a computer-controlled Sintech 20/G screw driven mechanical testing machine with a 20,000 lb (8,896 N) load cell at a constant cross-head speed of 0.005 in/s

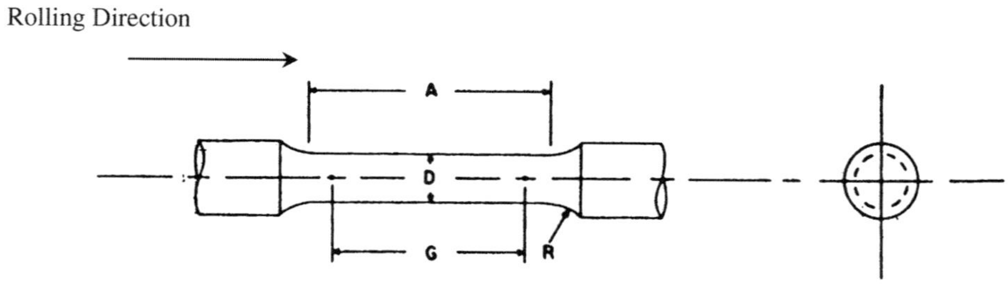

G (gage length)	30 ± 0.1 mm (1.18" ± 0.004")
D (gage diameter)	6 ± 0.1 mm (0.24" ± 0.004")
R (radius of fillet)	6 mm (0.24")
A (length of reduced section)	36 mm (1.42")

Fig. 1 Tensile test specimen dimensions (Standard ASTM E23)

(0.127 mm/s), using a calibrated extensometer of gage length 1 in. (25.4 mm). The load-time response was recorded using the TestWorks computer software package employing the actual cross-sectional areas and gage lengths of the specimens measured prior to testing. Area reduction and extension were also measured manually upon completion of the test. The ultimate tensile strengths, 0.2% offset yield strengths and total elongations were obtained based on two-sample averages.

Scanning Electron Microscopy. A Hitachi S-3500 scanning electron microscope (SEM) with a tungsten hairpin filament was used to investigate both the composition banding in the as-rolled samples and the fracture surfaces of the Charpy impact specimens. The microscope uses Quartz PCI Image Management Software for capturing images and for conducting quantitative analysis. For analysis, the samples were mounted on graphite tape and examined in the SEM with a 20 kV electron beam at a vacuum level of 10^{-4} torr. The secondary electron (SE) detector was used for imaging both the etched and fractured surfaces. The compositionally banded structure of the etched sample was characterized quantitatively near the metal-oxide interface using the PGT Energy Dispersive x-ray analyzer with digital pulse processing.

3D Atom Probe/Field ion Microscopy (AP-FIM). A three-dimensional atom probe microscope (MRC Atom-Probe Facility, Northwestern University) described in Ref 5 was used for characterizing the size, number-density and composition of nanoscale strengthening (Cu precipitates) and toughening (Ni-stabilized austenite) dispersions in the heat-treated samples. The atom probe, operated and maintained under an ultrahigh vacuum system (10^{-10}–10^{-11} torr) combined with a field ion microscope, operated with imaging gas at a pressure level of 10^{-5} torr, makes it an extremely high-resolution microscopy technique.

The specimens (atom probe tips) were prepared by a two-step electropolishing sequence of small rods (100 mm long with 200 × 200 μm square cross-section) cut from heat-treated hardness samples. Initial polishing used a solution of 10% perchloric acid in butoxyethanol at room temperature applying a DC voltage of 23 V until the square rods were shaped into long needles with a small taper angle. A solution of 2% perchloric acid in butoxyethanol at room temperature was used for necking and final polishing to produce a sharply pointed tip, with a radius of curvature less than 50 nm. The voltage was gradually decreased from 12 V DC to 5 V DC during the final stages of electropolishing.

FIM analysis was performed at temperatures between 50 and 80 K with a chamber pressure of 10^{-5} torr of neon gas. FIM images were captured during analysis using the Scion Image imaging software. Atom probe analysis was then conducted at temperatures 50 and 70 K under ultrahigh vacuum conditions (10^{-10} – 10^{-11} torr) for pulsed field-evaporation with a pulse fraction (pulse voltage/steady state DC voltage) of 20% at a pulse frequency of 1,500 Hz.

The standard error, σ for compositions was calculated using binomial statistics according to the equation (Ref 6):

$$\sigma = \sqrt{\frac{c_i(1-c_i)}{N}} \quad \text{(Eq 1)}$$

where c_i is the measured composition of element i and N is the total number of ions sampled. This standard error does not account for any overlapping mass to charge ratios between different elements. Systematic errors that may interfere with the collection of specific elements such as carbon may be an additional source of error.

The data from 3DAP were analyzed and visualized by the software *ADAM* developed by Hellman et al. (Ref 7), providing a range of tools to analyze 3DAP data (Ref 8, 9), including the ability to define planar or cylindrical regions of interest and perform analyses such as concentration profiles, ladder diagrams and composition maps with respect to a region of interest. Isoconcentration surfaces were employed to examine the size and shape of precipitates.

To analyze a larger volume of material a single sample was also run on a Local-Electrode Atom Probe (LEAP) at Imago LLC. The data were analyzed using the same analysis software.

Prototype Evaluation Results

The primary goal of prototype evaluation in this study is to experimentally verify the processing-structure and structure-property relationships quantified during the alloy design process. The analysis began with evaluation of the processability characteristics of the designed alloy at an experimental-heat scale. Experimental optimization of the multistep tempering treatment was then undertaken from the viewpoint of toughness/strength combination. Characterization of the strengthening and toughening dispersions provided the ultimate validation of the microstructural design.

Primary Processing Behavior

Microsegregation and Hot-Working Behavior. To study the microsegregation behavior in our cast prototype, the as-received material (homogenized for 8 h at 1204 °C, hot-rolled for 75% reduction to 0.45 in. or 4.5 cm thick plate and then annealed at 482 °C for 10 h) in the form of a 10 × 10 × 20 mm sample, was etched with 2% nital following standard metallographic polishing to 1 μm. Low-magnification transverse optical micrographs revealed both the banded structure oriented along the longitudinal rolling direction and the oxide-metal interface as shown in Fig. 2.

The centerline of the hot-rolled plate did not reveal as much of a banded structure as the surface region, as shown in Fig. 3. A higher magnification optical micrograph at the centerline of the plate presented in Fig. 4 shows an equiaxed microstructure, which is predominantly lath martensite in the form of packets

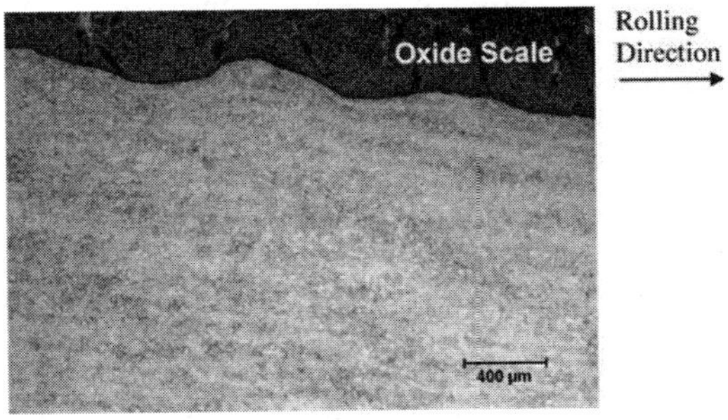

Fig. 2 Optical micrograph of the as-received plate viewed transverse to the rolling direction at the oxide-metal interface after etching with 2% nital

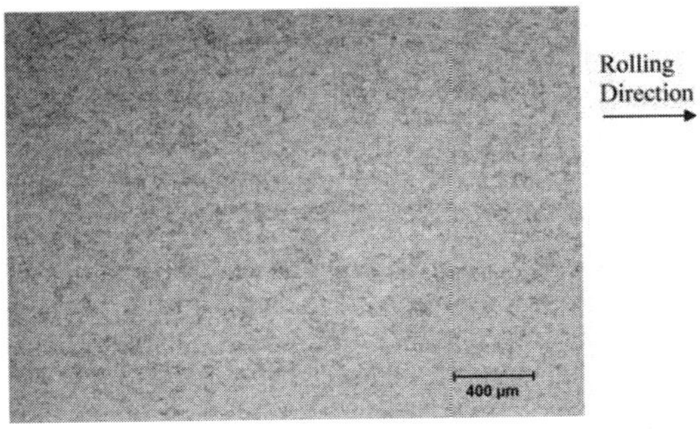

Fig. 3 Optical micrograph of the hot-rolled plate viewed transverse to the rolling direction at the centerline after etching with 2% nital

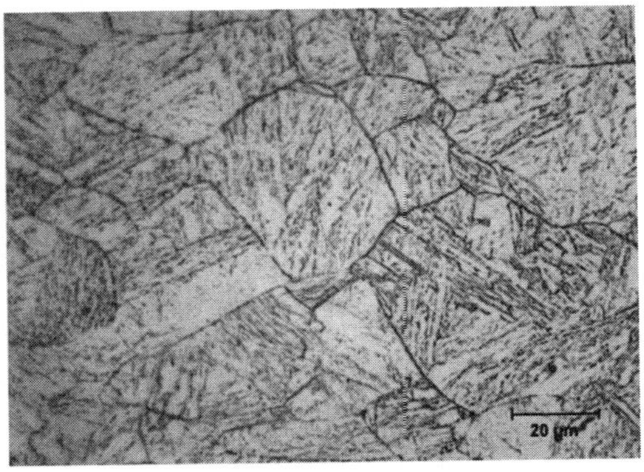

Fig. 4 Higher magnification optical micrograph of the hot-rolled plate at the centerline

within the prior austenite grains of an average size of 50 μm.

The composition bands revealed on etching in Fig. 2 were estimated to be of 40–50 μm thickness. The extent of microsegregation within these bands was determined by measuring the composition profile across the thickness of the plate near the oxide-metal interface. Composition data were collected every 4 μm starting from the metal-oxide interface and proceeding toward the center of the plate. The composition variation across the bands with respect to the major alloying elements Ni, Cu, Cr, and Mo is presented in Fig. 5. It was found that compositional banding in the plate was limited to an amplitude of approximately 6–7.5 wt% Ni, 3.5–5 wt% Cu, 1.6–2 wt% Cr, and 0.2–0.5 wt% Mo and agrees well with the microsegregation limits estimated from Scheil simulation in Ref 1. Based on the strength model, a variation in the level of Cu across the bands within 3.5–5 wt% corresponds to a predicted hardness variation of 30 VHN equivalent to 6.8 ksi (47 MPa) in yield strength. This will promote a smooth yielding behavior as confirmed by the measured tensile properties.

Investigations in the past (Ref 10–12) have shown that copper can be a particularly detrimental element with respect to hot shortness. At hot rolling temperatures above 1050 °C in an oxidizing atmosphere, iron is selectively oxidized leaving an enrichment of copper near the oxide-metal interface, whereby a copper-enriched liquid phase enters grain boundaries of the austenite causing intergranular fracture during hot rolling. The advantages of copper addition to steels for strengthening as well as improving atmospheric-corrosion resistance has led to extensive research (Ref 13–17) showing that the addition of nickel in an amount equal to 0.5–1 times that of copper eliminates surface cracking. Consistent with the high Ni/Cu ratio of 1.8 maintained in our design, no hot-shortness was encountered during processing. As further verification, the oxide layer of the as-received material shown in Fig. 6 was examined carefully. Composition analysis of various regions in the oxide layer did not reveal any Cu-rich phase but did show some Ni-enriched phases varying from 20% to 80% within the Fe-rich oxide. This study thus supports the ability of Ni to suppress Cu-enriched liquid during oxidation (Ref 14).

Evaluation of Allotropic Kinetics. A dilatometry study was next conducted to determine the allotropic kinetics of the prototype. Figure 7 presents a typical plot of the relative length change versus temperature for a fast cooled sample. Straight lines are fit to the single phase portions of heating and cooling curves, the full width between them defining full transformation. The series of dashed lines superimposed on the length and temperature trace represent varying degrees of partial martensitic transformation during rapid quench from an austenizing temperature of 1050 °C. The threshold for transformation defining M_S is taken as 1% (Ref 2).

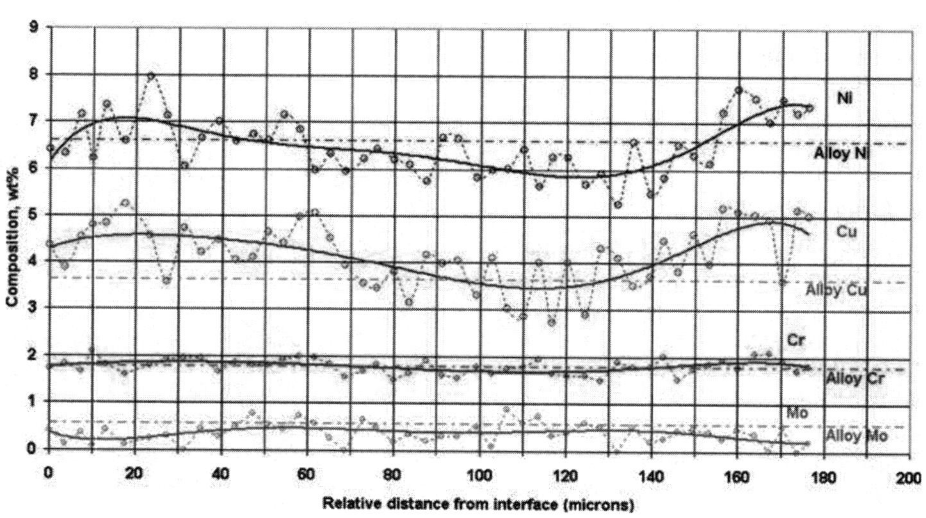

Fig. 5 Line profile compositions for as-received material from oxide-metal interface

The M_S temperature, averaged over 15 dilatometry runs, is 360 ± 8.4 °C. The predicted M_S temperature from the Ghosh-Olson model (18) using the SSOL database was 298 °C.

Since the alloy was designed to produce a bainite/martensite microstructure during air-cooling of plates, the bainite kinetics was determined by measuring the isothermal time-temperature-transformation diagram. The relative length change versus temperature dilatometry trace for a two-hour isothermal hold at 377 °C is presented in Fig. 8. In this case, the total bainitic transformation that took place after 2 h is 44.1%. The evolution of bainitic transformation with respect to time at 377 °C is shown in Fig. 9. It is apparent that the volume fraction of bainite is saturated after a 2 h isothermal hold. Similar analyses were carried out for each 2 h test performed at isothermal temperatures ranging from 362 to 407 °C. Table 2 summarizes the maximum transformation levels at all the temperatures. The TTT curve based on the data from all of the isothermal runs is presented in Fig. 10. It shows that we can achieve a 50% bainite/martensite mix in approximately 4 min at 360 °C. The experimental B_S temperature was determined to be 410 °C. The temperature interval between M_S and B_S is in good agreement with the model predictions, but both temperatures are about 50 K higher than the model predictions using the SSOL thermodynamic database.

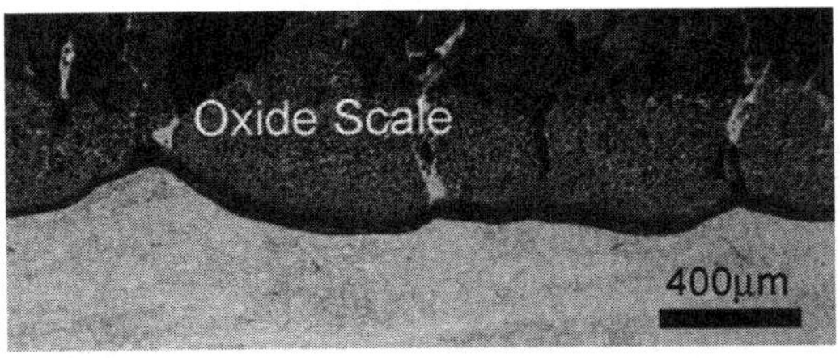

Fig. 6 Optical micrograph showing the oxide scale in the as-received plate.

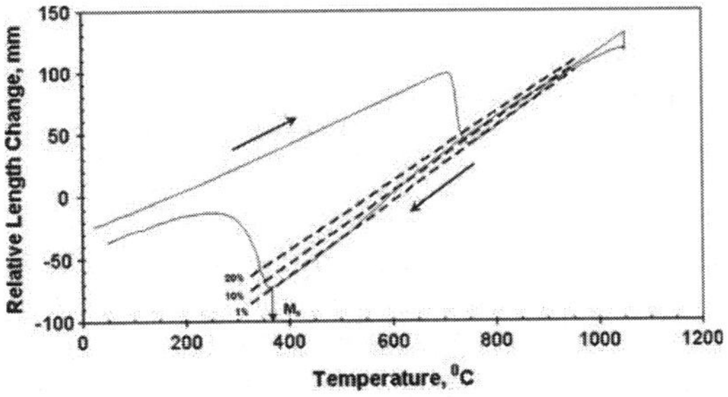

Fig. 7 Relative sample length change and temperature trace during heating, cooling (quench) cycle from dilatometry experiment

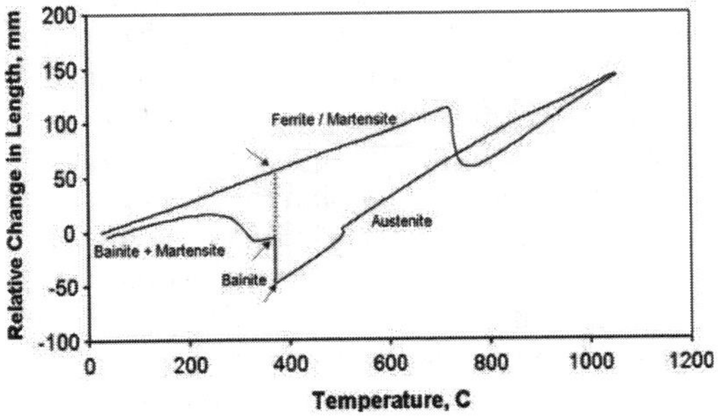

Fig. 8 Relative sample length change and temperature trace during heating, cooling and isothermal hold at 377 °C from dilatometry experiment

Thermal Process Optimization

Isochronal Tempering Response. An isochronal tempering study was conducted to evaluate the basic tempering characteristics of the prototype and provide a baseline for later

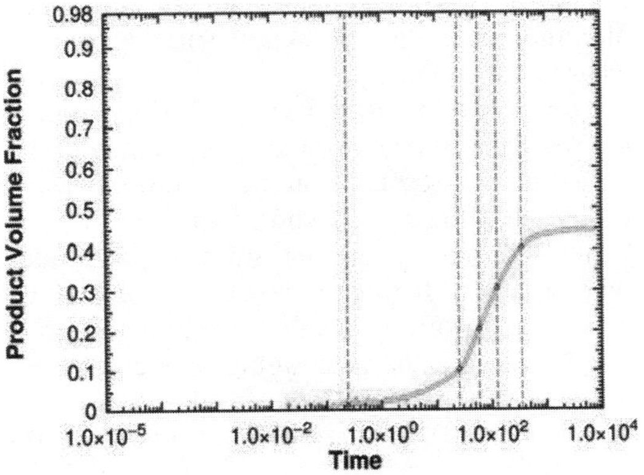

Fig. 9 Volume fraction evolution of bainite as a function of time for isothermal temperature of 377 °C

Table 2 Saturation volume fraction of bainite as a function of isothermal temperature

Temperature (°C)	Saturation volume fraction of bainite
362	0.61
367	0.53
372	0.50
377	0.44
382	0.24
387	0.27
392	0.10
402	0.02
407	0.01

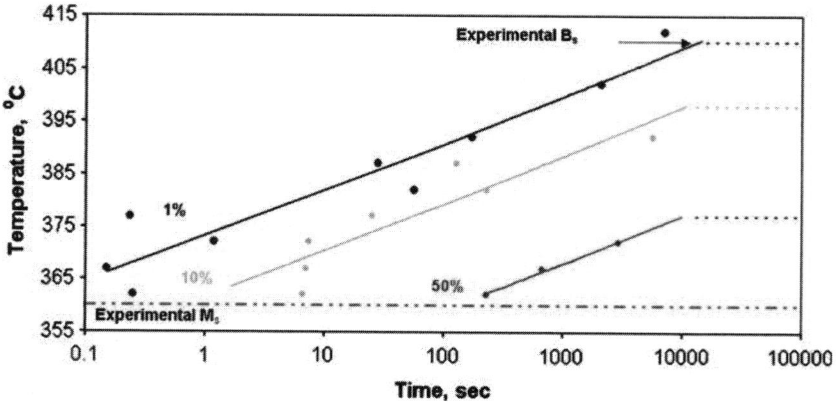

Fig. 10 Time-temperature-transformation (TTT) curve for bainite transformation reaction

studies of multistep tempering treatments. To simplify the optimization, the tempering response investigation was performed in a uniform martensite matrix to minimize retained austenite effects. After a solution treatment at 900 °C for 1 h followed by a water quench and liquid nitrogen cool, tempering was performed for 1, 5, and 10 h under vacuum. Samples were finish machined, notched and then tested at room temperature for Charpy impact toughness. Hardness measurements were taken directly from the polished surface of the Charpy specimens.

The tempering response for 1 h isochronal tempering was investigated over a temperature range of 200–600 °C in the solution-treated prototype alloy and is shown in Fig. 11. The 1 h isochronal tempering study demonstrates that a peak hardness level is reached at 420 °C followed by gradual overaging. This is consistent with the peak aging temperature for 30 minute tempering reported by Maruyama et al. (Ref 19) in martensitic steels strengthened by copper precipitation. The retention of high hardness even after the peak aging condition to 500 °C can be attributed to precipitation of M_2C carbides and a fine austenite dispersion observed at secondary hardening temperature of 482 °C in carbide strengthened ultra-high strength steels like AerMet100 and AF1410 (2). The hardness at 500 °C (represented by an arrow in Fig. 11) is in very good agreement with that predicted for the calculated final tempering temperature (490 °C) in Ref 1 to achieve the design objectives.

After confirming the basic secondary hardening characteristics of the prototype alloy, a series of isochronal tempering treatments of

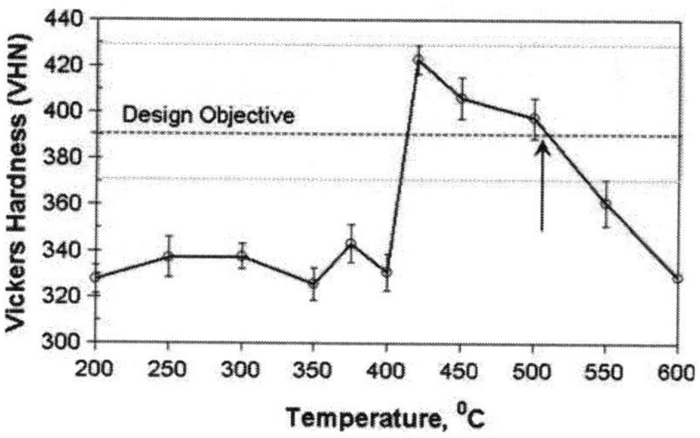

Fig. 11 Isochronal (1 h) tempering response of prototype alloy. The arrow superimposed on the plot shows that the design objective is achieved by tempering at 500 °C in agreement with design prediction

Charpy specimens were performed for 1, 5, and 10 h within a temperature range of 400–600 °C. Figure 12 illustrates the room temperature Charpy toughness (C_V)–Vickers hardness (VHN) trajectory for the indicated tempering temperatures. This establishes the baseline of the toughness-hardness (strength) combination in tempered martensitic microstructures. The shape of the trajectory is consistent with earlier studies of HSLA100 (Ref 20), AF1410, and AerMet100 (Ref 2) where the best combination of strength and toughness are obtained in slightly overaged condition corresponding to complete cementite dissolution.

At the shortest tempering time of 1 h we see from Fig. 12 that cementite formation limits toughness, and as Cu precipitates in its presence, strength increases from 400 °C to 450 °C tempering treatment while there is a sharp decline in toughness. With further tempering, cementite begins to dissolve as a result of M_2C carbide formation in combination with

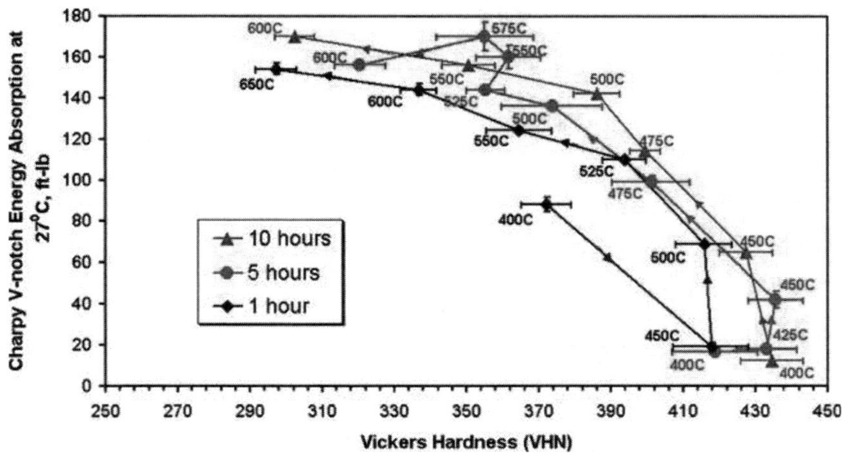

Fig. 12 Isochronal tempering response represented by Charpy toughness-Vickers hardness trajectory. The label corresponding to each data point indicates the tempering temperature

BCC copper precipitation at the peak aging condition. This results in an increase of both strength and toughness. The toughness-hardness trajectory takes a sharp turn thereafter, as the strengthening precipitates begin to coarsen exceeding their sizes predicted for peak strengthening effect and the strength continues to decrease with overaging. Figure 12 suggests that peak hardness occurs at 450 °C 5 h tempering and the corresponding toughness resides on an upper band suggesting full dissolution of paraequilibrium cementite by precipitation of a fine M_2C strengthening dispersion.

The highly overaged region is also likely associated with precipitation of a dispersion of austenite, which increases in stability due to Ni enrichment at higher tempering times. An interesting feature observed in the toughness-hardness trajectory for 5 h tempering in Fig. 12 between tempering temperatures of 525 and 575 °C is a toughness enhancement from the baseline toughness of 144 ft-lbs (195 J), to 170 ft-lbs (230 J), respectively, a toughness increment by 18% at a strength level corresponding to 355 VHN. This is characteristic of the transformation toughening phenomenon caused by the austenite reaching an optimal stability for the lower strength condition.

The tempering response of the hardness (strength) can be correlated to the Hollomon-Jaffe tempering parameter (Ref 21, 22), defined as $T(18 + \ln(t))$ where T is tempering temperature in K and t is the tempering time in minutes. Figure 13 presents the measured values of hardness for different tempering conditions as a function of this parameter. This provides a simple interpolation scheme to adjust tempering for a desired strength level.

The fracture surfaces of the broken Charpy impact testing samples were observed under SEM to characterize the mode of fracture. The fracture surface for the 450 °C 1 h tempering condition is presented in Fig. 14. The SEM micrograph reveals that the sample failed by quasi-cleavage fracture with signs of intergranular embrittlement. The fracture mode represents relatively brittle behavior attributed to the presence of undissolved cementite at short tempering times.

For higher tempering times and temperatures, ductile fracture occurred by microvoid nucleation and coalescence. Representative SEM micrographs showing a ductile mode of fracture for 5 h tempering marked by toughness enhancement due to transformation toughening in Fig. 12 are presented in order of increasing tempering temperature in Fig. 15–17. Figure 15 clearly shows that a completely ductile mode of fracture is achieved with 5 h 525 °C tempering and micrographs presented in Fig. 16 and 17 represent fracture surfaces with increased toughness due to transformation toughening, indicated by the relatively higher degree of primary void growth.

Toughness Optimization by Multistep Tempering. Heat treatment for stabilization of austenite for dispersed phase transformation toughening is directed towards combined size refinement and compositional enrichment of

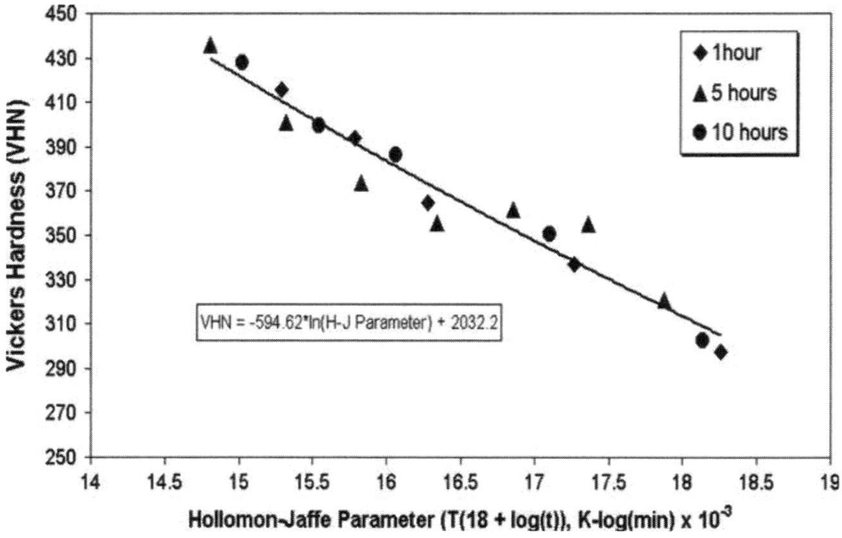

Fig. 13 Hollomon-Jaffe Parameter correlating the hardness data obtained for different tempering conditions in the overaged region

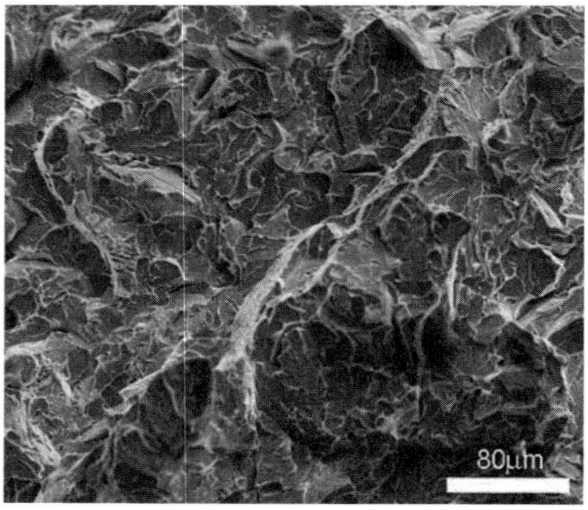

Fig. 14 SEM micrograph of quasi-cleavage fracture surface for prototype tempered at 450 °C for 1 h

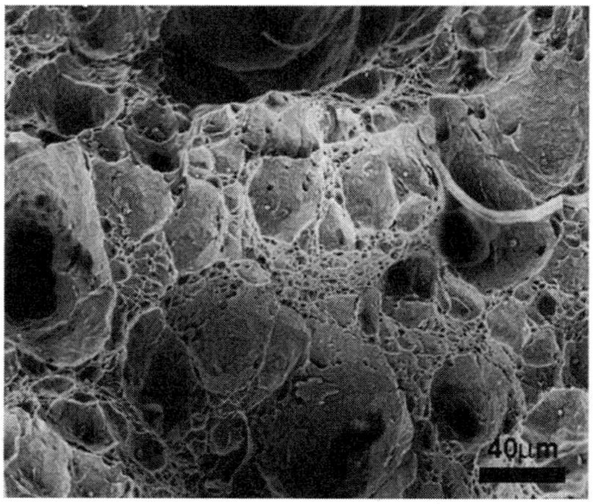

Fig. 15 SEM micrograph of ductile fracture surface for prototype tempered at 525 °C for 5 h

the austenite particles in order to achieve maximum transformation toughening at higher strength levels. A designed two-step tempering process consisting of an initial high-temperature, short time treatment followed by a standard isothermal tempering treatment is employed to achieve this goal. The first step is designed to nucleate a fine, uniform dispersion of intralath austenite and strengthening particles of suboptimal size formed directly by increasing the driving force for austenite precipitation. This is achieved by a short time, high-temperature tempering step designed to give an underaged state based on the isochronal tempering study. The second tempering step is optimized to enhance Ni-enrichment of the austenite particles coupled with completion of precipitation strengthening for peak aging condition involving enrichment of the 3 nm Cu precipitates and cementite conversion to 3 nm M_2C carbides. This is achieved by a longer-time final tempering at a lower temperature characterized by the peak strengthening condition. Thus, from the toughness-hardness trajectory for isochronal tempering presented in Fig. 12, an appropriate final stage tempering condition was determined to be 5 h at 450 °C, which produced a peak hardness of

436 VHN. The first step was then experimentally optimized by varying the tempering time from 5 to 90 min over a temperature range of 500 °C–575 °C in intervals of 25 °C.

Figure 18 shows the variety of two-step heat treatments investigated to maximize the toughness-strength combination in comparison with the HSLA100 alloy and is superimposed on the isochronal tempering plot. The labels in the plot represent the tempering time in min corresponding to the first step and the bold black arrow points to the condition for maximum strengthening, which is the final step in the tempering sequence. The short time, high-temperature austenite nucleation treatments were conducted in a molten salt-bath to reduce heating time followed by water quench to reduce cooling time.

The optimal combination of toughness and strength is determined from Fig. 18 to be a 550 °C 30 min + 450 °C 5 h heat treatment. The apparent achievement of optimal austenite stability by multistep tempering results in significant increase of impact toughness to

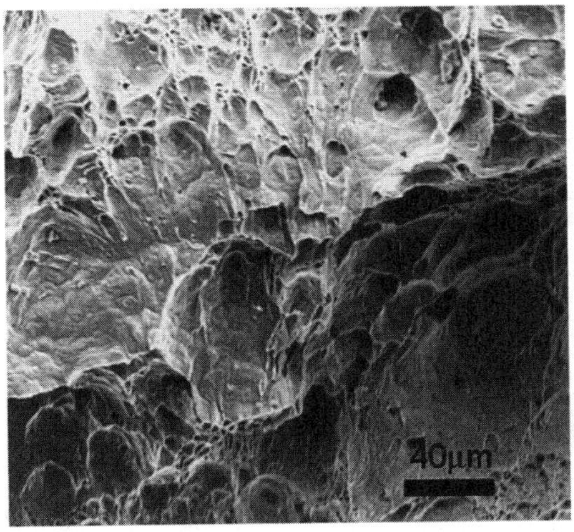

Fig. 16 SEM micrograph of ductile fracture surface representing toughness enhancement due to transformation toughening for prototype tempered at 550 °C for 5 h

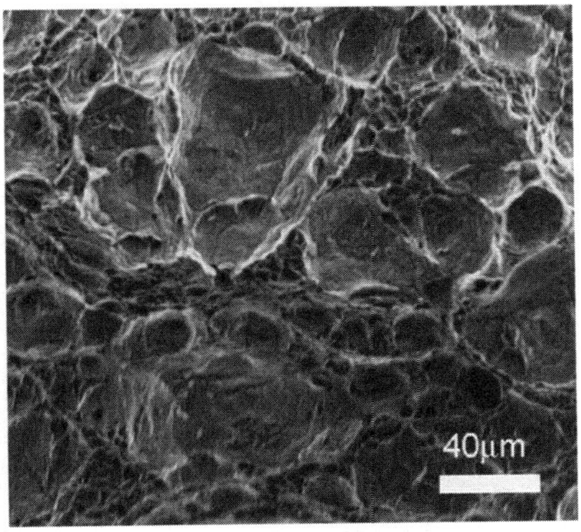

Fig. 17 SEM micrograph of ductile fracture surface representing toughness enhancement due to transformation toughening for prototype tempered at 575 °C for 5 h

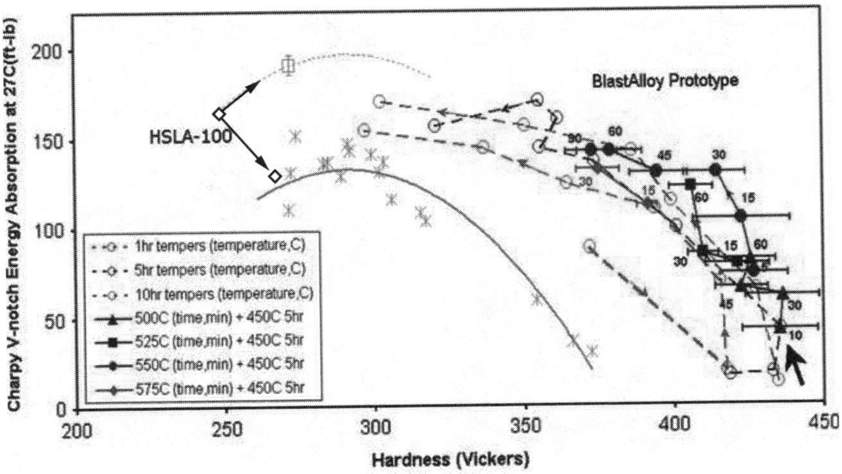

Fig. 18 Multistep tempering treatments designed to maximize transformation toughening response represented by Charpy toughness-Vickers hardness trajectory. The label corresponding to each data point indicates the tempering time during the first tempering step. The condition for the second step is listed on the legend

130 ft-lbs at a hardness level of 415 VHN. Comparing with the baseline toughness-strength combination from isochronal tempering data, a transformation toughening increment of 50% from 87 ft-lbs for the 10 h isothermal treatment and 70% from 77 ft-lbs for the 5 h isothermal treatment is observed at the same strength level. So an average of 60% toughness increment due to dispersed phase transformation toughening can be attributed to multistep tempering when compared to standard isothermal tempering at the same strength level. This toughness level exceeds the design goal of 85 ft-lbs by almost 55%.

SEM analysis of the fracture surfaces for the multistep treatment specimens indicate a transition from quasi-cleavage to ductile mode of failure as the time of initial tempering is increased, attributed to the transformation toughening increment as described in the previous section. Figure 19 presents a representative micrograph of the fracture surface for the optimal toughness-strength combination for the optimal tempering treatment of 550 °C 30 min + 450 °C 5 h. Figure 20 shows a higher magnification micrograph of a primary void in the same sample. The relatively higher degree of primary void growth is consistent with delayed microvoid instability, as expected for transformation toughening.

Mechanical Properties

Evaluation of Tensile Properties. An evaluation of the tensile properties was conducted to determine the actual yield strength of the prototype under the optimized tempering conditions and to provide a basis for closer hardness-strength correlation for this class of steels. Room-temperature tensile properties were assessed for the chosen heat treatment conditions based on the results of the toughness-hardness data from both isochronal and multistep tempering response. The tempering conditions were chosen to cover the full width of the toughness-strength combination plot (Fig. 18). Table 3 summarizes the results of the tensile testing for the solution treated and aged samples for each heat treatment condition and provides hardness values for comparison. Figure 21 presents the true stress versus true plastic strain curves for all the samples tested. The curves are represented as solid lines until the point of tensile instability (necking) or uniform elongation and by dotted lines thereafter. The tensile data presented in Fig. 21 and Table 3 confirms the design of a 160 ksi yield strength steel, which we here designate as "Blastalloy 160." The multistep tempering treatments helped to achieve the 160 ksi yield strength goal.

From data on the reduction in area at fracture and uniform elongation in Table 3, all the heat treatment conditions show reasonably high values of ductility. The ratio of YS/UTS (strength ratio) is a general measure of work hardening behavior. The low values of strength ratio for the "transformation

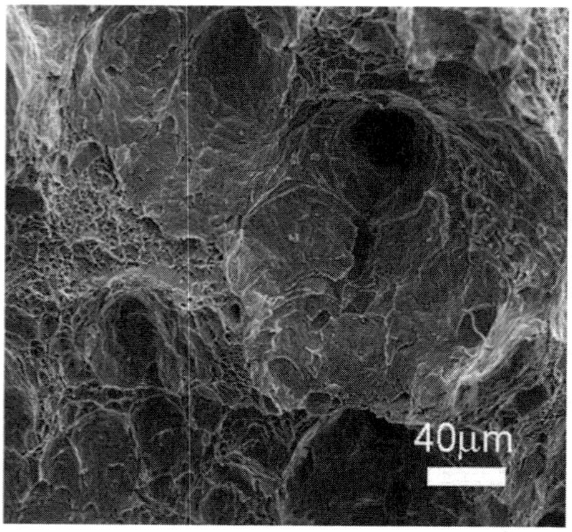

Fig. 19 SEM micrograph of ductile fracture surface representing toughness enhancement due to transformation toughening for the 550 °C 30 min + 450 °C 5 h multistep tempering treatment.

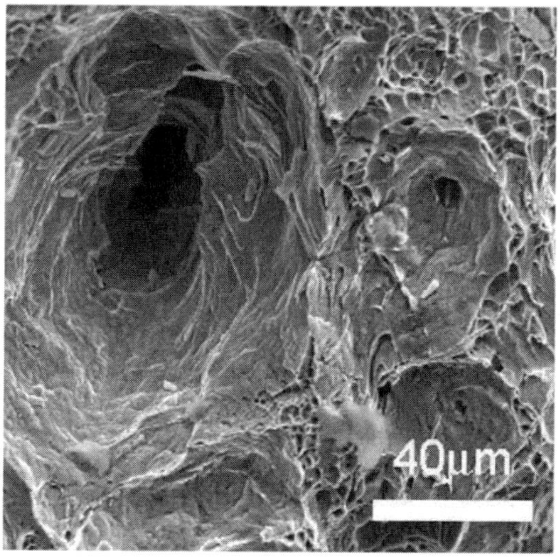

Fig. 20 SEM micrograph of a primary void in the fracture surface of prototype for 550 °C 30 min + 450 °C 5 h multistep tempering treatment

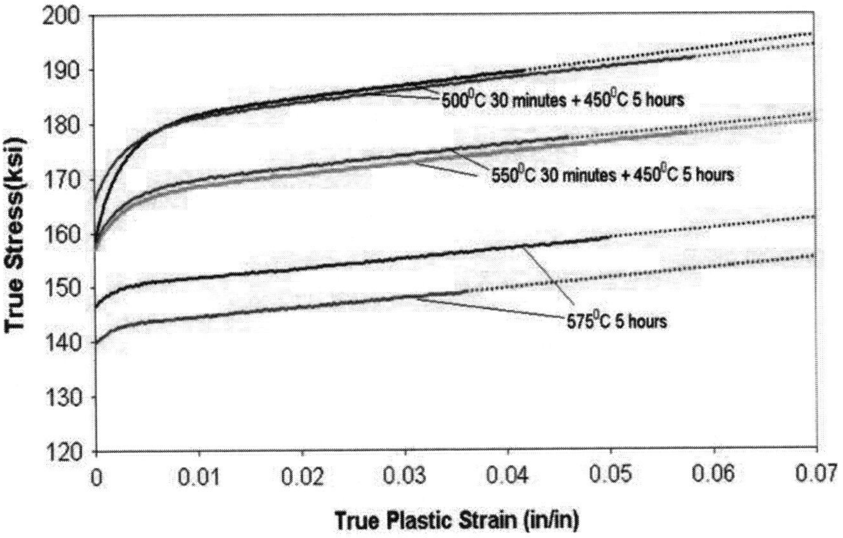

Fig. 21 True stress-true plastic strain response. The stress (σ)-plastic strain (ϵ_p) behavior is shown by solid lines until uniform elongation and by dotted line after necking

Table 3 Room temperature tensile properties of prototype

Tempering condition	0.2% off-set yield strength ksi (MPa)	Ultimate tensile strength ksi (MPa)	YS/UTS	Uniform elongation %	Reduction in area %	Hardness VHN
575 °C 5 h	142.12 (980)	146.45 (2019)	0.97	4.98	73.94	355.30
550 °C 30 min + 450 °C 5 h	156.35 (1078)	167.56 (1155)	0.93	4.89	64.60	414.70
500 °C 30 min + 450 °C 5 h	160.97 (1110)	180.16 (1242)	0.89	5.70	57.09	436.57

toughening optimized" multistep treatments compared to that for the single-step treatment condition suggests that the work hardening of the steel is appreciably improved by the optimal tempering treatments. The load-displacement curves for all the conditions showed smooth yielding without any distinguishable upper and lower yield points.

Toughness-Temperature Dependence. To characterize temperature dependence of toughness, Charpy V-notch impact tests were performed over temperatures ranging from −84 °C to 100 °C for the tempering condition that optimized the austenite for room-temperature dispersed phase transformation toughening. Thus, from Fig. 18 the tempering condition displaying the best toughness-strength combination, 550 °C 30 min + 450 °C 5 h was chosen.

Figure 22 shows the Charpy impact energy of the prototype as a function of test temperature. The corresponding impact energy values for 5 h and 10 h tempering treatments at room temperature are superimposed on the plot. Consistent with the concept that our composition and process design optimized the dispersed phase transformation toughening phenomenon at room temperature (Ref 1), the plot shows that there is a 30 ft-lbs toughness increment at 25 °C compared to the baseline ductile fracture toughness at lower and higher test temperatures. This provides further support for our transformation toughening design concept.

SEM micrographs of the fracture surfaces presented in Fig. 23–27 at each of the test temperatures establish the mode of fracture. Figure 23 shows that the fracture surface for the prototype tested at −84 °C is representative of quasicleavage fracture characterized by the array of flat facets with dimples and tear ridges around the periphery of the facets, indicating a brittle mode of failure. As the test temperature is increased to −40 °C, most of the fracture surface is characteristic of ductile mode of fracture. Closer investigation of Fig. 24 however shows that there are a few tear ridges with

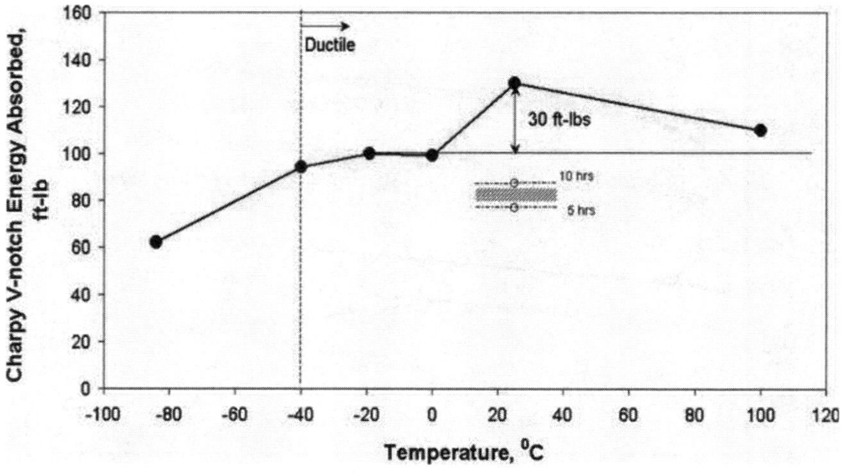

Fig. 22 Charpy impact energy absorbed as a function of testing temperature for prototype tempered at 550 °C 30 min + 450 °C 5 hr. Toughness increment of 30ft-lb due to dispersed phase transformation toughening is shown. The toughness baseline defined by 5 and 10 h single step tempering is superimposed

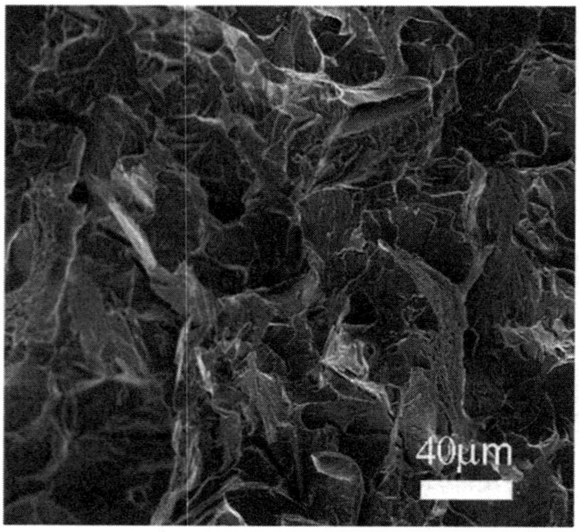

Fig. 23 SEM micrograph of quasicleavage fracture surface showing flat facets with dimples and tear ridges for the 550 °C 30 min + 450 °C 5 h multi-step tempering treatment tested at−84 °C

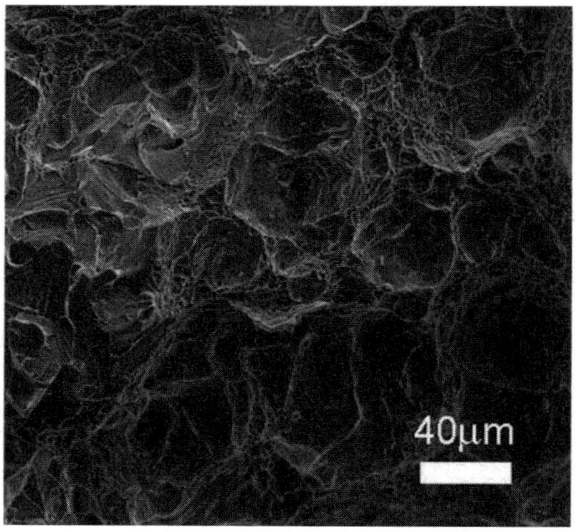

Fig. 24 SEM micrograph of mixed ductile/brittle mode fracture surface showing microvoids with some tear ridges for the 550 °C 30 min + 450 °C 5 h multi-step tempering treatment tested at−40 °C

facets, showing a slightly mixed fracture mode. Figures 25–27 are representative micrographs from fracture surfaces of prototypes tested at −20, 0, and 100 °C respectively, showing purely ductile mode of fracture characterized by primary voids and microvoids without any evidence of flat facets. The micrographs for the fracture surface of the prototype tested at room temperature are presented in Fig. 19–20, which contain larger primary voids with relatively fewer microvoids. This supports the expected effect of transformation toughening leading to more extensive growth of the primary voids before they coalesce by microvoiding. Transformation toughening studies by Leal (Ref 23) in fully austenitic steels indicate a toughness enhancement of 20–50% relative to the toughness of stable austenite. Figure 22 indicates that the impact toughness enhancement in the prototype is at least 30%.

Microstructural Validation

Optimization of the processing conditions of the prototype for dispersed phase transformation toughening in combination with a fine

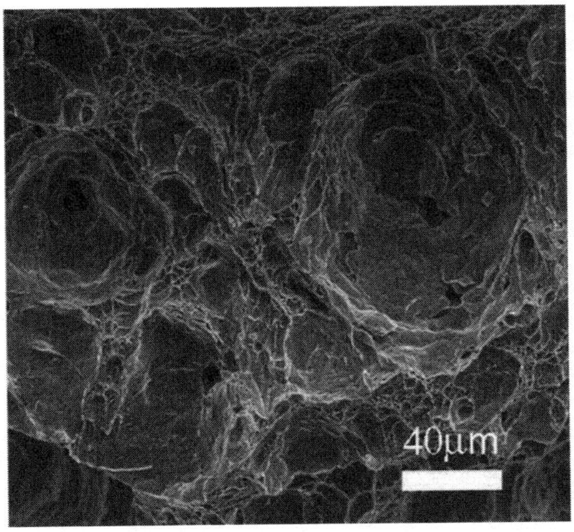

Fig. 25 SEM micrograph of purely ductile mode fracture surface showing primary voids and microvoids for the 550 °C 30 min + 450 °C 5 h multi-step tempering treatment tested at −20 °C

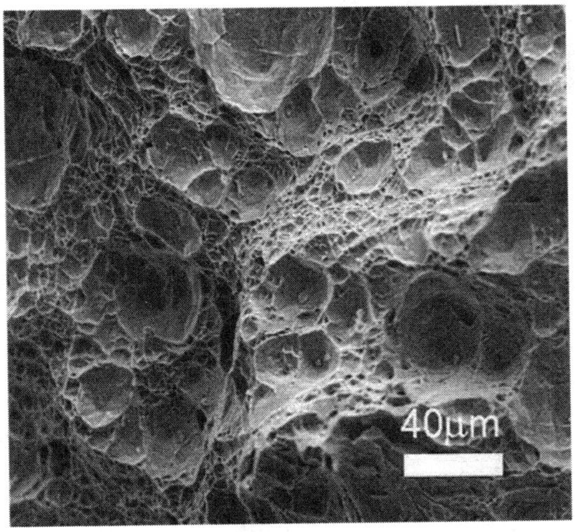

Fig. 26 SEM micrograph of purely ductile mode fracture surface showing primary voids and microvoids for the 550 °C 30 min + 450 °C 5 h multistep tempering treatment tested at 0 °C

dispersion of strengthening precipitates has been supported by property evaluation in the previous sections. Microanalytical characterization of the austenite dispersion and the strengthening precipitates and their interaction in the prototype was performed next to fully validate the design.

Three-Dimensional Atom Probe (3DAP) Microscopy. The choice of samples for 3DAP analysis was based on the condition of tempering treatment for the highest obtainable number density of the precipitates, determined

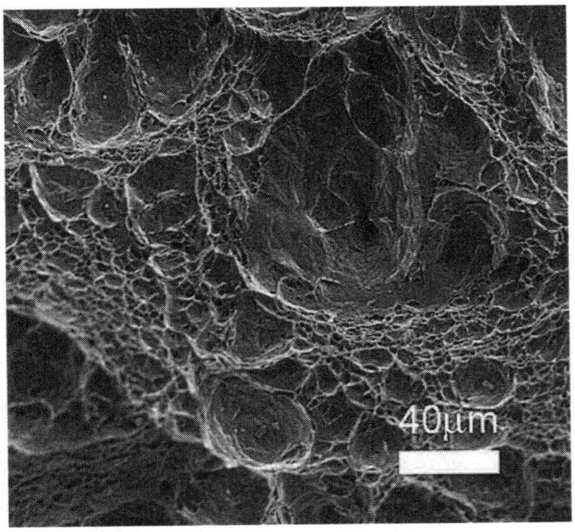

Fig. 27 SEM micrograph of purely ductile mode fracture surface showing primary voids and microvoids for the 550 °C 30 min + 450 °C 5 h multistep tempering treatment tested at 100 °C

from the assessed mechanical properties (Fig. 18). Thus, the tempering condition corresponding to the highest observed strength (Table 3), namely, 500 °C 30 min + 450 °C 5 h was chosen. The 450 °C 1 h tempering condition was also chosen as a reference for comparison with 3DAP data on similar Cu-strengthened steels (Ref 24). For simplification, the 450 °C 1 h tempering treatment specimen will be referred to as the "single-step temper" and the 500 °C 30 min + 450 °C 5 h tempering treatment specimen will be referred to as the "multistep temper" in this section. The data for both tempering conditions will be presented simultaneously for easier comparison.

The overall composition of the reconstructed volume from atom probe analysis was obtained and compared with the actual composition of the prototype as shown in Table 4. The measured overall compositions are within the range of variation observed in the SEM/EDS analysis of banding. The lower measured carbon content may indicate coarsely distributed carbides beyond the length scale of 3DAP analysis.

Atom probe analysis of the single-step temper was conducted at 50 K while that for the multistep temper at 70 K with a pulse fraction of 20% at a pulse frequency of 1.5 kHz from 7 to 10 kV steady state DC voltage. The complete analysis for the single-step temper contained a total of 751,608 atoms in a reconstruction volume of dimensions 13 × 13 × 84 nm. The

Table 4 Comparison between the actual overall composition of prototype and the overall compositions determined by 3DAP analysis.

	Actual overall composition		Overall composition from 3DAP	
			450 °C 1 hr	500 °C 30 min + 450 °C 5 h
Element	wt%	at.%	at.%	at.%
Fe	87.2	90	89.90 ± 0.08	88.58 ± 0.18
C	0.04	0.192	0.11 ± 0.24	0.12 ± 0.53
Cu	3.64	3.30	2.37 ± 0.23	1.13 ± 0.53
Ni	6.61	6.49	5.34 ± 0.23	7.01 ± 0.52
Cr	1.78	1.97	1.86 ± 0.23	2.1 ± 0.53
Mo	0.58	0.35	0.31 ± 0.24	0.89 ± 0.53
V	0.11	0.124	0.11 ± 0.24	0.16 ± 0.53

multistep temper analysis collected 254,917 atoms in a reconstruction volume dimension of 17 × 16 × 28 nm. Figures 28–29 show partial 3D reconstruction of the solute atoms detected for single-step and multistep tempering conditions, respectively. Iron is not shown in any reconstruction in this section for purpose of clarity, enabling larger microstructural features like precipitates to be seen distinctly.

The regions of high copper concentration are clearly noticeable in Fig. 28–29 confirming the presence of a nanometer-sized copper particle

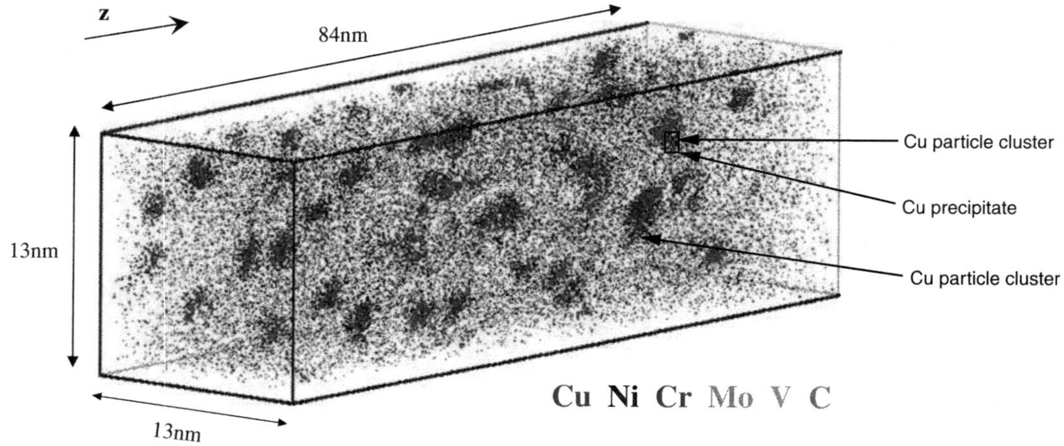

Fig. 28 3DAP reconstruction for prototype tempered at 450 °C for 1 h. The elements in the reconstruction are indicated by their shade of gray. Iron is not shown to provide more clarity in viewing the particles. z is the direction of analysis.

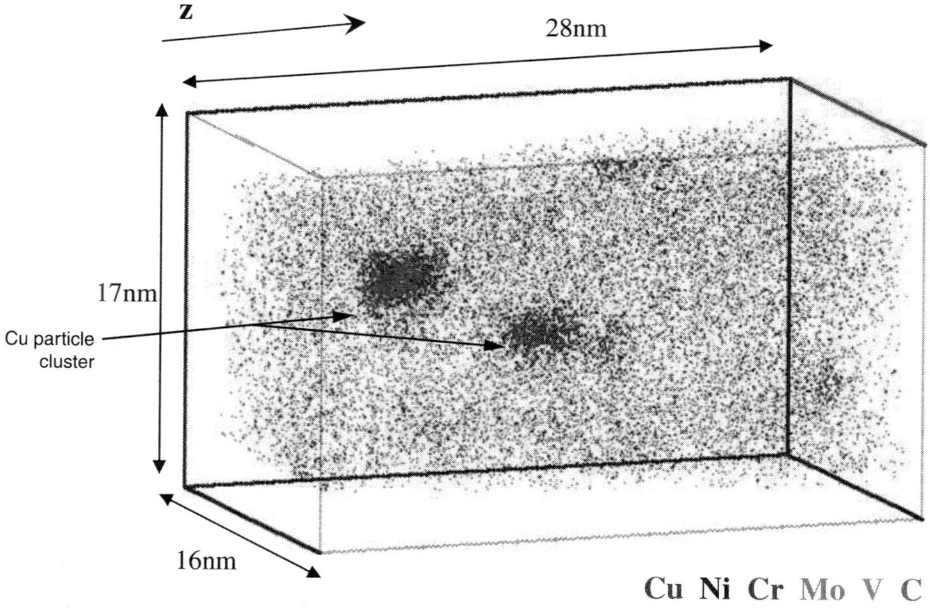

Fig. 29 3DAP reconstruction for prototype tempered at 500 °C 30 min + 450 °C 5 h. The elements in the reconstruction are indicated by their shade of gray. Iron is not shown to provide more clarity in viewing the particles **z** is the direction of analysis.

distribution in the microstructure. These copper-rich precipitates can be represented by an isoconcentration surface at 10 at.% copper level overlaid with the atomic positions of copper atoms as shown in Fig. 30–31. The isoconcentration surfaces more clearly outline the Cu-rich precipitates. The size of the copper precipitates for the single-step temper is relatively smaller than that for the multistep temper, while the number density of precipitates for the former is much higher.

The shape of the copper precipitates appears to be elliptical and stretched in the direction of analysis for both the tempering conditions. The apparent distortion is interpreted as an instrumental artifact due to a magnification effect caused by the difference in field evaporation of copper precipitates compared to the matrix. The precipitates are believed to be equiaxed in shape as evidenced by other researchers (Ref 24, 26),

Having defined the copper precipitates by the isoconcentration surface, the size, number densities, and compositions of these copper precipitates were determined with the help of the *ADAM* analysis software (Ref 7). Cross-

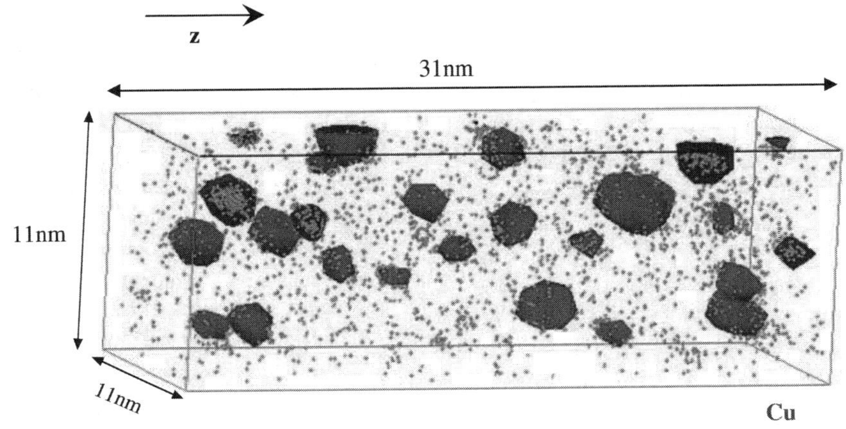

Fig. 30 3DAP reconstruction for prototype tempered at 450 °C for 1 h showing copper precipitates defined at 10 at.% isoconcentration surface overlaid on atomic positions of copper atoms. All other atoms in the reconstruction are not shown **z** is the direction of analysis

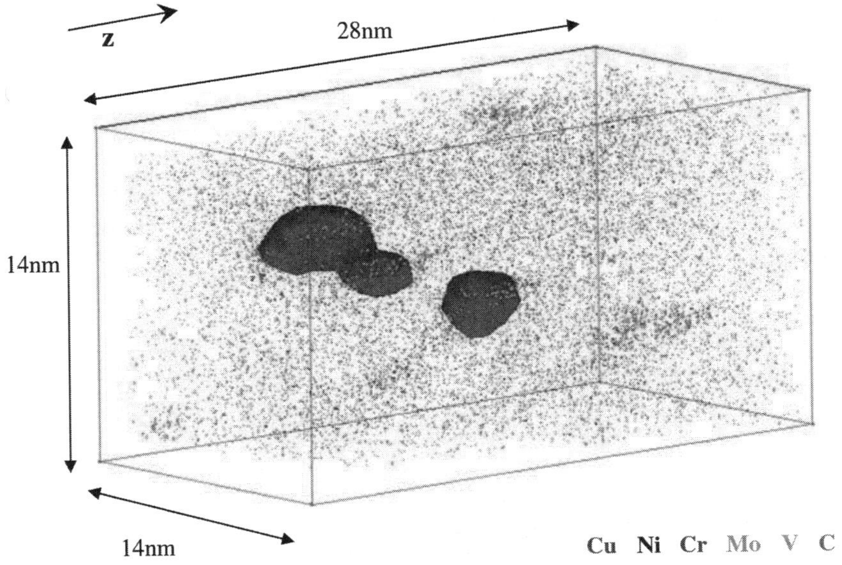

Fig. 31 3DAP reconstruction for prototype tempered at 500 °C 30 min + 450 °C 5 h showing copper precipitates defined by 10 at % isoconcentration surface overlaid on atomic positions of copper atoms. **z** is the direction of analysis

sectional views from an analyzed volume of the reconstruction were used to measure the size of the precipitates. For the single-step temper, the average diameter of the copper precipitates contained completely within the analysis volume was found to be 2.67 ± 0.57 nm while that for the multistep temper is 3.79 ± 0.13 nm. These measurements compare with the optimal particle size value of 2.9 nm obtained from atom-probe measurements by Isheim and Gagliano (Ref 24, 26) for copper-strengthened steel aged for 100 min at 490 °C and also with the peak aging size of 1–5 nm for BCC copper precipitates determined from previous literature as discussed in Ref 1.

The number density of strengthening Cu precipitates is significantly higher for the single-step temper than the multistep temper. The number density of the copper precipitates in the analyzed volume was estimated by Eq 2 (Ref 27):

$$N_V = \frac{N_p \zeta}{n\Omega} \qquad \text{(Eq 2)}$$

N_p and n are the number of particles and the total number of atoms detected in the volume, Ω is the average atomic volume, and ζ is the detection efficiency of a single ion detector, equal to 0.6 in this case. The number density of copper precipitates for the single-step temper was calculated to be 5.42×10^{18} precipitates/cm^3 while that for multistep temper was calculated to be 1.2×10^{18} precipitates/cm^3. The high number density measured for the single-step temper (4.5 times that for multistep temper) is consistent with the high Cu content of the alloy. Evidence for cementite dissolution in the toughness-hardness plots of Fig. 18 support the presence of M_2C carbides contributing to the strength of the multistep tempered material, while the single step represents a peak Cu strengthening condition without as significant carbide contribution.

The average matrix and precipitate compositions were determined from the analyzed volumes. To analyze the composition of the inner core of the precipitates, a higher threshold level of 15 at.% was set to isolate them. Tables 5 and 6 give the composition of the Cu precipitates and the matrix, respectively, with 2σ error bar limits for both the single-step and multistep conditions. Table 6 also compares alloy matrix composition with the equilibrium composition of the BCC matrix predicted at the optimal 490 °C tempering temperature for austenite stability.

The results of the 3DAP analysis indicate that the matrix compositions for both heat treatment conditions are near the predicted equilibrium values. The matrix Cu content is very near the predicted equilibrium composition at the earliest evolution stage, indicating a high degree of Cu precipitation and it remains at the equilibrium condition for the multistep temper composition analyzed. The relatively higher Ni level observed for both conditions suggest less complete austenite precipitation.

From Table 5, the Cu composition of the precipitates from 3DAP analysis is consistent with previous atom-probe results by Goodman (Ref 28) and Isheim-Gagliano (Ref 24, 26) during the early stages of evolution. They reported values ranging from 50 to 70% Cu in the precipitates during the initial stages of precipitation until the peak aged condition is reached. As

Table 5 Average copper precipitate compositions determined by 3DAP analysis for selected heat treatment compositions.

	BCC Cu precipitate composition from 3DAP analysis	
	450 °C 1 h	500 °C 30 min + 450 °C 5 h
Element	at.%	at.%
Fe	30.25 ± 3.53	43.79 ± 6.52
Cu	63.50 ± 2.55	46.69 ± 6.35
Ni	5.40 ± 4.11	8.76 ± 8.31
C	ND	ND
Cr	0.40 ± 4.21	0.57 ± 8.67
Mo	0.13 ± 4.22	ND
V	ND	0.19 ± 8.69

ND, not detected

Table 6 Average matrix compositions determined by 3DAP analysis for selected heat treatment compositions compared with equilibrium prediction from ThermoCalc

	BCC matrix composition from 3DAP analysis		Equilibrium prediction
	450 °C 1 h	500 °C 30 min + 450 °C 5 h	490 °C
Element	at.%	at.%	at.%
Fe	91.22 ± 0.49	92.01 ± 0.22	94.1
Cu	0.73 ± 0.66	0.22 ± 0.77	0.12
Ni	5.32 ± 1.62	6.33 ± 0.74	3.78
C	0.014 ± 1.67	0.041 ± 0.77	4.4×10^{-5}
Cr	2.18 ± 1.65	0.88 ± 0.76	1.88
Mo	0.44 ± 1.66	0.39 ± 0.77	0.10
V	0.09 ± 1.67	0.12 ± 0.77	0.02

suggested by Gagliano (Ref 24) thismay be a lower limit of the true Cu concentration value caused by aberrations of ion trajectories and local magnification effects in 3DAP, which limits the spatial resolution of this nanoanalytical technique to a few tenths of a nanometer (Ref 29). It is well established now (Ref 30, 31) that spatial overlap effects due to the difference in the field evaporation between the matrix and the precipitate in the Fe-Cu system leads to matrix atoms being projected into the precipitate, especially for particles smaller than 5 nm in size.

The average matrix and precipitate compositions and the concentration of the various solute atoms near the matrix/precipitate interface can be investigated by a proximity histogram, or "proxigram," available in *ADAM*, developed and implemented by Hellman et al. (Ref 9). The concentration values were determined by averaging the concentration in 0.2 nm peripheral shells around all the precipitates with respect to the 10 at.% copper isoconcentration surface, within and outside the precipitates. The negative values in abscissa represent the matrix composition while the positive values are indicative of the precipitate compositions. However, the zero point is not necessarily a correct estimate of the precipitate/matrix interface and serves as an approximate reference point (Ref 9). The proxigram obtained from analysis of copper precipitates in the single-step temper sample is presented in Fig. 32. The proxigram indicates that Ni shows considerable segregation to the precipitate/matrix interface while that for other solute atoms is within the error limit of estimation. The level of Ni enrichment at the interface is about 50% higher than the matrix Ni content for the single-step temper, which is similar to the results of Isheim and Gagliano (Ref 24, 26).

To investigate the expected Ni-rich austenite precipitation in the multistep tempered material, details of the Cu particle interface region were examined by varying Ni concentration threshold levels. Setting a 10 at.% level for Ni, the isoconcentration surface of a Ni-rich precipitate at the interface of the Cu-rich precipitates is shown in Fig. 33, overlaid with atomic positions of Cu and Ni from three different orientations. To confirm the Ni content of austenite, further investigation was done by a one-dimensional composition profile plotted along the atom-probe analysis direction in Fig. 34. This confirmed that the Ni content of austenite is 30 at.% and is consistent with our design value for stability optimization.

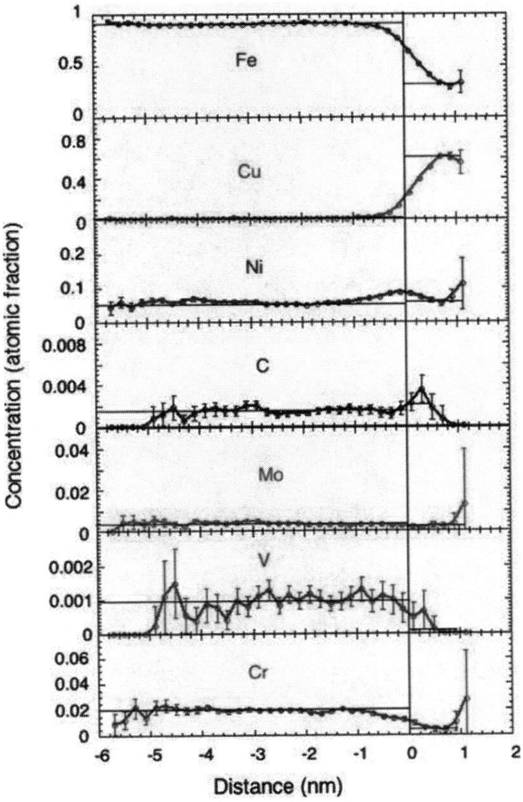

Fig. 32 Proxigram of all the solute species detected in the 450 °C 1 h temper specimen with respect to 10 at.% copper isoconcentration surface in the analysis volume

The compelling observation of a single Ni-rich particle in the conventional 3DAP was further supported by the opportunity to analyze a multistep tempered sample in the Imago Local Electrode Atom Probe. Analysis of a three-order-of-magnitude larger volume compared to Fig. 33 provides numerous examples of Ni-rich particles as shown in Fig. 35. The size and location of the austenite precipitates, measured as 5 nm from Fig. 34, confirms that it is intralath austenite nucleated on Cu precipitates. The size is comparable with that from dark-field TEM observation of intralath austenite of 5–10 nm by Lippard in multistep tempered AerMet100 alloy (Ref 3). Lippard also demonstrated austenite nucleation on M_2C carbide from evidence of high Cr signals associated with STEM EDS data gathered from dispersed austenite intralath precipitates. Our result provides more direct evidence of the heterogeneous nucleation of intralath austenite on

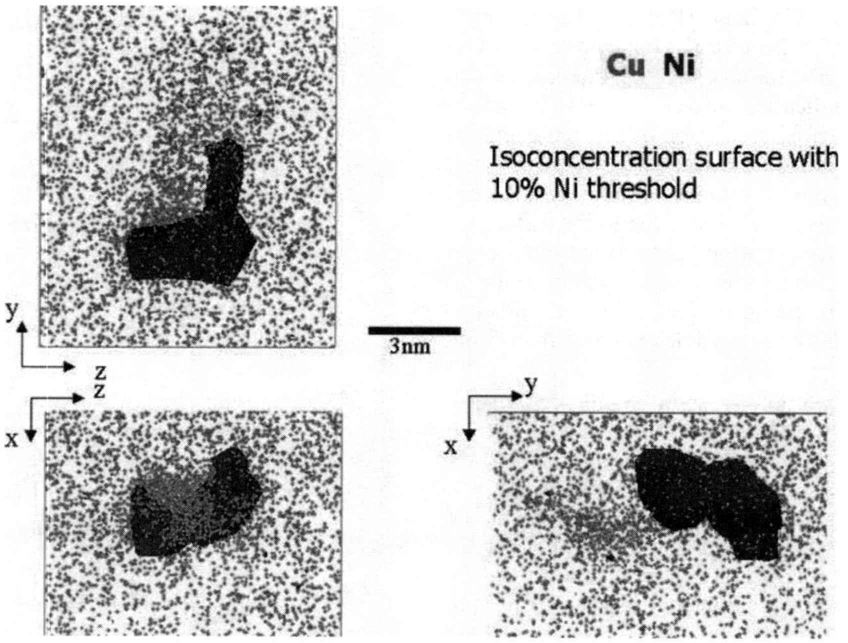

Fig. 33 3DAP reconstruction for prototype tempered at 500 °C 30 min + 450 °C 5 h showing austenite defined by 10 at.% Ni level isoconcentration surface overlaid on atomic positions of nickel and copper atoms **z** is the direction of analysis

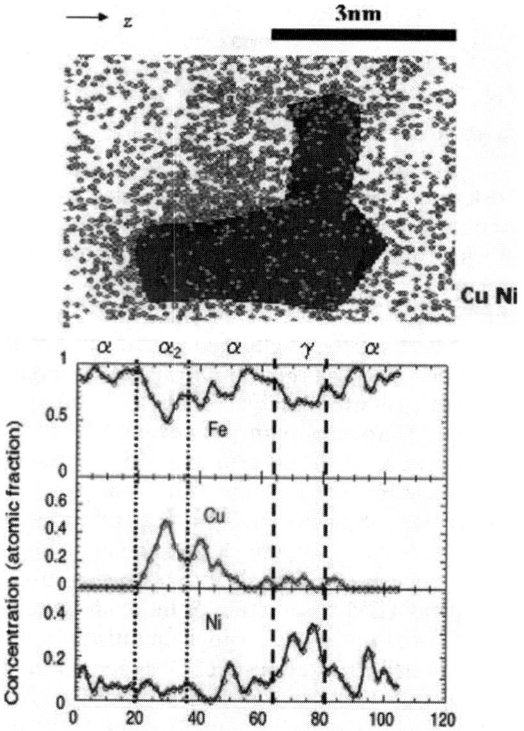

Fig. 34 One-dimensional composition profile along the atom-probe analysis direction in the 500 °C 30 min + 450 °C 5 h temper specimen with respect to 10 at.% Ni isoconcentration surface in the analysis volume. **z** is the direction of analysis

a fine dispersion of strengthening precipitates; in this case Cu precipitates. This finding also clearly validates the transformation toughening design concept of achieving an optimal stability austenite dispersion by employing a multi-step tempering treatment to nucleate the austenite in the first tempering step followed by a Ni-enrichment final tempering step.

Summary and Conclusions

The success of the prototype alloy reinforces the strengths of the design models and their integration. Evaluation of a hot-rolled slab-cast 34 lb VIM heat of the Blastalloy 160 design demonstrated microsegregation banding within the predicted limits and showed no evidence of oxidation-induced hot shortness. Dilatometric studies verified allotropic transformation kinetics allowing bainite/martensite microstructures in air-cooled plates. Thorough evaluation of multistep tempering optimization, mechanical properties and detailed microstructural characterization was then conducted in a fully martensitic condition. Hardness and tensile tests confirmed predicted precipitation strengthening behavior. Isochronal tempering studies at 1 h confirmed peak strengthening at 420 °C with gradual overaging, consistent with

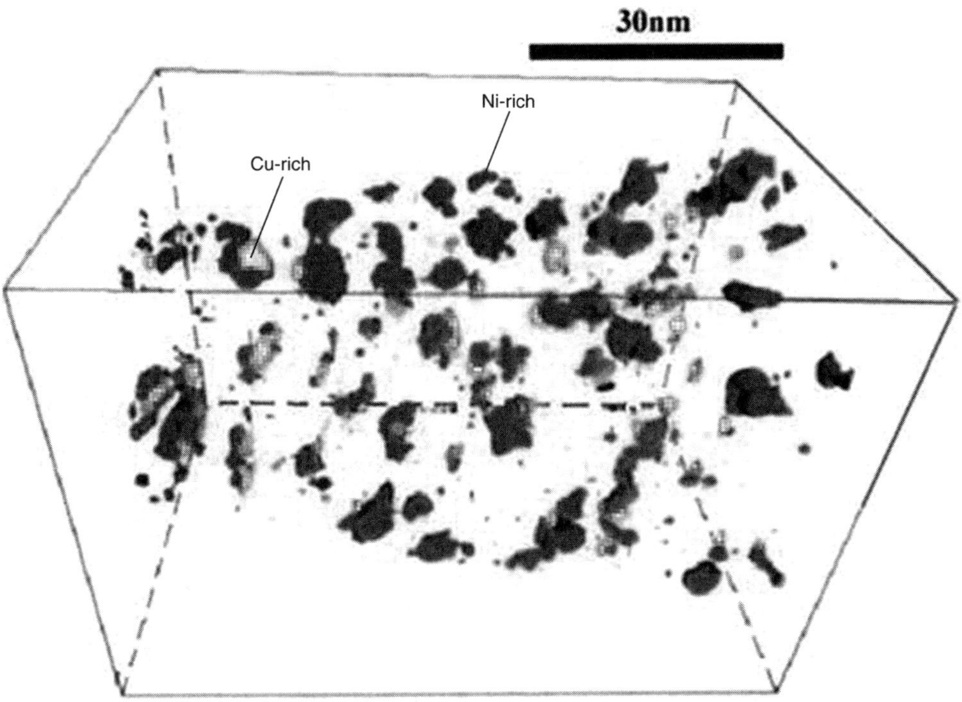

Fig. 35 Multistep tempered sample analyzed in the Imago Local Electrode Atom Probe.

literature findings for copper bearing systems. Multistep tempering was employed to optimize the austenite and a significant enhancement in toughness was observed with minimal loss in strength for a 550 °C 30 min + 450 °C 5 h tempering condition. An optimal austenite stability was indicated by a significant increase of impact toughness to 130 ft lb (176 J) at a strength level of 160 ksi (1100 MPa). Comparison with the baseline toughness-strength combination determined by isochronal tempering studies indicates a significant transformation toughening increment of 60% in Charpy energy, exceeding the actual toughness goal of 85 ft lb (115 J) by almost 55%. Tensile tests conducted on the optimum tempering conditions confirmed the predicted strength levels. Charpy impact tests and fractography demonstrate ductile fracture with $C_v > 80$ ft lbs (108 J) down to −40 °C, with a substantial toughness peak at 25 °C consistent with designed transformation toughening behavior. Predicted Cu particle number densities and the heterogeneous nucleation of optimal stability high Ni 5 nm austenite on nanometer-scale copper precipitates in the multistep tempered samples were confirmed using three-dimensional atom probe microscopy. The copper precipitate size was verified for peak strengthening at 2–3 nm and precipitate composition of 50–60% copper for short tempering times agreed with results from previous studies. Analysis of the fine austenite confirmed a Ni content near the theoretical prediction of 30%. The validated microstructure and properties of this prototype warrants evaluation of the alloy's response to larger scale processing and assessment of its designed weldability. Evaluation of 400 lb (180 kg) heats of the alloy is now under way.

Figure 36 graphically represents the toughness-strength combination of the Blastalloy 160 prototype for three different tempering conditions in comparison to other commercial and experimental alloys. The mechanical properties of the Blastalloy 160 prototype indicate significant improvement in strength-toughness combination compared to other commercial steels currently used by the Navy, and exceeds the performance of higher cost Co-Ni steels such as HY180. The success of the prototype alloy reinforces the strengths of the design models and their integration. Achieving these improvements in a single design and prototyping iteration is a significant advance in computational materials design capability.

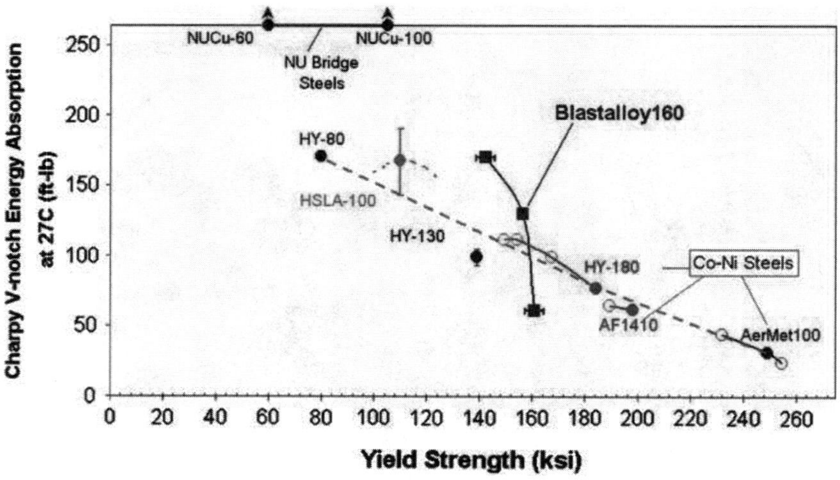

Fig. 36 Toughness-yield strength comparison plot of Blastalloy 160 with other commercial and experimental steels

ACKNOWLEDGMENTS

This research was carried out under the financial support by the Office of Naval Research (ONR) under the ONR Grand Challenge in Naval Materials by Design grant number N00014-01-1-0953, conducted as a part of the multi-institutional Steel Research Group program. The authors would like to acknowledge Dieter Isheim, Ph.D. for his time and his help during the operation of the atom probe and analysis of the data and to Imago for early use of their LEAP instrument. The authors are thankful to Jamie Heisserer for her assistance during the collection of the dilatometry data. The prototype was cast by Special Metals Corporation in New Hartford, New York and rolled by Huntington Alloys in Huntington, West Virginia.

REFERENCES

1. A. Saha and G.B. Olson, Computer-aided design of transformation toughened blast resistant naval hull steels: Part I. *J. Comput.-Aided Mat. Design* (2005) (Submitted)
2. C.J. Kuehmann and G.B. Olson, Gear steels designed by computer. *Adv. Mater. Proc.* 153, 40–43 (1998)
3. H.E. Lippard, Microanalytical investigations of transformation toughened Co-Ni steels. Doctoral dissertation. Northwestern University (1999)
4. G.N. Haidemenopoulos, G.B. Olson, and M. Cohen, Dispersed-Phase Transformation Toughening in Ultrahigh-Strength Steels. In G.B. Olson, M. Azrin, and E.S. Wright, (eds.) Innovations in *Ultrahigh-Strength Steel Technology* (34th Sagamore Army Materials Research Conference), p 549–593. US Government Printing Office, (1990)
5. J.T. Sebastian, Nanoscale three-dimensional studies of segregation at ceramic/metal interfaces. Doctoral dissertation. Northwestern University (2001)
6. M.K. Miller and G.D.W. Smith, *Atom Probe Microanalysis: Principles and Applications to Materials Problems*. MRS Publications (1977)
7. O.C. Hellman, J.A. Vandenbroucke, J. Blatz du Rivage, and D.N. Seidman, Application software for data analysis for three-dimensional atom probe microscopy. Mater. Sci. Eng. A 327, 29–33 (2002)
8. O.C. Hellman, J. Blatz du Rivage, and D.N. Seidman, Efficient sampling for three-dimensional atom-probe microscopy data. Ultramicroscopy 95, 199–205 (2003)
9. O.C. Hellman, J.A. Vandenbroucke, J. Rüsing, D. Isheim, and D.N. Seidman, Analysis of three-dimensional atom-probe data by the proximity histogram. Microsc. Microanal. 6, 437–444 (2000)
10. J.W. Halley, Effect of residual elements upon the properties of metals. ASM 71–87 (1957)
11. D.A. Melford, Surface hot shortness in mild steel. *J. Iron Steel Inst.*, 200, 290–299 (1962)
12. S. Seo, K. Asakura, and K. Shibata, Effects of 0.4% Si and 0.02% P additions on

surface hot shortness in 0.1%C-0.5%Mn steels containing 0.5% Cu. *ISIJ Int.*, 37, 240–249 (1997)

13. W.J.M. Salter, Effects of alloying elements on solubility and surface energy of copper in mild steel. *J. Iron Steel Inst.*, 204, 478–488 (1966)
14. G.L. Fisher, The effect of nickel on the high-temperature oxidation characteristics of copper-bearing steels. *J. Iron Steel Inst.*, 207, 1010–1016 (1969)
15. N. Imai, N. Komatsubara, and K. Kunishige, Effect of Cu and Ni on hot workability of hot-rolled mild steel, *ISIJ Int.*, 37, 224–231 (1997)
16. S. Akamatsu, T. Senuma, Y. Takada, and M. Hasebe, Effect of nickel and tin additions on formation of liquid phase in copper bearing steels during high temperature oxidation. *Mater. Sci. Tech.* 15, 1301–1307 (1999)
17. A. Nicholson and J.D. Murray, Surface hot shortness in low-carbon steel. *J. Iron Steel Inst.*, 203, 1007–1018 (1965)
18. G. Ghosh and G.B. Olson, Kinetics of FCC → BCC heterogeneous martensitic nucleation-I. the critical driving force for athermal nucleation. *Acta Mater.*, 42, 3361–3370 (1994)
19. N. Maruyama, M. Sugiyama, T. Hara, and H. Tamehiro, Precipitation and phase transformation of copper particles in low alloy ferritic and martensitic steels. *Mater. Trans.* JIM, 40, 268–277 (1999)
20. R.P. Foley, Normal and intercritical tempering behavior of a low-carbon, copper-bearing steel designed for ultraservice applications. Doctoral dissertation. Northwestern University (1992)
21. R.A. Grange and R.W. Baughman, Hardness of tempered martensite in carbon and low alloy steels. *Trans. Am. Soc. Met.* 48, 165–197 (1956)
22. J.H. Hollomon, Tensile deformation. Trans. Metall. Soc. AIME. 162, 268–290 (1945)
23. R.H. Leal, Transformation toughening of metastable austenitic steels. Doctoral dissertation. Massachusetts Institute of Technology (1984).
24. M.S. Gagliano, Co-precipitation of copper and niobium carbide in a low carbon steel. Doctoral dissertation. Northwestern University (2002). Atom-probe experiments by D. Isheim
25. K.C. Russell and L.M. Brown, A dispersion strengthening model based on differing elastic moduli applied to the iron-copper system. Acta Metall. 20, 969–974 (1972)
26. D. Isheim, Unpublished research. Northwestern University (2002)
27. M.K. Miller, *Atom Probe Tomography: Analysis at the Atomic Level.* Kluwer Academic/Plenum Publishers (1999)
28. S.R. Goodman, S.S. Brenner, and J.R. Low, An FIM-Atom probe study of the precipitation of copper from iron-1.4 At. Pct copper. Part II: Atom probe analyses. Metall. Trans. A, 4, 2371–2378 (1973)
29. F. Vurpillot, A. Bostel, and D. Blavette, Trajectory overlaps and local magnification in three-dimensional atom probe, *App. Phy. Lett.* 76, 3127–3129 (2000)
30. D. Blavette, F. Vurpillot, P. Pareige, and A. Menand, A model accounting for spatial overlaps in 3D atom-probe microscopy. Ultramicroscopy, 89, 145–153 (2001)
31. M.K. Miller, A. Cerezo, M.G. Hetherington, and G.D.W. Smith, In M.K. Miller, (ed.) *Atom Probe Field Ion Microscopy.* Clarendon, p 196–199 (1996)

Transformations—Selected Works of G.B. Olson on Materials, Microstructure, and Design
C.E. Campbell, M.V. Manuel, and W. Xiong, editors

CHAPTER 20

Design of Heusler Precipitation Strengthened NiTi- and PdTi-Base Shape Memory Alloys for Cyclic Performance*

Dana J. Frankel and Gregory B. Olson
Northwestern University

Abstract

For a wide range of actuation applications, the performance of NiTi-based shape memory alloys is limited by cyclic instability associated with accommodation slip. For medical applications, low-Ni compositions are also desirable. Increasing yield strength via precipitation of a coherent nanoscale Ni_2TiAl-type Heusler phase from a supersaturated B2 matrix is an effective approach for eliminating slip in order to improve the stability of the functional response and increase the structural fatigue life. Quaternary additions that partition into the $L2_1$ Heusler phase, such as Zr or Pd, are favorable for reducing interphase misfit and maintaining coherency during aging. Phase relations and precipitation kinetics in quaternary Ni(TiZrAl), low-Ni (PdNi)(TiAl), and Ni-free (PdFe)(TiAl) systems are summarized from TEM and atom probe tomography data in the literature. Strengthening behavior during isothermal aging is compared in the NiTiZrAl and PdNiTiAl systems, and recent work characterizing a high-strength, low-Ni "hybrid" (PdNi)(TiZrAl) alloy is presented. A systems design approach is taken in which an optimal microstructure for peak strengthening is identified while other property objectives such as transformation temperature, misfit, radiopacity, and biocompatibility are satisfied.

Introduction

Motivation and Background

Due to their unique thermomechanical properties, shape memory alloys (SMAs) are utilized across a vast array of industries including biomedical, aerospace, automotive, and civil applications. Near-equiatomic NiTi (or Nitinol) is one of the most widely used commercial SMAs. Large reversible strains associated with thermal recovery along with the highly compliant superelastic loading response makes NiTi-based SMAs ideal candidates for expandable transcatheter devices such as vascular stents and artificial heart valve frames. SMAs are also utilized as actuators in

*D.J. Frankel and G.B. Olson, "Design of Heusler Precipitation Strengthened NiTi-and PdTi-Base SMAs for Cyclic Performance," *Shape Memory and Superelasticity*, 1, 2 (2015) 1–18

automotive and aerospace applications because of their high power density and simple actuation mechanism, which allow for quiet, compact, and efficient actuator design.

In all of these applications, operational conditions include thermal and/or mechanical cyclic loading, often into the high or ultrahigh cycle fatigue (HCF/UHCF) regimes. Therefore, in the parametric design of materials for these applications we must focus on fatigue life as a key property objective in ensuring safe and reliable device performance. Increasing the yield strength of the austenite phase, σ_y, is important for improving both structural and functional fatigue resistance. Increasing σ_y (a) decreases the accumulation of plastic strain at stress raisers such as nonmetallic inclusions or other defects, (b) promotes deformation via reversible processes such as the thermoelastic martensitic transformation, and (c) helps minimize the buildup of accommodation slip at the austenite/martensite interface during phase transformation that can lead to instability and/or irreversibility during thermal or mechanical cycling. While cold working materials can lead to high-strength levels, it has been shown to degrade the superelastic response due to interference with interface movement during transformation (Ref 1). Therefore, utilizing a strengthening dispersion of coherent, nanoscale precipitates to increase the yield strength of the austenite phase is a promising method of improving fatigue life while maintaining cyclic stability. Controlled precipitation of an $L2_1$ (Ni_2TiAl-type) Heusler phase in a supersaturated B2 Ni(Ti,Al) matrix can be achieved by aging at intermediate temperatures (500–800 °C). Characterization of phase relations and kinetics of aging in this two-phase field is critical for designing processing parameters to achieve an optimized microstructure. Current progress in this regard is outlined in the "Process Optimization" section of this paper.

The $NiTi/Ni_2TiAl$ system is analogous to the gamma/gamma′ $NiAl/Ni_2TiAl$ system that has been studied extensively in the context of Ni superalloy development. Only recently has interest been developed in the $NiTi/Ni_2TiAl$ system for precipitation strengthened SMAs. This paper will focus on the recent research developments surrounding phase relations, precipitation and coarsening kinetics, phase stability, and strengthening behavior of the ternary $Ni_{50}(Ti,Al)_{50}$ system as well as selected higher-order quaternary and quinary systems of interest.

One system that has been studied is the $Ni_{50}(Ti,Zr,Al)_{50}$ system, which utilizes Zr addition for both its excellent solid solution strengthening capabilities and its beneficial effect on decreasing interphase lattice misfit. Based on ab initio calculations and experimental measurements, Zr has been found to preferentially inhabit Ti sites in the B2 and $L2_1$ phases (Ref 2). The ternary $NiTi/Ni_2TiAl$ system has a relatively large lattice misfit of −2.57%, which can cause loss of coherency during aging and is detrimental to precipitation strengthening. The larger Zr atoms partition preferentially into the Heusler phase, thus helping minimize interphase misfit. NiTi-based systems such as the NiTiZrAl system are of interest for automotive applications, where cost is a limiting factor and additions of precious metals such as Pd and Pt, as discussed in the following section, are prohibitive. A detailed review of this system will be discussed in the "Ni(Ti,Zr,Al) System" section.

When designing SMA systems for biomedical applications, biocompatibility is an important property objective to address. In individuals with Ni hypersensitivity, systemic allergic reactions to NiTi-based devices can cause inflammation near the implant and eventual device failure (Ref 3). Biocompatibility concerns thus drive the development of low-Ni and Ni-free alternatives to conventional NiTi-base SMAs. Preliminary cytology studies by Bender et al. (Ref 4) indicate that a low-Ni $Pd_{30}Ni_{20}Ti_{46}Al_4$ alloy shows improved cell viability over binary NiTi when live cell count was measured after exposure to metal alloy powders dispersed in growth media. Additionally, Pd-base alloys have increased radiopacity compared to binary NiTi allowing for better x-ray visibility of devices during placement procedures. The same low-Ni alloy studied by Bender et al. (Ref 4) exhibited a 141.9 % radiopacity improvement over binary NiTi. Therefore, the low-Ni and Ni-free PdTi-base systems show good promise for biomedical applications.

The PdTi system has been widely studied as a high-temperature SMA (Ref 5, 6) Hosoda et al. (Ref 7) found that ternary additions of 3d transition metals on the Pd sublattice can effectively stabilize the austenite phase down to room temperature or body temperature range, allowing for design of superelastic biomedical alloys. As with Zr, Pd has a larger atomic volume than Ni and partitions to the

Heusler phase, decreasing interphase misfit. An extensive characterization of transformation behavior and precipitation in low-Ni (Pd, Ni)$_{50}$(Ti,Al)$_{50}$ and Ni-free (Pd,Fe)$_{50}$(Ti,Al)$_{50}$ system is reviewed in the "PdTi-Base System" section of this paper.

Despite being of interest in many high-temperature actuator applications, binary PdTi SMAs demonstrate poor recoverable strain due to their low critical slip stress. Pd is a significant solution softener, and slip stress of the parent B2 phase decreases by more than 75% in $Ni_{49.5-x}Pd_xTi_{50.5}$ as Pd is increased from 15 to 46 at.% (Ref 8). Zirconium is of interest as an effective strengthening addition in the PdTi system, much like in the NiTi system. Kawakita et al. (Ref 9) note significant solid solution strengthening effects from the addition of 5% Zr, resulting in increased martensite yield strength, austenite transformation stress, and recoverable strain during compression. Models used to describe strengthening in these materials are outlined in the "Strengthening Models" section, while application of these models to prototype NiTiZrAl and PdNiTiAl systems is addressed in the "Calibrating Strength Models" section.

The preliminary investigation of the "hybrid" (Pd,Ni)$_{50}$ (Ti,Zr,Al)$_{50}$ system is also presented, which combines the biocompatibility of PdTi-base systems with the strengthening of Zr and the benefits of coherent Heusler precipitation to achieve a high-strength, fatigue-resistant, low-Ni SMA design. New data regarding phase relations and strengthening behavior in a non-transforming ($A_f \ll$ RT) alloy ($Pd_{30}Ni_{20}Ti_{38}Zr_8Al_4$ nominal) is characterized via atom probe tomography (APT) and Vickers microhardness. From APT data, average particle radius and phase composition can be measured. Using these measurements as model inputs, other material parameters such as phase fraction, interphase misfit, austenite finish temperature (A_f), and solid solution strengthening can be calculated, as described in the "Models" section.

Lastly, a brief overview of non-metallic inclusions in commercial NiTi is included in "Nonmetallic Inclusions" section. Nonmetallic inclusions such as micron-scale $Ti_4\,Ni_2O_x$-type oxides and Ti(C,O)-type carbides form due to oxygen and carbon impurities introduced during commercial melt processes. They can act as potent fatigue crack nucleation sites and are therefore important to address when considering the fatigue life of NiTi-based components. Both empirical and computational approaches have been taken to understand the role nonmetallic inclusions play in fatigue life. Recent advances in industrial processing technologies are also discussed.

Design Objectives

In an effective materials design, it is imperative to understand and define design objectives. These objectives will vary based on the application of interest. The operational temperature (T_{op}) and desired thermomechanical response for a given application will determine the optimal transformation temperature for an alloy. In the case of actuator applications such as SMA springs, valves, or switches, the application might require switching between two constrained/biased strain states (one-way shape memory effect) or two unconstrained/unbiased trained shapes (two-way shape memory effect) upon temperature cycling. Therefore in these applications the designed alloy must have $M_f > T_{low}$ and $A_f < T_{high}$ as well as minimal hysteresis. For applications which undergo mechanical loading in which a superelastic (SE) response is desired, the alloy should be austenitic at T_{op}, thus the design objective for transformation temperature is $A_f < T_{op}$. For SE biomedical applications such as vascular stents and artificial heart valve frames, A_f is targeted to be between 10 and 15 °C, sufficiently below body temperature to ensure a superelastic response during operation.

In both SE and shape memory applications, resistance to functional fatigue is critical for maintaining a stable cyclic response. Maximizing the yield strength of the B2 phase is important for improving both functional and structural fatigue resistance. The required yield strength goals will be determined by the specific application device design and loading conditions.

Minimizing interphase misfit is important for maximizing driving force for precipitation to increase fine-scale particle number density. Additionally, large coherency strains can be an obstacle for martensite interface motion during phase transformation, increasing interfacial frictional work and transformation hysteresis (Ref 10), which is detrimental to actuator applications with narrow temperature ranges. Lastly, early coherency loss during particle growth and coarsening would prohibit dislocation cutting and cause premature bypass, leading to nonoptimal peak strength values (Ref 11). For these reasons, the objective is set to reduce lattice misfit from $\delta = -2.57\%$ in the NiTiAl system to $|\delta| < 1\%$ in quaternary and higher-order system designs.

Cost is also an important design factor. For automotive applications, use of precious metals such as Pd or Pt may be prohibitive, while the cost-to-benefit ratio may be acceptable in aerospace or medical applications. A cost analysis of stent devices indicated that while replacing a binary NiTi alloy with a low-Ni $Pd_{30}Ni_{20}Ti_{46}Al_4$ alloy would increase the cost of raw materials by a factor of 3.7; the raw materials make up just a small portion of the device cost to hospitals, resulting in a projected device cost increase of only 8% (Ref 12). This cost increase might be outweighed by the benefit that a biocompatible PdTi-based device would provide to a severely Ni-hypersensitive patient with no treatment alternative.

In biomedical applications, biocompatibility and radiopacity must be considered. Cell cytology studies can be used to investigate relative cytotoxicity of different alloys, while attention must also be paid to surface finish and corrosion behavior. Additionally, the material for medical devices must be radiopaque, or visible under an x-ray fluoroscope, in order to aid device insertion and maintenance. Radiopacity may be quantified by calculating x-ray absorption as a function of composition as described in the "Radiopacity Model" section. The absorption of binary NiTi is considered the baseline and improvement over this level is desirable.

Systems Design of SMAs

Design Chart

The systems design chart provides a useful way to illustrate and elucidate the complex interrelationships between processing, structure, properties, and performance in a material system (Ref 13). This chart emphasizes the reciprocity of the relationships as well as the hierarchical nature of the microstructural subsystems. An example of a systems design chart for precipitation strengthened SMAs can be found in Fig. 1.

Developing computational and analytical models to describe the relationships between subsystem boxes, denoted with connector lines, allows for parametric design of high-performance materials. While a scientific analysis might focus on a smaller number of subsystems, understanding the interrelationships between each subsystem is crucial for a systems-based design.

Models

Strengthening Models. Yield strength can be considered a simple linear superposition

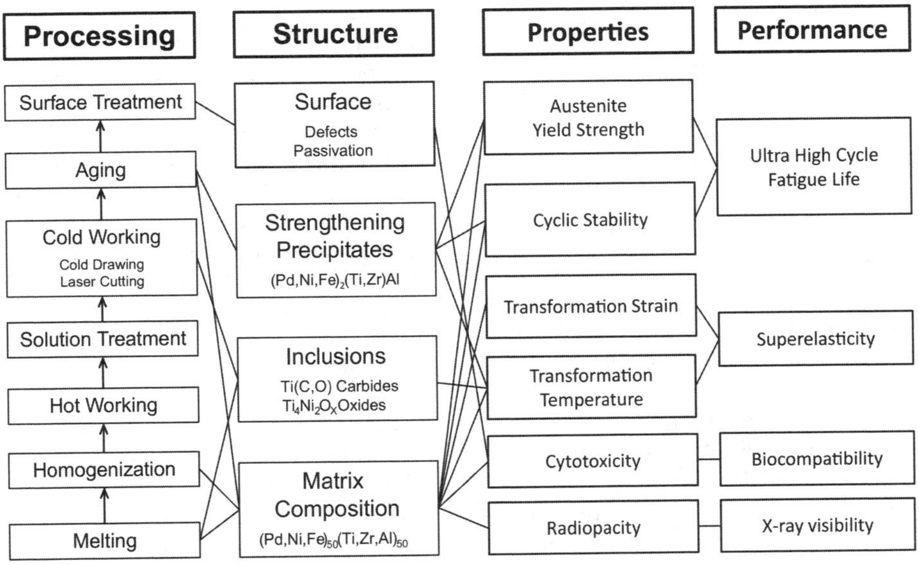

Fig. 1 Systems design chart for a high-performance precipitation strengthened SMA

of base strength (incorporating Hall-Petch strengthening at a constant grain size), solid solution strengthening, and precipitation strengthening contributions (Eq 1). Solid solution strengthening may be modeled using a sum of elemental strengthening or softening coefficients k_i multiplied by the mole fraction of the added component x_i raised to an empirical power coefficient $1/n_i$ ($n_i = 1,2,\cdots$), as shown in Eq 2. Strengthening and power coefficients used in this paper can be found in Table 1.

Precipitation strengthening contributions can be described by the classic shearing/looping interaction of a dislocation with precipitates in a matrix. At small particle sizes, a dislocation shearing mechanism dominates, in which it is energetically favorable for a dislocation to cut through a precipitate. Due to modulus, ordering, coherency, and antiphase boundary (APB) contributions, shearing resistance increases as a function of particle size and can be combined into a prefactor $K1$ with a dependence on r^n where $n = 1$ for these systems (Eq 3). At larger particle sizes and interparticle spacing distances, dislocation looping becomes the dominant interaction mechanism, in which dislocations bypass incoherent particles in a matrix and strengthening decreases as $1/r$ (Eq 4). Shearing behavior, that is, the magnitude of the prefactor $K1$, is dependent on precipitate composition, while looping behavior, that is, the value of $K2$, is independent of precipitate composition:

$$\sigma_{tot} = \sigma_{base} + \Delta\sigma_{ss} + \Delta\sigma_{ppt} \quad \text{(Eq 1)}$$

$$\Delta\sigma_{ss} = \sum_{i=1}^{n} k_i x_i^{\frac{1}{n_i}} \quad \text{(Eq 2)}$$

$$\Delta\sigma_{ppt}^{shear} = K1 \cdot f^{\frac{1}{2}} \cdot r^n \quad \text{(Eq 3)}$$

$$\Delta\sigma_{ppt}^{loop} = K2 \cdot f^{\frac{1}{2}} \cdot \frac{1}{r} \quad \text{(Eq 4)}$$

When experimental data are normalized by $f^{\frac{1}{2}}$ and the precipitation strengthening values are plotted versus the average particle radius, curves for Eqs 3 and 4 can be fit to the data and the analytically determined intersection corresponds to the peak-strengthened state and optimum particle radius.

Table 1 Fitted empirical solid solution strengthening constants and exponents for substitutional additions to NiTi.

i	k_i (MPa)	n_i	Fitted to data from
Al	2575	2	Ref 32
Zr	2730	2	Ref 19
Pd	−297	1	Ref 8

Kinetic Models. Lacking a completed kinetic database to provide a detailed calculation of phase diagrams (CALPHAD)-based computational model for particle growth trajectory as a function of aging time and temperature, the next best estimation is to determine rate constants for prototype systems as a function of aging time. Thus a clear processing/properties relationship can be modeled allowing the prediction of average particle radius and design of optimum processing method. For systems under high supersaturation it has been shown that nucleation is rapid and growth is suppressed (Ref 14), such that a simple Lifshitz-Slyozov-Wagner (LSW)-type coarsening model might adequately describe average particle size evolution as directly related to $t^{1/3}$ through a rate constant, k, in the pure coarsening regime (Ref 15):

$$\bar{r}^3 - \bar{r}_0^3 = kt \quad \text{(Eq 5)}$$

Here $\bar{r}$ is the average particle size at time t and $\bar{r}_0$ is the average radius at the onset of coarsening. For highly supersaturated systems in which precipitation is rapid and growth is suppressed, $\bar{r}_0$ can be approximated as r^*, the critical radius for precipitation (Ref 14). Jung et al. (Ref 2) found that microstructural dynamics in NiTiZrAl-base systems were governed by particle growth at 800 °C but by coarsening at 600 °C, which is consistent with the increased supersaturation of the matrix at lower aging temperatures.

Composition Trajectory Models. For supersaturated systems under evolving unstable equilibrium, phase composition can be modeled as a function of particle radius following Gibb-Thomson trajectories for dilute solutions with the following relation:

$$C_i^r = Ae^{\frac{b}{r}} \quad \text{(Eq 6)}$$

where C_i^r is the composition of component i in the phase of interest at radius r, and A and b are fitted constants where A corresponds to $C_i^{r=\infty}$, or the equilibrium solubility of component i the modeled phase. Fitting experimental data for phase composition and radius at selected aging conditions allows for parametric prediction of nonequilibrium phase relations at finite particle sizes, and also allows for extrapolation to $r = \infty$ to determine equilibrium solubilities.

Misfit Model. The unconstrained interphase misfit (in percent) between the B2 matrix phase and $L2_1$ precipitate phase is calculated by the relation:

$$\delta = \left(\frac{a_{L2_1} - 2a_{B2}}{2a_{B2}}\right) \times 100 \qquad \text{(Eq 7)}$$

For the ternary NiTiAl system, the lattice parameter for the B2 phase $a_{B2} = a_{NiTi} = 0.3010$ nm and the lattice parameter for the $L2_1$ phase $a_{L2_1} = a_{Ni_2TiAl} = 0.5865\,\text{nm}$ are known and the misfit can be calculated to be −2.57%. For higher-order systems, the composition-dependent lattice parameters may be calculated using a weighted sum of the known atomic volumes of the constituents in the B2 and $L2_1$ phases and the misfit can be calculated from the cube root of the ratio of the average atomic volume for each phase:

$$\delta = \left(\sqrt[3]{\frac{\bar{\Omega}_{L2_1}}{\bar{\Omega}_{B2}}} - 1\right) \times 100 \qquad \text{(Eq 8)}$$

$$\bar{\Omega}_{L2_1} = \sum_{i=1}^{n} x_i \Omega_i^{L2_1} \qquad \text{(Eq 9)}$$

$$\bar{\Omega}_{B2} = \sum_{i=1}^{n} x_i \Omega_i^{B2} \qquad \text{(Eq 10)}$$

Jung (Ref 2) utilized literature values along with x-ray diffraction data performed on experimental alloys to determine the average atomic volumes of Ni, Ti, Al, Zr, Hf, Pd, and Pt in the B2 and $L2_1$ phases. Atomic volume was derived based on lattice spacing by utilizing a simplifying assumption that atomic volume is not dependent on the lattice site that an atom occupies. The values calculated by Jung are found in Table 2.

Table 2 Measured and calculated atomic volumes at 25 °C for atomic species in the B2 and $L2_1$ phases (Ref 2)

Species, i	Ω_i^{B2} (nm³)	$\Omega_i^{L2_1}$ (nm³)
Ni	0.0123	0.0115
Ti	0.0151	0.0157
Al	0.0114	0.0124
Pd	0.0172	0.0155
Pt	0.0144	0.0110
Zr	0.0293	0.0305
Hf	0.0236	0.0245

A_f Model. The austenite finish temperature, A_f, can be approximated as T_0 for superelastic polycrystals (Ref 16) and modeled as a function of B2 composition using a Redlich-Kister polynomial for a two-sublattice system. This model is analogous to the substitutional regular solution model used in solution thermodynamics with a Redlich-Kister power series model for higher-order excess free energy contributions (Ref 17). The two-sublattice Redlich-Kister model for A_f temperature is given in Eq 11 for compounds $(A_1,\cdots,A_p)^{I}(B_1,\cdots,B_q)^{II}$ with sublattices I and II and elemental site fractions given by y_i^m for species A_i or B_i on sublattice m (m = I or II, respectively). The parameter $A_{f(i:j)}$ is the binary regular solution endpoint, corresponding to the A_f temperature of binary compound A_iB_j. The higher-order terms ^vL ($v = 0,\cdots,2$) represent the ternary interaction parameters which describe interactions between two species j and k on one sublattice in the presence of a third species i on the neighboring sublattice. The temperature-dependent term from solution thermodynamics is ignored:

$$A_f = \sum_{i=1}^{p}\sum_{j=1}^{q} y_i^{I} y_j^{II} A_{f(i:j)} + \sum_{i=1}^{p}\sum_{j=1}^{q}\sum_{k\neq j}^{q} y_i^{I} y_j^{II} y_k^{II} {}^{v}L_{(i:j,k)} \left(y_j^{II} - y_k^{II}\right)^{v} + \sum_{i=1}^{p}\sum_{j=1}^{q}\sum_{k\neq j}^{q} y_i^{I} y_j^{II} y_k^{I} {}^{v}L_{(j,k:i)} \left(y_j^{I} - y_k^{I}\right)^{v} \qquad \text{(Eq 11)}$$

Higher-order interaction parameters ^vL as well as some binary endpoints $A_{f(i:j)}$ which cannot be determined empirically must be fit to experimental data by minimizing the residual sum of squares. This fitting process is analogous to the process used for creating thermodynamic databases for CALPHAD techniques based on experimental data. A_f can be determined experimentally by differential scanning calorimetry (DSC) or differential thermal analysis (DTA) methods.

Data for many binary and ternary systems can be found in the literature. Hosoda et al. (Ref 7) summarized experimental M_s data for 3d transition metal additions in ternary $Pd_{(50-y)}X_yTi_{50}$ systems for X = Cr, Mn, Fe, Co, and Ni and found the B2 stabilizing effect to be significant and linear with respect to at.% addition and constant with respect to electron-to-atom (e/a) ratio, approximately 900 K/(e/a). Selected data from Hosoda et al. (Ref 7) is shown in Fig. 2 for ternary additions of Ni, Co, and Fe, which result in decreases in M_s temperature of -14.2, -25.5, and -36.2 °C/at.%, respectively.

While prominent transformation peaks in Ti-rich Ni(Ti,Zr) have been observed in alloys with up to 25 at.% Zr (Ref 18), only limited data on the effect of ternary additions of Zr in PdTi has been reported. Kawakita et al. (Ref 9) observed that in Pd(Ti,Zr) alloys, DSC peaks are prominent for 5 at.% Zr additions on the Ti sublattice, while at 10 at.% Zr the peaks become smaller and broader, and multiple peaks are exhibited upon cooling. In NiTi, Zr additions have a stabilizing effect on the martensite, while in PdTi systems Zr is found to decrease the transformation temperature (Fig. 3).

Radiopacity Model. The radiopacity, or extent to which a material can be seen in an x-ray fluoroscope, is quantified at a given x-ray energy by calculating the x-ray absorption, $1-\frac{I}{I_0}$, where I is the intensity of the transmitted x-ray and I_0 is the intensity of the incident x-ray. X-ray absorption is given by:

$$1-\frac{I}{I_0}=e^{-\left(\frac{\mu}{\rho}\right)\rho x} \quad \text{(Eq 12)}$$

where x is the strut thickness, ρ is the alloy density, and $\left(\frac{\mu}{\rho}\right)$ is the mass attenuation given by:

$$\left(\frac{\mu}{\rho}\right)=\sum_i w_i\left(\frac{\mu}{\rho}\right)_i \quad \text{(Eq 13)}$$

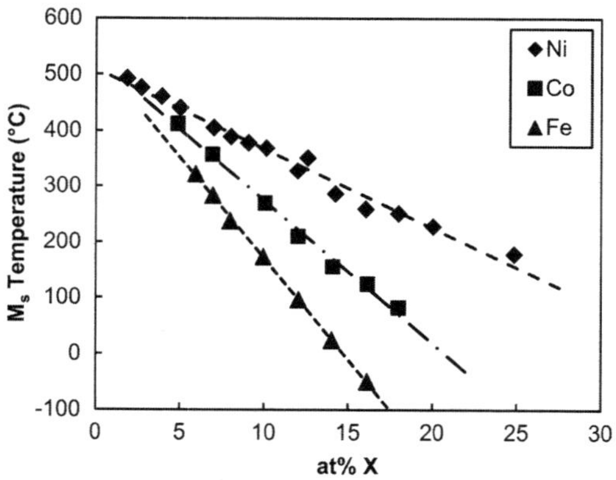

Fig. 2 M_s temperature dependence on ternary additions X = Ni, Co, Fe in a $Pd_{(50-y)}X_yTi_{50}$ alloy, from experimental data compiled in Ref 7.

where w_i is the mass fraction of component i and $\left(\frac{\mu}{\rho}\right)_i$ is the mass attenuation coefficient of the component i at the x-ray energy of interest, which for many medical x-ray fluoroscope applications is 30 keV. Radiopacity can be considered as the radiopacity improvement (RI) over binary NiTi, which has an x-ray absorption at 30 keV of 0.1863 or 18.6% for a part with a standard stent thickness of x = 40 μm (Ref 12).

Experimental Methods

Details regarding experimental procedures used in studying the NiTiAl-X, low-Ni, and Ni-free quaternary systems can be found in the literature (Ref 12, 19–21). The "hybrid" alloys presented in "Low-Ni Hybrid (Pd,Ni)(Ti,Zr,Al)" section were arc melted in Ar using >99% purity stock material and homogenized at 950 °C for 24 h. Samples were then cut into slices for mechanical, thermal, and structural characterization using an Accutom precision saw. Aging treatments were performed at various times at temperatures between 500 and 600 °C. All samples were encapsulated in evacuated quartz tubes with Ta foil getters to avoid surface oxidation and water quenched after each heat treatment.

Atom probe tomography (APT) was used to measure phase compositions and precipitate size after aging in a nontransforming

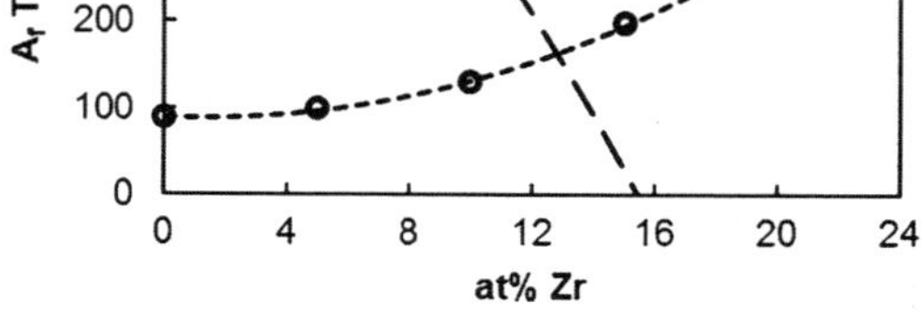

Fig. 3 A_f temperature as a function of Zr content in the $Ni_{50}Ti_{(50-x)}Zr_x$ (Ref 18) and $Pd_{50}Ti_{(50-x)}Zr_x$ (Ref 9) systems. Dotted lines indicate the predicted A_f temperatures in these ternary systems based on the Redlich-Kister A_f model with interaction parameters fitted to both data from the literature plotted here and experimental data.

$Pd_{30}Ni_{20}Ti_{38}Zr_8Al_4$ "hybrid" alloy prototype. For the isothermal study, samples were aged at 600 °C for 0.25, 1, 3, 5, 10, and 100 h. For the isochronal study, samples were aged for 3 h at 500, 530, and 600 °C. APT was performed on an imago local electrode atom probe (LEAP) at 80 K with a pulse frequency of 250 kHz and a laser energy of 20–30 pJ. Data analysis was performed using IVAS 6.3.2 software. Average precipitate radius and phase compositions were measured using proximity histograms (proxigrams) around Al isoconcentration surfaces representing the $L2_1$/B2 interface.

Vickers microhardness was used to characterize mechanical properties of the nontransforming "hybrid" sample as well as the peak-aged superelastic Ni-free $Pd_{42.2}Fe_{7.8}Ti_{44}Al_6$ alloy. Hardness tests were performed with 1000 g indent force/10 s dwell time and 20 indents per sample. For the precipitation hardening study, room temperature Vickers hardness was performed on hybrid alloy samples aged at 500 °C for 3, 5, 10, 24, and 50 and 360 h. Hot hardness measurements utilized a resistive ceramic heating stage in atmosphere and were performed on the PdFeTiAl alloy at T = 25, 40, 60, 80, 100, 125, 145, and 190 °C.

To measure M_s, M_f, A_s, and A_f temperatures in transforming alloys, differential scanning calorimetry (DSC) was performed using a Mettler Toledo DSC822 differential scanning calorimeter with a ramp rate of 10 °C/min for heating.

Process Optimization

To design effective aging treatments in order to obtain an optimized microstructure, it is imperative to develop descriptions of both particle growth kinetics and phase composition trajectory. In this section, experimental data for precipitate size and phase composition at various aging times and temperatures are used to fit the models described in the "Kinetic Models" and "Composition Trajectory Models" sections. Data from numerous prototype systems are collected and presented here, with the ultimate goal of facilitating the development of a thermodynamic database to predict phase relations in the multicomponent alloy system.

In addition to designing a peak-aged microstructure, much work has been done focusing on issues related to homogenization and the presence of metastable Ni-rich phases and Ti-rich intermetallics. Additionally, in Zr-containing systems the formation of brittle or low-melting phases that can cause workability or incipient melting issues must be addressed. Identification of appropriate solution and hot working temperatures to avoid these issues are discussed in "Ni(Ti,Zr,Al) System" section.

NiTi-Base Systems

NiTiAl-X (X = 5 at.% Pd, Pt, Zr, Hf) Systems. Early investigation by Jung (Ref 2, 19) of quaternary NiTiAl-X (X = Pd, Pt, Zr, Hf) systems examined elemental partitioning in highly overaged samples in order to evaluate the best candidate for reducing misfit. Early prototypes included small quaternary additions (5 at.%) of Pd and Pt on the Ni sublattice as well as Zr and Hf on the Ti sublattice, with 4–5 at.%Al on the Ti sublattice. A ternary alloy with 5 at.%Al was also studied. Average particle size was measured using centered dark field TEM imaging of samples aged at 600 and 800 °C for 100 and 1000 h. Phase compositions and partitioning coefficients were measured using AEM on highly overaged samples aged at 800 °C, and APT was used to measure phase relations in samples aged at 600 °C due to small precipitate sizes.

It was found that coarsening kinetics are retarded in the Zr- and Hf-containing systems when compared to particle growth in the ternary NiTiAl system, while the Pd- and Pt-containing systems had relatively faster kinetics. Therefore, the ternary, 5Zr, and 5Hf samples were aged for 1000 h while the 5Pd and 5Pt samples were aged for 100 h. After aging at 800 °C, the 5Zr alloy exhibited an average particle radius of 233 ± 81.3 nm but after aging at 600 °C the average particle size was only 5.65 ± 1.00 nm. Due to the extremely fast kinetics at 800 °C, 600 °C was chosen as a more appropriate aging temperature to achieve a finely dispersed array of nanoscale precipitates.

A coherent ⟨100⟩-oriented cuboidal microstructure with r_{avg} = 6.7 nm was seen in the Pt-containing system aged at 600 °C, but full coherency loss was observed at r_{avg} = 96.8 nm after aging at 800 °C. Coherency is observed in the Zr, Hf, and Pd systems after 600 °C aging treatments (r_{avg} = 5.65, 6.24, and 9.48 nm, respectively), although precipitates in the Zr and Hf systems are cuboidal while the Pd system maintains spherical precipitates.

Irregular and/or cuboidal semicoherent morphologies were observed in the Zr and Hf systems with average radii of 233 and 312 nm after aging at 800 °C, while precipitates in the Pd-containing system remained fully spherical indicating full coherency and very low misfit at an average radius of 34.6 nm. An additional 20Pd alloy studied exhibited fully coherent precipitates with an average radius as large as 75.5 nm after aging at 800 °C.

Key learnings from this initial research into quaternary NiTiAl-X systems were that while partitioning behavior is temperature dependent for Zr, Hf, and Pt, partitioning of Pd remained nearly constant across temperatures and compositions. Pt tends to strongly partition into the B2 matrix, making it an unfit candidate for low-misfit quaternary systems. Measured values for partitioning coefficients are summarized in Table 3.

Zr was shown to be more effective at decreasing misfit than Hf at 600 °C, and was therefore chosen for further investigation for strengthening studies for low-cost automotive applications. A preliminary $Ni_{50}Ti_{32}Zr_{15}Al_3$ prototype aged for 100 h at 600 °C demonstrated extremely effective strengthening with a yield strength of 2.1 GPa during compression at 180 °C (Ref 20). Further investigation into the precipitation behavior and strengthening in the NiTiZrAl system will be reviewed in "Ni (Ti,Zr,Al) System" and "Calibrating Strength Models" sections, respectively.

Pt-containing samples cannot be electropolished for APT sample preparation, making LEAP analysis for Pt-containing alloys prohibitive. Additionally, partitioning behavior of Pt at both 600 and 800 °C was detrimental to interphase misfit. Therefore, Pd was determined to be the most promising candidate for developing low-Ni biomedical alloys.

Ni(Ti,Zr,Al) System. Despite demonstrating excellent strengthening capabilities, the phase relations and processing issues involved with designing Zr-containing alloys must be further understood. There are a number of processing challenges that must be considered with Zr-containing alloy designs. Alloys with as low as 3 at.%Zr have shown brittleness in hot working processes due to the formation of the NiTiZr laves phase and $(Ti,Zr)_2Ni$ intermetallic (Ref 22). Hsieh and Wu (Ref 23) showed that in $Ni_{50}Ti_{(50-x)}Zr_x$ systems for $x \geq 10$ at.%, brittle $(Ti,Zr)_2Ni$ and NiTiZr were in equilibrium with the matrix after homogenizing at 900 °C (Ref 22).

Three scaled-up 5-lb NiTiZrAl prototypes were investigated by Bender (Ref 12): $Ni_{50}Ti_{32}Zr_{15}Al_3$, $Ni_{50}Ti_{38}Zr_8Al_4$, and $Ni_{50}Ti_{35.8}Zr_{8.7}Al_{5.5}$. The first two fractured during hot rolling at 950 °C, and the third was homogenized at 950 °C and then shattered during hot rolling at 825 °C. SEM/EDS investigation indicated the presence of both a NiTiZr laves phase and an metastable Ni-rich $Ni_3(Ti, Zr)_2$ T-phase in equilibrium with the B2 matrix. Scheil modeling was employed to investigate nonequilibrium cooling and the resulting effects on incipient melting at homogenization or hot working temperatures. In the three alloys listed above, liquid was predicted to be present down to 899, 961, and 894 °C, respectively, indicating that the first and third buttons probably experienced some melting during homogenization and/or hot rolling. In Fig. 4, the nonequilibrium cooling solidus lines are

Table 3 Experimentally measured partitioning coefficients X = Pd, Pt, Zr, and Hf in the NiTiAl-X system

$k_x^{B2/L2_1}$	600 °C	800 °C
Pd	0.60	0.64
Pt	3.71	3.72
Zr	0.75	1.35
Hf	0.87	2.17

Values at 800 °C were measured using AEM while values at 600 °C were measured using APT. Data taken from Ref 4.

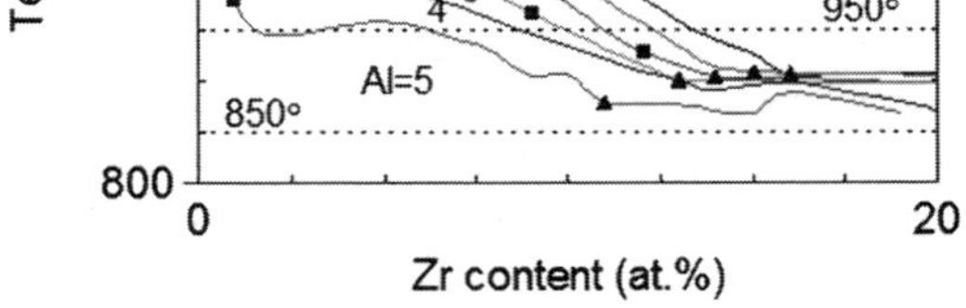

Fig. 4 Scheil simulations showing solidus as a function of Zr content for various Al contents. The square symbols indicate the composition at which the phases are in equilibrium with liquid at the solidus temperature shift from single-phase B2 to two-phase B2 + $L2_1$. The triangle indicates the Zr content above which the NiTiZr laves phase (C14) is present in equilibrium with liquid, B2, and $L2_1$. Figure reproduced with permission from Ref 12.

plotted for various Zr and Al compositions. In order to fully avoid incipient melting with compositions up to 5 at.%Al and 20 at.%Zr, both annealing and hot working temperatures must be limited to 850 °C. In addition it is clear that the C14 (laves) phase is stabilized at high Zr content levels, which decreases with added Al; the maximum Zr content limit as a function of Al composition should be carefully considered to avoid formation of the brittle laves phase.

With care taken toward avoiding incipient melting and formation of brittle intermetallic phases, Zr-based alloys show promise in reducing interphase misfit in low-cost NiTi-base designs for automotive applications. Atom probe tomography was performed on $Ni_{50}Ti_{38}Zr_8Al_4$ aged at 600 °C. Particle size evolution can be seen to follow a $t^{1/3}$ trend consistent with LSW-type coarsening behavior Fig. 5. The rate constant is smaller than in the ternary system, as Jung observed in his initial TEM analysis of overaged systems. Fitted Gibbs-Thomson constants for composition trajectory are presented in Tables 4 and 5 and are plotted in Fig. 6(a). The critical radius for nucleation, r_{crit}, can be identified by extrapolating the matrix composition trajectory to the bulk composition. The scaling of the x-axes in Fig. 6 have been adjusted to reflect the limits of $\left(\frac{1}{r_\infty}, \frac{1}{r_{crit}}\right)$.

APT reconstructions for the 15Zr3Al and 8Zr4Al systems aged at 600 °C for long times (100+ h) as well as 5Zr5Al starting even at shorter aging times (10 h) show the presence of a lenticular metastable Ni-rich Ni_4Ti_3 intermetallic phase which is not apparent in TEM micrographs at 800 °C (Ref 12). It should be noted when describing phase relations for these alloys during aging at 600 °C that they do not reflect a B2 + $L2_1$ 2-phase field.

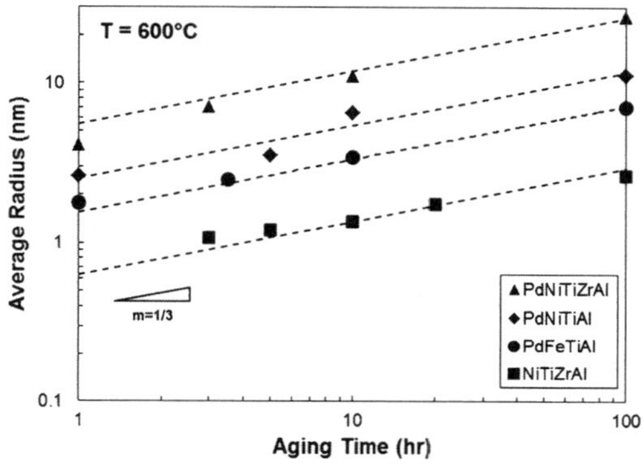

Fig. 5 Particle size trajectory versus aging time at 600 °C. Dotted lines represent fitted $t^{1/3}$ curves consistent with LSW-type coarsening. Experimental data are based on LEAP measurements of phase composition in four experimental prototypes: $Ni_{50}Ti_{38}Zr_8Al_4$ (Ref 12), $Pd_{(30-36)}Ni_{(14-20)}Ti_{(45-46)}Al_{(4-5)}$ (Ref 12), $Pd_{42.2}Fe_{7.8}Ti_{44}Al_6$ (Ref 21), and $Pd_{30}Ni_{20}Ti_{38}Zr_8Al_4$.

Table 4 Values for fitted constant A (at.%) from Gibbs-Thomson trajectory for phase composition fitted to experimental LEAP data.

		B2						$L2_1$					
Alloy	Temp (°C)	Ni	Pd	Fe	Ti	Zr	Al	Ni	Pd	Fe	Ti	Zr	Al
$Ni_{50}Ti_{38}Zr_8Al_4$	600	50.67	···	···	38.07	9.16	2.23	46.30	···	···	17.51	11.57	24.76
$Pd_{20}Ni_{30}Ti_{46}Al_4$	550	31.72	17.90	···	48.48	···	2.02	11.06	42.18	···	26.85	···	20.52
$Pd_{42.2}Fe_{7.8}Ti_{44}Al_6$	600	···	40.56	8.39	47.34	···	3.85	···	52.80	1.70	25.74	···	20.30
$Pd_{30}Ni_{20}Ti_{38}Zr_8Al_4$	600	20.40	29.05	···	41.22	7.35	2.00	7.74	44.61	···	13.06	11.65	24.34

The value of *A* corresponds to the equilibrium solubility of each component in the phase

Table 5 Values for fitted constant b (nm) from Gibbs-Thomson trajectory for phase composition fitted to experimental LEAP data.

		B2						$L2_1$					
Alloy	Temp (°C)	Ni	Pd	Fe	Ti	Zr	Al	Ni	Pd	Fe	Ti	Zr	Al
$Ni_{50}Ti_{38}Zr_8Al_4$	600	−0.0206	···	···	−0.0080	−0.0277	0.5000	0.0008	···	···	0.1647	−0.1176	−0.0854
$Pd_{20}Ni_{30}Ti_{46}Al_4$	550	−0.0257	0.1340	···	−0.0851	···	0.7765	0.6898	−0.4405	···	0.3320	···	−0.1726
$Pd_{42.2}Fe_{7.8}Ti_{44}Al_6$	600	···	0.0798	0.0033	−0.1480	···	0.6547	···	−0.2527	1.2631	0.1718	···	0.1590
$Pd_{30}Ni_{20}Ti_{38}Zr_8Al_4$	600	−0.0457	0.0715	···	−0.1310	0.1963	1.0626	0.3047	−0.1241	···	1.4595	−1.0414	−0.9341

PdTi-Base Systems

Low-Ni (Pd,Ni)(Ti,Al) System. A range of low-Ni prototypes have been investigated by Bender and Jiang using APT at 600 °C (Ref 12) and 550 °C (Ref 20). At 600 °C the investigated systems had compositions in the range of $Pd_{30-36}Ni_{14-20}Ti_{45-46}Al_4$. At 550 °C a non-transforming $Pd_{20}Ni_{30}Ti_{46}Al_4$ prototype was characterized, the results of which were utilized in calibrating a strengthening model in the "Calibrating Strength Models" section. The precipitate growth trajectory at 600 °C is plotted in Fig. 5 along with other systems characterized at that aging temperature. The kinetics are faster than in the NiTiZrAl system, consistent with TEM observations for the first quaternary prototypes presented in the "NiTi-Base Systems" section. The measured rate constants for systems aged at 600 °C can be found in Table 6.

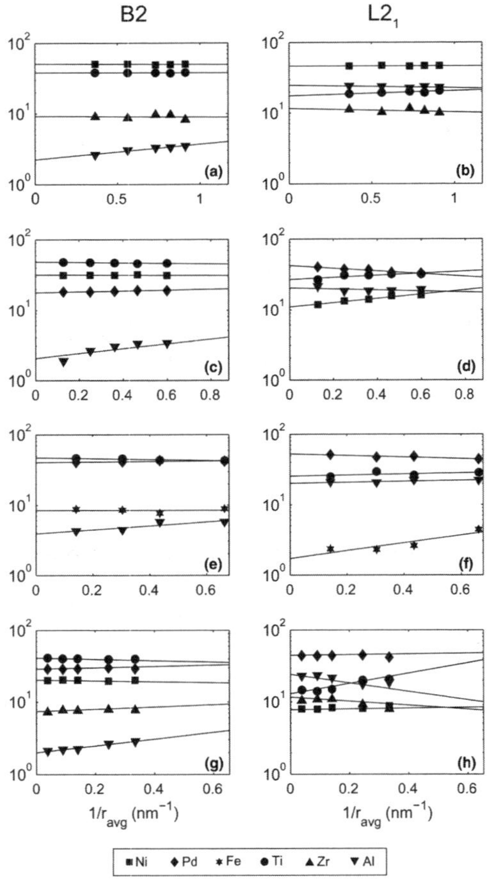

Fig. 6 Gibbs-Thomson plots for composition trajectory versus 1/*r*. Atomic composition of elemental components in the B2 phase (a,c,e,g) and L21 phase (b,d,f,h) is plotted versus the inverse of the measured average radius for various aging treatments, along with fitted Gibbs-Thomson curves (solid lines) for (a, b) $Ni_{50}Ti_{38}Zr_8Al_4$ at 600 °C (Ref 12), (c, d) $Pd_{20}Ni_{30}Ti_{46}Al_4$ at 550 °C (Ref 20), (e, f) $Pd_{42.2}Fe_{7.8}Ti_{44}Al_6$ at 600 °C (Ref 21), and (g, h) $Pd_{30}Ni_{20}Ti_{38}Zr_8Al_4$ at 600 °C. The x-axis is scaled such that $x_{max} = 1/r^*$, i.e., corresponding to the critical radius for nucleation.

Partitioning behavior during aging at 550 °C can be observed in the Gibbs-Thomson trajectories in Fig. 6, from which equilibrium solubility can be determined. From the proxigram in Fig. 7b, it is clear that in the PdNiTiAl system, Pd partitions into the $L2_1$ while Ni partitions into the B2 phase. A pseudoternary phase diagram in Fig. 8 shows tie lines for equilibrium solubilities at 600 °C based on the extrapolated Gibbs-Thomson trajectories for the NiTiZrAl, PdNiTiAl, and PdNiTiZrAl systems.

A secondary microscale interdendritic phase was observed in most low-Ni microstructures after homogenization. This phase has a Ti-rich $Ti_2(Ni,Pd)$-type composition and is thought to be a proto-oxide similar to the $Ti_4Ni_2O_x$ oxides seen in commercial arc-melted NiTi (Ref 24). Jiang (Ref 20) conducted a solution treatment study in which a 950 °C 20 h homogenization was followed by 100 h solution treatments at 900, 950, 1000, and 1050 °C. Solution treatments led to spheroidization and a slight reduction in phase fraction of the Ti-rich secondary phase, but full dissolution of the phase was not observed. The Ti_2Ni phase has been shown to be brittle in NiTi so attention must be paid to this phase despite the relatively low phase fraction and large interparticle spacing.

Ni-Free (Pd,Fe)(Ti,Al) System. A Ni-free $Pd_{42.2}Fe_{7.8}Ti_{44}Al_6$ prototype was aged at 600 °C and investigated using APT (Ref 21). Pd partitioning was not observed to be as strong as in low-Ni alloys because Ni was not present on the Pd sublattice in competition for B2 sites. Fe partitioned strongly out of the Heusler phase, with an equilibrium composition of only 1.70 at.% in $L2_1$. Al solubility in the B2 phase is greatly increased compared to

Table 6 Rate constants for systems characterized at 600 °C

System	Rate constant, k (nm/h$^{1/3}$)
NiTiZrAl	0.63
PdNiTiAl	2.52
PdFeTiAl	1.54
PdNiTiZrAl	5.44

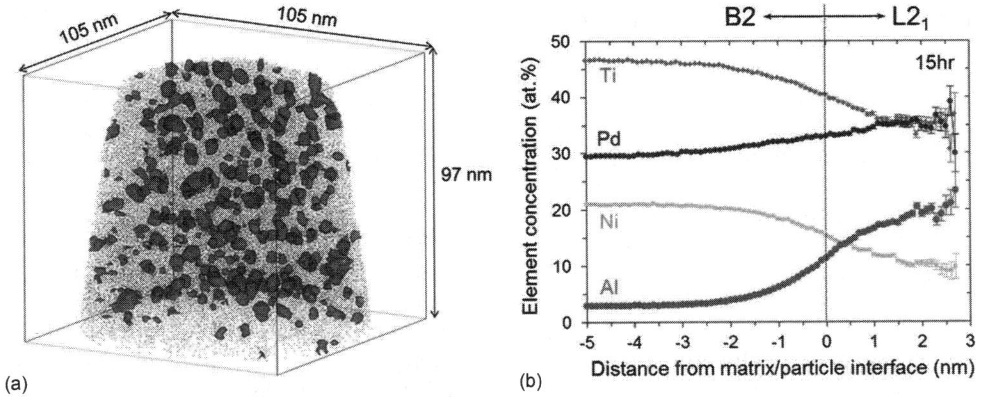

Fig. 7 $Pd_{30}Ni_{20}Ti_{46}Al_4$ alloy peak aged at 530 °C for 15 h. (a) IVAS reconstruction of APT data. Heusler phase is visualized using 12 %Al isoconcentration surfaces. (b) proxigram around B2/L2$_1$ interface shows compositional profiles. Data from Ref 20.

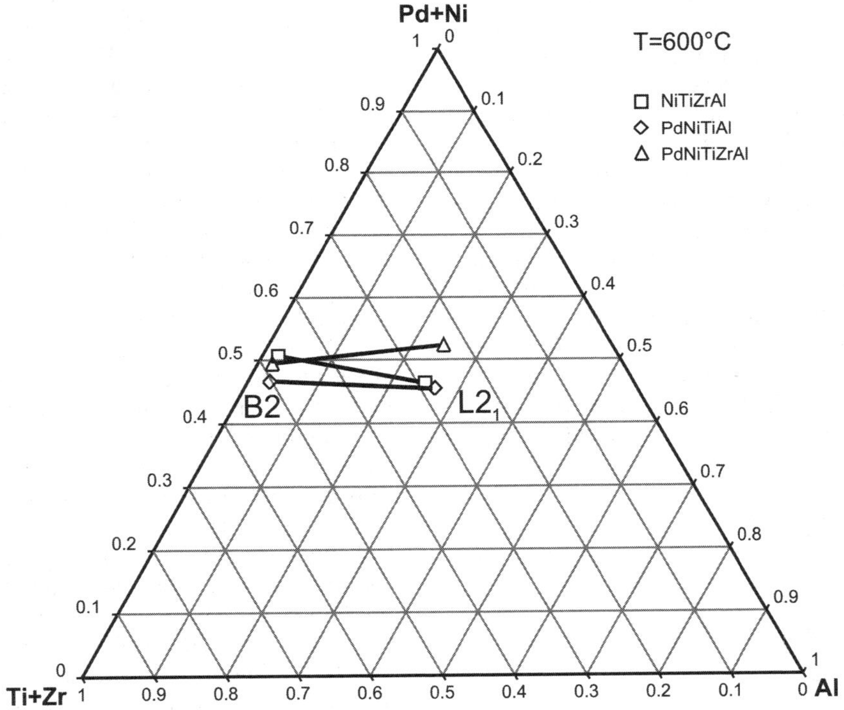

Fig. 8 Pseudoternary phase diagram for NiTiZrAl (squares), PdNiTiAl (diamonds), and PdNiTiZrAl (triangles). Tie lines represent equilibrium solubilities calculated from LEAP data for samples aged at 600 °C

low-Ni, with an equilibrium solubility value of 3.85%. Early prototypes with 4 at.%Al bulk compositions showed negligible precipitation due to lack of supersaturation. From Fig. 5 it is clear that the coarsening kinetics are faster than in the NiTiZrAl system but slower than in the low-Ni PdNiTiAl system.

Structure/Property Relations

Calibrating Strength Models

Hardness data were collected on nontransforming $Ni_{50}Ti_{38}Zr_8Al_4$ aged at 600 °C (Ref 12) and Pd_{20} $Ni_{30}Ti_{46}Al_4$ aged at 550 °C (Ref 20) for various times. An empirical power law fit to experimental data from Ref 25 relating hardness to yield strength in NiTiAl alloys was used to estimate yield strength values from Vickers hardness numbers.

Using Eq 1 to isolate the precipitation strengthening contribution and normalizing by $f^{1/2}$ based on phase fraction values measured by APT, the strengthening data from both systems can be plotted together (Fig. 9). The dashed lines represent the shearing curves (Eq 3) fit to the NiTiZrAl and PdNiTiAl data for $r \leq r_{opt}$, for which $K1$ was calculated to be 1419.1 and 491.3 MPa/nm, respectively. The much lower value for $K1$ in the Pd-containing system is most likely due to modulus softening of Pd in contrast to the modulus strengthening of Zr. The modulus hardening effect proves to be the most significant contributor to shear resistance, and the relationship between $L2_1$ composition and resulting $K1$ value, or shear resistance, can be quantified.

The dotted line is the looping curve (Eq 4) fit to both sets of experimental data for $r > r_{opt}$ ($K2$ = 3089.4 MPa-nm). Due to the different shear resistance values, each alloy has a different resulting value for r_{opt}. For the NiTiZrAl system, 1.5 nm is found to be the optimum particle size, corresponding to a 10 h 600 °C aging treatment. In the low-Ni PdNiTiAl alloy, peak strengthening occurs at r_{opt} = 2.5 nm. Previous curve fitting that constrained the value of n in Eq 3 to 0.5 resulted in an optimum radius value of 2.3, but a fit with n = 1 was found to describe

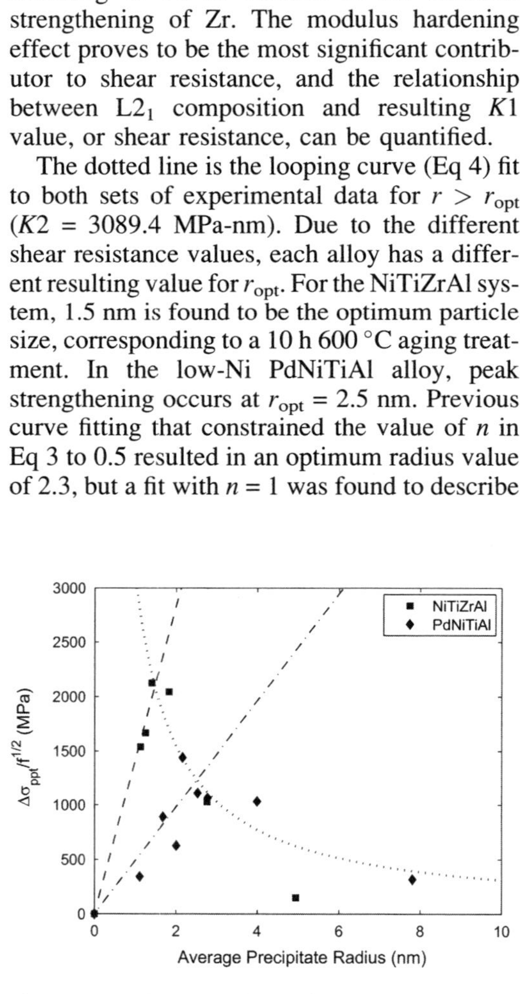

Fig. 9 Strengthening data from nontransforming $Ni_{50}Ti_{38}Zr_8Al_4$ aged at 600 °C (Ref 12) and $Ni_{30}Pd_{20}Ti_{46}Al_4$ aged at 550 °C (Ref 20). The dashed and dot-dashed lines represent the shearing curves (Eq 3) fit to the NiTiZrAl and NiPdTiAl data for $r \leq r_{opt}$, respectively. The dotted line is the looping curve (Eq 4) fit to both sets of experimental data for $r > r_{opt}$.

the behavior of these systems more accurately. The PdNiTiAl alloy exhibited a peak strength of 1.05 GPa with misfit value of −0.77% and radiopacity improvement (RI) of 129%, while the NiTiZrAl alloy exhibited a peak strength of 2.40 GPa, misfit of −0.67%, and RI of 35.6%.

Transformation Temperature

A_f values have been measured using DSC for a number of ternary and higher-order systems of interest here. Data for the ternary (Ni,Pd)Ti were reported by Linquist and Wayman as reproduced in Ref 26, from which values for the Ni:Ti and Pd:Ti binary regular solution endpoints $A_{f(Ni:Ti)}$ and $A_{f(Pd:Ti)}$ as well as higher-order ^0L and ^1L terms describing the ternary (Ni,Pd)Ti interaction can be calculated. As shown in Fig. 2, ternary data in (Pd,X)Ti systems have been studied by Hosoda et al. (Ref 7), including (Pd, Fe)Ti. Transformation temperatures in the Ni (Ti,Zr) (Zr = 0–20 at.%) system was characterized by Hsieh and Wu (Ref 18), while limited data for the Pd(Ti,Zr) (Zr = 0, 5, 10 at.%) have been reported by Kawakita et al. (Ref 9). Literature data with fitted Ni(Ti,Zr) and Pd(Ti,Zr) interaction parameters are plotted in Fig. 3.

Experimental data were collected on the quaternary low-Ni PdNiTiAl system (Ref 12, 20). While none of the Ni(Ti,Zr,Al) alloys characterized by Jung and Bender proved to be transformable, by switching Al to the Ni sublattice, transforming alloys can be designed. Data have been collected for transforming (Ni,Al)(Ti,Zr) prototypes (Ref 20), allowing for the calculation of Al:Ti, Al:Zr, and higher-order (Ni,Al):Ti parameters. In combination with the data from the Ni-free (Pd,Fe)(Ti,Al) system presented in the "Ni-Free (Pd,Fe)(Ti,Al) System" section, Fe:Ti and Fe:Al interaction parameters as well as higher-order (Pd,Fe):Ti have been calculated. The calculated interaction parameters are summarized in Tables 7 and 8. Contour plots containing A_f predictions using aforementioned fitted constants are calculated for four quaternary systems of interest: low-Ni PdNiTiAl, Ni-free PdFeTiAl, hybrid PdNiTiZr, and transforming NiAlTiZr (Fig. 10). The bold lines represent the contours at 10 and 15 °C, and the plotted points represent experimental data points used in fitting. Experimentally measured A_f values can be found in a supplementary Table A1 at the end of this chapter.

Table 7 Calculated interaction parameters for binary solution end-points $A_{f(i:j)}$ where i is the component on sublattice I and j is the component on sublattice II.

	Component j (sublattice II)		
Component i (sublattice I)	Ti	Zr	Al
Ni	89	1422.8	−3291.0
Pd	591	−2513.3	−1606.3
Fe	−2522.5	···	−4132.8
Al	7599.2	−3796.0	···

Table 8 Calculated ternary interaction parameters for the 0th (^0L)- and 1st (^1L)-order interactions

Ternary interaction	^0L	^1L
(Ni,Pd):Ti	−720.43	−507.46
Ni:(Ti,Zr)	−1400.8	0
Pd:(Ti,Zr)	982.2	1923.0
(Pd,Fe):Ti	1696.3	0
(Ni,Al):Ti	−9078.6	0

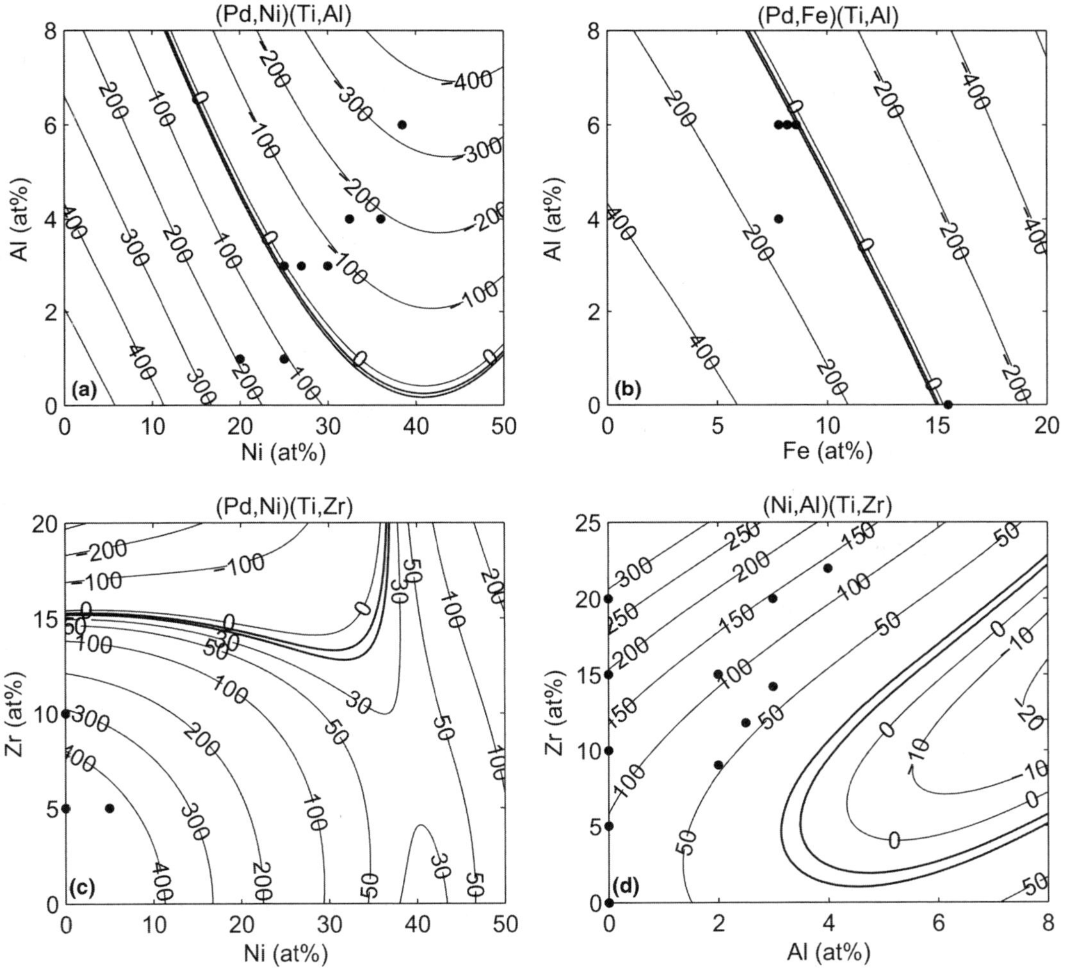

Fig. 10 Contour plots for A_f based on the fitted Redlich-Kister polynomial model for A_f temperatures. Bold contour lines correspond to A_f = 10 and 15 °C. Dots represent experimental data points used in the model fitting. Experimental A_f values can be found in supplementary Table A1

The parabolic shape of the ternary (Ni:Pd)Ti interaction in which the A_f temperature first decreases upon addition of Pd on the Ni sublattice and then increases with increasing Pd content (as seen in Fig. 11) is a result of the Pd stabilizing a B19 martensite instead of the B19′ martensite that is observed in binary NiTi systems. The ternary system could also be

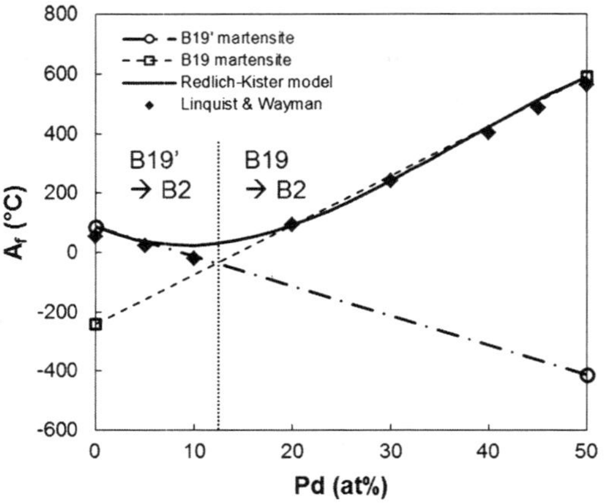

Fig. 11 A_f temperature in $Ni_{50-x}Pd_xTi_{50}$ for x = 0–50 at.%. Solid diamonds represent data from Linquist and Wayman as reported in Ref 26. The solid line represents the continuous Redlich-Kister function for predicting A_f, while the dotted lines represent the linear functions for B2 → B19′ (circles) and B2 → B19 (squares).

modeled by two linear models that have separate B2 → B19′ and B2 → B19 regular solution end-points for the two different martensites stabilized in that system. These endpoints are found to be $A^{B19}_{f(Ni:Ti)} = -240.0$ and $A^{B19'}_{f(Pd:Ti)} = -410.8$. Figure 11 plots experimental data points for A_f values in the (Ni,Pd)Ti system from Linquist and Wayman as reported in Ref 26 along with the discrete B19 and B19′ expressions (dashed lines) and continuous Redlich-Kister function (solid line) for the prediction of A_f temperature. It is clear that the Redlich-Kister function with higher-order parameters merely approximates the discrete function made up of two intersecting linear trends, leading to possible overestimations of A_f in the 10–15 at.% Pd compositional region.

Cyclic Stability Validation

Validation of cyclic stability performed on a transforming $Pd_{30}Ni_{20}Ti_{46}Al_4$ alloy aged at 600 °C for 5 h showed stable behavior in the DSC after 30 thermal cycles as well as good cyclic stability under compression cycles and reversible transformation strains up to 4% (Ref 12). However, when compared to samples aged at 550 and 530 °C, the best cyclic stability was found in samples aged at 530 °C (Ref 20). The increased stability at lower-temperature aging treatments is most likely due to an increased driving force for precipitation resulting in a finer dispersion of nanoscale precipitates with a larger number density. Jiang found that the temperature dependence of the rate constant in PdNiTiAl followed an Arrhenius trend, and used this to design a peak-strengthened alloy aged at 530 °C for 15 h. LEAP analysis of this peak-strengthened alloy shows precipitates with an average size of 2.76 nm, close to the targeted value of 2.5 nm. An A_f of approximately 20 °C was measured and cyclic compression performed at T = 36 °C showed superelastic loops exhibiting high cyclic stability and a recoverable superelastic strain of 5.5%.

An optimum heat treatment for the Ni-free $Pd_{42.2}Fe_{7.8}Ti_{44}Al_6$ of 3.5 h at 600 °C was designed to target an r_{opt} = 2.5 nm based on the strengthening study in the low-Ni system (Ref 21). LEAP analysis on a peak-aged alloy measured an average precipitate radius of 2.3 nm. The radiopacity improvement is calculated to be 217.2%. The A_f temperature of 52.3 °C was measured using DSC, which shifted by only 0.1 °C over 10 heating/cooling cycles. Cyclic compression showed stable loading behavior and recoverable strains on the order of 3%. Hardness values measured at T = 145 °C ($T = M_s^\sigma$) indicate a B2 yield strength of 423 MPa. The diminished strength of the Ni-free prototype is due to the solution and modulus softening effect of Pd. Cyclic stability is achieved despite relatively low strength due to the reduced transformation strain of the B2 → B19 transition such that lower strength levels are necessary to prevent accommodation slip. The addition of Zr in the "hybrid" design described in the "Low-Ni Hybrid (Pd,Ni)(Ti, Zr,Al)" section will attempt to mitigate this strength loss and prevent accommodation slip in low-Ni B19′ stabilized systems with larger transformation strains.

Low-Ni Hybrid (Pd,Ni)(Ti,Zr,Al)

Precipitation in a "hybrid" $Pd_{30}Ni_{20}Ti_{38}Zr_8Al_4$ prototype was characterized at 600 °C using APT. Phase composition trajectories can be observed in Fig. 6. Ni and Pd partitioning trends are similar to low-Ni systems, while Zr partitions to the $L2_1$ phase as in the NiTiZrAl system. Misfit values were calculated to be −0.02% after 100 h of aging, lower than any previously studied system. Kinetics at 600 °C in the hybrid system were found to be much faster than in any of the previously characterized systems (Fig. 5). For this reason, precipitation kinetics were also investigated at 530 and 500 °C, and 500 °C was determined to have sufficiently reduced kinetics such that a nanoscale precipitation study could be performed. Hardness measurements were performed on samples aged at 500 °C for various times, and peak strengthening was observed to occur at 50 h (Fig. 12). Peak strength was measured to be 550.8 VHN corresponding to an estimated yield strength of 1.8 GPa. This strength level is 70% higher than that observed in a low-Ni $Pd_{20}Ni_{30}Ti_{46}Al_4$ peak-aged system, demonstrating that even with higher Pd content, strength can still be improved significantly with relatively small Zr additions. However, at temperatures at or below 530 °C, a slightly Ni-group rich $(Pd,Ni)_x(Ti,Zr)$ metastable compound is found to be present in addition to the $L2_1$ phase. This third phase is Zr-rich and the compound stoichiometry deviates from Ni-rich Ni_4Ti_3 metastable intermetallic compounds commonly observed in binary NiTi as well as some NiTiZrAl alloys (Ref 12, 19). Characterizing this third phase is important in analyzing the strengthening contributions of precipitation in this alloy system at 500 °C and is the subject of continuing investigation.

To design a transforming hybrid alloy, prototypes needed to be made to fit interaction parameters in the hybrid system for the A_f model. Three transforming hybrid prototypes were fabricated: $Pd_{50}Ti_{41}Zr_5Al_4$, $Pd_{50}Ti_{39}Zr_5Al_4$, and $Pd_{45}Ni_5Ti_{41}Zr_5Al_4$. In combination with data on the ternary PdTiZr system from Ref 9, the A_f model was extended to include the (Pd,Ni)(Ti,Zr,Al) design space. A_f predictions based on this model can be observed in Fig. 10(c). Due to the limited, high-temperature data points used in fitting this model, its reliability in predicting A_f at lower temperatures is not optimized. More experimental data must be collected for transforming systems in the temperature region of interest to extend the reliability of the model for designing a superelastic hybrid alloy.

Nonmetallic Inclusions

In commercial NiTi melting techniques such as vacuum induction melting (VIM), vacuum arc remelting (VAR), and some combination of these processes (VIM/VAR), nonmetallic inclusion particles form during solidification due to C and O impurities introduced into the melt by source materials or the melt process itself. Both Ti(C,O)-type carbides or $Ti_4Ni_2O_x$-type oxide phases can form, the phase fractions of which are dependent upon the impurity concentration levels (Ref 24). Due to their brittle properties and incoherency with the surrounding matrix, inclusions can break up during cold drawing processes, forming long stringers of fractured particles in the order of 10–50 μm in length (Fig. 13). The fractured inclusions can partially delaminate from the surrounding matrix forming sharp crack tips or voids which act as potent fatigue crack nucleation sites. In order to characterize behavior in the UHCF regime, it is critical to be able to model the morphology, distribution, and potency of these fatigue nucleants.

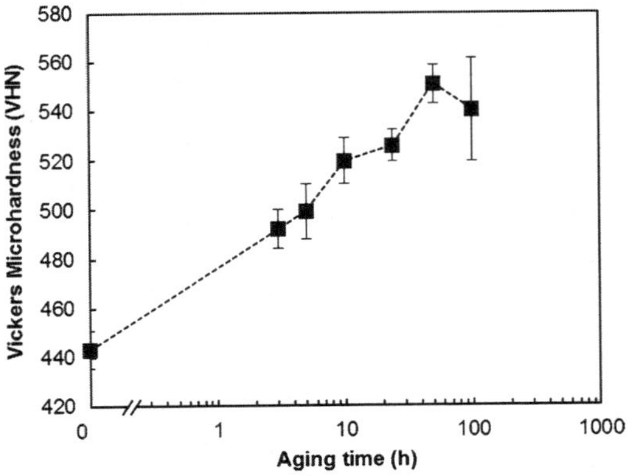

Fig. 12 Vickers microhardness versus aging time in $Pd_{30}Ni_{20}Ti_{38}Zr_8Al_4$ aged at 500 °C

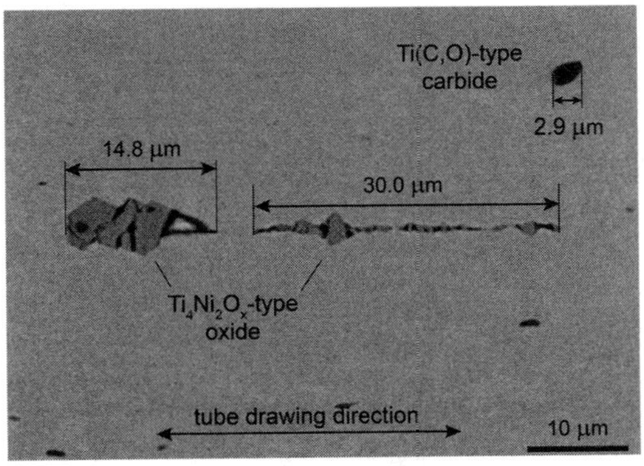

Fig. 13 Oxide and carbide inclusions in a cold-drawn NiTi tube. Broken and elongated stringers occur in the drawing direction (horizontal).

Moore (Ref 27) has utilized image-based meshes of a NiTi material with a realistic inclusion stringer to predict fatigue limit using finite element analysis. Employing crystal plasticity models, a fatigue indication parameter (FIP) was calculated at various applied strain amplitudes which can be correlated to Coffin-Manson type fatigue behavior (Ref 28) The FIP values were calibrated to the minimum fatigue life calculated from experimental NiTi diamond element fatigue data. The minimum (P = 0) fatigue life, i.e., fatigue limit, was calculated by fitting normalized experimental data to a three-parameter Weibull function as per the method outlined in Ref 29. The effect of matrix strength on the fatigue life was quantified using this technique. A 50% increase in NiTi yield strength over a base strength of 750 MPa resulted in a 44% increase in fatigue limit at 10^9 cycles, increasing from 0.27 to 0.39 % strain amplitude (Fig. 14).

A number of studies have attempted to quantify the effects of inclusion type and morphology on fatigue performance. Current consensus in the literature is that in high and ultrahigh cycle fatigue, the maximum inclusion size is a critical factor in determining fatigue behavior (Ref 30–32). Launey et al. (Ref 30) found that in a direct comparison between standard VAR material and high purity (HP) VAR material with reduced maximum inclusion size and area fraction, the HP VAR material exhibited a fatigue limit at 10^7 cycles five times higher than that of the standard VAR material. Rahim et al. (Ref 31) found that in bending rotation tests, HP material with a low volume fraction of oxides and extremely low volume fraction of carbides outperforms both oxide- and carbide-heavy samples in low cycle fatigue (LCF), while in the HCF and UHCF regimes fatigue performance in carbide-heavy samples outperformed both oxide-heavy and HP samples. This behavior was due to the fact that oxides form stringers both at a higher rate and of larger sizes than carbides during thermomechanical processing. Recent developments by G. Rau GmbH (Pforzheim, Germany) demonstrate the use of electron beam refining (EBR) techniques to reduce the maximum inclusion size. Compared to conventional VAR material, VAR samples treated with EBR showed a reduction in maximum inclusion size from >15 to 1.1 μm and area fraction from 0.73 to 0.06%, corresponding to an increase in fatigue strain limit at 10^7 cycles from 0.6 to 1.3% (personal communication from A. Pelton to D. Frankel). Process modifications such as EBR which reduce the nucleation potency of non-metallic inclusions are integral in addressing fatigue concerns in commercial NiTi. Calculations are under way to predict the fatigue performance potential of combining the inclusion size reduction with $L2_1$ nanodispersion strengthening.

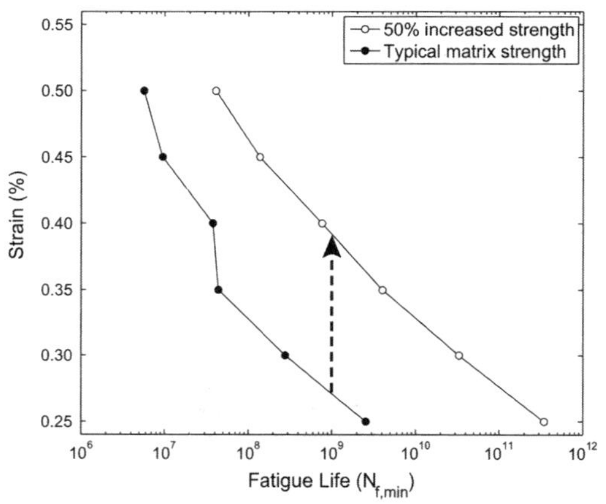

Fig. 14 Plot for applied strain amplitude versus minimum fatigue life ($N_{f,min}$) for a NiTi matrix with an inclusion stringer. Curves show fatigue limit for NiTi with typical B2 yield strength values (filled) and a NiTi matrix with 50 % enhanced B2 yield strength values (unfilled). The dashed arrow indicates the resultant increase in fatigue limit at 10^9 cycles. Results are based on calibrated FIP values computed from FEA simulations on an image-based mesh as reported in Ref 27.

Conclusions and Recommendations

$L2_1$ aluminide Heusler phase precipitation and strengthening behavior in NiTiZrAl, low-Ni PdNiTiAl, Ni-free PdFeTiAl, and hybrid PdNiTiZrAl have been characterized such that kinetic, thermodynamic, and strengthening models can be fit to experimental data and utilized in a parametric alloy design to achieve stated application-dependent property objectives. Zr and Pd are employed as quaternary alloying additions that reduce interphase misfit and help maintain coherency. Particle evolution is found to be dominated by particle coarsening at 500–600 °C and LSW-type $t^{1/3}$ rate constants can be used to predict particle size as a function of aging time. Compositional trajectories can be fit to Gibbs-Thomson equations allowing for prediction of phase composition at nonequilibrium (i.e., finite r) stages, enabling phase relations at peak strength to be determined. Additionally, Gibbs-Thomson trajectories can be used to calculate both equilibrium solubilities and the critical radius for nucleation.

A high-strength NiTiZrAl system for automotive actuator applications exhibited peak strengths of 2.40 GPa at an optimum precipitate radius of 1.5 nm. Lower peak strengths were achieved in the low-Ni PdNiTiAl system due to Pd softening. A peak strength of 1.05 GPa at r_{opt} = 2.5 nm was observed in a nontransforming PdNiTiAl alloy aged at 550 °C. A peak-strengthened superelastic PdNiTiAl alloy aged at 530 °C for 15 h exhibited excellent thermal and mechanical cyclic stability with reversible strains of 5.5% in compression. A superelastic Ni-free peak-aged alloy exhibited high thermal cyclic stability despite low strength values due to reduced transformation strain. However, low-Ni and Ni-free alloys exhibited significant improvements in radiopacity over binary NiTi, which is advantageous for biomedical applications. Low-Ni "hybrid" PdNiTiZrAl alloys demonstrate the capacity for combining high strength and biocompatibility, along with extremely low misfit and improved radiopacity. A nontransforming hybrid alloy aged at 500 °C for 50 h demonstrated a peak strength level of 1.8 GPa. Further development of an A_f model based on empirical data is necessary to achieve a superelastic design for biomedical applications.

Issues due to the presence of secondary intermetallic phases, low-melting phases, and large non-metallic inclusions were addressed as important aspects of process design. A Ti-rich interdendritic phase is observed to form during the melt process in low-Ni and Ni-free systems. Lenticular Ni_4Ti_3-type metastable intermetallic phases can be seen in many Zr-containing systems aged at or below 600 °C. Nonequilibrium solidification simulations predict that a low-melting NiTiZr laves phase may limit homogenization and hot working temperatures due to incipient melting concerns. Additionally, non-metallic phases such as oxides and carbides must be considered when addressing structural fatigue life in the UHCF regime. Advanced powder metallurgy techniques such as rapid solidification and thermal spray could be explored as fabrication routes to bypass solidification pathways that lead to the formation of undesirable secondary phases.

The development of these $L2_1$ precipitation strengthened SMA systems provides exciting avenues for future investigation. Most current methods for strengthening NiTi-base alloys rely on the use of cold working techniques. Utilizing work-free aging treatments to produce strengthening dispersions opens the door for many net shape fabrication techniques. With the rapid advances being made in the field of additive manufacturing, the ability to produce a device via 3D printing would afford radically new processing routes and provide an alternative to traditional cold drawing and laser cutting techniques, which can result in damaging inclusion stringers.

Moving forward, there is much work to be done in expanding the current thermodynamic and kinetic databases for Pd-containing systems so that computational CALPHAD methods can be fully utilized in the systems design of these alloys. The Redlich-Kister-based A_f model can be integrated into the thermodynamic database for prediction of transformation temperature. Capillary energy adjustments based on experimental APT data can be calibrated to allow for the calculation of peak-aged nonequilibrium phase relations using Thermo-Calc software. A complete database of mobility coefficients would allow for detailed DICTRA modeling of the homogenization and aging processes. With a CALPHAD framework in place, the fully parametric design of an optimized microstructure can be implemented.

ACKNOWLEDGMENTS

This material is based upon work supported by the National Science Foundation Graduate Research Fellowship under Grant No. DGE-1324585. Additionally, this work was performed under the following financial assistance award 70NANB14H012 from U.S. Department of Commerce, National Institute of Standards and Technology as part of the Center for Hierarchical Materials Design (CHiMaD). LEAP measurements were performed at the Northwestern University Center for Atom Probe Tomography (NUCAPT). The LEAP at NUCAPT was purchased and upgraded with funding from NSF-MRI (DMR-0420532) and ONR-DURIP (N00014-0400798, N00014-0610539, N00014-0910781) grants. NUCAPT received support through the Initiative for Sustainability and Energy at Northwestern and the

NSF's MRSEC program (DMR-0520513 and 1121262). This work also made use of the OMM Facility which receives support from the MRSEC Program (NSF DMR-1121262) of the Materials Research Center at Northwestern University. Lastly, the authors would like to acknowledge and thank Alan Pelton, Ph.D. of G. Rau GmbH (Germany) for helpful communications regarding EBR advances in industry, as well as Paul Adler, Ph.D. of Medtronic (USA) for fruitful discussion.

Table A1 Experimentally measured A_f temperatures from prototype alloys (nominal composition given)

Sublattice I				Sublattice II				
Pd	Ni	Fe	Al	Ti	Zr	Al	Measured A_f (°C)	Source
34.5	0	15.5	0	50	0	0	−11.35	1
42.2	0	7.8	0	46	0	4	142.45	1
42.2	0	7.8	0	44	0	6	47.88	1
41.8	0	8.2	0	44	0	6	34.25	1
41.4	0	8.6	0	44	0	6	16.325	1
0	48	0	2	41	9	0	65.37	1
0	47.5	0	2.5	38.2	11.8	0	55.33	1
0	47	0	3	35.8	14.2	0	61.63	1
0	48	0	2	35	15	0	106	2
0	47	0	3	30	20	0	143	2
0	46	0	4	28	22	0	141	2
50	0	0	0	41	5	4	327.37	1
45	5	0	0	41	5	4	190.72	1
50	0	0	0	39	5	6	211.55	1
20	30	0	0	49	0	1	74	3
25	25	0	0	49	0	1	147	3
25	25	0	0	47	0	3	−10	3
27	23	0	0	47	0	3	37	3
30	20	0	0	47	0	3	123	3
36	14	0	0	46	0	4	134	3
32.5	17.5	0	0	46	0	4	95	3
38.5	11.5	0	0	44	0	6	88.37	2
25	25	0	0	49	0	1	145	4
25	25	0	0	47	0	3	−10	4

Sources: 1. New data 2. [20] 3. [4] 4. [19].

REFERENCES

1. P.H. Adler, J. Allen, J. Lessar, and R. Francis Martensite transformations and fatigue behavior of nitinol, *Am. Soc. Test. Mater.*, 2007, 1481, 3–17
2. J. Jung, G. Ghosh, and G.B. Olson, A comparative study of precipitation behavior of Heusler phase (Ni_2TiAl) from B2-TiNi in Ni-Ti-Al and Ni-Ti-AI-X (X = Hf, Pd, Pt, Zr) alloys, *Acta. Mater.* 51, 2003, 6341–6357
3. G. Honari, S.G. Ellis, B.L. Wilkoff, M.A. Aronica, L.G. Svensson, and J.S. Taylor, Hypersensitivity reactions associated with endovascular devices. Contact Dermatitis 59, 2008, 7–22
4. M. Bender and G.B. Olson Designing a precipitation-strengthened, superelastic, TiNi-based alloy for endovascular stents, In: *ICOMAT 2008*
5. K. Otsuka, K. Oda, Y. Ueno, M. Piao, T. Ueki, and H. Horikawa, The shape-memory effect in a $Ti_{50}Pd_{50}$ alloy, *Scr, Metall, Mater.*, 29, 1993, 1355–1358
6. H.C. Donkersloot and J.H.N. Van Vucht, Martensitic transformations in gold-titanium, palladium-titanium a platinum-titanium alloys near equiatomic composition, *J. Less, Common Met.*, 20, 1970, 83–91
7. H. Hosoda, K. Enami, A. Kamio, and K. Inoue Alloys design of PdTi-based shape memory alloys based on defect structures and site preference of ternary elements, *J. Intell, Mater. Syst. Struct.*, 7, 1996, 312–320
8. G.S. Bigelow, S.A. Padula, A. Garg, and R.D. Noebe, Correlation between mechanical behavior and actuator-type performance of Ni-Ti-Pd high-temperature shape memory alloys, *Behav. Mech. Multifunct. Compos. Mater*, 6526, 2007, 1–12
9. M. Kawakita, M. Takahashi, S. Takahashi, and Y. Yamabe-Mitarai Effect of Zr on phase transformation and high-temperature shape memory effect in TiPd alloys, *Mater. Lett.*, 89, 2012, 336–338
10. M. Grujicic, G.B. Olson, and W.S. Owen, Mobility of martensitic interfaces, *Metall. Trans. A.* 16, 1985, 1713–1722
11. Z.L. Guo and W. Sha, Quantification of precipitation hardening and evolution of precipitates, *Mater. Trans.* 43, 2002, 1273–1282
12. M. Bender, Computational design of precipitation-strengthened TiNi-based shape memory alloys, Northwestern University, Evanston, 2008, p 202
13. G.B. Olson, Computational design of hierarchically structured materials, *Science* 277, 1997, 1237–1242
14. J.S. Langer and A.J. Schwartz Kinetics of nucleation in near-critical fluids, *Phys. Rev. A.* 21, 1980, 948–958
15. I.M. Lifshitz and V.V. Slyozov The kinetics of precipitation from supersaturated solid solutions, *J. Phys. Chem. Solids*, 19, 1961, 35–50
16. R.J. Salzbrenner and M. Cohen Thermodynamics of thermoelastic martensitic transformations, *Acta. Metall.*, 27, 1979, 739–748

17. H.L. Lukas, S.G. Fries, and B. Sundman *Computational thermodynamics: the CALPHAD method*. Cambridge University Press, Cambridge, 2007
18. S.F. Hsieh and S.K. Wu, A study on ternary Ti-rich TiNiZr shape memory alloys, *Mater. Charact*, 41, 1998, 151–162
19. J. Jung, Design of nanodispersion strengthened TiNi-base shape memory alloys. Northwestern University, Evanston, 2003, p 174
20. T. Jiang, Design of high performance NiTi-based shape memory alloys for automotive actuator applications. Northwestern University, Evanston, 2014, p 161
21. D.J. Frankel, T. Jiang, and G.B. Olson Design of a fatigue resistant Ni-free PdTi-base SMA, materials today, In: Proceedings—ICOMAT 2014, 25 (2015), 5801–5804
22. Z.W. Feng, B.D. Gao, J.B. Wang, D.F. Qian, and Y.X. Liu Influence of Zr additions on shape-memory effect and mechanical properties of Ni-rich NiTi alloys, *Shape Mem. Mater. Appl.*, 394–3, 2001, 365–368
23. S.F. Hsieh and S.K. Wu, Room-temperature phases observed in $Ti_{53-x}Ni_{47}Zr_x$ high-temperature shape memory alloys, *J. Alloys Compd*, 266, 1998, 276–282
24. A. Toro, F. Zhou, M. Wu, W. Van Geertruyden, and W. Misiolek, Characterization of non-metallic inclusions in superelastic NiTi tubes, *J. Mater. Eng. Perform.* 18, 2009, 448–458
25. Y. Koizumi, Y. Ro, S. Nakazawa, and H. Harada, NiTi-base intermetallic alloys strengthened by Al substitution, *Mater. Sci. Eng. A.* 223, 1997, 36–41
26. R.D. Noebe, T. Biles, and S.A. Padula, NiTi-based high-temperature shape-memory alloys, In: W.O. Soboyejo, and T.S. Srivatsan (ed.), *Advanced structural materials*. CRC Press, Boca Raton, FL, 2006, p 145–186
27. J. Moore, A Method for Multiscale Fatigue Prediction., Northwestern University, Evanston, 2015
28. D.L. McDowell, Simulation-based strategies for microstructure-sensitive fatigue modeling, *Mater. Sci. Eng. A*, 468, 2007, 4–14
29. D.G. Harlow, R.P. Wei, T. Sakai, and N. Oguma, Crack growth based probability modeling of S-N response for high strength steel, *Int. J. Fatigue*, 28, 2006, 1479–1485
30. M. Launey, S.W. Robertson, L. Vien, K. Senthilnathan, P. Chintapalli, and A.R. Pelton, Influence of microstructural purity on the bending fatigue behavior of VAR-melted superelastic Nitinol, *J. Mech. Behav. Biomed.*, 34, 2014, 181–186
31. M. Rahim, J. Frenzel, M. Frotscher, J. Pfetzing-Micklich, R. Steeg-muller, M. Wohlschlogel, H. Mughrabi, and G. Eggeler, Impurity levels and fatigue lives of pseudoelastic NiTi shape memory alloys, *Acta. Mater.*, 61, 2013, 3667–3686
32. M.M. Patel, D.L. Plumley, R.J. Bouthot, and J.L. Proft, The significance of melt practice on fatigue properties of superelastic NiTi fine diameter wire, SMST 2006. In: *Proceedings of the international conference on shape memory and superelastic technologies*, 2008, p 43–52

Transformations—Selected Works of G.B. Olson on Materials, Microstructure, and Design
C.E. Campbell, M.V. Manuel, and W. Xiong, editors

CHAPTER 21

Integrated Computational Materials Engineering—Success Stories and Cultural Barriers*

C.J. Kuehmann, QuesTek Innovations LLC
G.B. Olson, Northwestern University and QuesTek Innovations LLC

Introduction

Several national academy studies of materials and manufacturing (Ref 1–4) have emphasized the unique opportunity of computational materials engineering. While the breadth of the 2008 ICME study (Ref 1) creates an impression of a field in its infancy, a 2004 study (Ref 2) uniquely charged with identifying best industrial practices for accelerating transition of materials and process technology identified practices which are now mature and ready for broader dissemination. We here update the state of best practices and consider the principal cultural barriers to Integrated Computational Material Engineering (ICME) proliferation.

Computational Design Successes—Flying Cyber Steel

The first example of a commercial alloy created by computational design is the Ferrium C61 (AMS6517) high-durability gear steel now performing well in off-road racing applications (Ref 5). The first such designer alloy to reach flight qualification is the Ferrium S53 (AMS 5922) corrosion-resistant landing gear steel allowing a drop-in replacement for current nonstainless landing gear steels, eliminating the need for cadmium plating (Ref 6). S53 is a secondary-hardening steel strengthened by efficient M_2C carbide precipitates and contains sufficient Cr content to provide passivation against general corrosion (Ref 7) as shown in the flow-block diagram in Fig. 1. The sequential processing steps experienced by the alloy are depicted on the left-most column of the flow-block diagram and are constrained to existing processes for steels employed in current structural aircraft applications to maximize manufacturability. The subcomponents of the alloy system are connected by process-structure and structure-property relationships, representing the mechanistic modeling tools necessary to employ quantitative computational design.

The application of the system flow-block diagram in conjunction with the computational models begins with identification of key design trade-offs. Key to achieving the stated design

*C.J. Kuehmann and G.B. Olson, "ICME: Success Stories and Cultural Barriers," in *Models, Databases, and Simulation Tools Needed for the Realization of Integrated Computational Materials Engineering*, eds. S. Arnold and T. Wong, ASM International (2011), p 24–30

goals is the development of an efficient strengthening dispersion. M_2C carbide is an efficient strengthener in steels due to its high modulus misfit with BCC Fe and its ability to precipitate coherently at the nano scale. Models were utilized, as demonstrated in Fig. 2, to predict the overall precipitation driving force and the normalized coarsening rate constant of the M_2C carbide as a function of the Mo and V content in the S53 alloy. The approach maximizes the resulting strength through the precipitation driving force, while assuring that strength can be achieved with reasonable tempering parameters.

This design also maintains adequate martensite kinetics to ensure a lath martensitic alloy while achieving high strength as another key design trade-off. Quantitative martensite kinetic models (Ref 8) are used to predict the martensite start temperature along with the M_2C precipitation driving force in Fig. 3 as a function of Co and Ni content.

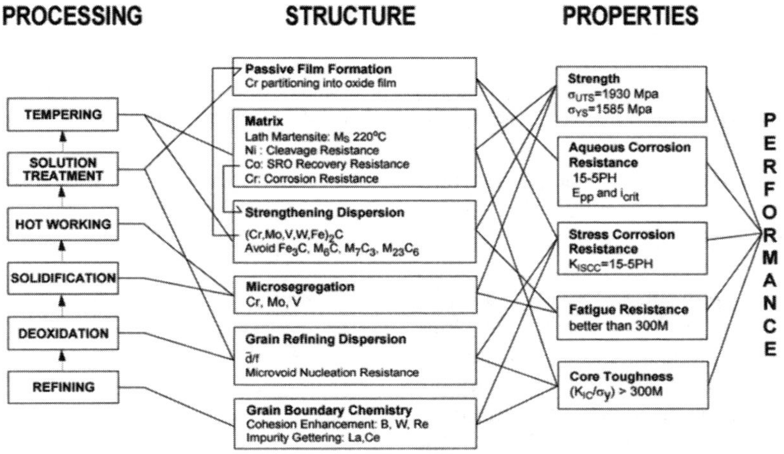

Fig. 1 The flow-block diagram for ultrahigh-strength corrosion-resistant steel indicates the desired property objectives, the microstructural subsystems, and sequential processing steps needed for design. Links between system blocks indicate quantitative models needed to effect the design via science-based computation.

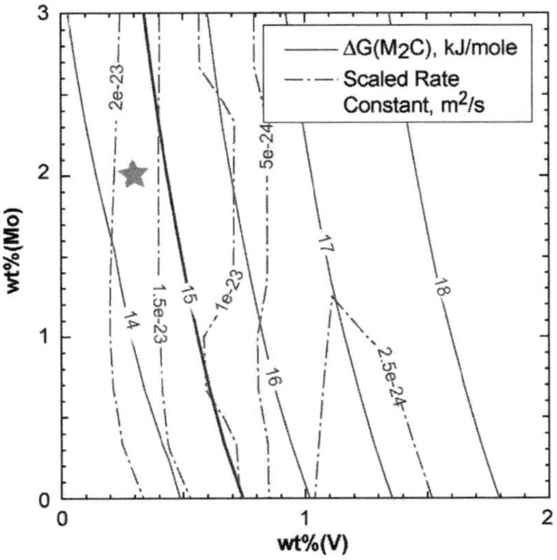

Fig. 2 Maximizing M_2C driving force while maintaining a normalized coarsening rate for adequate tempering kinetics provides a secondary-hardening alloy with highly efficient strengthening. The S53 alloy achieves greater than 1930 MPa tensile strength with only 0.2 wt% carbon.

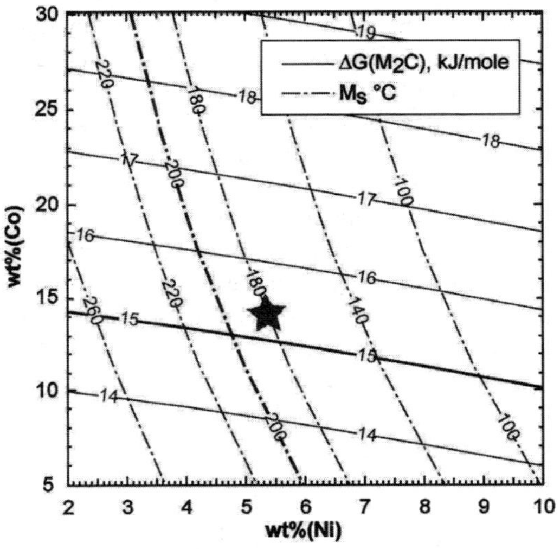

Fig. 3 Maintaining a sufficiently high M_S temperature while maximizing carbide driving force allows efficient strengthening in a fully martensitic alloy. Contours of M_2C precipitation driving force overlaid with the alloy M_S temperature determine optimal Co and Ni content.

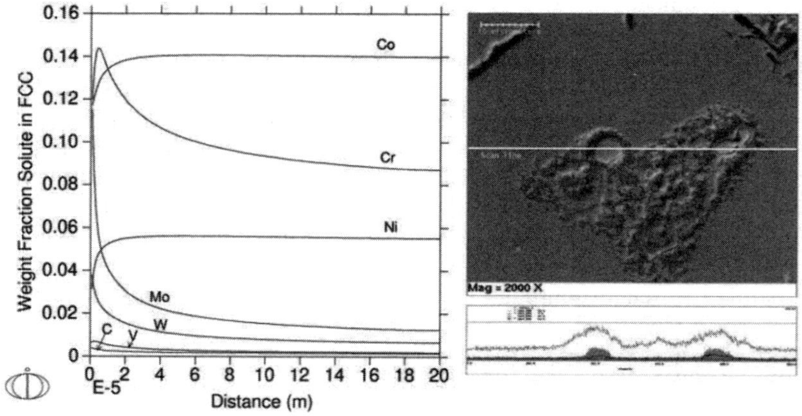

Fig. 4 One-dimensional multicomponent diffusion simulations of the solidification of S53 alloy indicate the extent of segregation that can be expected in final ingot. This segregation is validated by SEM observations of ingot material.

Additional constraints on ductile fracture, grain boundary chemistry, and grain pinning dispersions are used to complete the design optimization and uniquely identify the alloy composition that represents the best compromise of the diverse design goals and constraints. Grain boundary chemistry considerations are applied based on first principles calculations of impurity and alloying effects on the interfacial cohesion of BCC low angle grain boundaries (Ref 9). The grain pinning dispersion design was based on calculation of the TiC/Fe interfacial adhesion that demonstrated excellent resistance to microvoid formation (Ref 10).

Solidification simulations employing 1-D multicomponent diffusion calculations in DICTRA for candidate design compositions ensured production scale processability (Ref 11). Figure 4 demonstrates segregation profiles across secondary dendrite arms predicted for production scale ingots using VAR. These predictions were validated from actual production ingots. Figure 5 shows DICTRA predictions of homogenization Mo in the as-cast ingot of the S53 alloy. The simulation guides processing recommendations and indicates that sufficient homogenization can be achieved in production scenarios. Figure 6 demonstrates the validation of tensile properties in production ingots of up to 24,000 lb. size.

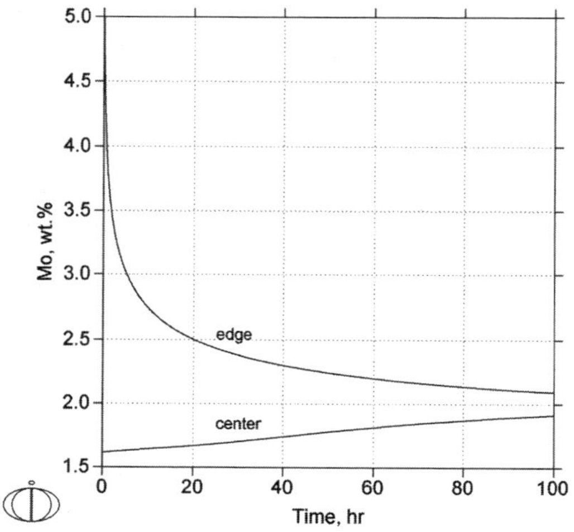

Fig. 5 The homogenization of Mo segregation in the S53 alloy as a function of time at 1350C. The two curves represent the composition of Mo at the dendrite center and edge, and, the initial profiles are provided by the solidification simulations from Fig. 4.

Extending Computation Design to Qualification and Implementation

Accelerated insertion of materials (AIM) methods (Ref 2) were applied to the prediction of the ultimate tensile strength, a primary design objective of the S53 alloy. Prediction of property design minimums, which normally require 10 production heats of alloy and 300 individual observations, were completed using only three heats and 30 individual observations. A large simulated data set with over 300 simulations was produced using expected process variation and computational mechanistic models of alloy strength. The simulated property distribution was scaled to fit 20 experimental observations from two production heats by a linear transformation function determined by the best fit. This analysis indicated the 1% minimum ultimate tensile strength was predicted to be below the 1930 MPa design goal by at least 10 MPa. This early indication

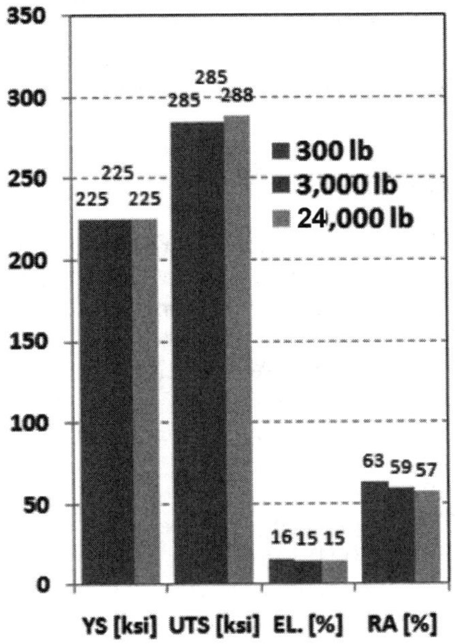

Fig. 6 Validation of scale-up results after the computational optimization of homogenization parameters. The chart represents typical tensile properties for wrought product produced at three production scales, including laboratory 300 lb ingots, intermediate 3000 lb ingots and full production 24,000 lb ingots. The similar tensile properties across all products indicate the ingots are homogeneous across all production size scales.

allowed additional process optimization to be completed, increasing the ultimate tensile strength by 10 MPa. The AIM analysis was repeated after a new temper was optimized, using 30 data produced from three heats of the alloy, and is shown by the solid curve of Fig. 7. The new property minimum estimate shows a prediction of 1930 MPa (280 ksi) meeting property objectives. The resulting full experimental data set of 10 heats and over 600 individual observations are represented by the square points of Fig. 7. The AIM prediction is within 7 MPa (1 ksi) of the 1% minimum tensile strength statistically defined by the 10 heats of data. In this AIM example, if data development had proceeded and the property deficit discovered when the full data set was complete, over a year of development and in excess of US $500,000 would have to be repeated to meet the property goals.

The first flight of a Ferrium S53 landing gear occurred on December 17, 2010.

Integrated Computational Material Engineering Cultural Barriers

Primary cultural barriers to the accelerated development and proliferation of ICME technology stem from the broader issue of the opposite nature of science and engineering. It has been

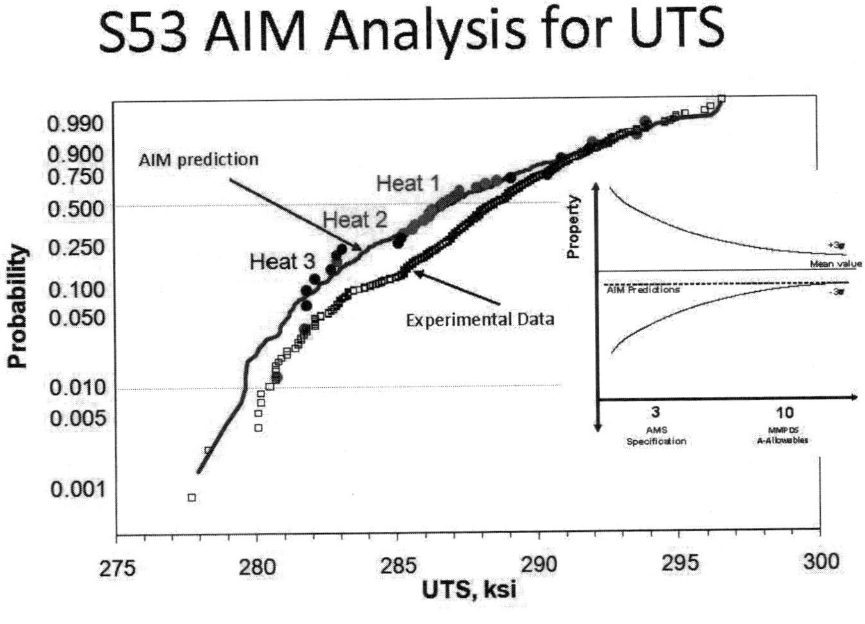

Fig. 7 An AIM analysis of the ultimate tensile strength of S53 using simulation results and 30 data from 3 individual heats agrees well with the completed data set from 10 heats and over 600 observations in determining the 1% minimum design allowable.

much discussed that science seeks the discovery of new knowledge of nature as it is, while engineering applies existing knowledge to bring new technology into existence. The unique position of materials at the interface of science and engineering intensifies the inherent conflict between these opposite philosophies.

With the glorification of basic science by the atom bomb, and the catalytic effect of Sputnik hysteria, it became temporarily fashionable for postwar American industry to invest directly in basic research. This created historic environments like Bell Labs (with achievements from this period still receiving Nobel Prizes) where science and engineering coexisted in an atmosphere of mutual respect and collaboration, enabling an unprecedented level of science-based technology creation. Concurrent with the development of MBA programs, major industry concluded that such basic research could not be justified in terms of short-term competitiveness, and largely dropped out of science. This in turn created a need for federally funded academic research to take full responsibility for basic science, with the unfortunate consequence that engineering at our research universities has been almost entirely replaced by engineering science. We are left with a system where universities do science and companies do engineering; probably the most egregious example of the much-discredited "over-the-wall" system. While limited technological competence at universities and limited scientific depth in industry offer an opportunity for unlimited finger pointing, the special case of university spinoff companies has provided the best exception to the prevalent systemic dysfunction, and it is no wonder that these have served as the principal source of the ICME methods, tools, and databases that are now available for broad dissemination.

While a few elite engineering schools have attracted private funding to conduct actual engineering on campus, there is a pressing need for federal funding agencies to take responsibility for restoration of engineering across our research university system. DoD agencies have provided the best example so far, but far more impact could be achieved by the formation of a National Engineering Foundation as an equal and opposite partner to NSF, with a mission of direct technology creation through design as the complement of discovery. Such an enterprise would powerfully demonstrate societal respect for engineering as the equal of science, would provide a fertile intellectual environment far more capable of achieving the goals of Bayhe-Dole legislation in a technological return on our vast public investment in science, and would create an educational environment for engineering equaling that in place for science.

REFERENCES

1. NRC 2008, Integrated Computational Materials Engineering: A Transformational Discipline for Improved Competitiveness and National Security
2. NRC 2004, Acceleration Technology Transition: Bridging the Valley of Death for Materials and Processes in Defense Systems
3. NRC 2003, Materials Research to Meet 21st Century Defense Needs
4. NRC 2004, Retooling Manufacturing: Bridging Design, Materials and Production
5. C.J. Kuehmann, and G.B. Olson, "GEARSTEELS: Designed by Computer," *Advanced Materials & Processes*, 153, 5 (1998), 40–43
6. C.J. Kuehmann, *et al.*, "Computational Design for Ultra High-Strength Alloy," *Advanced Materials and Processes*, 166 (2008), 37–40
7. C.E. Campbell and G.B. Olson, "Systems Design of High Performance Stainless Steels: I. Conceptual Design," *J. Computer-Aided Mat. Des.*, 7 (2001), 145–170
8. G. Ghosh and G.B. Olson, "Kinetics of FCC BCC Heterogeneous Martensitic Nucleation, Part I: The Critical Driving Force for Athermal Nucleation," *Acta Metall. Mater.*, 42 (1994), 3361–3370
9. W.T. Geng, *et al.*, "Influence of Alloying Additions on Grain Boundary Cohesion of Transition Metals: First-Principles Determination and its Phenomenological Extension," *Phys. Rev. B*, 63 (2001) 165415
10. J.-H. Lee, *et al.*, "Strong Interface Adhesion in Fe/TiC," *Philosophical Magazine*, 85, 31 (2005), 3683–3697
11. H.E. Lippard, *et al.*, "Microsegregation Behavior during Solidification and Homogenization of AerMet 100® Steel," *Met. Trans B*, 29B (1998) 205–210

CHAPTER 22

Computer Simulations for the Prediction of Microstructure/Property Variation in Aeroturbine Disks*

Herng-Jeng Jou, QuesTek Innovations LLC
Peter Voorhees and Gregory B. Olson, Northwestern University

Abstract

In support of a new methodology of accelerated development and qualification of new materials, the PrecipiCalc software has been developed as an efficient, high-fidelity 3D multiphase precipitation simulator grounded in computational thermodynamics and multi-component diffusional nucleation and growth models incorporating interfacial dissipations. With extensive validation under a major federally sponsored program, the software has been successfully applied to the accurate and rapid simulation of multiscale (from nm to μm), multimodal γ' microstructure development in IN100 Ni-based aeroturbine disk alloys under complex commercial heat-treatment processing. Integration of PrecipiCalc with FEM heat transfer simulation and a strength model has demonstrated accurate prediction of property variation for accelerated process optimization and probabilistic modeling of design minimum properties.

Introduction

The recently completed three-year DARPA-AIM program on Accelerated Insertion of Materials has successfully demonstrated a new methodology of integrated computational simulation and focused testing, to rapidly optimize materials processing at the component level and efficiently predict design minimum properties through mechanistically based probabilistic methods. As a central tool of this methodology, the PrecipiCalc software was developed under the AIM initiative to provide rapid high-fidelity simulation of microstructural evolution during thermal processing. To support efficient property prediction, the code is structured to predict the microstructural parameters necessary for state-of-the-art structure/property relations, incorporating only the physics necessary for required accuracy.

Under the AIM initiative, a high degree of simulation fidelity was achieved through intensive iterative interaction of theory and experiment through microstructural characterization of Ni-based IN100 aeroturbine disks over a wide range of scales. Various experimental techniques were used to characterize the IN100 samples taken from the aeroturbine disks after commercial heat treatment, including TEM for microstructural morphology of precipitations and EDS/3DAP for compositions of both precipitates and matrix. Major

*H.-J. Jou, P. Voorhees, and G.B. Olson, "Computer Simulations for the Prediction of Microstructure/Property Variation in Aeroturbine Disks," *Superalloys*, TMS, (2004) 877–886

microstructural constituents in IN100 are γ (FCC) matrix and γ' precipitates, as well as small amounts of carbide and boride phases. For this study, we focused on a three-step heat treatment cycle for the IN100 aeroturbine disk, and the cycle includes:

1. 1143 °C solution (subsolvus) treatment, followed by a transfer and oil quench;
2. 982 °C heat treatment followed by fan cool;
3. 732 °F heat treatment followed by fan cool.

The three-step heat treatment cycle produces a final microstructure with three distributions of γ' precipitates at distinct sizes, named primary (large), secondary (intermediate), and tertiary (small), with a total fraction of about 60%. Figure 1 provides two example micrographs of IN100 γ' microstructure. Primary γ' particles are formed in the powder extrusion and forging process, and are coarsened at subsolvus solution temperature, providing grain pinning to matrix grain structure. The secondary (tertiary) γ' develops from the quench after the first (second) treatment temperature and coarsens in the second (third) heat treatment step. Fine γ' particles have a much higher number density than the coarse ones. As shown in Fig. 1, each primary (and secondary) particle is surrounded by many secondary (tertiary) particles.

This article discusses the effort under the AIM program to mechanistically capture the γ' precipitation microstructure in the IN100 disk alloy. We will cover the PrecipiCalc multicomponent precipitation models, software implementation techniques, and the applications to the IN100 alloys.

Multicomponent Precipitation Models

Precipitation is one of the most important types of phase transformation processes in material systems. It changes material properties in many ways and is often critical to the material performance. The ability to control the microstructure of precipitates provides a powerful tool in material development. Due to its importance, precipitation modeling has been an active research area in materials science. Precipitation generally includes three physical mechanisms: nucleation, growth, and coarsening. Each of them has received extensive research in the past. In the continuum modeling arena, different approaches to the precipitation model have been proposed, involving different levels of approximation and computational complexity. From the Avrami-type single state variable (usually volume fraction) models to complex phase-field models, the computational expense scales exponentially with the details captured by the models.

PrecipiCalc precipitation models are based on the foundation of the double state variable approach, following the framework of Langer-Schwartz (LS) (Ref 1), which provided an integrated approach of nucleation, growth and coarsening, and Kampmann-Wagner (KW) (Ref 2), which provided a numerical solution to LS model. The two state variables are particle size and number density, and together they define the particle size distribution (PSD). With the PSD, higher level quantities such as volume fractions, mean sizes, and interparticle distances can be calculated. Within this framework, the key models are nucleation (source of new particles), growth (particle size changes) under mean field, continuity (time

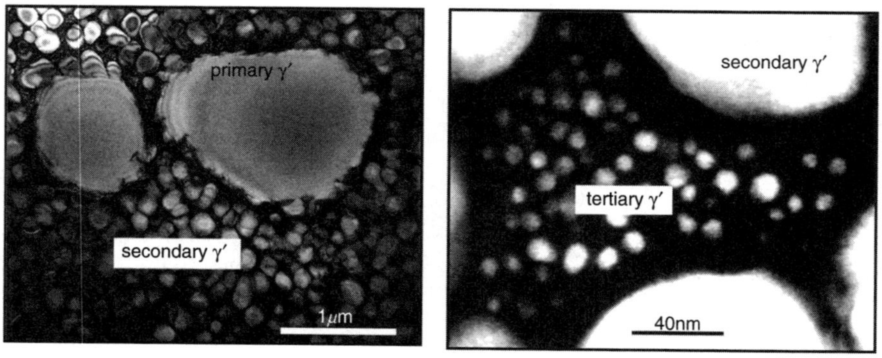

Fig. 1 Experimental micrographs of IN100 at two resolutions clearly show three distinct size distributions of γ'. Courtesy of Gautam Ghosh, Northwestern University

evolution of PSD), and mass balance. Coarsening comes out naturally within this framework. Due to the mechanistic integration of physical mechanisms in LS framework, the transitions between mechanisms are captured without ad hoc treatments. For example, high supersaturation precipitation in an alloy system can show only nucleation and coarsening, without an intermediate growth stage, which has been captured by the LS model.

The adoption of numerical LS-KW framework in modeling precipitation has been successfully applied to many alloy systems (Ref 3–6). However, those work have only provided the models for binary systems, and the applications to multicomponent systems require the reduction to pseudo-binary systems. The true multicomponent treatment within the LS-KW framework is limited. The work of Popov (Ref 8, 9) dealt with the multicomponent growth and evolution of PSD, without nucleation. In the coarsening region, several multicomponent models (Ref 9–12), have been proposed. The complete model and implementation of integrated nucleation, growth and coarsening for multicomponent systems are not available. Furthermore, many of the mentioned literature work focused on isothermal conditions. In complex commercial heat treatment of alloys, it is important to model precipitation under nonisothermal conditions.

In the following subsections, detailed PreciPiCalc multicomponent precipitation models are presented.

Continuity Equation

The particle size distribution, $f(R, t)$ is defined such that $f(R, t)dR$ is the number of particles per unit volume of size R to $R + dR$. Thus, the number of particles per unit volume in the sample, N_v, is:

$$N_v = \int_0^{\infty} f(R,t)dR \quad \text{(Eq 1)}$$

In our case, the volume fraction of particles, ϕ, is:

$$\phi = \int_0^{\infty} \frac{4\pi}{3} f(R,t)R^3 dR \quad \text{(Eq 2)}$$

The particle size distribution evolves in time according to:

$$\frac{\partial f}{\partial t} + \frac{\partial (f dR/dt)}{\partial R} = I(R,t) \quad \text{(Eq 3)}$$

where $I(R, t)$ is the nucleation rate.

Growth Model

The growth rate is determined by the diffusion field surrounding the particle within the approximation of local equilibrium at the particle-matrix interface. We employ the quasistationary approximation and assume that the diffusion coefficient is independent of composition. Unlike binary alloys, the compositions at the interface are not fixed by the phase diagram, but are functions of the diffusion coefficients and supersaturation. We determine these compositions self consistently within the small supersaturation approximation. We also allow for capillarity, interfacial kinetics, and nonzero volume fraction correction. The growth rate of a particle of size R is given by:

$$\frac{dR}{dt} = \left(1 + R\sqrt{4\pi N_v \bar{R}}\right) \frac{\overline{\Delta C}_i G^{\alpha}_{ij} \Delta C^{\infty}_j + \bar{C}^{\beta}_{\delta}\left(\bar{\mu}^{\alpha}_{\delta} - \bar{\mu}^{\beta}_{\delta}\right) - 2V^{\beta}_m \sigma/R}{R\overline{\Delta C}_i G^{\alpha}_{ij} D^{-1}_{jk} \mathcal{B}_k + 1/M} \quad \text{(Eq 4)}$$

where the sum over repeated Latin indices from $2 \cdots N$ and repeated Greek indices from $1 \cdots N$ is assumed, $\overline{\Delta C}_i = \bar{C}^{\beta}_i - \bar{C}^{\alpha}_i$, $\bar{C}$ is the mole fraction of the denoted phase about which the free energy function is expanded, $G_{ij} = \partial^2 G/\partial C_i \partial C_j$ evaluated at $\bar{C}_i$, G is the molar Gibbs free energy of the noted phase, $\Delta C^{\infty}_j = C^{\infty}_j - \bar{C}^{\alpha}_j$, C^{∞}_j is the composition of component j at an infinite distance from the spherical particle, $\bar{\mu}_i$ is the chemical potential of component i evaluated at $\bar{C}_i i > 1$, V^{β}_m is the molar volume of the particle phase β, $\sigma(R)$ is the size-dependent interfacial energy, D_{jk} is the diffusion tensor, R_g is the gas constant, $\bar{R}$ is the mean size, M is the interfacial mobility, and:

$$\mathcal{B}_k = L_k + \overline{\Delta C}_k + \left(\delta_{kj} - A_{kj}\right)A^{-1}_{jl}\Delta C^{\infty}_l - \frac{y\left(\delta_{kj} - A_{kj}\right)A^{-1}_{jh}D^{-1}_{hl}\overline{\Delta C}_l}{\left(\overline{\Delta C}_j G^{\alpha}_{ji} D^{-1}_{ik} \overline{\Delta C}_k\right)} \quad \text{(Eq 5)}$$

$$L_k = G^{\beta-1}_{jk}\left(\tilde{\mu}^{\alpha}_j - \tilde{\mu}^{\beta}_j\right) - 2G^{\beta-1}_{jk}\overline{\Delta V}_j \sigma/R \quad \text{(Eq 6)}$$

$$A_{kj} = G^{\alpha-1}_{kl} G^{\beta}_{lj} \qquad \text{(Eq 7)}$$

$$y = \overline{\Delta C_i} G^{\alpha}_{ij} \Delta C^{\infty}_j + \bar{C}^{\beta}_{\delta}\left(\bar{\mu}^{\alpha}_{\delta} - \bar{\mu}^{\beta}_{\delta}\right) - 2V^{\beta}_m \sigma / R \qquad \text{(Eq 8)}$$

$\tilde{\mu}_i = \bar{\mu}_i - \bar{\mu}_1$, $i = 2 \cdots N$, and $\overline{\Delta V_i} = \bar{V}_i - \bar{V}_1$, $i = 2 \cdots N$.

Interfacial kinetics is included by assuming the interfacial velocity is proportional to the interfacial mobility M and total chemical driving force on the interface. M is expressed as $M_0 \exp(-Q/R_gT)$ where Q is the activation energy for interfacial kinetics, R_g is the gas constant, and M_0 is the prefactor.

The composition at which the free energy is expanded is chosen such that the resulting expression for the growth rate is accurate. In a typical run this involves changing this composition, but not for every time step. Unlike a binary, where the growth rate is a function solely of the supersaturation, in a multicomponent alloy, the growth rate depends on the supersaturation and a thermodynamic factor, G_{jk}. This is because the concentrations at the interface cannot be specified by the local equilibrium conditions, but are functions of the mass balance equations at the interface.

When $y = 0$ the growth rate is zero, and thus this gives an expression for the critical radius of nucleation in a multicomponent alloy:

$$R_c = \frac{2V^{\beta}_m \sigma}{\Delta G} \qquad \text{(Eq 9)}$$

where

$$\Delta G = \overline{\Delta C_i} G^{\alpha}_{ij} \Delta C^{\infty}_j + \bar{C}^{\beta}_{\delta}\left(\bar{\mu}^{\alpha}_{\delta} - \bar{\mu}^{\beta}_{\delta}\right) \qquad \text{(Eq 10)}$$

the free energy change on forming β of the composition of the critical nucleus from α of composition C^{α}_k where both phases are assumed incompressible at the pressure of the matrix is given by the denominator in Eq 5. This agrees with the result obtained by the usual approach.

Homogeneous and Transient Nucleation Models

Isothermal Nucleation. The nucleation rate is given by:

$$I(R) = e^{-\tau/t} j(R) \qquad \text{(Eq 11)}$$

where $j(R)$ is the nucleation rate, τ is the incubation time and t is time. Assume that the nucleation rate is a function of particle size:

$$j(R) = \frac{J_s}{\delta\sqrt{2\pi}} \exp\left[-\frac{(R - R_c)^2}{2\delta^2}\right] \qquad \text{(Eq 12)}$$

where J_s is the steady state nucleation rate, δ is the variance of the Gaussian distribution, R_c is the critical radius of nucleation. $j(R)$ is chosen such that:

$$J_s = \int_{R_c}^{\infty} j(R) dR \qquad \text{(Eq 13)}$$

Thus only particles greater than the critical are considered part of the precipitate phase.

The steady state nucleation rate is given by:

$$J_s = Z\beta^* N_o e^{-W^*_R/k_BT} \qquad \text{(Eq 14)}$$

where Z is the Zeldovitch factor, β^* is the rate of atom impingement, N_o is the number of atoms per volume, W^*_R is the work on forming the critical nucleus, k_B is Boltzmann's constant, and T is temperature. The work on forming the critical nucleus is:

$$W^*_R = \frac{16\pi\sigma^3}{3\left(\Delta G / V^{\beta}_m\right)^2} \qquad \text{(Eq 15)}$$

where σ is the interfacial energy, ΔG is the volume free energy change on forming a nucleus of critical composition, and V^{β}_m is the molar volume of the β phase. The Zeldovitch factor is:

$$Z = \left[\frac{\omega^2_a \sigma}{4\pi^2 k_B T R^4_c}\right]^{1/2} \qquad \text{(Eq 16)}$$

where ω_a is the volume per atom in α, and k_B is Boltzmann's constant.

The form of the expression for the impingement rate depends on the rate-limiting mass transport process. One limit is where the impingement rate is controlled by the rate of atom motion across the interface. The other extreme is where the impingement rate is controlled by long-range diffusion to a cluster. Here we examine the limit where long-range diffusion of atoms to a cluster sets the impingement rate. Thus the impingement rate can be determined by solving a diffusion equation for the flux of solute atoms to a cluster. This will allow us to determine, for example, the effects of slow diffusing species on the kinetics of cluster growth. Once this is known, the growth rate of the cluster is specified and thus the impingement rate is given by:

$$\beta^* = 4\pi R^2 N_a \rho dR/dt \qquad \text{(Eq 17)}$$

where R is the critical radius, ρ is the constant density of the solid (moles/volume), N_a is Avogadro's number, dR/dt is the growth rate of a cluster of the critical size. In this case we neglect the corrections to dR/dt due to capillarity and obtain using Eq 4:

$$\beta^* = 4\pi R N_a \rho \frac{\overline{\Delta C}_i G_{ij}^{\alpha} \Delta C_j^{\infty} + \bar{C}_{\delta}^{\beta}\left(\bar{\mu}_{\delta}^{\alpha} - \bar{\mu}_{\delta}^{\beta}\right)}{\overline{\Delta C}_i G_{ij}^{\alpha} D_{jk}^{-1} \mathcal{B}_k} \qquad \text{(Eq 18)}$$

where $\mathcal{B}_k$ is given in Eq 5 with:

$$L_k = G_{jk}^{\beta-1}\left(\tilde{\mu}_j^{\alpha} - \tilde{\mu}_j^{\beta}\right) \qquad \text{(Eq 19)}$$

In the limit that the growth of the phase is limited by long-range diffusion (interface kinetics limited nucleation will be included in future extension), The incubation time is:

$$\tau = \frac{1}{\theta \beta^* Z^2} \qquad \text{(Eq 20)}$$

where $2 \leq \theta \leq 4\pi$.

Nonisothermal Nucleation. An exact calculation of the incubation time for nucleation requires a solution to the Fokker-Planck equation. In the spirit of our past work, we shall develop an approximate theory. For isothermal nucleation the incubation time τ is given by Eq 20. The incubation time can be computed by considering the characteristic time required for a nucleus to random walk in cluster size space a distance given by the number of atoms in a cluster 1 kT from the top of the W_R curve.

Since a random walk is described by a diffusion equation, it is not surprising that the Fokker-Planck equation in the region where $\partial W_R/\partial n$ is small (near the top of the W_R curve where the derivative of W_R is small) is in fact a diffusion equation. Using a solution to a diffusion equation where we allow for temperature dependent β^* and Z, we find that the incubation time during a continuous cool, where β and Z will vary with temperature and thus time is given by:

$$\int_0^{\tau} \beta^*(t')dt' = \frac{1}{\pi Z^2(\tau)} \qquad \text{(Eq 21)}$$

This is an integral equation for the incubation time τ. The Zeldovitch factor varies with temperature and hence time. By design, Eq 16 will not yield the steady-state nucleation current as $t \rightarrow \infty$, since we have neglected terms related to the gradient of W_R, but only an estimate of τ.

Mass Conservation

The atoms of component i are divided between the β and α phases. Thus:

$$C_i^o = (1-\phi)C_i^{\alpha} + \phi\tilde{C}_i^{\beta} \qquad \text{(Eq 22)}$$

where C_i^o is the mole fraction of i in the alloy, $\phi = n_i^{\beta}/n_i^o$, n_i^o is the number of atoms of i in the alloy, n_i^{β} is the number of atoms of i in the β phase, C_i^{α} is the mole fraction of i in the α phase, and $\tilde{C}_i^{\beta}$ is the mole fraction of i in the β phase. We shall approximate ϕ as the volume fraction of β, which is exact if the densities of α and β are equal. Within the quasistationary approximation $C_i^{\alpha} = C_i^{\infty}$. Thus the mass conservation equation is:

$$C_i^o = (1-\phi)C_i^{\infty} + \phi\tilde{C}_i^{\beta} \qquad \text{(Eq 23)}$$

The composition in the precipitate at the matrix-precipitate interface is a function of the radius of the particle due to capillarity and, since the alloy contains more than two components, the diffusion coefficient tensor and the supersaturation:

$$\tilde{C}_i^{\beta} = \bar{C}_i^{\beta} + G_{ik}^{\beta-1} G_{kl}^{\alpha} \left[C_l^{\infty} - \bar{C}_l^{\alpha} - \frac{D_{ik}^{-1}\overline{\Delta C}_k}{\overline{\Delta C}_i G_{ij}^{\alpha} D_{jk}^{-1} \mathcal{B}_k} \left(\overline{\Delta C}_i G_{ij}^{\alpha} \Delta C_j^{\infty} + \bar{C}_{\delta}^{\beta}\left(\bar{\mu}_{\delta}^{\alpha} - \bar{\mu}_{\delta}^{\beta}\right) - \frac{8\pi\sigma V_m^{\beta} f_2}{3\phi} \right) \right] - \frac{8\pi\sigma f_2}{3\phi} G_{ik}^{\beta-1} \overline{\Delta V}_k \qquad \text{(Eq 24)}$$

where f_2 is the second moment of the particle size distribution and $\mathcal{B}_k$ is given in Eq 5 with L_k given in Eq 19 and y given in Eq 8.

Software Implementation

Numerical Schemes

There is no general analytical solution available to the mathematical equations of PrecipiCalc's multicomponent precipitation models described in previous sections. Utilization of numerical methods within computer software is hence necessary. PSD of particle size and number density distribution is discretized in both particle size and time. The continuity equation (Eq 3) describing the evolution of PSD is solved in PrecipiCalc with the characteristic methods (Ref 13), and an adaptive discretization framework to provide accurate and efficient computation. To provide efficiency, the thermodynamics/diffusivity quantities are not calculated at every timestep, and are done only when the departure of current temperature and matrix composition exceeds certain limits from the state where expansion is performed.

CALPHAD Tool Integration

A unique feature of PrecipiCalc software is the tight integration of CALPHAD thermodynamics and mobility data and tools. In PrecipiCalc software, TCAPI (Thermo-Calc Advanced Programming Interface) (Ref 14), library is selected to efficiently provide all the necessary thermodynamics and diffusivity calculations. Though there are other CALPHAD software tools available, they do not currently meet the PrecipiCalc model requirements, in particular the needs of driving forces, second derivative matrices of free energies, and diffusivity matrices for multicomponent systems. Before the availability of TCAPI library, an earlier version of PrecipiCalc relied on a system-level layer based on UNIX IPC (Inter-Process Communication) to communicate with an external DICTRA process. However, this approach involves the system level communication overhead, and produces non-standalone user programs. In literature, Serriere et al., (Ref 6) have utilized Thermo-Calc for the precipitation simulation of binary system, while using simple representation for diffusivity without using DICTRA and atomic mobility.

Software Engineering and AIM Integration

PrecipiCalc software was mostly written in C++ programming language to take advantage of many modern object-oriented features. In addition, several C and FORTRAN libraries are used, such as BLAS (Ref 15) and LAPACK (Ref 16) for linear algebra calculation, and Hierarchical Data Format (HDF) library (Ref 17) for data storage. PrecipiCalc is available for both Linux and MS Windows platforms. The user interface of PrecipiCalc calculation program and its postprocessor are text-based, and currently without a graphical user interface (GUI). The postprocessor can be used to quickly produce plots for fast visualization of calculation results. The text-based PrecipiCalc interface, though not as friendly for interactive operation as a GUI, does provide a simple mechanism for higher level integration discussed later.

One of the critical components in the AIM initiative was the ability to perform the mechanistic computations of spatial variation and uncertainty/error propagation analysis, which relies heavily on extensive process-microstructure-property calculation. For the applications to aeroturbine disk alloy, the PrecipiCalc software acted as a process-microstructure simulation tool. Combined with SFTC's FEM-based DEFORM-HT (Heat Treat module) (Ref 18) as the heat treatment process simulation tool to determine the temperature evolution history at every location in a part, and the strength model developed by University of Michigan (Ref 19) for calculation of yield strength at room and elevated temperatures, these tools complete the process-microstructure-property calculation cycles. Furthermore, a CALPHAD-based APB (Anti-Phase Boundary) energy model (Ref 20) (needed by the strength model) was used to calculate APB energies for three different distributions of γ' particles based on the compositions output from PrecipiCalc software. In addition, a grain pinning model (Ref 21) is also used to calculate the grain size of γ matrix based on the γ' particle size after the solution treatment.

Under the AIM program, a high level multisite distributed integration was achieved using the iSIGHT (Ref 22) engineering design/optimization/automation tool and secure shell utilities for networking communication. The text-based interface of PrecipiCalc allows easy integration into a script and iSIGHT to facilitate the massive automation. The overall integration frame work is shown in Fig. 2.

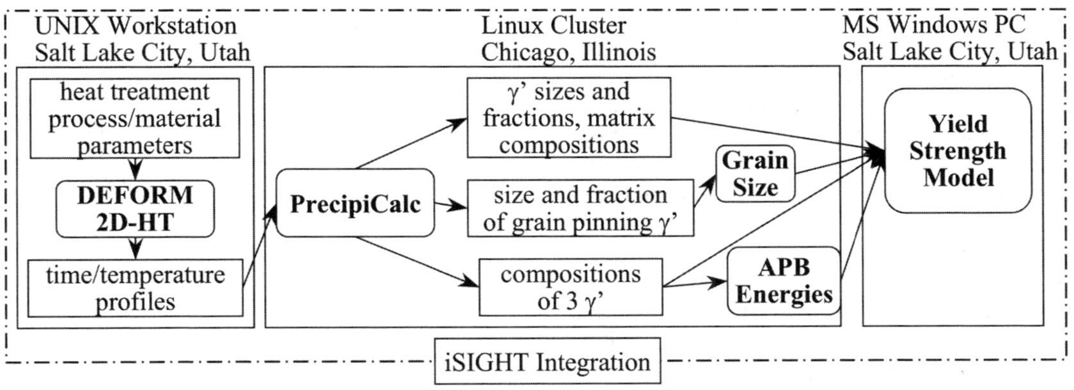

Fig. 2 Distributed iSIGHT integration for mechanistic process-microstructure-property simulations, performed on different computer hardware and operating systems across network.

Applications to IN100 Aeroturbine Disk Alloy

Applications of PrecipiCalc to the γ' microstructure simulation for heat treatment of Ni-based IN100 utilize several sources of data described as:

- Multicomponent thermodynamics database is based on ThermoTech's Ni-Data version 4.0 (Ref 23) with additional assessment of vanadium (Ref 24) under the AIM program. The IN100 compositions for γ/γ' simulation were obtained by excluding higher temperature carbides and borides from overall compositions, based on the equilibrium calculation at solution temperature. The resulting overall γ/γ' composition is Ni-13.02Cr-17.03Co-4.55Ti-10.0Al-1.86Mo-0.778V in at%. Based on this composition, the three-step sequential equilibria calculation (Ref 24) yields 19.1%, 27.6%, and 13.6% for primary, secondary, and tertiary γ', respectively, with a total of about 60%.
- Multicomponent mobility database is based on NIST Ni-based mobility database (Ref 25) for γ matrix. All the relevant components of IN100 alloy are available in this database, and the full 6×6 (without Ni) diffusivity matrices for γ matrix at different temperatures and matrix compositions were used in the PrecipiCalc simulation.
- Molar volume of γ' is based on a multicomponent and temperature-dependent model (Ref 26) established under the support of the AIM program.
- Interfacial mobility of low-coherency interface is described based on massive transformation kinetics with an activation energy scaled to that for solvent self diffusion. This was applied only to primary γ'.
- γ' microstructure before the heat treatment cycle is a necessary input to start the PrecipiCalc heat treatment simulation. Experimental information from forging was employed which gave a starting γ' distribution with mean diameter of 0.4 μm and fraction of 43%.

Instead of applying PrecipiCalc to the entire three-step heat treatment cycle with one single calculation, a sequential three-step simulation is used for IN100 applications. The first (and second) step ends when the secondary (and tertiary) γ' starts to come out during the quench right after the first (and second) treatment, and calculation restarts with primary (and secondary) γ' particles removed. This is necessary for two reasons. First, the IN100 PrecipiCalc calculations used uniform particle compositions in the mass balance calculation and the precipitate compositions automatically follow the thermodynamics description. However, during the fast quench after the first and second heat treatments and with relatively slow diffusion in γ' particles, primary and secondary γ' particles do not have enough time to adjust and relax their compositions. Removing primary and secondary γ' particles during the first and second quenches prevent incorrect particle (and hence matrix) compositions for the subsequent calculations. Second, the mean field approximation utilized by PrecipiCalc particle growth model assumes every particle interacts with all others through a single composition pool. This is a good assumption for unimodal particle size distribution where no significant size difference exists. For multiple distributions where a large particle is surrounded by many small particles, the large particle only interacts with a layer of smaller ones around it. Applying mean field approximation hence overestimates the growth of large particles and the dissolution of small particles, yielding faster coarsening kinetics to this multimodel distribution than reality. As γ' particles in IN100 have three distinct size scales, with many secondary (tertiary) particles around a single primary (secondary) particle, utilizing the multistep simulation prevents the over-coarsening problem.

Simulations for the Disk Center Location

The microstructure of a center location of a full scale disk has been fully characterized at University of Connecticut under the AIM program and is used to calibrate four selected PrecipiCalc tuning parameters. An optimization to the above PrecipiCalc parameters based on available experimental information of γ' sizes and fractions was performed with the help of iSIGHT. Lower optimization weighting was placed for tertiary microstructure than those for primary and secondary microstructure. These tuning parameters and their determined values are:

- Surface energy of primary γ' (0.12 J/m^2) represents the incoherent (due to large primary size) γ/γ' interface.
- Surface energy of secondary/tertiary γ' (0.061 J/m^2) represents the coherent γ/γ' interface. The determined value is on the high side compared to other literature studies based on coarsening kinetics of γ' particles.
- Diffusivity scaling factor (0.964) is a positive value multiplied to all diffusivity matrix components to account for possible inaccuracy in the mobility database. A number close to unity was found, indicating the NIST mobility database provides accurate predictions and needs little correction.
- Interfacial mobility prefactor for primary γ' (0.546 m/sec) determines the level of interfacial dissipation due to the low-coherency interface between primary γ' and matrix. For coherent secondary and tertiary γ', interfacial dissipation is considered insignificant.

Using these optimized tuning parameters, the results of γ' simulation for the disk center are shown in Fig. 3, where time evolutions of both mean γ' sizes and fractions of individual distributions are shown along with the heat treatment temperature profile overlapped. Labels ‘P’, ‘S’, and ‘T’ are placed to denote primary, secondary, and tertiary γ' kinetics, respectively.

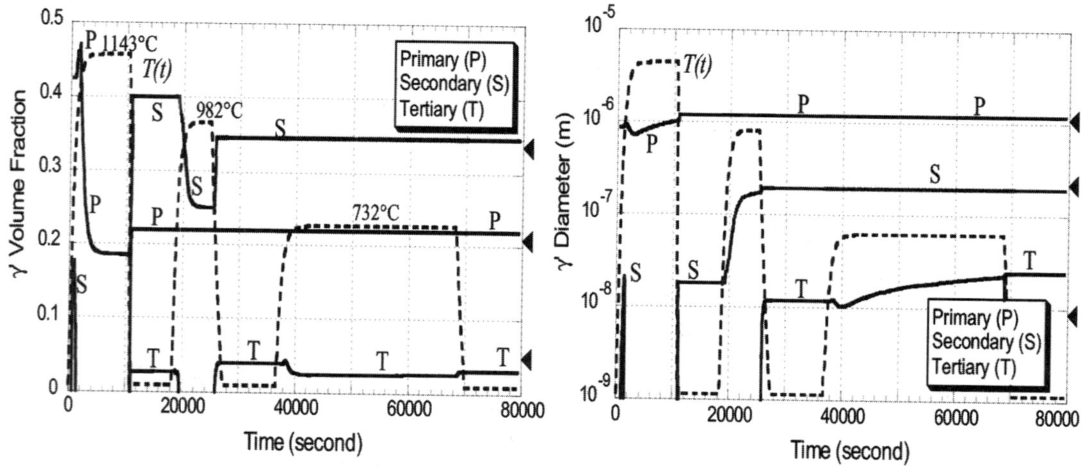

Fig. 3 PrecipiCalc simulation of a center location of an aeroturbine disk under the three-step heat treatment cycle. Time evolutions of mean sizes and fractions of three γ' distributions are shown with overlapped temperature profile. Experimental γ' microstructural results after heat treatment are shown as triangles on the right of both plots.

The major observations of this PrecipiCalc simulation are summarized as:

- Multimodal γ' distributions are resulted from the cyclic heat treatment. The high-temperature aging puts back some γ' formers to solution and coarsens up existing particles. The subsequent quench releases remaining supersaturation into a new nucleation, which at lower temperature with lower diffusivity, requires higher driving force to activate, leading to a small size scale in the new precipitation.
- The final γ' microstructure after the heat treatment observed experimentally are shown as triangles in Fig. 3. PrecipiCalc simulation results with optimized parameters has provided very good agreement with experimental results except tertiary microstructure, where the calculated mean size (20 nm) is twice larger than that determined experimentally (8.6 nm). Note that experimental observation has shown a big spread of tertiary size with maximum size found to be 17 nm. The experimentally estimated tertiary fraction (5% which is back determined from overall fraction and the observed primary/secondary fractions), has high uncertainty and is slightly higher than the calculated value (3%).
- The equilibrium fractions of γ' are reached at three holding temperatures, comparing to the sequential equilibria calculation. For example, at second heat treatment, a total (primary + secondary) of about 57% of γ' is found for both PrecipiCalc and sequential equilibria results. However, after isothermal treatments, the temperature drop pushes coarsening γ' particles into growing particles during the quench. This introduced additional fraction increases, and led to higher primary and secondary fractions than those calculated from sequential equilibria. The final tertiary fraction from PrecipiCalc simulation ended up smaller than that from sequential equilibria results, as the total γ' fraction is still around 60%.
- With further undercooling and more accumulated γ' supersaturation during the quench, new nucleation occurs and finer secondary (tertiary) γ' particles started to come out. The amount of finer γ' can achieve depends on how much γ' has formed previously.
- The simulated results have shown several transient and temporal γ' precipitations that may come out at low temperature during heat up, due to remaining and unused supersaturation of γ'. Those transient γ' particles become subcritical and disappear when temperature goes up further.

The calculated compositions of γ and γ' are also compared with the experimental measurements with EDS and 3DAP (3D Atomic Probe) (Ref 27). The 3DAP removed a total of 3 million atoms, in a box size of $12 \times 12 \times 476\ nm^3$ along the direction of <110>. The box cuts through two secondary and one γ' particles, and two matrix channels between secondary and tertiary particles. The results are summarized in Table 1. The calculated γ' compositions captured the correct trends for most alloying components from high-temperature γ' (primary) to low-temperature γ' (tertiary), when compared to the experiments. However, deviations in some elements are evident. For the compositions of γ' particles, PrecipiCalc underestimated the amounts of Mo, V, and Cr, and overestimated the amounts of Al and Co. This is caused by the inaccuracy of the thermodynamics information used in simulation.

Table 1 Comparison of compositions from PrecipiCalc (PpC) and experimental results for a disk central location

at.%		Ni	Co	Mo	V	Cr	Al	Ti
Primary	EDS	63.1	11.06	0.84	0.74	3.73	12.4	8
	PpC	61.1	12.1	0.38	0.27	3.14	14.14	8.83
Secondary	3DAP	62.83	10.06	0.945	0.725	2.82	14.65	7.88
	PpC	63.22	10.66	0.507	0.568	2.72	15.25	7.06
Tertiary	3DAP	63.12	9.1	1.885	1.712	2.49	16.31	5.27
	PpC	64.61	9.32	1.104	1.14	2.99	17.18	3.65
Matrix	3DAP	32.03	30.78	3.415	0.86	30.53	2.09	0.21
	PpC	37.51	26.14	3.841	1.3	28.38	2.71	0.11

Spatial Calculations for IN100 Disk

Based on the experience of IN100 simulation of one location, we applied PrecipiCalc to compute the spatial γ' microstructure distribution of an entire heat-treated minidisk (a scaled-down aeroturbine disk for testing purposes), including the integral which will be machined off after heat treatment. Using the DEFORM calculated time-temperature profiles of IN100 minidisk, we applied PrecipiCalc calculations to every FEM node. The results of PrecipiCalc were used to calculate γ grain sizes, APB energies and yield strength, based on the software integration shown in Fig. 2. All the input and parameters to PrecipiCalc remain the same as the previous disk center calculation, except for different thermal profiles depending on location, and the initial γ' size distribution to start the heat treat simulation has 40% larger mean size than that used in previous single location calculation. The adjustment to the initial size distribution is necessary to yield reasonable final primary γ' sizes observed experimentally. This adjustment falls within the microstructural variations caused by the variations of extrusion/forging conditions.

In the plot of a cross section of the minidisk with integral shown in Fig. 4, the calculated secondary γ' size, which is the most sensitive γ' microstructure feature with respect to the thermal profile (particularly quench rate), is presented as a contour plot. The thick dashed line overlaid on the contour plot represents the minidisk shape machined out after the heat treatment. As indicated from the contours, the secondary γ' size varies from 40 nm at the vertical surface to 160 nm at the bottom center portion of the disk. This is expected, as there is fastest cooling (smallest γ') at the vertical surface, and slowest cooling (largest γ') at the bottom surface with heat trap. The simulations were then compared to the experimental measurements at three areas: bore, rim and attachment, indicated in Fig. 4 with shaded rectangles. Based on PrecipiCalc results, the secondary γ' sizes are fairly uniform in the vertical direction within the bore and attachment. In the rim area, however, the secondary γ' sizes can vary dramatically along the y-direction. To provide the comparison of simulation and experiments, microstructure results are summarized in Table 2. Multiple experimental values are given when different minidisks were used in characterization. A large variation in the experimentally measured secondary size was observed between different disks. Multiple PrecipiCalc values at rim location were from three points with the same x-coordinate, but different y-coordinate, to represent the range of variation at that location. Overall, the PrecipiCalc calculations captured the correct trend in secondary size at different locations. The volume fractions of primary and secondary γ' are within 2.5% of measurements, within the experimental accuracy. All other calculated microstructural features compare reasonably well with experiments, including tertiary sizes.

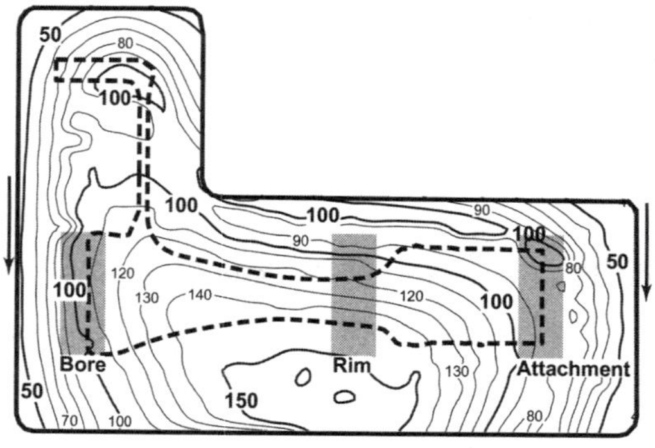

Fig. 4 Calculated PrecipiCalc spatial contours of secondary γ' size (in nm) for a radially symmetric cross-section of the minidisk. The left and right sides of the cross sections represent the inside and outside diameters, respectively. During the oil quench, the disk is immersed into quench media indicated by the arrows.

Using the strength model (Ref 19) developed under the AIM program, we also calculated the 260 °C (500 °F) yield strength distribution as shown in the contour plot of Fig. 5. The experimental tensile data at five selected locations are shown as five circles on the plot, representing the grip locations of the tensile bars. The experimental yield strengths are around 1166 MPa, in good agreement with predictions.

Additional Applications of PrecipiCalc in AIM Program

In addition to the previous applications to the IN100 heat treatment simulation, PrecipiCalc was utilized to model the variation of γ' microstructure in extrusion and forging, based on available information of processing variations in time and temperature. The results agreed very well with experiments. Furthermore, we also had preliminary simulation of both carbides and γ' precipitation at different

Table 2 Minidisk microstructure prediction from PrecipiCalc and experimental results

		Bore		Rim		Attachment	
Minidisk comparison		Exp.	PpC	Exp.	PpC	Exp.	PpC
Primary γ'	Fraction (%)	24 25.2	22.6	23.5 25	23.5	23.1 25.7	23.3
	Size(μm)	1.28	1.29	1.23 1.27	1.32	1.18 1.2	1.31
Secondary γ'	Fraction (%)	32.4	35		34		34.6
	Size(nm)	109 129	107.9	132 157	120 135 146	103 114	84.2
Tertiary γ'	Size (nm)	18 20.8	21.5	19.7 21.8	21.4	21.4	20.7

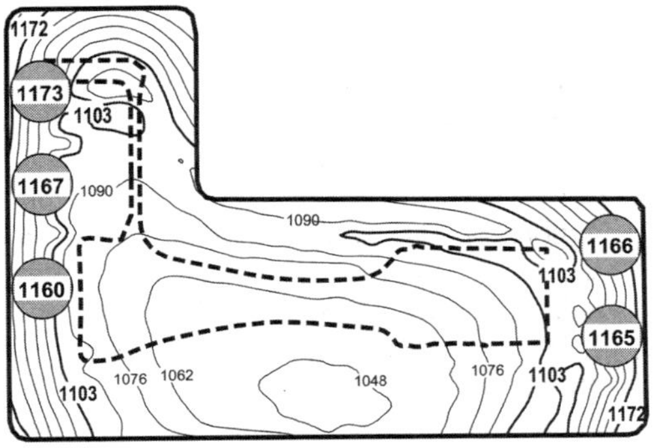

Fig. 5 Calculated spatial contour of yield strength (in MPa) at 260°C for minidisk.

solution treatment temperatures using the multiphase feature of PrecipiCalc. Combining with the distributed grain size model, we computed the grain sizes and clearly demonstrated the grain pinning control from γ' particles to carbides with increasing solution temperatures (Ref 21) when passing through the γ' solvus temperature. The high-fidelity PrecipiCalc γ' precipitation simulations and structure/property provided a mechanistic base for probabilistic methodology of accelerated prediction of minimum property, which was demonstrated in the AIM program.

Conclusions

Under the AIM initiative, PrecipiCalc has been greatly improved and enhanced. With the mechanistic multicomponent precipitation models of integrated treatment of nucleation, growth and coarsening, nonisothermal treatment and multiphase capability, PrecipiCalc has shown to be a useful tool for the simulations of commercial alloy processing. The successful application of PrecipiCalc to the γ/γ' simulation of the IN100 Ni-based aeroturbine disk alloy under commercial heat-treatment processing is discussed in this article. The particular emphasis here is the spatial microstructure and yield strength prediction in the entire disk, with the integration of processing and property models and tools. With this mechanisitc microstructure prediction tool, it is possible to address the property variation with other statistical models and tools developed under the AIM initiative.

Though PrecipiCalc was originally developed to capture the γ/γ' precipitation in Ni-based superalloys, it was built with mechanistic CALPHAD-based models, allowing quick applications to other alloy systems. For example, in other government-funded projects, PrecipiCalc has been used to simulate the strengthening dispersion such as Heusler phase in NiTi alloys and carbides/intermetallics in steels, the grain pinning dispersion in heat affected zone of welding, and the devitrification kinetics from amorphous iron/aluminum alloys.

Many modeling enhancements for PrecipiCalc are currently under development, including heterogeneous nucleation (on different types of defects and catalyst particles), denuded zone to provide an estimate of the size of solute depleted zones associated with growth, precipitation under constrained equilibrium.

ACKNOWLEDGMENTS

This work was supported by DARPA/USAF under Contract No. F33615-00-2-5216 and under subcontract from Pratt & Whitney. The authors acknowledge Pratt & Whitney, Martin Blackburn of University of Connecticut, and Gautam Ghosh of Northwestern University for providing the experimental results, Tresa Pollock of University of Michigan for the strength model, and Mike Long of Engineous Software for performing the iSIGHT integrated calculations. The authors would also like to thank Ryan Rathbun, Lei Li, and Christina Scharer of QuesTek for their support in software implementations and applications, and additional support provided by State of Illinois Technology Challenge Grant, No. 02-49106.

REFERENCES

1. J.S. Langer and A.J. Schwartz, *Physical Review A*, 21 (1980) 948–958
2. R. Kampmann and R. Wagner, *Materials Science and Technology*, vol. 5. VCH Weinheim (1991) 213–303
3. A. Deschamps and Y. Brechet, *Acta Mater.*, 47 (1999) 293–305
4. J.D. Robson, M.J. Jones, and P.B. Prangnell, *Acta Materialia* 51 (2003) 1453–1468
5. R. Kampmann and R. Wagner, presented at *Decomposition of Alloys: The Early Stages*, Sonnenberg, Germany, 1983
6. M. Serriere et al., *Materials Science Forum*, 396–402 (2002) 747–752
7. V.V. Popov, *Phys. Met. Metallogr.*, 93 (2003) 303–309
8. V.V. Popov and I.I. Gorbachev, *The Physics of Metals and Metallography*, 95 (2003), 417–426
9. C.J. Kuehmann and P.W. Voorhees, *Metal. Trans. A.*, 27 (1996) 937–943
10. J. Morrel and G. Purdy, *Scripta Mater.*, 30 (1994) 905–908
11. J. Ågren et al., *CALPHAD*, 204 (2000) 41–54
12. A. Umantsev and G.B. Olson, *Scripta Mater.*, 29 (1993) 1135–1140
13. S. Kumar and D. Ramkrishna, *Chemical Engineering Science*, 52 (1997) 4659–4679
14. Thermo-Calc Application Programing Interface, Thermo-Calc AB, Stockholm, Sweden, 2003, www.thermocalc.com
15. J.J. Dongarra et al., *ACM Trans. Math. Soft.*, 16 (1990) 18–28
16. E. Anderson et al., *LAPACK Users' Guide*, 3rd ed., Society for Industrial and Applied Mathematics, (1999), Philadelphia, PA
17. http://hdf.ncsa.uiuc.edu/HDF5
18. http://www.deform.com
19. R. Kozar, J. Schirra, W. Milligan and T.M. Pollock, submitted to *Metall. Trans.*
20. A.P. Miodownik and N. Saunders, *Applications of Thermodynamics in the Synthesis and Processing of Materials*, ed. Nash and Sundman, TMS (1995) 91
21. J. Wright and R. Rathbun, to be published
22. http://www.engineous.com
23. N. Saunders, M. Fahrmann and C.J. Small, in *Super-alloys 2000*, eds. K.A. Green, T.M. Pollock, and R.D. Kissinger (TMS, 2000), 803
24. A.M. Wusatowska-Sarnet et al., *J. Mater. Res.*, 18 (2003) 2653–2663
25. C.E. Campbell, W.J. Boettinger, and U.R. Kattner, *Acta Materialia*, 50 (2002) 775–792
26. H.-J. Jou, unpublished work performed under DARPA-AIM program, 2000
27. D. Isheim, to be published

CHAPTER 23

Precipitation Model Validation in Third-Generation Aeroturbine Disc Alloys*

G.B. Olson, QuesTek Innovations LLC and Northwestern University
H.-J. Jou, J. Jung, J.T. Sebastian, and A. Misra, QuesTek Innovations LLC
I. Locci and D. Hull, NASA Glenn Research Center

Abstract

In support of application of the DARPA-AIM methodology to the accelerated hybrid thermal process optimization of third-generation aeroturbine disc alloys with quantified uncertainty, equilibrium and diffusion couple experiments have identified available fundamental thermodynamic and mobility databases of sufficient accuracy. Using coherent interfacial energies quantified by Single-Sensor DTA nucleation undercooling measurements, PrecipiCalc simulations of nonisothermal precipitation in both supersolvus and subsolvus treated samples show good agreement with measured γ' particle sizes and compositions. Observed long-term isothermal coarsening behavior defines requirements for further refinement of elastic misfit energy and treatment of the parallel evolution of incoherent precipitation at grain boundaries.

Introduction

An important ongoing application of predictive science-based computational materials engineering is the accelerated hybrid process optimization of dual microstructure aeroturbine discs fabricated from third-generation high-refractory Ni-base superalloys. In support of higher fidelity application of the DARPA-AIM methodology previously demonstrated on the IN100 and R88DT disc alloys (Ref 1, 2), a collaborative model and database validation study is supported by the NASA Aviation Safety Program to better quantify uncertainty and improve prediction accuracy of the QuesTek PrecipiCalc multiphase/multicomponent precipitation simulation code developed as the primary engine of the AIM methodology.

Four third-generation nickel-based disk superalloys, including ME3 (also called René104), LSHR (Low-Solvus, High-Refractory alloys developed by NASA), Alloy 10 (developed by Honeywell), and RR1000 (developed by Rolls-Royce) are being studied under this program, and their nominal compositions are listed in Table 1. In the following sections, key results from the first year of this NASA/QuesTek collaboration are provided.

*G.B. Olson, H.-J. Jou, J. Jung, J.T. Sebastian, A. Misra, I. Locci, and D. Hull, "Precipitation Model Validation in 3rd Generation Aeroturbine Disc Alloys," *Superalloys*, TMS, (2008) 923–932

Table 1 Nominal compositions (balance Ni) of the four third-generation disc alloys studied under this program

		Cr	Co	Mo	W	Al	Ti	Nb	Ta	Hf	C	B	Zr
ME3	wt%	13.1	20.0	3.8	1.9	3.5	3.6	1.1	2.3	...	0.04	0.03	0.05
	at.%	14.5	19.5	2.3	0.6	7.5	4.3	0.7	0.7	...	0.19	0.16	0.03
LSHR	wt%	13.0	21.0	2.7	4.3	3.5	3.5	1.5	1.6	...	0.03	0.03	0.05
	at.%	14.5	20.7	1.6	1.4	7.5	4.2	0.9	0.5	...	0.15	0.16	0.03
Alloy10	wt%	10.2	14.9	2.7	6.2	3.7	3.9	1.9	0.9	...	0.03	0.03	0.10
	at.%	11.5	14.8	1.6	2.0	8.0	4.8	1.2	0.3	...	0.15	0.16	0.07
RR1000	wt%	15.0	18.5	5.0	...	3.0	3.6	...	2.0	0.5	0.03	0.015	0.06
	at.%	16.5	17.9	3.0	...	6.4	4.3	...	0.6	0.2	0.13	0.080	0.04

PrecipiCalc Calibration and Validation Protocol

An initial objective of this NASA program is to develop a standard PrecipiCalc calibration and validation protocol for intragranular γ' precipitation. The protocol needs to employ independent experimental measurements to decouple and/or minimize the cross interaction between model parameters, allowing the determination of the model parameters with high fidelity and minimum overfitting. The protocol must sequentially address the foundational databases and model parameters of the PrecipiCalc method, which include:

- CALPHAD fundamental databases. PrecipiCalc relies on CALPHAD-based databases to capture fundamental mechanistic features of multicomponent alloys. These databases and associated tuning parameters include:
  - Thermo-Calc (Ref 3) compatible thermodynamic databases, representing bulk free energy with state variables such as composition and temperature;
  - ΔE, a phase free energy shift in Thermo-Calc to locally tune equilibrium phase fractions;
  - DICTRA (Ref 4) compatible mobility databases — representing atomic mobility and allowing calculation of diffusivity when combined with the thermodynamic database;
  - D_{scale}, diffusivity correction factor used by PrecipiCalc to easily rescale the diffusivity matrix for local fitting;
  - Molar Volume, preliminary multicomponent molar volume models were developed for both γ and γ' under the AIM program.
- Material kinetic model parameters:
  - σ_{coh}, coherent surface energy, which is the key parameter affecting the nucleation barrier when particles are small and coherent with the surrounding matrix;
  - G_{el}, elastic coherency (misfit) energy adds additional energy penalty to the precipitation when the particles are coherent with the matrix;
  - σ_{incoh}, incoherent surface energy captures the increased surface energy when the particles lose coherency with the matrix;
  - M_o, prefactor for the interfacial mobility term to describe incoherent interfaces, while the corresponding activation energy is scaled to that for solvent self diffusion;
  - $R_{coh \rightarrow incoh}$, the characteristic particle size for coherency transition.

The current PrecipiCalc calibration and validation protocol is summarized in Table 2. High-temperature long aging is applied to Ni-based superalloys to produce near equilibrium microstructures, and matrix/particle compositions are measured by APT and EDS, (see the section "Thermodynamics"). The results help down-select a candidate thermodynamic database and determine potential simple energy corrections ΔE. Next, diffusion couples between Ni-based superalloys and pure Ni with high-temperature aging are used to determine if there is a need to correct mobility of individual components or scale with a simple scaling factor (D_{scale}) to the calculated diffusivity from databases. Ultimately, XRD can be used to evaluate the importance of misfit and modify the molar volume models if needed.

For material-specific model parameters of precipitation kinetics, nonisothermal nucleation experiments using SSDTA (see later discussion) were identified to determine coherent γ' surface energy (σ_{coh}) and estimated elastic coherency energy (G_{el}). Next, later stage coarsening experiments are used to determine the incoherent γ' surface energy (σ_{incoh}) and possible interfacial mobility term (M_o). Finally,

TEM study can be used to determine the coherency transition size ($R_{coh \rightarrow incoh}$), and to assist the calculation of elastic coherency energy (G_{el}).

Thermodynamics

To assess equilibrium phase fractions and compositions, water quenched supersolvus treated samples of ME3, LSHR, Alloy10, and RR1000 disc alloys are given 1000hr treatments at temperatures of 1093 °C (2000 °F), 927 °C (1700 °F), and 760 °C (1400 °F). Commercial thermodynamic databases compared in this study, using Thermo-Calc (Ref 3) software, include Thermotech Ni-DATA versions 4 to 7 (Ref 5), Computherm PanNickel (Ref 6), Ni-NIST (Ref 7) (not including Nb), and TCNI1 (Ref 8) (not including Mo, Nb, and Ta).

Alloy 10/ME3 Equilibrium Studies

Comparison in equilibrium phase compositions was undertaken with evaluation of the Alloy10 and ME3 samples quenched from the 1093 °C 1000 hr age treatment. The high-temperature matrix composition of ME3 was determined by both Energy Dispersive Spectroscopy (EDS) analysis conducted at NASA GRC and Atom-Probe Tomography (APT) using an Imago Local Electrode Atom Probe (LEAP) with new larger-FOV LEAP detector at the Northwestern University Center for Atom Probe Tomography (NUCAPT). In addition, γ' composition was also successfully measured by EDS for both alloys.

A LEAP data set with 2.2 million atoms and dimensions of 36 × 37 × 60 nm^3 was analyzed for the ME3 sample. No large isothermally aged γ' was detected in LEAP (the fine γ' formed during quench will be discussed later), and thus the entire LEAP data set represents the matrix during the high temperature age. Table 3 summarizes both the quantitative APT and semi-quantitative EDS results, and a comparison with the predicted high temperature matrix compositions using different thermodynamics databases. Based on the root mean square (RMS) error of solute content relative to the APT measurement, the Ni-DATA 7 database gives the best agreement. In the last column of Table 3, an experimental phase fraction of 24.8% was estimated by mass balance using the APT measured matrix composition and the Ni-DATA 7 predicted γ' composition, and again the Ni-DATA 7 prediction provides the best agreement. The

Table 2 Summary of calibration and validation protocol

Experiments	CALPHAD Fundamental Databases	Material Kinetic Model Parameters
Equilibrium Age + APT and EDS Compositions	Thermodynamics, ΔE	
Diffusion Couple + Microanalysis	Mobility, D_{scale}	
SSDTA + APT		σ_{coh}, G_{el} (est.)
Coarsening Age + SEM/TEM for γ' size and fraction		σ_{incoh}, M_o
XRD, TEM for misfit	Molar volume	G_{el}, $R_{coh \rightarrow incoh}$

Table 3 Comparison of the predicted γ and γ′ compositions and γ′ phase fraction in ME3 at 1093 °C (2000 °F) with the experimental APT γ and EDS γ and γ′ compositions. Ni-DATA version 5 (or 7) is abbreviated as Ni 5 (or 7) in the table

		Composition at 1093 °C, at.%											
		Ni	Al	Cr	Co	Ti	Mo	W	Nb	Ta	C	RMS	% γ' fraction
γ	APT	44.9	6.7	17.5	21.7	3.5	3.4	1.0	0.5	0.5	0.04	...	24.8
	2σ	0.04	0.02	0.03	0.03	0.01	0.01	0.01	0.01	0.01	0.002	...	1.5
	EDS	46.1	6.6	16.6	22.7	3.3	3.0	1.3*	0.6	...	...	...	...
	Ni 5,6	45.7	5.6	19.7	22.1	2.4	3.1	0.7	0.3	0.3	0.01	0.93	30.7
	Ni 7	**46.4**	**6.0**	**18.5**	**21.8**	**2.7**	**3.0**	**0.7**	**0.4**	**0.5**	**0.01**	**0.54**	25.5
	Pan-Nickel	44.3	5.3	20.3	23.0	2.4	2.9	0.8	0.4	0.5	0.05	1.19	32.3
	NIST-Ni	44.7	5.1	19.9	23.8	2.4	3.2	0.8	...	0.2	...	1.44	29.4
γ'	EDS	58.4	12.9	4.1	12.8	8.9	0.5	0.3	0.6	1.4	...	...	...
	Ni 5,6	61.5	5.7	2.6	14.4	7.1	0.5	0.9	2.4	4.9	...	3.09	...
	Ni 7	**60.2**	**12.0**	**2.8**	**13.0**	**8.8**	**0.3**	**0.3**	**1.3**	**1.2**	...	**0.63**	...
	Pan-Nickel	61.8	12.3	2.2	12.5	8.2	0.4	0.1	1.2	1.1	...	0.81	...
	NIST-Ni	68.3	6.3	1.3	9.7	7.6	0.3	0.3	...	6.2	...	3.27	...

* Value includes both W and Ta due to peak overlaps in EDS.

comparison of predicted versus EDS measured γ′ compositions also confirm the best agreement from the Ni-DATA 7 database.

High-temperature γ and γ′ compositions of Alloy 10 (1093 °C 1000 hr) were successfully measured by EDS, and compared with several thermodynamic databases in Table 4. Combined W+Ta composition is reported for EDS due to peak overlap. Overall, Ni-DATA 7 agrees the best with the semi-quantitative EDS measurement.

A systematic underestimation of Ti in the calculated γ matrix was also noted in QuesTek's earlier simulations in the DARPA-AIM initiative on commercial Ni-base superalloys (Ref 1) and may have to be addressed later in this program.

Ni-Al-Cr(-X) (X=Re,W) Model Alloys

Further comparison of thermodynamic databases is enabled by previous APT and ICP (Inductively Coupled Plasma atomic-emission spectroscopy measurement) studies at Northwestern University of isothermal γ′ precipitation in simple Ni-Al-Cr-X model alloys (Ref 9–13). The measured time evolution of particle size suggests the precipitation composition trajectory can be modeled as an unstable equilibrium with capillarity. Comparison of phase compositions with the database predictions for these alloys is made in Tables 5–8. For the base Ni-Al-Cr alloy at 600 °C, measurement of the composition trajectory includes the initial critical nucleus composition. As metrics of relative impact of phase

Table 4 Comparison of the predicted γ and γ′ compositions at 1093 °C and EDS measurement for Alloy 10 aged at 1093 °C for 1000 hr.

		Compositions, at.%									
		Ni	Al	Cr	Co	Ti	Mo	W	Ta	Nb	RMS
γ	EDS	53.8	7.0	13.4	17.5	3.2	2.3	1.6	...	1.13	...
	Ni 5,6	50.8	5.74	17.60	17.76	2.61	2.30	2.54	0.14	0.50	1.75
	Ni 7	**51.1**	**5.97**	**16.97**	**17.72**	**2.76**	**2.21**	**2.48**	**0.20**	**0.63**	**1.49**
	Ni-NIST	51.3	5.44	17.00	18.60	2.58	2.31	2.66	0.10	...	1.68
γ′	EDS	62.3	13.0	4.0	10.9	7.2	0.6	0.9	...	1.3	...
	Ni 5,6	62.2	12.02	2.82	11.01	8.42	0.24	0.93	0.49	1.89	0.80
	Ni 7	**64.1**	**12.11**	**2.80**	**10.53**	**8.64**	**0.23**	**0.91**	**0.40**	**1.73**	**0.83**
	Ni-NIST	66.8	13.5	1.35	7.54	9.62	0.13	0.37	0.68	...	1.93

Table 5 Phase composition comparison of APT (Ref 9) and the database predictions for a Ni-5.2Al-14.2Cr (at.%) alloy at 600 °C

			at.%						
	600 °C		Ni	Al	Cr	RMS	APB (J/m²)	600 °C misfit %	% γ′ fraction
γ	Equilibrium matrix composition	APT	81.26	3.13	15.61	...	...	...	...
		2σ	0.18	0.08	0.18	0.14	...	...	...
		Ni 5,6	81.40	3.64	14.96	0.58	...	...	...
		Ni 7	**81.34**	**3.62**	**15.04**	**0.53**	...	...	...
		TCNI1	81.17	3.93	14.90	0.76	...	...	...
		Ni-NIST	80.60	5.20	14.20	1.77	...	...	...
γ′	γ′ equilibrium composition	APT	76.53	16.69	6.77	...	0.19	−0.72	15.4
		2σ	0.50	0.44	0.30	0.38	0.01	...	0.4
		Ni 5,6	74.98	16.23	8.80	1.47	0.19	−0.6	12.4
		Ni 7	**74.97**	**16.21**	**8.82**	**1.49**	0.19	−0.6	12.4
		TCNI1	75.59	14.55	9.57	2.49	0.18	−0.8	12.3
		Ni-NIST	75.96	12.25	11.79	4.74	0.15	−0.9	0.0
	Critical γ′-nucleus composition	APT	72.40	18.30	9.30	...	0.16	−0.2	...
		2σ	2.20	1.80	1.40	1.61	0.03	...	...
		Ni 5,6	74.91	17.37	7.71	1.30	0.20	−0.5	...
		Ni 7	**74.91**	**17.40**	**7.69**	**1.30**	0.20	−0.5	...
		TCNI1	75.36	16.43	8.21	1.53	0.20	−0.6	...
		Ni-NIST	No precipitation			...	...	...	...

composition error, Tables 5–8 include not only the RMS solute concentration error and γ′ phase fraction, but the corresponding predicted error in thermodynamically computed γ′ APB enthalpy (Ref 13), and the at-temperature interphase lattice-parameter misfit based on QuesTek's molar volume model. These comparisons again support the Ni-DATA 7 database as most accurate, and suggest the corresponding error in APB enthalpy and misfit is acceptable within current structure/property model uncertainty.

While predicted phase fractions show good agreement with experiment for the complex superalloy of Table 3, we note a significant discrepancy for the simpler model alloys of Tables 5–8. For these alloys a ΔE shift in the relative free energy of γ′ could be appropriate for further modeling of precipitation dynamics.

Table 6 Phase composition comparison of ICP (Ref 10) and the database predictions for a Ni-10Al-8.5Cr (at.%) alloy at 800 °C

800 °C		at.% Ni	Al	Cr	RMS	% γ′ fraction
γ equilibrium matrix composition	ICP	82.71	8.43	8.86		
	2σ	0.24	0.16	0.18	0.17	
	Ni 5,6	82.93	8.32	8.75	0.11	
	Ni 7	**82.93**	**8.32**	**8.75**	**0.11**	
	Ni-NIST	81.90	9.76	8.34	1.01	
γ′ equilibrium precipitate composition	ICP	76.60	17.41	5.99		14.6
	2σ	0.36	0.34	0.12	0.25	0.3
	Ni 5,6	75.52	18.64	5.84	0.88	13.9
	Ni 7	**75.52**	**18.64**	**5.84**	**0.88**	13.9
	Ni-NIST	76.47	17.40	6.13	0.10	0.0

Misfit Estimates

Comparison of relative interphase misfit also provides a useful metric of database accuracy for the complex disc alloys. Figure 1 summarizes the predicted equilibrium at-temperature γ/γ′ misfit for disc alloys, taking into account the presence of equilibrium carbides, using phase compositions predicted by the Ni-DATA 7 database. The alloys fall into two groups of relatively high and low misfit. Of the third-generation disc alloys, only the RR1000 alloy is predicted to maintain the low misfit of the IN100 and René88DT alloys. In

Table 7 Phase composition comparison of APT [11] and the database predictions for a Ni-Al-Cr-Re alloy at 800 °C

800 °C			at.% Ni	Cr	Al	Re	RMS	APB (J/m²)	800 °C misfit %	% γ′ fraction
γ	Equilibrium matrix composition	APT	81.07	10.04	6.74	2.15				
		2σ	0.16	0.04	0.08	0.02	0.05			
		Ni 7	**80.55**	**8.94**	**8.35**	**2.16**	**1.13**			
		Ni-NIST	80.22	8.79	8.87	2.12	1.42			
γ′	Equilibrium composition	APT	76.17	4.97	18.05	0.81		0.19	−0.71	24.8
		2σ	0.26	0.10	0.18	0.04	0.12	0.00		0.2
		Ni7	**75.29**	**5.68**	**18.80**	**0.23**	**0.68**	0.20	−0.68	15.9
		Ni-NIST	75.82	6.13	17.65	0.39	0.75		−0.77	7.2
	composition at 80% completion	APT	76.33	5.46	16.92	1.29		0.17		20.3
		2σ	0.30	0.16	0.26	0.08	0.18	0.00		0.2
		Ni7	**75.22**	**5.56**	**18.99**	**0.23**	**1.35**	0.20		12.8

Table 8 Phase composition comparison of APT (Ref 12) and the database predictions for a Ni-Al-Cr-W alloy at 800 °C.

800 °C		at.% Ni	Cr	Al	W	RMS	APB (J/m²)	800 °C misfit %	% γ′ fraction
γ equilibrium matrix composition	APT	81.31	5.83	11.52	1.34				
	2σ	0.07	0.04	0.05	0.02	0.04			
	Ni 7	**81.44**	**6.22**	**10.53**	**1.80**	**0.67**			
	Ni-NIST	80.48	8.57	8.84	2.11	2.26			
γ′ equilibrium precipitate composition	APT	76.30	17.00	3.91	2.80		0.21	−0.49	37.9
	2σ	0.08	0.07	0.04	0.03	0.05	0.00		0.0
	Ni7	**75.39**	**18.00**	**4.20**	**2.42**	**0.64**	0.22	−0.36	32.1
	Ni-NIST	75.87	16.61	6.42	1.09	1.77		−0.61	18.7

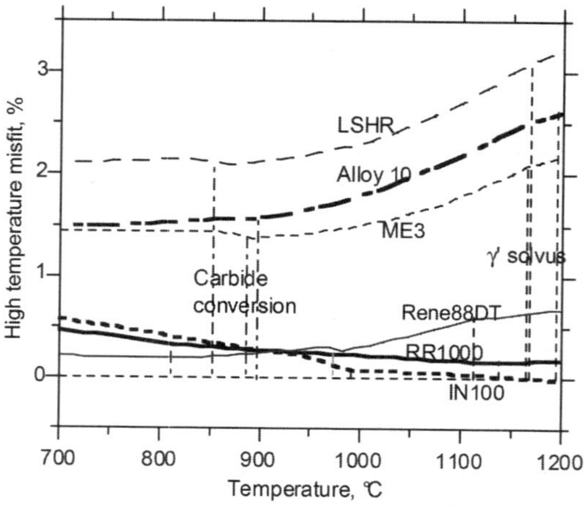

Fig. 1 Summary of predicted equilibrium at-temperature γ/γ' misfit for disc alloys, taking into account the presence of equilibrium carbides, using phase compositions predicted by the Ni-DATA 7 database and the preliminary molar volume model.

support of this prediction, Fig. 2 compares the γ' microstructures of supersolvus treated (1193 °C/1 hr) Alloy10 and RR1000 after long-term aging at 1093 °C for 1000 hr. While the fine intragranular precipitates in Alloy10 show the cuboidal morphology and ordered arrays consistent with the predicted higher misfit, the γ' in RR1000 shows a spheroidal morphology and disordered spatial distribution consistent with its predicted low misfit. The combination of thermodynamic database and misfit model thus appears to give a reasonable ranking of relative misfit between alloys. However, the higher misfit values appear susprisingly large and XRD studies are planned to calibrate the molar volume model. Elastic energy based misfit estimates from relative nucleation undercoolings are discussed later.

Mobility

To assess accuracy of available atomic mobility databases for combination with thermodynamic databases in order to predict multicomponent diffusivities, a linear diffusion "multiple" was prepared at NASA GRC. A 5 mm (0.2-inch) thick disc of pure Ni was diffusion bonded by hot pressing to a 3.2 mm (0.125-inch) thick disc of ME3 on one side and to a 3.2 mm (0.125-inch) thick disc of Alloy 10 on the other, for 4 hr at 877 °C (1610 °F) in vacuum. After the initial bond was formed, the diffusion couple was subjected to additional annealing in a horizontal tube furnace in an argon atmosphere at two temperatures, 1093 °C or 927 °C for 100 or 300 hr.

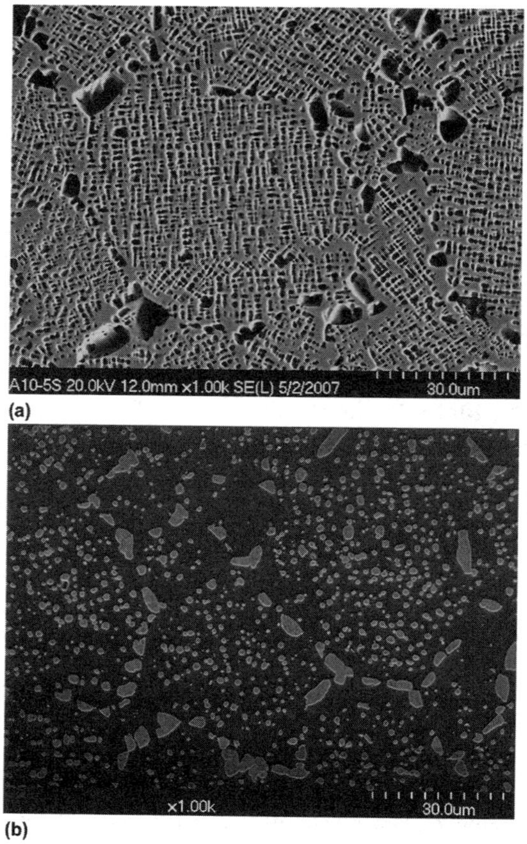

Fig. 2 Comparison of the γ' microstructures of (a) Alloy 10 and (b) RR1000. Courtesy of L. Birrell at University of Cambridge

DICTRA (Ref 4) simulations with Ni-DATA 7 and the NIST mobility database (Ref 14) of the diffusion multiple at the hot-pressing temperature 877 °C predicted no significant diffusion. Therefore, subsequent simulations ignored the hot-pressing step.

Next, the NIST mobility database employed in the AIM project was compared with the Thermo-Calc MOB1 database in combination with various thermodynamic databases. In order to compare microprobe analyses of the diffusion multiple with DICTRA simulations, the average Matano interface (defined as the interface across which equal number of atoms have crossed in both directions) was equated to the origin of the calculated profiles, and simulations assumed that γ' precipitates act only as sink or source of solute for diffusion, that is, no diffusion through γ'.

The agreement with the experimental microanalysis results is best using Ni-DATA 7 and NIST mobility databases. However, there are two discrepancies: (a) with the exception of Al, the experimental data does not confirm the predicted strong nonmonotonic profiles near the interface, and (b) the actual diffusion of Nb, Ta, Al, and Ti is less than predicted. Using the NIST mobility database with minor adjustments to the diffusivity prefactors of Cr, Nb, and Ti (by factors of 1.4, 0.53, and 0.70, respectively), the predicted DICTRA composition profiles agree reasonably well with the microanalysis, as shown in two of the selected examples in Fig. 3. However, comparative simulations reveals that these adjustments do not affect the γ' precipitation simulation with PrecipiCalc. Hence, we conclude there is no correction needed to the NIST mobility database and thus D_{scale} is set to 1.

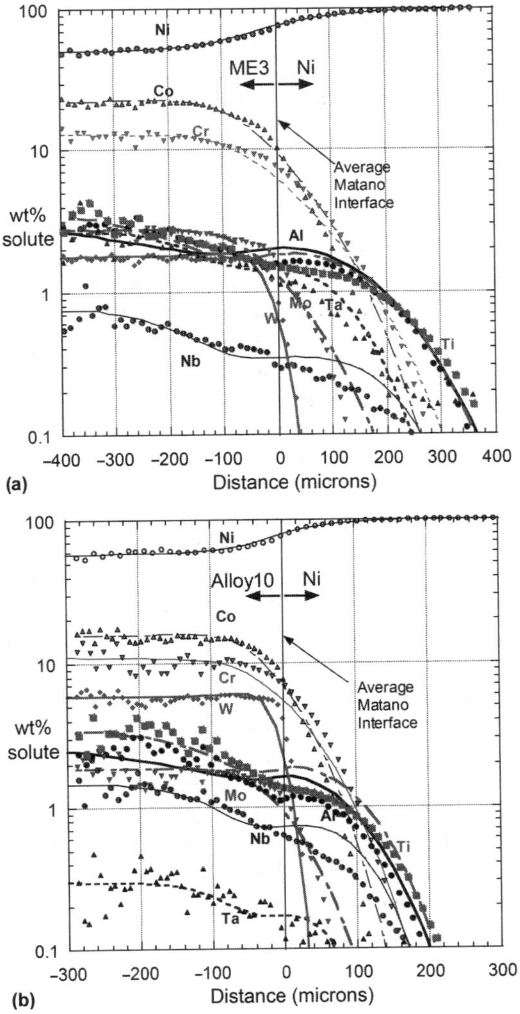

Fig. 3 Comparison of microanalysis (discrete points) and DICTRA (curves) predicted composition trace of (a) ME3/Ni diffusion couple aged at 1093 °C for 300 hr, and (b) Alloy10/Ni diffusion couples aged at 1093 °C for 100 hr.

Furthermore, as seen in the micrograph of Fig. 4(b), the width of significant grain growth in ME3 due to MC/γ' dissolution also agrees well with the DICTRA predicted phase fraction distribution in Fig. 4(a).

Nonisothermal Precipitation

Nucleation Onset Temperatures and SSDTA

As an efficient experimental check of nucleation behavior, transformation onset temperatures were measured in rapidly cooled arc melted samples of ME3 and Alloy10 using a Single Sensor Differential Thermal Analysis (SSDTA) technique developed at Ohio State University (Ref 15). From nucleation theory, the most important material parameters affecting the nucleation rate are driving force (which is determined by the bulk thermodynamics), surface energy and elastic coherency energy. In this section, utilizing the Ni-DATA 7 and NIST mobility database, we present the results of calibrating the surface energy, while initially ignoring the coherency elastic energy.

The transformation onset temperatures (T_{onset}) from the SSDTA experiments were determined from SSDTA data processing software, which takes the measured temperature versus time, fits the baseline prior to the transformation with a choice of function suitable for describing cooling, and then determines the departure point of measured data from the

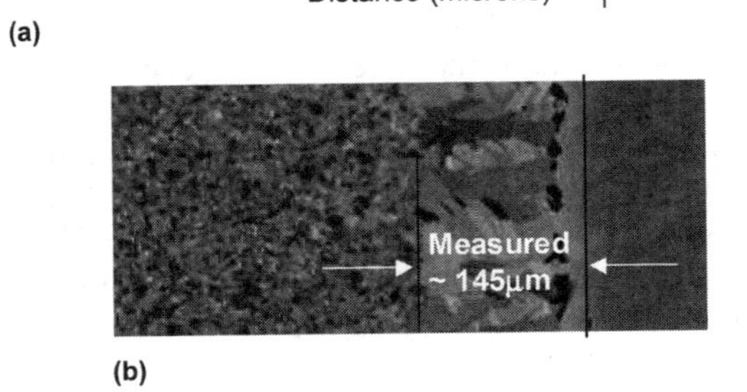

Fig. 4 Comparison of (a) predicted equilibrium phase fractions, and (b) measured phase boundaries with adjusted Matano interface, for ME3/Ni 1093 °C/100 hr diffusion couple sample.

fitted function. The calibration of the SSDTA thermocouples has shown an accuracy of ±4.5 °C (Ref 16) in the relevant temperature range. To determine the onset temperature from PrecipiCalc simulation with physically equivalent interpretation as closely as possible to the SSDTA results, the following summarizes our procedure.

- Process SSDTA measured temperature profiles to remove the latent heat contribution;
- Calculate material compositions relevant to the γ' precipitation from matrix γ phase—the high-temperature phases (borides, carbides, and undissolved γ') are removed with equilibrium calculations at highest SSDTA measured temperatures using the Ni-DATA 7 database.
- Perform PrecipiCalc simulations using Ni-DATA 7 and NIST mobility databases, with estimated surface energy;
- Collect time (or temperature) evolution results of γ' volume fraction, and compositions of matrix and γ', compute the time (or temperature) evolution of molar enthalpy;
- Compute temperature derivatives of the molar enthalpy with respect to temperature, dH/dT;
- Determine the transformation onset temperature where dH/dT changes by more than 10% of average dH/dT values, with decreasing temperature.

We note the difference of calculated T_{onset} values with the above enthalpy change approach and a simple approach of using a small γ' volume fraction of 0.1% is within the SSDTA accuracy.

Figure 5 summarizes the interfacial energy calibrated PrecipiCalc results and comparison to the SSDTA measured T_{onset}, for all four alloy samples. We varied the surface energy within 0.0225 to 0.0325 J/m^2, and Fig. 5 shows the best fit surface energy and T_{onset} results with minimum RMS of the difference between prediction and measurement. Except LSHR (which will be further discussed later), the optimized interfacial energies of the remaining three alloys are similar and agrees with the value of interfacial energy, 0.022~0.023 J/m^2, reported by Sudbrack *et al.*, (Ref 9) for the Ni-Al-Cr ternary model alloys.

The experimental uncertainty of T_{onset} associated with SSDTA is represented by error bars in Fig. 5. The PrecipiCalc predictions with optimized interfacial energy agree with

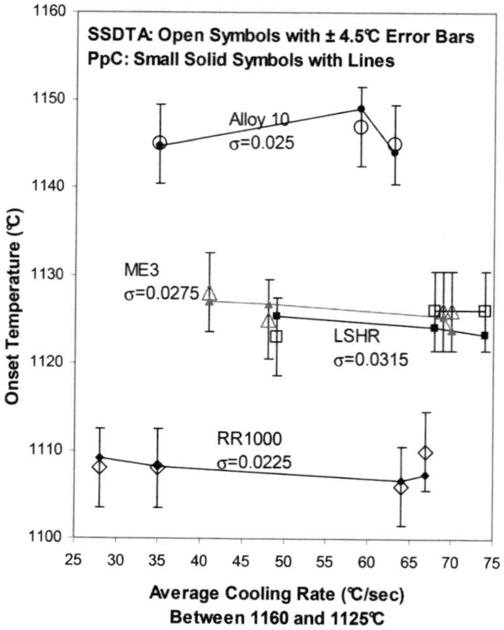

Fig. 5 Summary of SSDTA results and PrecipiCalc (PpC) predictions with calibrated coherent interfacial energy σ_{coh} (values in J/m^2). The data are plotted with respect to average cooling rate from measured temperature profiles.

SSDTA results well, and mostly fall within the experimental uncertainty. Note that the T_{onset} temperatures, both measured and predicted, do not monotonically decrease with the average cooling rate between 1125 to 1160 °C. If the temperature followed linear cooling, T_{onset} should have decreased monotonically with higher cooling rate. The actual local temperature profile clearly affected the measured T_{onset}, and PrecipiCalc predictions capture the nonmonotonic behavior very well.

Due to a furnace temperature limitation of the SSDTA experiment, Alloy10, having the highest predicted γ' solvus temperature, contains undissolved large γ' as confirmed by SEM (Fig. 6). The observed amount is in good agreement with predicted equilibrium γ' fraction at the highest sample temperature in the furnace. The calculated T_{onset} of Alloy10 presented in Fig. 5 was based on the assumption that undissolved γ' does not grow during the quench, allowing us to remove the large γ' completely in the PrecipiCalc simulations. To test this assumption, we performed an Alloy10 simulation allowing primary γ' growth during the quench, which resulted in a small reduction of nucleation onset temperature by only 1.6 °C. Hence, we conclude the assumption to remove large undissolved γ' in SSDTA Alloy 10 quench simulation is reasonable.

The relative ranking of best-fitting interfacial energies in Fig. 5 ranks with the relative experimental nucleation undercoolings. This in turn ranks with the relative predicted misfit (RR1000 lowest, LSHR highest). Incorporation of a coherency elastic energy G_{el} in further modeling will likely reduce the range of the coherent σ values. Using the surface energy of 0.0225 J/m^2, found for RR1000 and assumed to correspond to zero misfit, we reran PrecipiCalc for ME3, LSHR, and Alloy 10 for SSDTA simulations with the same surface energy and then estimated the coherency elastic energy G_{el} by fitting to the measured nucleation onset temperature. Table 9 summarizes the results, and also the corresponding at-temperature misfits based on the theoretical elastic strain energy of inclusions (Ref 17). According to this analysis, LSHR corresponds to the highest misfit, followed by ME3 and Alloy 10, in a similar ranking but at about 1/3 of the misfit values of the model prediction in Fig. 1. These misfit estimates will be validated against ongoing XRD experiments. In subsequent PrecipiCalc simulation to be discussed here, the surface energies in Fig. 5 are used.

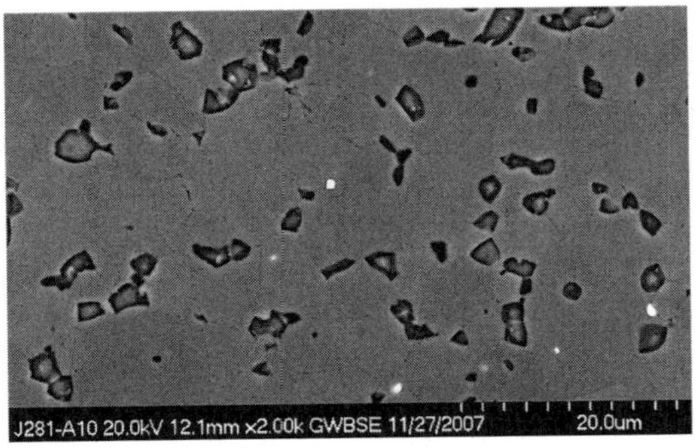

Fig. 6 NASA SEM observation of an Alloy10 SSDTA sample. γ' volume fraction was estimated to be around 9.5%, which is compatible with Ni-DATA 7 equilibrium prediction of 8.4% γ' phase.

Table 9 Elastic energy based misfit estimate from SSDTA nucleation onset temperatures and RR1000 surface energy.

	Estimated G_{el} (J/mole)	Misfit (%)
RR1000	0	0
Alloy 10	30	0.5
ME3	50	0.61
LSHR	80	0.8

γ' *Microstructure in SSDTA Samples*

The well-defined thermal history of the SSDTA samples also provides an excellent opportunity for experimental validation of PreciCalc simulations of nonisothermal precipitation. Using LEAP APT, we analyzed the ME3 Ar gas-quenched sample (ME3-1Ar) from the SSDTA experiment. In total, 32 million atoms were analyzed. The overall bulk composition is very close to the expected composition. Larger γ' precipitates, about 40–60 nm in diameter, were observed together with ultrafine γ' precipitates, about 5 nm or less in diameter.

High-resolution SEM (HR SEM) analysis was also conducted on the same ME3-1Ar SSDTA sample as well as an LSHR Ar gas-quenched sample (LSHR-2Ar). Consistent results between the APT and SEM for ME3-1Ar are shown in Fig. 7. As shown by the

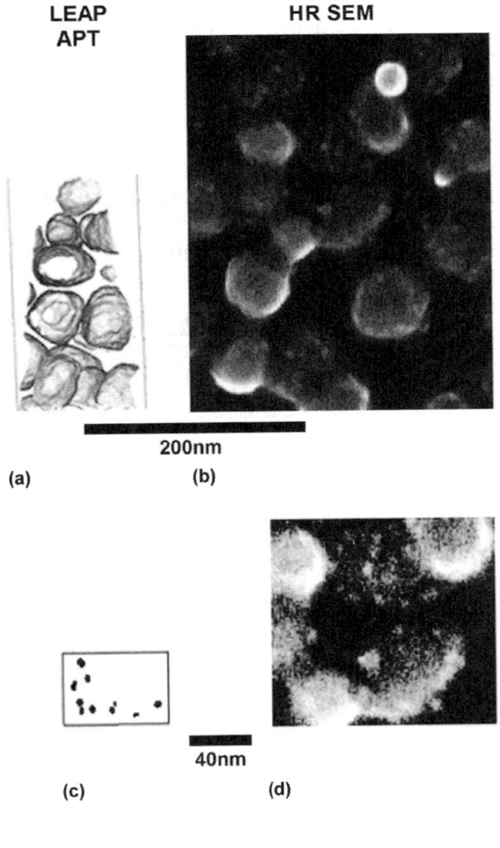

Fig. 7 Characterization of γ' microstructure in ME3-1Ar SSDTA sample. Left (a, c) figures from LEAP APT and right (b, d) figures from HR-SEM are compared at two different length scale: (a) and (b) at smaller magnification, showing larger γ', and (c) and (d) at slightly higher magnification showing very fine γ'

higher magnification views at the bottom of Figure 7, fine γ' formed at lower temperature during SSDTA cooling was also observed in both ME3-1Ar LEAP APT and HR SEM.

Table 10 shows a summary of measured larger γ' size and also PrecipiCalc predictions using the measured SSDTA temperature profiles. PrecipiCalc provides reasonable estimates with the approximations employed thusfar. Figure 8 shows reasonable agreement in particle size histograms between HR SEM and simulation for the LSHR-2Ar sample. In RR1000-3He (He gas-quenched) LEAP study, fraction of 35% was determined for largest γ', which agrees well with PrecipiCalc's 37.6%.

As in the case of the model alloy of Table 8, the APT measurements in the ME3-1Ar sample show W partitioning in the direction of thermodynamic prediction for the larger γ' population. However, the W profiles observed for the ultrafine population suggest solute trapping of the slowest diffusing components may be occurring during precipitation at the lowest temperatures.

Table 10 Comparison of the largest γ' mean diameters between two SSDTA samples and PrecipiCalc simulation using the coherent interfacial energies discussed previously.

	LEAP APT	HR SEM	PrecipiCalc Predictions
ME3-1Ar	40–60 nm	53.8 nm	41 nm
LSHR-2Ar	...	50.1 nm	45.6 nm
RR1000-3He	20–45 nm	31.8 nm	20.5 nm

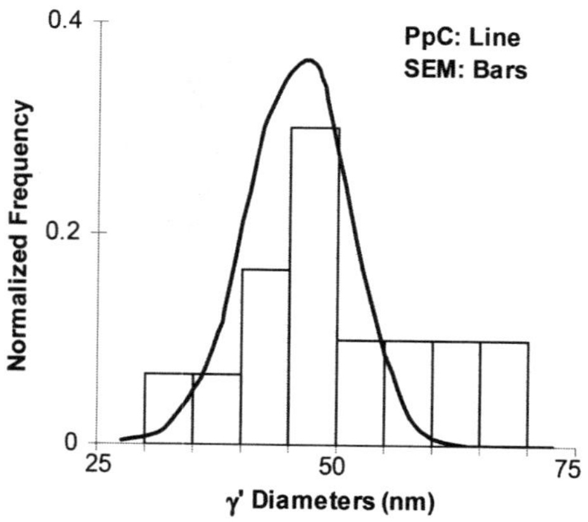

Fig. 8 Comparison of particle size histogram of PrecipiCalc (3D results) and HR SEM with an observation of 30 γ' particles in an LSHR-2Ar SSDTA sample.

γ' *Microstructure in ME3 Water-Quenched Samples*

LEAP APT microanalysis was conducted on ME3 samples water quenched from both subsolvus treatment at 1093 °C (1000 hr) and supersolvus treatment at 1200 °C (1 hr). For the supersolvus treated sample, a LEAP data set containing 8.7 million atoms within a box of $128 \times 55 \times 54\ nm^3$ was collected. The LEAP data analysis shows nanoscale γ' precipitates of about 15–30 nm in diameter, which is in good agreement with our PrecipiCalc prediction of 30 nm with an estimated cooling temperature profile. TEM work at NASA GRC indicates a larger mean diameter of γ' at 44 nm.

For the 1093 °C subsolvus treated ME3, the LEAP experiment used to obtain 1093 °C matrix composition (Table 3) was further analyzed for fine quench γ'. An estimated 10–20 nm effective diameter from LEAP APT is in reasonable agreement with the predicted 29 nm diameter from PrecipiCalc simulation with an estimated cooling temperature profile. A proximity histogram of the subsolvus LEAP sample summarizes the measured partitioning of solutes between the γ and γ' phases in Fig. 9. As summarized in Table 11 for both subsolvus and supersolvus treated samples, the partitioning is again in the direction of thermodynamic prediction, with the notable exception of the flat W profile offering further evidence that the slowest diffusing component

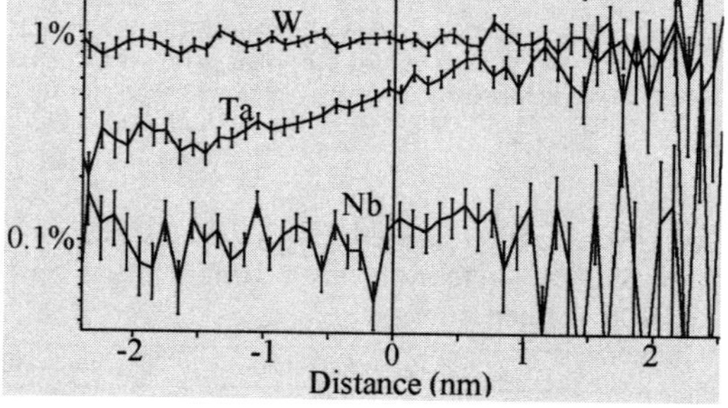

Fig. 9 γ and fine quench γ' compositions (in atomic %) analysis of LEAP experiment on ME3 1093 °C/1000 hr isothermal hold sample.

Table 11 Comparison of measured nanoscale γ′ particle and matrix compositions in ME3 with corresponding PrecipiCalc simulations.

		Compositions, at.%									
ME3		Ni	Al	Cr	Co	Ti	Mo	W	Nb	Ta	RMS
Subsolvus											
γ	APT	42.5	4.9	21.3	23.7	2.3	3.6	1.0	0.4	0.3	
	PpC	35.0	2.82	27.3	27.7	0.51	5.28	1.30	0.07	0.08	2.80
γ′	ATP	55.6	9.3	7.3	17.5	6.1	2.1	0.8	0.4	0.8	
	PpC	61.3	12.8	2.1	12.2	8.2	0.42	0.52	1.2	1.2	3.07
Supersolvus											
γ	APT	40.5	4.8	22.5	23.8	2.0	4.2	0.8	0.6	0.5	
	PpC	39.2	3.00	24.6	27.5	0.74	3.94	0.87	0.08	0.13	1.71
γ′	APT	56.5	12.0	3.5	13.6	8.9	1.9	0.5	1.0	1.6	
	PpC	62.2	12.4	1.94	11.9	8.50	0.32	0.34	1.0	1.3	1.01

exhibits solute trapping at extreme cooling conditions. Further study will address whether solute trapping need be considered for industrially relevant processing conditions.

γ′ Microstructure in Furnace-Cooled Samples

Supersolvus treated ME3, LSHR and Alloy10 samples followed by instrumented furnace cooling were characterized under SEM. The cooling rate between 1160 °C (2120 °F) and 1125 °C (2057 °F) is around 0.54 °C (0.97 °F)/sec, which is about two orders of magnitude slower than the SSDTA and water quench experiments. These experiments offer growth-dominated conditions for calibration of interfacial mobility. PrecipiCalc simulations were conducted using the coherent surface energies determined from the SSDTA study and the individually measured temperature cooling profiles for each samples. The PrecipiCalc 3D results converted to 2D are in reasonable agreement with the SEM results, as shown in Table 12. Unlike the finer γ′ particles formed during SSDTA fast cooling (see Fig. 8) which were etched to reveal 3D sizes in SEM, γ′ particles here are much larger and should be compared to the 2D results. As shown in Fig. 10, we observe reasonable agreement between simulation and experimental size distribution for these growing supercritical γ′ particles, showing sharper size distribution than that from coarsening theory. Note the truncation of the calculated 2D distribution is 10 nm, which is about the resolution of SEM and image analysis.

These PrecipiCalc simulations were carried out with coherent γ/γ′ surface energy and without interfacial dissipation (M_o term in Table 2). Though these γ′ particles are significantly larger than those found in SSDTA and water quench experiments, there is no indication that they are incoherent with the matrix. However, it is possible that they are semi-incoherent and are in the process of losing coherency.

Figure 11 compares the PrecipiCalc simulated γ′ size and growth rate evolution of fast-cooled SSDTA and slow furnace-cooled LSHR samples. In these two very distinct cooling rates, γ′ nucleates at similar sizes within a narrow temperature window (1140~1142 °C). In the slower furnace-cooled case (LSHR-FC), nucleated γ′

Table 12 Comparison of the largest γ′ mean sizes (diameters) between furnace cooled samples and PrecipiCalc simulations

	HR SEM	PrecipiCalc results converted to 2D
ME3	430 nm	348 nm
LSHR	408 nm	402 nm
Alloy 10	376 nm	396 nm

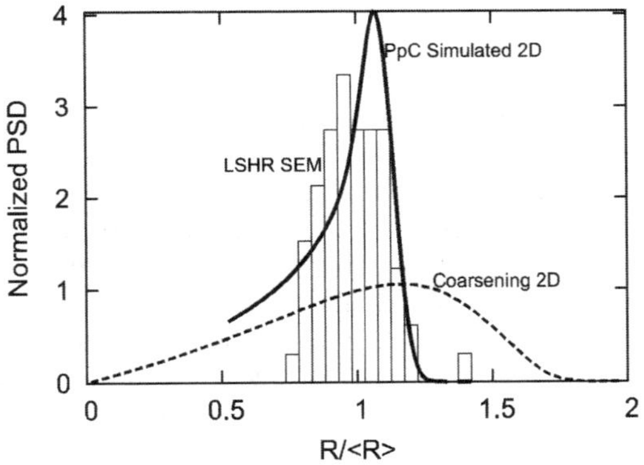

Fig. 10 Comparison shows good agreement between experimental SEM (bars) and PrecipiCalc simulated (converted to 2D, solid line) and normalized particle size distributions for LSHR furnace cooled sample. 2D coarsening size distribution is shown in a dash line.

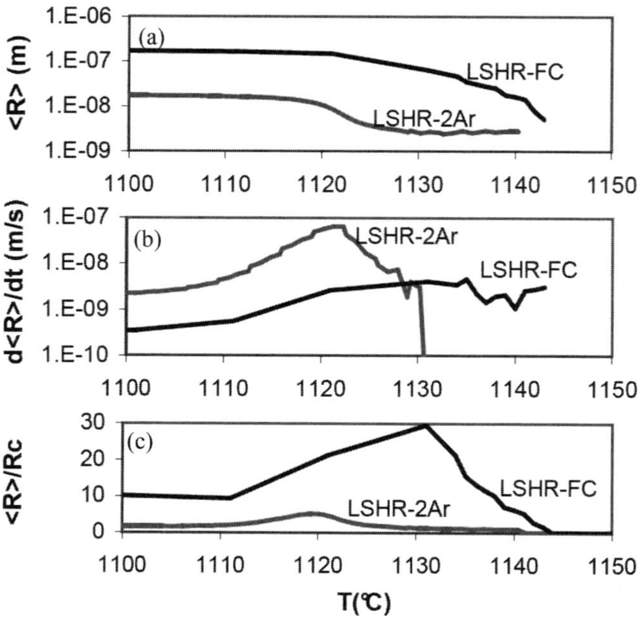

Fig. 11 Comparison of PrecipiCalc simulated γ' (a) mean radius (<R>), (b) growth rate and (c) mean radius over critical radius (R_C) ratio with respect to temperatures for both LSHR furnace cooled (FC) and SSDTA (2Ar). In this temperature range, only the first γ' distribution is formed in these two calculations.

particles can grow to much higher multiples of critical size (Fig. 11c), despite the slower growth rate (due to lower supersaturation in slower cooling) compared to the fast cool at temperatures below 1130 °C (Fig. 11b). This dominance of growth leads to the sharper size distribution of Fig. 10. The good agreement with average size in Table 12 suggests that interfacial dissipation can be negliected for particle size up to the 500 nm scale.

Coarsening Experiments

ME3, LSHR, and Alloy10 samples were supersolvus treated at 1193 °C (2180 °F) for 1 hr, water quenched, and then aged at different conditions, followed by a water quench. The aging conditions selected are 1093 °C (2000 °F) for 20 and 1000 hr, 927 °F (1700 °F) for 1000 hr, and 760 °C (1400 °F) for 1000 hr. The samples were then analyzed under TEM or SEM, depending on the size scale of γ' particles. For higher temperature aging, both intragranular γ' and grain boundary γ' particles are monitored as represented in Fig. 12. Similar to Alloy 10 in Fig. 2(a), the intragranular γ' in ME3 shows an ordered alignment, indicating coherency based elastic

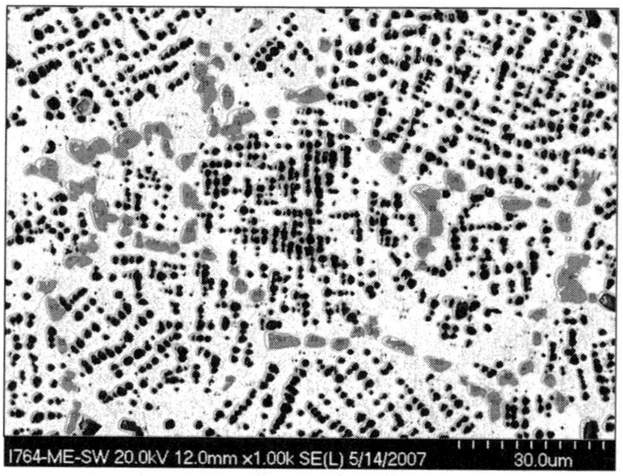

Fig. 12 SEM image (processed to show different features) of ME3 supersolvus treated and then aged at 1093 °C for 1000 hr. Intragranular and grain boundary γ' are represented as black and grey, respectively.

interaction. A consistent image processing technique was used to identify the grain boundary γ' and to obtain the sizes and fractions of both types of γ' particles for all ME3, LSHR, and Alloy 10 samples. All of these heat treatment and experimental analysis were conducted at NASA GRC.

Table 13 summarizes the measured γ' sizes in these heat treatment conditions. The grain boundary γ' particles are typically larger than

Table 13 Summary of experimental measured γ' mean equivalent diameters (μm) of isothermally aged samples. Intragranular (G) γ' and grain boundary (GB) γ', when observed, are reported separately

Aging conditions	Exp	ME3 G	ME3 GB	LSHR G	LSHR GB	Alloy 10 G	Alloy 10 GB
Before aging	TEM	0.044	...	0.028	...	0.044	...
1093C-20 h	SEM	0.510	0.810	0.504	0.984	0.546	1.084
1093C-1000 h	SEM	1.380	2.840	1.420	2.996	0.894	2.618
927C-1000 h	SEM	0.584	0.758	0.598	0.954	0.670	1.224
760C-1000 h	TEM	0.058	...	0.062	...	0.062	...

the intragranular γ' particles by a factor of 1.5 to 2. Between ME3, LSHR, and Alloy 10, there are no consistent trends in the relative γ' sizes. We note that the TEM and SEM measured sizes correspond to the mean 3D and 2D sizes, respectively.

The PrecipiCalc modeling of coherent intragranular γ' in these isothermally aged coarsening experiments is ongoing. Program plans include the modeling of incoherent grain boundary γ' using PrecipiCalc's heterogeneous grain boundary nucleation model. The grain boundary γ' growth simulations will include effects of grain boundary diffusion and possibly interfacial mobility to account for incoherent interfaces.

Conclusion and Summary

Measured γ- γ' phase compositions in equilibrated third-generation disc alloys and reported phase compositions in simple Ni-Cr-Al-X model alloys show best agreement with the Thermotech Ni-Data version 7 thermodynamic database. In the content of multicomponent disc alloys, the assessed sensitivity of predicted phase fractions, APB enthalpy and interphase misfit to measured composition error is judged as acceptable within current structure/property model uncertainty. Prediction of a relatively high misfit in the high-refractory third-generation alloys is supported by microstructural evidence, but precise predictions will require further misfit model calibration. Comparison of measured composition profiles in diffusion couples of ME3 and Alloy 10 with pure Ni against DICTRA multicomponent diffusion simulations indicates the combination of the Ni-DATA version 7 thermodynamic database with the NIST mobility database gives sufficient accuracy to support PrecipiCalc simulation of diffusional precipitation without diffusivity modification. The SSDTA measurement of critical nucleation undercoolings defines reasonable values of coherent interfacial energies. Incorporation in PrecipiCalc simulation of nonisothermal precipitation for both supersolvus and subsolvus treated materials gives good agreement with observed particle sizes and compositions. APT microanalysis shows some evidence of solute trapping of W, the slowest diffusing component, in fine precipitates forming at extreme cooling rates. Further modeling of measured long-term coarsening behavior requires refinement of coherency elastic energy estimates and treatment of the parallel evolution of incoherent precipitation at grain boundaries. The accuracy of predictions validated so far indicates that available fundamental databases and precipitation models offer sufficient fidelity for effective application of the AIM methodology to the accelerated optimization of hybrid thermal processing of third-generation aeroturbine disc alloys.

ACKNOWLEDGMENT

This work is supported by NASA Aviation Safety Program and Glenn Research Center (GRC) under Contract No. NNC07CB01C. The authors acknowledge John Gayda and Tim Gabb of NASA GRC for project collaboration and experimental work, Jeff Simmons and Chris Woodward of the Air Force Research Laboratory for serving as science advisors of this program, David Furrer, Rob Mitchell, and Mark Hardy of Rolls-Royce Corporation for serving as industrial advisors of this program, Carelyn Campbell of NIST for providing the mobility database, Boian Alexandrov of The Ohio State University for SSDTA work, Chris Kern and Sean Backs of QuesTek for SEM characterization and analysis, S.-L. Chen of CompuTherm and Paul Mason of Thermo-Calc AB for assisting thermodynamics calculation, Howard Stone of Univeristy of Cambridge for RR1000 data, Eugine Kang of Northwestern University for SEM data analysis, and David Seidman of Northwestern University for discussion and suggestions on LEAP APT results.

REFERENCES

1. H.-J. Jou, P. Voorhees, and G.B. Olson, "Computer Simulations for the Prediction

of Microstructure/Property Variation in Aeroturbine Disks," *Superalloys 2004*, eds. K.A. Green, T.M. Pollock, H. Harada, T.E. Howson, R.C. Reed, J.J. Schirra, and S. Walston, 2004, 877–886
2. A.M. Wusatowska-Sarnek, G. Ghosh, G.B. Olson, M.J. Blackburn, and M. Aindow, "A Characterization of the Microstructure and Phase Equilibria Calculations for the Powder Metallurgy Superalloy IN100," *J. Materials Research*, 18 (2003), 2653–2663
3. Thermo-Calc, Thermo-Calc Software AB, Stockholm, Sweden, 2007: www.thermo-calc.com/Products/TCC.html
4. DICTRA, Thermo-Calc Software AB, Stockholm, Sweden, 2007: www.thermo-calc.com/Products-services/software/diffusion-module-(dictra)/
5. N. Saunders, M. Fahrmann, and C.J. Small, "The Application of CALPHAD Calculations to Ni-Based Superalloys," *Superalloys 2000*, eds. K.A. Green, T.M. Pollock and R.D. Kissinger (TMS, Warrendale, 2000), 803
6. Pan Nickel database; CompuTherm LLC, Madison, WI, 2007: www.computherm.com/
7. U.R. Kattner, Construction of a thermodynamic database for ni-base superalloys: A case study, in: Turchi P.E.A., A. Gonis and R.D. Shutt (Ed.), TMS, Warrendale, PA, 2002, 147–164
8. TCS Ni-based Superalloys Database, v1, Thermo-Calc Software, 2007: www.thermo-calc.com/Products-services/databases/thermodynamic/
9. C.K. Sudbrack, D. Isheim, R.D Noebe, N.S. Jacobson, and D.N. Seidman, "The Influence of Tungsten on the Chemical Composition of a Temporally Evolving Nanostructure of a Model Ni-Al-Cr Superalloy," *Microsc. Microanal*, 10 (2004), 355–365
10. C.K. Sudbrack, T.D. Ziebell, R.D. Noebe, D.N. Seidman, "Effects of a Tungsten Addition on the Morphological Evolution, Spatial Correlations and Temporal Evolution of a Model Ni-Al-Cr Superalloy," *Acta Materialia*, 56 (2008) 448–463
11. C.K. Sudbrack, R.D. Noebe, and D.N. Seidman, "Compositional Pathways and Capillary Effects during Isothermal Precipitation in a Nondilute Ni-Al-Cr Alloy," *Acta Materialia*, 55 (2007), 119–130
12. K.E. Yoon, R.D. Noebe, and D.N. Seidman, "Effects of Rhenium Addition on the Temporal Evolution of the Nanostructure and Chemistry of a Model Ni-Cr-Al Superalloy. II: Analysis of the Coarsening Behavior," *Acta Materialia*, 55 (2007), 1145–1157
13. A.P. Miodownik and N. Saunders, *Applications of Thermodynamics in the Synthesis and Processing of Materials*, ed. Nash and Sundman, TMS (1995) 91
14. C.E. Campbell, W.J. Boettinger, and U.R. Kattner, *Acta Materialia*, 50 (2002) 775–792
15. B.T. Alexandrov and J.C. Lippold, "Single Sensor Differential Thermal Analysis of Phase Transformations and Structural Changes during Welding and Postweld Heat Treatment," *Welding in the World*, 51, 11/12 (2007) 48-59 (Doc. IIW 1843–07)
16. Private communication with B. Alexandrov, Ohio State University, 2008
17. T. Mura, *Micromechanics of Defects in Solids*, 2nd ed., Kluwer Academic, 1987

Transformations—Selected Works of G.B. Olson on Materials, Microstructure, and Design
C.E. Campbell, M.V. Manuel, and W. Xiong, editors

CHAPTER 24

Multiscale Ductile Fracture Integrating Tomographic Characterization and 3D Simulation*

S.C. O'Keeffe, S. Tang, A.M. Kopacz, J. Smith, and G.B. Olson, Northwestern University
D.J. Rowenhorst, Naval Research Laboratory
W.K. Liu, Northwestern University and King Abdulaziz University

Abstract

Ductile fracture in alloys is a multiscale process in which primary voids formed at micron-scale particles coalesce by a zig-zag pattern of shear localization driven by finer-scale micro-voiding at submicron-scale secondary particles. Employing the method of serial sectioning, unprecedented 3D microstructural reconstructions of steel crack-tip process zones are obtained and implemented into a large-scale simulation for ductile fracture analysis. A quantitative understanding of the microvoid sheeting mechanism and mixed-mode failure controlling the zig-zag fracture surface are presented using the modeling technique utilized herein. We define and quantify metrics of fracture by analyzing the crack opening distance, process zone size, zig-zag wavelength, and void growth ratios in the crack tip reconstructions. The quantitative agreement of these metrics between experiment and simulation supports a new and developing predictive structure/property theory to enable materials design.

Introduction

The US Materials Genome Initiative (Ref 1) acknowledges the new opportunity of integrated computational tools grounded in fundamental databases to greatly accelerate the full materials development cycle, as already demonstrated by a new generation of "designer steels" (Ref 2) taken to full flight qualification for aircraft. While the structure-property relations enabling computational materials design (Ref 3) are well developed for material strength, the equally important property of fracture toughness, defining resistance to material fracture, has not been well represented in terms of quantitative predictive capability (Ref 4). To address the need for such capabilities, an integrated approach of tomographic characterization and 3D fracture simulation has been undertaken in order to bring the maturity of the structure-property relations of toughness to the same level as that of strength. Ductile fracture in steel is a process consisting of nucleation, growth and coalescence of voids and microvoids on at least two microscopic levels (Ref 5). On the micron scale, primary voids (approximately 1–10 μm in diameter in this work) nucleate from primary inclusions that have cracked and/or debonded from the matrix. On the secondary (submicron) scale, microvoids nucleate from submicron precipitates, such as grain-refining carbides (approximately 100–200 nm in diameter in this work). The primary voids continue to grow

*S.C. O'Keeffe, S. Tang, A.M. Kopacz, J. Smith, D.J. Rowenhorst, G. Spanos, W.K. Liu, and G.B. Olson, "Multiscale Ductile Fracture Integrating Tomographic Characterization and 3D Simulation," *Acta Mater* (2015) 503–510

until a critical strain level in the intervoid ligaments, where a process of microvoid softening driven by the submicron secondary microvoids acts to link the primary voids to one another by a pattern of shear localization. Rice and Johnson (Ref 6) have shown that the critical crack-tip opening displacement for crack propagation directly scales with the spacing of primary inclusions (thus, $J_{IC} \propto X_0$). This has motivated the technology of "clean steels" in which the content of inclusions governing this spacing has been substantially reduced. The linking of the primary voids by a pattern of zig-zag shear localization corresponding to micro-void sheets along paths of maximum shear strain has been well established (Ref 7, 8), and the critical crack opening displacement has been shown to scale with the wavelength of this pattern (Ref 9); however, the quantitative role of the submicron secondary particles in this pattern of shear localization has remained elusive. Our goal is to create a comprehensive model which envelops microstructure features at multiple length scales to predict fracture toughness. Based on the aforementioned characteristics of ductile fracture, toughness and shear instability can be determined from an examination of microstructural quantities, such as inclusion spacing, critical primary void growth ratio (final void size at fracture/inclusion size upon which the void nucleated), zig-zag wavelength of crack morphology, and crack-tip opening displacement. Many of these quantities have traditionally been experimentally measured in only two dimensions (Ref 6, 7, 10–15) but now with advanced tomographic capabilities developed as part of a multi-institutional effort (Ref 16), we can obtain data that is more precise, of greater statistical volume, and of 3D character. This, in turn, supports 3D simulation with improved quantitative validation capabilities. Previous work by Tian et al., (Ref 17) utilized the multiresolution continuum theory with purely phenomenological models used for the constitutive behavior in order to obtain an accurate depiction of a crack tip during ductile fracture. While this work did capture fracture at two length scales and included strong multiple void interaction, intervoid ligament strain analysis, a very coarse crack wavelength measurement, and an effort to relate microstructure evolution to macroscale properties, it was unable to capture the finer details of the crack both experimentally and through simulation, and lacked an explanation of the underlying mechanisms that drive the evolution of the fracture path. Thus, the present work is a notable improvement directed at determining the underlying mechanisms which govern the fracture properties and evolution of hierarchically structured alloy materials. There are a number of key features which distinguish this work as an improvement over that which was previously conducted. Experimentally, the resolution of the digital data set is seven times greater in-plane and almost two times greater section-to-section, the volume of material captured in the reconstruction is almost three times greater, and the segmentation methods are vastly improved with the utilization of color imaging and development of more advanced algorithms for automatic detection of microstructural features. This results in a more accurate depiction of the microstructure that better represents the actual process zone and provides more data for implementation and comparison with simulation (e.g., we integrated eight times more primary particles into our simulation than that used in the previous work; Ref 17). For the simulation, we implemented the multiresolution theory and adopted two mechanics-based constitutive models for both the macro- and microzones, each of which was selected based on the mechanics known to occur during the ductile fracture process at their respective scales. A modified Gurson-Tvergaard-Needleman model is applied to the macrozone, accounting for primary void growth and shape change, and a modified Fleck-Hutchinson model is used to govern the elastic-plastic behavior in the microzone, accounting for shear localization in the microvoid region. The mechanistic nature of the models incorporated within the multiresolution theory for the current work, in combination with the vastly more extensive microstructural data available from the improved experimental techniques, yields a unified framework that is much better suited to quantify the fundamental driving forces behind the evolution of the ductile fracture process. The resulting implementation of the 3D microstructural data into the fully mechanics-based 3D multiscale model, combined with large-scale parallel computing capabilities, has revealed a comprehensive understanding of the fracture process including detailed micron-scale crack paths and crack branching that have not been previously captured by existing models. A summary of this implementation is given in the Supplementary Information (see Appendix A).

Experimental Reconstruction of a Crack

Studying ductile fracture processes (void nucleation, growth, and coalescence) under the conditions of engineering toughness measurement standards requires a specific mechanical test in which a plane-strain condition is achieved. This means that material must be constrained in the plane normal to the crack such that the investigated region is not subjected to shear lip deformation. We performed a modified-J_{IC} fracture toughness test on fatigue precracked miniature compact-tension specimens of a titanium-modified 4330 high-strength steel; the alloy composition and processing are described elsewhere (Ref 18). The test was halted near an estimated critical point corresponding to toughness measurement standards, to produce a specimen exhibiting the full stages of the ductile fracture process.

The crack was reconstructed in three dimensions through serial sectioning, the process of collecting a series of consecutive, closely spaced 2D layers of images and assembling them to form a 3D volume. Each thin layer or "slice" of material is imaged by light optical microscopy and is subsequently removed by metallographic polishing. This process of imaging and removal is repeated until over 200 sections of images are collected. To simultaneously acquire images with high resolution and with large fields of view, photomontaging is employed. Aided by an automatic stage on an optical microscope, an array of 8 × 8 images is collected for each slice. These overlapping images are carefully photostitched together to create one large image for each section, with an in-plane resolution of 0.133 μm per pixel and a total area of 800 × 1000 μm^2. The depth of material removal for each section is 1.5 μm, and is tracked by fiducial marks in the form of an array of Vickers hardness indents surrounding the crack-tip region. This slice thickness was chosen as an appropriate compromise between obtaining fine microstructure detail and obtaining a large volume of data in a timely manner and with a reasonable data size. The repetitive serial sectioning process required approximately 3–4 weeks of man hours to complete. Although this is cumbersome, the more time-consuming step lies in the image processing. The photostitching (in-plane image-to-image alignment), registration (section-to-section alignment), and segmentation (identification of each microstructural feature) steps, though seemingly straightforward, required automatic and semi-automatic methods. Further details can be found in the Supplementary Information.

The crack-tip data set was reconstructed in three dimensions utilizing IDL and Amira commercial software. IDL was used to visualize 3D reconstructions and perform the quantitative measurements presented here, and Amira was used to create the video data included with this publication (refer to Appendix A for the online Supplementary Data, Movie S1). A snapshot of the fully reconstructed crack tip is represented in Fig. 1, showing the distribution of primary inclusions and the topography of the associated fracture process.

Simulation of Ductile Fracture

Taking the digital data sets from the experiments for the initial microstructure configuration, the fracture process was simulated using

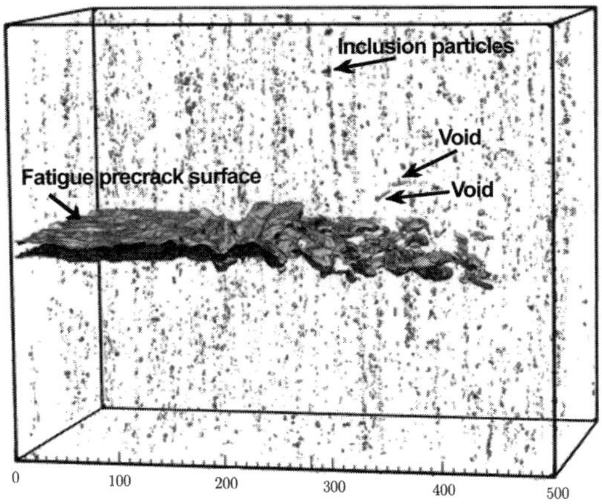

Fig. 1 Experimental 3D reconstruction of a steel crack tip via serial sectioning. The surfaces of the crack/voids and inclusion particles are shown. The fatigue precrack region is on the left side and the crack progresses rightward. At the far right are isolated voids. In between these regions are primary voids connected by shear bands that comprise the zig-zag pattern, indicative of shear localization and a microvoid sheeting mechanism. Scale is in μm. The total volume shown is 500 μm × 400 μm × 250 μm. (For more information regarding a color version of this figure, see the section, "Appendix A: Supplementary Data," in this chapter.)

a multiresolution framework in which the multiscale microstructure was modeled concurrently. Because we have a representative 3D microstructure from an experimental reconstruction, we can directly input the position, size, and even general morphology of the inclusions (primary particles). The primary particles found during collection of the 3D microstructure data were explicitly modeled in the simulation. It is important to note that for this simulation, primary voids were input in place of the primary particles, that is, inclusions were assumed to be fully debonded from the surrounding matrix such that particle-matrix debonding was not modeled. The additional morphological components of the microstructure were modeled implicitly via the mechanics-based constitutive models. The secondary submicron particles (titanium carbides) were modeled using statistical descriptors gathered from a submicron level focused serial sectioning experiment described elsewhere (Ref 19), and represented by a homogenized constitutive law, which is the macrozone constitutive law embedded within the multiresolution framework for this material, that accounts for microvoiding and shear localization (described in the Supplementary Information, see Appendix A). Because the number density of the secondary particles is so large (1.7×10^{17} m^{-3}), it is not feasible to explicitly represent them since the computational costs would be insurmountable. Instead, our new model is based on direct simulation of microvoid growth and coalescence underlying strain softening, and is calibrated to shear localization experiments on the same steel. The mechanics of grain boundary fracture are neglected for the simulation and thus the damage within the material is assumed to result only from the evolution of voids. The microzone constitutive law embedded in the multiresolution theory accounts for strain gradient effects and thus represents an interaction length scale for the secondary inclusions.

On the outer boundary of the simulated domain, we imposed the elastic K-fields corresponding to Mode I and plane-strain conditions (Ref 20). The displacements in the plane normal to the z-direction are set to zero to satisfy the plane-strain condition. It should be noted that the far-field solution where the boundary condition is applied is sufficiently far enough away from the fracture process zone, the area of interest and the location where the multiresolution theory is applied that the solution is assumed to have only elastic effects. Therefore, the boundary condition related to the K-field described by linear elastic fracture mechanics can be applied (Ref 20). The K-field was applied by varying the K_I value in Eq 35 (in the Supplementary Information, see Appendix A) from 0.0 to just above the reference K_{IC} fracture toughness value obtained experimentally, K_{ref}, at fracture of the modified 4330 steel in order to create the best possible comparison between the experimental and numerical fracture results. This variation was discretized such that it was fixed at each time step. Thus, the displacement, u_x and u_y, given by Eq 35 for a specific node on the boundary, located a distance R from the crack tip and an angle θ from the plane of the notch, can be determined for each discrete K_I value. Once the peak K_I value was reached, the specimen was relaxed to an unloaded state. A video of this simulation is included in the online Supplementary Data, Movie S2 (see Appendix A).

The multiresolution framework that is described in detail in the Supplementary Information was implemented with large-scale high-performance computing in the commercial software LS-DYNA. Our code handles the computing and memory requirements of a large domain size with up to 2.1×10^8 degrees of freedom. The simulations were carried out on Linux clusters at Northwestern University.

Integration of Experiment with Simulation

From qualitative and quantitative comparisons, the experimental data corroborate the simulation results, indicating that the models used to drive the simulation are accurately providing the first steps in the full prediction of microstructure evolution for the ductile fracture process. Optical micrographs from experiment showing slices of the 3D data set are displayed above comparable slices of the final unloaded state of the simulated crack tip in Fig. 2. We define a coordinate system with the x-axis representing the crack propagation direction, the y-axis representing the crack opening and loading direction, and the z-axis representing the specimen thickness direction. The fatigue precrack (shown extending from

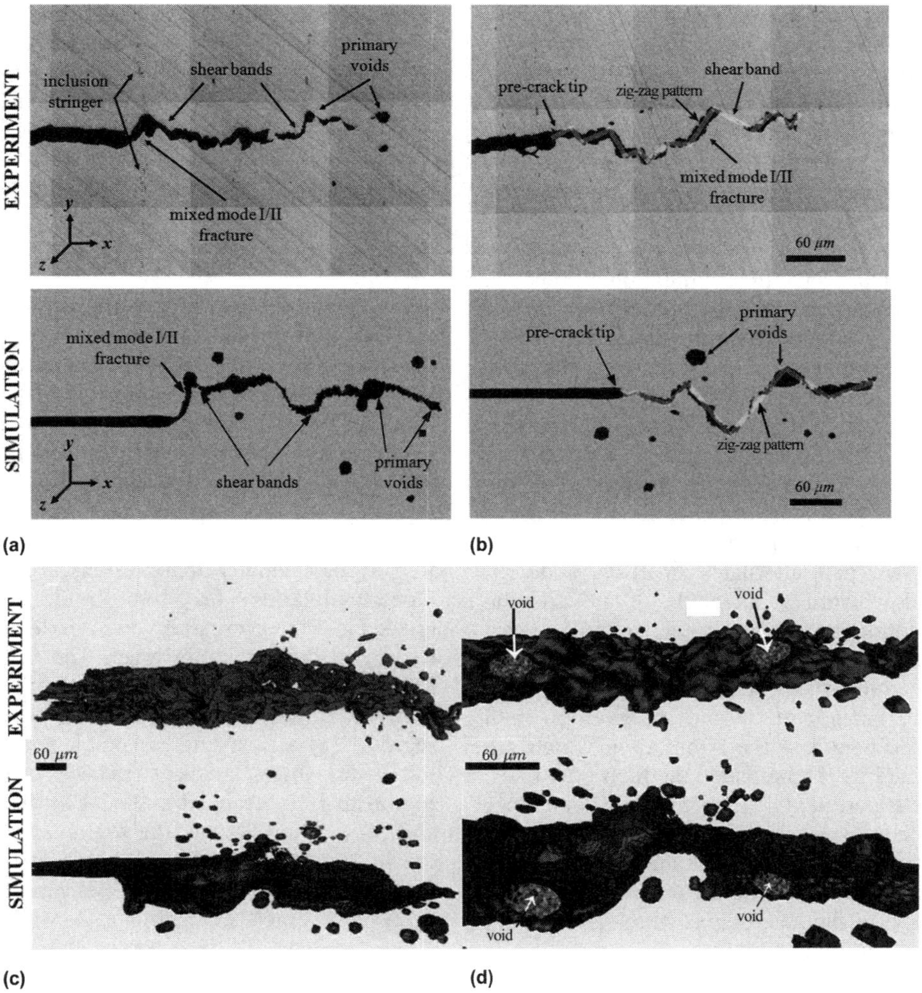

Fig. 2 (a,b) Two selected planar views of the reconstructed cracks from experiment (top) and simulation (bottom) with comparable crack morphology, including primary voids, shear bands and zig-zagging from mixed Mode I/II fracture. The zig-zag pattern is highlighted by the medium gray line in (b) with the experimental zig-zagging being finer than that captured in the simulation. (c) Both the experiment and simulation show evidence of a zig-zag fracture pattern in this zoomed in side view of the respective three-dimensionally reconstructed cracks. (d) Both the experiment and simulation show connection of the crack to a preexisting primary void, as evidenced by regions of the crack with faces of opposite curvatures identified by the arrows. Although qualitative comparisons can be visually observed in this figure, quantitative comparisons explained in the text further support these similarities and differences. (For more information regarding a color version of this figure, see the section, "Appendix A: Supplementary Data," in this chapter.)

the left side of the micrograph to the right in the positive x-direction) has a large crack opening displacement when compared to the remainder of the crack and is mostly flat in morphology. At the end of the precrack is a region of crack-tip blunting, where the stretched faces of the crack taper off, and where monotonic loading first initiated the ductile fracture process. At this point, the crack deviates ~45 degrees from the growing reference planar crack, indicative of a shear band. The whole crack has a zig-zag-like appearance, which was previously observed and described by Rogers (Ref 7), referring to the pattern formed when the primary voids are linked by shear bands of alternating directions and in a periodic fashion. In Fig. 2(a) (top), there are two obvious inclusion "stringers," corresponding to clusters of inclusion particles aligned with the y-axis as a result of prior deformation processing of the material. One of the stringers extends through the crack.

Interrupting this inclusion stringer line is a large primary void, likely nucleated from one of the inclusion stringer particles. It is notable that not all particles within the stringer have contributed to coalesced voids, suggesting that inclusion stringer spacing on the fracture plane is more relevant in defining and predicting fracture than the individual particle spacing.

The slice from the simulation shown in Fig. 2(a) (bottom) exhibits strikingly similar features. The flatness of the fatigue precrack region is idealized, but immediately at the original crack tip is the same abrupt deviation away from the main straight crack line. This portion of the crack makes a sharp detour upward at an angle greater than 60° to connect with the first primary void. Immediately after this to the right, the next primary void is connected as the crack path alternates directions and continues downward at an angle of 45°. As the crack continues to the right, more primary voids are connected through shear bands. For another comparison shown in Fig. 2(b), very similar zig-zag patterns are observed in both the experiment and the simulation, which are highlighted by the indicated medium gray lines, with the experimental zig-zag pattern being finer than in the simulation. Even though the nature of the zig-zag pattern cannot be simply or formally deduced from a continuum theory, large-scale simulation of the heterogeneous material shows a distinct periodic pattern.

Three-dimensional reconstructions of the crack tips for both experiment and simulation are compared in Fig. 2(c). Following the precrack region, the crack has a clear wave-like appearance, with fine ripples that correspond to the zig-zag patterns of the shear bands linking the primary voids. At the front of the process zones are single voids that appear somewhat equiaxed. In Fig. 2(d), reconstructions from both the experiment and the simulation contain oblate-shaped voids (indicated by the arrows), indicating more void growth in the transverse plane than in the loading direction. The multiple stages of the ductile fracture process—void nucleation, growth and coalescence—can all be observed within the reconstructed process zone. One main difference lies in the lack of fine corrugations in the simulated crack tip, which is a result of the homogenized representation of the secondary particle microvoiding rather than the explicit representation of them. Although this would be ideal, current computational limitations prevent the possibility of performing a simulation utilizing a volume of data this immense.

Metrics of Fracture

Crack Opening Displacement

For quantitative comparison, we plotted the crack opening displacement (COD), the distance between the top and bottom faces of the crack, in contour maps as shown in Fig. 3. The first requirement for this analysis was proper segmentation, or identification and separation of the top and bottom faces of the crack. On the experimental side, this was simply achieved by noting the sharp change in grayscale values of the pixels at the crack face boundaries. For the simulation, new crack faces were created when the local strain reached a critical strain level. The COD contour plots are quite similar for the experiment and the simulation. There are three distinct regions: (a) a fairly flat precrack zone on the left side, (b) a crack extension zone, the region of fully cracked material in the center, and (c) a process zone, the region of partially cracked material on the right side of the plot. Quantitatively, the boundary between the precrack and crack extension zone lies at the x-position where the average COD changes abruptly from 14 to 10 μm. The boundary between the crack extension zone and process zone is denoted as the x-position where the first COD values of 0 appear as one progresses along the crack in the positive x-direction. The end of the process zone is demarcated by the x-position in which no COD or surrounding voids are present. There is also a transition region between the precrack and crack extension regions, corresponding to the crack-tip blunting region as denoted by region D in Fig. 3 where the faces of the crack are curved and stretch towards each other (Ref 21). The average values for crack extension and process zone size are comparable for both the simulation and experimental results, as shown in Table 1. Process zone size as a function of crack extension is plotted in Fig. 3 on the right, showing that process zone size increases with crack extension. The simulation result lies on this experimentally derived curve,

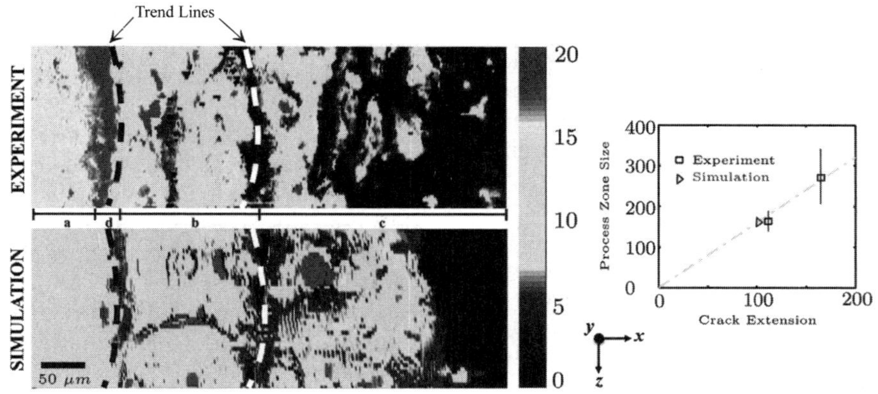

Fig. 3 These contour plots reveal crack opening displacement on the *xz*-plane in μm. Both experiment (top) and simulation (bottom) show three distinct regions of the crack: fatigue precrack (region a), full cracking (region b), and partial cracking (region c). Region d depicts the crack-tip location. Crack extension (Δa, μm) and process zone width (μm) are measured and correlated with one another on the right. The average process zone size (region of partially cracked material) is 160 and 164 μm for simulation and experiment, respectively. These have corresponding average crack extension zone sizes (region of fully cracked material) of 100 and 113 μm for simulation and experiment, respectively. The other experimental point is from a similar, but lower-resolution serial sectioning data set from Ref 17. The dashed trend line runs through the origin since it assumed that there is no process zone if there is no crack extension.

Table 1 Experiment versus simulation process zone measurements and input.

	Experiment	Simulation
Experiment and simulation fatigue precrack opening displacement (post-test, after unloading) (μm)	14	14.5
Critical crack-tip opening displacement (δ_{IC}) (μm)	N/A	12.43
Residual crack-tip opening displacement (δ^*_{IC}) (μm)	10.2	7.83
Crack extension zone width (Δa) (μm)	113	100
Process zone width (x-direction) (μm)	164	160
Primary zig-zag wavelengths (μm)	92, 50, 33	100, 54, 32
Primary zig-zag amplitude (μm)	3.4	5.0
Void growth ratio (transverse)		
COD	5.94	...
Adjusted COD	4.51	...
Stereomicroscopy	5.78	3.3
Void growth ratio (longitudinal)		
COD	4.55	...
Adjusted COD	2.64	...
Stereomicroscopy	2.43	2.18
Projected areal void density (on fracture surface) (m^{-2})	9.6×10^8	11.0×10^8
Effective critical strain		
Rice-Tracey	0.13–0.27	0.08–0.30
Direct measurement	N/A	0.05–0.35

corresponding to the degree of extension specified by engineering toughness measurement standards, and exemplifying further validation of the simulation with experiment.

Critical crack-tip opening displacement, δ_{IC}, is one measure of plane-strain fracture toughness (Ref 22, 23). Traditionally, the direct measurement of δ_{IC} is accomplished through 2D micrographic examination and involves some user discretion in delineating the point of crack initiation. Pardoen and Delannay (Ref 24) developed a method for defining a critical crack-tip opening distance, δ_{IC}, in the presence of crack-tip blunting. From experiments, only the unloaded value δ^*_{IC} is measured. From simulation, one can obtain both the total δ_{IC} (under load) and residual δ^*_{IC} (after unloading) (see Table 1). The ratio of these two parameters ($\delta^*_{IC}/\delta_{IC}$) was theoretically computed as 0.67 using the model of Ref 24. The ratio obtained from our detailed simulation is 0.63, in good agreement with the estimate in Ref 24. The quantification of this ratio thus allows more precise measurement of toughness from the measured residual COD in sectioned crack tips.

Shear Localization Wavelength

To quantitatively characterize the periodic shear instability behavior, we plot average COD versus x-position by taking the average of all COD values within a given x-coordinate and plotting this at the respective position on the x-axis (see Fig. 4(a). For convenience of observing the zig-zag pattern, the adjusted average COD was calculated by subtracting the background linear variation along the process zone from the average COD (also shown in Fig. 4a). From the adjusted curves, a sinusoidal pattern is clearly observed with a measurable amplitude and wavelength (see Table 1). A quantitative wavelength is deduced from the Fourier transform of the adjusted COD waveforms (see Fig. 4b). Three similar primary wavelengths exist, confirming that on the primary scale there are similarities in crack morphology between the experimental and simulated crack. The largest primary wavelength is in good agreement with previous 2D measurements in a similar steel of equivalent strength level (Ref 14).

Void Growth Ratio

Previous studies have examined the correlation of toughness with the extent of primary void growth. The void growth ratio is defined as R_v/R_0, where R_v is the final void radius and R_0 is the radius of the particle upon which the void nucleated. Here R_T/R_0 refers to the void growth ratio in the transverse direction or the plane of the crack, while R_L/R_0 refers to the void growth ratio in the longitudinal or loading direction. For measurements from the experimental COD plots, R_0 is taken as the average radius of the inclusions reconstructed in an equivalent, randomly selected region of undeformed material. For measurements from the simulation plots, R_i is taken as the initial implanted void radius. For the measurement of final void diameters on the fracture surface, a few analytical methods were utilized and compared for the experimental data; measurements were taken from the COD contour plot, the adjusted COD contour plot, and from scanning electron stereomicroscopy of the fracture surface (Ref 19). On the COD contour plot, voids are defined as local regions of large COD relative to surrounding material. Because this plot also accounts for overall crack opening (including the shear band contribution), an adjusted COD contour plot is formed that virtually closes the crack faces. By subtracting this overall crack opening, we can now observe the true dimple depth of each primary void

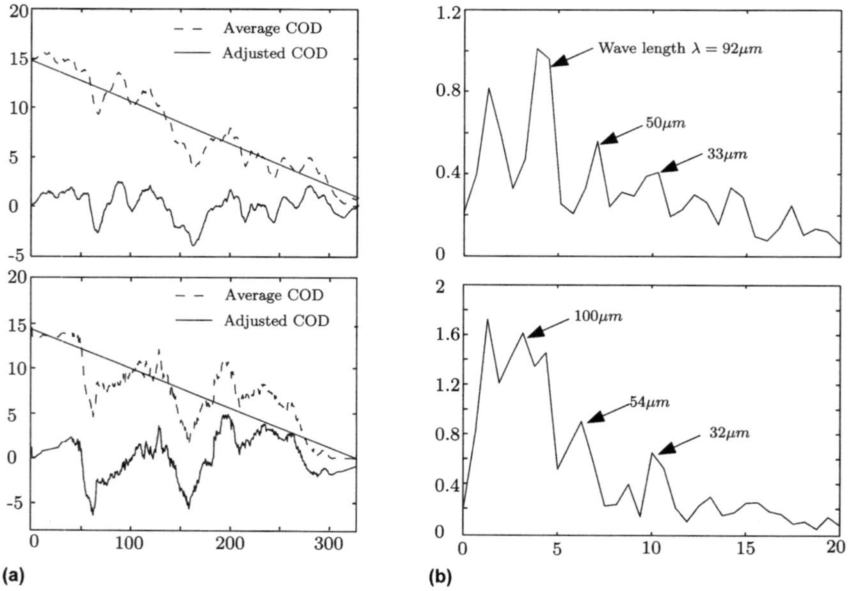

Fig. 4 Crack opening distance (μm) is plotted (a) as a function of x (distance along the crack, μm). Spectral density versus wavenumber is plotted (b) for both experiment (top) and simulation (bottom), which is deduced from the Fourier transformation to reveal three similar primary characteristic wavelengths. Both the experiment and simulation show evidence of a sinusoidal pattern.

prior to the coalescence by the shear bands. From the COD contour plots, a total of 80 voids were identified with most of the voids being somewhat equiaxed in the transverse xy-plane. From stereomicroscopy, a total of 28 voids were identified and measured. In addition, the inclusion diameters upon which the voids nucleated were measured and recorded for each void/inclusion pair. These are used in the void growth ratio calculations and are compared with those from the simulation since they utilize individual particle measurements rather than overall averages. Further details about the stereomicroscopic measurements can be found in Ref 19.

For the simulation, the voids were identified and measured by virtual fractographic examination. Because voids were explicitly input beforehand, the starting positions and dimensions of the voids were known exactly and void growth could be measured for each void individually. Due to increased mesh node density at the boundaries, void boundaries could be easily identified manually and from these, void diameters in the transverse and longitudinal direction were measured for 32 voids on the fracture surface. Only the voids modeled explicitly in the simulation were considered for the void growth ratio calculations, not those contained in the modified Gurson-Tvergaard-Needleman model.

The void growth ratios are presented in Table 1. Between the three analytical methods employed for the experiment, the void growth ratios are all comparable for the transverse void growth ratio, though the ratios from COD and stereomicroscopy show slightly more similarity. For the longitudinal void growth ratio, the COD-adjusted contour plot and stereomicroscopy ratios are very similar, confirming that the measurement from the unadjusted COD plot is overestimating the ratio, and that the COD-adjusted contour plot is adequately estimating the longitudinal ratio. The transverse and longitudinal void growth ratios from simulation are both slightly less than those of experiment, though they reflect the same void growth shape. Taking the ratio of the transverse ratio versus longitudinal ratio gives an average value of 2.05 for the experiment and 1.51 for the simulation. The voids are thus oblate ellipsoids with more void growth in the plane of the crack than in the loading direction. This result shows a close correspondence to that reported by Bray (Ref 12) in a study of five ultrahigh-strength steels of similar yield strengths. In that study, R_T/R_0 ranges from 3.7 to 7.55 with an average of 5.88, while one measurement is provided for R_L/R_0 of 2.88, again reflecting the oblate void shape.

The final void radii are plotted versus initial nucleation particle radius in Fig. 5. While final void size increases with initial size, they do not directly scale, indicating a smaller void growth ratio for larger inclusions. Between simulation and experiment, both display the same trends, though the slopes are different. The trend observed in the void growth ratio vs. initial nucleation particle size may be related to a strain gradient effect. Gao et al. (Ref 25) suggested that the higher strain gradient will increase the critical ligament strain for shear localization, thus allowing a larger void growth ratio prior to shear coalescence.

In plots of void growth ratio and final void size versus x-position (shown in the Supplementary Information), void shape change is observed in the various zones of fracture—full cracking, late process zone, and early process zone. While void shape is highly oblate at the

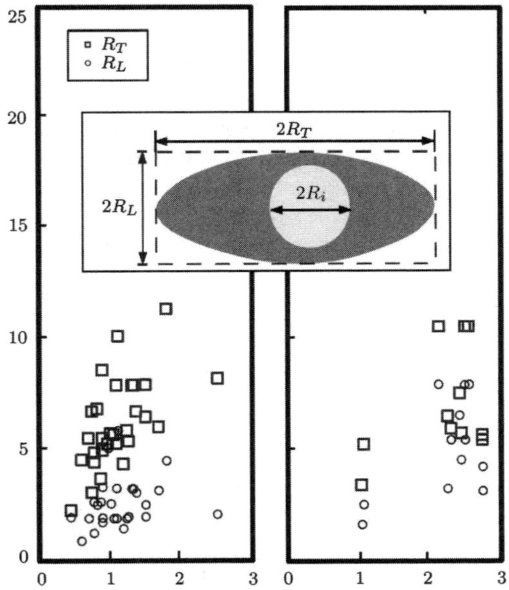

Fig. 5 Final void radius (μm) at fracture in the transverse and longitudinal directions versus initial nucleation particle radius (μm) is plotted for both experiment (left) and simulation (right). In general, final void radius increases as a function of the initial nucleation particle radius (larger particles result in larger voids). The transverse final void radius is also larger than the longitudinal radius, and also increases more rapidly as a function of initial nucleation particle radius than for the longitudinal radii.

earliest stages of simulation, the shape becomes closer to spherical at final fracture in both simulation and experiment. For the experimental void measurements, observed void growth ratios can be related to an estimated critical instability strain at the ligament, using the Rice-Tracey model relating growth of a single void to an associated remote plastic strain: $\varepsilon_c = ln(R_c/R_0)exp(-3\sigma_m/2\sigma_0)/C$, where $C = 0.283$ as derived in Ref 10 and $\sigma_m/\sigma_0 = 2.1$, a typical value at the onset of void coalescence in our simulations (Ref 26). The average effective critical strain from experimental void growth ratios is shown in Table 1. From simulation, individual critical strains for adjacent ligaments surrounding each void were measured directly, a unique capability of simulation relative to experiment. The average effective critical strains from simulation (Table 1) compare well with those estimated from the Rice-Tracey relationship.

In parallel with the detailed analysis presented here, the mechanistic insights gleaned from the synthesis of experiment and simulation motivated a thorough reanalysis (Ref 27) of available literature relating measured fracture toughness in high-strength steels to primary inclusion spacing and transverse final void growth ratios measured by frac-tography. This new analysis (Ref 27) showed a stronger sensitivity of toughness to the critical void growth ratio and the associated critical instability strain controlling it. Our present work has provided a new quantitative structure-property relation for ductile fracture toughness to support materials design, and has identified control of the finer-scale microvoid-induced shear localization as the highest priority for further toughness modeling improvement.

Conclusion

We have shown that the integration of advanced tomographic techniques with high-performance computing and multiscale simulation provide the foundation for predictive capabilities for determining ductile fracture behavior in high-strength alloys. While 3D microstructure data from experiment was explicitly input into the primary level of the multiresolution simulation, a homogenized constitutive law accounted for secondary-scale microvoiding and associated shear localization. The metrics of fracture that were measured and compared for both experiment and simulation included void growth ratios, shear localization wavelength and crack opening distance distributions. We also gathered unprecedented quantitative measurements of the crack extension zone (~100 μm), process zone (~160 μm), critical instability strain (0.05–0.35) and the ratio of crack-tip opening distance for unloaded to loaded specimens (0.63), all obtained from specimens loaded to the onset of fracture. This work supports new quantitative structure-property relations, enabling new directions in science-based materials design for high-strength alloys with greatly enhanced fracture toughness.

ACKNOWLEDGMENTS

The authors would like to acknowledge the financial support from the ONR/DARPA-sponsored Dynamic 3-Dimensional Digital Structures (‘D3D’) Program (Program Manager: Julie Christodoulou, Ph.D.) and the National Science Foundation Graduate Research Fellowship Program.

APPENDIX A: SUPPLEMENTARY DATA

Supplementary data associated with this article can be found, in the online version, at http://dx.doi.org/10.1016/j.actamat.2014.09.016.

REFERENCES

1. In: Materials Genome Initiative for Global Competitiveness, OSTP 2011, Washington, USA, 2011
2. C.J. Kuehmann, G.B. Olson. In: Models, Databases, and Simulation Tools Needed for the Realization of Integrated Computational Materials Engineering: Proceedings of the Symposium Held at Materials Science & Technology 2010, Houston, Texas, USA. ASM International, vol. 24, 2011
3. G.B. Olson, Science 288 (2000) 993
4. B. Boyce, S. Kramer, H. Fang, T. Cordova, M. Neilsen, K. Dion, et al., *Int. J. Fract.* 186 (2014) 5
5. P.F. Thomason, Ductile Fracture of Metals, Pergamon Press, UK, 1990
6. J.R. Rice, M.A. Johnson, McGraw-Hill, New York, 1970, p 641
7. H.C. Rogers, *AIME Trans.* 218 (1960) 498

8. J.G. Cowie, M. Azrin, G.B. Olson, *Metall. Mater. Trans. A* 20 (1989) 143
9. C.A. Berg, M.F. Kanninen, McGraw-Hill, New York, 1970, p. 171
10. J.R. Rice, D.M. Tracey, *J. Mech. Phys. Solids* 17 (1969) 201
11. W.M. Garrison, A.L. Wojcieszynski, *Mater. Sci. Eng.* A 464 (2007) 321
12. J.W. Bray, K.J. Handerhan, W.M. Garrison, A.W. Thompson, *Metall. Mater. Trans.* A 23 (1992) 485
13. A.W. Thompson, M.F. Ashby, *Scr. Metall.* (1984) 18
14. J.A. van den Avyle, Correlation of fractography, microstructure and fracture toughness behavior of high strength alloys Ph.D. thesis, Massachusetts Institute of Technology, 1975
15. M.F. Ashby, R. Ebeling, *AIME Met. Soc. Trans.* 236 (1966) 1396
16. J. Christodoulou, *J. Mater.* 61 (2009) 21
17. R. Tian, S. Chan, S. Tang, A.M. Kopacz, J.S. Wang, H.J. Jou, et al., *J. Mech. Phys. Solids* 58 (2010) 1681
18. C.L. Briant, E. Sylven, M.N. Shabrov, D.H. Sherman, L. Chuzhoy, A. Needleman, in: Mechanisms and Mechanics of Fracture, John Knott Symposium, 2002, p 169–174
19. S.C. Chan, Multi-scale tomographic analysis of ductile fracture in ultrahigh strength steels, (Ph.D. thesis), Northwestern University, 2012
20. M.F. Kanninen, C.L. Popelar, Advanced Fracture Mechanics, Oxford University Press, New York, 1985
21. R.M. McMeeking, *J. Eng. Mater. Technol.* (1977) 99
22. C.F. Shih, *J. Mech. Phys. Solids* 29 (1981) 305
23. C.F. Shih, Tables of Hutchinson-Rice-Rosengren singular field quantities, Division of Engineering, Brown University, 1983
24. T. Pardoen, F. Delannay, *Eng. Fract. Mech.* 65 (2000) 455
25. H. Gao, Y. Huang, W.D. Nix, J.W. Hutchinson, *J. Mech. Phys. Solids* 47 (1999) 1239
26. C. McVeigh, F. Vernerey, W.K. Liu, B. Moran, G.B. Olson, *J. Mech. Phys. Solids* 55 (2007) 225
27. J.S. Wang, G.B. Olson, in: 12th International Conference on Fracture, Ottawa, Ontario, Canada, 2009

SUPPLEMENTARY REFERENCES*

- F.J. Vernerey, C. McVeigh, W.K. Liu, B. Moran, D. Tewari, D.M. Parks, et al., JOM 58 (2006) 45
- F. Vernerey, W.K. Liu, B. Moran, J. Mech. Phys. Solids 55 (2007) 2603
- F.J. Vernerey, W.K. Liu, B. Moran, G.B. Olson, J. Mech. Phys. Solids 56 (2008) 1320
- C. McVeigh, W.K. Liu, Comput. *Methods. Appl. Mech. Eng.* 197 (2008) 3268
- C. McVeigh, W.K. Liu, *J. Mech. Phys. Solids* 57 (2009) 244
- C. McVeigh, W.K. Liu, *J. Mech. Phys. Solids* 58 (2010) 187
- S. Tang, A.M. Kopacz, S.C. O'Keeffe, G.B. Olson, W.K. Liu, *J. Mech. Phys. Solids* 61 (2013) 2108
- S. Tang, S.M. Greene, W.K. Liu, *Comput. Methods Appl. Mech. Eng.* 217 (2012) 213
- N. Aravas, *Int. J. Numer. Methods Eng.* 24 (1987) 1395
- K. Nahshon, J.W. Hutchinson, *Eur. J. Mech. A-Solids* 27 (2008) 1
- N.A. Fleck, J.W. Hutchinson, *J. Mech. Phys. Solids* 41 (1993) 1825

*Please see Appendix A for access to the supplementary information http://dx.doi.org/10.1016/j.actamat.2014.09.016.

Transformations—Selected Works of G.B. Olson on Materials, Microstructure, and Design
C.E. Campbell, M.V. Manuel, and W. Xiong, editors

CHAPTER 25

3D-Modeling of Subsurface Fatigue Crack Nucleation Potency of Primary Inclusions in Heat-Treated and Shot Peened Martensitic Gear Steels*

R. Prasannavenkatesan and D.L. McDowell, Georgia Institute of Technology
J. Zhang, University of Nevada
G.B. Olson, Northwestern University and QuesTek Innovations LLC
H.-J. Jou, QuesTek Innovations LLC

Abstract

A computational strategy is developed to characterize the driving force for fatigue crack nucleation at subsurface primary inclusions in carburized and shot peened C61 martensitic gear steels. Experimental investigation revealed minimum fatigue strength to be controlled by subsurface fatigue crack nucleation at inclusion clusters under cyclic bending. An algorithm is presented to simulate residual stress distribution induced through the shot peening process following carburization and tempering. A methodology is developed to analyze potency of fatigue crack nucleation at subsurface inclusions. Rate-independent 3D finite element analyses are performed to evaluate plastic deformation during processing and service. The specimen is subjected to reversed bending stress cycles with $R = 0.05$, representative of loading on a gear tooth. The matrix is modeled as an elastic-plastic material with pure nonlinear kinematic hardening. The inclusions are modeled as isotropic, linear elastic. Idealized inclusion geometries (ellipsoidal) are considered to study the fatigue crack nucleation potency at various subsurface depths. Three distinct types of second-phase particles (perfectly bonded, partially debonded, and cracked) are analyzed. Parametric studies quantify the effects of inclusion size, orientation, and clustering on subsurface crack nucleation in the high cycle fatigue (HCF) or very high cycle fatigue (VHCF) regimes. The nonlocal average values of maximum plastic shear strain amplitude and Fatemi-Socie (FS) parameter calculated in the proximity of the inclusions are considered as the primary driving force parameters for fatigue crack nucleation and microstructurally small crack growth. The simulations indicate a strong propensity for crack nucleation at subsurface depths in agreement with experiments in which fatigue cracks nucleated at inclusion clusters, still in the compressive residual stress field. It is observed that the gradient from the surface of residual stress distribution, bending stress, and carburized

*R. Prasannavenkatesan, J. Zhang, D.L. McDowell, G.B. Olson, and H-J. Jou, "3D Modeling of Subsurface Fatigue Crack Nucleation Potency of Primary Inclusions in Heat Treated and Shot Peened Martensitic Gear Steels," *International Journal of Fatigue*, 31, 7 (2009), 1176–1189

material properties play a pivotal role in fatigue crack nucleation and small crack growth at subsurface primary inclusions. The fatigue potency of inclusion clusters is greatly increased by prior interfacial damage during processing.

Introduction

Although much progress has been made, development of quantitative descriptions of potency of primary inclusions for fatigue crack nucleation and early stages of growth in high strength steels requires further understanding to support quantitative materials design. Carburization and shot peening are well established techniques for improving high cycle fatigue strength of steel components (Ref 1–7). The fact that bending fatigue strength of steel gear teeth is increased considerably by carburization and shot peening has been extensively utilized in industry (Ref 8–11). With carburization and shot peening, crack initiation (nucleation and early growth at the scale of the primary inclusion) is observed to shift from the surface toward the subsurface of the carburized gears; this is primarily due to presence of high compressive residual stress and comparatively high hardness at the surface. It is widely observed that cracks in unnotched specimens nucleate at non-metallic inclusions and other microstructure inhomogeneities. Such second-phase particles often serve as fatigue crack starters in a variety of metallic materials (Ref 12–18). Extensive experimental observations have been reported earlier on subsurface crack initiation (Ref 19–23).

Toyada et al. (Ref 21) performed a series of rotating-bending fatigue tests on Cr-Mo steels and estimated the influence of the inclusions on the fatigue strength by employing the projected $\sqrt{area}$ approach suggested by Murakami and Endo (Ref 13), with necessary modifications to incorporate the residual stress effects. They argued that there exists a critical inclusion size that initiates subsurface fatigue failure within the specimen. The semi-empirical Murakami and Endo model has successfully predicted upper and lower bounds on service lives for specimens containing inclusions. The fracture mechanics approach has also been widely applied for bending fatigue analysis of carburized steel components (Ref 24–26). Kim and co-workers (Ref 24) studied the effect of surface residual stress on fatigue life by analyzing the change in stress intensity factor due to residual stress. Sharma and co-workers (Ref 25) presented a detailed study of various factors influencing K_{1c} of case steels including alloying elements, retained austenite and carbon content. Lin et al. (Ref 26) investigated the general methodology of bending fatigue life prediction using strain-life and linear elastic fracture mechanics (LEFM) approaches for SAE 8620 steel. Both approaches were reported to predict the fatigue lives very well, although the strain-life approach proved to be more conservative in its predictions. Life prediction schemes based on nominal strain-life or LEFM are insufficient to fully reflect the mechanisms for crack nucleation at nonmetallic inclusions; as such, they cannot serve to quantify the microstructure-level details of threshold condition on cyclic stress amplitude to form a fatigue crack which can propagate away from the inclusion. More detailed analyses of the inclusion size and shape, interface characteristics, local residual stresses, contrast of elastic and thermal properties, and even surrounding microstructure of the matrix are necessary to explore this issue in more detail.

Nomenclature

σ_{ys}	uniaxial cyclic yield strength
σ_{ij}	stress tensor components
ε_{ij}	strain tensor components
S_{ij}	deviatoric stress tensor components
σ^h	hydrostatic stress
α_{ij}	back stress tensor components
$\bar{\sigma}$	Mises uniaxial effective stress
α'_{ij}	deviatoric back stress tensor components
α^h	hydrostatic part of the back stress tensor
$\dot{\underset{\sim}{\varepsilon}}^e$	elastic strain rate tensor
$\dot{\underset{\sim}{\varepsilon}}^p$	plastic strain rate tensor
$\dot{\bar{\varepsilon}}^p$	equivalent plastic strain rate
σ^{res}	residual stress magnitude
ε_a	applied strain amplitude
E	Young's modulus
ν	Poisson's ratio
γ	engineering shear strain
$\Delta\gamma^{p*}_{max}$	nonlocal maximum plastic shear strain range
$\Delta\Gamma$	Fatemi-Socie (FS) parameter
σ^n_{max}	maximum stress normal to plane of maximum shear strain range in a loading cycle

Mackaldener and co-workers (Ref 27) developed a two-dimensional plane strain gear tooth FE model and studied the stress state history at every point within the tooth during a single contact load cycle. The residual stress induced through case hardening was simulated by applying a thermal load cycle and the McDiarmid critical plane approach (Ref 28) was employed to predict fatigue crack initiation at various depths with necessary modifications to incorporate shot peening effects. They reported that the tensile residual stresses due to case hardening led to high risk of fatigue crack initiation in the interior of the tooth. Although the method was effective computationally and considered various aspects of surface treatment in the macroscopic analysis, it was not able to describe the root cause for subsurface crack initiation where the residual stresses are still compressive and the role of microstructural inhomogeneity in such failure scenarios. Borbely et al. (Ref 29) attempted to study the cause for subsurface failures initiating from pores in AZ91 magnesium alloys through 3D FE simulations. Results of local stresses and plastic strains in the vicinity of the pores showed a critical size for pores just beneath the surface with a maximum tendency to initiate cracks. Wang and co-workers (Ref 16) developed a deterministic model to correlate the total fatigue life and subsurface failure for six different ultrahigh-strength steels. Although the method was successful in obtaining a good correlation between inclusion size and fatigue life, it is insufficient to address many other factors which play a pivotal role, for example, particle clustering, orientation with respect to principal stresses, and residual stresses.

Residual stresses play a pivotal role in affecting crack initiation and propagation. Considerable effort has been made to accurately model the residual stresses and account for their influence on service life of the component (Ref 30–37). Some previous work has considered the effects of damaged particles or inclusions on fatigue crack formation and early growth in the micronotch root field of cracked or debonded inclusions (Ref 17, 18, 38–41). As mentioned above, a simple assumption that fatigue cracks form immediately at such inclusions as a precursor to application of LEFM ignores the complexities of the details of inclusion geometry, damage state, interface condition, and microstructure, all of which can play a role in the near-threshold regime of fatigue strength of these materials. This is important for HCF or very high cycle fatigue (VHCF) applications.

In support of a new generation of high-fidelity materials design tools, the aim of the present work is to develop a 3D modeling strategy for a comparative parametric study of the potency for nucleation of fatigue cracks at damaged primary inclusions in carburized and shot peened martensitic gear steels. Specific attention is devoted to the case of Al_2O_3 inclusions in carburized and shot peened C61 alloy gear steel (Ref 42). A new algorithm is presented to model representative residual stresses induced via plastic deformation due to shot peening. Case hardening (carburization and tempering) is considered by varying the material properties from surface to core. To determine quantitative relationships between local plastic strains and far-field loading, representative 3D finite element meshes are constructed which contain inclusions surrounded by carburized matrix. Both cracked and debonded inclusions are considered, consistent with experimental findings. We focus on processes of crack formation and early growth in the matrix under the influence of the strain concentration of the inclusion notch root field, and do not include the fracture mechanics problem of physically small and long crack growth away from the particle. The former (nucleation and early growth) establish the crack origin and often a majority of the fatigue life in HCF and VHCF. We define nucleation as the formation of a crack within the matrix in the region of highest cyclic plastic strain localization near the matrix-inclusion interface; this is considered heuristically as a zone of subcritical crack formation and growth to a stable nucleus that would be amenable to propagation analysis via fracture mechanics.

Relative fatigue crack nucleation potency at various case depths is compared with experimental findings regarding subsurface depths at which cracks form. Nonlocal average values of plastic shear strain range (Ref 18, 38, 41) and the Fatemi-Socie parameter (Ref 43) are considered as the candidate driving forces for fatigue crack nucleation. All simulations were performed using the ABAQUS (Ref 44) finite element code.

Experimental Observations of Fatigue in Case-Hardened Steel

The material of investigation is commercial C61 high-performance gear steel subjected to carburization, tempering, and shot peening. This is a martensitic steel with predominantly tempered lath martensite microstructure as confirmed by x-ray diffraction. The experimentally measured profiles of microhardness after heat treatment and residual stress distribution after shot peening are shown in Fig. 1 (a) and 1(b), respectively (Ref 1). Microhardness measurements were made using Buehler Micromet II Micro Hardness Tester following ASTM standard E92 (Ref 45). Prior to measurement, specimens were mounted in Bakelite and then ground and polished up to 1 μm finish. The procedure to measure the residual stress profile following shot peening is explained elsewhere (Ref 46). With the increase of case depth, the carbon content decreases, as does the microhardness. As measured by x-ray diffraction, initial compressive residual stress exists after shot peening within a thickness of 0.5 mm in the surface layer and reaches a maximum value of −1400 MPa at a depth of around 75–100 μm. Under reversed bending stress with $R = 0.05$, it is observed that fatigue cracks defining minimum properties form at a depth of approximately 200–270 μm below surface (Fig. 2a) in spur gear teeth. Similar observations on subsurface fatigue crack initiation have been reported earlier by Shiozawa and Lu (Ref 23). Scanning electron microscopy (SEM) was employed to characterize the fracture surface of failed spur gear teeth as well as the topography and features of various spur gear surface conditions. A Hitachi S-3500 microscope with tungsten hairpin filament was used with a 20 kV electron beam while the specimen chamber maintained a vacuum level of 10^{-4} torr. A secondary electron detector was used to image topographic features while a backscatter electron detector was used to achieve chemical contrast of inclusion phases located on fracture surfaces. Qualitative chemical analysis and mapping of inclusion phases was performed using a PGT energy dispersive x-ray analyzer. The site of crack nucleation is not in the tensile residual stress region. Analysis of the fracture surface and its mating surface indicates that the cracks are formed at clusters of inclusions. A single row of Al_2O_3 inclusions (particle 1, 2, 3, and 4) on the tooth fracture surface and the corresponding location of cavities on the mating surface are shown for clarity in Fig. 2(b) (Ref 1). The large particle (#3), about 10 μm in size, could be the favorable site for fatigue crack nucleation (Ref 13). Debonding of particles was observed, with uncracked particles on the fracture surface and corresponding cavities on the mating surface. Inclusion clustering observed at the initiation site indicates a potential role for inclusion interaction. Furthermore, the variation in material properties, distribution of residual stress, and magnitude and gradient of applied stress affect the fatigue crack nucleation process in the subsurface in a cooperative manner. The aforementioned brief introduction of the essential experimental results on microhardness, XRD residual stress measurements, fatigue tests, and SEM micrographs of fracture surface should suffice to convey the problem we intend to address.

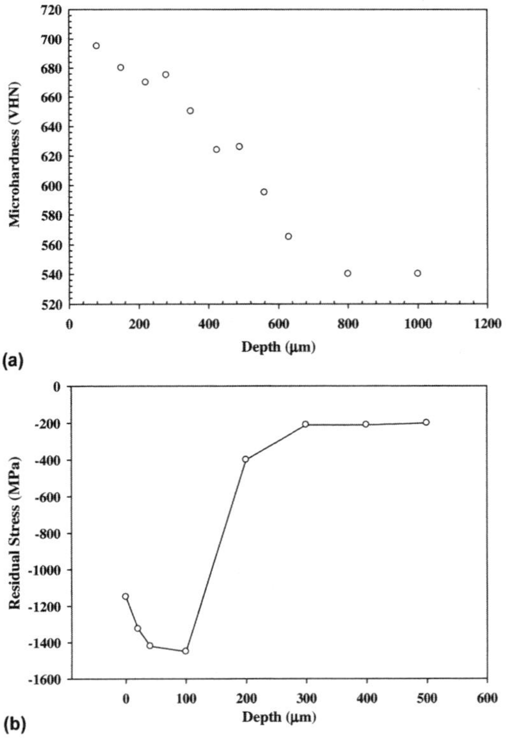

Fig. 1 Measured hardness and residual stress profiles (Ref 1): (a) variation of Vickers microhardness after heat treatment (carburization and tempering) with depth and (b) variation of residual stress with depth after shot peening.

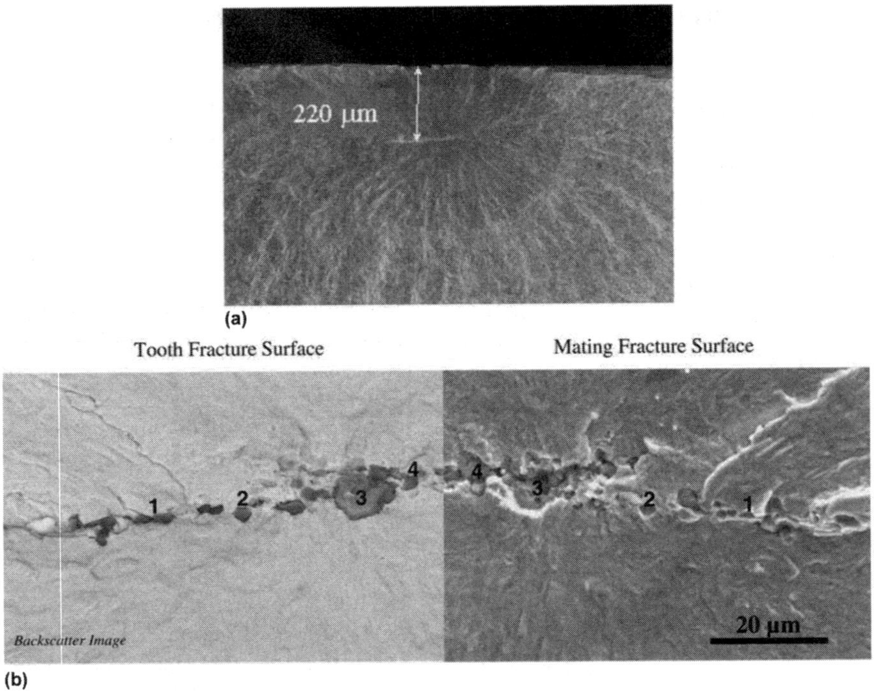

Fig. 2 (a) SEM micrograph of fracture surface indicating the depth of fatigue crack origin and (b) backscatter SEM image of crack initiation zone showing characteristic inclusion sizes, clustering and debonding of inclusion-matrix interfaces. Source: Ref 1

Characterization of Mechanical Properties of Matrix and Inclusions

Three-dimensional FE analyses were performed to understand the mechanisms of fatigue crack nucleation within the tempered martensite matrix. Detailed description of heat treatment, surface treatment, and composition of the material is presented by Tiemens (Ref 1, 2). The inclusions were assumed to be isotropic linear elastic with Young's modulus E = 400 GPa and Poisson's ratio ν = 0.2. The matrix was modeled as a rate-independent elastic-plastic material with a nonlinear kinematic hardening law cast in a format of hardening with dynamic recovery. The variation of carbon content from the surface to core following carburization has significant influence on local mechanical properties of the material. Elastic modulus, yield strength, and work hardening rate depend on carbon content. Donzella et al. (Ref 47) and Pedersen et al. (Ref 48) presented an empirical relation between yield strength and microhardness for case-hardened steel. However, for carbon steel subjected to cyclic loading, softening occurs and thus the yield strength may be scaled down by 15–35% from the initial value (Ref 49). Considering the experimental observations and applying the empirical relations, the material parameters of C61 steel were estimated for purposes of FE analyses. It is acknowledged that an accurate measurement of material property gradients in the carburized layer is tedious; in view of the desire to account for the effect of carburization we adopt the aforementioned strategy. In this work, we present some primary results and focus more on the development of the methodology. The variations of material parameters with depth adopted in the current investigation are listed in Table 1. Figure 3 shows the simulated monotonic tension stress-strain curves at various depths for the material considered.

General Outline of Elasto-Plastic Framework

The rate-independent plasticity model with combined kinematic-isotropic hardening for the matrix is based on the simple Mises yield surface:

$$F = f\left(\underset{\sim}{\sigma} - \underset{\sim}{\alpha}\right) - \sigma_{ys}^{2} \qquad \text{(Eq 1)}$$

Table 1 Variation of material properties with the depth from surface.

Depth (m)	*E* (MPa)	$_{ys}$ (MPa)	*c* (MPa)	*r*
0 (surface)	193,000	998	112,000	200
110	193,550	970	112,428	200
150	193,750	942	113,083	200
170	193,850	927	113,517	200
190	193,950	912	114,018	200
200	194,000	904	114,292	200
220	194,100	890	114,883	200
250	194,250	867	115,872	200
275	194,375	860	116,780	200
300	194,500	850	117,756	200
325	194,625	840	118,790	200
350	194,750	830	119,875	200
1000 (core)	198,000	800	120,000	200

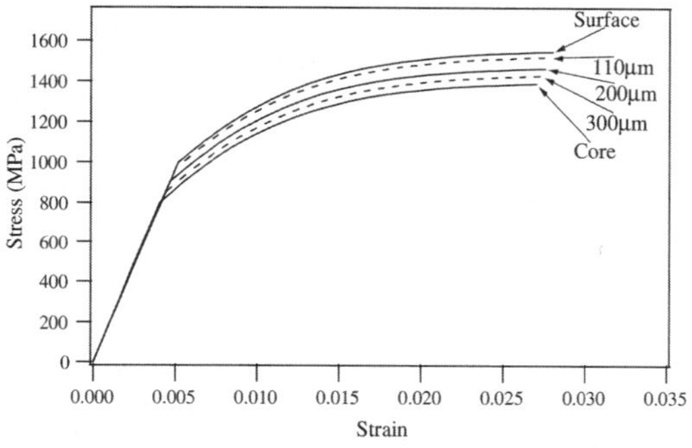

Fig. 3 Simulated monotonic tension stress-strain curves as a function of depth from surface. Depth of core is 1.0 mm below surface.

with $F = 0$ during plastic flow. Here, $\underset{\sim}{\sigma}$ and $\underset{\sim}{\alpha}$ are the stress and back stress tensors, respectively, and σ_{ys} is the uniaxial yield strength. The function f in Eq 1 is defined by:

$$f\left(\underset{\sim}{\sigma} - \underset{\sim}{\alpha}\right) = \frac{3}{2}\left(\underset{\sim}{S} - \underset{\sim}{\alpha'}\right) : \left(\underset{\sim}{S} - \underset{\sim}{\alpha'}\right) = \frac{3}{2}\left(S_{ij} - \alpha'_{ij}\right)\left(S_{ij} - \alpha'_{ij}\right) \quad \text{(Eq 2)}$$

The associative plastic flow rule is given by:

$$\underset{\sim}{\dot{\varepsilon}^p} = \sqrt{\frac{3}{2}}\dot{\bar{\varepsilon}}^p \frac{\partial F}{\partial \underset{\sim}{\sigma}} \left\| \frac{\partial F}{\partial \underset{\sim}{\sigma}} \right\|^{-1} = \sqrt{\frac{3}{2}}\dot{\bar{\varepsilon}}^p \underset{\sim}{N} = \sqrt{\frac{3}{2}}\dot{\bar{\varepsilon}}^p \frac{\left(S_{ij} - \alpha'_{ij}\right)}{\left\|\left(S_{ij} - \alpha'_{ij}\right)\right\|} \quad \text{(Eq 3)}$$

where $\dot{\bar{\varepsilon}}^p$ is defined by:

$$\dot{\bar{\varepsilon}}^p = \sqrt{\frac{2}{3}\underset{\sim}{\dot{\varepsilon}^p} : \underset{\sim}{\dot{\varepsilon}^p}} \quad \text{(Eq 4)}$$

The evolution equation for the back stress tensor $\underset{\sim}{\alpha}$ is expressed as (Ref 50):

$$\underset{\sim}{\dot{\alpha}} = \frac{c}{\sigma_{ys}}\left(\underset{\sim}{\sigma} - \underset{\sim}{\alpha}\right)\dot{\bar{\varepsilon}}^p - r\,\underset{\sim}{\alpha}\,\dot{\bar{\varepsilon}}^p \quad \text{(Eq 5)}$$

where c and r are material parameters. Here, c is the initial kinematic hardening modulus and r determines the rate of dynamic recovery of the back stress with increasing plastic deformation. Isotropic hardening is neglected in view of the desire to simulate cyclically stable response in parametric studies with pure kinematic hardening. The elastic response is given by:

$$\underset{\sim}{\sigma} = \underset{\sim}{C} : \underset{\sim}{\varepsilon}^e \quad \text{(Eq 6)}$$

where $\underset{\sim}{\varepsilon}^e$ denotes the elastic strain tensor and $\underset{\sim}{C}$ is the isotropic elastic stiffness tensor.

Imposition of Initial Residual Stress: Modeling the Effect of Shot Peening Process

During processing, heat treatment (carburization and tempering), and surface treatment (shot peening) introduce a lateral compressive residual stress field at and near the surface. Comparatively, the residual stress introduced by carburization and tempering (−100 to −300 MPa) is much less than that introduced through shot peening, especially close to the surface layer of the specimen (Ref 1, 51). A series of nonequilibrium thermo-mechanical processes, such as diffusion, temperature-induced phase transformation, dynamic elastic-plastic deformation, and deformation-induced phase transformation are potentially involved during carburization, tempering and shot peening. As a result, it is very complex to simulate the evolution of residual stress based on the physical mechanisms. Although mechanism-based FE simulation of residual stress is effective for two-dimensional calculations, three-dimensional simulations become computationally more demanding. Alternatively, a simplified scheme is followed to impose the residual stresses.

After carburization, the mechanical properties vary only with the depth. Kobayashi and co-workers (Ref 52) considered that the shot peening process effectively induces equi-biaxial residual compressive stresses within the specimen by virtue of plastic deformation. It is reasonable to assume that the residual stress in the steel matrix near the surface after shot peening is also equi-biaxial; of course, very near the surface the material may undergo tribological transformation (phase transformation, extensive shear, refinement of structure, etc.), but this zone is typically small (<30 μm or so) compared to the depths at which fatigue cracks form. Figure 4(a) shows a schematic of the shot peening process. Figure 4(b) shows an equivalent volume element of the plastic zone (subsurface affected by shot peening) divided into multiple "subsurface elements," each corresponding to a depth *d* below surface. The residual elastic strains in the subsurface correspond to a self-equilibrating residual stress field, with compression near the surface and a tensile stress state deeper into the subsurface. During shot peening the principal stress components are compressive. With these observations and idealizations, a methodology is developed to simulate residual stresses within the material induced by shot peening with respect to the surface coordinate system *XYZ* (Fig. 5). The procedure is summarized as:

1. Consider a homogeneous matrix subsurface element of the specimen at a depth *d* below surface, initially without inclusions, as shown in Fig. 5(a). The carburized material properties corresponding to the depth *d* is assigned to the subsurface element.
2. Imposing the boundary conditions shown in Fig. 5, the subsurface element is subjected to uniaxial compression in direction Y to strain levels calculated from the elasto-plastic constitutive relations, in connection with the following step (ε_{yy}^{load} in Fig. 5b). The total strain components (elastic plus plastic) in the *X* and *Z* directions are assumed as zero at each stage of the process, in accordance with overall constraints.
3. The compression is followed by unloading in direction Y (ε_{yy}^{final} in Fig. 5b) to elastic strain levels that match the measured macroscopic residual stress distribution in the material as a function of depth (Fig. 5c).
4. The resulting strain-time history is then applied as boundary conditions on the same subsurface element with embedded inclusions that are intact, debonded or cracked to assess the local residual stresses and strains in the vicinity of inclusions as initial conditions for subsequent fatigue simulations. This is repeated for each subsurface depth of interest.

When considering the embedded inclusion(s) in Step 4, the dimensions of the subsurface element are chosen to be sufficiently large that the inclusion experiences negligible boundary interaction effects (i.e., boundary conditions obtained from Steps 1–3 are adequate). There is an additional consideration that the gradient of the residual stress state cannot be too large compared to the thickness of the subsurface volume element selected in Step 1; this is dictated by inclusion size. A detailed derivation of the approach outlined in Steps 1–3 above appears in Appendix A. The following direction convention holds for all subsequent 3D calculations:

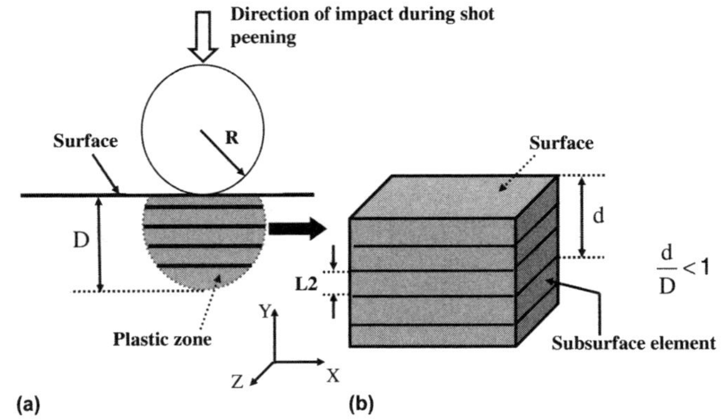

Fig. 4 (a) A schematic showing a metallic ball impacting the specimen surface during shot peening, inducing constrained plastic deformation to a depth D below surface and (b) an equivalent volume element of the plastic zone divided into several subsurface layers ("elements"), each corresponding to a depth d below the surface at which shot peening and fatigue simulations are performed.

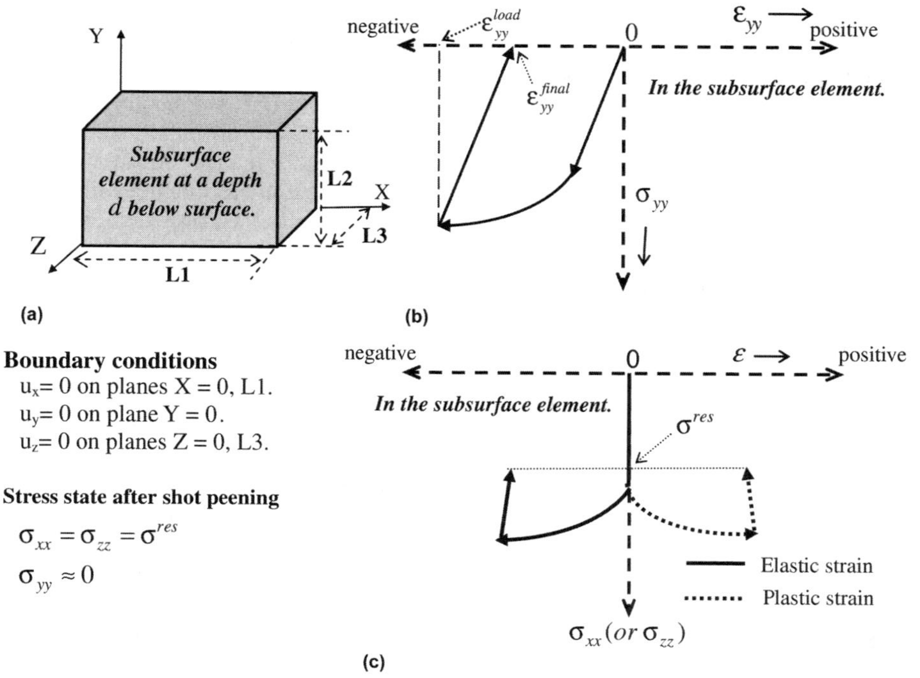

Fig. 5 Methodology to simulate the shot peening process: (a) schematic showing a subsurface element on which the strains are imposed, (b) σ_{yy} versus ε_{yy} response in the subsurface element during shot peening simulation, and (c) variation of elastic and plastic strain in X and Z directions in the subsurface element during shot peening.

- Direction of impact (shot peening) –*Y* direction
- Equibiaxial residual stress state –*X* and *Z* directions
- Cyclic loading-bending stress along *Z* direction

It is noted that damage (cracking or debonding) can occur to inclusions during primary forming or shot peening. We therefore pursue the strategy of simulating inclusions that are assumed to be already cracked or debonded at the beginning of the shot peening process as a means of assessing their influence on the local residual stress state and the driving forces for fatigue crack nucleation at the inclusion.

Fatigue Crack Nucleation Analysis under Cyclic Loading

Three-dimensional FE simulations were performed to parametrically explore the relative potency for fatigue crack nucleation at depths ranging from 75 μm to 300 μm below the surface. Idealized shapes of inclusions (ellipsoidal) were employed. Figure 6 shows a cross-section of a

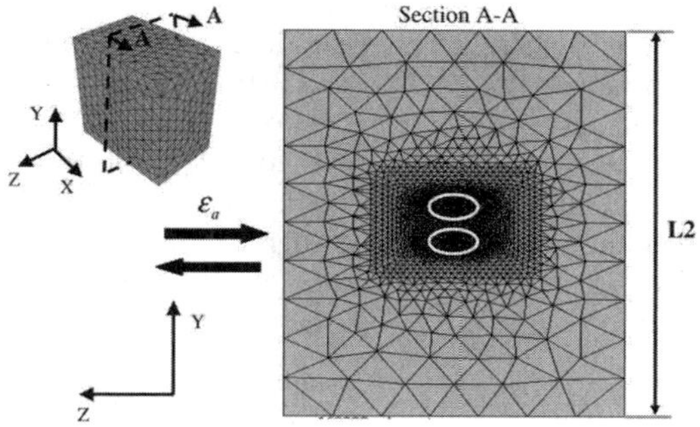

Fig. 6 Cross-section of FE mesh through the center of inclusion with refinement close to the inclusion.

3D FE mesh with closely spaced inclusions. A fine mesh is employed close to the inclusion(s) for detailed investigation of deformation and stress state in the proximity of the inclusions. All simulations were performed using 4-node 3D tetrahedral elements.

It was assumed that the inclusion sizes are small enough that the gradient of applied stress and residual stress over the scale of an inclusion is negligible. With this assumption, the stress state is imposed at discrete depths within the subsurface inclusion-matrix volume element and the stress and plastic strain components around the inclusion(s) are computed. Figure 7 shows the variation of peak applied uniaxial compressive strain, $\varepsilon_{yy}^{\text{load}}$, and strain after unloading, $\varepsilon_{yy}^{\text{final}}$, with depth for C61 gear steel during the shot peening simulation. The numerical estimate of strain levels to be imposed to mimic shot peening effect utilizes experimental residual stress value corresponding to the depth of interest as input. For pure matrix without inclusions, the numerical and experimental residual stress values coincide. While simulating the residual stress distribution corresponding to various case depths, we interpolate for the depth where the residual stresses were not measured, with the limited available experimental data. It is noted that when the interval between two neighboring measuring points is smaller, the confidence in the interpolated result is greater. Additionally, between 200 and 300 μm below the surface, the variation in the residual stress magnitude is relatively small, and the interpolation scheme is reasonable. The simulation that imposes residual stress is followed by three cycles of strain-controlled loading with $R_\varepsilon = 0.05$ (applied strain ratio). Figure 8 shows the variation with depth of the peak applied strain during cyclic bending. The remote applied stress pertains to HCF loading conditions and residual stress relaxation may not be significant; however, the local plasticity in the vicinity of primary inclusions can alter the residual stress state and thereby influence the driving force for fatigue crack nucleation. Driving forces for crack nucleation are calculated during the third load cycle. Two parameters, namely the nonlocal average maximum plastic shear strain range, $\Delta\gamma_{\max}^{p*}$, and Fatemi-Socie (FS) parameter, $\Delta\Gamma$, are considered as possible driving force parameters for fatigue crack nucleation at various depths. The latter parameter includes limited consideration of small crack propagation (Ref 53). The FS parameter considers the effect of the maximum normal stress $\sigma_n^{\max}$ perpendicular to the plane of maximum cyclic plastic shear strain, and has proven to be a very effective correlative parameter for multiaxial fatigue (Ref 54). It has been generalized by McDowell (Ref 55) and Shenoy et al. (Ref 56) to form the basis for microstructure-sensitive small crack growth laws, with the latter study using computational micromechanics to compute the FS parameter within individual grains in a polycrystalline Ni-base superalloy. Good correlations with experiments have been obtained in terms of fatigue lives of different metallic materials using the FS critical plane approach (Ref 57–62). The FS parameter is defined by:

$$\Delta\Gamma = \frac{\Delta\gamma_\theta^p}{2}\left(1 + k\frac{\sigma_n^{\max}}{\sigma_{ys}}\right) \qquad \text{(Eq 7)}$$

where $k = 0.6$ is a material parameter (Ref 43).

In local FE analyses, the micronotch root maximum plastic shear strain amplitude is mesh sensitive. Furthermore, fatigue cracks form over some finite region of intense shear, typically on the order of a micron in scale. Moreover, for cracks growing in a notch root field, a transition crack length on the order of 10–20% of the notch root radius is often invoked, below which it is assumed that the

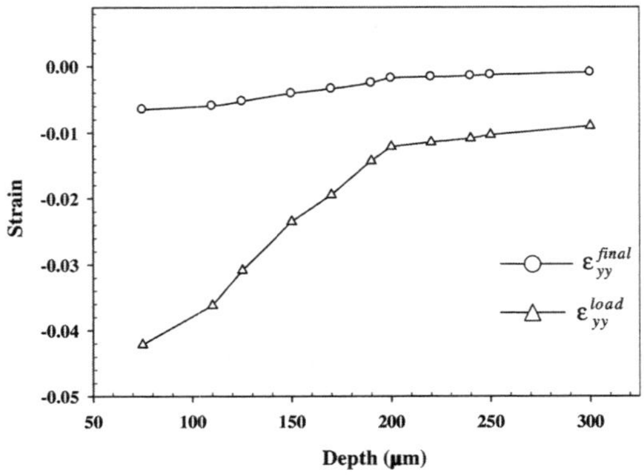

Fig. 7 Variation of the peak compressive strain $(\varepsilon_{yy}^{load})$ and the strain after unloading $(\varepsilon_{yy}^{final})$ along surface normal direction Y during shot peening simulation with depth.

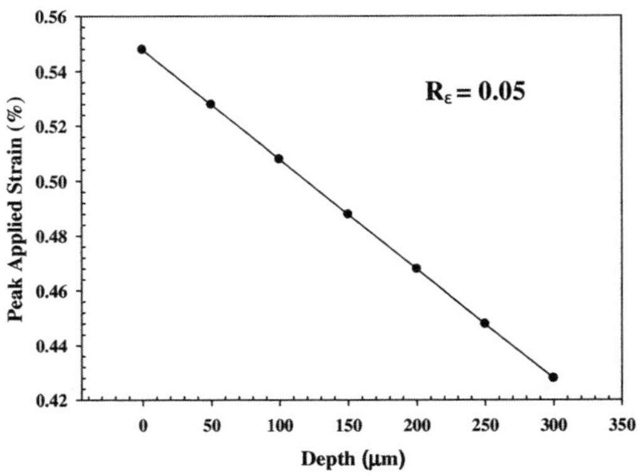

Fig. 8 Peak applied bending strain as a function of depth during cyclic loading.

local notch root field controls behavior and above which the notch becomes part of the crack. It is therefore necessary to introduce a nonlocal volume averaging procedure over integration points in the mesh to (a) effectively remove mesh dependence and (b) accord with the physical process zone for subcritical crack formation processes. Motivated by earlier work (Ref 18, 38), the maximum plastic shear strain amplitude is averaged over 1 μm^3 (unit cube), which corresponds to about 10% of the volume of an inclusion of minimum size in the present study, and in each simulation is selected as that particular volume of matrix material at the notch root that maximizes the nonlocal plastic shear strain amplitude. We employ this nonlocal volume for all the cases examined in this parametric computational study. Naturally, such a small volume is considered in terms of evaluating the probability to form a crack, and larger domains would be necessary to consider propagation resistance.

At every integration point lying within the nonlocal averaging region, the plastic shear strain γ_θ^p is computed on the plane-θ by projecting onto it the plastic strain tensor ε_{ij}^p at each integration point, that is:

$$\frac{\gamma_\theta^p}{2} = n_i \varepsilon_{ij}^p t_j \qquad \theta = 1 \ldots N \qquad \text{(Eq 8)}$$

where n_i and t_j are the unit normal and tangent vectors, respectively, on the plane-θ, and N is the number of discrete planes. The nonlocal average plastic shear strain is then calculated by averaging over the volume V of the nonlocal region, that is:

$$\gamma_\theta^{p*} = \frac{1}{V}\int_V \gamma_\theta^p dV \qquad \text{(Eq 9)}$$

The nonlocal cyclic plastic shear strain range for every plane is calculated during the third load cycle of the simulation. The nonlocal maximum cyclic plastic shear strain range, is defined by:

$$\Delta\gamma_{max}^{p*} = \max_\theta\left(\Delta\gamma_\theta^{p*}\right) \qquad \text{(Eq 10)}$$

Once the critical plane and corresponding nonlocal maximum plastic shear strain range are determined, $\Delta\Gamma$ is calculated using Eq 7 at every integration point on the critical plane and a volume average is performed in the nonlocal region similar to Eq 9; this effectively results in a nonlocal FS parameter. Alternatively, direct maximization of $\Delta\Gamma$ could be considered instead of first maximizing $\Delta\gamma_{max}^{p*}$. A simple local form of the Coffin-Manson law has been used to correlate the number of cycles for nucleation of fatigue cracks with the candidate driving forces (Ref 41, 63), but of course more complex analyses can be conceived that reflect lower scale details of microstructure and slip irreversibility.

Results and Discussion

Cracked Inclusion Analyses

Subsurface layers with cracked ellipsoidal inclusions, embedded in the elasto-plastic matrix, were subjected to boundary conditions during shot peening based on the analyses conducted without inclusions, as described earlier. Ellipsoidal inclusions were chosen to facilitate parametric studies; however, inclusion morphology can influence the local stresses and cyclic plastic strain distribution. Investigation of realistic particle shape and distribution will be performed in future. Each layer was then subjected to three cycles of strain-controlled cycling with amplitude appropriate for each depth, and an R ratio of 0.05. Simulations were also performed with intact particles and results showed negligibly small driving force in this study for all case depths below the surface. Similar observations were reported by Gall and co-workers (Ref 18) for A356 aluminum alloy. This may be due to the fact that the peak applied bending stress is less than $\leq 0.8\sigma_{ys}$ and hence the stress amplitude is well below the elastic limit of C61 gear steel. Also, localization of stresses and plastic strains around intact particles are not as intense as that observed near damaged particles. In addition to the aforementioned observations, it is important to note that fatigue crack nucleation in the high cycle fatigue regime is highly heterogeneous at the grain scale (Ref 63). Investigation of the potency of intact particles to nucleate fatigue cracks while operating close to a fatigue limit associated with nonpropagating cracks demands more detailed scale-dependent, discrete dislocation plasticity simulations; J_2 plasticity assumption may not be sufficient to reflect such failure scenarios. Accordingly, this ideal case study focuses on cracks nucleating from damaged particles (cracked and partially debonded inclusions) for which plastic strain localization in the matrix is of sufficient magnitude to merit treatment with simple J_2 plasticity to facilitate parametric comparisons.

Figure 9 illustrates the three-dimensional boundary conditions and orientation of a cracked particle with respect to the direction of applied bending stress and the surface. It was assumed that inclusions tend to crack when the major axis of the ellipsoidal particle is oriented parallel to the applied stress direction. Table 2 summarizes the list of cases considered while analyzing the effect of cracked particles on fatigue crack nucleation in the matrix at various subsurface depths. In Tables 2 and 3, 2a and 2b represent the major and minor axis dimensions of the ellipsoidal inclusion, respectively, 2c and 2d represent those of the neighboring inclusion, and L is the longitudinal end-to-end spacing between the inclusions, chosen to be 1 μm in this limited parametric study. For convenience, all cases and results presented in the subsequent sections will be addressed with the "case pointer" listed in Table 2. For example, Fig. 10(a) shows the equivalent plastic strain contour plot around the cracked inclusions located at 250 μm below surface, at the end of the third loading cycle for case R1. The study cases listed in Tables 1 and 2 were chosen in order to examine the factors that affect the severity of plastic strain localization at nonmetallic primary inclusions which include inclusion spacing, size, and orientation. The aforementioned factors are important in HCF (Ref 63). A frictionless surface-to-surface contact is used to model all debonded interfaces in this study.

Plastic strain intensification is observed to occur in the matrix near the intersection of the cracked particle with the matrix. The

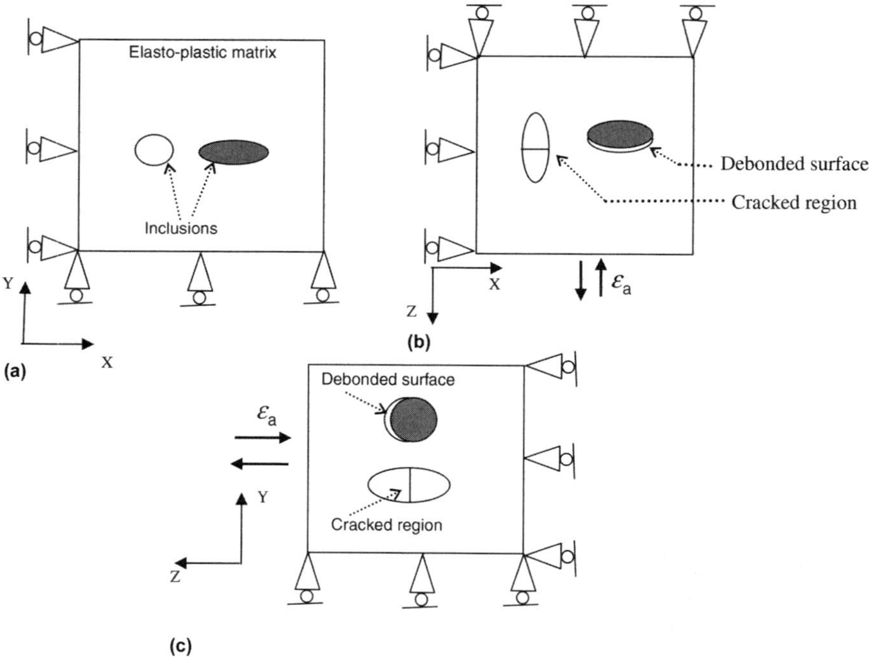

Fig. 9 Cross-sectional views of the 3D section with embedded inclusions, showing the debonded and cracked surfaces, boundary conditions for bending, and cyclic loading direction. Views on (a) *XY*, (b) *XZ*, and (c) *YZ* cutting planes through the center of the inclusion.

Table 2 List of cases analyzed in the study of cracked inclusions.

Case pointer	Number of inclusions	Shape of inclusion	2a (μm)	2b (μm)	2c (μm)	2d (μm)	L (μm)
P1	1	Ellipsoidal	5	2	...	...	...
Q1	2	Ellipsoidal	5	2	5	2	1
R1	2	Ellipsoidal	5	2	10	4	1

Table 3 List of cases analyzed in the study of partially debonded inclusions.

Case pointer	Number of inclusions	Shape of inclusion	2a (μm)	2b (μm)	2c (μm)	2d (μm)	L (μm)	ϕ_1	ϕ_2
A1	1	Ellipsoidal	5	2	...	...	...	0°	...
B1	2	Ellipsoidal	5	2	5	2	1	0°	0°
C1	2	Ellipsoidal	5	2	5	2	1	90°	90°
D1	2	Ellipsoidal	5	2	10	4	1	0°	0°
E1	2	Ellipsoidal	5	2	10	4	1	90°	90°

Note: ϕ_1 and ϕ_2 represent the orientation of inclusion 1 and inclusion 2, respectively).

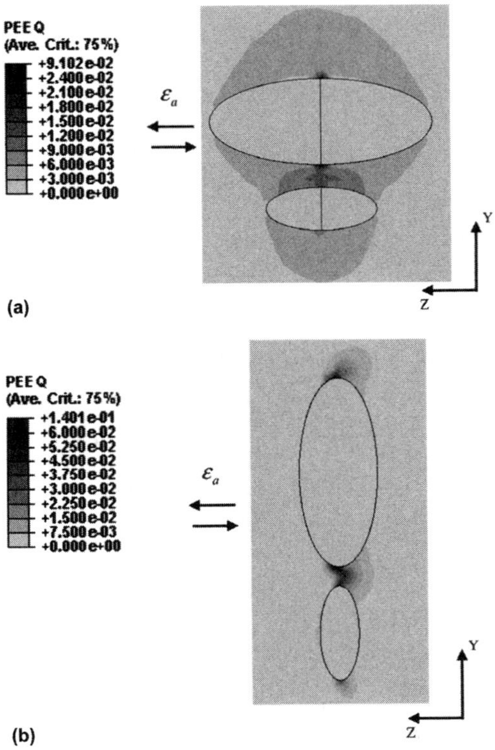

Fig. 10 Contour plot of the equivalent plastic strain at the end of third loading cycle at a depth of 250 μm below surface: (a) case R1 and (b) case E1.

severity of stresses and plastic strain localization was observed to depend on the size of the inclusion and inclusion spacing. The highest values of stresses and plastic strains were observed in case R1 as compared to cases P1 and Q1 in the vicinity of cracked particle. Stress intensification arising due to interaction of inclusions could magnify the crack driving force and thereby reduce the fatigue crack nucleation and early growth life relative to the case of a single inclusion. An example of a detailed investigation of the effect of particle shape, alignment, spacing and configuration on local stress distribution under cyclic loading can be found elsewhere (Ref 64).

Figure 11 shows the variation of the nonlocal maximum plastic shear strain amplitude, $\Delta\gamma^{p*}_{max}/2$, and nonlocal FS parameter, $\Delta\Gamma$, with depth for case P1. Interestingly, the maximum value of these parameters corresponds to a depth of about 250 μm below the surface, which falls within the range of depths observed in experiments for C61 gear steels. Note that this depth is still within the regime of initial compressive residual stress following shot peening. The driving forces are negligibly small at depths between 75 and 150 μm even though the applied strain amplitude is higher close to the surface. Shot peening suppresses the fatigue crack nucleation at the surface and shifts the crack nucleation site to subsurface depths. The critical depth depends on residual stress distribution, applied stress, and property gradients. Presence of other inhomogeneities such as pores and soft inclusions (sulfide particles) could alter the favorable site for crack nucleation depending on the intensity of localization around such inclusions.

The nonlocal maximum plastic shear strain amplitude and FS parameter were also calculated for cases with two cracked inclusions closely spaced (case Q1and R1); results are shown in Fig. 12 and 13, respectively. The trends are similar to that of the single cracked inclusion (case P1) but the values are an order of magnitude higher due to size and interaction of inclusions. Increase of the fatigue crack nucleation driving force with increase in inclusion size (case R1 compared to case Q1) supports the observation of Toyoda and coworkers (Ref 21) relating the inclusion size to threshold fatigue strength. The sensitivity of potency of fatigue crack nucleation to the size and number of particles (clustering) is most pronounced at lower stress levels than at stress levels above macroscopic yielding (Ref 41). Several other factors affect the HCF crack nucleation life, including inclusion shape and orientation (Ref 17, 40, 64). However, detailed investigation of all the factors influencing

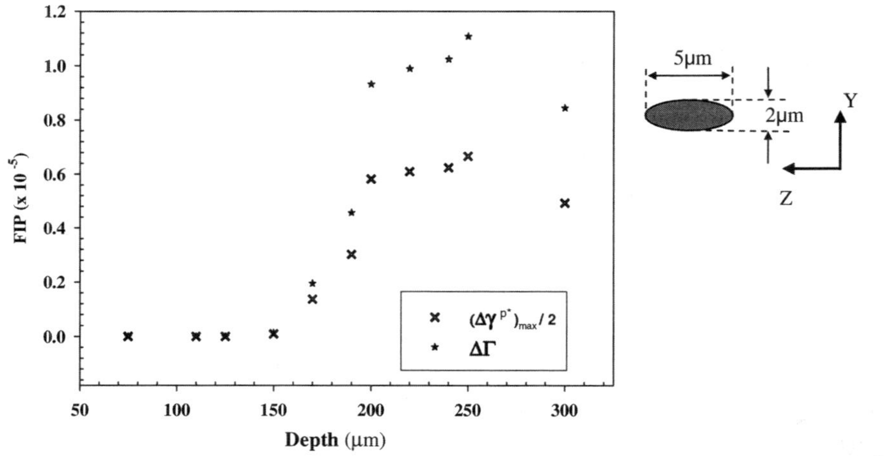

Fig. 11 Variation of nonlocal maximum cyclic plastic shear strain amplitude and FS parameter with depth for case P1.

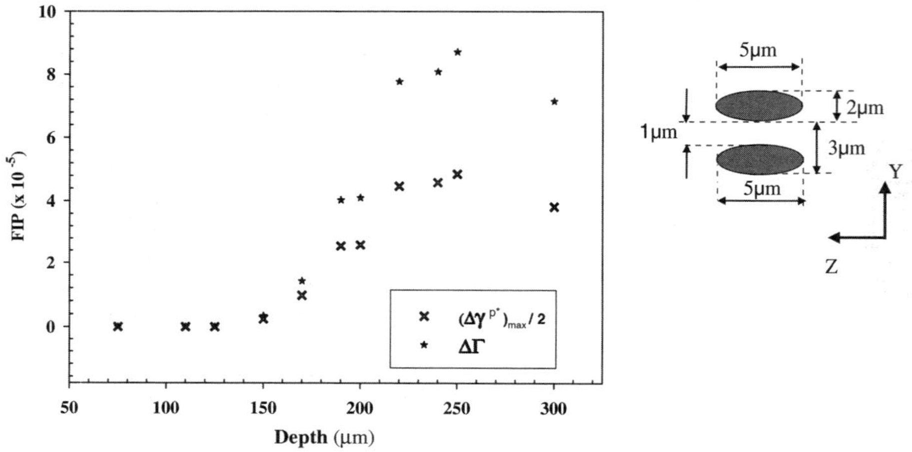

Fig. 12 Variation of nonlocal maximum cyclic plastic shear strain amplitude and FS parameter with depth for case Q1.

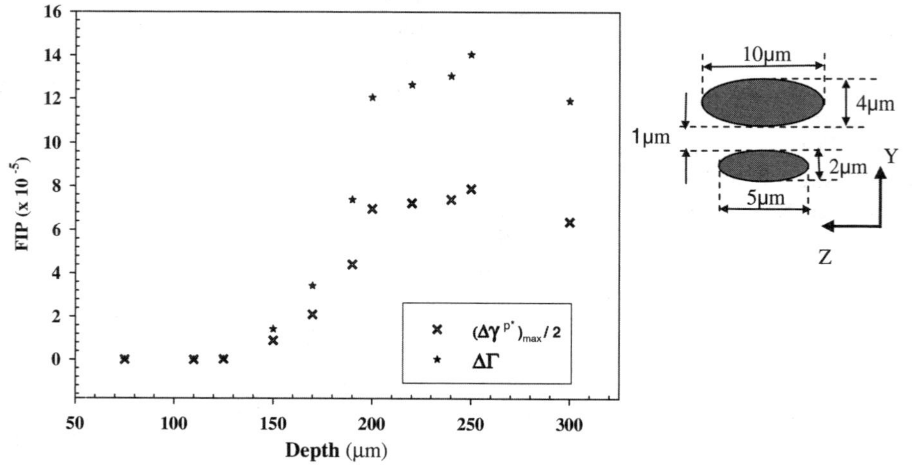

Fig. 13 Variation of nonlocal maximum cyclic plastic shear strain amplitude and FS parameter with depth for case R1.

subsurface fatigue crack nucleation in shot peened C61 steel is beyond the scope of the present parametric study.

Partially Debonded Inclusion Analyses

Three-dimensional FE analyses were performed involving partially debonded inclusions (ellipsoidal) embedded in an elasto-plastic matrix. McDowell et al. (Ref 17) showed for A356-T6 Al alloy that partially debonded inclusions offer the most conservative (i.e., severe) scenario for assessment of fatigue potency. Figure 14 describes the convention followed to represent ellipsoidal inclusion orientation in XY plane denoted by angle φ. In all cases, the ellipsoidal inclusions were aligned such that the major axis of the inclusion was perpendicular to the direction of applied bending stress. Figure 9 shows cross-sectional views of the 3D domain, elaborating on the boundary conditions, loading direction, and debonded surface of the inclusion considered in the parametric study.

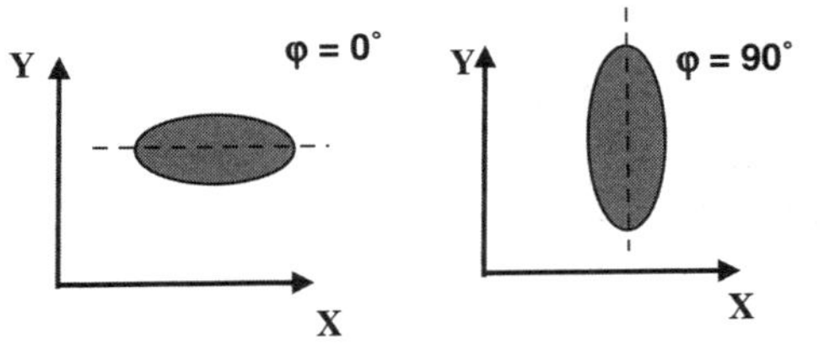

Fig. 14 Convention followed to represent the orientation of inclusion. Angle φ is measured with respect to *X* axis (axis of applied bending strain along *Z* direction). Direction *Y* is normal to the surface.

Each subsurface volume element was subjected to similar loading conditions as described in cracked inclusion analyses. Table 3 summarizes the different cases of a partially debonded inclusion analyzed in the parametric study. All the cases presented in the following plots and discussion will be addressed with the "case pointer" listed in Table 3. For example, Fig. 10(b) shows the equivalent plastic strain contour plot for case E1 at the end of the third loading cycle. Intensified plastic strain at the intersection of the debonded and bonded surfaces of the matrix can be observed in the contour plot. The intensification is due to the crack-like defect present at the tip of the debond seam coupled with the constraint and contact of embedded inclusions. It was mentioned by Gall et al. (Ref 18) that high stresses around cracked inclusions can facilitate debonding at the particle matrix interface. It was also shown that once a cracked particle begins to debond, the local plastic strains are intensified in the matrix with the same order of magnitude as a crack-free but partially debonded inclusion. Similar to cracked inclusion analyses, the severity of stress and plastic strain localization was influenced by particle size and spacing. The magnitudes of plastic strains and stresses in the matrix encompassing the partially debonded alumina inclusion were highest in cases D1 and E1 compared to all other cases listed in Table 3.

Figure 15 shows the variation of nonlocal maximum plastic shear strain amplitude and nonlocal FS parameter with depth for case A1. Similar to the results obtained in the

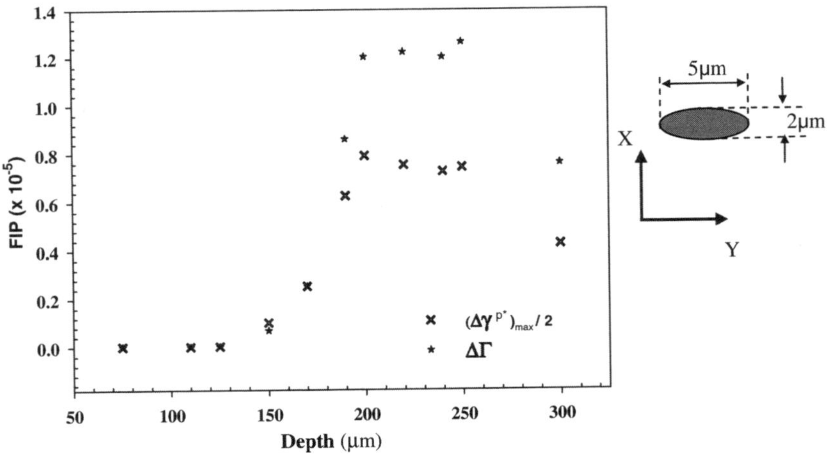

Fig. 15 Variation of nonlocal maximum cyclic plastic shear strain amplitude and FS parameter with depth for case A1.

cracked inclusion analyses, the critical depth for fatigue crack nucleation trends toward 250 μm. The debonded inclusion is observed to facilitate higher plastic strains under monotonic loading, which in turn leads to higher local stress states. Consequently, reversed yielding occurs more readily near the debonded inclusion. Consistent results were obtained for all the cases listed in Table 3 (see Fig. 16–19) where the predicted critical depth for subsurface fatigue crack nucleation was in accordance with experimental observations. Size and spacing of partially debonded inclusions are observed to play a significant role in fatigue crack nucleation potency at subsurface depths. Case E1 is observed to be most detrimental under cyclic loading among all cases analyzed in this parametric study. This is due to the presence of a large partially debonded inclusion coupled with intensification arising due to interaction with neighboring inclusion. The variations arising due to spatial alignment of inclusions in a cluster (comparing case D1 and case E1) will be addressed later. The predictions and observations reported in this parametric study emphasize the importance of inclusion-matrix interface character on fatigue crack nucleation potency. Comparing the single cracked particle results (case P1) in Fig. 11 and single partially debonded inclusion results (case A1) in Fig. 15 for which the inclusion size was equal, the driving force for fatigue crack nucleation is relatively higher in presence of partially debonded inclusions.

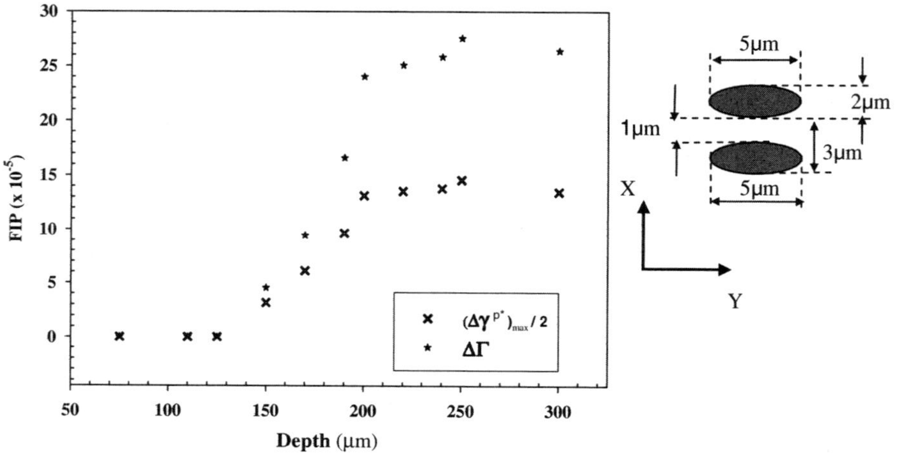

Fig. 16 Variation of nonlocal maximum cyclic plastic shear strain amplitude and FS parameter with depth for case B1.

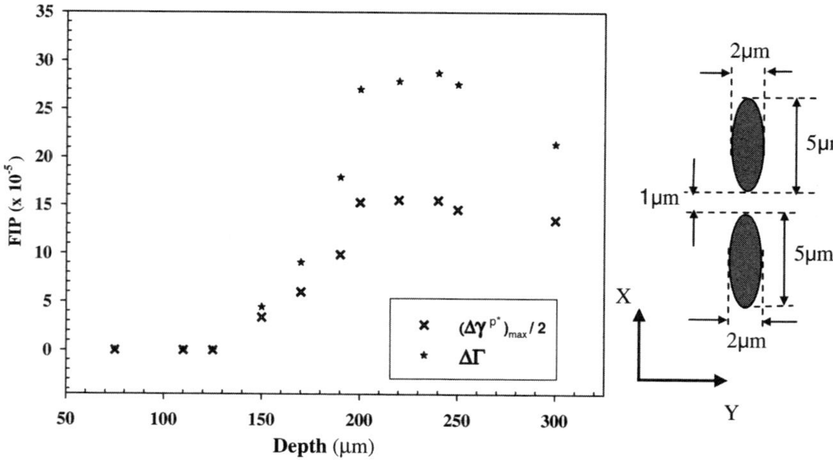

Fig. 17 Variation of nonlocal maximum cyclic plastic shear strain amplitude and FS parameter with depth for case C1.

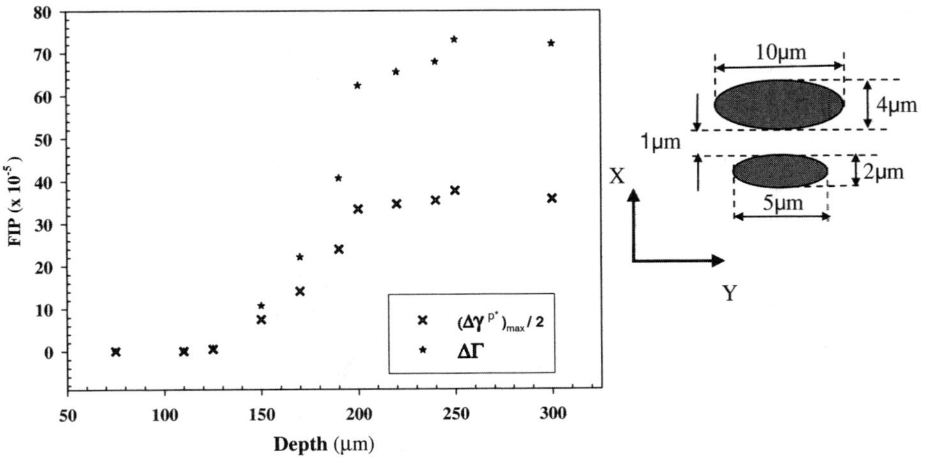

Fig. 18 Variation of nonlocal maximum cyclic plastic shear strain amplitude and FS parameter with depth for case D1.

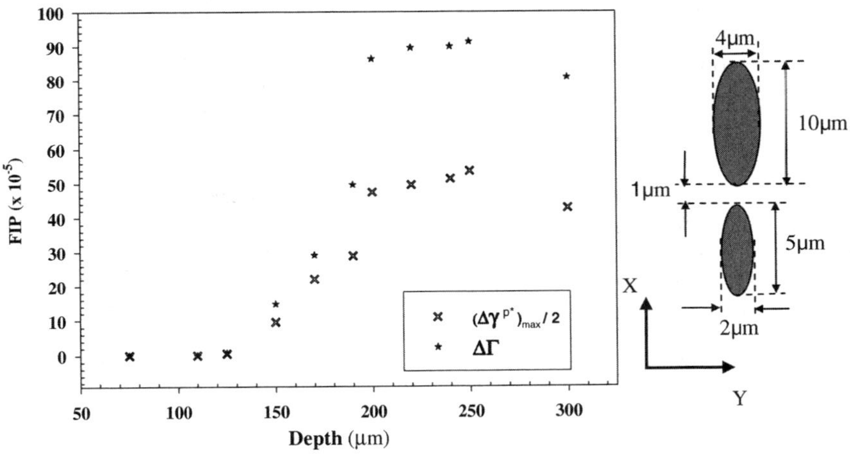

Fig. 19 Variation of nonlocal maximum cyclic plastic shear strain amplitude and FS parameter with depth for case E1.

Similar observations can be made by comparing different cases of partially debonded and cracked inclusions with equal size and spacing; it is evident that cracked inclusions do not localize plastic strains and stresses in the matrix as severely as partially debonded inclusions.

The use of ellipsoidal inclusions in this parametric study facilitates assessment of the effect of inclusion alignment in a cluster with respect to loading direction on fatigue crack nucleation. Although partially debonded inclusions were aligned such that the major axis of the ellipsoid was perpendicular to loading direction, some disparity in the driving force for fatigue crack nucleation arises by varying the relative alignment of inclusions in the *XY* plane. The differences were significant between case D1 (Fig. 18) and case E1 (Fig. 19). The driving force parameters are relatively higher at subsurface depths for case E1. Since the plastic strain localization is most severe at the inclusion notch root in ellipsoidal particles, the magnitudes of these parameters are expected to be relatively higher when the inclusions are aligned such that the major axis of individual inclusions coincides (as in case E1). Comparing cases B1 and C1 in Fig. 16 and 17, respectively, the difference in the magnitudes of FIPs are relatively small. This observation suggests that the effect of relative spatial alignment of inclusions in a cluster on fatigue crack nucleation is more pronounced for large inclusions ($\geq$10 μm) and can be regarded as a second-order effect. The disparity in fatigue response arising due to variations in spatial arrangement of inclusions in a cluster can be a major contributor to the scatter in HCF lives observed in experiments. In addition to the intensification of the driving force for nucleation, the size of the embryo for long crack growth is on the order of the cluster size, which can be substantially larger than the size of individual primary inclusions (see Fig. 2). Study of subcritical crack formation and growth within inclusion clusters of more complex character is a topic of ongoing work.

It is evident from cases D1, E1 and R1 that the presence of large inclusions has more detrimental effect under cyclic loading. This is widely acknowledged from the perspective of initial flaw size, treating the inclusion as an initial crack starter for application of fracture mechanics. In this work, we undertake a different approach that focuses more on the details of inclusion size, spacing, interface character and residual stresses on HCF response.

Summary

Finite element modeling is used to parametrically study the effects of process history and and nonmetallic inclusions on the high cycle fatigue of a martensitic gear steel. Significant contributions and findings of this work are.

- A simplified algorithm is presented to simulate the residual stress distribution induced by shot peening prior to applying a fatigue loading history, including effects of cracked and debonded subsurface inclusions on local residual stresses.
- An integrated three-dimensional methodology is developed to analyze potency of fatigue crack nucleation at subsurface primary inclusions in carburized and shot peened gear steels. The methodology was applied to predict the favorable depth for fatigue crack nucleation in C61 alloy gear steel, with predicted depths in accordance with experimental observations (Ref 1). Fatigue crack nucleation occurs in a depth range where the initial residual stresses are compressive (not tensile) following shot peening.
- Under the given external loading, material model, and the selected fatigue crack nucleation driving force parameter, intact inclusions were observed to show negligible driving force for crack nucleation in the HCF regime. Preexisting damage from primary forming or from shot peening processes, such as partially debonded inclusion-matrix interfaces, promoted fatigue crack nucleation.
- The driving force for subsurface fatigue crack nucleation depends on size and clustering of inclusions. Relative alignment of inclusions in a cluster was observed to influence the driving force for fatigue crack nucleation in presence of sufficiently large inclusions ($\geq$10 μm). Presence of large inclusions in addition to clusters has detrimental effects under cyclic loading.

Future work will address mechanisms specific to the martensitic microstructure in the matrix with regard to crack formation and early growth to refine the framework presented here. In addition, relaxation of residual stress during fatigue cycling can be very important even in the HCF regime (Ref 65–67). and must be studied using more refined material models such as polycrystal plasticity. The framework developed here can be readily extended to consider this effect.

ACKNOWLEDGMENTS

The authors are grateful for the support of the ONR/DARPA D3D tools consortia (J. Christodoulou, monitor), under contract # N00014-05-C-0241. The authors also thank Mahesh Shenoy for the helpful discussions during the course of this work.

APPENDIX A

The stress state at any subsurface depth after shot peening is given by:

$$\sigma_{xx} = \sigma_{zz} = \sigma^{res}$$
$$\sigma_{yy} = 0$$
$$\tau_{xy} = \tau_{xz} = \tau_{yz} = 0$$
$$\alpha_{xy} = \alpha_{xz} = \alpha_{yz} = 0 \quad \text{(Eq A.1)}$$

During plastic deformation (shot peening), the following quantities and their differential increments always satisfy:

$$\varepsilon^p_{xy} = \varepsilon^p_{xz} = \varepsilon^p_{yz} = 0 \quad d\varepsilon^p_{xy} = d\varepsilon^p_{xz} = d\varepsilon^p_{yz} = 0$$

$$\varepsilon^p_{xx} + \varepsilon^p_{yy} + \varepsilon^p_{zz} = 0 \text{ and } d\varepsilon^p_{xx} + d\varepsilon^p_{yy} + d\varepsilon^p_{zz} = 0$$
(incompressibility)

$$\varepsilon^p_{xx} = \varepsilon^p_{zz} = -\frac{1}{2}\varepsilon^p_{yy} \quad d\varepsilon^p_{xx} = d\varepsilon^p_{zz} = -\frac{1}{2}d\varepsilon^p_{yy} \quad \text{(Eq A.2)}$$

During the shot peening process the shear strain components are zero in the XYZ principal coordinate frame, that is:

$$\varepsilon_{xy} = \varepsilon_{yz} = \varepsilon_{zx} = 0 \qquad \text{(Eq A.3)}$$

Expanding the yield function in Eq 1 gives:

$$\begin{aligned} F = &\left[2(\sigma_{xx} - \alpha_{xx}) - (\sigma_{yy} - \alpha_{yy}) - (\sigma_{zz} - \alpha_{zz})\right]^2 \\ &+\left[2(\sigma_{zz} - \alpha_{zz}) - (\sigma_{xx} - \alpha_{xx}) - (\sigma_{yy} - \alpha_{yy})\right]^2 \\ &+\left[2(\sigma_{yy} - \alpha_{yy}) - (\sigma_{xx} - \alpha_{xx}) - (\sigma_{zz} - \alpha_{zz})\right]^2 \\ &-6\sigma_{ys}^2 \\ = &\, 0 \end{aligned} \qquad \text{(Eq A.4)}$$

Due to equi-biaxial deformation, the following relationships hold during the entire process of shot peening:

$$\begin{aligned} &\sigma_{xx} - \alpha_{xx} = \sigma_{zz} - \alpha_{zz} \\ &S_{xx} - \alpha'_{xx} = S_{zz} - \alpha'_{zz} \\ &\varepsilon^p_{xx} = \varepsilon^p_{zz}, \quad \dot{\varepsilon}^p_{xx} = \dot{\varepsilon}^p_{zz} \end{aligned} \qquad \text{(Eq A.5)}$$

Inserting Eq A.5 into Eq A.4, the yield function reduces to:

$$\begin{aligned} &F = \left[(\sigma_{xx} - \alpha_{xx}) - (\sigma_{yy} - \alpha_{yy})\right]^2 - \sigma_{ys}^2 = 0 \text{ or} \\ &\left|(\sigma_{xx} - \alpha_{xx}) - (\sigma_{yy} - \alpha_{yy})\right| = \sigma_{ys} \end{aligned} \qquad \text{(Eq A.6)}$$

Using the relation in Eq A.2 the increment of equivalent plastic strain is given by:

$$d\bar{\varepsilon}^p = \left|d\varepsilon^p_{yy}\right| = 2\left|d\varepsilon^p_{xx}\right| \qquad \text{(Eq A.7)}$$

Using the three-dimensional Hooke's law, the elastic and plastic strains at the end of shot peening process can be obtained by applying the boundary conditions shown in Fig. 5, that is:

$$\begin{aligned} &\varepsilon^{e,final}_{xx} = \frac{\sigma^{res}}{E}(1 - \nu) = -\varepsilon^{p,final}_{xx} \\ &\varepsilon^{e,final}_{zz} = \frac{\sigma^{res}}{E}(1 - \nu) = -\varepsilon^{p,final}_{zz} \\ &\varepsilon^{e,final}_{yy} = \frac{-\nu}{E}(2\sigma^{res}) \end{aligned} \qquad \text{(Eq A.8)}$$

Here, the superscript *final* corresponds to final strain state at the end of shot peening. Coupling Eqs A.2, A.7, and A.8 the final strain component along direction Y after unloading can be obtained as:

$$\varepsilon^{final}_{yy} = \frac{-\nu}{E}(2\sigma^{res}) + 2\varepsilon^{e,final}_{xx} \qquad \text{(Eq A.9)}$$

In order to obtain the peak uniaxial compressive strain to be imposed along direction Y $\left(\varepsilon^{load}_{yy}\right)$ to induce the target residual stress level, the above equations coupled with the nonlinear kinematic hardening relation described in Eq 5 are solved numerically. We employ the method of successive substitution along with an interval-halving scheme to iteratively solve the set of equations. The return mapping scheme to solve the elasto-plastic relations (Eq 1–6) is implemented along the lines proposed by Kobayashi and Ohno (Ref 68).

REFERENCES

1. B.L. Tiemens, Performance optimization and computational design of ultra-high strength gear steels. Department of materials science and engineering. PhD thesis, Evanston (IL): Northwestern University, 2006
2. B.L. Tiemens, A.K. Sachdev, and G.B. Olson, Performance of computationally designed ultra-hard steel for transmission gear applications. Collaborate report CL - 04/41/MPL, GM Research and Development Center, August 2004
3. L. Ceschini and G. Minak, Fatigue behaviour of low temperature carburised AISI 316L austenitic stainless steel. *Surf. Coat. Technol.* 2008, 202(9), 1778–84
4. K. Tokaji and M. Akita, Effect of carburizing on notch fatigue behaviour in AISI 316 austenitic stainless steel. *Surf. Coat Technol*, 2006, 200(20–21), 6073–8
5. S. Farfan, C. Rubio-Gonzalez, T. Cervantes-Hernandez, and G. Mesmacque, High cycle fatigue, low cycle fatigue and failure modes of a carburized steel. *Int. J. Fatigue*, 2004, 26(6), 673–8
6. N. Kawagoishi, T. Nagano, and M. Moriyama, Effect of shot peening on fatigue strength of maraging steel, Sixth international conference on computer methods and experimental measurements for surface treatment effects, Crete, Greece 2003, 7, 99–108
7. N. Iwata, Y. Tomota, K. Katahira, and H. Suzuki, Effect of shot peening on fatigue fracture for an as quenched martensitic steel. *Mater. Sci. Technol.* 2002, 18(6), 629–32

8. J. Brown, Shot peening increases gear life. Power Transmission Des 1996, 38(5), 61–4
9. M. Benedetti, V. Fontanari, B.R. Hohn, P. Oster, and T. Tobie, Influence of shot peening on bending tooth fatigue limit of case hardened gears. *Int. J. Fatigue.* 2002, 24(11), 1127–36
10. J.O. Almen, Fatigue weakness of surfaces. *Prod. Eng.* 1950, 21(11), 117–40
11. R. Sieber, Bending fatigue performance of carburized gear steels. SAE International congress and exposition Detroit, MI. Warrendale (PA), SAE, 1992, p 1–18
12. T. Kunio, M. Shimizu, K. Yamada, K. Sakura, and T. Yamamoto, Early stage of fatigue crack growth in martensitic *steel. Int. J. Fracture*, 1981, 17(2), 111–9
13. Y. Murakami and T. Endo, Effects of small defects on fatigue strength of metals. *Int. J. Fatigue* 1980, 2(1), 23–30
14. Y. Murakami and M. Endo, Quantitative evaluation of fatigue strength of metals containing various small defects or cracks. *Eng. Fracture Mech.* 1983, 17(1), 1–15
15. O. Umezawa, and K. Nagai, Deformation structure and subsurface fatigue crack generation in austenitic steels at low temperature. *Metall. Mater. Trans. A (Phys. Metall. Mater. Sci.)* 1998, 29A(3), 809–22
16. Q.Y. Wang, C. Bathias, N. Kawagoishi, and Q. Chen, Effect of inclusion on subsurface crack initiation and gigacycle fatigue strength, *Int. J. Fatigue.* 2002, 24(12), 1269–74
17. D.L. McDowell, K. Gall, M.F. Horstemeyer, and J. Fan, Microstructure-based fatigue modeling of cast A356-T6 alloy. *Eng. Fract. Mech.*, 2003, 70(1), 49–80
18. K. Gall, M.F. Horstemeyer, B.W. Degner, D.L. McDowell, and J. Fan, On the driving force for fatique crack formation from inclusion and voids in a cast A356 aluminum alloy. *Int J Fracture* 2001, 108(3), 207–33
19. M.A.S. Torres and H.J.C. Voorwald, An evaluation of shot peening, residual stress and stress relaxation on the fatigue life of AISI 4340 steel, *Int. J. Fatigue.*, 2002, 24(8), 877–86
20. B.A. Shaw, C. Aylott, P. O'Hara, and K. Brimble, In: M.A. Hyannis, ed., The role of residual stress on the fatigue strength of high performance gearing, Vol 25. Elsevier; 2003, p 1279–83
21. T. Toyoda, T. Matsui, and Y. Murakami, A study of inclusions causing fatigue cracks in steels for carburized and shot peened gears., *JSAE* Review 1990, 11, 50–4
22. S. Nishijima and K. Kanazawa, Stepwise S-N curve and fish-eye failure in gigacycle fatigue, *Fatigue Fracture Eng. Mater. Struct.* 1999, 22(7), 601–7
23. K. Shiozawa and L. Lu, Very high-cycle fatigue behaviour of shot-peened high-carbon-chromium bearing steel, *Fatigue Fracture Eng. Mater. Struct.* 2002, 25(8–9), 813–22
24. C. Kim, D.E. Diesburg, and G.T. Eldis, Effect of residual stress on fatigue behavior of carburized steels—an analytical model. In: ASTM STP 776, Residual stress effect in fatigue; 1982, p 224–234
25. V.K. Sharma, G.H. Walter, and D.H. Breen, Factors influencing fracture toughness of high carbon martensitic steels. *Gear Technol.*, 1989, 7–18
26. H. Lin, G.A. Fett, and R.R. Binoniemi, Bending fatigue life analysis of carburized components using strain life and fracture mechanics approaches., *J. Mater. Manuf.* 2004, 112, 154–64
27. M. MackAldener and M. Olsson, Tooth interior fatigue fracture—computational and material aspects, *Int. J. Fatigue* 2001, 23(4), 329–40
28. D.L. McDiarmid, General criterion for high cycle multiaxial fatigue failure. *Fatigue Fracture Eng. Mater. Struct.* 1991, 14(4), 429–53
29. A. Borbely, H. Mughrabi, G. Eisenmeier, and H.W. Hoppel, A finite element modelling study of strain localization in the vicinity of near-surface cavities as a cause of subsurface fatigue crack initiation, *Int. J. Fracture*, 2002, 115(3), 227–32
30. K. Ogawa, and T. Asano, Theoretical prediction of residual stress produced by shot peening and experimental verification for carburized steel, *Mater. Sci. Res. Int.*, 2000, 6(1), 55–62
31. K. Inoue and M. Kato, Crack growth resistance due to shot peening in carburized gears. *J. Propul. Power* 1995, 11(5), 973–9
32. A. Blarasin, M. Guagliano, and L. Vergani, Fatigue crack growth prediction in specimens similar to spur gear teeth. *Fatigue Fracture. Eng. Mater. Struct.* 1997, 20(8), 1171–82

33. J.L. Chaboche and O. Jung, Application of a kinematic hardening viscoplasticity model with thresholds to the residual stress relaxation, *Int. J. Plasticity*, 1997, 13(10), 785–807
34. J.K. Li, M. Yao, and D. Wang, Mechanical approach to the residual stress field induced by shot peening, *Mater. Sci. Eng. A (Struct Mater: Properties, Microstruct. Process)* 1991, A147(2), 167–73
35. S.A. Meguid, G. Shagal, J.C. Stranart, K.M. Liew, and L.S. Ong, Relaxation of peening residual stresses due to cyclic thermo-mechanical overload. *J. Eng. Mater. Technol, Trans ASME*, 2005, 127 (2), 170–8
36. R. Fathallah, A. Laamouri, H. Sidhom, and C. Braham, High cycle fatigue behavior prediction of shot-peened parts, *Int. J. Fatigue*, 2004, 26(10), 1053–67
37. L. Yanhuai, L. Jian, and X. Kewei, Calculation of relaxation of residual stress and change of yield strength in shot peened layer, *Mater. Sci. Forum*, 2005, 490–491, 396–403
38. M.M. Shenoy, R.S. Kumar, and D.L. McDowell, Modeling effects of nonmetallic inclusions on LCF in DS nickel-base superalloys, *Int. J. Fatigue*. 2005, 27(2), 113–27
39. A. Melander, Finite element study of short cracks with different inclusion types under rolling contact fatigue load, *Int. J. Fatigue*. 1997, 19(1), 13–24
40. K. Gall, N. Yang, M. Horstemeyer, D.L. McDowell, and F. Jinghong, The debonding and fracture of Si particles during the fatigue of a cast Al-Si alloy, *Metall. Mater. Trans A (Phys. Metall. Mater. Sci.)* 1999, 30A(12), 3079–88
41. J. Fan, D.L. McDowell, M.F. Horstemeyer, and K. Gall, Cyclic plasticity at pores and inclusions in cast Al-Si alloys, *Eng. Fract. Mech.*, 2003, 70(10), 1281–302
42. C.J. Kuehmann and G.B. Olson, Advanced gear steels by computer-aided design, *Adv. Mater. Process*, 1998, 153, 40–3
43. A. Fatemi and D.F. Socie, A critical plane approach to multiaxial fatigue damage including out-of-phase loading, *Fatigue Fracture Eng. Mater. Struct.*, 1988, 11(3), 149–65
44. ABAQUS (Ver 6.5) Hibbitt, Karlsson, and Sorensen Inc., Providence, RI; 2005
45. ASTM E92-82 standard test method for Vickers hardness of metallic materials, 2003
46. Y. Qian, J. Almer, U. Lienert, B.L. Tiemens, and G.B. Olson, Nondestructive residual stress distribution measurements in nanostructured ultrahigh-strength gear steels. In: 5th International conference on synchrotron radiation in materials science, 2006
47. G. Donzella, A. Pola, L. Solazzi, and G.P. Marconi, Effect of shot peening on carburized surfaces, *Int. J. Mater. Prod. Technol.*, 2000, 15(1–2), 117–30
48. R. Pedersen and S.L. Rice, "Case crushing" of carburized and hardened gears, Society of Automotive Engineers (SAE); 1960
49. C. Boller and T. Seeger, Materials data for cyclic loading, part b: low alloy steels. Elsevier, 1987
50. ABAQUS/Standard User's Manual (Ver. 6.5). HKS, 2005
51. H. Xiao, Q. Chen, E. Shao, D. Wu, Z. Chen, and Z. Wang, The effect of shot peening on rolling contact fatigue behaviour and its crack initiation and propagation in carburized steel, *Wear*, 1991, 151 (1), 77–86
52. M. Kobayashi, T. Matsui, and Y. Murakami, Mechanism of creation of compressive residual stress by shot peening, *Int. J. Fatigue.*, 1998, 20(5), 351–7
53. D.L. McDowell, and J.Y. Berard, A ΔJ based approach to biaxial fatigue, *Fatigue. Fracture. Eng. Mater. Struct.*, 1992, 15(8), 719–41
54. D.L. McDowell, Multiaxial Fatigue Strength, *ASM Handbook*, 1996, p 263–273
55. D.L. McDowell, Multiaxial small fatigue crack growth in metals, *Int. J. Fatigue*. 1997, 19, 127–35
56. M. Shenoy, J. Zhang, and D.L. McDowell, Estimating fatigue sensitivity to polycrystalline Ni-base superalloy microstructures using a computational approach, *Fatigue Fracture Eng. Mater. Struct.*, 2007, 30 (10), 889–904
57. Y. Jiang, O. Hertel, and M. Vormwald, An experimental evaluation of three critical plane multiaxial fatigue criteria, *Int. J. Fatigue*, 2007, 29(8), 1490–502
58. K.S. Kim, and J.C. Park, Shear strain based multiaxial fatigue parameters applied to variable amplitude loading, *Int. J. Fatigue*, 1999, 21(5), 475–83
59. K.O. Findley and A. Saxena, Low cycle fatigue in Rene 88DT at 650 C: crack

nucleation mechanisms and modeling, *Metall. Mater. Trans. A (Phys. Metall. Mater. Sci)* 2006, 37A(5), 1469–75

60. X. Chen, Q. Gao, and X.F. Sun, Damage analysis of low-cycle fatigue under non-proportional loading, *Int. J. Fatigue*. 1994, 16(3), 221–5
61. J.D. Kwon and J.C. Park, Multiaxial fatigue life prediction of duplex stainless steels with thermal aging at 430 °C under axial-torsional load, *Key Eng. Mater*. 2004, 270–3, 1183–8
62. L. Reis and M. De Freitas, Analytical and experimental studies on fatigue crack path under complex multi-axial loading, *Fatigue Fracture Eng. Mater. Struct*. 2006, 29(4), 281–9
63. D.L. McDowell, Simulation-based strategies for microstructure-sensitive fatigue modeling, *Mater. Sci. Eng. A*, 2007, 468–70, 4–14
64. K. Gall, M. Horstemeyer, and D.L. McDowell, J. Fan, Finite element analysis of the stress distributions near damaged Si particle clusters in cast Al–Si alloys, *Mech. Mater*., 2000, 32(5), 277–301
65. J.R. Teodosio, M.C. Fonseca, and P.D. Pedrosa, Relaxation of residual stresses during fatigue cycles in steels, *Mater. Sci. Forum*, 2003, 426–432, 3981–6
66. W.Z. Zhuang and G.R. Halford, Investigation of residual stress relaxation under cyclic load, *Int. J. Fatigue*, 2001, 23, S31–7
67. J.D. Almer, J.B. Cohen, and B. Moran, The effects of residual macrostresses and microstresses on fatigue crack initiation, *Mater. Sci. Eng. A*, 2000, 284, 268–79
68. M. Kobayashi and N. Ohno, Implementation of cyclic plasticity models based on a general form of kinematic hardening. *Int. J. Numer. Methods. Eng*., 2002, 53(9), 2217–38

Transformations—Selected Works of G.B. Olson on Materials, Microstructure, and Design
C.E. Campbell, M.V. Manuel, and W. Xiong, editors

CHAPTER 26

Genomic Materials Design—The Ferrous Frontier*

G.B. Olson, QuesTek Innovations LLC and Northwestern University

Abstract

Our unique depth of scientific knowledge has allowed ferrous metallurgy to lead the development of a successful computational materials design methodology grounded in a system of fundamental databases. Expansion of this approach to embrace acceleration of the full development and qualification cycle has resulted in the flight of the first fully computationally designed and qualified material. The example of ferrous alloys provides a template for a general methodology of computational design for all materials.

Introduction

Several recent US national academy studies of materials and manufacturing (Ref 1–5) have emphasized the unique opportunities of computational materials engineering. While the breadth of the best known 2008 Integrated Computational Material Engineering (ICME) study (Ref 2) created an impression of a field in its infancy, the 2004 Accelerated Technology Transition (ATT) study (Ref 3), uniquely charged with identifying best industrial practices to accelerate the transition of materials and process technology, elucidated practices which are now mature and ready for broader dissemination. The achievements of these practices have now been highlighted in a recent lightweighting technology study (Ref 1), and the importance of the underlying fundamental databases, now known as the Materials Genome, have been heralded in a recent report of the US President's Office of Science and Technology Policy (Ref 6).

Ferrous metallurgy has played a leading role in the creation of this technology (Ref 7). While virtually every microstructural phenomenon occurring in structural materials can be found in ferrous alloys, ferrous systems enjoy the unique position that these phenomena have been studied long enough and deeply enough that theories for them exist, with quantitative validation by fundamental experiments. We here review the development of this technology with an emphasis on the leading examples of computationally designed "cybersteels." European efforts have made significant contributions to this technology, notably including computational thermodynamics and their application (Ref 8), computational quantum mechanics applications (Ref 9), microstructural simulation tools (Ref 10), computational materials design (Ref 11), and through process modeling (Ref 12, 13). This overview focuses on developments in the USA, which remain unique in their scope of integration.

*G.B. Olson, "Genomic Materials Design: The Ferrous Frontier," *Acta Mater.* Diamond Jubilee Issue, 61 (2013) 771–81

Background—Emergence of a Design Discipline

Founded in 1985 as the Steel Research Group, a multi-institutional materials design consortium has undertaken a quarter century of development of an interdisciplinary systems approach to the science-based computational design of hierarchically structured multiphase, multicomponent materials (Ref 7, 14, 15), grounded in the system of fundamental databases now known as the Materials Genome (Ref 6). Integrating materials science, continuum mechanics and quantum physics, the approach has featured a suite of validated computational tools enabled by an iterative interaction of theory, simulation and experiment, spanning the full range of process/structure/property/performance relations. An initial focus on high-performance steels, exploiting not only their unique depth of predictive scientific understanding but also their level of fundamental database development, has demonstrated the capabilities of the approach, providing a range of ultrahigh-strength ferrous alloys (Ref 16–20). now in use in commercial applications. The founding of QuesTek in 1997 as the first computational materials design company has provided both commercial materials design services as well as successful licensing of the designed alloys (Ref 21–23), with broadened application of the approach to Al-, Ni-, Ti-, and Cu-based alloys. Recognizing the central role of chemical thermodynamics in the science of all materials (as reflected in the structure of undergraduate materials curricula) a fundamental genomic database system centered on CALPHAD thermodynamics (Ref 24) has provided a framework for computational materials design with broad applicability, as attested by university projects in the 1990s demonstrating its application to polymers, ceramics and composites (Ref 25–27).

The successes in computational materials design in the 1990s helped to bring about the seminal DARPA-AIM (US Defense Advanced Research Projects Agency: Accelerated Insertion of Materials) initiative of the 2000s, expanding the scope of science-based computational materials engineering to embrace the full component level process optimization and material qualification scheme in order to significantly compress the full materials development cycle (Ref 28). Under this initiative the PrecipiCalc precipitation simulator (Ref 29, 30), grounded in the CALPHAD thermodynamics and mobility database and software system, was efficiently linked to macroscopic component-level process simulation tools to successfully accelerate process optimization and accurately forecast manufacturing variation with calibration using minimal data sets for the test case of aeroturbine disc manufacturing. Designed concurrently with the development of the AIM process, the historic first flight of QuesTek's Ferrium S53 landing gear steel in December 2010 represented the first fully computationally designed and qualified material (Ref 31), fully exploiting the AIM methodology. As the principal follow-on to the DARPA-AIM successes, the recent 5-year "D3D" Digital Structure consortium program (Ref 32), jointly supported by the US Office of Naval Research (ONR) and DARPA, extended the fidelity of AIM modeling tools by integrating a multiscale suite of three-dimensional (3-D) tomographic microstructural characterization tools with a new generation of 3-D microstructural simulators. Combining local electrode atom probe (LEAP), focused ion beam (FIB) and metallographic serial sectioning techniques to span the nanometer, micron and millimeter microstructural scales, respectively, computer-aided tomographic reconstructions have provided a new level of spatial characterization informing novel numerical simulators addressing strength, toughness, and fatigue resistance, using advanced steels as the test cases. This has notably included several generations of multiscale ductile fracture simulators (Ref 33–36), as well as the direct micromechanical simulation of fatigue nucleation processes at observed nucleants (Ref 37). Extending the 3-D approach to bond topological analysis of quantum mechanical calculations has also provided new heuristics to guide materials design component selection (Ref 38, 39). Seeking to accelerate the development of fundamental databases supporting computational materials engineering, the design of niobium-based alloys under the MEANS (Materials Engineering for Affordable New Systems) initiative of the US Air Force Office of Sponsored Research (AFOSR) demonstrated efficient integration of first principles predictions with rapid experimental validation of phase relations in small scale samples (Ref 40–42).

The Predictive Science Foundation

Early design-driven research to establish a predictive science of precipitation strengthening began with a thorough coordinated multi-technique evaluation of the precipitation hardening behavior of the commercial secondary hardening martensitic steel AF1410, as summarized in Fig. 1 (Ref 43). Field ion microscopy (FIM), transmission electron microscopy (TEM), small angle neutron scattering (SANS), atom probe field ion microanalysis (APFIM), and scanning transmission electron microanalysis (STEM) were combined to quantify the

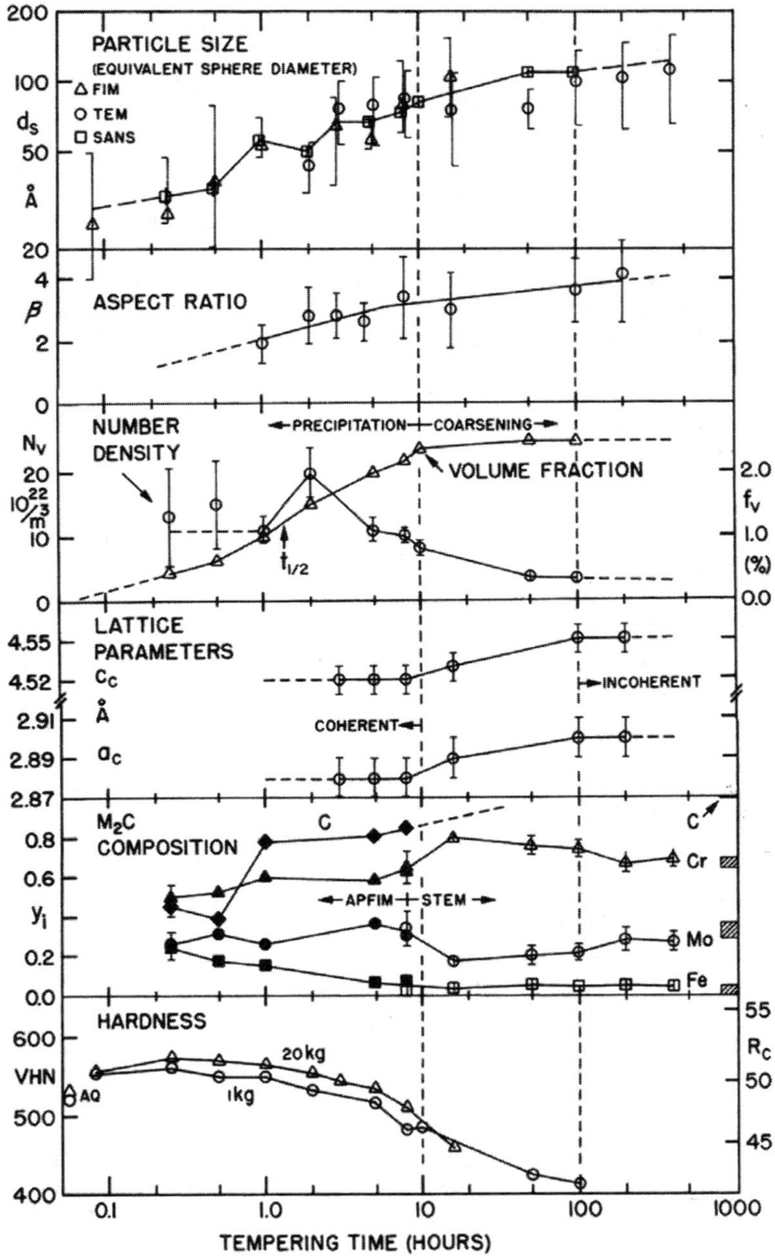

Fig. 1 Multitechnique quantification of the M_2C carbide precipitation hardening evolution in a commercial secondary hardening martensitic AF1410 Co-Ni steel at the standard tempering temperature of 510 °C (Ref 43).

detailed temporal evolution of the M_2C alloy carbide strengthening dispersion in terms of particle size, shape, number density, phase fraction, lattice parameter, and chemical composition, correlated with the evolution of alloy hardness. The evolution of size validates the prediction of the Langer-Schwartz theory of precipitation at high supersaturation (Ref 44), where the system evolves along a trajectory of unstable equilibrium, enabling space-time separation in its description and control. Particle size can then be controlled through the thermodynamic driving force governing the initial critical nucleus size, while the timescale of precipitation can be independently controlled through a multicomponent coarsening rate constant (Ref 45, 46) predictable from mobility databases. Observed shifts in lattice parameters and composition of the M_2C phase are consistent with initial coherent precipitation, and x-ray line broadening measurements in related model alloys give supporting evidence of associated matrix coherency strains (Ref 47). The prediction of thermodynamic driving forces controlling particle size has required extension of the CALPHAD chemical thermodynamic databases to incorporate the elastic energy of the coherent state (Ref 48, 49). Calibrating analytical strength models to the measurements in AF1410 and related model alloys (Ref 50), parametric design controlling the length scales and timescales through the fundamental databases to place the carbide size at an optimal diameter of 3 nm has allowed a 50% strength increase in secondary hardening martensitic steels over previous steels of a given carbon content (Ref 51).

Constraining the alloy processability at the ingot stage employs multicomponent diffusion simulation of solidification microsegregation and the subsequent response to homogenization heat treatment (Ref 52). Control of the martensitic transformation temperatures employs the Olson-Cohen nucleation theory (Ref 53) supported by a kinetic database describing solution hardening effects in martensitic interfacial mobility (Ref 54–56). Prediction of alloy hardenability employs a related model of bainitic transformation as a coupled displacive/diffusional mechanism (Ref 57). Predictive control of austenite grain refinement employs a calibrated Zener pinning model (Ref 58). Control of cleavage fracture resistance uses a "master curve" model of the ductile-brittle transition in which the transition temperature is expressed as a function of matrix composition, grain size, and alloy hardness. Improved resistance to ductile fracture employs insights from microvoid softening-induced plastic shear localization simulation (Ref 35) to optimize the grain refining dispersion, including phase selection for enhanced interfacial adhesion (Ref 38). Further enhancement of the ductile fracture toughness has been demonstrated using transformation plasticity theory (Ref 59) to optimize the mechanical stability of dispersed austenite precipitated during secondary hardening, aided by two step tempering treatments (Ref 14).

A major barrier to the advance of ultrahigh-strength steels has historically been their extreme sensitivity to segregation-induced intergranular embrittlement, as exacerbated by the effect of internally and externally induced hydrogen. Perhaps the most challenging science problem addressed, an early significant achievement of the SRG design consortium, was the Rice-Wang thermodynamic theory of intergranular embrittlement, predicting the potency of a segregant from the difference in its segregation energy in the grain boundary and free surface environments (Ref 60). While this theory is of universal applicability to grain boundaries in all materials, a unique feature of ferrous alloys is that the embrittlement potency of segregants has been measured along with the segregation energies in the two environments to validate the theory. With such limited thermodynamic data available, and recognizing the relative difficulty of surface thermodynamic measurements, the capability of predicting the underlying thermodynamics was developed through application of the highly precise all electron density functional theory full potential augmented plane wave (FLAPW) quantum mechanical method (Ref 61). Figure 2

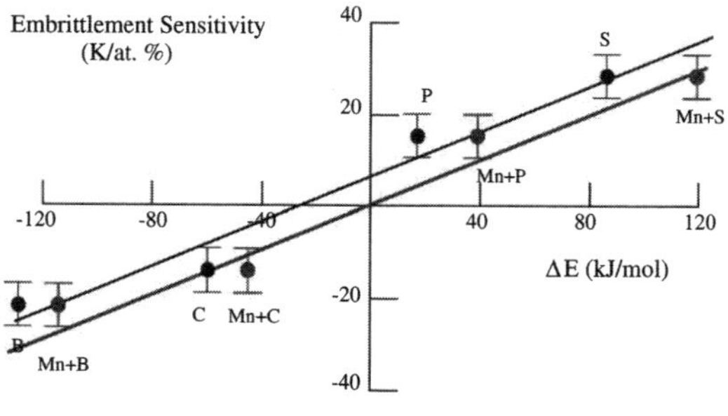

Fig. 2 Measured grain boundary embrittlement potency of B, C, P, and S in steels versus the FLAPW-computed boundary/surface segregation energy difference, with and without prior segregation of Mn. Source: Ref 14

summarizes the correlation of measured embrittlement potency (as a shift in the intergranular ductile-brittle transition temperature with amount of boundary segregant) with the calculated boundary-surface segregation energy difference. The two curves depict predictions with and without the prior segregation of Mn at the boundary, as Mn is a common alloying element in many of the steels for which the embrittlement potencies were measured. The first principles values generally agree with the measured thermodynamic quantities within 0.1 eV $atom^{-1}$, and show a strong correlation with the relative embrittling potencies of P and S, and the relative cohesion enhancing potencies of B and C. With the predictive capability thus validated, attention turned to the less well-characterized effects of substitutional alloying elements. After rigorous FLAPW calculations for Mn, Mo and Pd a simplified model was developed using boundary parameters (such as site size) from the FLAPW calculations together with handbook quantities such as elemental cohesive energies to make the prediction of potencies represented in Fig. 3 (Ref 62). After screening for strongly negative values to identify potent cohesion enhancers, these estimates were followed by rigorous FLAPW calculations confirming the high potency of elements such as W and Re. In this way a genomic fundamental surface thermodynamic database was assembled to enable design of the boundary composition to offset the effect of H (Ref 63). Integration into alloy design led to the first "quantum steels" in which new directions in alloy composition led by first principles calculations allowed complete elimination of the intergranular mode of stress corrosion cracking in ultrahigh strength steels (Ref 14, 64).

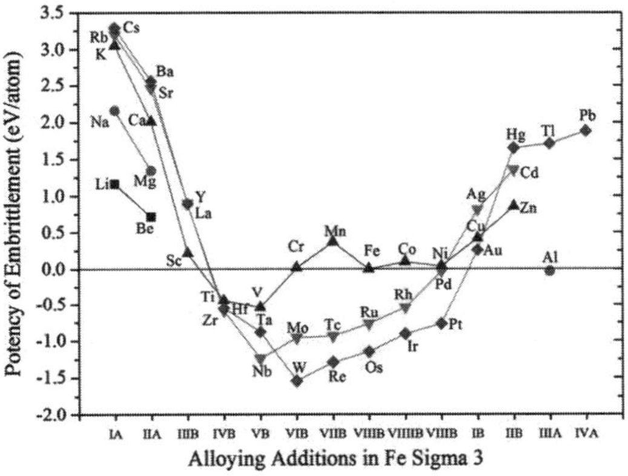

Fig. 3 FLAPW-calibrated model prediction of the embrittlement potency of substitutional elements in Fe grain boundaries. Source: Ref 62

The resulting suite of computational models supporting materials design is summarized in Fig. 4, ordered according to length scale. At

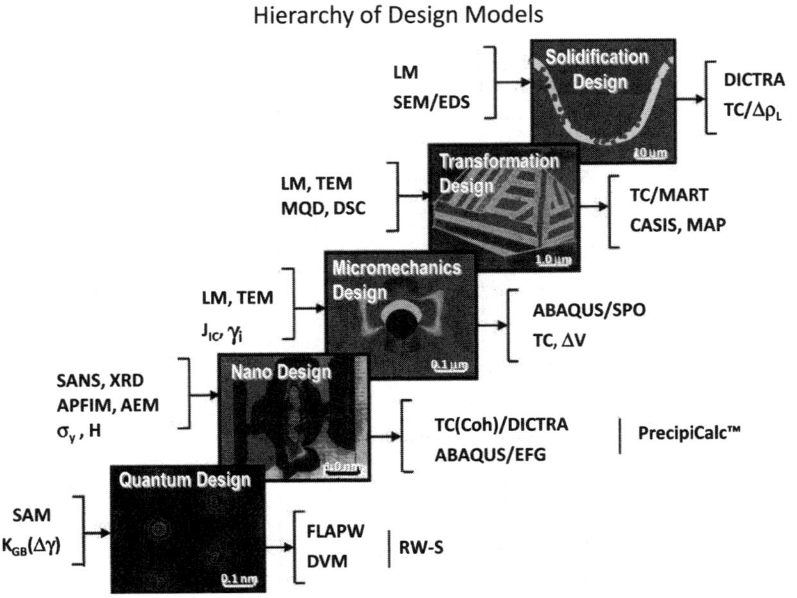

Fig. 4 Multiscale hierarchy of interdisciplinary mechanistic models supporting computational materials design. The acronyms are as described in Olson (Ref 14).

the electronic level quantum mechanical calculations predict surface thermodynamics, at the nano-scale materials science continuum models control precipitation strengthening, at the submicron scale numerical mechanics addresses unit processes of ductile fracture, while at higher scales allotropic transformation and solidification models address processability. Acronyms on the right denote the design models and their software platforms, while acronyms on the left denote the equally important advanced instrumentation allowing calibration and validation for quantified predictive accuracy, which are described in detail elsewhere (Ref 14).

Computational Design Successes—Flying Cybersteel

The first example of a commercial alloy created by computational design using these tools is Ferrium C61 (AMS6517) high-durability gear steel, now performing well in off-road racing applications (Ref 22, 51). The first such designer alloy to reach flight qualification is Ferrium S53 (AMS 5922), a corrosion-resistant landing gear steel allowing drop-in replacement for current nonstainless landing gear steels, eliminating the need for cadmium plating (Ref 23, 31, 65). S53 is a secondary hardening martensitic steel strengthened by efficient M_2C carbide precipitates which contains sufficient Cr to provide passivation against general corrosion. Employing a systems approach to materials design integration (Ref 14) the full structural hierarchy of the alloy is represented by the flow-block diagram in Fig. 5. The sequential processing steps experienced by the alloy are depicted in the leftmost column of the flow-block diagram and are constrained to existing processes for steels employed in current structural aircraft applications to maximize manufacturability. The subcomponents of the alloy system are connected by process-structure and structure-property relationships, as addressed by the suite of mechanistic modeling tools just described.

Application of the system flow-block diagram in conjunction with the computational models begins with the identification of key design trade-offs motivated by the full set of quantitative property objectives represented by the right column. The early conceptual design then defines the specific microstructural subsystems represented in the central column. To address the strong conflict between corrosion resistance and strengthening efficiency a thermodynamic model for the metastable spinel

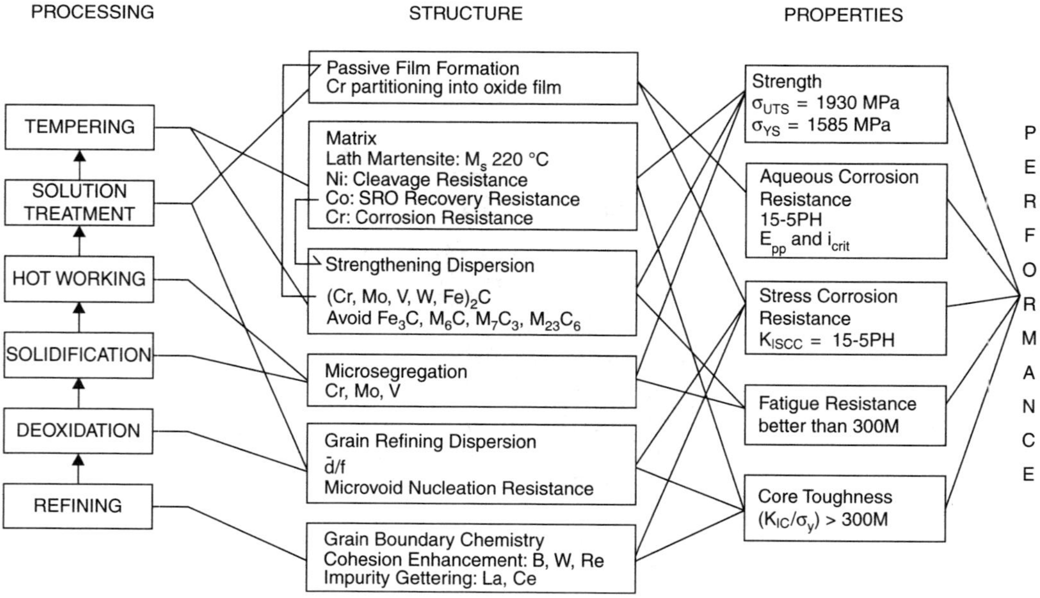

Fig. 5 Flow-block system diagram for ultrahigh-strength corrosion-resistant steels, indicating the desired property objectives, the microstructural subsystems, and the required sequence of processing steps. Links between system blocks indicate quantitative models needed to effect the design via science-based computation. Source: Ref 31

oxide was developed to precisely control the passive film defining corrosion resistance (Ref 17). The multicomponent thermodynamics of the bcc alloy matrix were then exploited to achieve a Cr potential providing a passive film equivalent to a 15Cr alloy using less than 12Cr in the design. In support of this a detailed reassessment of the fcc/bcc thermodynamics of the base Fe-Cr-Co-Ni system was undertaken to simultaneously maintain accurate control of the allotropic transformation behavior.

Integration of the new information with the models in the section, "The Predictive Science Foundation," was undertaken by graphical parametric design in which mechanistic understanding motivated the definition of key scalar parameters computable from phase composition vectors and diffusion matrices via the fundamental CALPHAD databases. An example is presented in Fig. 6. With other composition variables fixed by the initial estimates, Fig. 6(a) depicts the dependencies on two composition variables (Mo and C) of primary importance to the nanoscale carbide strengthening dispersion. Contours of the computed solution temperature T_s and martensite start temperature M_s are superimposed as processability constraints. With a secondary hardening tempering temperature of 482 °C selected from the initial rate constant estimates the contours are shown for the precipitation driving force for coherent M_2C carbides (taking into account prior precipitation of metastable Fe_3C carbides under "paraequilibrium" constraints against substitutional partitioning). This driving force, together with a total phase fraction defined by the C content, predicts the contours of peak hardness (VHN) defining the outlined region meeting the property requirements within constraints. Similarly, Fig. 6(b) depicts attributes of the grain refining MX carbonitride dispersion versus a composition variable (V) and the process temperature (T). Attributes of importance here are the MX phase fraction (for Zener pinning) and component site fraction (for interfacial adhesion). Process temperature windows define constraints of full solubility during homogenization, precipitation during hot work, and stability during austenitizing/solution treatment. In addition to meeting specified parameter values, the slopes defined by the contour plots support sensitivity assessment for robust design.

Using similar cross plots, additional constraints on ductile fracture and grain boundary chemistry are used to complete the design optimization and uniquely identify the alloy composition and process temperatures (each with specified tolerances) that represent the best compromise of the diverse design goals and constraints. By iterative design and prototyping

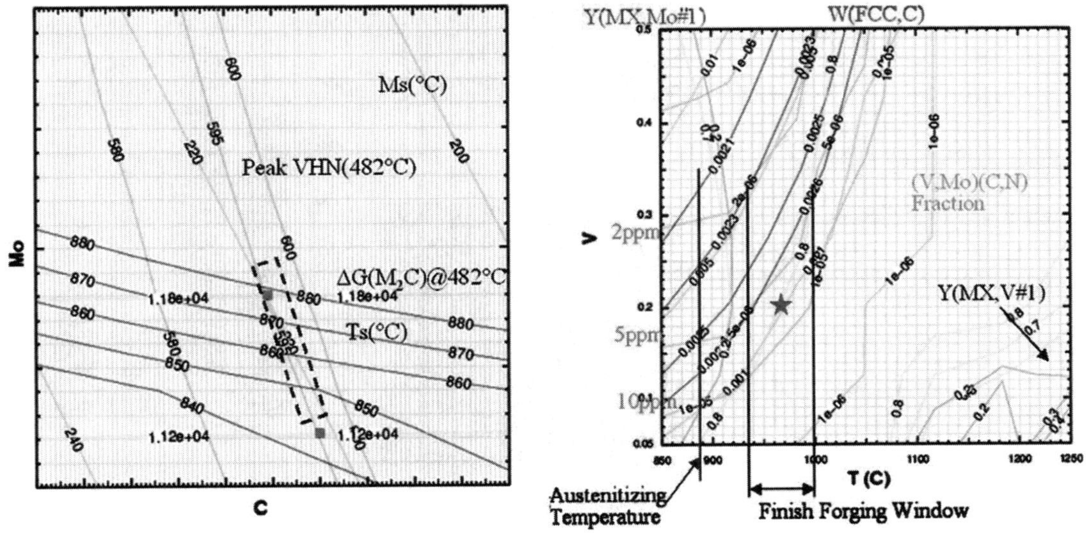

(a) Strengthening Dispersion Design

(b) Grain Pinning Dispersion Design

Fig. 6 Parametric design cross plots employed to achieve the required microstructural attributes within the processing constraints for (a) strengthening dispersion and (b) grain refining dispersion. Source: Ref 31 with full-scale microstructural evaluation to test all models local correction factors are passed back to design models with increased accuracy in the new composition region of interest. In this way accurate design solutions are typically achieved within three iterations of prototyping. For the very challenging objectives of Ferrium S53 design successful achievement of all objectives required five iterations spanning a 3-year period. Experience has shown that the number of required iterations and the associated timescale will steadily decrease as the accuracy of the general CALPHAD databases continues to increase.

Extending Computational Design to Qualification and Implementation

The design of Ferrium S53 steel ran concurrently with the DARPA-AIM initiative on accelerated materials technology transition. The steel was included in the AIM program to serve as a first demonstrator of the new methodology. This began by running the design models within the iSIGHT design integration software (Ref 28) to apply Monte Carlo simulation of parametric sensitivities in support of a higher level of robust design practice for reduced intrinsic variation within allowed composition and process tolerances. To circumvent the traditional problem of process scale-up in empirically developed materials Ferrium S53 was the first example of a "design for scale" approach in which process simulation was used to constrain the up-front alloy design to be compatible with full scale processing. As summarized in Fig. 7, solidification simulations employed the DICTRA multicomponent diffusion code (Ref 52) for candidate design compositions under the cooling rates of full scale production ingots. The acceptable level of microsegration represented in the first panel was defined by simulating its response to homogenization treatment, as represented in the second. As the production scale was increased sections of as-solidified ingots were obtained to measure the microsegregation for

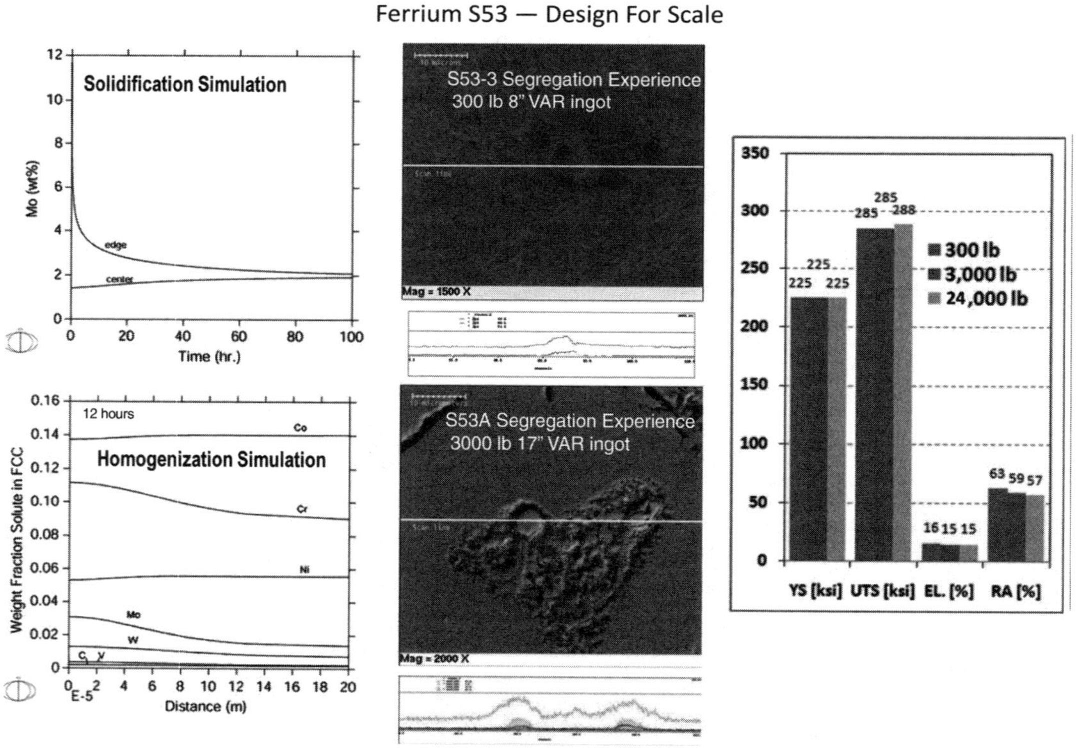

Fig. 7 Simulation of the solidification microsegregation and homogenization response in Ferrium S53, validated by ingot microanalysis, to constrain the design for full-scale production, as demonstrated by no loss of tensile properties with increasing scale. Source: Ref 31

comparison with model predictions, as depicted in the middle panels. The summary of tensile properties in the last panel demonstrates no significant loss of properties (notably including the reduction in area, %RA) with increasing scale from the 300 to the 24,000 lb. level, successfully eliminating scale-up from alloy development.

The greatest challenge of the DARPA-AIM program was forecasting the manufacturing variation to arrive at the minimum design-allowable properties aided by the limited data. Using a new method developed under the AIM initiative, prediction of Ferrium S53 property design minimums, which normally require 10 production heats of alloy and 300 individual observations, were completed using only three heats and 30 individual observations. A large simulated data set with over 300 simulations was produced using the expected process variation and mechanistic models of alloy strength. The simulated property distribution was calibrated for only 20 experimental observations from two production heats by a linear transformation method. This analysis indicated that the 1% minimum ultimate tensile strength was below the 1930 MPa design goal by at least 10 MPa. This early indication allowed additional process optimization to be completed, increasing the ultimate tensile strength by 10 MPa. The AIM analysis was repeated after a new temper was optimized, using 30 data produced from three heats of the alloy, and is shown by the solid curve in Fig. 8. The new property minimum estimate is 1930 MPa (280 ksi), meeting the property objectives. The resulting full experimental data set of 10 heats and over 600 individual observations are represented by the square points in Fig. 8. The AIM prediction is within 7 MPa (1 ksi) of the 1% minimum tensile strength statistically defined by the 10 heats data. In this AIM example over a year of development and in excess of US $500,000 would have to be expended to meet the property goals if data development had proceeded and the property deficit discovered when the full data set was complete. The first flight of a Ferrium S53 landing gear occurred on December 17, 2010.

Building on the success of Ferrium S53 a second landing gear steel design for the high toughness levels required for carrier-based aircraft has been undertaken, designated Ferrium M54. Figure 9 summarizes the timeline of technology readiness levels for these two steels at the landing gear level, along with their corresponding materials development milestones. While the timescale to qualified design allowable for Ferrium S53 was 8.5 years, Ferrium M54 met its objectives in the first iteration of design, and is on track to reach materials qualification within 5 years.

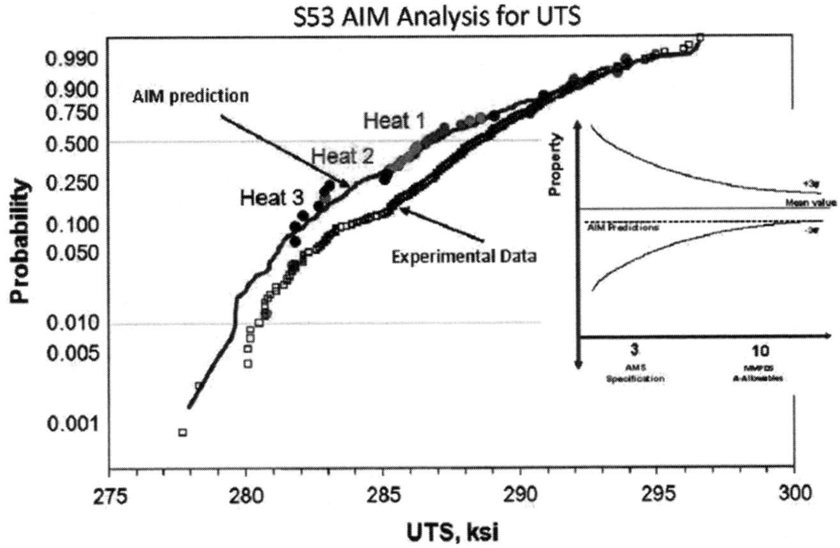

Fig. 8 AIM analysis of the ultimate tensile strength of S53 using simulation results and 30 data from three individual heats agrees well with the completed dataset from 10 heats and over 600 observations in determining the 1% minimum design allowable. Source: Ref 31

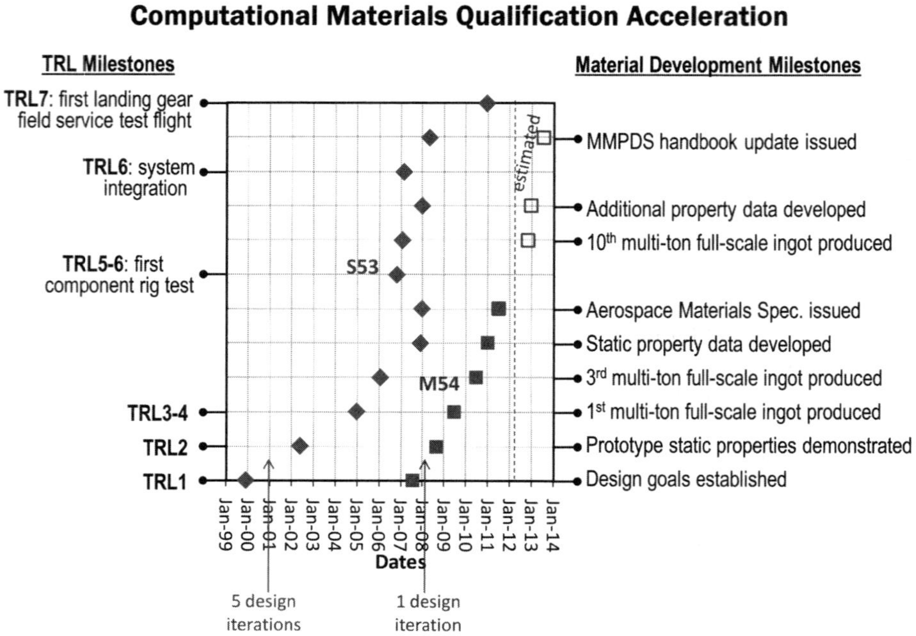

Fig. 9 Time line of landing gear level technology readiness levels (TRL) and the corresponding materials development milestones for computationally designed S53 and M54 alloys.

Research/Education Integration

Recognizing that engineering disciplines are defined by what can be practiced with a bachelor's degree, it was the vision of the founders of the SRG design consortium in 1985 to develop a methodology of computational design that can be taught to undergraduates. This has motivated a series of materials design education innovations at Northwestern University, beginning with the teaching of an undergraduate Materials Design course in 1989 (Ref 66). In a unique integration of research and education annual iterations of theoretical design optimization techniques employing the newest experimental measurements and model/simulation predictions are conducted by teams of materials science undergraduates in this novel class. The course features a series of computational sessions teaching the suite of computational design tools grounded in the materials genome fundamental databases (Ref 67) and the graphical parametric design integration approach of Fig. 6. Fostering a high level of technical design, the undergraduate teams are coached by doctoral students participating in funded design projects (Ref 68, 69). These doctoral students are in turn assisted by a broader group of graduate students contributing to projects under a special interdisciplinary doctoral cluster program in Predictive Science and Engineering Design (PSED) supported by Northwestern's Graduate School. A seminar course under the PSED program explores the role of computational tools in the interdisciplinary integration of predictive science and engineering design. A central outreach activity to introduce the new technology to a broader audience is a new master's (MS) certificate program in Integrated Computational Materials Engineering, in which first-year MS students also participate in the interdisciplinary PSED seminars, leading to a culminative integrative project within the Materials Design course. Undergraduates taking the Materials Design course in their third year can participate in the experimental validation of their design prototypes in their senior projects in the following year. To enhance recruitment to the materials program teams of students from a special "Murphy Scholars" section of a freshman-level Engineering Design and Communication course also collaborated with the undergraduate design teams, adding exploration of device applications of the new materials (Ref 70). As summarized in Fig. 10, the design, development, and accelerated flight qualification of Ferrium S53 as the first fully

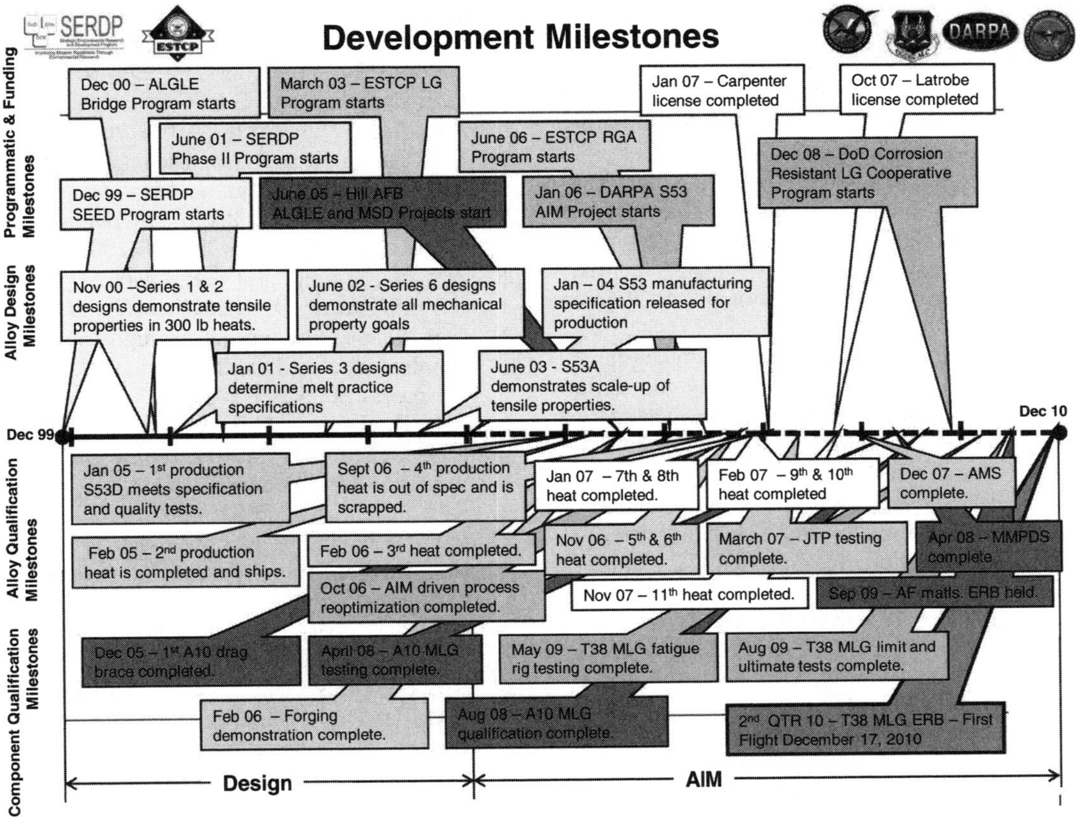

Fig. 10 Case study summary chart of the computational design and qualification of Ferrium S53 landing gear steel. The top level represents the programmatic and funding milestones of the multi-agency effort. The second level represents iterative alloy design milestones leading to the composition and processing specifications. The bottom levels represent alloy and component level AIM-based qualification, leading to the first flight in December 2010.

computationally designed and qualified material now serves as the central case study for all of these education initiatives.

Conclusions

Moving beyond a system of "technology by accident" grounded in near-random walk scientific discovery to a new system of tightly integrated science-based engineering constitutes a true revolution in materials technology, offering substantial societal benefit as well as a new justification for scientific investment. Its emergence has reestablished ferrous metallurgy as the true intellectual frontier of the materials enterprise. Our unique depth of scientific understanding in ferrous alloys will continue to serve as the central template for a general methodology of computational design for all materials, fully grounded in fundamental genomic databases.

ACKNOWLEDGMENTS

Materials design research by the SRG design consortium and its spin-off company QuesTek was supported by the US Office of Naval Research, the Defense Advanced Research Projects Agency, the Air Force Office of Sponsored Research, the Army Research Office, the Department of Energy, the National Aeronautics and Space Agency, and the National Science Foundation, with supplemental industry support. Design and qualification of the Ferrium S53 alloy was principally supported by the US joint agency Strategic Environmental Research and Development Program and Environmental Security Technology Certification Program.

REFERENCES

1. National Research Council, Application of lightweighting technology to military aircraft, vessels, and vehicles, Washington, DC, National Academies Press, 2011
2. National Research Council. Integrated computational materials engineering, a transformational discipline for improved competitiveness and national security, Washington, DC, National Academies Press, 2008
3. National Research Council, Accelerating technology transition: bridging the valley of death for materials and processes in defense systems, Washington, DC, National Academies Press, 2004
4. National Research Council, Retooling manufacturing: bridging design, materials and production, Washington, DC, National Academies Press, 2004
5. National Research Council, Materials research to meet 21st century defense needs, Washington, DC, National Academies Press, 2003
6. OSTP, Materials genome initiative for global competitiveness, Washington, DC, Office of Science and Technology Policy, 2011
7. G.B. Olson, M. Azrin, and E.S. Wright, ed., Innovations in ultrahigh-strength steel technology, In: Proceedings of the 34th Sagamore army research conference, Washington, DC, US Government Printing Office, 1990
8. J. Ågren, *Metall. Mater. Trans. A*, 2012, 43, 3453–61
9. L. Vitos, H.L. Zhang, N. Al-Zoubi, S. Lu, J-O. Nilsson, S. Hertzman, et al. *La Metall. Ital.*, 2012, 5, 19–27
10. I. Steinbach, *Mod. Sim. Mater. Sci. Eng.*, 2009, 17, 073001–31
11. H.K.D.H. Bhadeshia, *Mater. Sci. Technol.*, 2008, 24, 128–36
12. J. Hirsch, ed., Virtual fabrication of aluminum products. Weinheim: Wiley-VCH; 2006
13. G. Gottstein, ed., Integral materials modeling: towards physics-based through-process models. Weinheim: Wiley-VCH; 2007
14. G.B. Olson, *Science*, 1997, 277(5330), 1237–42
15. G.B. Olson, *Science*, 2000, 288, 993–8
16. T.A. Stephenson, C.E. Campbell, and G.B. Olson, In: R.J. Richmond, and S.T. Wu, editors. Advanced Earth-to-orbit propulsion technology, NASA conference, publication 3174, Vol 2, 1992. p 299–307
17. C.E. Campbell and G.B. Olson, *J. Comput Aid Mater. Des.*, 2001, 7, 145–70
18. C.E. Campbell and G.B. Olson, *J. Comput Aid Mater. Des.*, 2001, 7, 171–94
19. A. Saha and G.B. Olson, *J. Comput Aid Mater. Des.*, 2007, 14(2), 177–200
20. A. Saha, J. Jung, and G.B. Olson, *J. Comput. Aid Mater. Des.*, 2007, 14(2), 201–33
21. G.B. Olson, Fracture-tough high hardness stainless steel and method of making same, US patent no. 5,221,372, 1993
22. C.J. Kuehmann, G.B. Olson, J.P. Wise, and C. Campbell, Advanced case carburizing secondary hardening steels, US patent no 6,176,946 B1; 2001
23. C.J. Kuehmann, G.B. Olson, and H-J. Jou, Nanocarbide precipitation strengthened ultrahigh-strength, corrosion resistant, structural steels, US patent no. 7,160,399, 2007
24. N. Saunders and A.P. Miodownik, CALPHAD (calculation of phase diagrams), a comprehensive guide, New York, Elsevier Science, 1998
25. S.H. Carr, R. D'Oyen, and G.B. Olson, In: D. Hui, editor. Proceedings of the fourth international conference on composites engineering: international community for composites engineering, Big Island of Hawaii, 1997, p 205
26. C.M. Neubauer, J. Thomas, M. Garci, K. Breneman, G.B. Olson, and H.M. Jennings, In: Proceedings of the 10th international congress on the chemistry of cement, Gothenburg, Sweden, 2–6 June, Göteborg, Sweden, Amarkai AB and Congrex Göteborg AB, 1997
27. B. Files and G.B. Olson, In: SMST-97: Proceedings of the 2nd international conference on shape memory and superelastic technologies, Santa Clara, CA, SMST, 1997, p 281–6
28. D.S. Stolz and J.J. Schirra, AIAA 2011-1740, Proceedings of the 52nd AIAA/ASME/ASCE/AHS/ASC structures, structural dynamics and materials conference, 4–7 April 2011, Denver, Colorado, p 1–8
29. H-J. Jou, P. Voorhees, and G.B. Olson, *Superalloys*, 2004, 877–86
30. G.B. Olson, H-J. Jou, J. Jung, J.T. Sebastian, A. Misra, I. Locci, et al. *Superalloys*, Warrendale, PA, TMS, 2008, p 923–32

31. C.J. Kuehmann and G.B. Olson, In: S. Arnold, and T. Wong, ed., Models, databases, and simulation tools needed for the realization of integrated computational materials engineering, Materials Park, OH, ASM International, 2011
32. J.A. Christodoulou, In: AIAA 2012-1405, Proceedings of the 53rd AIAA/ASME/ASCE/AHS/ASC structures, structural dynamics and materials conference, 23–26 April 2012, Honolulu, Hawaii, p 1–4
33. S. Hao, B. Moran, W-K. Liu, and G.B. Olson, *J. Comput. Aid Mater. Des.*, 2003, 10(2), 99–142
34. F.J. Vernerey, C. McVeigh, W.K. Liu, B. Moran, D. Tewari, D. Parks, et al. *J. Miner. Met. Mater. Soc.*, 2006, 58, 45–51
35. F.J. Vernerey, W-K. Liu, B. Moran, and G.B. Olson, *J. Mech. Phys. Solids.*, 2008, 56, 1320–47
36. R. Tian, S. Chan, S. Tang, A.M. Kopacz, J-S. Wang, H-J. Jou, et al. *J. Mech. Phys. Solids.*, 2010, 58, 1681–700
37. R. Prasannavenkatesan, J. Zhang, D.L. McDowell, G.B. Olson, and H-J. Jou, *Int. J. Fatigue.*, 2009, 31(7), 1176–89
38. J-H. Lee, T. Shishidou, Y-J. Zhao, A.J. Freeman, and G.B. Olson, *Philos. Mag.*, 2005, 85(31), 3683–97
39. T.E. Jones, M.E. Eberhart, S. Imlay, C. Mackey, and G.B. Olson, *Phys. Rev. Lett.*, 2012, 109, 125506
40. G.B. Olson, A.J. Freeman, P.W. Voorhees, G. Ghosh, J. Perepezko, M. Eberhart, et al. In: Y-W. Kim, T. Carneiro, ed., International symposium on niobium for high temperature applications, Warrendale, PA, TMS, 2004, p 113–22
41. M. Kim, A.J. Freeman, S. Kim, J.H. Perepezko, and G.B. Olson, *Appl. Phys. Lett.*, 2005, 87
42. G. Ghosh and G.B. Olson, *Acta. Mater.*, 2007, 55(10), 3281–303
43. J.S. Montgomery and G.B. Olson, In: G. Krauss, P.E. Repas, ed., Proceedings of the R. Gilbert, Speich symposium: fundamentals of aging and tempering in bainitic and martensitic steel products, Warrendale, PA, ISS-AIME, 1992, p 177–214
44. J.S. Langer and A.J. Schwartz, *Phys. Rev. A.* 1980, 21, 948–58
45. A. Umantsev and G.B. Olson, *Scripta. Metall.*, 1993, 29, 1135–40
46. C.J. Kuehmann and P.W. Voorhees, *Metall. Mater. Trans. A.*, 1996, 27, 937–43
47. Y. Nagataki, J.B. Cohen, and G.B. Olson, In: Proc advanced materials for the 21st century: the Julia R. Weertman symposium, Warrendale, PA, TMS, 1999
48. K.C. King, P.W. Voorhees, and G.B. Olson, *Metall. Trans.*, 1991, 22A, 2199–210
49. K.C. King, G.B. Olson, and T. Mura, In: J.J. Wu, T.C.T. Ting, D.M. Barnett, ed., Modern theory of anisotropic elasticity and applications, Philadelphia PA, SIAM, 1991, p 103–21
50. J. Wise, Doctoral thesis, Northwestern University, 1998
51. C.J. Kuehmann and G.B. Olson, *Adv. Mater. Process.*, 1998, 153(5), 40–3
52. H.E. Lippard, C.E. Campbell, T. Bjorklind, U. Borggren, P. Kellgren, V.P. Dravid, et al. *Met. Trans. B.*, 1998, 29B, 205–10
53. G.B. Olson and M. Cohen, In: F.R.N. Nabarro, ed., Dislocations in solids, Vol 7, Amsterdam, North-Holland, 1986, p 295–407
54. G. Ghosh and G.B. Olson, *Acta. Metall. Mater.*, 1994, 42, 3361–70
55. G. Ghosh and G.B. Olson, *Acta. Metall. Mater.*, 1994, 42, 3371–9
56. G. Ghosh and G.B. Olson, *J. Phase. Eq.*, 2001, 22(3), 199–207
57. G.B. Olson, H.K.D.H. Bhadeshia, and M. Cohen, *Metall. Trans. A.*, 1990, 21, 805–9
58. G.B. Olson, In: G.B. Olson, M. Azrin, and E.S. Wright, ed., 34th Sagamore army materials research conference proceedings, innovations in ultrahigh-strength steel technology, 1990, p 3–66
59. G.B. Olson and Z.D. Feinberg, In: S. Arnold, and E. Pereloma, ed., Phase transformations in steels: Fundamentals and diffusion controlled transformations, Vol 1, Cambridge, Woodhead Publishing Ltd., 2012
60. J.R. Rice and J-S. Wang, *Mater. Sci. Eng. A.*, 1989, 107, 23
61. R. Wu, A.J. Freeman, and G.B. Olson, *Science* 1994, 265, 376–80
62. W.T. Geng, A.J. Freeman, and G.B. Olson, *Phys. Rev. B.*, 2001, 63, 165415
63. L. Zhong, A.J. Freeman, R. Wu, and G.B. Olson, *Phys. Rev. B.*, 2000, 62, 13938–41
64. C. Kantner, Doctoral thesis, Northwestern University, 2002
65. C.J. Kuehmann, B.E. Tufts, and P. Trester, *Adv. Mater. Process.*, 2008, 166, 37–40
66. G.B. Olson, In: P.K. Liaw, J.R. Weertman, H.L. Markus, and J.S. Santner, ed.,

E. Morris, fine symposium. Warrendale, PA, TMS-AIME, 1991, p 41–8
67. G.B. Olson, *J. Int. Eng. Ed.*, 2001, 17(4/5), 468–71
68. A.F. McKenna, J.E. Colgate, and G.B. Olson, Proceedings of the American Society for Engineering (ASEE) annual conference, Chicago, Illinois, Washington D.C, ASEE, 2006
69. M.V. Manuel, A.F. McKenna, and G.B. Olson, IJEE, 2008, 24(2), 260–5
70. A.F. McKenna, J.E. Colgate, S.H. Carr, and G.B. Olson, *Int. J. Eng. Educ.*, 2006

Transformations—Selected Works of G.B. Olson on Materials, Microstructure, and Design
C.E. Campbell, M.V. Manuel, and W. Xiong, editors

CHAPTER 27

Materials Genomics—From CALPHAD to Flight*

G.B. Olson, Northwestern University and QuesTek Innovations LLC
C.J. Kuehmann, QuesTek Innovations LLC

Abstract

Unique depth of scientific knowledge has allowed physical metallurgy to lead the development of a successful computational materials design methodology grounded in a system of fundamental databases. Expansion of the approach to embrace acceleration of the full development and qualification cycle has resulted in the first fully computationally designed and qualified material reaching the flight stage. The example of high-performance alloys provides the template for a general methodology of computational design for all materials.

Introduction

Several recent US national academy studies of materials and manufacturing (Ref 1–4) have emphasized the unique opportunity of computational materials engineering. While the breadth of the best-known 2008 Integrated Computational Material Engineering (ICME) study (Ref 3) creates an impression of a field in its infancy, the 2004 Accelerated Technology Transition (ATT) study (Ref 1), uniquely charged with identifying best industrial practices for accelerating transition of materials and process technology, elucidated practices which are now quite mature and ready for broader dissemination. The achievements of these practices have now been highlighted in a recent Lightweighting Technology study (Ref 4), and the importance of the underlying fundamental databases now known as the Materials Genome have been heralded in a recent report of the US President's Office of Science and Technology Policy (Ref 5).

As part of a Materials Genome viewpoint set, this article is an updated and condensed version of a recent overview (Ref 6) which includes discussion of the underlying scientific foundation of the engineering achievements discussed here. Our emphasis in this article is the overarching engineering architecture of computational materials design and accelerated qualification, and its level of achievement relative to stated acceleration goals of the ongoing US research initiative (Ref 5). We note that European efforts have made significant contributions to this technology, notably in computational thermodynamics and its application (Ref 7), computational quantum mechanics applications (Ref 8), microstructural simulation tools (Ref 9), computational materials design (Ref 10), and through-process modeling (Ref 11, 12). This viewpoint paper focuses on developments in the United States which underly the ongoing Materials Genome

*G.B. Olson and C.J. Kuehmann, "Materials Genomics: From CALPHAD to Flight," *Scripta Mater.*, 70 (2014) 25–30

initiative, and remain unique in their scope of engineering integration. The new architecture of science-based computational materials engineering is here illustrated with the case study of the first flight-qualified designer alloy.

Background—Emergence of a Materials Genome Infrastructure

A timeline of the evolution of ICME technology is summarized in Fig. 1. While broad debate continues over what a "materials genome" could be, it is clear that the genome we already have is the CALPHAD (CALculation of PHAse Diagrams) fundamental database and software system (Ref 13), whose origins can be traced back to Kaufman and Cohen in 1956 (Ref 14). Becoming an organized international collaboration in the 1970s, the system has broadened to encompass a range of fundamental phase-level thermochemical and physical properties spanning solution thermodynamics, molar volumes, atomic mobilities and elastic constants, treating phases as the core building blocks of multiphase materials. Originally based entirely on empirical thermodynamic measurements, the improved accuracy of atomistic predictive methods such as density functional theory (DFT) has more recently made them an equal partner in the development of this fundamental database system.

Recognizing the potential of these fundamental databases to make predictions from scientific understanding highly quantitative, the emergence of a second generation of higher-accuracy CALPHAD tools in the 1980s inspired the founding of the SRG design consortium in 1985 (Ref 15) with the goal of a general systems-based methodology of computational materials design, using metal alloys as first examples (Ref 15–17)—represented here as the second line in Fig. 1. Integrating materials science, continuum mechanics and quantum physics, the systems design approach has featured a suite of validated computational tools enabled by an iterative interaction of theory,

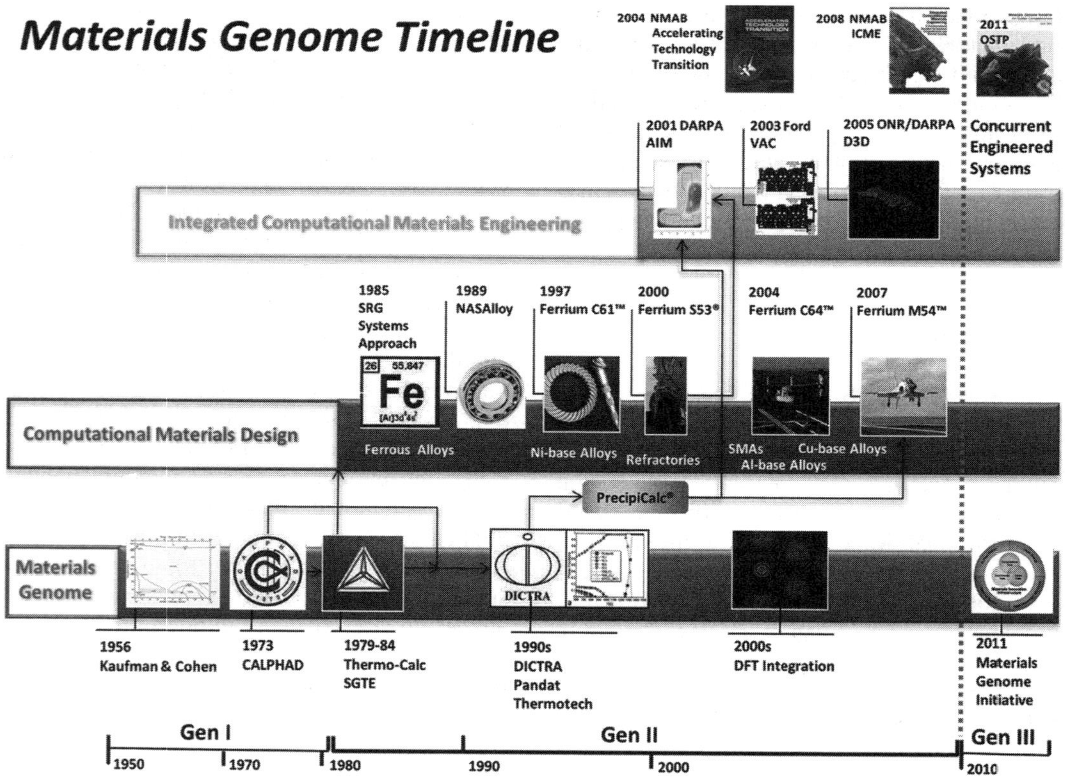

Fig. 1 Materials Genome technology timeline.

simulation and experiment, spanning the full range of process/structure/property/performance relations, as described in detail elsewhere (Ref 15–17). An initial focus on high-performance steels exploiting not only their unique depth of predictive scientific understanding but also their level of fundamental database development has demonstrated the capabilities of the approach, providing a range of ultrahigh-strength ferrous alloys (Ref 18–22) now in commercial use. Building on an established small business CALPHAD supply chain network (Ref 1), the founding of a computational materials design company in 1997 provided commercial materials design services and successful licensing of designed alloys [23–25], with broadened application of the approach to Al-, Ni-, Ti-, and Cu-based alloys. Recognizing the central role of chemical thermodynamics in the science of all materials (as reflected in the structure of undergraduate materials curricula), the fundamental genomic database system centered on CALPHAD thermodynamics has provided a framework for computational materials design with broad applicability, as attested by university projects in the 1990s demonstrating its application to polymers, ceramics and composites (Ref 26–28).

As represented by the third line in Fig. 1, the successes in computational materials design in the 1990s helped to bring about the seminal DARPA-AIM (US Defense Advanced Research Projects Agency: Accelerated Insertion of Materials) initiative of the 2000s, expanding the scope of science-based computational materials engineering to embrace the full component-level process optimization and material qualification scheme in order to significantly compress the full materials development cycle (Ref 29). Under this initiative, the PrecipiCalc precipitation simulator (Ref 30, 31), grounded in the CALPHAD thermodynamics and mobility database and software system, was efficiently linked to macroscopic component-level process simulation tools to successfully accelerate process optimization and accurately forecast manufacturing variation with calibration by minimal datasets for the test case of aeroturbine disc manufacturing.

As the principal follow-on to the DARPA-AIM successes, the recent five-year "D3D" Digital Structure consortium program (Ref 32), jointly supported by the US Office of Naval Research (ONR) and DARPA, extended the fidelity of the AIM modeling tools by integrating a multiscale suite of 3-D tomographic microstructural characterization tools with a new generation of 3-D microstructural simulators. Combining local electrode atom probe (LEAP), focused ion beam (FIB), and metallographic serial sectioning techniques to span the nanometer, micron, and millimeter microstructural scales, respectively, computer-aided tomographic reconstructions provided a new level of spatial characterization informing novel numerical simulators addressing strength, toughness, and fatigue resistance, using advanced steels as test cases. This has notably included several generations of multiscale ductile fracture simulators (Ref 33–36) as well as the direct micromechanical simulation of fatigue nucleation processes at observed nucleants (Ref 37). Extending the 3-D approach to bond topological analysis of quantum mechanical calculations has also provided new heuristics to guide materials design component selection (Ref 38, 39).

Computational Design Successes—Flying Cybersteels

The first designer alloy to demonstrate the full cycle of computational design and AIM qualification is the Ferrium S53 (AMS 5922) corrosion-resistant landing gear steel allowing a drop-in replacement for current nonstainless landing gear steels, eliminating the need for cadmium plating (Ref 25, 40, 41). S53 is a secondary-hardening martensitic steel efficiently strengthened by M_2C carbide precipitates and containing sufficient Cr content to provide passivation against general corrosion. Employing the systems approach to materials design integration (Ref 16), the full structural hierarchy of the alloy is represented by the flow-block diagram in Fig. 2. The sequential processing steps experienced by the alloy are depicted on the left-most column of the flow-block diagram and are restricted to existing processes for steels employed in current structural aircraft applications to maximize manufacturability. The subcomponents of the alloy system are connected by process-structure and structure-property relationships, as addressed by the suite of mechanistic modeling tools just described.

The application of the system flow-block diagram in conjunction with the computational models begins with identification of key design trade-offs motivated by the full set of quantitative property objectives represented by the

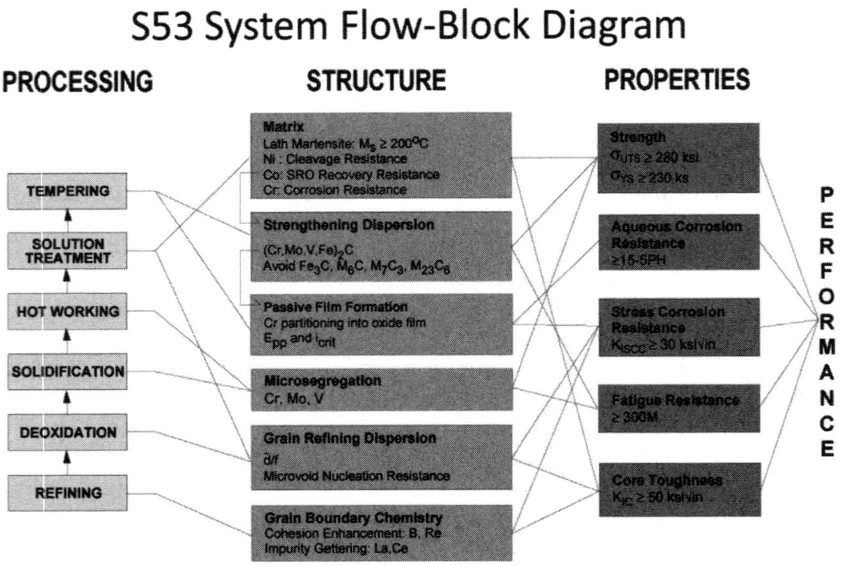

Fig. 2 System chart representing corrosion-resistant martensitic landing-gear steel.

right column. Early conceptual design then defines the specific microstructural subsystems represented in the central column. To address the strong conflict between corrosion resistance and strengthening efficiency, a thermodynamic model for the metastable spinel oxide was developed to precisely control the passive film defining corrosion resistance (Ref 19). The multicomponent thermodynamics of the body-centered cubic alloy matrix was then exploited to achieve a Cr potential providing a passive film equivalent to a 15Cr alloy using less than 12Cr in the design.

Integration of the design models was undertaken by graphical parametric design in which mechanistic understanding motivated the definition of key scalar parameters computable from phase composition vectors and diffusion matrices via the fundamental CALPHAD databases.

By iterative design and prototyping with full-scale microstructural evaluation to test all models, local correction factors are passed back to design models for increased accuracy in the new composition region of interest. In this way, accurate design solutions are typically achieved within three iterations of prototyping. For the very challenging objectives of the Ferrium S53 design, successful achievement of all objectives required five iterations spanning a three-year period. Experience has shown that the number of required iterations and associated time scale will steadily decrease as the accuracy of general CALPHAD databases continues to increase.

Extending Computational Design to Qualification and Implementation

The design of the Ferrium S53 steel ran concurrently with the DARPA-AIM initiative on accelerated materials technology transition. The steel was brought into the AIM program to serve as a first demonstrator of the new methodology. This began by running the design models within the iSIGHT design integration software (Ref 29) to apply Monte Carlo simulation of parametric sensitivities in support of a higher level of robust design practice for reduced intrinsic variation within allowed composition and process tolerances. To circumvent the traditional problem of process scaleup in empirically developed materials, Ferrium S53 was the first example of a "design for scale" approach in which process simulation was used to constrain the up-front alloy design to be compatible with full-scale processing. Solidification simulations employed the DICTRA multicomponent diffusion code (Ref 42) for candidate design compositions under the cooling rates of full-scale production ingots. An acceptable level of microsegration was defined by simulating its response to homogenization treatment. As production scale increased, sections of as-solidified ingots were obtained to measure the microsegregation for comparison with model prediction. Tensile properties demonstrated no significant loss of properties (notably including the reduction-in-area, %RA) with increasing scale from 300 to the 24,000 lb. level, successfully eliminating scaleup from the alloy development.

The greatest challenge of the DARPA-AIM program was the forecast of manufacturing variation to arrive at minimum design-allowable properties aided by limited data. Using a new linear transformation method developed under the AIM initiative, prediction of Ferrium S53 property design minima, which normally require 10 production heats of alloy and 300 individual observations, were completed using only 3 heats and 30 individual observations. A large simulated data set with over 300 simulations was produced using expected process variation and the mechanistic models of alloy strength. The simulated property distribution was then calibrated by linear transformation using only 30 data produced from three heats of the alloy. The full experimental data set of 10 heats and over 600 individual observations confirmed the AIM prediction was within 7 MPa (1 ksi) of the 1% minimum tensile strength statistically defined by the 10 heats of data.

Building on the success of Ferrium S53, a second landing gear steel design meeting the high toughness levels required for carrier-based aircraft has been undertaken, designated Ferrium M54. Figure 3 summarizes for these two steels the timeline of technology readiness levels (TRL) at the landing-gear level, along with their corresponding materials development milestones. While the timescale to qualified design allowables for Ferrium S53 was 8.5 years, the Ferrium M54 met its objectives in the first iteration of design, and is on track to reach materials qualification within 5 years, corresponding to the stated acceleration goal of the US Materials Genome Initiative (Ref 5).

Materials Genome Future

Figure 4 summarizes the current set of fundamental databases, models and methodologies of the Materials Genome system described here. New opportunities to significantly expand capabilities are represented by dashed boxes. At the most foundational level represented at the bottom of the figure is the opportunity for a large, publically shared database system of "Proto-Data" corresponding to a central repository of evolving raw fundamental data such as measured phase relations, calorimetric measurements and DFT predictions. These can be periodically queried during database assessments at the "Processed Data" level in support of both commercial and open CALPHAD databases. While the current output of the AIM

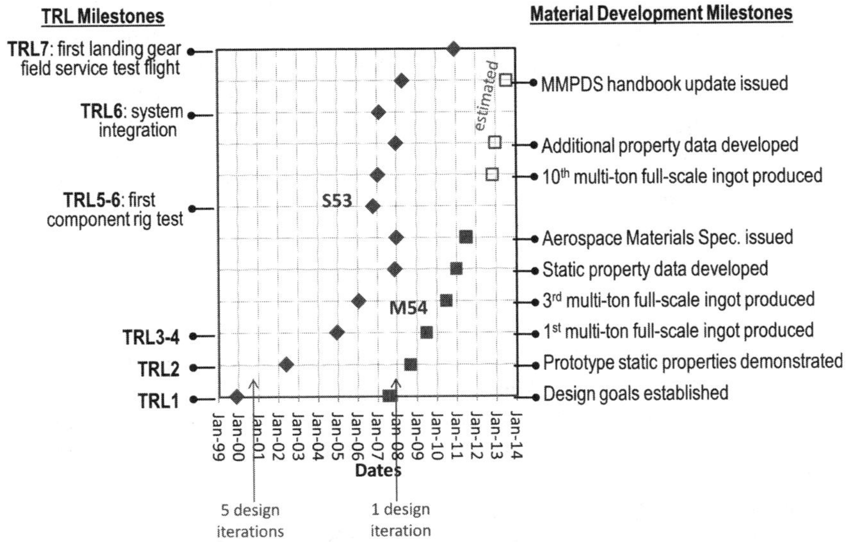

Fig. 3 Timeline of landing-gear-level technology readiness levels (TRLs) and corresponding materials development milestones for computationally designed S53 and M54 alloys.

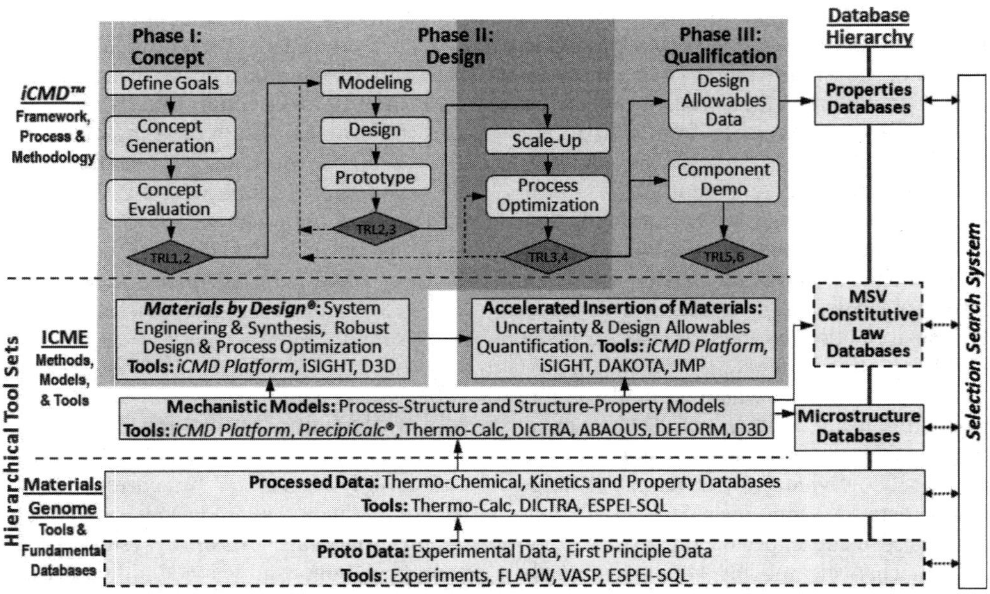

Fig. 4 Hierarchy of present and future Materials Genome methods, tools, and databases.

methodology is a fixed dataset of material minimum properties that can be queried by materials selection search systems, a new opportunity offered by the inherent predictability of designed materials is another evolving database of model-generated "Microstructural State Variable" constitutive laws (dashed box at middle right) to integrate in the adaptive tuning of microstructural evolution in manufacturing and service, to rapidly respond to new experience in materials application. The opportunity also exists to apply the same selection search systems developed for material selection to component and phase selection in microstructure design by directly linking to the full set of genomic fundamental databases.

Conclusions

Moving beyond a system of "technology by accident" grounded in near-random-walk scientific discovery to a new system of tightly integrated science-based engineering constitutes a true revolution in materials technology, offering substantial societal benefit as well as a new justification for scientific investment. Clear opportunities exist for the further expansion of the existing Materials Genome system to further enhance its value creation potential.

ACKNOWLEDGMENTS

Materials design research of the SRG design consortium and its spinoff company QuesTek have been supported by the US Office of Naval Research (ONR), Defense Advanced Research Projects Agency (DARPA), Air Force Office of Sponsored Research (AFOSR), Army Research Office (ARO), Department of Energy (DOE), National Aeronautics and Space Agency (NASA), and the National Science Foundation (NSF), with supplemental industry support. Design and qualification of the Ferrium S53 alloy was principally supported by the US joint agency Strategic Environmental Research and Development Program (SERDP) and Environmental Security Technology Certification Program (ESTCP).

REFERENCES

1. National Research Council, Accelerating Technology Transition: Bridging the Valley of Death for Materials and Processes in Defense Systems, National Academies Press, Washington, DC, 2004
2. National Research Council, Retooling Manufacturing: Bridging Design, Materials and Production, National Academies Press, Washington, DC, 2004

3. National Research Council, Integrated Computational Materials Engineering: A Transformational Discipline for Improved Competitiveness and National Security, National Academies Press, Washington, DC, 2008
4. National Research Council, Application of Lightweighting Technology to Military Aircraft, Vessels, and Vehicles, National Academies Press, Washington, DC, 2011
5. OSTP, Materials genome initiative for global competitiveness, Office of Science and Technology Policy, Washington, DC, 2011
6. G.B. Olson, *Acta. Mater.*, 61, (2013), 771–781
7. J. Ågren, *Metall. Mater. Trans. A*, 43, (2012), 3453–3461
8. L. Vitos, H.L. Zhang, N. Al-Zoubi, S. Lu, J.-O. Nilsson, S. Hertzman, G. Nilson, and B. Johansson, *La Metallurgia Italiana* 5, (2012), 19–27
9. I. Steinbach, *Modell. Simul. Mater. Sci. Eng.*, (2009), 073001–73031
10. H.K.D.H. Bhadeshia, *Mater. Sci. Technol.*, 24, (2008), 128–136
11. J. Hirsch (Ed.), Virtual Fabrication of Aluminum Products, Wiley-VCH, Weinheim, 2006
12. G. Gottstein (Ed.), Integral Materials Modeling: Towards Physics–Based Through-Process Models, Wiley-VCH, Weinheim, 2007
13. N. Saunders and A.P. Miodownik, CALPHAD (Calculation of Phase Diagrams): A Comprehensive Guide, Elsevier, New York, 1998
14. L. Kaufman and M. Cohen, *Trans. AIME*, 206 (1956), 1393
15. G.B. Olson, M. Azrin, and E.S. Wright (Ed.), Innovations in ultrahigh-strength steel technology, in: Proceedings of the 34th Sagamore Army Research Conference, US Government Printing Office, Washington, DC, 1990
16. G.B. Olson, *Science* 277, (1997), 1237–1242
17. G.B. Olson, *Science* 288, (2000), 993–998
18. T.A. Stephenson, C.E. Campbell, and G.B. Olson, sdnbn, in: R.J. Richmond, and S.T. Wu (Ed.), Advanced Earth-to-Orbit Propulsion Technology, NASA Conference, Publication 3174, Vol 2, 1992, p 299–307
19. C.E. Campbell and G.B. Olson, *J. Comput. Aided Mater. Des.*, 7 (2001), 145–170
20. C.E. Campbell and G.B. Olson, *J. Comput. Aided Mater. Des.*, 7 (2001), 171–194
21. A. Saha and G.B. Olson, *J. Comput. Aided Mater. Des.*, 14 (2007), 177–200
22. A. Saha, J. Jung, and G.B. Olson, *J. Comput. Aided Mater. Des.*, 14 (2007), 201–233
23. G.B. Olson, Fracture-tough high hardness stainless steel and method of making same, US patent no. 5,221,372, 1993
24. C.J. Kuehmann, G.B. Olson, J.P. Wise, and C. Campbell, Advanced case carburizing secondary hardening steels, US patent no. 6,176,946 B1, 2001
25. C.J. Kuehmann, G.B. Olson, and H.-J. Jou, Nanocarbide precipitation strengthened ultrahigh-strength, corrosion resistant, structural steels, US patent no. 7,160,399, 2007
26. S.H. Carr, R. D'Oyen, and G.B. Olson, in: D. Hui (Ed.), Proceedings of the Fourth International Conference on Composites Engineering: International Community for, Composites Engineering, 1997, p 205
27. C.M. Neubauer, J. Thomas, M. Garci, K. Breneman, G.B. Olson, and H.M. Jennings, in: Proceedings of the 10th International Congress on the Chemistry of Cement, Gothenburg, Sweden, 2–6 June, Amarkai AB and Congrex Goteborg AB, Goteborg, 1997
28. B. Files and G.B. Olson, in: SMST-97: Proceedings of the Second International Conference on Shape Memory and Superelastic Technologies, *SMST*, Santa Clara, CA, 1997, p 281–286
29. D.S. Stolz and J.J. Schirra, in: AIAA 2011-1740, Proceedings of the 52nd AIAA/ASME/ASCE/AHS/ASC Structures, Structural Dynamics and Materials Conference, 4–7 April 2011, Denver, CO, p 1–8
30. H.-J. Jou, P. Voorhees, and G.B. Olson, *Superalloys*, (2004), 877–886
31. G.B. Olson, H.-J. Jou, J. Jung, J.T. Sebastian, A. Misra, and I. Locci, et al., *Superalloys*, TMS, Warrendale, PA, 2008, p 923–932
32. J.A. Christodoulou, in: AIAA 2012–1405, Proceedings of the 53rd AIAA/ASME/ASCE/AHS/ASC Structures, Structural Dynamics and Materials Conference, 23–26 April 2012, Honolulu, HI, p 1–4
33. S. Hao, B. Moran, W.-K. Liu, and G.B. Olson, *J. Comput. Aided Mater. Des.*, 10 (2), (2003), 99–142

34. F.J. Vernerey, C. McVeigh, W.K. Liu, B. Moran, D. Tewari, and D. Parks, et al., *JOM* (2006), 45–51
35. F.J. Vernerey, W.-K. Liu, B. Moran, and G.B. Olson, *J. Mech. Phys. Solids.*, 56 (2008), 1320–1347
36. R. Tian, S. Chan, S. Tang, A.M. Kopacz, J.-S. Wang, and H.-J. Jou, et al., *J. Mech. Phys. Solids.*, 58 (2010), 1681–1700
37. R. Prasannavenkatesan, J. Zhang, D.L. McDowell, G.B. Olson, and H.-J. Jou, *Int. J. Fatigue.*, 31 (7) (2009), 1176–1189
38. J.-H. Lee, T. Shishidou, Y.-J. Zhao, A.J. Freeman, and G.B. Olson, *Philos. Mag.*, 85 (31), (2005), 3683–3697
39. T.E. Jones, M.E. Eberhart, S. Imlay, C. Mackey, and G.B. Olson, *Phys. Rev. Lett.*, 109 (2012), 125506
40. C.J. Kuehmann and G.B. Olson, in: S. Arnold, and T. Wong (Ed.), Databases, and Simulation Tools Needed for the Realization of Integrated Computational Materials Engineering, ASM, 2011
41. C.J. Kuehmann, B.E. Tufts, and P. Trester, *Adv. Mater. Processes.*, 166 (2008), 37–40
42. H.E. Lippard, C.E. Campbell, T. Bjorklind, U. Borggren, P. Kellgren, and V.P. Dravid, et al., *Metall. Mater. Trans. B*, 29B (1998), 205–210

CHAPTER 28

Cybermaterials—Materials by Design and Accelerated Insertion of Materials*

W. Xiong, Northwestern University
G.B. Olson, Northwestern University and QuesTek Innovations LLC

Abstract

Cybermaterials innovation entails an integration of Materials by Design and accelerated insertion of materials (AIM), which transfers studio ideation into industrial manufacturing. By assembling a hierarchical architecture of integrated computational materials design (ICMD) based on materials genomic fundamental databases, the ICMD mechanistic design models accelerate innovation. We here review progress in the development of linkage models of the process-structure-property-performance paradigm, as well as related design accelerating tools. Extending the materials development capability based on phase-level structural control requires more fundamental investment at the level of the Materials Genome, with focus on improving applicable parametric design models and constructing high-quality databases. Future opportunities in materials genomic research serving both Materials by Design and AIM are addressed.

Background—ICMD Blueprint

The far-reaching multi-agency enterprise, Materials Genome Initiative (MGI) (Ref 1, 2), highlights computational materials design techniques grounded in fundamental databases, which can support an ambition of decreasing the full development cycle of new materials from the present 10–20 years to ≤ 5 years. As a subfield of the broader field of integrated computational materials engineering (ICME), which includes modeling of existing materials (Ref 3, 4), the MGI centers on design of new materials and their accelerated qualification through the inherent predictability of designed systems. Creation of the infrastructure of this technology has been a global activity as summarized in a recent series of viewpoint papers on materials genomics (Ref 5–10). Particularly notable has been the design work of Bhadeshia (Ref 8) at the University of Cambridge and his former students including Harada (Ref 11, 12) and Reed. (Ref 13) The highest achievements of full-cycle compression have been demonstrated in US research, which will be the focus of this paper.

In the development of applied engineering materials design powered by fundamental thermodynamic and kinetic genomic databases, a hierarchical infrastructure called ICMD presents a proven scenario of materials genomic design for accelerated engineering innovation. As indicated in Fig. 1, the backbone of the

*W. Xiong and G.B. Olson, "Cybermaterials: Materials by Design and Accelerated Insertion of Materials," *npj Computational Materials*, 2016 (published online February 12, 2016), doi: 10.1038/npjcompumats.2015.9

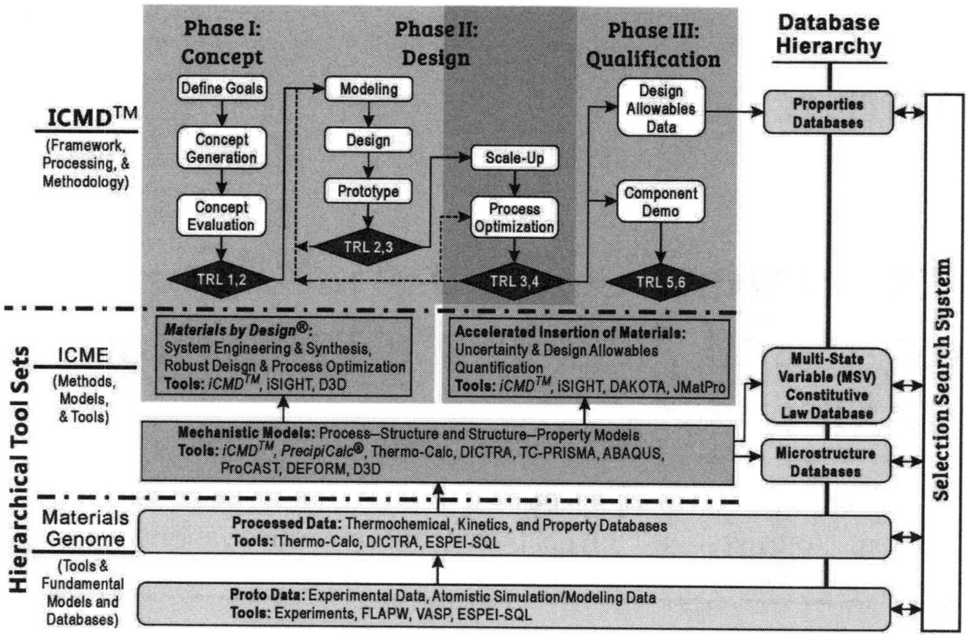

Fig. 1 Overall hierarchical architecture of ICMD methods, tools and techniques are highlighted in green and databases are marked in yellow. Please note that iCMD is the toolkit used during application of ICME methods based on Materials Genome. Figure is reprinted with permission from Elsevier.

ICMD infrastructure is composed of Materials by Design and accelerated insertion of materials (AIM) techniques. The application of both techniques spans the entire course of materials innovation, which can be divided into three phases: concept implementation, materials and processing design, and material qualification. The quality of the materials innovation is determined by the mechanistic models applied in the ICMD framework, which follow the universal process-structure-property-performance paradigm in materials science (Ref 14). In Materials by Design, both process-structure and structure-property models are evaluated and refined to maximize the model-predictability grounded in the Materials Genome, which is powered by fundamental research on high-quality databases. Although the AIM technique also takes mechanistic models as its basis, it requires extra effort in uncertainty quantification, design sensitivity analysis, integrated component-level process simulation and probabilistic prediction of manufacturing variation.

A recent review of ICMD application highlights notable achievements made in ferrous materials (Ref 15). This review provides further detail on the linkage models for Materials by Design, and methods of qualification adopted in AIM for a technology transfer from lab-scale materials innovation to industrial commercial practice. It is formulated as a snapshot of materials innovations on selected alloy designs for improving mechanical performance. Its foundation is sufficiently general that all levels of materials design should share the same principles, not only for metal and alloys, but also for oxides and polymers (Ref 14, 16, 17).

Here we revisit some key process-structure and structure-property linkage models applied in Materials by Design followed by a technical review of the AIM method application in projects supported by DARPA (Defense Advanced Research Projects Agency) and the National Aeronautics and Space Administration. Although the above linkage models are applied in AIM technology as well, related discussion on AIM will be more about its unique efforts on model integration, uncertainty quantification, and design sensitivity control. Future research opportunities and perspectives are presented as concluding remarks in the last section.

Materials by Design—Design Engine

The paradigm of materials design guided by governing processing-structure-properties-performance is a universal principle for developing advanced materials (Ref 14). Predictive-science-based computational modeling reinforced by experimental calibration allows us to perform multiscale modeling for various properties with temperature/time dependence. Therefore, the development of the process-structure and structure-property models will be beneficial to Materials by Design, and boost AIM applications.

As shown in Fig. 2, process-structure models empower two levels of modeling: phase stability and microstructural evolution, which establish a fundamental basis for structure-property design models. With particular engineering design purpose, high-quality engineering models are integrated with science-based mechanistic models forming a hierarchical model system.

Process–Structure Relation and Design Models

In the course of systems design for advanced materials, a thorough understanding of thermodynamics and kinetics enables processing design, which often starts with a parametric design for chemical composition. As shown in Fig. 3, intrinsic design parameters for materials processing encompass chemical composition, temperature, pressure and external force fields, on which even a small adjustment may lead to significantly different evolution of microstructure driven by the thermodynamics and kinetics of the system. The CALPHAD (calculation of phase diagrams) method (Ref 18) and its correlated modeling techniques quantitatively express our mechanistic understanding of the process-structure linkage relations.

Thermodynamics is the core of the precipitation simulation. Considerable efforts have been made on integration of density functional theory (DFT)-based atomistic modeling, CALPHAD and phase equilibrium measurements for assembling materials genomic databases. DFT calculations for zero-kelvin thermal-chemical properties, such as enthalpy of formation, are often used as protodata to support CALPHAD thermodynamic modeling. This is now considered as a routine method to perform physically sound thermodynamic modeling, which can sometimes circumvent the restriction of calorimetry (e.g., radiative

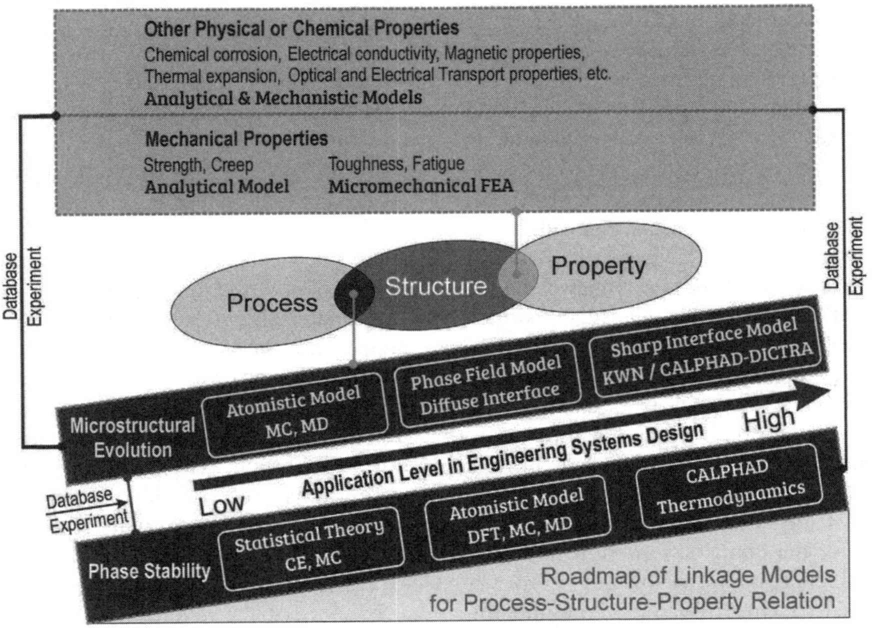

Fig. 2 Materials design roadmap for linkage models of process-structure and structure-property relations. CE, cluster expansion; MC, Monte Carlo; MD, molecular dynamics.

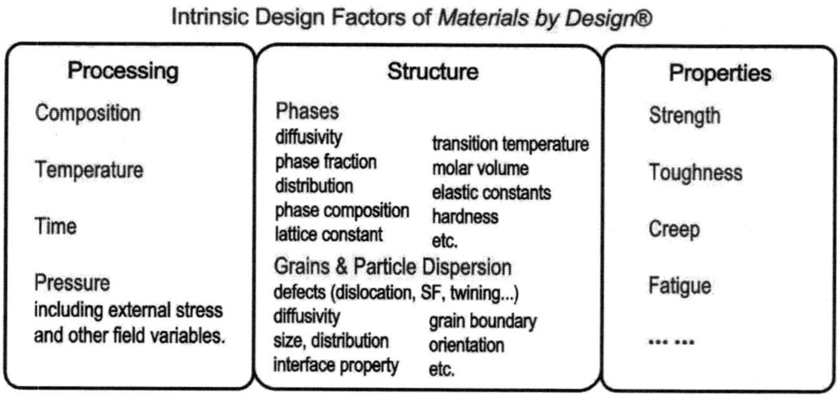

Fig. 3 Intrinsic design factor of the process-structure-property-performance paradigm. The most common attributes of design factors for microstructure control are presented.

alloy systems) (Ref 19, 20), and assist in determining reliable experimental data sets (Ref 19, 21–23). At present, there are many commercial software packages available based on the CALPHAD technique, for example, Thermo-Calc, Pandat, FactSage, MTDATA (Ref 24, 25). However, not until the release of the DICTRA (diffusion-controlled transformations) package by the Thermo-Calc software company in early 1990s (Ref 26, 27) for diffusion kinetic modeling, did the CALPHAD approach start to demonstrate its capability of handling sophisticated process-structure modeling. The significance of the CALPHAD method nowadays has exceeded its original definition as phase diagram computation. It is indisputable that the modern CALPHAD approach (Ref 28) serves as a fundamental technique for establishing the materials genomic databases, and is widely recognized as a foundational tool for Materials by Design.

When related CALPHAD genomic databases are available, although thermodynamic and diffusion calculations can be readily performed, one should be aware of application limits and evaluate the quality of database predictions prior to a pragmatic design, or before adopting CALPHAD as an input for explicit microstructural simulation. One should also bear in mind that a fundamental CALPHAD database generally provides the Gibbs energy of bulk phases by neglecting the contribution from interfaces to the total energy of the system. However, when modeling precipitation strengthening, the Gibbs-Thompson effects, that is, size-dependent capillary effects, need to be considered. When experimental data determined by atom probe tomography (APT) is available, a semi-log plot of composition versus inverse of particle size can often generate a good trajectory defining equilibrium tie-lines and initial critical nucleus size defining interfacial energy (Ref 29, 30). In addition, elastic coherency (misfit) energy, G_{el}, can be evaluated for a further energy penalty to the precipitation when particles are coherent. In the work by Olson *et al.*, (Ref 31) the method of estimating coherent interfacial energy (G_{coh}) and elastic misfit energy was demonstrated using single sensor differential thermal analysis (Ref 32) combined with APT.

Another issue related to phase stability should be addressed when applying CALPHAD thermodynamics is evaluation of competitive growth behavior of different phases. As CALPHAD thermodynamic database calculates phase equilibrium as a default option, some phases with sluggish formation should be carefully evaluated in a quantitative comparison with experimental data determined from materials processing, which are generally nonequilibrium. For example, in the work performed by Wusatowska-Sarnek *et al.*, (Ref 33) experimental phase fraction were initially compared with calculated results under full equilibrium. Good agreement was ultimately obtained when the calculation was performed using constrained equilibria, under which topologically close-packed phases are excluded because of their sluggish formation.

In view of the above, for the sake of parametric design, experimental validation is essential to assess the accuracy of a CALPHAD model-prediction. This can in turn contribute to database refinement in support of higher fidelity applications of Materials by Design and AIM methodologies. Similar to the case in thermodynamics, diffusional kinetic databases also require adjustments for the constrained case in a quantitative comparison with experimental results measured after specific processing. For developing dual microstructure processing turbine disk in a project supported by National Aeronautics and Space Administration Glenn Research Center (Ref 31), a multicomponent CALPHAD diffusion database was constructed by National Institute of Standards and Technology, which demonstrates a good predictability after calibration on the atomic mobility of Nb, Al, Ti, and Cr in fcc using multicomponent diffusion couples made between disc alloys and pure Ni. A need for calibration of CALPHAD diffusivity databases originates from the simplifications made by omitting the influence of fast diffusion path, for example, dislocations, phase/grain boundaries. A diffusion scaling factor, D_{scale}, can be applied to the diffusivity matrix for calibration, particularly for a precipitation simulation.

A well-calibrated CALPHAD thermodynamic and kinetic database is a prerequisite of accurate parametric control of microstructural length and time scales. For example, effective precipitation strengthening relies on the size control of the particle precipitation, and thus requires predictive models for the kinetic process of particle size evolution. A good example is the extensive application of the coarsening model represented by the Lifshitz-Slyozov-Wagner (LSW) theory (Ref 34–38) for precipitation size evolution in a dilute binary system. To couple with the CALPHAD thermodynamic/kinetic databases for multicomponent and multiphase simulation, several models (Ref 39, 40) were proposed based on extensions of the LSW theory. One of the simplified models was proposed by Lee *et al.*, (Ref 39) based on the dilute solution case. A later work performed by Umantsev and Olson (Ref 40) removed the constraint of dilute solution thermodynamics, but neglected effects of capillarity on the precipitation composition. Further improvements have been made by Morral and Purdy (Ref 36, 37) with a more general description of the coarsening in multicomponent alloys. Kuehmann and Voorhees (Ref 38) considered flux balance boundary conditions on the interface rather than local equilibrium, but off-diagonal terms in the diffusion matrix were omitted for simplification. For practical design purposes, it is found that the simplified dilute solution model proposed by Lee *et al.*, (Ref 39) may be directly applied, if only an underdeveloped database is available. However, the Morral-Purdy model (Ref 36, 37) overall provides coarsening rate with highest accuracy.

To maximize hardening effects by reducing initial strengthening particle size, a designed alloy usually preserves a high-solute supersaturation, that is, high-driving force for nucleation of a second-phase precipitate particle, which can allow a bypass of the growth regime from nucleation directly to the coarsening regime. Therefore, the coarsening rate predicted by the aforementioned multicomponent coarsening models is often adopted as the principle design factor to control overall precipitation rate. Such a concept can be confirmed using a more rigorous approach based on the Kampmann and Wagner numerical (KWN) model (Ref 41) for concomitant nucleation, growth, and coarsening of precipitates, extended to multicomponent alloy systems. A detailed discussion on the influence of supersaturation was made by Robson, (Ref 42) analyzed different scenarios with overlapping of nucleation, growth, and coarsening during precipitation. Another key factor to control particle size is interfacial energy, which also markedly influences kinetic behaviour of precipitation. As shown in Fig. 4(a) and (b) in the KWN modeling, both interfacial energy and supersaturation driving force can bring significant impact and cause the overlap of nucleation and coarsening regime. However, in the design process, supersaturation expressed as the thermodynamic driving force is the most accessible parameter using the CALPHAD genomic databases. Figure 4(c) is a kinetic map calculated by Robson (Ref 42) indicating the correlation between supersaturation and interface energy where nucleation and coarsening can be controlled by both parameters. Experiments are from studies carried out on Fe-Cu (Ref 43), Cu-Co (Ref 44), Cu-Ti (Ref 45, 46), and Ni-Al (Ref 47) alloys.

Although a design model-chain from CALPHAD thermodynamics and kinetics to phase-level microstructural information is developed and can be directly used as a robust tool for Materials by Design, some constraints require further development of the sharp-

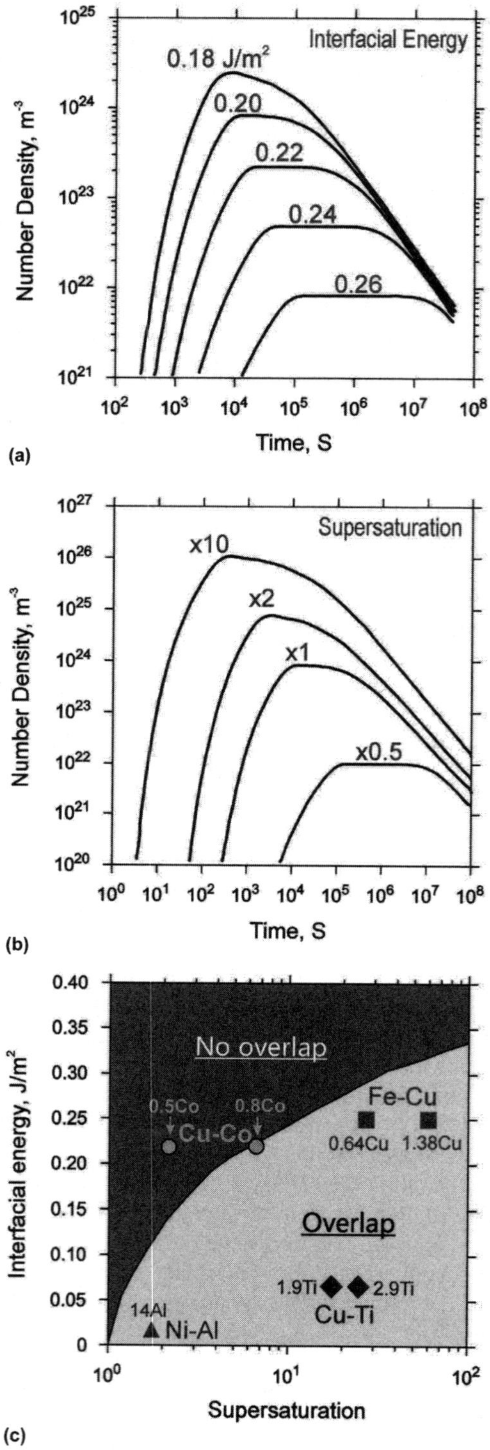

Fig. 4 Predicted evolution of number density with time for a selected range of (a) interfacial energies and (b) supersaturation. (c) Kinetic map showing the overlap region of nucleation and coarsening influenced by supersaturation and interfacial energy. Simulations are all based on the work performed by Robson (Ref 42). Alloy compositions are given in atomic percentage. Figure is reprinted with permission from Elsevier.

interface model. For example, the current KWN model-based simulations assume a spherical geometry for the second-phase particle precipitation, which is an oversimplification for rod or plate shape precipitates, for example, Q phase in the Al-Cu-Mg-Si-based alloy systems. A compromising method is to simulate precipitate kinetics by assuming spherical particles of equivalent particle volume. However, such a compromise could deviate significantly from experimental observation, and should be applied with caution for design purposes.

A less efficient but sometimes useful method which may circumvent the above difficulties by considering geometrical/crystal anisotropy is the so-called phase-field model (PFM) or diffuse interface model. For the sake of the process-structure modeling, the governing functionals of PFM are equivalent to the time-dependent Ginzburg-Landau equation (Ref 48) and Cahn-Hilliard equation (Ref 49). Similar to the case of sharp-interface KWN model, both approaches take fundamental thermodynamic and kinetic parameters as the input, which can be read from the CALPHAD databases. In addition, PFM also experiences the difficulties of determining several key input parameters, such as interfacial energy, gradient energy term, and interfacial mobility, which are generally hard to measure, without even further considering their time and temperature dependence. It is noteworthy, even though PFM of anisotropic structure is feasible, the anisotropic input parameters, for example, interfacial energy, often require a good quantitative estimation, which is still a daunting task for both atomistic modeling and experiments. Therefore, more often, these input parameters are still considered as the fitting parameter to perform the simulation to reach a good agreement with experimental microstructural evolution.

Figure 5 shows the relation among CALPHAD, experiments, sharp-interface (KWN), and diffuse interface models (PFM) for a typical process-structure simulation of solid-state precipitation phenomena. Although PFM can also be performed based on the CALPHAD method, nowadays it is still a significant challenge for effective coupling with CALPHAD database for multicomponent and multiphase simulations (Ref 50, 51) owing to the multiplicity of both bulk and interfacial properties. Another distinctive disadvantage of PFM is its limitation on simulating nucleation stage by generating nucleation sites using some

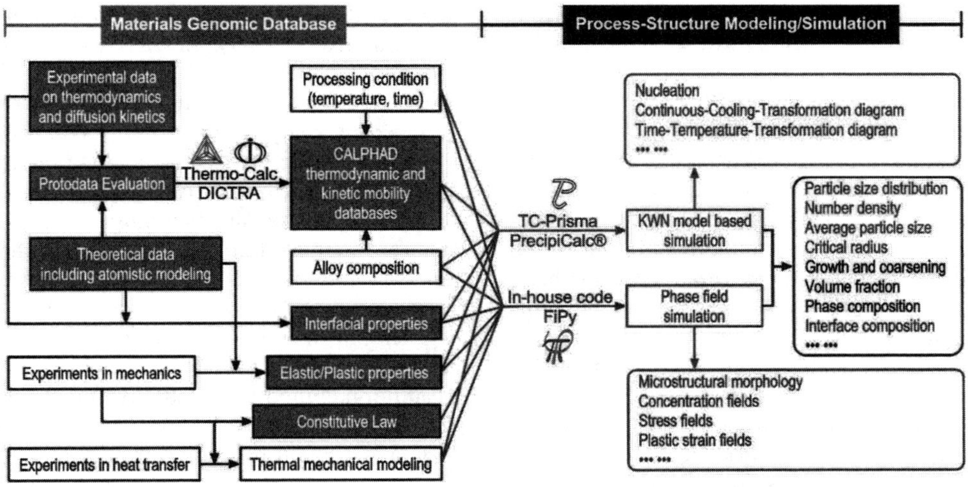

Fig. 5 Flow chart of process-structure modeling of solid-state transformation for second-particle growth. Both KWN model-based simulation and phase-field simulation take materials genomic database for the model input. The unique outputs of these two different simulation techniques are KWN (top) and phase-field (bottom).

ad hoc methods. Overall, in a practical simulation, comparing with KWN methods, PFM usually contains more adjustable parameters, which are not easy to determine, despite its capability of handling more complex phenomena. Nevertheless, PFM is often a useful science tool to enhance understanding of mechanism, it is generally less useful as an engineering tool than KWN-based methods that also provide better-defined statistical quantities.

A parametric design often relies on efficient computational design toolkits. Because the KWN model-based simulations can be implemented in an intuitive way by coupling with fundamental CALPHAD thermodynamic and kinetic parameters, they are widely adopted in the process-structure design. The PrecipCalc software developed by the QuesTek Innovations is a good example utilizing the sharp-interface precipitation model for alloy design and process optimization, which led the commercializing of several Ferrium high-performance alloys, for example, S53, M54, C61, and C64 (Ref 52).

Structure–Property Relation and Design Models

As shown in Fig. 3, strength, toughness, creep, and fatigue are four elementary mechanical properties of importance in structural materials design. Although corrosion resistance is also important, it is commonly treated as a composition constraint while improving these four elementary properties. It is a formidable task to cover all of the structure-property linkage models in this review. For elucidating the application of structure-property models to linkage design, in this section, we selectively discuss some well-developed mechanical models, which are available for application, by citing some practical structural alloy designs. As indicated in Fig. 2, whereas analytical models are available for improving strength and creep, toughness and fatigue are optimized by performing micromechanical finite element analysis (FEA).

Structure-Strength. One of the most well-developed structure-property analytical models is probably the one for predicting mechanical strength based on alloy hardening mechanism. Table 1 shows a summary of elementary hardening mechanism of materials strength by Hornbogen. (Ref 53) Therefore, the microstructure from zero- to three-dimensional (3D) discontinuities are defined as the source of elementary hardening mechanism. Excellent examples of applying such rules for materials strengthening can be found in the hardening enhancement design for high strength steels (Ref 54, 55). Typically, overall yield strength is described using the superposition law:

$$\sigma_Y = M\left[\tau_0 + \sum_i \tau_i\right] + \sigma_{HP} \qquad \text{(Eq 1)}$$

where M represents the Taylor factor, τ_0 is the Peierles-Nabarro stress contribution (Ref 56–58), which often adopts the intrinsic value of the pure base alloy. τ_i takes into account contributions from work hardening τ_W, solid solution strengthening τ_{SS}, and precipitation strengthening τ_P. The Hall-Petch effect σ_{HP} is the strengthening caused by grain refinement. Different contributions to the yield strength model is listed in Table 2. On the basis of the alloy composition and processing time, σ_P in Table 2 representing the dispersed particle strengthening contribution can be modeled based on the size of obstacle precipitates. As illustrated in Fig. 6(a), as a start, the small shearable particles follow the Friedel stress; with particle growth, the Orowan bypass stress becomes dominant, decreasing with the incremental particle size. However, in the case of ordered precipitate strengthening for superalloys, due to the antiphase boundary behaviour with paired dislocations, the peak hardening usually arrives earlier than the case with single-dislocation interaction with particle obstacles as shown in Fig. 6(a) (Ref 59–61). In this case, the peak-aged state corresponds to the intensity of pair dislocation interaction with obstacle particles.

A recent study performed by Wang *et al.*, (Ref 54) demonstrated the application of the above superposition law on predicting the yield strength of a secondary-hardening steel Blastalloy 160. A new transformation strengthening model accounting for Cu particle phase transformation was developed by associating with dislocation interaction with Cu of different crystal structure (bcc, 9R martensite, and fcc). In addition, a modified predictive yield strength model was developed based on 3D APT of M_2C and bcc Cu-hardening precipitates. Especially, to directly link quantitative grain structure characterisation and yield strength, 3D precipitate size distribution and characterisation parameters (e.g., phase composition, phase fraction) measured by 3D APT were adopted in the strengthening model for predicting yield strength within 10% uncertainty. Such a

Table 1 Summary of elementary hardening mechanism of crystalline phases

Symbol	Obstacles	Hardening mechanisms
σ_0	Lattice friction, order	Order hardening
$\Delta\sigma_s$	Solute atoms, vacancies	Solid solution hardening
$\Delta\sigma_d$	Dislocations	Work hardening
$\Delta\sigma_b$	Grain- or phase-boundaries	Grain boundary hardening
$\Delta\sigma_p$	Particles	Precipitation hardening
$\Delta\sigma_{ac}$	Crystal anisotropy	Texture hardening
$\Delta\sigma_{am}$	Microstructure anisotropy	Fibre hardening
$\Delta\sigma_m$	Crystal metastability	Premartensitic softening, transformation hardening

Source: Ref 53

Table 2 Contributions to yield strength by considering effects of solution strengthening, precipitation strengthening, work hardening, Hall-Petch effects based on intrinsic strength due to Peierles-Nabarro stress

Numerical expression		Constants and variables	
σ_0	$K_0 \cdot \left(\frac{2G}{1-\nu}\right) \cdot \exp\left(\frac{-4\pi w}{b}\right)$	G: shear modulus w: dislocation width K_0: Taylor coefficient	b: Burgers vector ν: Poisson's ratio
σ_W	$K_W \cdot \frac{\sqrt{3}}{60}\alpha_0\left[3\alpha_0\vartheta k_1 + (3\alpha_0\vartheta k_1)^2 + \beta b\rho^{1/2}\right]$	ϑ: misorientation angles between two subgrains α_0: Taylor factor, ρ: dislocation density β: elastic modulus coefficient k_1: subgrain interaction constant K_W: Taylor coefficient	
σ_{HP}	$K_{HP} \cdot d^{-1/2}$	d: grain size or martensite packet size K_{HP}: Hall-Petch coefficient	
σ_{SS}	$K_{SS} \cdot \left(\sum_i k_{SS,i}^2 C_i\right)^{1/2}$	$k_{SS,i}$: strengthening coefficient c_i: mole fraction of element in matrix K_{SS}: model coefficient	
σ_P	$K_P \cdot \tau_P = K_P \cdot \min(\tau_{PF}, \tau_{PO})$	K_P: model coefficient τ_P: Precipitation stress by taking minimum value between Friedel τ_{PF} and Orowan stress τ_{PO}	

Source: Ref 54

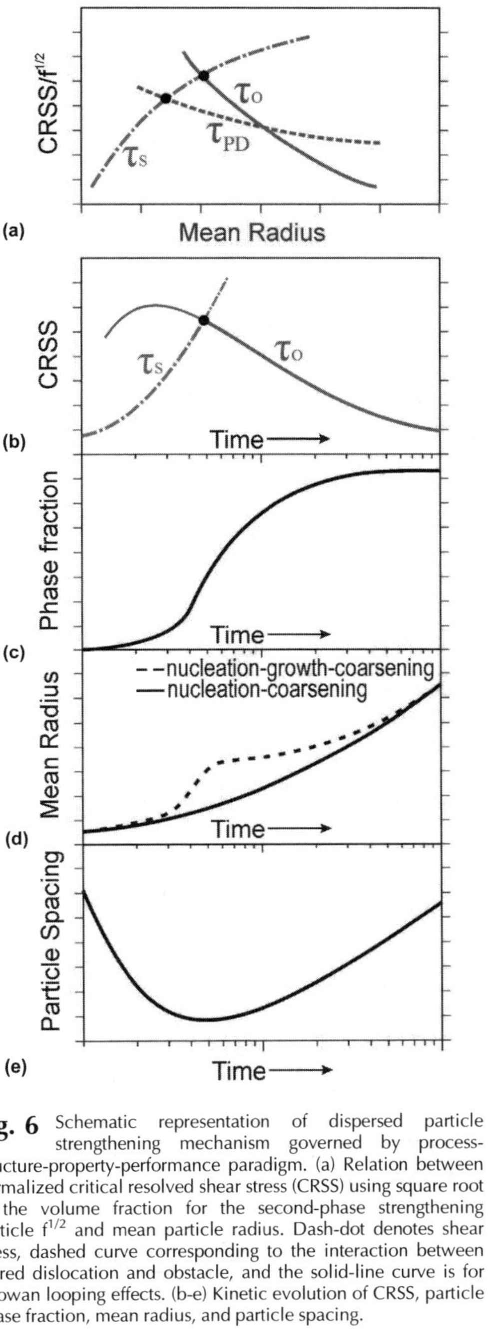

Fig. 6 Schematic representation of dispersed particle strengthening mechanism governed by process-structure-property-performance paradigm. (a) Relation between normalized critical resolved shear stress (CRSS) using square root of the volume fraction for the second-phase strengthening particle $f^{1/2}$ and mean particle radius. Dash-dot denotes shear stress, dashed curve corresponding to the interaction between paired dislocation and obstacle, and the solid-line curve is for Orowan looping effects. (b-e) Kinetic evolution of CRSS, particle phase fraction, mean radius, and particle spacing.

method can directly bridge the aforementioned process-structure and structure-property models for improving yield strength as long as all input physical parameters can be well determined with high accuracy. Therefore, a comprehensive scenario for modeling dispersed particle-hardening uses the process-structure model to obtain desired kinetic behaviour of the particle dispersion (see Fig. 6c-e). The structure-strength model can determine the optimal particle radius to reach the maximum strengthening as shown in Fig. 6(a) and (b) providing a clear microstructural objective for design.

Structure-Toughness. Materials design is clearly a delicate compromise of a number of controlling parameters, which may be beneficial to one type of mechanical properties, but detrimental to the other. A classic example is the intrinsic inverse proportionality of toughness to materials strength. A specific example of designing Co-Ni ultrahigh-strength (UHS) steels is illustrated in Fig. 7 and 8, which summarize a systems design effort on improving both strength and toughness synchronously. Refining grain size can be beneficial to both strength and toughness, but it is not always achievable. Therefore, understanding multiple ways to maximize both strength and toughness using structure-toughness models is crucial to a successful design.

Much research on toughness mechanism of high-strength steels is devoted to avoidance of brittle fracture failure modes such as intergranular fracture. In a systems design for Co-Ni UHS steel, several methods were applied for improving both toughness and strength synergistically in the steel research group (SRG) consortium centred at Northwestern University (Ref 15, 62–68). The related design efforts for enhancing both strength and toughness are summarised in Fig. 8, which focuses on (i) ductile-brittle transition temperature (DBTT) (Ref 69, 70), (ii) enhancement of interphase chemical bonding to delay microvoid softening, and (iii) transformation toughening. Further, as hydrogen embrittlement can promote brittle intergranular fracture, grain boundary (GB) chemistry optimization is also a primary requirement as indicated in Fig. 7 and 8.

On the heuristic level, DBTT is often a primary design parameter to avoid brittle fracture at desired service temperature. An analytical "master curve" model based on the shift of DBTT was constructed for designing the SRG UHS steels (Ref 69) by considering the influence of weight percent of key components, hardness and grain size as primary model variables. Such a model directly correlating with alloy composition made it possible to create a criteria for optimizing Ni and Co content in the Co-Ni UHS steel design by making tradeoff among toughness, strength, and Martensite start temperature.

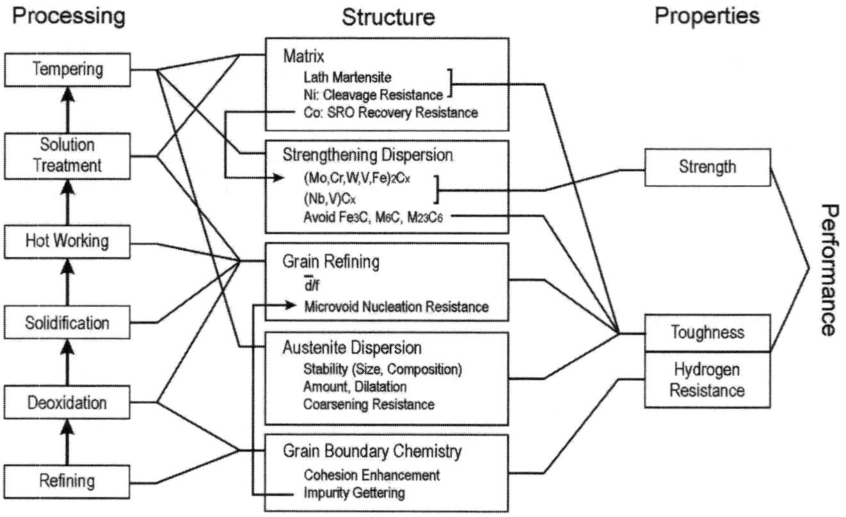

Fig. 7 Systems design chart of the Co-Ni UHS steel

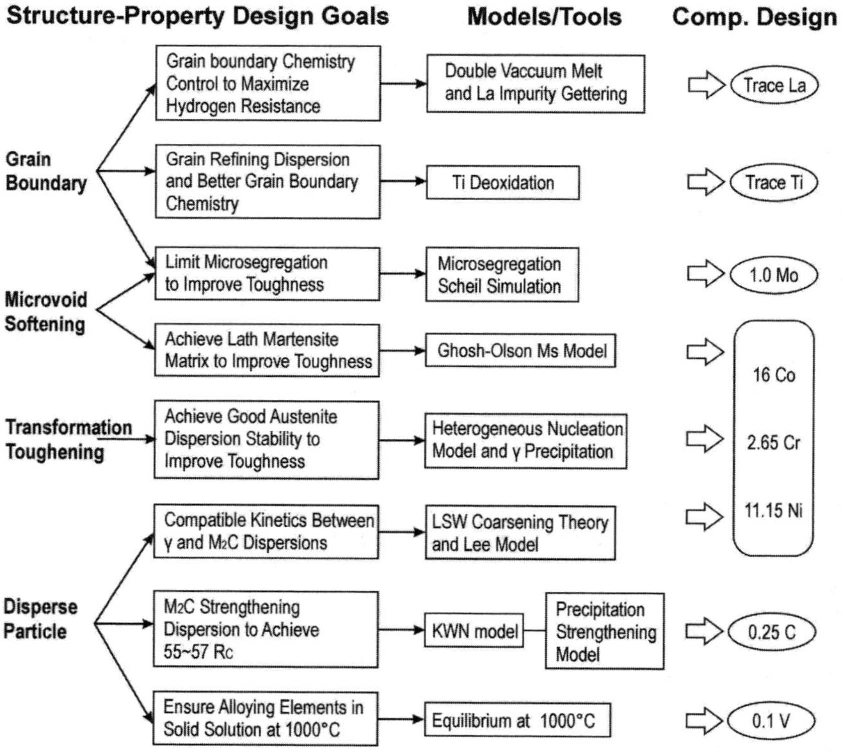

Fig. 8 Summary of design goals and models for the Co-Ni UHS steel based on an inversely inductive design approach

In the course of the UHS SRG steel designs, the submicron microstructure design for improving ductile toughness and retain high strength is based on suppressing microvoid softening and introducing dispersed-austenite transformation toughening. It is well established that microvoid nucleation during plastic deformation accelerates ductile fracture via a process of shear localisation, whereby sudden microvoid nucleation destabilizes plastic flow by strain location into shear bands (Ref 64, 71). However, the quantitative role of fine ~0.1 μm scale secondary particle dispersion which nucleate the microvoids is less well understood. In an optimized microstructure, these small particles correspond to the Zener-pinning particles (Ref 72–74) acting as the necessary grain refiners. Considerable efforts have been made by the SRG consortium developing several generations of multiscale ductile fracture simulators (Ref 75–79) to quantify the role of this level of microstructure. Extensive efforts were made on micromechanical FEA combined with the DFT calculation of interfacial strength (Ref 75, 76, 80–83), which can evaluate ideal work of adhesion. For example, in the work by Hao *et al.* (Ref 75, 76) a hierarchical constitutive model was derived with the plastic potential by considering DFT-calculated interfacial debonding and microvoid softening effects. Such an effort directly bridges atomistic modeling and continuum finite element simulation, which offers deeper insights into correlation between strength and fracture toughness. Furthermore, the interfacial adhesion energy can be enhanced by controlling dispersed phase compositions and geometric features in the design process using process-structure models.

On the basis of these simulations, a phenomenological model of toughness related to microvoid nucleation (Ref 84–86) was developed to aid parametric design of ductile toughness. According to the model developed by Wang and Olson, (Ref 84) the limit of ductile fracture toughness can be quantitatively related to the critical strain for shear localisation governing primary void coalescence.

As illustrated in Fig. 8, another interface-level design concept for avoiding environmental GB embrittlement is the GB cohesion enhancement, which is mainly applied to minimize hydrogen-induced intergranular embrittlement (Ref 68, 87, 88). As a basis for establishing an understanding on the electronic level, the full-potential augmented planewave (FLAPW) method was applied in a series studies on UHS steel (Ref 68, 87–91) based on a thermodynamic theory developed by Rice and Wang, (Ref 92). Their theory describes the mechanism of impurity-induced intergranular embrittlement via the competition between plastic crack blunting and brittle boundary separation. The derived "embrittlement potency" is defined as the segregation energy difference between segregant species along the GB and free surface (FS), $\Delta E_{GB} - \Delta E_{FS}$ (Ref 80, 82, 92), which indicates a species as a cohesion enhancer when this quantity is negative. Rice and Wang (Ref 92) directly correlated this quantity with measured embrittlement sensitivity quantities, relating intergranular DBTT and interfacial segregant atomic fraction. Accurate prediction of this quantity based on FLAPW were then performed by the SRG at the Northwestern University (Ref 89, 91). Combining initial FLAPW calculations and handbook quantities, such as elemental cohesive energies, a simplified model was developed to predict potencies represented in Fig. 9 (Ref 89). To quantify the accuracy of this simplified model, several rigorous validation FLAPW calculations were performed (solid points in the inset of Fig. 9), and confirmed that W and Re exhibit the highest cohesion enhancement. Eventually, a genomic surface thermodynamic database was constructed to enable design of the GB composition to entirely eliminate the inter-granular mode of hydrogen embrittlement (Ref 55, 87, 88).

In addition to these methods of increasing GB cohesion and interfacial bonding between matrix and particles, transformation plasticity was further introduced to improve toughness during plastic deformation in the mechanical Co-Ni UHS steels as shown in Fig. 8. A key design parameter is the transition temperature called M_S^σ, below which stress-assisted martensitic transformation controls materials yielding. As quantified by the Olson-Cohen theory (Ref 93, 94), below M_S^σ, the transformation-controlled yield stress rises due to the stability of the parent austenite increasing with temperature. Above M_S^σ, the slip-controlled yield

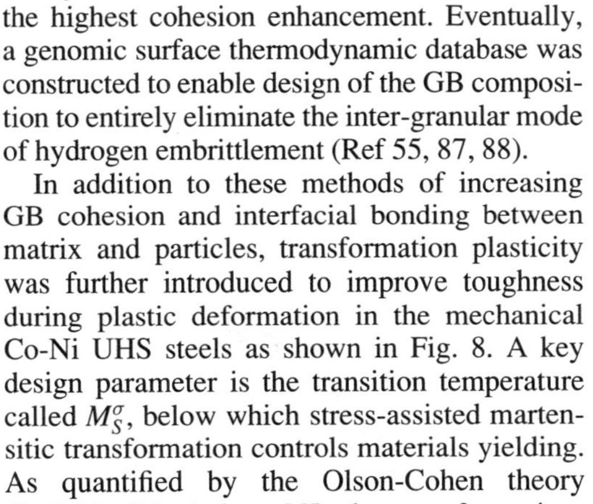

Fig. 9 Embrittlement potency prediction made for substitutional elements in Fe grain boundaries (Ref 89). Solid points in inset represent FLAPW calibration/validation results (Ref 55).

stress decreases with temperature through thermal activation of dislocation motion. Controlling the stability of the austenite phase qualified by M_S^σ is the key to optimizing transformation toughening. In the studies by Leal (Ref 95) and Stavehaug, (Ref 96) the M_S^σ temperature was further characterized by distinguishing between that defined by uniaxial tension $M_S^\sigma(ut)$ and by the crack-tip stress state denoted as $M_S^\sigma(ct)$. As shown in Fig. 10, the uniform ductility reaches a maximum between these two temperatures, and maximum fracture toughness occurs at $M_S^\sigma(ct)$. To quantitatively design transformation toughness in UHS steel, a model linked with thermodynamic database is proposed by Olson and his colleagues (Ref 67, 97–101) in a series of works. The underlining approach is based on an austenite stability parameter (ASP) (Ref 97, 98), which can be expressed as:

$$\Delta G_{ch}(x_i,T) + W_f(x_i,T) = -G_n - \Delta G_\sigma(\sigma, \Delta V/V) \quad \text{(Eq 2)}$$

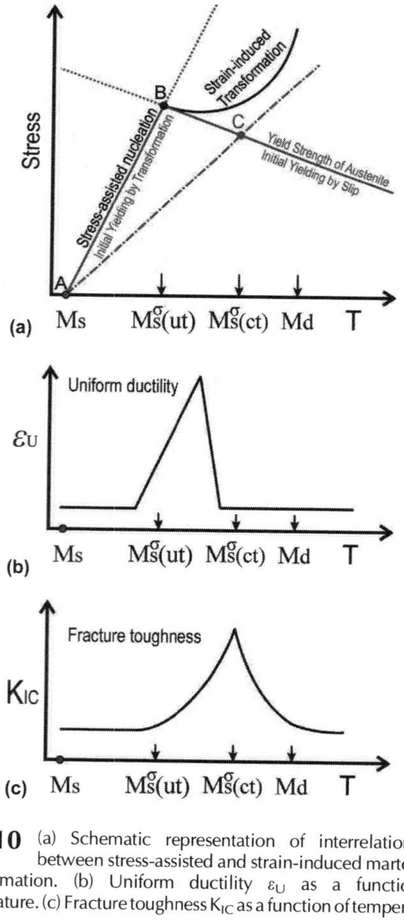

Fig. 10 (a) Schematic representation of interrelationships between stress-assisted and strain-induced martensitic transformation. (b) Uniform ductility ε_U as a function of temperature. (c) Fracture toughness K_{IC} as a function of temperature.

where the first term on the left side of equation is chemical contribution of the total energy, the second term is the interfacial friction energy, where both are functions of temperature T and composition x_i. The right side of Eq 2 is composed of defect potency, G_n, which can be approximated as a constant value for a certain material (Ref 97, 98), and mechanical driving force, $\Delta G\sigma$, which is a function of stress state and strength goal of the material. According to the Olson-Cohen model (Ref 93), given the stress state of the material and desired M_S^σ temperature, it is possible to provide a critical value for the austenite stability parameter, and thus allow dispersed-austenite composition to be designed through the CALPHAD genomic databases.

Structure-Creep/Fatigue. Unlike strength and toughness, creep and fatigue are dynamically structure-sensitive properties, where predictive models are more complex. For creep modeling, many available phenomenological models require numerous empirical parameters, and thus generate a heavy load for experiments. However, there is a promising method for parametric design purposes when dynamic creep models are correlated to diffusion kinetic modeling. For example, in single-crystal Ni superalloys, under a high-temperature, low-stress condition, where dislocation climb at the matrix γ-precipitate γ' interface becomes the dominant rate-controlling step in the kinetic creep process, modeling vacancy flux can be directly employed for creep resistance design. Studies published by Dyson (Ref 102) and Zhu *et al.*, (Ref 103) provide such promising models with vacancy diffusivity as a key design parameter, which can be calculated using the CALPHAD diffusion database for multicomponent alloy design. Further at very high temperatures, where rafting affects creep, structure-property models can also be related to the CALPHAD genomic database. In the viscoplastic regime described by the Socrate-Parks model (Ref 104), the thermodynamic driving force for rafting is primaryily controlled by the lattice mismatch of the matrix γ and the precipitate γ', which can be predicted through models (Ref 105) linking to the CALPHAD molar volume databases (Ref 106, 107).

With regard to structure-property models for fatigue, as indicated in Fig. 2, besides conventional statistical analysis-based phenomenological models, more attention is now paid to micromechanical FEA as a foundation for a predictive probabilistic approach. The state-of-the-art micromechanical FEA on fatigue is comprehensively reviewed by Pineau *et al.* (Ref 108) highlighting the seminal contributions of McDowell and his former students. Fatigue is an intrinsically multiscale and multistage phenomenon, and thus highly sensitive to microstructure level design. Conventional fatigue modeling separates the total life into two stages: initiation life N_i and crack propagation life defined by number of cycles N_p. For low cycle fatigue, N_p dominates, and thus can be modeled by statistical methods via some parametric scaling law (e.g., the Coffin-Manson law; Ref 109). In contrast, fatigue crack initiation dominates the total life of (ultra) high cycle fatigue. Therefore, it should be predictable by the micromechanical model with process-microstructure sensitivity. A thesis work by Moore (Ref 110) further developed the micromechanics-based method for multiscale fatigue nucleation prediction, which successfully predicted the fatigue behaviour of the Ni-Ti shape memory alloy. The 3D image-based micromechanical FEA demonstrates that an alloy design with enhanced yield strength can significantly increase fatigue strength in addition to the effects of minimizing inclusion size. It should be emphasized that such micromechanical FEA for ultrahigh cycle fatigue simulation directly drove a shape memory biomedical alloy design (Ref 111, 112) performed synchronously. It was determined that a 50% increase in Ni-Ti yield strength resulted in a 44% increase in ultra-high cycle fatigue limit at 10^9 cycles.

We note that, in a practical design, although a completely deductive process-structure-property model focusing on sequential cause and effects is difficult to obtain, an inversely inductive design approach directed by moving fundamental design parameters in the right direction is usually sufficient to implement a hierarchical design using specific means based on the available mechanistic models of Materials by Design.

AIM—Accelerating Technology Transition

As indicated in the ICMD hierarchical architecture (see Fig. 1), the AIM method completes the process of component-level materials development. Here, final design allowables bounded by the process-structure-property models are evaluated deliberately. To fully implement AIM, an integrated product team consisting of original equipment manufacturer, small company, university, and government laboratory are often assembled, determining design goals, sharing information, and integrating available design models/tools within the framework now known as ICME.

Although the above described process-structure-property models in Materials by Design is utilized extensively in AIM as a basis, there are three additional unique thrusts in the AIM method, which includes (i) multidisciplinary engineering model integration, (ii) location-specific modeling throughout a component, and (iii) uncertainty quantification and management.

Standard statistical methods are naturally vital for uncertainty quantification, which guides design tolerances. Such techniques were applied in the aforementioned design of the UHS steel, Ferrium S53 by the SRG consortium, run concurrently with the DARPA-AIM project. Design sensitivity analysis was performed using a combined method of Monte Carlo simulation and iCMD (an integrated computational materials design software package created by QuesTek Innovations LLC, Evanston, IL, USA), implemented through the commercial iSIGHT design optimization platform. For example, within 12 min on a Pentium IV 2.2 GHZ CPU (QuesTek Innovations LLC), a thousand parametric of structure-property iCMD calculations using the three-σ limits of the composition and process temperature tolerances generated the probability distribution of Charpy V-Notch toughness, hardness, Martensite start temperature and DBTT.

Under the DARPA-AIM 3-year project starting from 2000 using the example of IN100 superalloy turbine disks, the iSIGHT design integration system was adopted as the cross-disciplinary platform to link distributed software capabilities in Utah and Illinois as shown in Fig. 11(a). The turbine disk heat treatment process modeling performed using a FEA-based DEFORM Heat Treatment module (Scientific Forming Technologies Corporation, Columbus, OH, USA) determined the history of temperature profile at every location in a part, which were taken as the input of the precipitation simulator, PrecipiCalc (QuesTek Innovations), for the process-structure

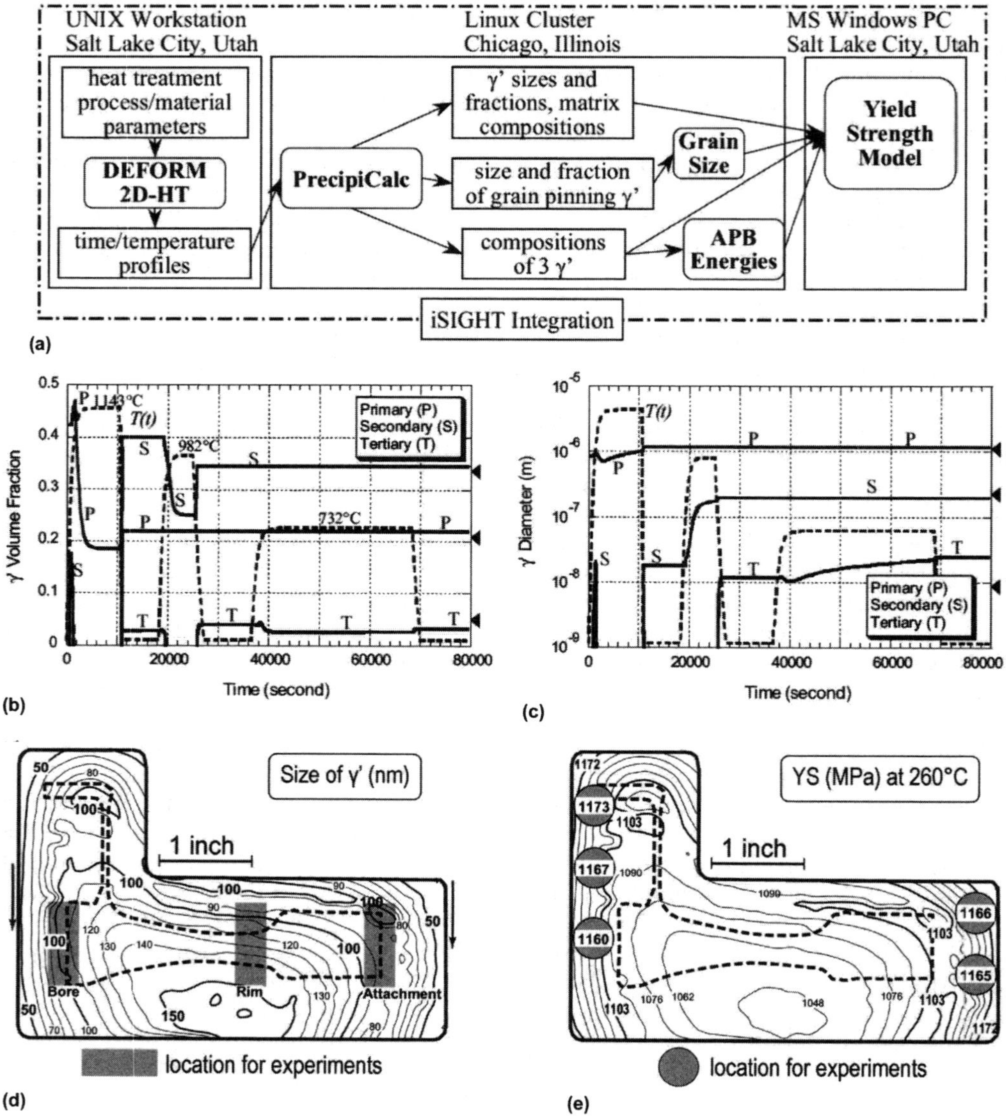

Fig. 11 (a) Distribution iSIGHT integration for mechanistic process-structure-property simulations in the DARPA-AIM project, performed on different computer hardware and operating systems across network. (b,c) Kinetic simulation performed using PrecipiCalc for a center location of an aero-turbine disk under the three-step heat treatment cycle. Time evolutions of mean sizes and fractions of three γ' distributions are shown in solid curves overlapping with temperature profile as dashed curve. Experimental γ' microstructural results are indicated by solid triangles on the right. (d) Calculated spatial contours of secondary γ' size (nm) for a radially symmetric cross-section of the minidisk. The disk is immersed into quenching media from bottom to top denoted by side-arrows. (e) Calculated spatial contour of yield strength (MPa) at 260 °C for minidisk. Figure is reproduced based on the work performed by Jou *et al.* (Ref 113) Figure is reprinted with permission. Copyright 2004 by The Minerals, Metals & Materials Society.

modeling. In the final step, a yield strength structure-property model was applied to predict the strength at both room temperature and elevated temperatures by taking results of PrecipiCalc and a CALPHAD-prediction of antiphase boundary energies as the input. The efficiency and high fidelity inherited from the Materials by Design models for the AIM methodology can be reflected by the quantitative simulation of structure and yield strength for the disk center location. By tuning parameters of interfacial property and diffusivity through model-calibration performed at the University of Connecticut, the dynamic behavior of the strengthening particle can be quantitatively captured using PrecipiCalc simulation with a high accuracy (Ref 113). Figure 11(b) and (c) shows the time evolution of mean sizes and fractions of three γ' populations under a three-step heat treatment cycle. To further perform a location-specific design, process-structure-property models were applied to an entire minidisk as shown in Fig. 11(d) and (e). The calculated spatial contours of secondary γ' size show high consistency with experimental observation on bore, rim, and attachment parts indicated with shaded rectangles in Fig. 11(d). Using the structure-strength model, the tensile yield strength at five different locations was predicted within quantified uncertainty ranges by comparing with experimental results on specific locations (see circles highlighted in Fig. 11d).

As a follow-on to the DARPA-AIM program, a National Aeronautics and Space Administration-funded project at QuesTek Innovations in 2005–2008 further extended the above AIM techniques for a hierarchical calibration/validation uncertainty management approach for all available process-structure models (Ref 31) for a Dual Microstructure Heat Treatment (DMHT) process design indicated in Fig. 12. As different location-specific properties require different microstructures, the design sensitivity of the process-structure model is crucial to meet property requirements. Table 3 summarizes the sequential characterization protocol performed for standard uncertainty quantification of individual inputs in the process-structure models. Subsequently, a hierarchical uncertainty management strategy was adopted by balancing model sensitivity against database intrinsic accuracy. Rigid shifts to second-phase free-energy functions according to Gibbs-Thompson effects were applied for a precise description of the secondary-strengthening particles. Meanwhile, DICTRA atomic mobility databases were recalibrated by diffusion couple experiments, which achieved an improved accuracy of diffusivity prediction. Regarding the high sensitivity of interfacial energy in precipitation simulation, PrecipiCalc simulations were carried out by adjusting interfacial energy to match the critical nucleation undercooling accurately determined by the single sensor differential thermal analysis technique, which measures a precise thermal history of a rapidly quenched 3 mm diameter

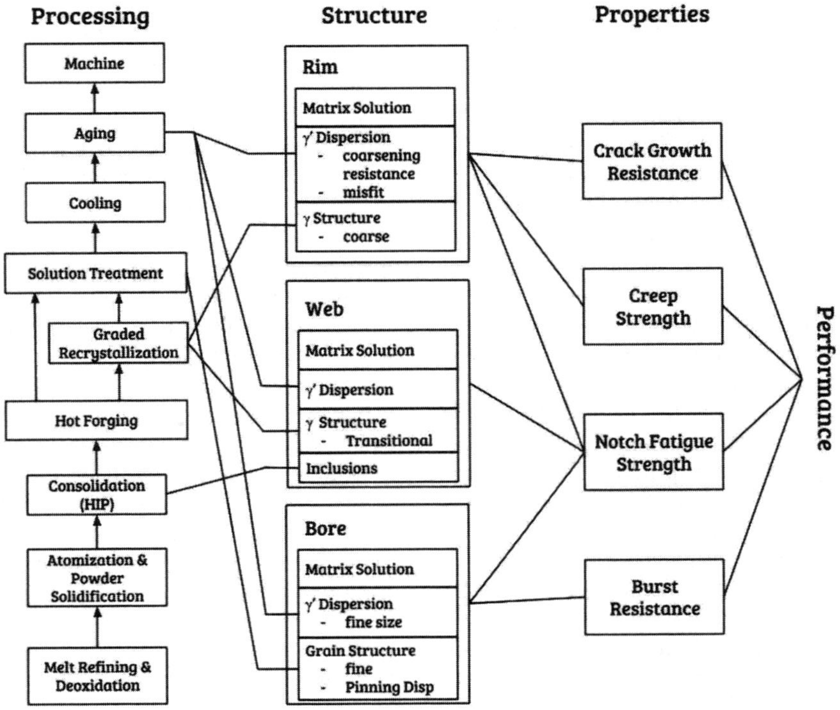

Fig. 12 Systems design chart of Ni turbine disk alloy using the AIM method. Hierarchical microstructure are designed for specific locations in the turbine disk.

Table 3 Summary of database, model, calibration, and validation protocol for precipitation modeling adopted in the AIM method.

Database	Quantity	Model	Model parameters	Calibration and validation
Thermodynamics	Bulk energy	CALPHAD	Free energy	XRD/EDS/TEM/APT
	Interfacial energy	Gibbs-Thompson	G_{el}, σ_{coh}	APT/SSDTA
Molar volume	Lattice mismatch	CALPHAD	G_{el}, R	XRD/TEM/DFT
Kinetics	Atomic mobility	CALPHAD	Mobility	Diffusion couple
	D_{scale}		D_{scale}	EPMA or WDS

Abbreviations: APT, atom probe tomography; DFT, density functional theory; EDS, energy dispersive spectroscopy; G_{el}, estimated elastic coherency energy; R, coherency transition size from coherent to incoherent state; SSDTA: single sensor differential thermal analysis (Ref 32); TEM, transmission electron microscopy; XRD, x-ray diffraction.

pins (Ref 31, 32). Afterwards the PrecipiCalc simulated microstructural attributes were further compared with the APT characterisation to evaluate uncertainty limits.

It should be noted that, having demonstrated the importance of the AIM technique, a suite of advanced 3D tomographic multiscale characterization tools were further developed under the support of ONR and DARPA. Thus, the AIM technique is now supported by both novel experiments and 3D process-structure-property models reaching a new level of fidelity, which is greatly beneficial to the enterprise of materials cycle compression.

Summary—Cybermaterials Innovation Based on Materials Genome

The continued development of the ICMD techniques, Materials by Design and AIM, is already accelerating materials innovation, thus leveraging the current design techniques in materials science and mechanical engineering, which foster synergistic interactions among disciplines.

Materials by Design and AIM techniques serve as the two major pillars in the ICMD hierarchical architecture. As reviewed in the foregoing sections, sustained efforts have made grand achievements in transferring science-based ideas to industrial production. However, a sustainable development strategy is still needed to fulfil the potential of the ICMD blueprint in Fig. 1 for cybermaterials innovations. It is clear that Materials Genome database as the cornerstone has a decisive role.

First, the systems design models of process-structure-property-performance paradigm can all benefit from further improvement. For instance, the preceding discussion on Materials by Design has highlighted needs of model development for nonspherical particles in the framework of the KWN model-based simulation. In view of the high sensitivity of nucleation to interfacial energy, more rigorous atomistic modeling of interfacial thermodynamics would enhance predictive capability.

Second, because process-structure linkage design relies heavily on the quality of the fundamental materials genomic databases assembled by computational thermodynamics and kinetics, more dedicated research on constructing high-quality genomic database should be motivated and promoted. Importantly, these research activities also include experiments providing protodata. The so-called high-throughput experiment represented by diffusion multiples (Ref 114) is a good example to make such contributions. Meanwhile, more efforts should be made on bridging the current DFT-based atomistic models and CALPHAD method by extending their temperature scales (Ref 115). A recent work on the Fe-Cr system (Ref 22) demonstrated that the CALPHAD models extended to 0 K will garner valuable insights into anomalous phenomenon, that is, negative enthalpy of mixing on Fe-rich side, found in enthalpy of formation generated by DFT calculations (Ref 116).

Expanding the impact of the structure-property model development, genomic databases of mechanistic constitutive laws for mechanical FEA should be initiated. Such efforts need intimate collaborations between experimentalists and modeling experts in both materials science and mechanical engineering. As mentioned in the above discussion, the state-of-the-art materials design usually considers experimental calibration as an important part in design iterations. Therefore, technical improvement of such experiments for fundamental model-calibration purposes should be a primary effort.

Materials genomic database construction requires corresponding efforts on standardisation and database management as indispensable support. Taking DBTT as an example, although DBTT can be evaluated using Charpy impact energy tests, precise modeling of DBTT is currently data intensive (Ref 117, 118). If a standard of efficient DBTT measurement is available, limitations affecting the accuracy of DBTT will be significantly diminished, and thus improve the DBTT model predictability. Further, we note that related topics are often much broader than the research field of materials science, driving the call for collaborative efforts by various scientific and engineering communities.

Last but not least, perhaps the greatest promise of materials development cycle compression is the new opportunity of concurrent design of materials and devices (Ref 119). Notable achievements have already been demonstrated in the field of consumer electronics which would never have been possible under trial-and-error empirical materials development.

Overall, it is evident that the next generation of cybermaterials innovation applicable across all materials classes will be grounded in ICMD hierarchical architecture powered by an expanded system of materials genomic databases.

ACKNOWLEDGMENTS

This work was performed within the CHiMaD Center for Hierarchical Materials Design at the Northwestern Univeristy, funded by the National Institute of Standards and Technology, USA. ICMD and Materials by Design are trademarks of QuesTek Innovations. We are grateful to Professor John Ågren (KTH, Sweden), Qing Chen, Ph.D. (Thermo-Calc Software AB, Sweden), Ikumu Watanabe, Ph.D. (NIMS, Japan), and Abhinav Saboo (Northwestern University) for stimulating discussions.

REFERENCES

1. D. Apelian, *et al. Accelerating Techology Transition: Bridge the Valley of Death for Materials and Processes in Defense Systems*. National Research Council of The National Academies, National Academies Press, (2004). http://www.nap.edu/catalog.php?record_id = 11108
2. J.P. Holdren, *Materials Genome Initiative for Global Competitiveness*. (Executive office of the president national science and technology council, 2011)
3. T.M. Pollock, *et al. Integrated Computational Materials Engineering: A Transformational Discipline for Improved Competitiveness and National Security*. (National Research Council of the National Academies, National Academies Press, 2008) http://www.nap.edu/catalog/12199.html
4. M.F. Horstemeyer, *An Introduction to Integrated Computational Materials Engineering (ICME). Integr Comput Mater Eng Met*. (John Wiley, 2012)
5. G.B. Olson, Preface to the viewpoint set on: the Materials Genome. *Scr. Mater*. 70, 1–2 (2014)
6. L. Kaufman and J. Ågren, CALPHAD, first and second generation—birth of the materials genome. *Scr. Mater*. 70, 3–6 (2014)
7. C.E. Campbell, U.R. Kattner, and Z.K. Liu, File and data repositories for Next Generation CALPHAD. *Scr. Mater*. 70, 7–11 (2014)
8. H.K.D.H. Bhadeshia, Computational design of advanced steels. *Scr. Mater*. 70, 12–17 (2014)
9. S. Christensen and R. D'Oyen, Computational formulation of a new composite matrix. *Scr. Mater*. 70, 18–24 (2014)
10. G.B. Olson and C.J. Kuehmann, Materials genomics: from CALPHAD to flight. *Scr. Mater*. 70, 25–30 (2014)
11. Y. Yuan, *et al*. A new method to strengthen turbine disc superalloys at service temperatures. *Scr. Mater*. 66, 884–889 (2012)
12. Y. Yuan, *et al*. Deformation mechanisms in a new disc superalloy at low and intermediate temperatures. *Scr. Mater*. 67, 137–140 (2012)
13. R.C. Reed, T. Tao, and N. Warnken, Alloys-By-Design: Application to nickel-based single crystal superalloys. *Acta. Mater*. 57, 5898–5913 (2009)
14. G.B. Olson, Computational design of hierarchically structured materials. *Science* 277, 1237–1242 (1997)
15. G.B. Olson, Genomic materials design: the ferrous frontier. *Acta. Mater*. 61, 771–781 (2013)
16. L. Martinetti, *et al*. A critical gel fluid with high extensibility: the rheology of

chewing gum. *J. Rheol.* 58, 821–838 (2014)
17. R.M. D'Oyen, *Systems Design of Case Hardened Polymers for Gears*, Doctoral thesis, Northwestern University (1997)
18. N. Saunders and A.P. Miodownik, CALPHAD (Calculation of Phase Diagrams): *A Comprehensive Guide Pergamon Mater Ser.* Pergamon, (1998). http://dx.doi.org/10.1016/S1470-1804(98)80019-9
19. W. Xiong, W. Xie, C. Shen, and D. Morgan, Thermodynamic modelling of the U-Zr system—a revisit. *J. Nucl. Mater.* 443, 331–341 (2013)
20. W. Xiong, W. Xie, and D. Morgan, Thermodynamic evaluation of the Np-Zr system using CALPHAD and ab initio methods. *J. Nucl. Mater.* 452, 569–577 (2014)
21. W. Xiong, Y. Du, X. Lu, J.C. Schuster, and H. Chen, Reassessment of the Ce-Ni binary system supported by key experiments and ab initio calculations. *Intermetallics* 15, 1401–1408 (2007)
22. W. Xiong, *et al.* An improved thermodynamic modelling of the Fe-Cr system down to zero kelvin coupled with key experiments. *CALPHAD* 35, 355–366 (2011)
23. W. Xiong, *et al.* Construction of the Al-Ni-Si phase diagram over the whole composition and temperature ranges: thermodynamic modelling supported by key experiments and first-principles calculations. *Int. J. Mater. Res.* 99, 598–612 (2008)
24. B. Hallstedt and Z.-K. Liu, Software for thermodynamic and kinetic calculation and modelling. *CALPHAD* 33, 265 (2009)
25. H. Lukas, S.G. Fries, and B. Sundman, *Computational Thermodynamics: The Calphad Method* (Cambridge University Press, 2007)
26. J.-O. Andersson, L. Höglund, B. Jönsson, and J. Ågren, in *Proc Metall Soc Can Inst Min Metall* (ed. Purdy G.) 153–163 (Pergamon, 1990). http://dx.doi.org/10.1016/B978-0-08-040412-7.50023-2
27. A. Borgenstam, A. Engstrom, L. Hoglund, and J. Agren, DICTRA, a tool for simulation of diffusional transformations in alloys. *J. Phase Equilibria.* 21, 269–280 (2000)
28. W. Xiong, CALPHAD-based integrated computational materials engineering research for materials genomic design. *JOM* 67, 1864–1865 (2015)
29. M.D. Bender and G.B. Olson, Designing a precipitation-strengthened, super-elastic, TiNi-based alloy for endovascular stentsin *Int. Conf. Martensitic Transform* 159–166 (John Wiley, 2010)
30. M.D. Bender and G.B. Olson, Computational thermodynamics-based design of nanodispersion-strengthened shape memory alloys. in *SMST-2007—Proc. Int. Conf. Shape Mem. Superelastic Technol.* 115–122 (2008)
31. G.B. Olson, *et al.* Precipitation model validation in 3rd generation aeroturbine disc alloysin *11th Int. Symp. Superalloys* 923–932 (Minerals, Metals and Materials Society, 2008)
32. B.T. Alexandrov and J.C. Lippold, Single sensor differential thermal analysis of phase transformations and structural changes during welding and postweld heat treatment. *Weld World* 51, 48–59 (2007)
33. A.M. Wusatowska-Sarnek, G. Ghosh, G.B. Olson, M.J. Blackburn, and M. Aindow, Characterization of the microstructure and phase equilibria calculations for the powder metallurgy superalloy IN100. *J. Mater. Res.* 18, 2653–2663 (2003)
34. I.M. Lifshitz and V.V. Slyozov, The kinetics of precipitation from supersaturated solid solutions. *J. Phys. Chem. Solids* 19, 35–50 (1961)
35. C. Wagner, Theorie der Alterung von Niederschlägen durch Umlösen (Ostwald-Reifung). *Zeitschrift für Elektrochemie, Berichte der Bunsengesellschaft für Phys Chemie* 65, 581–591 (1961)
36. J.E. Morral and G.R. Purdy, Particle coarsening in binary and multicomponent alloys. *Scr. Metall. Mater.* 30, 905–908 (1994)
37. J.E. Morral and G.R. Purdy, Thermodynamics of particle coarsening. *J. Alloys Compd.* 220, 132–135 (1995)
38. C.J. Kuehmann and P.W. Voorhees, Ostwald ripening in ternary alloys. *Metall. Mater. Trans. A* 27, 937–943 (1996)
39. H.M. Lee, S. Allen, and M. Grujicic, Stability and coarsening resistance of M2C carbides in secondary hardening steelsin *Innov Ultrahigh-Strength Steel Technol Sagamore Army Mater. Res. Conf.* (eds G.B. Olson, M. Azrin, and E.S. Wright)

127–146 (U.S. Government Printing Office, 1990)

40. A. Umantsev and G.B. Olson, Ostwald ripening in multicomponent alloys. *Scr. Metall. Mater*. 29, 1135–1140 (1993)
41. R. Wagner, R. Kampmann, and P.W. Voorhees, in *Phase Transform Mater* (ed. G. Kostorz) 309–407 (Wiley-VCH and Co. KGaA, 2005)
42. J.D. Robson, Modelling the overlap of nucleation, growth and coarsening during precipitation. *Acta. Mater*. 52, 4669–4676 (2004)
43. R. Kampmann and R. Wagner, in *At Transp Defects Met by Neutron Scatt SE—12* (eds C. Janot, W. Petry, D. Richter, and T. Springer) 10, 73–77 (Springer, 1986)
44. H.I. Aaronson and F.K. LeGoues, Assessment of studies on homogeneous diffusional nucleation kinetics in binary metallic alloys. *Metall. Trans. A, Phys. Metall. Mater. Sci*. 23A, 1915–1945 (1992)
45. L.V. Alvensleben and R. Wagner, FIM-atom probe studies of early stage decomposition in Cu-Ti alloysin *Proc 2nd Acta-Scripta Metall. Conf.* (eds P. Haasen, V. Gerold, R. Wagner, and M.F. Ashby) 143–148 (Pergamon, 1984) http://dx.doi.org/10.1016/B978-0-08-031651-2.50026-4
46. H. Eckerlebe, R. Kampmann, and R. Wagner, in *At Transp Defects Met by Neutron Scatt SE—11* (eds C. Janot, W. Petry, D. Richter, and T. Springer) 10, 66–72 (Springer, 1986)
47. H. Wendt and P. Haasen, Nucleation and growth of γ'-Precipitates in Ni-14 at.% Al. *Acta. Metall*. 31, 1649–1659 (1983)
48. S.M. Allen and J.W. Cahn, A microscopic theory for antiphase boundary motion and its application to antiphase domain coarsening. *Acta. Metall*. 27, 1085–1095 (1979)
49. J.W. Cahn and J.E. Hilliard, Free energy of a nonuniform system. I. interfacial free energy. *J. Chem. Phys*. 28, 258–267 (1958)
50. T. Kitashima, Coupling of the phase-field and CALPHAD methods for predicting multicomponent, solid-state phase transformations. *Philos. Mag*. 88, 1615–1637 (2008)
51. H. Larsson and L. Höglund, A scheme for more efficient usage of CALPHAD data in simulations. *CALPHAD* 50, 1–5 (2015)
52. J. Sebastian and G.B. Olson, Examples of QuesTek Innovations' Application ICME to Materials Design, Development, and Rapid Qualificationin *55th AIAA/ASMe/ASCE/AHS/SC Struct Struct Dyn Mater Conf* 1–7 (American Institute of Aeronautics and Astronautics, 2014)
53. E. Hornbogen, Design alloys for > 3 GPa steel in *Innov Ultrahigh-Strength Steel Technol Sagamore Army Mater Res Conf* (eds G.B. Olson, M. Azrin, and E.S. Wright) 113–126 (U.S. Government Printing Office, 1990)
54. J.S. Wang, M.D. Mulholland, G.B. Olson, and D.N. Seidman, Prediction of the yield strength of a secondary-hardening steel. *Acta. Mater*. 61, 4939–4952 (2013)
55. C.D. Kantner, *Designing Strength, Toughness, and Hydrogen Resistance*, Doctoral thesis, Northwestern University (2002)
56. F.R.N. Nabarro, Fifty-year study of the Peierls-Nabarro stress. *Mater Sci Eng A* 234–236, 67–76 (1997)
57. F.R.N. Nabarro, Dislocations in a simple cubic lattice. *Proc. Phys. Soc*. 59, 256 (1947)
58. J.N. Wang, A new modification of the formulation of Peierls stress. *Acta. Mater*. 44, 1541–1546 (1996)
59. E. Nembach, *Particle Strengthening of Metals and Alloys*. (John Wiley, 1997)
60. E. Nembach, S. Schänzer, W. Schröer, and K. Trinckauf, Hardening of nickel-base superalloys by high volume fractions of γ'-precipitates. *Acta. Metall*. 36, 1471–1479 (1988)
61. W. Huether and B. Reppich, Interaction of dislocations with coherent, stress-free, ordered particles. *Zeitschrift fuer Met. Res. Adv. Tech*. 69, 628–634 (1978)
62. B. Tiemens, A. Sachdev, and G.B. Olson, Cu-precipitation strengthening in ultra-high-strength carburizing steels. *Metall. Mater. Trans. A* 43, 3615–3625 (2012)
63. B.L. Tiemens, A.K. Sachdev, R.K. Mishra, and G.B. Olson, Three-dimensional (3-D) atom probe tomography of a Cu-precipitation- strengthened, ultrahigh-strength carburized steel. *Metall. Mater. Trans. A Phys. Metall. Mater. Sci*. 43, 3626–3635 (2012)
64. J.G. Cowie, M. Azrin, and G.B. Olson, Microvoid formation during shear deformation of ultrahigh strength steels.

Metall. Trans. A, Phys. Metall. Mater. Sci. 20A, 143–153 (1989)

65. H.E. Lippard, *et al.* Microsegregation behavior during solidification and homogenization of AerMet100 steel. *Metall. Mater. Trans. B Process Metall. Mater. Process. Sci.* 29, 205–210 (1998)
66. G.B. Olson, Transformation plasticity and toughening. *J. Phys. IV JP* 6, C1-407–C1-418 (1996)
67. G.N. Haidemenopoulos, M. Grujicic, G.B. Olson, and M. Cohen, Thermodynamics-based alloy design criteria for austenite stabilization and transformation toughening in the Fe-Ni-Co system. *J. Alloys Compd.* 220, 142–147 (1995)
68. R. Wu, A.J. Freeman, and G.B. Olson, First principles determination of the effects of phosphorus and boron on iron grain boundary cohesion. *Science* 265, 376–380 (1994)
69. G.B. Olson, *Corrosion Resistant Steels for Structural Applications in Aircraft* (Strategic Environmental Research and Development Program (SERDP), Department of Defense, 2005). https://www.serdp-estcp.org/content/download/6302/84489/file/PP-1224-FR-01.pdf
70. R. Moskovic and P.E.J. Flewitt, An overview of the principles of modeling charpy impact energy data using statistical analyses. *Metall. Mater. Trans. A* 28, 2609–2623 (1997)
71. J.G. Cowie and F.R. Tuler, The influence of second-phase dispersions on shear instability and fracture toughness of ultrahigh strength AISI 4340 steel. *Mater. Sci. Eng. A* 141, 23–37 (1991)
72. M. Hillert, On the estimation of the Zener drag on grain boundaries. *Scr. Metall.* 18, 1431–1432 (1984)
73. M. Hillert, Inhibition of grain growth by second-phase particles. *Acta. Metall.* 36, 3177–3181 (1988)
74. C.H. Wörner, A. Cabo, and M. Hillert, On the limit for particle attachment in Zener drag. *Scr. Metall.* 20, 829–831 (1986)
75. S. Hao, B. Moran, W. Kam Liu, and G.B. Olson, A hierarchical multi-physics model for design of high toughness steels. *J. Comput. Mater. Des.* 10, 99–142 (2003)
76. S. Hao, W.K. Liu, B. Moran, F. Vernerey, and G.B. Olson, Multi-scale constitutive model and computational framework for the design of ultra-high strength, high toughness steels. *Comput. Methods Appl. Mech. Eng.* 193, 1865–1908 (2004)
77. F. Vernerey, *et al.* The 3-D computational modeling of shear-dominated ductile failure in steel. *JOM* 58, 45–51 (2006)
78. F.J. Vernerey, W.K. Liu, B. Moran, and G.B. Olson, A micromorphic model for the multiple scale failure of heterogeneous materials. *J. Mech. Phys. Solids* 56, 1320–1347 (2008)
79. R. Tian, *et al.* A multiresolution continuum simulation of the ductile fracture process. *J. Mech. Phys. Solids* 58, 1681–1700 (2010)
80. S. Zhang, O.Y. Kontsevoi, A.J. Freeman, and G.B. Olson, First principles investigation of zinc-induced embrittlement in an aluminum grain boundary. *Acta. Mater.* 59, 6155–6167 (2011)
81. N.I. Medvedeva, Y.N. Gornostyrev, O.Y. Kontsevoi, and A.J. Freeman, Ab-initio study of interfacial strength and misfit dislocations in eutectic composites: NiAl/Mo. *Acta. Mater.* 52, 675–682 (2004)
82. S. Zhang, O.Y. Kontsevoi, A.J. Freeman, and G.B. Olson, Cohesion enhancing effect of magnesium in aluminum grain boundary: a first-principles determination. *Appl. Phys. Lett.* 100, 231904 (2012)
83. J.H. Lee, T. Shishidou, Y.J. Zhao, A.J. Freeman, and G.B. Olson, Strong interface adhesion in Fe/TiC. *Philos. Mag.* 85, 3683–3697 (2005)
84. J.-S. Wang and G.B. Olson, Prediction of ductile fracture toughness. in *12th Int. Conf. Fract. 2009, ICF-12* 1, 145–154 (2009)
85. T. Pardoen and J.W. Hutchinson, An extended model for void growth and coalescence. *J. Mech. Phys. Solids* 48, 2467–2512 (2000)
86. T. Pardoen and J.W. Hutchinson, Micromechanics-based model for trends in toughness of ductile metals. *Acta. Mater.* 51, 133–148 (2003)
87. W.-T. Geng, A.J. Freeman, G.B. Olson, Y. Tateyama, and T. Ohno, Hydrogen-promoted grain boundary embrittlement and vacancy activity in metals: Insights from ab initio total energy calculatons. *Mater. Trans.* 46, 756–760 (2005)
88. L. Zhong, R. Wu, A.J. Freeman, and G.B. Olson, Charge transfer mechanism of

hydrogen-induced intergranular embrittlement of iron. *Phys. Rev. B* 62, 13938–13941 (2000)
89. W.T. Geng, A.J. Freeman, and G.B. Olson, Influence of alloying additions on grain boundary cohesion of transition metals: first-principles determination and its phenomenological extension. *Phys. Rev. B* 63, 165415 (2001)
90. R. Wu, A.J. Freeman, and G.B. Olson, On the electronic basis of the phosphorus intergranular embrittlement of iron. *J. Mater. Res.* 7, 2403–2411 (1992)
91. R. Wu, A.J. Freeman, and G.B. Olson, Effects of carbon on Fe-grain-boundary cohesion: first-principles determination. *Phys. Rev. B* 53, 7504–7509 (1996)
92. J.R. Rice and J.-S. Wang, Embrittlement of interfaces by solute segregation. *Mater. Sci. Eng. A* 107, 23–40 (1989)
93. G.B. Olson and M. Cohen, Martensitic transformations as a deformation processin *Mechanical properties and phase transformations in engineering materials-Earl R Parker Symposium on Structure Property Relationships* (eds S.D. Antolovich, R.O. Ritchie, and W.W. Gerberich) 367–390 (The Metallurgical Society, Inc., 1986)
94. G.B. Olson and Z.D. Feinberg, in *Phase Transform Steels Fundam Diffus Control Transform Vol 2* (eds S. Arnold and E. Pereloma) 59–82 (Woodhead Publishing, 2012)
95. R.H. Leal, *Transformation Toughening of Metastable Austenitic Steels*, Doctoral thesis, MIT (1984)
96. F. Stavehaug, *Transformation Toughening of gamma-strengthened Metastable Austenitic Steels*, Doctoral thesis, MIT (1990)
97. G. Ghosh and G.B. Olson, Kinetics of f.c.c. → b.c.c. heterogeneous martensitic nucleation—I. The critical driving force for athermal nucleation. *Acta. Metall. Mater.* 42, 3361–3370 (1994)
98. G. Ghosh and G.B. Olson, Kinetics of f.c.c. → b.c.c. heterogeneous martensitic nucleation—II. Thermal activation. *Acta. Metall. Mater.* 42, 3371–3379 (1994)
99. G.B. Olson and M. Cohen, A general mechanism of martensitic nucleation: Part I. General concepts and the FCC → HCP transformation. *Metall. Trans. A* 7, 1897–1904 (1976)
100. G.B. Olson and M. Cohen, A general mechanism of martensitic nucleation: Part II. FCC → BCC and other martensitic transformations. *Metall. Trans. A* 7, 1905–1914 (1976)
101. G.B. Olson and M. Cohen, in *Dislocations Solids* (ed. F.R.N. Nabarro) (Elsevier Science, 1986)
102. B.F. Dyson, Microstructure based creep constitutive model for precipitation strengthened alloys: theory and application. *Mater. Sci. Technol.* 25, 213–220 (2009)
103. Z. Zhu, H. Basoalto, N. Warnken, and R.C. Reed, A model for the creep deformation behaviour of nickel-based single crystal superalloys. *Acta. Mater.* 60, 4888–4900 (2012)
104. S. Socrate and D.M. Parks, Numerical determination of the elastic driving force for directional coarsening in Ni-superalloys. *Acta. Metall. Mater.* 41, 2185–2209 (1993)
105. J. Jung, G. Ghosh, and G.B. Olson, A comparative study of precipitation behavior of Heusler phase (Ni2TiAl) from B2-TiNi in Ni-Ti-Al and Ni-Ti-Al-X (X = Hf, Pd, Pt, Zr) alloys. *Acta. Mater.* 51, 6341–6357 (2003)
106. X.-G. Lu, M. Selleby, and B. Sundman, Theoretical modeling of molar volume and thermal expansion. *Acta. Mater.* 53, 2259–2272 (2005)
107. X.-G. Lu, M. Selleby, and B. Sundman, Assessments of molar volume and thermal expansion for selected bcc, fcc and hcp metallic elements. *CALPHAD* 29, 68–89 (2005)
108. A. Pineau, D.L. McDowell, E.P. Busso, and S.D. Antolovich, Failure of metals II: Fatigue. *Acta. Mater.* http://dx.doi.org/10.1016/j.actamat.2015.05.050 (in press)
109. D. Sornette, T. Magnin, and Y. Brechet, The physical origin of the Coffin-Manson law in low-cycle fatigue. *Europhys. Lett.* 20, 433 (1992)
110. J.A. Moore, *A Micromechanics-Based Method for Multiscale Fatigue Prediction*, Doctoral thesis, MIT (2015)
111. D.J. Frankel, *Design of Fatigue Resistant Heusler-strengthened PdTi-based Shape Memory Alloys for Biomedical Applications*, Doctoral thesis, MIT (2015)

112. D. Frankel and G. Olson, Design of Heusler precipitation strengthened NiTi- and PdTi-Base SMAs for cyclic performance. *Shape Mem. Superelasticity* 1, 162–179 (2015)
113. H.-J. Jou, P.W. Voorhees, and G.B. Olson, Computer simulations for the prediction of microstructure/property variation in aeroturbine disks. in *Superalloys* (eds K.A. Green, *et al.*) 877–886 (2004)
114. J.-C. Zhao, M.R. Jackson, L.A. Peluso, and L.N. Brewer, A diffusion multiple approach for the accelerated design of structural materials. *MRS Bull.* 27, 324–329 (2002)
115. M. Palumbo, *et al.* Thermodynamic modelling of crystalline unary phases. *Phys. Status Solidi.* 251, 14–32 (2014)
116. W. Xiong, M. Selleby, Q. Chen, J. Odqvist, and Y. Du, Phase equilibria and thermodynamic properties in the Fe-Cr System. *Crit. Rev. Solid State Mater. Sci.* 35, 125–152 (2010)
117. D.R. Squires and E.A. Wilson, Effect of cobalt on impact toughness of steels. *Mater. Sci. Technol.* 10, 52–55 (1994)
118. T. Shibayama, I. Yamagata, H. Kayano, and C. Namba, Effect of small additional elements on DBTT of V-4Cr-4Ti irradiated at low temperatures. *J. Nucl. Mater.* 258–263, 1361–1368 (1998)
119. D.L. McDowell and G.B. Olson, Concurrent design of hierarchical materials and structures. *Sci. Model. Simul. SMNS* 15, 207–240 (2008)

CHAPTER 29

Hierarchical Model for Coaching Technical Design Teams*

M.V. Manuel, University of Florida
A.F. McKenna and G.B. Olson, Northwestern University

Abstract

A new design education model has been developed to address the issue of technical depth in undergraduate design education by utilizing graduate students as coaches. This model has been implemented in a junior-level capstone materials science and engineering design course. The model allows the graduate student coaches to experience supervising other researchers while serving as a technical resource and role model for the undergraduate design team. Meanwhile, the instructor serves as an expert who oversees the instruction of the course and the graduate student coaches. With reduced responsibility, the instructor can focus on the quality of research performed while influencing a greater number of design teams. This paper provides an overview of the model, its characteristics, and execution, and presents findings from interviews with the graduate students coaches and undergraduate design team members. Results from our study suggest that the model provides benefits for all participants in the hierarchy.

Coaching Model

In industry, design work is typically done in teams. However, some speculate whether work done in undergraduate design teams embodies the same level of technical depth that one would see in industry. We implemented a design education model to address the issue of technical depth by utilizing graduate students as undergraduate design team coaches. This model aims to bring undergraduate design education to the level of graduate engineering research.

The hierarchical model used for coaching design teams can be found in Fig. 1. The hierarchical arrangement of the instructor, coaches, and design teams acknowledges the hierarchical nature of the knowledge levels (expert, intermediate, and novice). Those with the greatest technical expertise are placed at the top of the hierarchy. Initially, information

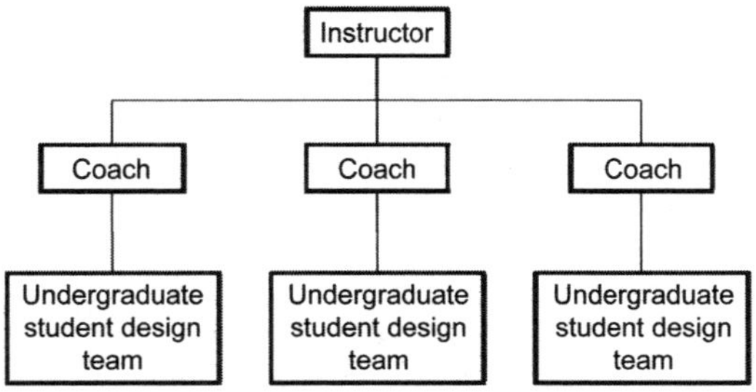

Fig. 1 Hierarchical model for coaching design teams.

*M.V. Manuel, A.F. McKenna, and G.B. Olson, "Hierarchical Model for Coaching Technical Design Teams," *IJEE*, 24, 2, (2008) 260–65

flows in a downward direction to the undergraduate design team. As the design team progresses through the project they feed information back up through the branches of the hierarchy, reporting to the coaches and in turn the coaches relay the information to the instructor. The complex and iterative exchange of information that occurs between all the levels of the hierarchy leads to a bidirectional movement of information, consequently the structure of the hierarchy is no longer rigid but becomes more compliant as research progresses. As the course progresses, the hierarchal arrangement shifts as the technical level of the undergraduate design teams approaches that of the coaches.

At the beginning of the course, design projects are posed to the undergraduate students by the instructor. Students form four- to five-member teams in which each team is coached by a graduate student who is selected prior to the start of the course. The design projects relate directly to the thesis work of the graduate student. This allows the team to be coached by someone who has technical expertise in the project and helps guide the team through the design process. It is this interaction which facilitates the elevation of undergraduate design education to a higher technical level similar to that of funded graduate student research.

The coaching model was used in a junior-level capstone course on materials design (Ref 1). A similar model was also used in an upper-level interdisciplinary design project course (Ref 2) where the faculty serve as "mentors" and coaching occurs concurrently, as opposed to the hierarchical manner discussed in this paper. The details and characteristics of the coaching model used in the interdisciplinary design course are reported elsewhere (Ref 3). All students enrolled in both of these courses have been exposed to engineering design through a firstyear required design and communication course which they take in their freshman year (Ref 4).

The hierarchical coaching model can be related to instructional theories such as Schon's teaching model of "reflection-in-action" (Ref 5) and the "cognitive apprenticeship" model of instruction (Ref 6). That is, Schon's model suggests that teaching consists of a dialogue between the coach and student where understanding is developed through communication and reflection about the design itself. In addition, the cognitive apprenticeship model of instruction suggests a process of modeling, coaching, scaffolding, reflection, and exploration. In cognitive apprenticeship, the teacher models effective practice then observes and coaches students while they perform a similar task. The coaching model described in this paper presents a multiplicative approach of these two educational theories. The involvement of graduate student coaches enables the instructor to influence a greater number of undergraduate students. These theories highlight the advantages of using a hierarchical coaching model in engineering design education.

Model Implementation

The materials design course is composed of lectures, computational labs, homework, a midterm, and a final design project. The homework and midterm evaluates the students understanding of what has been taught in the lectures and labs, and constitutes 40% of the final grade. The primary goal of the course is to teach design practices grounded in materials science through active learning in a group design project and therefore 60% of the final grade is derived from a composite final design project grade. To determine this grade, students submit a project proposal midway through the course and deliver a final project presentation and report at the end of the course.

All undergraduate students participating in the class must be enrolled in the lecture. The lecture provides fundamental technical understanding of the concepts and principles of the materials design methodology used by practicing engineers. It also serves to engage and prepare students for the computational labs and design project. Case studies are presented to analyze the process of arriving at possible design solutions in the context of the undergraduate design projects. The case studies highlight industrial-oriented design projects and enhance the relevancy of the scientific principles learned in the lecture to solving real-world design problems. The lecture is prepared and taught by the instructor, and classroom assessment is performed through the use of a midterm. The lecture emphasizes key concepts and enables interaction between the instructor and undergraduate design teams, although the principles learned in lecture are commonly reinforced by the graduate student coaches.

The computational laboratory allows students to apply theoretical concepts learned in lecture in an active, hands-on environment. The goal of each laboratory session is to familiarize students with computer applications commonly used in materials design for future use in their respective projects. Software packages include the Cambridge Engineering Selector (CES) developed by Ashby (Ref 7, 8) ThermoCalc software system developed by the Royal Institute of Technology (Ref 9), and Computational Materials Dynamics (CMD) developed by QuesTek Innovations. Lab work is done individually by the student and the level of understanding is evaluated using homework. Laboratory instruction is primarily provided by a graduate student fulfilling the dual role of a teaching assistant and graduate student coach. It is the course instructor who is charged with the responsibility of coordinating coherency between the lectures and laboratory sessions.

During the quarter, the instructor assesses the progress of each design team in meetings with the graduate student coaches. Meetings provide a venue where the technical details of the project can be discussed between the instructor and graduate student coaches. The coaches in turn meet with their design teams to discuss the technical details, thereby following the multiplicative/hierarchical nature of the model. Meetings between the coaches and design teams take place in and outside of class.

Approximately one class a month is devoted to in-class coaching sessions where the undergraduate students are separated into their respective design teams with their coach. These sessions provide an informal, supportive atmosphere where information from the instructor can be passed to the design teams via the coach. The coach provides a clear assessment of the team's progress and encourages questions, round-table discussions, and participation by all members of the undergraduate design team. The coach also sets clear, attainable goals and deadlines to keep the team on schedule. This helps the team to set a good pace so that the team does not run out of time at the end of the course. In-class meetings are typically followed by a series of outside meetings among the undergraduate team members.

The final assessment of design teams occurs at the end of the quarter. Each team presents a 30-minute presentation which is accompanied by a final report. The instructor evaluates the team's performance and issues a team grade.

Design Projects

Descriptions of two design projects used in the materials design course are listed in Table 1. These projects also represent ongoing government and industry funded programs. Each team was coached by a graduate student who is performing research in that area. This allows the course to leverage the existing computational and human infrastructure available at the university.

The goals of the Blastalloy project were based on the CyberSteel 2020 project focused on developing ultrahigh-toughness plate steels for blast-resistance naval hull applications. The project was funded by the Office of Naval Research (ONR) "Naval Materials by Design" Grand Challenge. In collaboration with freshmen and upper-level interdisciplinary students, an undergraduate design team coached by a graduate student designed and built a pressurized water blast testing chamber based on numerical blast simulations performed by a design team in the previous quarter.

The Terminator 4 project leveraged an ongoing research project utilizing a graduate student coach funded by the National Aeronautics and Space Administration (NASA) to develop a high specific-strength magnesium-based alloy composite for aerospace applications that can self-heal. Utilizing computation thermodynamic software, the team designed a high-performance magnesium-zinc-aluminum alloy with thermodynamically compatible shape memory alloy (SMA) wires that demonstrates yield strengths greater than a commonly used magnesium-based alloy, AZ91. The team performed interfacial testing to characterize the effects of reinforcement surface modifications and demonstrated the feasibility of adding yttrium to the matrix for self-fluxing.

Table 1 Examples of design projects.

Project name	Project goal
Blastalloy	Design a high performance steel for anti-terrorism blast protection.
Terminator 4	Design a high specific-strength self-healing magnesium-based alloy composite for UAV applications.

This successful implementation of this model in the undergraduate curriculum has led to national recognition. The students participating in the Blastalloy and Terminator 4 design projects were the winners of the 2004 and 2005 national TMS-AIME (The Minerals, Metals and Materials Society of the American Institute of Mining, Metallurgical and Petroleum Engineering) undergraduate design competition. These experiences reflect and represent the elevated knowledge level and problem-solving skills that can be developed with the implementation of this model.

Method

The purpose of our data collection was to understand the project and learning needs of upper-level undergraduate design teams and to characterize the nature of guidance necessary to effectively implement the hierarchical coaching model. We used a qualitative approach for collecting data to determine the impact of our coaching model, and to understand the mentoring experience from both the students' and coaches' perspective.

Specifically, we interviewed six of the graduate student coaches and two undergraduate students to obtain feedback on the role of the coach, the challenges that each side faced, and strategies that were utilized to move the project forward. The coaches were asked the following questions:

- Describe your experience as a coach.
- What has been challenging about coaching the design teams?
- What help do students seek from you?
- How might coaches be better prepared to coach the undergraduate teams?

The undergraduates were asked similar questions but they were tailored to the design team experience. The questions were selected to elicit general details about the mentoring/coaching process and to reveal any issues faced by graduate/undergraduate students in order to better inform our coaching model. Responses from this exploratory study provide insights into how the hierarchical model was implemented, and how it may be improved for future applications.

The responses to each of the questions were reviewed in order to characterize the types of responses students and coaches generated. All responses were read to identify common themes. This method of analyzing qualitative data reflects the iterative tradition of qualitative research corresponding with the grounded theory approach of Glaser and Strauss (Ref 10) and the strategies for analysis of Miles and Huberman (Ref 11).

Results

The study presented in this paper is preliminary, focusing mainly on responses from graduate student coaches on their coaching experience in the materials design course during the Spring 2007 quarter or earlier.

The following subsections provide representative responses regarding two broad categories that are directly relevant to the model. The first category, "mentoring to achieve a high technical level," presents general comments about the mentoring process specific to our model and the second category, "motivation," is a main theme that emerged from the data. The quotes preceded by a "C" are from the coaches while those preceded by a "U" are from individual undergraduate materials design team students.

Mentoring to Achieve a High Technical Level

It has been previously stated that the high technical level of undergraduate design team is fostered by the graduate student coach who is considered to have technical expertise on the project. This can be seen in the following responses by undergraduate design team members and a graduate student coach:

(U) "[Projects in the materials design course] which are very sophisticated and hard to understand, it is really necessary to have a mentor [coach] because most of us [the undergraduate design team] ... didn't know what to do. The mentors [coaches] really helped us try to figure out the whole goal of this project."

(U) "At first, because we [the undergraduate design team] didn't know anything, our mentor [coach] guided us... but once we knew what we were trying to do, everyone [the undergraduate design team] was self-motivated to figure out the models... equations... and data for the system, and now it seems like the mentor [coach] is almost useless..."

(C) "By the end, they know as much as we do... their learning curve is always very steep... by the time they [the undergraduate design team] end their project, they are at such a good level. They are so confident about the project that is it amazing at how much they learned in such a short period of time."

Furthermore, while commenting about the elevation of the undergraduate design team experience to graduate level research, a graduate student coach indicated that the flow of information during the project is bidirectional, benefiting multiple participants in the hierarchy.

(C) "A lot of times the students [undergraduate design teams] are able to work on smaller side projects that you [the coach] may not necessarily have time to work on or it may not be the main objective of your thesis research. Sometimes on the smaller projects, the undergraduates find interesting phenomenon or some scientific thing that maybe needs to be further looked into and that helps out the coaches in their research and kind-of provides feedback on where they should go with their [the coaches] own research."

One interesting finding is the critical role that the instructor plays in the hierarchical model. Our original assumption was that the instructor's role is minimized through the use of graduate student coaches. The data reveal that the instructor plays a critical role in the mentoring needs of the graduate student coaches. To advance the project, quite often the existing technical knowledge threshold must be increased. This forces the graduate student coaches to expand their knowledge and expertise. As one coach stated,

(C) "[The projects] push the boundaries of the coaches' knowledge."

Our findings suggest that the coaches look to the instructors, who are considered technical experts, as mentors for technical advice on how to meet the specific project goals that are outside of the area of their expertise. This need for mentorship is captured in these comments by graduate student coaches, who stated:

(C) "[The instructor] has a vision that is not available to the graduate students."
(C) "There are a lot of things I need to know. It is all in [the instructor's] head, I feel like [the instructor] knows it's possible but I don't... I need to get to that point before I could explain to other people what it is all about."

This technical knowledge is then translated and transferred to the undergraduate design teams which is reflected in the following statements made by four graduate student coaches in response to question about the needs of the undergraduate design teams.

(C) "I think that they [the undergraduate design team] really just want to know step-by-step, what do I do, where do I start, what am I looking for?"
(C) "They [the undergraduate design team] know that coaches are pretty much critical to the success of your project."
(C) "We [the coach] tell them how, what machines you [the undergraduate design team] need to use and how to use them. We teach them, we train them on that 'cause they have to use it."
(C) "I [the coach] think a lot of it relies on coaches to explain new concepts."

Motivation

In addition to the instructor's critical role in the hierarchy, motivation also plays a vital role and occurs through several mechanisms. One mechanism is the connection of project goals with a societal need. Project goals demonstrate the significance and relevance of the project in a broader context and shows that what is being learned is worthwhile and intrinsically valuable. The following quotation was from an undergraduate materials design student working on a project to improve the fatigue performance in shape memory alloy stents, their comment is in reference to team trip to a local hospital to see a surgeon deploy a stent in a patient.

(U) "We [the team] were all standing around and we could see like x-ray vision, how the stent alloy goes in and expands and like it opens up the blood vessel... First of all it was cool, second of all, wow, we are actually working on a project that could be used in a surgical room. That motivated all of us to like just work on this and try to improve the fatigue limit which is our project for this quarter."

Another mechanism is the relationship of the project goals to the thesis research of the graduate student coach. This provides motivation for the graduate student coach to advance the knowledge level of the project which will in turn help them to complete their thesis. The following quotations are from graduate student coaches who have served as coaches on several design teams. This statement was in response to a question asking them to explain their coaching experience.

(C) "We [the coach and the undergraduate design team] have the project that I'm working... is directly related to my own PhD thesis so it is much easier for me."
(C) "It is really easy for me to guide students [the undergraduate design teams] because I'm more knowledgeable, I guess, at different things they need to know for their project."

Meanwhile, graduate students interested in pursuing academic careers are motivated by the teaching experience as seen in the following quotations by two graduate student coaches.

(C) "I think it is a good teaching experience, mentoring experience, it does take a significant amount of time but if that is your goal and your are interested in teaching and mentoring, I think it is a satisfying experience."
(C) "It's what I want to do after research, after industry, is working like, be a high school teacher, just to be like 50-year-old crazy high school teacher that blows up things in the chemistry lab. ... I [the coach] find myself teaching a lot, I think a lot of it like intimate teaching experience."

On the other hand, this experience provides an impression of graduate school for undergraduate students. This was observed in the following two quotations from a graduate student coach and an undergraduate materials design team member, both of whom participated on separate design teams.

(C) "A lot of times they [the undergraduate design team members] ask me about grad school, what is grad school like, is this typical of a grad student that they have to learn with undergrads"
(U) "One of the girls [an undergraduate design team member] was talking about how it must be hard for the grads [the coaches], like she doesn't want to be a graduate student because it is like, they can try all these different directions and you work on a project for four years and like it is not even complete"

The graduate student coaches also play a critical role by motivating the undergraduate design team through the use of constant encouragement. Encouragement can occur through several mechanisms such as recognition and establishing confidence but also through a personal connection between the coach and the team. This sentiment is reflected in the following comments by an undergraduate design team member and various graduate student coaches:

(U) "[The coach] is needed there for guidance issues and trying to figure out what we [the undergraduate design team] need to do but to do that effectively, I think... for the mentor [coach] to be more friendly to us and like act as a friend versus a grad student who is just doing this for like payroll or something like that... I think that personal connection is really important for a mentor [coach] to have...
"We [the coach and undergraduate design team member] play soccer together... that is pretty cool."
(C) "Getting them [the undergraduate design team] passionate and interested in what is going on and actually understand what is going on is really the difficult part."
(C) "I think the most important thing, the part of success, is to bond really well with the team members and from the very first I [the coach] make sure that they [the undergraduate design team] are comfortable..."
(C) "A lot of it is motivation because I [the coach] think... people work better, design better when they understand the motivation of their project... because motivation itself is a way to get people excited about doing it [the project]... it is a way to get them [the undergraduate design team] to feel like they are doing something worthwhile."

Discussion

Benefits occur at each level in the model. Minimizing the interaction between the instructor and undergraduate students gives the instructor more time to focus on the lecture and thorough assessment of design projects, meanwhile influencing a greater number of students. The undergraduate students benefit by having a technically knowledgeable project coach whose expertise directly relates to the design project. The coach can provide greater individualized attention: teaching hands-on experimental techniques, providing access to resources and a sounding board for design ideas. Meanwhile, the coach can help the team set realistic goals and help the students with time management. Undergraduate students can feel free to express their ideas, concerns, and questions while receiving feedback and instruction in an informal and supportive environment. Since work of the undergraduate design team is closely tied to the thesis work of the graduate student, progress of the project can provide valuable information of specific aspects of their research. Graduate student coaching also provides the opportunity for modeling academic career paths for those undergraduate students interested in pursuing a graduate degree.

Coaching provides a means of professional development for the graduate student. Graduate students receive sound mentoring and feedback from the instructor on the technical aspects of the project and developing effective coaching skills. Furthermore, coaching can provide graduate students experience in supervising researchers. To enhance the professional development of gradate student coaches, workshops focusing on the skills needed to be an effective coach could be implemented. In addition to providing support, the workshop would encourage and reinforce the use of sound coaching and teaching practices.

Conclusions and Future Work

The hierarchical coaching model presented in this paper suggests a method to increase the level of technical depth in undergraduate design teams through the use of graduate student coaches. A graduate student coach whose thesis research directly relates to the design project creates successful design teams while minimizing the need for personalized attention from the instructor. While providing a good role model for undergraduate students, graduate student coaches gain valuable experience in teaching and supervising researchers. The successful use of the hierarchical coaching model can be beneficial to faculty, students, and institutions.

This paper provides a basic introduction to the coaching model used in an upper-level capstone materials science and engineering design course. Since the involvement of graduate student coaches is vital to the effectiveness of the hierarchy, we will study the potential impact of including features such as coaching workshops for graduate student coaches. This will provide a solid framework to build a coaching model which will benefit all participants in the hierarchy.

ACKNOWLEDGMENTS

The authors appreciate the helpful and insightful discussions with the graduate student coaches and undergraduate design team members who participated in our study. We would also like to acknowledge the transcript support from the Wilson-Cook Endowment. In addition, we would like to thank the Northwestern Center for Engineering Education Research (NCEER) for their feedback and guidance on the current work.

REFERENCES

1. G.B. Olson, *Materials Design: An Undergraduate Course*, in *Morris E. Fine Symposium.* 1991. PA Warrendale,: TMS-AIME
2. A.F. McKenna, J.E. Colgate, S.H. Carr, and G.B. Olson, *IDEA: Formalizing the Foundation for an Engineering Design Education.* International Journal of Engineering Education, 2006. 22(3): p 671–678
3. A.F. McKenna, J. Colgate, and G.B. Olson, *Characterizing the Mentoring Process for Developing Effective Design Engineering*, in *American Society for Engineering Education Annual Conference Proceedings*, paper no. 2006–1378, 2006. Chicago, IL
4. P.L. Hirsch, B.L. Shwom, C. Yarnoff, J.C. Anderson, D.M. Kelso, G.B. Olson, and J.E. Colgate, *Engineering Design and Communication: The Case for Interdisciplinary Collaboration.* International Journal of Engineering Education, 2001. 17 (4–5): p 342–348
5. D.A. Schön, *Educating the Reflective Practitioner: Toward a New Design for Teaching and Learning in the Professions.* 1st ed., The Jossey-Bass Higher Education Series. 1987, San Francisco: Jossey-Bass. xvii, p 355

6. A. Collins, J.S. Brown, and A. Houm, *Cognitive Apprenticeship: Making Thinking Visible*. American Educator, 1991. 6 (11): p 38–46
7. M.F. Ashby, *Materials Selection in Mechanical Design*. 3rd ed. 2005, Elsevier Butterworth-Heinemann
8. M.F. Ashby, *Materials Selection in Conceptual Design*. Materials Science and Technology, 1989. 5(6): p 517–525
9. B. Sundman, B. Jansson, and J. Andersson, *The Thermo-Calc Databank System*. CALPHAD, 1985. 9(2): p 153–190
10. B.G. Glaser and A.L. Strauss, *The Discovery of Grounded Theory: Strategies for Qualitative Research*. 1967, Chicago: Aldin Publishing Co.
11. M.B. Miles and A.M. Huberman, *Qualitative Data Analysis*. 1994, Thousand Oaks, CA: Sage Publications

APPENDIX

Patents and Publications

Patents

- G.B. Olson, J.F. Watton, and M. Cohen, Hydrogen-Resistant Ultrahigh-Strength Steel, U.S. Patent 4,836,869, 6 June 1989
- G.B. Olson, Fracture-Tough High Hardness Stainless Steel and Method of Making Same, U.S. Patent 5,221,372, 22 June 1993
- C.J. Kuehmann, G.B. Olson, J.P. Wise, C.E. Campbell, and E. Carelyn, Advanced Case Carburizing Secondary Hardening Steels, Australian Patent 742,183, 28 Jan 1999
- C.J. Kuehmann, G.B. Olson, J.P. Wise, and C. Campbell, Advanced Case Carburizing Secondary Hardening Steels, U.S. Patent 6,176,946 B1, 23 Jan 2001
- C.J. Kuehmann, P.A. Bernikowicz, G.B. Olson, J.P. Wise, and C.E. Campbell, Case Hardened Steel Blades for Sports Equipment and Method of Manufacture, U.S. Patent 6,458,220 B1, 1 Oct 2002
- C.J. Kuehmann, G.B. Olson, J.P. Wise, C.E. Campbell, and E. Carelyn, Advanced Case Carburizing Secondary Hardening Steels, U.S. Patent 3,487,825, 31 Oct 2003
- C.J. Kuehmann, G.B. Olson, J.P. Wise, C.E. Campbell, and E. Carelyn, Advanced Case Carburizing Secondary Hardening Steels, Republic of Korea Patent 422,416, 2 Mar 2004
- C.J. Kuehmann, G.B. Olson, J.P. Wise, C.E. Campbell, and E. Carelyn, Advanced Case Carburizing Secondary Hardening Steels, U.S. Patent 6,464,801, 15 Oct 2002
- C.J. Kuehmann, G.B. Olson, J.P. Wise, C.E. Campbell, and E. Carelyn, Advanced Case Carburizing Secondary Hardening Steels, U.S. Patent 6,485,582, 26 Nov 2002
- G.B. Olson, W. Tang, C. Qiu, and H.-J. Jou, Nanophase Precipitation Strengthened Al Alloys Processed through the Amorphous State, U.S. Patent Pending, filed 24 Apr 2003
- C.J. Kuehmann, G.B. Olson, J.P. Wise, C.E. Campbell, and E. Carelyn, Advanced Case Carburizing Secondary Hardening Steels, U.S. Patent 6,635,126, 21 Oct 2003
- C.J. Kuehmann, G.B. Olson, J.P. Wise, C.E. Campbell, and E. Carelyn, Process for Case Hardened Dies for Improving Die Life, U.S. Patent 6,695,930, 24 Feb 2004
- G.B. Olson, A.J. Freeman, W.-T. Geng, and C. Kantner, Method for Enhancement of Grain Boundary Cohesion in Crystalline Materials and Compositions of Matter Therefore, U.S. Patent Pending, filed 21 June 2004
- G.B. Olson, D.J. Bryan, and A. Misra, Oxidation Resistant Niobium Based Alloys, U.S. Patent Pending, filed 30 July 2004
- C.J. Kuehmann, G.B. Olson, and H.-J. Jou, Nanocarbide Precipitation Strengthened Ultrahigh-Strength, Corrosion Resistant, Structural Steels, U.S. Patent 7,160,399, 9 Jan 2007
- J.-W. Jung and G.B. Olson, Coherent Nanodispersion-Strengthened Shape-Memory Alloys, U.S. Patent 7,316,753, 8 Jan 2008
- G.B. Olson, W. Tang, C. Qiu, and H.-J. Jou, High-Temperature High-Strength Aluminum Alloys Processed through the Amorphous State, U.S. Patent 20080138239 A1, 12 June 2008
- J.A. Wright, G.B. Olson, and W. Tang, Martensitic Stainless Steel Strengthened by Copper-Nucleated Nitride Precipitates, U.S. Patent 20110094637 A1, 28 Apr 2011
- G.B. Olson and A. Saha, Ultratough High-Strength Weldable Plate Steel and Method of Manufacture Thereof, U.S. Patent 8,016,954, 13 Sep 2011

- G.B. Olson and P. Sadhukhan, High Strength Austenitic TRIP Steel, U.S. Patent 8,092,620 B2, 10 Jan 2012
- A.K. Sachdev, B.L. Tiemens, and G.B. Olson, Copper Precipitate Carburized Steels and Related Method, U.S. Patent 8118949 B2, 21 Feb 2012
- Y. Qian and G.B. Olson, Case Hardenable Nickel-Cobalt Steel," U.S. Patent 8,157,931, 17 Apr 2012
- J. Yan and G.B. Olson, Computationally-Designed Transformation-Toughened Near-Alpha Titanium Alloy, U.S. Patent 20,160,060,742, 3 Mar 2016

Publications

Books

Editor

- G.B. Olson, M. Azrin, and E.S. Wright, Eds., *Innovations in Ultrahigh-Strength Steel Technology*, Proc. 34th Sagamore Army Research Conf. (1990)
- G.B. Olson and W.S. Owen, Eds., *Martensite*, ASM International (1992)
- G.B. Olson, D.S. Lieberman, and A. Saxena, Eds., *Proc. ICOMAT-08*, TMS, (2009)

Chapter Author

- G.B. Olson, Transformation Plasticity and the Stability of Plastic Flow, in *Deformation, Processing and Structure*, 1982 ASM Materials Science Seminar, ed. G. Krauss, ASM International (1984), p 391–424
- G.B. Olson and M. Cohen, Principles of Martensitic Transformations, in *Frontiers in Materials Technologies*, ed. M.A. Meyers and O.T. Inal, Elsevier (1985), p 43–87
- G.B. Olson, Mechanically-Induced Phase Transformations in Alloys, *Encyclopedia of Materials Science and Engineering*, ed. M.B. Bever, Pergamon Press (1986), p 2929–32
- G.B. Olson and M. Cohen, Dislocation Theory of Martensitic Transformations, in *Dislocations in Solids* vol 7, ed. F.R.N. Nabarro, North-Holland (1986), p 295–407
- G.B. Olson, Introduction: Martensite in Perspective, in *Martensite*, ed. G.B. Olson and W.S. Owen, ASM International, (1992), p 1–10
- G.B. Olson and A.L. Roitburd, Martensitic Nucleation, in *Martensite*, ed. G.B. Olson and W.S. Owen, ASM International (1992), p 149–74
- G.B. Olson, Mechanically Induced Phase Transformations in Alloys: An Update, in *Encyclopedia of Materials Science and Engineering*, supp. vol 3, Pergamon Press (1993), p 1786–89
- G.B. Olson, Systems Design of Materials, in *Encyclopedia of Materials Science and Engineering*, Pergamon (1993), p 2041–45
- G.B. Olson, Effects of Stress and Deformation on Martensitic Formation, *Encyclopedia of Materials: Science and Technology*, ed. H. Aaronson, Elsevier (2002) p 2381–84
- G.B. Olson, Designing a New Material World," in *Science Pathways of Discovery*, ed. I. Amato, John Wiley & Sons (2002), p 91–108
- G.B. Olson and A. Saxena, "Models for Martensitic Transformations," in *ASM Handbook, Volume 22A: Fundamentals of Modeling for Metals Processing*, ASM International (2009), p 191–202
- C.J. Kuehmann and G.B. Olson, "ICME: Success Stories and Cultural Barriers," in *Models, Databases, and Simulation Tools Needed for the Realization of Integrated Computational Materials Engineering*, ed. S. Arnold & T. Wong, ASM International (2011)
- G.B. Olson, Z.D. Feinberg, "Kinetics of Martensite Transformations in Steels," in *Phase Transformations in Steels: Fundamentals and diffusion controlled transformations*, vol 2, ed. S. Arnold and E. Pereloma, Woodhead Publishing (2012), p 59–82

Journals and Proceedings

- B. Sadagopan, H.C. Gatos, and G.B. Olson, Superconducting Properties of BCC Alloys in the Nb-Ti-Mo System, *J. Appl. Phys.*, 41, 4 (1970), 1874
- G.B. Olson and M. Cohen, A Mechanism for the Strain-Induced Nucleation of Martensitic Transformations, *J. Less-Common Metals.*, 28 (1972), 107
- G.B. Olson and M. Cohen, Kinetics of Strain-Induced Martensitic Nucleation, *Metall. Trans.* 6A (1975), 791
- G.B. Olson and M. Cohen, Thermoelastic Behavior in Martensitic Transformations, *Scripta. Metall.*, 9 (1975), 1247
- M. Azrin, G.B. Olson, and R.A. Gagne, Inhomogeneous Deformation and Strain-rate

Effects in High-Strength TRIP Steels, *Mat. Sci. and Eng.*, 23 (1976), 33
- G.B. Olson and M. Cohen, A General Mechanism of Martensitic Nucleation: Part I. General Concepts and the FCC-HCP Transformation, *Metall. Trans.*, 7A (1976), 1897–904
- G.B. Olson and M. Cohen, A General Mechanism of Martensitic Nucleation: Part II. FCC-BCC and Other Martensitic Transformations, *Metall. Trans.*, 7A (1976), 1905–14
- G.B. Olson and M. Cohen, A General Mechanism of Martensitic Nucleation: Part III. Kinetics of Martensitic Nucleation, *Metall. Trans.*, 7A (1976), 1915–23
- M. Cohen and G.B. Olson, Martensitic Nucleation and the Role of the Nucleating Defect, *New Aspects of Martensitic Transformations*, Proc. First JIM International Symposium (JIMIS-1) Kobe, Japan. Suppl. *Trans. JIM.*, 17 (1976), 93
- G.B. Olson and W.S. Owen, The Stress Field of a Martensitic Particle and the Conditions for Thermoelastic Behavior, *New Aspects of Martensitic Transformations*, Proc. First JIM International Symposium (JIMIS-1) Kobe, Japan. Suppl. *Trans. JIM*, 17 (1976), 105
- M. Azrin, G.B. Olson, and R.A. Gagne, Austenite Stability and Tensile Properties of Warm-Extruded TRIP Steels, *Proc. Fourth North American Metalworking Research Conference*, ed. T. Altan, (1976), 25
- G.B. Olson and M. Cohen, Reply to "On the Equilibrium Temperature in Thermoelastic Martensitic Transformations," *Scripta Metall.*, 11 (1977), 345
- G.B. Olson and M. Azrin, Transformation Behavior of TRIP Steels, *Metall. Trans. A.*, 9A (1978), 713
- M. Azrin, G.B. Olson, and R.A. Gagne, Warm Extrusion of TRIP Steels: Process Control and Tensile Properties, *Mat. Sci. and Eng.*, 40 (1979), 175
- G.B. Olson and M. Cohen, Interphase Boundary Dislocations and the Concept of Coherency, *Acta Metall.*, 27 (1979), 1907
- M. Cohen, G.B. Olson, and P.C. Clapp, On the Classification of Displacive Phase Transformations, *Proc. ICOMAT-79*, Intl. Conf. on Mart. Transf., MIT (1979), 1
- G.B. Olson and M. Cohen, Kinetics of Isothermal FCC-HCP Martensitic Nucleation, *Proc. ICOMAT-79*, Intl. Conf. on *Mart. Transf.*, MIT (1979), 310
- M. Azrin, G.B. Olson, E.B. Kula, and W.F. Marley Jr., Soviet Progress in Thermomechanical Treatment of Metals, *J. Appl. Metalworking*, 2 (1980), 5–34
- G.B. Olson, R. Chait, M. Azrin, and R.A. Gagne, Fatigue Strength of TRIP Steels, *Metall. Trans. A*, 11A (1980), 1069
- M. Suga, J.L. Goss, G.B. Olson, and J.B. Vander Sande, Austenitic Grain Growth Characteristics in Rapidly Solidified Martensitic Steels, *Proc. 2nd Intl. Conf. on Rapid Solidification Processing Principles and Technologies*, Reston, VA (1980), 364
- G.B. Olson and M. Cohen, A Perspective on Martensitic Nucleation, *Ann. Rev. Mat. Sci.*, 11 (1981), 1
- G.B. Olson, Fine Structure of Interphase Boundaries, *Acta Metall.*, 29 (1981), 1475
- M.L. Green, M. Cohen, and G.B. Olson. The Pseudoelastic Force Balance and Its Application to βFe-Be Alloys, *Mat. Sci. and Eng.*, 50 (1981), 109
- G.B. Olson, J.F. Mescall, and M. Azrin, Adiabatic Deformation and Strain Localization, Keynote paper, *Shock Waves and High Strain-Rate Phenomena in Metals: Concepts and Applications*, ed. M.A. Meyers and L.E. Murr, Plenum (1981), 221–47
- G.B. Olson and M. Cohen, Stress-Assisted Isothermal Martensitic Transformation and Transformation Plasticity, *Proc. US/Japan Seminar on Mechanical Behavior of Metals and Alloys Associated with Displacive Phase Transformations*, RPI (1982), 7
- G.B. Olson and M. Cohen, Theory of Martensitic Nucleation: A Current Assessment, *Proc. Intl. Conf. Solid-Solid Phase Transf.*, (Carnegie-Mellon Univ.) AIME (1982), 1145
- G.B. Olson and M. Cohen, Dislocation Structure of Martensitic Interfaces, *Proc. Intl. Conf. Solid-Solid Phase Transf.*, (Carnegie-Mellon Univ.) AIME (1982), 1209
- H.C. Ling and G.B. Olson, Strain Energy of a Martensitic Nucleus, *Proc. Intl. Conf. Solid-Solid Phase Transf.*, (Carnegie-Mellon Univ.) AIME (1982), 1171
- G.B. Olson, H.C. Ling, J.S. Montgomery, J.B. Vander Sande, and M. Cohen, Grain-Coarsening Resistance and the Stability of Second-Phase Dispersions in Rapidly Solidified Steels, *Proc. MRS Symp. Rapidly Solidified Amorphous and Crystalline Alloys*, Elsevier (1982) 355
- T.F. Kelly, G.B. Olson, and D.J.B. Vander Sande, Analysis of a Rapidly Solidified High-Phosphorus Austenitic Steel Containing an Amorphous Phase, Proc. *MRS Symp.*

Rapidly Solidified Amorphous and Crystalline Alloys, Elsevier (1982), 343

- G.B. Olson and M. Cohen, Stress-Assisted Isothermal Martensitic Transformation: Application to TRIP Steels, *Metall. Trans. A*, 13A (1982), 1907
- G.B. Olson and H. Hartman, Martensite and Life: Displacive Transformations as Biological Processes, Proc. ICOMAT-82 *J. de Physique*, 43 (1982), C4-855
- G.B. Olson and M. Cohen, Classical and Non Classical Mechanisms of Martensitic Transformations, Proc. ICOMAT-82 *J. de Physique*, 43 (1982), C4-75
- M. Grujicic, G.B. Olson, and W.S. Owen, Kinetics of Martensitic Interface Motion, Proc. ICOMAT-82 *J. de Physique*, 43 (1982), C4-173
- T. Narutani, G.B. Olson, and M. Cohen, Constitutive Flow Relations During Strain-Induced Martensitic Transformation of Austenitic Steels, Proc. ICOMAT-82 *J. de Physique*, 43 (1982), C4-429
- G.B. Olson, A.A. Anctil, T.S. De Sisto, and E.B. Kula, Anisotropic Embrittlement in High Hardness ESR 4340 Steel Forgings, *Metall. Trans. A*, 14A (1983), 1661
- G.B. Olson and M. Cohen Early Stages of Aging and Tempering of Ferrous Martensites, *Metall. Trans. A*, 14A (1983), 1057
- G.B. Olson, M. Cohen, and G.J. Yurek, Structure-Property Relationships in RSP Steels, *Proc. 3rd Conf. Rapid Solidification Processing: Principles & Technologies*, ed. R. Mehrabian, NBS (1983), 171
- G.B. Olson and P.H. Adler, On the Lattice Correspondence of the $\delta \rightarrow \alpha$ Displacive Transformation in Plutonium Alloys, *Scripta Metall.*, 18 (1984), 401
- G.B. Olson, Interfacial Dislocations and Coherency, *Surf. Sci.*, 144 (1984), 302
- G.B. Olson, M. Azrin, and N. Tsangarakis, Plastic Shear Instability in 4340 Steel, *Materials Behavior Under High Stress and Ultrahigh Loading Rates*, Proc. 29th Sagamore Army Materials Research Conference, Part II (AMMRC MS 84-2), ed. J. Mescall and V. Weiss (1984), 139
- L. Chang, A. Cerezo, G.D.W. Smith, M.K. Miller, M.G. Burke, S.S. Brenner, K.A. Taylor, T. Abe, and G.B. Olson, Ageing of Fe-Ni-C Martensite, Proc. 31st Field-Emission Symposium *J. de Physique*, (1984), 1713
- R.W. Balluffi and G.B. Olson, On the Hierarchy of Interfacial Dislocation Structure, *Metall. Trans. A*, 16A (1985), 529
- M. Grujicic, G.B. Olson, and W.S. Owen, Mobility of Martensitic Interfaces, *Metall. Trans. A*, 16A (1985), 1713
- M. Grujicic, G.B. Olson, and W.S. Owen, Mobility of the $\beta_1 - \gamma_{1'}$ Martensitic Interface in Cu-Al-Ni. I. Experimental Measurements, *Metall. Trans. A*, 16A (1985), 1723
- M. Grujicic, G.B. Olson, and W.S. Owen, Mobility of the $\beta_1 - \gamma_{1'}$ Martensitic Interface in Cu-Al-Ni. II. Model Calculations, *Metall. Trans. A*, 16A (1985), 1735
- G.B. Olson and R. G. Bourdeau, Rapidly Solidified Ferrous Alloys, *Rapidly Solidified Crystalline Alloys*, ed. S.K. Das, B.H. Kear, and C. M. Adam, TMS-AIME, (1985), 185
- L. Chang, G.D.W. Smith, and G.B. Olson, Ageing and Tempering of Ferrous Martensites, Proc. 32nd Intl. Field Emission Symposium *J. de Physique*, 47 (1986), C2-265
- P.H. Adler, G.B. Olson, and D.S. Margolies, Kinematics of $\delta \rightarrow \alpha$ and $\alpha \rightarrow \delta$ Martensitic Transformation in Pu Alloys, *Acta Metall.*, 34 (1986), 2053
- M. Azrin, J.G. Cowie, and G.B. Olson, Shear Instability Mechanisms in High-Hardness Steels, *Annals of the Israel Physical Society*, 8 (1986), 409
- P.H. Adler, G.B. Olson, and W.S. Owen, Strain Hardening of Hadfield Manganese Steel, *Metall. Trans. A*, 17A (1986), 1775
- J.A. Liddle, G.D.W. Smith, and G.B. Olson, Alloy Carbide Precipitation in a High Cobalt-Nickel Secondary Hardening Steel, Proc. 33rd Intl. Field-Emission Symposium *J. de Physique*, 47 (1986), C7-223
- G.B. Olson, Structure/Property Relationships in High-Strength Steels, *Proc. 31st Sagamore Army Materials Research Conference*, ed. J.W. McCauley and V. Weiss, Plenum (1986), 109
- G.B. Olson and M. Cohen, Martensitic Transformation as a Deformation Process, *Proc. Earl R. Parker Symposium on Structure/Property Relationships*, ed. S.D. Antalovich, W.W. Gerberich, and R.O. Ritchie, TMS-AIME (1986), 367
- G.B. Olson, Lattice Stability and the Mechanism of Martensitic Transformations, *Proc. ICOMAT 86*, Japan Institute of Metals (1986), 25–34
- P.H. Adler and G.B. Olson, Crystallographic Considerations of the $\delta \leftrightarrow \alpha$ Displacive Transformation in Plutonium Alloys, *Proc. ICOMAT 86*, Japan Institute of Metals (1986), 1095–99

- G.B. Olson, Interphase Kinematics and the Roles of Structure and Composition in Solid-State Transformations, *Scripta Metall.*, 21 (1987), 1023
- G.B. Olson, K. Tsuzaki, and M. Cohen, Statistical Aspects of Martensitic Nucleation, Turnbull Symposium: *Phase Transitions in Condensed Systems*, ed. G.S. Cargill, F. Spaepen, and K.N. Tu, MRS (1987), 129
- M.R. Libera, G.B. Olson, and J.B. Vander Sande, Nucleation during the Solidification of Atomized Droplets Catalyzed by Spherically-Shaped Oxide Particles, *Science and Technology of Rapidly Quenched Alloys*, ed. M Tenhover, MRS (1987)
- G.B. Olson, H.K.D.H. Bhadeshia, and M. Cohen, Coupled Diffusional and Displacive Transformations, *Proc. Phase Transformations '87*, ed. G.W. Lorimer, Institute of Metals, (1988), 322–25
- G.B. Olson, Alloy Design Principles for RSP Steels, *Proc. 4th Conf. Rapid Solidification Processing: Principles and Technologies*, ed. R. Mehrabian and P.A. Parrish, Claitor's (1988), 82–98
- J.H. Moll, W. Stasko, A.T. Galbato, G.B. Olson, and R. Ray, New Developments in Rapidly Solidified Iron-Base Materials, *Proc. 4th Conf. Rapid Solidification Processing: Principles and Technologies*, ed. R. Mehrabian and P.A. Parrish, Claitor's (1988) 174–89
- M. Grujicic and G.B. Olson, Ferrite-M2C Coherent Phase Equilibrium in an AF1410 Ultrahigh-Strength Steel, *Calphad* 12, 4 (1988), 403–12
- P.H. Adler and G.B. Olson, Thermodynamics and Kinetics of $\delta \rightarrow \alpha$ Martensitic Transformation in Plutonium Alloys, *Metall. Trans.*, 19A (1988), 2705–11
- G.M. Carinci, M.G. Hetherington, and G.B. Olson, M2C Carbide Precipitation in AF1410 Steel, *J. de Physique* 49 (1988), C6-311
- G.N. Haidemenopoulos, G.B. Olson, M. Cohen, and K. Tsuzaki, Transformation Plasticity of Retained Austenite in Stage I Tempered Martensitic Steels, *Scripta Metall.*, 23 (1989) 207–11
- G.B. Olson, H.K.D.H. Bhadeshia, and M. Cohen, Coupled Diffusional/Displacive Transformations, *Acta. Metall.*, 37, (1989) 381–89
- K. A. Taylor, L. Chang, G.B. Olson, G.D.W. Smith, M. Cohen, and J.B. Vander Sande, Spinodal Decomposition during Aging of Fe-Ni-C Martensites, *Metall. Trans.*, 20A (1989), 2717–37
- K.A. Taylor, G.B. Olson, M. Cohen, and J.B. Vander Sande, {011} Twinning in Fe-Ni-C Martensites, *Metall. Trans.*, 20A (1989), 2739–47
- K.A. Taylor, G.B. Olson, M. Cohen, and J.B. Vander Sande, Carbide Precipitation during Stage I Tempering of Fe-Ni-C Martensites, *Metall. Trans.*, 20A (1989), 2749–65
- M.J. Gore, M. Grujicic, G.B. Olson, and M. Cohen, Thermally Activated Grain Boundary Motion through a Dispersion of Particles, *Acta Metall.*, 37 (1989), 2849–54
- G.N. Haidemenopoulos, M. Grujicic, G.B. Olson, and M. Cohen, Transformation Microyielding of Retained Austenite, *Acta Metall.*, 37 (1989), 1677–82
- J.G. Cowie, M. Azrin, and G.B. Olson, Microvoid Formation during Shear Deformation of Ultrahigh Strength Steels, *Metall. Trans. A*, 2A (1989), 143–53
- G.N. Haidemenopoulos, M, Grujicic, G.B. Olson, and M. Cohen, On the Optimization of Precipitated Austenite for Transformation Toughening of Steels, *Calphad*, 13 (1989), 215–33
- G.L. Krasko and G.B. Olson, Energetics of BCC-FCC Lattice Deformation in Iron, *Phys. Rev. B*, 40 (1989), 536–44
- G.L. Krasko and G.B. Olson, Ferromagnetism and Crystal Lattice Stability of BCC and FCC Iron, *J. Appl. Phys.*, 67, 9 (1990), 4570–72
- G.L. Krasko and G.B. Olson, Energetics of BCC-FCC Lattice Deformation in Iron, in Atomic Scale Calculations in Materials Science, ed. J. Tersoff, D. Vanderbilt, V. Vitek, *MRS Symp. Proc.*, vol 141 (1989), 135
- G.B. Olson, New Steels by Design, *J. Materials Education*, 11 (1989), 515–28
- G.B. Olson, Science of Steel, *Innovations in Ultrahigh-Strength Steel Technology*, ed. G.B. Olson, M. Azrin, and E.S. Wright, 34th Sagamore Army Materials Research Conference Proceedings, (1990), 3–66
- J.S. Montgomery and G.B. Olson, Kinematics of M2C Carbide Precipitation, *Innovations in Ultrahigh-Strength Steel Technology*, ed. G.B. Olson, M. Azrin, and E.S. Wright, 34th Sagamore Army Materials Research Conference Proceedings, (1990), 147–78
- G. Carinci, J.A. Liddle, G.D.W. Smith, and G.B. Olson, APFIM Study of Multicomponent M_2C Carbide Precipitation, *Innovations in Ultrahigh-Strength Steel Technology*, ed.

G.B. Olson, M. Azrin, and E.S. Wright, 34th Sagamore Army Materials Research Conference Proceedings, (1990), 179–208

- J.G. Cowie, M. Azrin, and G.B. Olson, Microvoid Formation during Shear Deformation of UHS Steels, *Innovations in Ultrahigh-Strength Steel Technology*, ed. G.B. Olson, M. Azrin, and E.S. Wright, 34th Sagamore Army Materials Research Conference Proceeding, (1990), 357–81
- M.J. Gore, G.B. Olson, and M. Cohen, Grain Refining Dispersions and Properties in Ultrahigh-Strength Steels, *Innovations in Ultrahigh-Strength Steel Technology*, ed. G.B. Olson, M. Azrin, and E.S. Wright, 34th Sagamore Army Materials Research Conference Proceedings, (1990), 425–41
- M. Grujicic, M. Buonanno, S.M. Allen, G.B. Olson, and M. Cohen, Heterogeneous Precipitation of Austenite for Stabilization, *Innovations in Ultrahigh-Strength Steel Technology*, ed. G.B. Olson, M. Azrin, and E.S. Wright, 34th Sagamore Army Materials Research Conference Proceedings, (1990), 527–47
- G.N. Haidemenopoulos, G.B. Olson, and M. Cohen, Dispersed-Phase Transformation Toughening in UHS Steels, *Innovations in Ultrahigh-Strength Steel Technology*, ed. G.B. Olson, M. Azrin, and E.S. Wright, 34th Sagamore Army Materials Research Conference Proceedings, (1990), 549–93
- G.L. Krasko and G.B. Olson, Evaluation of Decohesion Mechanisms by LMTO-ASA Cohesive Energy Calculations, *Innovations in Ultrahigh-Strength Steel Technology*, ed. G.B. Olson, M. Azrin, and E.S. Wright, 34th Sagamore Army Materials Research Conference Proceedings, (1990), 677–92
- J.F. Watton, G.B. Olson, and M. Cohen, A Novel Hydrogen-Resistant UHS Steel, *Innovations in Ultrahigh-Strength Steel Technology*, ed. G.B. Olson, M. Azrin, and E.S. Wright, 34th Sagamore Army Materials Research Conference Proceedings, (1990), 705–37
- G.B. Olson, H.K.D.H. Bhadeshia, and M. Cohen, Coupled Diffusional/Displacive Transformations: Part II. Solute Trapping, *Metall. Trans.*, 21A (1990), 805–9
- F-R. Chen and G.B. Olson, Transmission Electron Microscopy Studies of the Structure of Martensitic Interfaces in Cu-Zn: HRTEM, *Proc. XIIth Intl. Conf. Electron Microscopy*, (1990)
- F-R. Chen and G.B. Olson, Transmission Electron Microscopy Studies of the Structure of Martensitic Interfaces in Cu-Zn: Diffraction Contrast, *Proc. XIIth Intl. Conf. Electron Microscopy*, (1990)
- T.G. Zocco, M.F. Stevens, P.H. Adler, R.I. Sheldon, and G.B. Olson, Crystallography of the $\delta \rightarrow \alpha$ Phase Transformation in a Pu-Ga Alloy, *Acta Metall. Mater.*, (1990),, 2275–82
- G.B. Olson, K.A. Taylor, and M. Cohen, Reply to Discussion of "Spinodal Decomposition during Aging of Fe-Ni-C Martensites,"*Metall Trans.*, 21A (1990) 2083–86
- G.L. Krasko and G.B. Olson, The Effect of Boron, Carbon, Phosphorus, and Sulfur on Intergranular Cohesion in Iron, *Solid State Comm.*, 76 (1990), 247–51
- G.B. Olson, Mechanisms of First-Order Lattice-Distortive Transformations, Proc. ICOMAT'89 *Mater. Sci. Forum*, Vol 56–58, Trans. Tech (1990), 89–100
- G.B. Olson, Materials Design: An Undergraduate Course, *Morris E. Fine Symposium*, ed. P.K. Liaw, J.R. Weertman, H.L. Markus, and J.S. Santner, TMS-AIME, (1991), 41–48
- M. Libera, G.B. Olson, and J.B. Vander Sande, Heterogeneous Nucleation of Solidification in Atomized Liquid Metal Droplets, *Mater. Sci. Eng.*, A132 (1991), 107–18
- G.B. Olson, T.J. Kinkus, and J.S. Montgomery, APFIM Study of Multi-Component M_2C Carbide Precipitation in AF1410 Steel, *Surf. Sci.*, 246, (1991) 238
- K.C. King, P.W. Voorhees, and G.B. Olson, Solute Distribution around a Coherent Precipitate in a Multicomponent Alloy, *Metall., Trans.*, 22A (1991), 2199–210
- R.G. Stringfellow, D.M. Parks, and G.B. Olson, A Constitutive Model for Transformation Plasticity Accompanying Strain-Induced Martensitic Transformations in Metastable Austenitic Steels, *Acta Metall.*, 40, 7 (1992), 1703–16
- G.L. Krasko, R.J. Harrison, and G.B. Olson, Effect of H on the Electronic Structure of Iron, *MRS Symp. Proc.*, 209 (1991), 238
- G.L. Krasko and G.B. Olson, Effect of Hydrogen on the Electronic Structure of a Grain Boundary in Iron, *Solid State Comm.*, 79 (1991), 113–17
- K.C. King, G.B. Olson, and T. Mura, Elastic Energy of Coherent Precipitation at Dislocations in an Anisotropic Matrix, *Modern Theory of Anisotropic Elasticity and*

Applications, ed. J.J. Wu, T.C.T. Ting, and D.M. Barnett, SIAM, (1991), 103–21

- P.H. Adler, G.B. Olson, M.F. Stevens, and G.F. Gallegos, On the Constitutive Relations for $\delta \rightarrow \alpha$ and $\alpha \rightarrow \delta$ Martensitic Transformation Plasticity in Plutonium Alloys, *Acta Metall.*, 40, 5 (1992), 1073–82
- T.J. Kinkus and G.B. Olson, Microanalytical Evaluation of a Prototype Stainless Bearing Steel, *Surf. Sci.*, 266 (1992), 391–96
- R. Wu, A.J. Freeman, and G.B. Olson, On the Electronic Basis of the Phosphorus Intergranular Embrittlement of Iron, *J. Mater. Res.*, 7, 9 (1992), 2403–11
- T.A. Stephenson, C.E. Campbell, and G.B. Olson, Systems Design of Advanced Bearing Steels, *Advanced Earth-to-Orbit Propulsion Technology 1992*, ed. R.J. Richmond and S.T. Wu, NASA Conference Publication 3174, 2, (1992), 299–307
- A. Umantsev and G.B. Olson, Modulation Mechanism for First-Order Transformations with Nonconserved Order Parameters, *Phys. Rev. A*, 46 (1992), R6132-35
- M. Lin, G.B. Olson, and M. Cohen, Distributed-Activation Kinetics of Heterogeneous Martensitic Nucleation, *Metall. Trans.*, 23A (1992), 2987–98
- M.L. Brandt and G.B. Olson, Bainitic Stabilization of Austenite in Low Alloy Steels, *Proc. Gilbert R. Speich Symposium: Fundamentals of Aging and Tempering in Bainitic and Martensitic Steel Products*, ed. G. Krauss and P.E. Repas, ISS-AIME, (1992), 257–65
- J.S. Montgomery and G.B. Olson, M_2C Carbide Precipitation in AF1410, *Proc. Gilbert R. Speich Symposium: Fundamentals of Aging and Tempering in Bainitic and Martensitic Steel Products*, ed. G. Krauss and P.E. Repas, ISS-AIME, (1992), 177–214
- S. Tang, A.J. Freeman, and G.B. Olson, Phosphorus-Induced Relaxation in an Iron Grain Boundary: A Cluster Model Study, *Phys. Rev. B*, 47 (1993), 2441–45
- M. Lin, G.B. Olson, and M. Cohen, Homogeneous Martensitic Nucleation in Fe-Co Precipitates Formed in a Cu Matrix, *Acta Metall. Mater.*, 41 (1993), 253–63
- P. Hong and G.B. Olson, Estimation of the Domain-Boundary Energy by Landau-Ginzburg Model for Cubic to Tetragonal Transformations, *Solid-State Comm.*, 85 (1993), 681–83
- M.L. Brandt and G.B. Olson, Bainitic Stabilization of Austenite in Low Alloy Steels, *Iron and Steel Maker*, ISS, (1993), 55–60
- R. Wu, A.J. Freeman, and G.B. Olson, Effects of P and B Adsorbates on the Fe (111) Surface, *PRB Rapid Comm.*, 47, 11 (1993), 6855–58
- G. Ghosh and G.B. Olson, Impurity Gettering in Rapidly Solidified Advanced Steels, *Proc. 51st Annual Mtg. Microscopy Society of America*, ed. G.W. Bailey and C.L. Rieder, San Francisco Press (1993)
- G. Ghosh, G.B. Olson, and M.E. Fine, Coherency Induced Morphological Instability in NiAl-Type Precipitate, *Proc. 51st Annual Mtg. Microscopy Society of America*, ed. G.W. Bailey and C.L Rieder, San Francisco Press (1993)
- A. Umantsev and G.B. Olson, Ostwald Ripening in Multicomponent Alloys, *Scripta Metall.*, 29 (1993), 1135–40
- A. Umantsev and G.B. Olson, Phase Equilibria and Transformations in Adiabatic Systems, *Phys. Rev. E*, 48 (1993), 4229–49
- G. Ghosh and G.B. Olson, Solid Solution Strengthening and Kinetics of Martensitic Transformation in Fe-Base Alloys, *Proc. ICOMAT-92, Intl. Conf. Martensitic Transformations*, ed. C.M. Wayman and D.J. Perkins, Monterey Inst. Adv. Studies, (1993), 353–58
- M. Lin, G.B. Olson, and M. Cohen, Homogeneous Martensitic Nucleation in Fe-Co Precipitates Formed in a Cu Matrix, *Proc. ICOMAT-92, Intl. Conf. Martensitic Transformations*, ed. C.M. Wayman and D.J. Perkins, Monterey Inst. Adv. Studies, (1993), 191–96
- A. Umantsev and G.B. Olson, Modulation Mechanism for Displacive Transformations, *Proc. ICOMAT-92, Intl. Conf. Martensitic Transformations*, ed. C.M. Wayman and D.J. Perkins, Monterey Inst. Adv. Studies, (1993), 215–20
- P. Hong and G.B. Olson, "A Phenomenological Model for Magnetically-Driven Lattice-Distortive Transformations in Fe-Pd Alloys," *Proc. ICOMAT-92, Intl. Conf. Martensitic Transformations*, ed. C.M. Wayman and D.J. Perkins, Monterey Inst. Adv. Studies, (1993), 161–66
- R. Wu, A.J. Freeman, and G.B. Olson, First Principles Determination of the Effects of Phosphorus and Boron on Iron Grain-Boundary Cohesion, *Science*, (July 15, 1994), 376–80

- C.J. Kuehmann, J. Cho, T.A. Stephenson, and G.B. Olson, Systems Design of High Performance Steels, *Metallic Materials for Lightweight Applications*, 40th Sagamore Army Materials Research Conference, ed. E.B. Kula and M.G.H. Wells, U.S. Government Printing Office, (1994), 337–55
- R. Wu, A.J. Freeman, and G.B. Olson, The Nature of Phosphorus Embrittlement of the Fe$\sum$3[110](111) Grain Boundary, *Phys. Rev. B*, 50 (1994), 75
- P. Hong and G.B. Olson, Magnetic Origin of the Lattice Instability of FePd Alloys, *J. Mag. & Mag. Matls.*, 129 (1994), 191–99
- S. Tang, A.J. Freeman, and G.B. Olson, Local Density Studies of the Structure and Electronic Properties of B and S in an Fe Grain Boundary, *Phys. Rev. B*, 50 (1994), 1
- G. Ghosh and G.B. Olson, Kinetics of FCC → BCC Heterogeneous Martensitic Nucleation, Part I: The Critical Driving Force for Athermal Nucleation, *Acta Metall. Mater.*, 42 (1994), 3361–70
- G. Ghosh and G.B. Olson, Kinetics of FCC → BCC Heterogeneous Martensitic Nucleation, Part II: Thermal Activation, *Acta Metall. Mater.*, 42 (1994), 3371–79
- G.B. Olson and H. Ohtsuka, Kinetics of Thermoelastic Martensitic Transformation, *Proc. Intl. Symp. on Shape Memory Materials*, Intl. Acad. Publishers, (1994), 17–23
- G.B. Olson, K.C. Hsieh, and H.K.D.H. Bhadeshia, On the Nucleation Driving Force for Displacive Transformations in Steels, *Microstructures LCS '94*, Iron and Steel Inst. of Japan (1994)
- W.B. Lee, F-R. Chen, S.K. Chen, G.B. Olson, and C.M. Wan, Transmission Electron Microscopy Studies of the Crystallography of B.C.C./18R Martensite in Fe-Mn-Al-C, *Acta Metall. Mater.*, 43, 1 (1994), 21–30
- Y.A. Chu, B. Moran, P. Hong, and G.B. Olson, Nonclassical Nucleation of Displacive Phase Transformations, *Proc. Intl. Conf. Solid → Solid Phase Transform. Inorg. Min.*, (1994), 655–60
- G. Ghosh, G.B. Olson, and M.E. Fine, Coherency Induced Morphological Evolution, *Proc. Intl. Conf. Solid → Solid Phase Transform. Inorg. Min.*, (1994), 611–16
- G. Ghosh, G.B. Olson, T.J. Kinkus, and M.E. Fine, Phase Separation in Fe-Ni-Al and Fe-Ni-Al-Cr Alloys, *Proc. Intl. Conf. Solid → Solid Phase Transform. Inorg. Min.*, (1994), 359–64
- G.N. Haidemenopolous, M. Grujicic, G.B. Olson, and M. Cohen, Thermodynamics-Based Alloy Design Criteria for Austenite Stabilization and Transformation Toughening in the Fe-Ni-Co System, *J. Alloys Compd.*, 220 (1995), 142–47
- J.W. Christian, G.B. Olson, and M. Cohen, Classification of Displacive Transformations: What Is a Martensitic Transformation?, Proc. ICOMAT 95, supplement to *J. de Phys. III*, 5 (1995), Coll. C8, 3–10
- G.B. Olson, Opportunities in Martensite Theory, Proc. ICOMAT 95, supplement to *J. de Phys. III*, 5 (1995), Coll. C8, 31–40
- C.J. Kuehmann and G.B. Olson, Computer-Aided Systems Design of Advanced Steels, *Proc. International Symposium of Phase Transformations During the Thermal/ Mechanical Processing of Steel—Honoring Professor Jack Kirkaldy*, ed. E.B. Hawbolt et al., Metallurgical Society of the Canadian Institute of Mining, Metallurgy and Petroleum, (1995), 345–56
- G.B. Olson, Transformation Plasticity and Toughening, *J. de Phys. V*, Colloque C1, (1996), C1-407
- R. Wu, A.J. Freeman, and G.B. Olson, Effects of Carbon on Fe Grain Boundary Cohesion: First Principles Determination, *Phys. Rev. B*, 53, 11 (1996), 7504–9
- B. Moran, Y.A. Chu, and G.B. Olson, Homogeneous Nucleation of a Solid-Solid Dilatational Phase Transformation, *Int. J. Solids Struct.*, 33, 13 (1996), 1903–19
- J.J. Felten, T.J. Kinkus, A.C.E. Reid, J.B. Cohen, and G.B. Olson, Solid-Solution Structure and the Weakly First-Order Displacive Transformation in Fe-Pd Alloys, *Metall. Trans.*, 28A (1997), 527–36
- C.A. Knepfler, K.T. Faber, J. Weertman, G.B. Olson, C.R. Hubbard, O.B. Cavin, and N. Packen, High Temperature Stability and Thermal Expansion Behavior of Molybdenum-Chromium M_2C Carbides, *J. Alloys Comp.*, 248 (1997), 139–42
- H. Ohtsuka, K. Takashima, and G.B. Olson, Nonthermoelastic and Thermoelastic Martensitic Transformation Behavior Characterized by Acoustic Emission in an Fe-Pt Alloy, *Mat. Res. Soc. Symp. Proc.*, Vol 459 (1997), 407–12
- G.B. Olson, Systems Design of Hierarchically Structured Materials: Smart Steels, *Proc. Intl. Forum on Creation of Super Metallic Materials Consisting of Amorphous,*

Nanoscale and Mesoscopic Structures, R&D Institute of Metals and Composites for Future Industries, (RIMCOF) (1997)

- L. Zhong, R. Wu, A.J. Freeman, and G.B. Olson, Effects of Mn Additions on the P Embrittlement of the Fe Grain Boundary, *Phys. Rev. B*, 55, 11 (1997), 133–37
- G.B. Olson, Computational Design of Hierarchically Structured Materials, *Science*, 277, 5330 (1997), 1237–42
- G.B. Olson, Brains of Steel: Designing Materiallurgists, *Adv. Mat. & Proc.*, 152, 1 (1997), 72–79
- L. Zhong, R. Wu, A.J. Freeman, and G.B. Olson, Effects of Interfacial Relaxation on the Magnetic Coupling of Mn/Fe (111), *J. Appl. Phys.*, 81 (1997), 4479–81
- S.H. Carr, R. D'Oyen, and G.B. Olson, Design of Thermoset Resins with Optimal Graded Structures, *Proc. Fourth Intl. Conf. on Composites Engineering*, ed. David Hui, International Community for Composites Engineering, (1997), 205
- C.M. Neubauer, J. Thomas, M. Garci, K Breneman, G.B. Olson, and H.M. Jennings, Cement Hydration, *Proc. 10th Intl. Congress on the Chemistry of Cement*, Amarkai AB and Congrex Göteborg AB, (1997)
- B. Files and G.B. Olson, Terminator 3: Biomimetic Self-Healing Alloy Composite, *SMST-97: Proc. 2nd Intl. Conf. Shape Memory & Superelastic Technologies*, SMST, (1997), 281–86
- C.J. Kuehmann, J.P. Wise, C.E. Campbell, and G.B. Olson, Simulation of Carburization in Secondary Hardening Steels, *Heat Treating*, Proc. 17th Conf., ASM International (1997), 313–19
- G.B. Olson, Systems Design of Hierarchically Structured Materials: Advanced Steels, *J. Comput. Aided Mater. Des.*, 4 (1997), 143–56
- D.E. Ellis, X. Chen, and G.B. Olson, Effects of Impurities and Alloying Elements on Iron Grain Boundary Cohesion, *MRS Sympos. Proc.*, 475 (1997), 499–506
- X. Chen, D.E. Ellis, and G.B. Olson, Effects of Alloying Elements on Iron Grain Boundary Cohesion, *MRS Sympos. Proc.*, 492 (1998), 243–48
- H.E. Lippard, C.E. Campbell, T. Bjorklind, U. Borggren, P. Kellgren, V.P. Dravid, and G.B. Olson, Microsegregation Behavior during Solidification and Homogenization of AerMet 100® Steel, *Metall. Trans. B*, 29B (1998), 205–10
- L. Sagert, G.B. Olson, and D.E. Ellis, Chemical Embrittlement of Iron Grain Boundaries: P and the P/Mo Couple, *Philos. Mag. B*, 77 (1998), 871–89
- M. Grujicic and G.B. Olson, Dynamics of Martensitic Interfaces, *Interface Sci.*, 6, 155 (1998), 155–64
- G.B. Olson, Materials Design: Building a Better Martensite, *Proc. Intl. Conf. Displacive Phase Transformations and Their Applications in Materials Engineering*, TMS-AIME, (1998), 15–26
- G.B. Olson and C.J. Kuehmann, Materials Design: Weld Metals for Elevated Temperature Service, *Mathematical Modelling of Weld Phenomena 4*, Proc. 4th Intl. Seminar on Numerical Analysis of Weldability, IOM Communications, (1998), 290–301
- C.J. Kuehmann and G.B. Olson, Gear Steels Designed by Computer, *Adv. Mat. & Proc.*, 153, 5 (1998), 40–43
- G. Ghosh, C. Kantner, and G.B. Olson, Thermodynamic Modeling of the Pd-*X* (*X* = Ag, Co, Fe, Ni) Systems, *J. Phase Eq.*, 20, 3 (1999), 295–305
- G. Ghosh, C.E. Campbell, and G.B. Olson, An Analytical Electron Microscopy Study of Paraequilibrium Cementite Precipitation in Ultra-High-Strength Steel, *Metall. Mater. Trans. A*, 30A (1999), 501–12
- A.C.E. Reid, G.B. Olson, and B. Moran, Dislocations in Nonlinear Nonlocal Media: Martensitic Embryo Formation, *Phase Transitions*, 69 (1999), 309–28
- V.I. Levitas, A.V. Idesman, and G.B. Olson, Continuum Modeling of Strain-Induced Martensitic Transformation at Shear-Band Intersections, *Acta Mater.*, 47, 1 (1999), 219–33
- G.B. Olson, New Directions in Martensite Theory, Proc. ICOMAT 98, *Mater. Sci. Eng.* A, 273–75 (1999), 11–20
- A.C.E. Reid and G.B. Olson, Martensitic Embryo Formation at Dislocations Near Surfaces, Proc. ICOMAT 98, *Mater. Sci. Eng.* A, 273–75 (1999), 257–61
- Y. Nagataki, J.B. Cohen, and G.B. Olson, X-Ray Line Broadening Analysis of Coherent M2C Precipitation in Ni-Co Secondary Hardening Steels, *Proc. Advanced Materials for the 21st Century: The Julia R. Weertman Symposium*, MRS (1999)
- P. Hirsch, B. Shwom, C. Yarnoff, J. Anderson, D. Kelso, G.B. Olson, and J.E. Colgate, Engineering Design and Communication: The Case for Interdisciplinary Collaboration, *Intl. J. Eng. Ed.*, 17, 4–5 (2000), 342–48

- G.B. Olson, Designing a New Material World, *Science*, 288 (2000), 993–98
- L. Zhong, A.J. Freeman, R. Wu, and G.B. Olson, Charge Transfer Mechanism of Hydrogen-Induced Intergranular Embrittlement of Iron, *Phys. Rev.* B, 21 (2000), 938–41
- G. Ghosh and G.B. Olson, Thermodynamic Modeling of the Cr-Pd and Mo-Pd Systems, *J. Phase Eq.*, 21, 1 (2000), 32–39
- W.T. Geng, A. J. Freeman, R. Wu, and G.B. Olson, Effect of Mo and Pd on the Grain-Boundary Cohesion of Fe, *Phys Rev. B*, 62, 10 (2000), 6208–14
- Y.A. Chu, B. Moran, A.C.E. Reid, and G.B. Olson, A Model for Non-Classical Nucleation of Solid-Solid Structural Phase Transformations, *Met. Mater. Trans. A*, 31A (2000), 1321
- W.T. Geng, A.J. Freeman, and G.B. Olson, Influence of Alloying Additions on Grain Boundary Cohesion of Transition Metals: First-Principles Determination and its Phenomenological Extension, *Phys. Rev. B*, 63 (2001), 165415
- G. Ghosh and G.B. Olson, Simulation of Paraequilibrium Growth in Multicomponent Systems, *Metall. Mater. Trans. A*, 32A (2001), 455–67
- A.C.E. Reid and G.B. Olson, A Continuum Description of Dislocations in Elastically Nonlinear Media: Martensitic Embryo Formation, *Mater. Sci. Eng. A*, 309–10 (2001), 370–76
- G.B. Olson, Brains of Steel: Mind Melding with Materials, *Intl. J. Eng. Ed.*, 17, 4–5 (2001), 468–71
- G.B. Olson, Beyond Discovery: Design for a New Material World, *Calphad*, 25, 2 (2001), 175–90
- C.E. Campbell and G.B. Olson, Systems Design of High Performance Stainless Steels: I. Conceptual Design, *J. Comput. Aided Mater. Des.*, 7 (2001), 145–70
- C.E. Campbell and G.B. Olson, Systems Design of High Performance Stainless Steels: II. Prototype Characterization, *J. Comput. Aided Mater. Des.*, 7 (2001), 171–94
- G.B. Olson, New Age of Design, *J. Comput. Aided Mater. Des.*, 7 (2001), 143–44
- G. Ghosh and G.B. Olson, Computational Thermodynamics and the Kinetics of Martensitic Transformation, *J. Phase Eq.*, 22, 3 (2001), 199–207
- G. Ghosh and G.B. Olson, Precipitation of Paraequilibrium Cementite: Experiments, and Thermodynamic and Kinetic Modeling, *Acta. Mater.*, 50 (2002), 2099–119
- G. Ghosh and G.B. Olson, The Isotropic Shear Modulus of Multicomponent Fe-Base Solid Solutions, *Acta Mater.*, 50 (2002), 2655–75
- V.I. Levitas, A.V. Idesman, G.B. Olson, and E. Stein, Numerical Modeling of Martensitic Growth in an Elastoplastic Material, *Philos. Mag. A*, 82, 3 (2002), 429–62
- G. Ghosh, A. van de Walle, M. Asta, and G.B. Olson, Phase Stability of the Hf-Nb System: From First-Principles to CALPHAD, *Calphad*, 26, 4 (2002), 491–511
- D. Isheim, D.N. Seidman, J.H. Perepezko, and G.B. Olson, Nanometer-Scale Solute Clustering in Aluminum-Nickel-Ytterbium Metallic Glasses, *Mater. Sci. Engin.*, A353 (2003), 99–104
- W.K. Liu, E.G. Karpov, S. Zhang, and H.S. Park, An Introduction to Computational Nanomechanics and Materials, *Comput. Meth. Appl. M.*, 193 (2004), 1529–78
- J. Jung, G. Ghosh, D. Isheim, and G.B. Olson, Precipitation of Heusler Phase (Ni_2TiAl) from B2-TiNi in Ni-Ti-Al and Ni-Ti-Al-X (X=Hf, Zr) Alloys, *Met. Mater. Trans. A*, (2003), 1221–35
- T. Shishidou, J-H. Lee, Y-J. Zhao, A.J. Freeman, and G.B. Olson, Overlayer and Superlattice Studies of Metal/Ceramic Interfaces: Fe/TiC, *J. Appl. Phys.*, 93, 10 (2003), 6876–78
- A. Misra, R. Bishop, G. Ghosh, and G.B. Olson, Phase Equilibria in Prototype Nb-Pd-Hf-Al Alloys, *Met. Mater. Trans. A*, 34A (2003), 1771–82
- J. Jung, G. Ghosh, and G.B. Olson, A Comparative Study of Precipitation Behavior of Heusler Phase (Ni_2TiAl) from B2-TiNi in Ni-Ti-Al and Ni-Ti-Al-X (X=Hf, Pd, Pt, Zr) Alloys, *Acta Mater.*, 51 (2003), 6341–57
- A.M. Wusatowska-Sarnek, G. Ghosh, G.B. Olson, M.J. Blackburn, and M. Aindow, Characterization of the Microstructure and Phase Equilibria Calculations in Powder Metallurgy Superalloy IN100, *J. Mater. Res.*, 18 (2003), 2653–63
- G. Ghosh and G.B. Olson, The Kinetics of Lath Martensitic Transformation, *J. de Physique IV*, 112 (2003), 139–42
- S. Hao, B. Moran, W-K. Liu, and G.B. Olson, A Hierarchical Multi-Physics Model for Design of High Toughness Steels, *J. Comput. Aided Mater. Des.*, 10, 2 (2003), 99–142

- G.B. Olson and C.J. Kuehmann, Transformation Toughening in Dispersed-Phase Systems, *Symposium on the Thermodynamics, Kinetics, Characterization and Modeling of: Austenite Formation and Decomposition*, (2003), 493–504
- G.B. Olson, A.J. Freeman, P.W. Voorhees, G. Ghosh, J. Perepezko, M. Eberhart, and C. Woodward, Quest for Noburnium: 1300C Cyberalloy, *International Symposium on Niobium for High Temperature Applications*, ed. Y-W. Kim and T. Carneiro, TMS, (2004), 113–22
- S. Hao, W-K. Liu, B. Moran, F. Vernerey, and G.B. Olson, Multi-Scale Constitutive Model and Computational Framework for the Design of Ultra-High Strength, High Toughness Steels, *Comput. Methods Appl. Mech. Engrg.*, 193 (2004), 1865–908
- A. Saha, G. Ghosh, and G.B. Olson, An Assessment of Interfacial Dissipation Effects at Reconstructive Ferrite-Austenite Interfaces, *Acta Mater.*, 53 (2004), 141–49
- G.B. Olson, Dynamics-Based Microstructural Design in Multicomponent Alloys, *JOM*, 56, 11 (2004), 264
- H.-J. Jou, P. Voorhees, and G.B. Olson, Computer Simulations for the Prediction of Microstructure/Property Variation in Aeroturbine Disks, *Superalloys 2004*, TMS, (2004), 877–86
- J.-H. Lee, T. Shishidou, Y.-J. Zhao, A.J. Freeman, and G.B. Olson, Strong Interface Adhesion in Fe/TiC, *Phil. Mag.*, 85, 31 (2005), 3683–97
- W.-T. Geng, J.-S. Wang, and G.B. Olson, Comment on "Grain Boundary Decohesion by Impurity Segregation in a Nickel-Sulfur System," *Science*, 309 (2005), 1677
- G.B. Olson, Materials by Design, *Handbook of Materials Modeling*, (2005), 2667–69
- M.E. Eberhart, T.E. Jones, M.A. Batchelder, and G.B. Olson, Structure Property Relationships in the Design of Alloy Composition: Moving beyond Electron to Atom Ratios, *J. Mater. Res.*, 20, 5 (2005), 1330–35
- M. Kim, A.J. Freeman, S. Kim, J. H. Perepezko, G.B. Olson, Phase Stability of the Intermetallic $L2_1$ Heusler Alloys of $A_2(Hf_{1-x}Zr_x)Al$ (Where A=Pd and Pt) for an Nb-Based High-Temperature Materials Design, *Appl. Phys. Lett.*, 87 (2005)
- W.-T. Geng, A. J. Freeman, G.B. Olson, Y. Tateyama, and T. Ohno, Hydrogen-Promoted Grain Boundary Embrittlement and Vacancy Activity in Metals: Insights from *Ab Initio* Total Energy Calculations, *Mater. Trans.*, 46, 4 (2005), 756–60
- A.F. McKenna, J.E. Colgate, S.H. Carr, and G.B. Olson, IDEA: Formalizing the Foundation for an Engineering Design Education, *Int. J. Eng. Educ.*, (2006)
- A.F. McKenna, J.E. Colgate, G.B. Olson, and S.H. Carr, Exploring Adaptive Expertise as a Target for Engineering Design Education, ASME 2006 International Design Engineering Technical Conferences and Computers and Information in Engineering Conference, *Proc IDETC/CIE*, (2006), 1–6
- A.F. McKenna, J.E. Colgate, and G.B. Olson, Characterizing the Mentoring Process for Developing Effective Design Engineers, *Proc. the American Society for Engineering (ASEE) Annual Conference*, (2006)
- C. Qiu, S.M. Opalka, G.B. Olson, and D.L. Anton, The Na—H System: From First-Principles Calculations to Thermodynamic Modeling, *Int. J. Mat. Res.*, 97 (2006), 845
- C. Qiu, S.M. Opalka, G.B. Olson, and D.L. Anton, Thermodynamic Modeling of the Sodium Alanates and the Na-Al-H System, *Int. J. Mat. Res.*, 97 (2006), 1484–94
- W.-T. Geng, A.J. Freeman, and G.B. Olson, Atomic Size Effect in Impurity Induced Grain Boundary Embrittlement, *Mater. Trans.*, 47, 8 (2006), 2113–14
- E.L. Thomas and G.B. Olson, Morris Cohen, *Physics Today* 59 (2006), 86
- G.B. Olson, "Morris Cohen: A Memorial Tribute," *Mater. Sci. Eng. A*, 438–40 (2006), 2–11
- G.B. Olson, Advances in Theory: Martensite by Design, *Mater. Sci. Eng. A*, 438 (2006), 48–54
- F.J. Vernerey, C. McVeigh, W.K. Liu, B. Moran, D. Tewari, D. Parks, and G. Olson, 3D Computational Modeling of Shear Dominated Ductile Failure of Steel, *JOM*, (2006), 45–51
- G. Ghosh and G. Olson, Integrated Design of Nb-Based Superalloys: Ab Initio Calculations, Computational Thermodynamics and Kinetics, and Experimental Results, *Acta Mater.*, 55, 10 (2007), 3281–303
- C. McVeigh, F. Vernerey, W-K. Liu, B. Moran, and G.B. Olson, An Interactive Micro-Void Shear Localization Mechanism in High Strength Steels, *J. Mech. Phys. Solids*, 55 (2007), 225–44
- A. Saha and G.B. Olson, Computer-Aided Design of Transformation Toughened Blast Resistant Naval Hull Steels: Part I, *J.*

Computer-Aided Mater. Des., 14, 2 (2007), 177–200
- A. Saha, J. Jung, and G.B. Olson, Prototype Evaluation of Transformation Toughened Blast Resistant Naval Hull Steels: Part II, *J. Computer-Aided Mater. Des.*, 14, 2 (2007), 201–33
- M.V. Manuel and G.B. Olson, Biomimetic Self-Healing Metals, *Proc. 1st Intl. Conference on Self-Healing Materials*, (2007)
- A.C. To, W.-K. Liu, G.B. Olson, T. Belytschko, W. Chen, M.S. Shephard, Y.-W. Chung, R. Ghanem, P.W. Voorhees, D.N. Seidman, C. Wolverton, J.S. Chen, B. Moran, A.J. Freeman, R. Tian, X. Luo, E. Lautenschlager, and A.D. Challoner, Materials Integrity in Microsystems: A Framework for a Petascale Predictive-Science-Based Multiscale Modeling and Simulation System, *Comput Mech.*, 42, 4 (2008), 485–510
- M. Bender and G.B. Olson, Computational Thermodynamics-Based Design of Nanodispersion-Strengthened Shape Memory Alloys, *SMST-2007: Proceedings of the International Conference on Shape Memory and Superelastic Technologies*, (2008), 115–22
- F. Vernerey, W.-K. Liu, B. Moran, and G.B. Olson, A Micromorphic Model for the Multiple Scale Failure of Heterogeneous Materials, *J. Mech. Phys. Solids*, 56 (2008), 1320–47
- S. Zhang and G.B. Olson, Local-Electrode Atom-Probe Tomographic Investigation of Strengthening Precipitates in a Mg-7Zn-3Al Alloy at Peak Hardness, Magnesium Technology Symposium 2008, *Magnesium Technology 2008*, (2008), 113–16
- M.V. Manuel, A.F. McKenna, and G.B. Olson, Hierarchical Model for Coaching Technical Design Teams, *Int. J. Engin. Edu.*, 24, 2 (2008), 260–65
- D.L. McDowell and G.B. Olson, Concurrent Design of Hierarchical Materials and Structures, *Scientific Modeling and Simulation SMNS*, (2008), 1–34
- G.B. Olson, H.-J. Jou, J. Jung, J.T. Sebastian, A. Misra, I. Locci, and D. Hull, Precipitation Model Validation in 3rd Generation Aeroturbine Disc Alloys, *Superalloys 2008*, TMS, (2008), 923–32
- C. Qiu, S.M. Opalka, O.M. Løvvik, and G.B. Olson, Thermodynamic Modeling of the Na-Al-Ti-H System and Ti Dissolution in Sodium Alanates, *Calphad*, 32, 4 (2008), 624–36
- K. Thornton, S. Nola, R.E. Garcia, M. Asta, and G.B. Olson, Computational Materials Science and Engineering Education: Survey of Trends and Needs, *J. Mater.*, 61 (2009), 12–17
- R. Prasannavenkatesan, D.L. McDowell, G.B. Olson, and H-J. Jou, Modeling Effects of Compliant Coatings on HCF Resistance of Primary Inclusions in High Strength Steels, *J. Eng. Mater. Technol.*, 131, 1 (2009), 011012
- R. Prasannavenkatesan, J. Zhang, D.L. McDowell, G.B. Olson, and H-J. Jou, 3D Modeling of Subsurface Fatigue Crack Nucleation Potency of Primary Inclusions in Heat Treated and Shot Peened Martensitic Gear Steels, *Int. J. Fatigue*, 31, 7 (2009), 1176–89
- W-K. Liu, L. Siad, R. Tian, S. Lee, D. Lee, X. Yin, W. Chen, S. Chan, G.B. Olson, L-E. Lindgen, M.F. Horstemeyer, Y-S. Chang, J-B. Choi, and Y.J. Kim, Complexity Science of Multiscale Materials via Stochastic Computations, *Int. J. Numer. Meth. Engng*, 80, 6–7 (2009), 932–78
- C.J. Kuehmann and G.B. Olson, Computational Materials Design and Engineering, *Mater. Sci. Technol.*, 25, 4 (2009), 472–78
- M. Bender and G.B. Olson, Designing a Precipitation-Strengthened, Superelastic, TiNi-Based Alloy for Endovascular Stents, *Proc. ICOMAT '08*, TMS, (2009), 159–66
- F. Vernerey,W-K. Liu, B. Moran, and G.B. Olson, Multi-Length Scale Micromorphic Process Zone Model, *Comput. Mech.*, 44, 3 (2009), 433–45
- R. Tian, S. Chan, S. Tang, A.M. Kopacz, J-S. Wang, H-J. Jou, L. Siad, L-E. Lindgren, G.B. Olson, and W-K. Liu, A Multiresolution Continuum Simulation of the Ductile Fracture Process, *J. Mech. Phys. Solids*, 58 (2010), 1681–700
- J-S. Wang and G.B. Olson, Prediction of Ductile Fracture Toughness, *Proc. of the 12th International Conference on Fracture, 2009*, Curran Associates, (June 2010)
- F. Latourte, Z. Feinberg, L.F. Mori, G.B. Olson, and H.D. Espinosa, Shear and Tensile Plastic Behavior of Austenitic Steel TRIP-120 Compared with Martensitic Steel HSLA-100, *Int. J. Fract.*, 162 (2010), 187–204
- S. Zhang, O.Y. Kontsevoi, A.J. Freeman, and G.B. Olson, Sodium-Induced Embrittlement of an Aluminum Grain Boundary, *Phys. Rev. B*, 82 (2010), 224107
- G. Spanos, D.J. Rowenhorst, S. Chan, and G.B. Olson, 3D Characterization of

Microstructures in Metallic Systems by Serial Sectioning, *Proceedings of the 31st Risoe International Symposium on Materials Science: Challenges in Materials Science and Possibilities in 3D and 4D Characterization Techniques*, ed. N. Hansen, D. Juul Jensen, H. Poulsen, and B. Ralph, Risoe National Laboratory for Sustainable Energy, Technical University of Denmark, (2010), 159–70

- I.G. Kim, G. Rahman, J.H. Jang, Y.Y. Song, S.W. Seo, H.K.D.H. Bhadeshia, A.J. Freeman, and G.B. Olson, A Systematic Study on Iron Carbides from First-Principles, *Mater. Sci. Forum*, 47–50 (2010), 654–56
- S. Zhang, O.Y. Kontsevoi, A.J. Freeman, and G.B. Olson, First Principles Investigation of Zinc-Induced Embrittlement in an Aluminum Grain Boundary, *Acta Materialia*, 59, 15, (2011), 6155–67
- S. Zhang, O.Y. Kontsevoi, A.J. Freeman, and G.B. Olson, First Principles Determination of the Effect of Boron on Aluminum Grain Boundary Cohesion, *Phys. Rev. B*, 84, 13 (2011), 134104
- S. Zhang, O.Y. Kontsevoi, A.J. Freeman, and G.B. Olson, Cohesion Enhancing Effect of Magnesium in Aluminum Grain Boundary: A First-Principles Determination, *Appl. Phys. Lett.*, 100 (2012), 23
- M.V. Manuel, A.F. McKenna, and G.B. Olson, Supporting Students' Technical Innovation in Capstone Design: Insights into the Human Connection, *Int. J. Engin. Educ.*, 28, 2 (2012), 310–16
- F. Latourte, X. Wei, Z.D. Feinberg, A. de Vaucorbeil, P. Tran, G.B. Olson, and H.D. Espinosa, Design and Identification of High Performance Steel Alloys for Structures Subjected to Underwater Impulsive Loading, *Int. J. Solids Struct.*, 49, 13 (2012), 1573–87
- G.B. Olson, B.L. Tiemens, and A.K. Sachdev, Cu-Precipitation Strengthening in Ultrahigh-Strength Carburizing Steels, *Metall. Mater. Trans. A*, 43 (2012), 3615
- G.B. Olson, B.L. Tiemens, A.K. Sachdev, and R.K. Mishra, Three-Dimensional (3-D) Atom Probe Tomography of a Cu-Precipitation-Strengthened, Ultrahigh-Strength Carburized Steel, *Metall. Mater. Trans. A*, 43 (2012), 3626
- S. Zhang, O.Y. Kontsevoi, A.J. Freeman, and G.B. Olson, Aluminum Grain Boundary Decohesion by Dense Sodium Segregation, *Phys. Rev. B*, 85, 21 (2012), 21409
- G.J. Fraley, G.B. Olson, F. Mazzolani, and R. Herrera, Concurrent Design of High-Strength TRIP Steels and Passive Damping Devices, *Stessa 2012: Proceedings of the 7th International Conference on Behavior of Steel Structures in Seismic Areas*, (2012)
- T.E. Jones, M.E. Eberhart, S. Imlay, C. MacKay, and G.B. Olson, Better Alloys with Quantum Design, *Phys. Rev. Lett.*, 109 (2012), 125506
- H.-J. Jou, G.B. Olson, T. Gabb, A. Garg, and D. Miller, Characterization and Computational Modeling of Minor Phases in Alloy LSHR, *Superalloys 2012*, ed. E.S. Haron, R.C. Reed, M.C. Hardy, M.J. Mills, R.E. Montero, P.D. Portella, and J. Telsman, TMS, (2012), 893–902
- S. Tang, A.M. Kopacz, S. Chan O'Keeffe, G.B. Olson, and W-K. Liu, Concurrent Multiresolution Finite Element: Formulation and Algorithmic Aspects, *Comput. Mech.*, 52, 6 (2013), 1255–79
- G.B. Olson, Genomic Materials Design: The Ferrous Frontier, *Acta Mater.*, Diamond Jubilee Issue, 61 (2013) 771–781
- J. Grabowski, J. Sebastian, G. Olson, A. Asphahani, and R. Genellie, Integrated Computational Materials Engineering Helps Successfully Develop Aerospace Alloys, *Adv. Mater. Proc.*, 171, ASM International (2013), 17–19
- G.B. Olson, Integrated Computational Materials Design: From Genome to Flight, Proc. AIAA SDM, (2013)
- G.B. Olson and D. Gupta, Tribute to Professor D.S. Lieberman: On the Importance of Being Invariant, *J. Alloys Compd.*, 577 (2013), S9–12
- J. Sebastian and G.B. Olson, Examples of QuesTek Innovations' Application ICME to Materials Design, Development, and Rapid Qualification, 55th AIAA/ASME/ASCE/AHS/SC Structures, Structural Dynamics, and Materials Conference, (2014)
- G.B. Olson, Preface to Viewpoint Set On: The Materials Genome, *Scripta Mater.*, 70 (2014) 1–2
- G.B. Olson and C.J. Kuehmann, Materials Genomics: From CALPHAD to Flight, *Scripta Mater.*, 70 (2014), 25–30
- S. Chan, S. Tang, A.M. Kopacz, D. Rowenhorst, G. Spanos, W.K. Liu, and G.B. Olson, Multiscale Ductile Fracture Integrating Tomographic Characterization and 3-D Simulation, *Acta Mater.*, 82 (2015), 503–10

- J. Yan and G.B. Olson, Cahn-Rosenberg Distortion and the Orthorhombicity of Titanium Martensites, *Proc. PTM2015, International Conference on Solid-Solid Phase Transformation in Inorganic Materials*, TMS, (2015), 477–84
- N. Salajegheh, R. Prasannavenkatesan, D.L. McDowell, G.B. Olson, and H.J. Jou, Finite Element Simulation of Shielding/Intensification Effects of Primary Inclusion Clusters in High Strength Steels under Fatigue Loading, *J. Eng. Mater. Technol. Tran. ASME*, 136, 3 (2014), 031003
- N.J. Wengrenovich and G.B. Olson, Optimization of a TRIP Steel for Adiabatic Fragment Protection, Materials Today: Proc ICOMAT 2014, *Materials Today: Proc.* 2, supplement 3, (2015), S639–42
- D. Frankel, T. Jiang, and G.B. Olson, Design of a Fatigue Resistant Ni-free PdTi-Base SMA, Materials Today Proceedings ICOMAT (2014), *Materials Today: Proc.* 2, supplement 3, (2015), S801–4
- D.J. Frankel and G.B. Olson, Design of Heusler Precipitation Strengthened NiTi- and PdTi-Base SMAs for Cyclic Performance, *Shape Memory Superelas.*, 1 (2015), 1–18
- J. Zhou, J. Odqvist, J. Ågren, A. Ruban, M. Thuvander, W. Xiong, G.B. Olson, and P. Hedström, Direct Atom Probe Tomography Observations of Concentration Fluctuations in Fe-Cr Solid Solution, *Scripta Mater.*, 98 (2015), 13–15
- S.C. O'Keeffe, S. Tang, A.M. Kopacz, J. Smith, D.J. Rowenhorst, G. Spanos, W.K. Liu, and G.B. Olson, Multiscale Ductile Fracture Integrating Tomographic Characterization and 3-D Simulation, *Acta Mater.*, 82 (2015), 503–10
- W. Xiong and G.B. Olson, Integrated Computational Materials Design for High-Performance Alloys, *MRS Bull.*, 40 (2015), 1035–44
- Y. Idell, L.E. Levine, A.J. Allen, F. Zhang, C.E. Campbell, G.B. Olson, J. Gong, D.R. Snyder, and H.Z. Deutchman, Unexpected Delta-Phase Formation in Additive-Manufactured Ni-Based Superalloy, *JOM*, 68, 3 (2016), 950–59
- J.A. Moore, D. Frankel, R. Prasannavenkatesan, A.G. Domel, G.B. Olson, and W.K. Liu, A Crystal Plasticity-Based Study of the Relationship between Microstructure and Ultra-High-Cycle Fatigue Life in Nickel Titanium Alloys, *Int. J. Fatigue*, 91 (2016), 183–94
- J. Weertman, J. Weertman, and G.B. Olson, Member News, *JOM* 68, 1 (2016), 2–3
- J.Y. Yan and G.B. Olson, Computational Thermodynamics and Kinetics of Displacive Transformations in Titanium-Based Alloys, *J. Alloys Compd.*, 673 (2016), 441–54
- J.-Y. Yan and G.B. Olson, Molar Volumes of bcc, hcp, and Orthorhombic Ti-Base Solid Solutions at Room Temperature, *Calphad*, 52 (2016), 152–58
- W. Xiong and G.B. Olson, Cybermaterials: Materials by Design and Accelerated Insertion of Materials, *Comput. Mater.*, 2 (2016), 15009-–32
- J. Zhou, J. Odqvist, A. Ruban, M. Thuvander, W. Xiong, J. Agren, G.B. Olson, and P. Hedstrom, Effect of Solution Treatment on Spinodal Decomposition during Aging of an Fe-46.5 at.% Cr Alloy, *J. Mater. Sci.*, 52, 1 (2017), 326–35

Index

B

C

D

E

F

G

H

L

M

N

O

P

T

U

V

X

Y

Z